HANDBOOK
of DETERGENTS

Part B: Environmental Impact

SURFACTANT SCIENCE SERIES

ADDITIONAL VOLUMES IN PREPARATION

HANDBOOK of DETERGENTS

Editor-in-Chief
Uri Zoller
Haifa University–Oranim
Kiryat Tivon, Israel

Part B: Environmental Impact

edited by
Uri Zoller
Haifa University–Oranim
Kiryat Tivon, Israel

CRC Press
Taylor & Francis Group
Boca Raton London New York

CRC Press is an imprint of the
Taylor & Francis Group, an **informa** business

CRC Press
Taylor & Francis Group
6000 Broken Sound Parkway NW, Suite 300
Boca Raton, FL 33487-2742

First issued in paperback 2019

© 2004 by Taylor & Francis Group, LLC
CRC Press is an imprint of Taylor & Francis Group, an Informa business

No claim to original U.S. Government works

ISBN-13: 978-0-8247-0353-0 (hbk)
ISBN-13: 978-0-367-39395-3 (pbk)

Visit the Taylor & Francis Web site at
http://www.taylorandfrancis.com

and the CRC Press Web site at
http://www.crcpress.com

Handbook Introduction

The battle cry for sustainable development is persistent in all circles, gaining acceptance, worldwide, as the guiding rationale for activities or processes in the science–technology–environment–economy–society interfaces targeting improvement and growth. Such activities are expected to result in higher standards of living, leading eventually to a better quality of life for our increasingly technology-dependent modern society. Models of sustainable development and exemplary systems of sustainable management are continually being developed and/or adapted and creatively applied, taking into consideration human needs versus wants on the one hand, and long- versus short-term benefits and tradeoffs on the other.

"Detergents" constitute a classic case study within this context: this is a multi-dimensional systemic enterprise, operating within complex sociopolitical/technoeconomical realities, locally and globally, reflecting in its development and contemporary "state-of-affairs" the changing dynamic equilibria and interrelationships between demands/needs, cost/benefits, gains/tradeoffs, and social preferences. Interestingly, it is not surprising, despite the overall maturity of the consumer market, that detergents continue to advance more rapidly than population growth.

The soap and detergent industry has seen great change in recent years, responding to the shifts in consumer preferences, environmental pressures, the availability and cost of raw materials and energy, demographic and social trends, and the overall economic and political situation worldwide. Currently, detergent product design is examined against the unifying focus of delivering to the consumer performance and value, given the constraints of the economy, technological advancements, and environmental imperatives. The annual 2–3% growth of the detergent industry and a higher growth in personal care products reflect impressive developments in formulation and application. The detergent industry is thus expected to continue steady growth in the near future.

For the detergent industry, the last decade of the twentieth century has been one of transformation, evolution, and even some surprises (e.g., the increase of heavy-duty liquid detergents at the expense of powder detergent products). On both the supplier and consumer market sides (both remain intensely competitive), the detergent industry has undergone dramatic changes, with players expanding their offerings, restructuring divisions, or abandoning the markets altogether. This has resulted in the consolidation of the market, especially in the last several years, and this trend appears to be gaining momentum. The key concepts have been and still are innovation, consumer preferences, needs, multipurpose products, cost/benefit, efficiency, emerging markets, partnership–cooperation–collaboration–merging (locally, regionally, and globally), and technological advancements. Although substantial gains and meaningful rapid changes with respect to the preceding concepts have been experienced by the surfactants/detergents markets, the same cannot be said for detergent/surfactant technology itself. The $9-billion-plus detergent ingredients market has many entrenched workhorse products. This may suggest that the supply of "solutions" to most cleaning "problems" confronted by consumers in view of the increasing global demand for a full range of synergistic, multifunctional detergent formulations having high performance and relatively low cost, and the need for compliance with environmentally oriented (green) regulation, may be based on modifications of existing technologies. What does all this mean for the future of the detergent enterprise? How will advances in research and development affect future development in detergent production, formulation, applications, marketing, consumption, and relevant human behavior as well as short- and long-term impacts on the quality of life and the environment? Since new findings and emerging technologies are generating new issues and questions, not everything that can be done should be done; that is, there should be more response to real *needs* rather than *wants*.

Are all the questions discussed above reflected in the available professional literature for those who are directly involved or interested engineers, scientists, technicians, developers, producers, formulators, managers, marketing people, regulators, and policy makers? After a thorough examination of the literature in this and/or related areas, I came to the conclusion that a comprehensive series was needed that focuses on the practical aspects of the topic and provides the detergent industry perspective to all those involved and interested. The *Handbook of Detergents* is an up-to-date compilation of works written by experts each of whom is heavily engaged in his or her area of expertise, emphasizing the practical and guided by a common systemic approach.

The aim of this six-volume handbook (Properties, Environmental Impact, Analysis, Formulation, Applications, and Production) is to reflect the above and to provide readers who are interested in any aspect of detergents a state-of-the-art comprehensive treatise, written by expert practitioners (mainly from industry) in the field. Thus, various aspects involved—raw materials, production, economics, properties, formulations, analysis and test methods, applications, marketing, environmental considerations, and related research problems—are dealt with, emphasizing the practical in a shift from the traditional or mostly theoretical focus of most of the related literature currently available.

The philosophy and rationale of the Handbook of Detergents series are reflected in its title, its plan, and the order of volumes and flow of the chapters (within each volume). The various chapters are not intended to be and should not necessarily be considered mutually exclusive or conclusive. Some overlapping facilitates the presentation of the same issue or topic from different perspectives, emphasizing different points of view, thus enriching and complementing various perspectives and value judgments.

There are many whose help, capability, and dedication made this project possible. The volume editors, contributors, and reviewers are in the front line in this respect. Many others deserve special thanks, including Mr. Russell Dekker and Mr. Joseph Stubenrauch, of Marcel Dekker, Inc., as well as my colleagues and friends in (or associated with) the detergent industry, whose dedication and involvement facilitated this work. My hope is that the final result will complement the tremendous effort invested by all those who contributed; you the reader, will be the ultimate judge.

Uri Zoller
Editor-in-Chief

Preface

Regardless of the state-of-the-art and affairs in the detergent industry worldwide, with respect to scientific, technological, economic, safety, and regulatory aspects of detergent production, formulation, application, and consequently consumption, their environmental impact constitutes and will continue to be an issue of major concern. This is particularly so given the operating global free-market economy, which is supposed to, and is expected to, ensure sustainable development. Avoidance of detrimental environmental impact primarily requires *prevention* rather than *correction*, which in turn should dictate what the detergent industry should do and what needs to be accomplished in the future to ensure environmentally-oriented sustainable development, given contemporary shifts in consumer preferences, the availability and cost of raw materials and energy, demographic and social trends, and the overall economical/political situation worldwide.

This second volume (Part B) of the six-volume series *Handbook of Detergents* deals with the potential environmental impact of detergents—surfactants, builders, and sequestering/chelating agents—as well as other components of detergent formulations as a result of their production, formulation, usage/consumption, and ultimate disposal into the various compartments of the environment, particularly the aquatic compartment. Since commercial detergent formulations comprise many homologs, oligomers/polymers, and isomers, their identification, quantification, distribution, and persistence as well as specific and/or synergistic environmental impact (toxicity, estrogenicity, health risk, exotoxicity, and other factors) should be assessed, to be used as the solid and reliable scientific basis for action.

This volume is a comprehensive treatise on the multidimensional issues involved, and represents an international industry–academia collaborative effort of over 50 experts and authorities worldwide.

Part I, "The Multidimensionality of Detergent-Related Environmental Impact and Its Assessment" contains:

- A historical review of detergents and the environment
- A critical review and discussion of the distribution rate, effects, biodegradation, toxicology, and ecotoxicology of surfactants and other components of detergent formulations
- Risk assessment, life-cycle assessment, biodegradability, toxicity, and structure–activity relationships of detergent components and their evaluation
- An examination of the environmental impact of detergent packaging
- Environmental safety legislation on detergents

The topics addressed in Part II, "Environmental Behavior, Effects, and Impact of Detergent Components," include:

- The fate, effects, safety, survival, distribution, biodegradability, biodegradation, ecology, and toxicology of anionic, cationic, and nonionic surfactants
- Environmental impact and ramifications of inorganic detergent builders, chelating agents, bleaching activators, perborates, and other components of detergent formulations
- Toxicology and ecotoxicology of minor components in personal care detergent formulations
- Biodegradation of surfactants in sewage treatment plants and in the natural environment
- Science versus politics in the environment-related regulatory process

All the above are accompanied and supported by extensive research-based data, occasionally accompanied by a specific "representative" case study, the derived conclusions of which are transferable.

This resource contains more than 2300 cited works and is aimed to serve as a practical reference for environmental, surfactant, chemical/biochemical, toxicological/ecotoxicological scientists and engineers, regulators, and policy makers associated with the detergent industry. I thank all the contributors who made the realization of this volume possible.

Uri Zoller

Contents

Contents

Contributors

Walter Aulmann Cognis Deutschland GmbH & Co. KG, Düsseldorf, Germany

I. J. A. Baker Kodak (Australasia) Pty Ltd., Coburg, Victoria, Australia

Harald P. Bauer Division of Pigments and Additives, Clariant GmbH, Hürth, Germany

Scott Belanger Miami Valley Laboratories, The Procter & Gamble Company, Cincinnati, Ohio, U.S.A.

José Luis Berna Research and Development, Petresa, Madrid, Spain

Geert Boeije European Technical Center, The Procter & Gamble Company, Brussels, Belgium

Luciano Cavalli* Sasol Italy, Milan, Italy

Y. Chen Department of Soil and Water Sciences, The Hebrew University of Jerusalem, Rehovot, Israel

Christina Cowan-Ellsberry Miami Valley Laboratories, The Procter & Gamble Company, Cincinnati, Ohio, U.S.A.

Vince Croud† Advanced Technologies Division, Warwick International Ltd., Holywell, Flintshire, United Kingdom

Andreas Domsch Degussa Goldschmidt Personal Care, Essen, Germany

---

Current affiliation: UNICHIM, Milan, Italy.
†*Current affiliation*: Antec International Ltd., Sudbury, Suffolk, United Kingdom.

xiii

C. J. Drummond CSIRO Molecular Science, Clayton South, Victoria, Australia

Shinya Ebata Human Safety Evaluation Center, LION Corporation, Kanagawa, Japan

Tom Feijtel European Technical Center, The Procter & Gamble Company, Brussels, Belgium

K. K. Fox Department of Civil and Structural Engineering, University of Sheffield, Sheffield, United Kingdom

Marcel Friedman Research and Development, Neca-Agis Group, Petach-Tikva, Israel

D. N. Furlong Applied Chemistry, RMIT University, Bundoora, Victoria, Australia

Ester Gorelik Zohar Dalia, Kibbutz Dalia, Israel

F. Grieser School of Chemistry, University of Melbourne, Melbourne, Victoria, Australia

Otto Grundler BASF Aktiengesellschaft, Ludwigshafen, Germany

Bent Halling-Sørensen Department of Analytical Chemistry, The Danish University of Pharmaceutical Sciences, Copenhagen, Denmark

John E. Heinze Executive Director, Environmental Health Research Foundation, Manassas, Virginia, U.S.A.

Louis Ho Tan Tai Consultant, Lambersart, France

M. S. Holt European Center for Ecotoxicology and Toxicology of Chemicals, Brussels, Belgium

Hans-Ulrich Jäger BASF Aktiengesellschaft, Ludwigshafen, Germany

Klaus Jenni Degussa Goldschmidt Personal Care, Essen, Germany

Kristine A. Krogh Department of Analytical Chemistry, The Danish University of Pharmaceutical Sciences, Copenhagen, Denmark

Zenon Łukaszewski Institute of Chemistry, Poznan University of Technology, Poznan, Poland

Torben Madsen Department of Ecotoxicology, DHI Water and Environment, Hørsholm, Denmark

Horst Messinger   Cognis Deutschland GmbH & Co. KG, Düsseldorf, Germany

Betty B. Mogensen   Department of Atmospheric Chemistry, National Environmental Research Institute, Roskilde, Denmark

Carter G. Naylor*   Consultant, Austin, Texas, U.S.A.

D. Prats   Department of Chemical Engineering, University of Alicante, Alicante, Spain

David W. Roberts   Safety and Environmental Assurance Center (SEAC), Unilever R&D, Sharnbrook, Bedford, United Kingdom

M. Rodriguez   Department of Chemical Engineering, University of Alicante, Alicante, Spain

Erwan Saouter   The Procter & Gamble Company, Petit Lancy, Switzerland

Susan E. Selke   School of Packaging, Michigan State University, East Lansing, Michigan, U.S.A.

John Solbé   Consultant, Denbighshire, North Wales, United Kingdom

Yutaka Takagi   Human Safety Evaluation Center, LION Corporation, Kanagawa, Japan

Toshiharu Takei   Human Safety Evaluation Center, LION Corporation, Kanagawa, Japan

J. Tarchitzky   Israeli Ministry of Agriculture, Bet-Dagan, Israel

Karen L. Thorpe[†]   School of Biological Sciences, The Hatherley Laboratories, Exeter University, Exeter, Devon, United Kingdom

Charles R. Tyler   School of Biological Sciences, The Hatherley Laboratories, Exeter University, Exeter, Devon, United Kingdom

C. G. van Ginkel   Akzo Nobel Chemicals Research Arnhem, Arnhem, The Netherlands

Gert van Hoof   The Procter & Gamble Company, Strombeek-Bever, Belgium

Oded Vashitz   Zohar Dalia, Kibbutz Dalia, Israel

---

*Retired from Huntsman Corporation, Austin, Texas, U.S.A.
†*Current affiliation*: AstraZeneca UK Limited, Brixham, Devon, United Kingdom.

Karl V. Vejrup   Department of Atmospheric Chemistry, National Environmental Research Institute, Roskilde, Denmark

Donald Versteeg   Miami Valley Laboratories, The Procter & Gamble Company, Cincinnati, Ohio, U.S.A.

Peter White   The Procter & Gamble Company, Newcastle upon Tyne, United Kingdom

Andreas Willing   Cognis Deutschland GmbH & Co. KG, Düsseldorf, Germany

Helmut Witteler   BASF Aktiengesellschaft, Ludwigshafen, Germany

Guang-Guo Ying   Adelaide Laboratory, CSIRO Land and Water, Glen Osmond, Australia

Uri Zoller   Faculty of Science and Science Education-Chemistry, Haifa University–Oranim, Kiryat Tivon, Israel

1

Sustainable Relationships: Raw Materials–Surfactant/ Detergent Production/ Formulation–Usage/ Consumption–Environment

A Systemic Challenge for the New Millennium

URI ZOLLER   Haifa University–Oranim, Kiryat Tivon, Israel

I. SUSTAINABLE DEVELOPMENT: DETERGENT INDUSTRY–ENVIRONMENT RELATIONSHIPS

Sustainable development is a key demand in our world of finite resources and endangered ecosystems. Given the environmental imperatives, the potential ecotoxicological/health risks of anthropogenic chemicals/man-produced formulations, and the limited economic feasibility of large-scale treatment and remediation technologies, the currently emerging corrective-to-preventive paradigm shift in the exploitation of raw materials, production/formulation, usage/consumption and disposal, as well as in the conceptualization of future developments in these and related activities is unavoidable [1–3].

The role of science and technology in meeting the sustainable development challenge is obvious and is recognized worldwide by all "stake holders." In this context, environmental sciences are emerging as a new multidimensional, cross-interdisciplinary scientific discipline and beyond. They draw on all the basic sciences to explain the working of the entire complex and dynamic earth system—the environment—which is constantly changing by natural causes and under human impact [4]. At present, they are in a process of moving from a specialized, compartmentalized, (sub-) disciplinary, unidimensional enterprise into a multidimensional, cross-boundary endeavor in the context of the science–technology–environment–society (STES) interfaces [5–7]. This poses new challenges with respect to both the intrinsic science and technology organization and performance and the way the relevant generated and acquired knowledge and accompanying processes will be put into action, guided by the superordinate idea(l)s of social responsibility and sustainability. Ultimately, this would require all involved to operate within an open-ended ideas-oriented culture [8].

In view of the fact that the public, many policymakers, some scientists and engineers, and even some environmental professionals believe that science and technology can solve most pollution problems, prevent future environmental impact, and (should) pave the way for sustainable development, it is of the utmost importance to recognize the limits of environmental science and technology (alone) to meet the challenge of sustainable development [9]. This is because science and technology are useful in establishing what we can do. However, neither of them, or both, can tell us what we should do [1,6,7]. The latter requires the application of evaluative thinking [7,10] by socially responsible, reflective, and active individual, group, and organizational participants in the STES-economic-political decision-making process [1–2,6–7,10], particularly in the context of the contemporary "stressed ecology" imperative.

The detergent industry is deliberate, steady, and mature, so its pattern of change is evolutionary, avoiding drastic step changes. In spite of gloomy economic forecasts, detergent sales are expected to continue increasing, both in dollar and physical volume, in the first decade of the new millennium, as new formulations providing better convenience for customers improve the value-added component of the products. Anionic surfactants still dominate world output and consumption, accounting in the United States for about two-thirds of the total, compared with about one-fourth of the nonionic detergents. The combined production of the United States, western Europe, and Japan, which amounts to about 60% of that of the entire world, is shared almost equally (~30%) between the first two, the rest being produced by Japan (about 6%). Although some differences–as far as market share is concerned–are apparent, the general pattern is quite similar worldwide. The laundry detergent segment dominates the market by far, comprising, together with the segment of dishwashing products, more than 75% of the market in the United States and western Europe and about two-thirds of the market in Japan. Not surprising, the annual growth in production/consumption in the last decade of the twentieth century (~2–3%) followed that of the GNP in these countries, and world production has been more or less stable in recent years at a level of around 22×10^6 tons. These facts are very pertinent to the issue of sustainable development and relationships, since, one way or another, following their use all kinds of detergent formulations' components and/or their degradation products find their way into man-made sewage systems and/or soils, natural surface water, and groundwater. Since (1) both world detergent production and consumption are expected to grow in the years to come and (2) world population relies on both surface water and groundwater (primarily the latter) as its primary sources of drinking water, detergent/surfactant distribution, persistence, and survival as well as their potential health risk in the environment constitute issues of major concern.

From the raw materials–natural resources perspective, detergent formulations are based on surfactants derived from petrochemicals and/or fats and oils. Additionally, they contain builders: sequestrants such as carbonates, phosphates, silicates, as well as oxidants and other ingredients.

Compared to other industries, the detergent industry recognized rather early the ecological challenge. Its voluntary (for the most part) switch, in the 1960s from the nonbiodegradable ('hard') anionic, branched-chain dodecylbenzene sulfonate (DDBS or ABS: alkyl-benzene sulfonate) to the substituting biodegradable linear alkylbenzene sulfonate (LABS or LAS) is remarkable. The subsequent large-scale (preregulation) switch from polyphosphates to, mainly, zeolites in laundry and dishwashing formulations is no less impressive. Since then, the detergent industry worldwide has been constantly confronted by one demand, the "minimization commandment": that detergent formulation be the very best that yields the desired effect with the least amount. This demand is quite obvious in view of the fact that surfactants and other components of detergent formulations constitute a significant portion of municipal sewage water profiles.

Ultimately, positive feedback–type relationships have been developed between environmental concerns and detergent formulation: The higher the public awareness of the former, the higher the "environmental acceptability" of the latter. Indeed, the current reformulation of detergent products reflects the response of the detergent industry to environmental regulatory as well as economic-technological and demographic social factors in an attempt to cope with the increased awareness of environmental concerns, the upswing in action against phosphate builders and the unclear future of other builder systems, the tight sewage treatment requirements, the higher demand for cost performance and added-value compositions associated with lowering of washing temperatures, the increasing share of washload held by synthetic textiles, the increasing demand for liquid formulations, "supereffective" or powdery "concentrates," and the pressure of the change in customer habits requiring efficient and convenient multipurpose time-saving processes.

The appropriate response of the detergent industry to these pressures required (1) an overall increase of surfactants at the expense of builders in formulations, with the nonionics gaining most of the increased share; (2) substitution of polyphosphates mainly by zeolites as well as other effective sequestering agents; and (3) higher concentrations of active components in multifunctional formulations effective in low-temperature processes [14].

A major outcome of the foregoing was a three-fold development:

1. A dramatic switch from heavy-duty powder laundry formulations to heavy-duty liquid (HDL) formulations, particularly in the United States, where the latter accounts for more than 40% of heavy-duty laundry detergent sales
2. A substantial reduction in the use of polyphosphates in detergent formulations, with concomitant replacement, partially or totally, by zeolites
3. The development and introduction, to the markets, of concentrated and/or multifunctional heavy-duty, low-suds laundry formulations for use at low temperatures and having extra detergency.

Currently, concern about the environment is leading the detergent industry to develop "environmentally friendly" products, which are increasingly being sold in recycled packaging material to meet regulatory requirements and satisfy customer demand. A case in point: the concentrated detergent formulations that are both more powerful and require less packaging material.

Thus, in the final analysis, the new products and modifications made within the basic formulations did make a difference as far as the environment is concerned. What is in store for us concerning the sustainable relationships—raw materials–surfactant/detergents production/formulation–usage/consumption environment—is contingent on the way that the relevant "guiding models" are conceptualized and ultimately implemented.

II. PARADIGM SHIFTS IN ENVIRONMENTAL SCIENCE, TECHNOLOGY RESEARCH, AND SUSTAINABILITY

Our so-called "global village/free global market/man-made world" requires a new type of flexible, contextually relevant, adaptive knowledge, followed by evaluative and decision-making action, in accord, in order to sustainably cope with complexity and the fragility of multidimensional socioeconomic technological/environmental systems. This implies the importance of inter- and transdisciplinarity in environmental research [5,15,16], appropriate research methodologies [16,17], as well as strategies for technology assessment in the context of sustainable action.

From the perspective of the environmental impact of human activity–sustainable action/development relationships, selected paradigm shifts that are currently taking place in environmental science, technology, research, and consequent action have been identified and are summarized in Table 1 [1]. In-depth systematic examination of these shifts reveals their pertinence and relevance to the systemic challenge of maintaining sustainable relationships as far as detergents and their environmental impact are concerned.

What are the implications with respect to sustainable detergent production and consumption–environmental relationships? Given environmental concerns such as the exploitation/use/consumption of natural resources and their environmental impact, particularly the potential health risk associated with persistent pollutants and/or their metabolites [3,12,17], it appears that sustainable development as well as maintaining a sustainable detergent–environmental relationship are imperative.

Being led by the sustainable development imperative, Table 1 provides the essence of the required shifts (some of them already implemented) in subjects, objectives, and methodologies of the environment-related scientific and technological research and development associated with detergent production, use/consumption, and disposal. Ensuring sustainable development requires, to begin with, a radical change in the environmental behavior (as well as 'thinking Environment') of individuals, institutions, industry, social organizations, politicians, and governments. This, in turn, requires reconceptualization of long-accepted relevant concepts and beliefs [9,13,14,18]. Thus, for example, the shift from the acceptance of new technologies to facilitating sustainable technologies in responding to society needs is substantially dependent on the shift from people's or customers' "wants" to people's needs. On the other hand, the technological feasibility of the economically and socially healthy shift may carry the seeds of contradiction with the shift from people's "wants" to people's needs if economics is the governing criteria. Similarly, a shift from the conceptualization of environmental science and technology as omnipotent to a recognition of their limits in solving pollution problems, preventing future environmental impact, and paving the way for sustainable development through appropriate design [9] has its clear implications (and consequences) as far as detergents and their environmental impact are concerned.

If the foregoing imperative paradigm shifts are about to be realized, then different quality criteria for research and practice in the sustainable development–environmental context become necessary. This is because not only do methodological disciplinary aspects have to be rethought and reevaluated, but critical questions or issues arise, such as societal and practical relevance as well as external validity, particularly with respect to the risks and potentials of only partly controllable

TABLE 1 Selected Paradigm Shifts in Environmental Science, Technology Research, and Consequent Action

From	To
A. Sustainable development–environment interrelationship	
• Technological, economic, and social growth at all cost	Sustainable development
• Increase in the competitive gap between countries, nations, societies	Increase in collaboration/cooperation and decrease in polarization
• People's "wants"	People's needs
• Passive consumption of "goods," culture, and education	Active participation/social action in the decision-making process
• Selection from among available alternatives	Generation of alternatives
• Selected environmental improvement on the local level at all costs	"Globalization" in ecoeffective/ efficient action
• Environmental ethics	Environmental sustainability-oriented "pragmatism"
B. Scientific and technological research and development	
• Corrective	Preventive
• Reductionism, i.e., dealing with in vitro isolated, highly controlled, decontextualized components	Uncontrolled, in vivo complex systems
• Compartmentalization	Comprehensiveness, "holism"
• Descriptive, as it is—"here and now"	(Attempted) Predictive models/modeling
• Disciplinarity	Problem-solving oriented, systemic, inter-/cross-/transdisciplinarity
• Technological feasibility	Economic-social feasibility
• Scientific inquiry (per se)	Social accountability and responsible and environmental soundness
• Technological development per se	Integrated technological development and assessment
• Convergent, self-centered	Divergent, interactive/reflective/ adaptive and related to different frames of reference

Source: Ref. 1.

variables and data of "in vivo" studies [1]. Further, different stakeholders, values, and perspectives will be involved and, consequently, integrated into the relevant science [13], technology, and development as well as the production/formulation/use and disposal processes related to the environment [19]. Clearly, the application of the identified paradigm shifts in the context of the environmental impact of detergents requires a corresponding paradigm shift in the related conceptualization.

III. RECONCEPTUALIZATION OF CONSENSUSLY ACCEPTED CONCEPTS

Given (1) the environmental imperatives and (2) the limited economic feasibility of many of even the most innovative/advanced technologies, the switch from the currently dominant corrective paradigm to the emerging preventive practice in production–development–consumption–disposal is unavoidable. This requires a revolutionary change in the guiding philosophy, rationale, and models of the surfactant/detergent industry concerning:

Customers wants or needs, at what cost?

The more producing/selling the better?

Raising the standard of living equals raising the quality of life?

Responding to market trends or generating/leading them?

Relying on disciplinary or transdisciplinary science research–based technology for rational management of the environment and sustainable development.

A case in point, to serve as an example: "We are committed to meet our customers' needs" is a currently dominant central concept. A clear distinction between customers' "wants" and customers' "needs" has to be made. The first led to overconsumption, which is not necessarily beneficial to the consumer and, in fact, is perpetually and aggressively being promoted an industry motivated by "growth and profits at all cost," with all the uncontrolled socioenvironmental consequences involved. The latter, however, should be targeted and responded to by a responsible, environmentally concerned detergent industry. Only an orientation to people's needs has the chance (albeit not guaranteed) to meaningfully contribute to sustainable development, not only in developing countries ("emerging markets") but also in highly developed Western countries. A "needs" orientation is, mainly, a promoter of quality of life, with a consumption-limiting potential. In contrast, a "wants" orientation is a promoter of standard of living, which is not only inconsistent with the existing trend of ever-increasing overconsumption, but in most cases further accelerates the pace of this trend. The environmental consequences of over-consumption are apparent [1,14,18].

With respect to science and technology, virtually any discussion concerning the current and future states of scientific and technological research and problem solving is typified by statements about the importance of enabling researchers and engineers to work seamlessly across disciplinary boundaries and by declarations that some of the most exciting problems, particularly the complex systemic environmental ones, span the disciplines. Moreover, transdisciplinary applied research evolves from real, complex problems in the interdisciplinary STES context, which are relevant to societies living in different environments. Such problems have no disciplinary algorithmic solutions or even resolutions. It is growing increasingly difficult to establish the transdisciplinary basis necessary for addressing complex environmental problems [1,13,17]. Therefore, the challenge for this kind of target-oriented research and technology development is to develop problem-solving methodologies that not only integrate different qualities and types of knowledge, but also envision researchers and engineers as an integral (nonobjective "insiders") part of the investigated, or to be remediated ("corrected"), system. Sustainable development via appropriate environmental management and industrial production, formulation, marketing, and business policies is, thus, highly dependent on transdisciplinary research and development in the STES context. This will facilitate transfer beyond the specific subject(s) or discipline(s) and, consequently (hopefully), a higher success in coping with previously unencountered complex problem situations [13,17].

In view of the compartmentalized disciplinary orientation in science and technology research and development (R&D) and the corrective approach in dealing with point and diffuse pollution problems in the different parts of the environment, the failure to strengthen the links between the social/behavioral sciences and advances in science and technology applied in different socioeconomic, cultural, and environmental contexts is of no surprise. However, the prevention approach to ensure environmental quality and restoration of ecosystems requires, most of all, appropriate and responsible environmental behavior and action on the part of both producers and customers, which, in turn, is contingent on an adequate environmental education [20]. This implies an urgent need for strengthening the social and educational components within the corrective-to-preventive paradigm shift process concerning the sustainable management of, and maintaining system-sustainable relationships in, our environment. Therefore, a major goal of sustainable development–promoting science and technology, research, education, and training at all levels should be the development of students' higher-order cognitive skills (HOCS) in the context of both the specific content and processes of science and the processes/interrelationships related to interdisciplinary societal, economic, scientific, technological, and environmental issues. The expected resultant critical thinking and interdisciplinary transfer capabilities mean rational, logical, reflective, and evaluative thinking in terms of what to accept (or reject) and what to believe in, followed by a decision—what to do (or not to do) about it and responsible action-taking. Thus, any meaningful response to meet the challenge of sustainable development requires transdisciplinarity, essentially by definition, that is, the development and implementation of policies and cross-disciplinary methodologies, which can lead to the changes in behavior–of individuals, industries, organizations, and governments–that will allow development and growth to take place within the limits set by ecological imperatives. The educational challenge is rather clear. It is a precondition for the required reconceptualization, which, in turn, will ensure sustainable development and growth.

The detergent industry is a representative case in point; e.g., the phosphates-euthrophication issue (transdisciplinarity prevention) and the recent growth of the personal care ethnic markets in the United States: "These ethnic consumers not only can afford to buy their share . . . of cosmetics . . . but they also want products formulated to meet their needs" [21]. The sociobehavioral consumption economics and environmental links are apparent.

Four recent pertinent publications, two more general and two more specific, deal with the surfactants–environment–health relationship issue and can serve to illustrate the importance of reconceptualization in the context of the environmental system's challenge that we are confronting.

1. It is claimed that since major environmental pollutants are coming under the control of regulatory authorities, this part of ecotoxicology is more or less completed, although there is still work, not expected to call for major scientific innovation and discovery, remaining to be done. It is concluded that the merger between ecotoxicology and ecology would give rise to a new science, *stress ecology*, at the crossroads of ecology, genomics, and bioinformatrics [13].
2. Given that the public, many policymakers, and some environmental professionals believe that science and technology can solve most pollution problems,

prevent future (undesirable) environmental impacts, and pave the way for sustainable development, this paper makes the claim that since science and technology alone cannot meet the challenge of sustainable development, we all should recognize the limits (reconceptualize the problem-solving capability) of environmental science and technology [9].

3. A recent publication by CLER (Council for LAB/LAS Environmental Research) and ECOSOL (European Council on Studies on LAB/LAS) [21] reports that their revised dossier and assessment report on LAS, which reviews the considerable data on LAS, showing that LAS is safe for human health and the environment, was submitted to the U.S. EPA and recommends that no further testing is needed and that the EPA agreed that there is no need for further studies [21].

4. In contrast, following the completion of a comprehensive interdisciplinary, longitudinal EU-sponsored research program (COMPREHEND), on environmental hormones and endocrine disruptors, including APEOs, the investigators conclude that, although their studies did demonstrate impacts and endocrine disruption in fish exposed to environmentally realistic levels of estrogenic substances, the question of deleterious impacts of estrogenic effluents on fish populations is, as yet, one of the most important remaining to be answered [17].

Do the apparent different approaches to a similar (although, obviously, not identical) environmental issue in the detergent–environment–sustainable relationships context represent different (contradictory?) conceptualizations of the issue(s) at hand? This question and the response to it remain open.

IV. SUMMARY AND CONCLUSIONS: MEETING THE CHALLENGE OF DETERGENT–ENVIRONMENT– SUSTAINABLE RELATIONSHIPS

Sustainability is an enormous challenge, particularly in the STES context. Most problems and issues boil down to: Who does what, for what, at what price, at the expense of whom (or what), and in what order of priorities? The widely agreed-upon call for sustainable development requires rational hard choices to be made between either available or to-be-generated options [7]. This poses an even greater challenge to science, technology, and education for sustainability, whatever that means. This is so because dealing effectively and responsively with complex interdisciplinary problems within complex systems in the context of STES interfaces requires evaluative thinking and the application of value judgment by technologically, environmentally, and sociologically (i.e., STES-) literate, rational scientists, engineers, and citizens within a continuous process of critical thinking, problem solving, and decision making [6,10,22].

This implies an urgent need to strengthen the HOCS-promoting components of STES-oriented education within the corrective-to-preventive paradigm shift process concerning the sustainable management of our environment [1,10,22].

The expected resultant critical thinking and interdisciplinary transfer capabilities mean a rational, logical, reflective, and evaluative thinking in terms of what to accept (or reject) and what to believe in, followed by a decision — what to do (or not to do) about it and taking responsible action accordingly. Thus, any meaningful response to the current leading challenge of sustainability requires transdisciplinarity in environmental science, technology, research, and action [1].

It follows, then, that what we are dealing with is not just a simple matter of economics that the free "market forces" (which, incidentally, are not God's creation but, rather, changeable, people-made, and people-controlled) will take care of. Rather, we are dealing with an array of very complicated problems within a complex system, the components of which are natural, man-made, and human environments and their related subsystems. Most of these problems have no "right" solutions (definitely not algorithmic), but rather resolutions that can be worked out via the use of appropriate methodologies, simultaneously guided by a sustainable development–oriented value system. The disciplinary/correction–transdisciplinary/prevention paradigm shift concerning environmental issues, the initial steps of which we are currently witnessing, is crucial for both sustainable development and our survival on planet Earth. As far as the chemical and detergent industries are concerned, this reconceptualization-based paradigm shift has to be translated into "sustainable action," in terms of raw materials to be used, "green" production in accord with needs rather than wants, economic feasibility/cost–benefit–profit with respect to all involved, marketing (where, when, and to whom), given the local particular realities of constraints, risk assessment (methodologies and criteria applied), and environmental compatibility [14,18].

Can we meet the systemic challenge of sustainable detergents–environment relationships?

The evolutionary pattern of change in the deliberate and steady detergent industry can serve as a test case for a reasonable response, by taking a historical perspective: the switch from DDBS to LABS, the continuing use of the (potentially estrogenic?) branched-chain nonylphenol-based nonionic ethoxylates, the polyphosphates-to-zeolites switch, the recent extensive use of enzymes, the development of the activators/perborate bleach systems, the current switch (in the United States) from horizontal to vertical drums in washing (laundry) machines, the proliferation of personal-care products, and many other innovations. Whether or not each of these is consonant with the new sustainable development–oriented criteria and in line with the paradigm shift in the STES context remains an open question. It is up to each of us, following our own a evaluative thinking, conceptualization, and assessment process, to respond. Can we meet the challenge? Are we getting it right? Then we should act accordingly and take responsibility, each in her or his environmentally related milieu.

This Part B of the *Handbook of Detergents*: *Environmental Impact* deals, from different perspectives, with the relevant issues involved.

REFERENCES

1. Zoller, U.; Scholz, R.W. Environ. Sci. Technol. 2003. *submitted.*
2. Zoller, U. Environ. Sci. Pollut. Res. 2000, *7*, 63–65.
3. Zoller, U.; Plaut, I.; Hushan, M. Wa. Sci. Technol. 2003 *submitted.*
4. Glaze, W.H. Environ. Sci. Techno. 2002, *36* (23), 438A–439A.
5. Gibbons, M.; Nowotny, H. In *Transdisciplinarity*: *Joint Problem Solving Among Science, Technology, and Society. An Effective way for Managing Complexity*; Klein, J.T., Grossenbacher-Mansuy, W., Haberli, R., Bill, A., Scholz, R.W., Welti, M., Eds.; Birkhäuser: Basel, 2001; 67–80.

6. Zoller, U. Higher Educ. Europ. 1990, *15* (4), 195–197.
7. Zoller, U. Environ. Sci. Pollut. Res. 2001, *8* (1), 1–4.
8. Negroponte, N. Tech. Rev. 2003, *106* (1), 33–35.
9. Huesemann, M.H. Environ. Sci. Technol. 2003, *37* (13), 259A–261A.
10. Zoller, U. J. Chem. Educ. 1993, *70* (3), 195–197.
11. Glaze, W.H. Environ. Sci. Technol. 2002, *36* (17), 337A pp.
12. Schnoor, J.L. Environ. Sci. Technol. 2003, *37* (7), 119A pp.
13. Van Straalen, N.M. Environ. Sci. Technol. 2003, *37* (17), 324A–330A.
14. Zoller, U. Chem. Educ. 1993, *70* (3), 195–197.
15. Zoller, U. Environ. Sci. Pollut. Res. 1999, *7* (2), 63–65.
16. Scholz, R.W., Marks, D., Klein, J.T., Grossenbacher-Mansuy, W., Häberli, R., Bill, A., Scholz, R.W., Welti, M., Eds.; *Transdisciplinarity: Joint Problem Solving among Science, Technology and Society*; Birkjhäuser: Basel, 2001; 236–252.
17. Pikering, A.S.; Sumpter, J.P. Environ. Sci. Technol. 2003, *37* (17), 331A–336A.
18. Zoller, U. Proceedings of the 5th World Surfactants Congress, Cesio, Firenze, May 29–June 2, 2000; 2, 15766–1571.
19. Bill, A.; Oetliker, S.; Thompson-Klein, J. In *Transdisciplinarity: Joint Problem Solving among Science, Technology and Society. An Effective Way for Managing Complexity*; Thompson-Klein, J., Grossenbacher-Mansuy, W., Häberli, R., Bill, A., Scholz, R.W., Welti, M., Eds.; Birkhäuser: Basel, 2001; 25–34.
20. Keiny, S., Zoller, U., Eds.; *Conceptual Issues in Environmental Education*; Peter Lang: New York, 1991.
21. Reisch, M.S. Chem. Eng. News, Apr 12; 19–24.
22. Zoller, U. J. Coll. Sci. Teach. 1999, *26* (9), 409–414.

2

Detergents and the Environment Historical Review

MARCEL FRIEDMAN   Neca-Agis Group, Petach-Tikva, Israel

I. INTRODUCTION

Soaps and detergents are essential products that safeguard our health. They belong to that group of consumer products that are indispensable for the maintenance of cleanliness, health, and hygiene. It has been said that the amount of soap consumed in a country is a reliable measure of its civilization. The increase in per capita consumption of soap and detergents in various countries was found to correlate well with life span.

Cleanliness is essential to our well-being. A clean body, a clean home, and a clean environment are the norm of today and a general concern shared by everyone. "Cleanliness is next to godliness" was the ultimate historical religious praise of physical cleanliness leading to spiritual purity. Paraphrasing it, cleanliness was, throughout history, next to environment. For thousands of years soaps, and, in the last century the synthetic detergents, followed by more complex washing and cleaning products, were the blessed way to get it. The use of soaps and detergents always led to a significant contribution to the modern quality of life, the close environment always being part of this. However, in the past 50 years a new dimension to this obvious positive symbiosis has been imposed, and a long-unfinished detergents–environment debate opened.

The detergent industry has faced dramatic changes since the early 1980s, developing into an interdisciplinary, multidimensional enterprise trying to cope with scientific/technological/ecological/toxicological/social/economic/political constraints. At the outset of the new millenium, the detergent industry is focused on coping with four challenges: economics, safety and environment, technology, and consumer requirements. The products must not just meet consumer needs for quality and efficacy, but must be dangerous neither to manufacture nor to use and must in no way have a detrimental impact on the user's health. The products and their packaging should not accumulate in the environment and shift or harm the ecological balance.

Not only the products but also washing habits are changing. In an age of growing environmental concern, a change of attitude toward the washing process has taken place in many countries. Nowadays, the consumption of energy, water, and chemicals is faced not only from economic but also from ecological angles. As a consequence, new raw materials, washing processes, laundry practices, and cleaning technologies have been developed, with the common challenge to use carefully the limited resources of the earth, to exploit the renewable ones, and to prevent environment pollution as much as possible.

The development of surfactants and detergents over the past decades has been affected tremendously by their environmental acceptability. The challenge for the future is to meet the most modern risk assessment approaches.

The aim of this chapter is to review this detergents–environment interrelated development frame throughout history.

II. DETERGENTS, WASTEWATER, ENVIRONMENT

The complex relationship between surfactants/detergents/cleaning products on one hand and wastewater, sewage treatment, surface water, and the environment on the other became the basis for most of the ecological issues, leading to a twofold approach toward their solution.

First, surfactants and phosphates, as the main components of detergent formulations and cleaning products, have been the subject of longstanding and ongoing detergent regulation and legislation. Second, efficient management has been imposed on sewage treatment, which is being updated continuously. A better understanding of this multidimensional input/output interaction is needed.

During the washing process, detergent components are released to the wastewater stream, to become a potentially undesirable troublemaker in sewage treatment plants and in the environment. Wastewaters vary considerably in composition and concentration and hence in their environmental impact. These differences are geographically dependent and arise partly due to differences in laundry habits and soil levels and partly to the composition and amount of the detergent used. The environmental impact further depends on the specific ecological requirements and public awareness in each particular geographical area. Also, household wastewater, generated mainly from laundry and personal care products, differs from that from industrial and institutional outlets.

Typical sewage treatment systems in countries with developed environmental protection include intensive biochemical and physical degradation processes that bring about the elimination of the pollutants. The extent of elimination depends on sludge levels, aeration efficacy, and residence time. In addition to bacterial metabolic reactions, physicochemical processes, such as adsorption on sewage sludge, contribute to the reduction of pollutant levels [1].

Variations in sludge loading and in peak loads of the wastewater are moderated considerably by the predilution of household wastewater in the public sewage system. Wastewater from industrial sources is generally pretreated by pH adjustment and physical separation in on-site sewage plant prior to discharge to municipal sewage treatment plants. Sometimes a well-designed treatment plant on an industrial site may permit direct discharge.

The sewage treatment effluents are discharged in the surface waters. Quality requirements for these effluents depend upon the intended use of the surface waters. The ultimate use, such as for drinking water, agriculture, or recreation, also governs the technical basis of the wastewater treatment as well as the intensity of the sewage treatment process. Phosphorous and nitrogen elimination normally require additional treatment steps.

The sewage treatment effluent is diluted in surface waters. The dilution factor varies according to the geographical place. Human waste and some remaining traces of surfactants in the surface waters are further biologically treated by a self-cleaning process [2].

The efficacy of a sewage treatment is evaluated by different parameters, such as

- BOD (biochemical oxygen demand), which registers the biodegradable organic matter present
- COD (chemical oxygen demand)
- TOC (total organic carbon)

However, the most significant parameter is the direct measurement of the pollutant levels, mainly surfactant and phosphate concentrations.

Highly sensitive analytical test methods have been developed for the accurate determination of surfactant concentration [3,4]. Anionic surfactants are determined as "methylene blue–active substances" (MBAS) by a method based on a modified Epton two-phase titration. Nonionic surfactants are determined as bismuth-active substances (BiAS) after passage through cation and anion exchange columns.

Cationic surfactants are determined as "disulfide blue–active substances" (DSBAS).

Strict implementation of detergent regulations combined with effective sewage treatment hes led to low surfactant concentrations in large rivers, such as those in the Rhine, which currently are as follows [2]:

Anionic surfactants, about 0.05 mg/L MBAS (linear alkylbenzene sulfonates)
Anionic surfactants, less than 0.01 mg/L LAS
Nonionic surfactants, less than 0.01 mg/L BiAS
Cationic surfactants, less than 0.01 mg/L DSBAS

The sewage treatment and surfactant levels just referred to represent, more or less, the present state of the art, which should come close to an appropriate resolution to the ecological threat of surface contamination. It has taken some 40 years to reach the present state from the time when rivers all over began to foam and gave rise to the First Detergent Law.

III. SURFACTANTS REGULATION (1950–1980)

A. First Surfactants' Environmental Complication

As early as 1952, sewage treatment problems were observed in the UK. Foam on rivers was increasing, and tap water, drawn from wells located close to household discharge points, also tended to foam. In 1959, Germany encountered similar difficulties when foam formed on German rivers and stable foam layers developed downstream from dams.

The sewage treatment of the time, based on physicochemical separations and some biological treatment, was not able to cope with the surfactant load. The impact on sewage treatment was immediate and significant. The efficacy of the sedimentation process was reduced, leaving a high dispersion of suspended solids in the treatment plant. The few biological sewage treatment plants operating with an activated sludge process collapsed, producing foam layers several meters high above the aeration tanks.

Increased surfactant concentrations were found not only in sewage waters and rivers, but also, as a result of soil infiltration, in the groundwater. As a result, the drinking water supply was contaminated. The anionic surfactant content of drinking water from the Ruhr River increased to as much as 1.73 mg/L, while the required limit was nil [2]. During 1958–1960, increasing surfactant consumption in the United States led to similar ecological problems. The German market of 1960 consumed about 80,000 tons/year of surfactants. DDBS accounted for more than 80%, while nonionics and cationics added up to 15% and 5%, respectively, of market volume. Municipal sewage treatment plants could not cope with typical influent concentrations of about 20 mg/L MBAS on peak washing days. At a maximum reduction of 25% of the influent, effluents were released into surface waters with concentrations as high as 16 mg/L [2].

It was soon understood that these serious problems for water management were due to the poor biodegradation profile of DDBS. The abnormal quantities of foam were attributed to the presence of DDBS, which, in turn, was the result of incomplete biodegradation of propylene-based alkylbenzenesulfonates by the natural bacteria present in effluents. The branched-chain structure of alkylbenzene seemed to hinder attack by the bacteria. Supporting evidence for this judgment was provided by the facile degradation of fatty alcohol sulfates and soap. Both are derived from straight-chain fatty acids, suggesting that a straight-chain, linear alkylbenzene might also be degradable.

B. First Detergent Law on Surfactants in Detergents and Cleaning Products (1961)

In 1960, Germany established the Main Committee on Detergents, which elaborated in September 5, 1961, the first piece of detergent legislation, known as *First Detergent Law* [1,5]. This law imposed a strict requirement of a minimum of 80% biodegradability for all the surfactants (anionic, nonionic, cationic, amphoteric). However, a 1962 directive provided a dynamic test method according to a specified test protocol only for control of anionic surfactants.

By this biodegradability test, the straight-chain, linear alkylbenzenesulfonates (LAS or LABS) were found to be readily biodegradable and, as a result of the strict legislation, replaced totally the branched-chain dodecylbenzenesulfonates (ABS or DDBS):

DDBS (ABS): $CH_3[CH(CH_3)CH_2]_3CH(CH_3)_4C_6H_4SO_3Na$
LABS (LAS): $CH_3(CH_2)_{11}C_6H_4SO_3Na$

Beginning October 1, 1964, only those surfactants with a biodegradability of over 80% were allowed to be incorporated into detergents. During the short time between mid-1964 and 1965, known as the "conversion period," the foaming problems in surface waters and sewage treatment were solved. The degradation/elimination rate in sewage treatment plants, which was 19–25% before the First Detergent Law, increased to 60%. The effects of the conversion to biodegradable surfactants were similar on surface waters and rivers, with a decrease in surfactant concentration of up to 77% having been measured [2].

The conversion from "hard" to "soft" surfactants was also legislated during 1963–1964 in the United States, as amendments to the Federal Water Pollution Control Act, creating new water pollution control standards [6]. However, none of these measures was implemented. The problem was solved in 1966 by the detergent industry, which agreed voluntarily to switch over from ABS to LAS. Similar regulations and voluntary agreements were put into place during the late 1960s and early '70s in several countries of Western Europe and in Brazil and Japan.

LABS was made commercially available in 1966. The manner in which the DDBS problem was solved is an excellent example of environmental improvements achieved by cooperation of government, industry, and science. However, the detergent industry faced advantages and disadvantages. The change to LABS offered better detergency in heavy-duty formulations and lower cloud points and viscosities in pastes and slurries. But, on the other hand, while a lower viscosity in slurries offered an advantage for a spray-dry process, the liquid and paste LABS detergent of lower viscosity looked less appealing to the consumer. Also, the LABS powders became sticky and were less free flowing [7].

It was found that the actual isomer distribution of the linear alkylate has an effect on the stickiness of the powder, identifying the 2-phenyl isomer as giving the greatest tendency to stickiness. The different phenyl isomers are obtained when, during alkylation, the benzene molecules attach to the different carbons along the alkyl chain. For instance, an attachment at the second carbon of the alkyl chain gives a 2-phenyl isomer. It was found that the phenyl isomer distribution depends on the catalyst used during alkylation. Aluminum chloride ($AlCl_3$) alkylation gives a "high 2-phenyl" distribution (29%), while hydrofluoric acid (HF) catalysis results in a "low 2-phenyl" distribution (19%) [8].

This catalytic versatility in LAB production, as well as additives further developed, overcame most of the formulation problems.. However, in the case of solid laundry bars the lower viscosity and the less bulky molecular structure of LABS provided a softer bar hardness and a stickier appearance. This disadvantage could hardly be overcome for highly concentrated (above 30% active) bars.

C. Surfactant Biodegradation

Biodegradation is the process by which microorganisms in the environment convert complex materials into simpler compounds that are used as food for energy and growth. Biodegradation of the surfactants used in detergents is important because of the large volumes used worldwide and, of course, the detrimental toxic effects on the aqueous and soil environments.

Biodegradation is a multistep process that starts with the transformation of the parent compound into a first degradation product (primary degradation) and leading, ultimately, to mineralization products (carbon dioxide, water) and bacterial biomass (ultimate or total degradation). A typical surfactant biodegradation is illustrated by the linear alkylbenzenesulfonate (LAS) biodegradation path in Figure 1 [9].

A good understating of past and present biodegradation issues requires precise definitions of biodegradability terms [5,10,11].

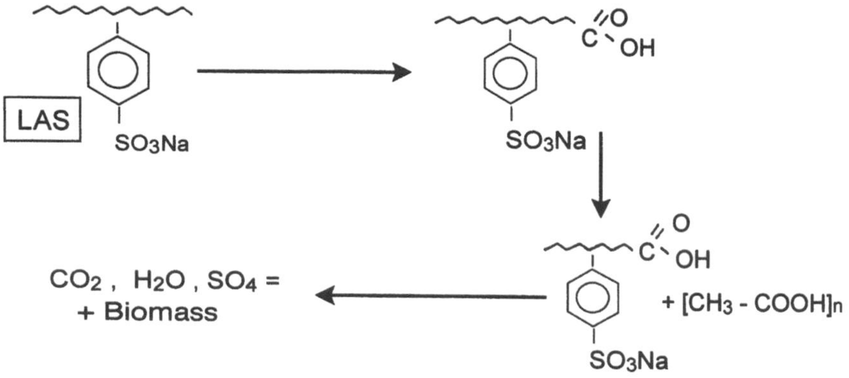

FIG. 1 Surfactant (LAS) biodegradation. (From Ref. 9.)

Primary biodegradability is the change in the chemical structure of an organic substance, resulting from a biological action that causes the loss of the specific chemical and physical properties of the substance. When this stage of biodegradation is reached, the remaining material is no longer a surfactant; it no longer has any surface-active properties, including the ability to foam.

Ultimate biodegradability in the presence of oxygen (aerobic conditions) represents the total level of degradation by which a test substance is consumed by microorganisms to produce carbon dioxide, water, mineral salts, and constituents of microbe cells (biomass).

Ready biodegradability is an arbitrary classification for chemical compounds that satisfy immediate biodegradability tests. The severity of the tests (biodegradation and acclimation time) ensures that such compounds will degrade quickly and completely in an aquatic environment under aerobic conditions.

Intrinsic/inherent biodegradability relates to chemical compounds that unambiguously undergo primary or ultimate biodegradation in any biodegradability test.

Anaerobic biodegradability: Most biodegradation processes take place in the presence of oxygen (aerobic conditions). However, biodegradation also proceeds in the absence of oxygen in anaerobic environments, albeit at slower rate. Anaerobic media are known as either anoxic (in which the rate of oxygen consumption exceeds the rate of oxygen diffusion) or strictly anaerobic (in which the oxygen is totally absent).

Because of concerns about the presence of detergent ingredients in all parts of the environment, anaerobic biodegradability has been proposed as the criterion for several Eco-label requirements. Experimentally, LAS was found to pose no risk to anaerobic environments [12].

D. Biodegradability Requirements and Test Methods (1973–1982)

Germany's First Detergent Law required a minimum of 80% primary biodegradability for the total separated anionic and/or nonionic surfactants present in a commercial detergent. Separation of anionics and nonionics from a detergent composition is required prior to biological testing to eliminate artifacts or interfering interactions. The results are reported as % (loss) MBAS for anionics and % (loss) BiAS for nonionic. The disappearance of the specific analytical species corresponds to the loss of signficant ecological surface activity.

Concurrently with the legal biodegradability requirements, specific test methods for measurement of biodegradability of synthetic anionic and nonionic surfactants in laundry detergents and cleaners were elaborated and approved. They are based mainly on the quantity of consumed oxygen and the disappearance of dissolved organic carbon (DOC).

In 1973, two biological tests were approved and mandated by the OECD (Organization for Economic Cooperation and Development) for establishing biodegradability [13,14]:

1. OECD Screening Test
2. OECD Confirmatory Test

The detergents and/or anionic and nonionic surfactants are considered biodegradable "only if the analysis shows a level of biodegradability higher than 80%" as given under laboratory conditions of the Screening Test. This equates to a real-life biodegradability of at least 90%. If the level of biodegradability is lower or if the results are in about, a subsequent Confirmatory Test is required. The results of this are decisive and definitive. The Confirmatory Test is specified as a reference test in the 73/405 Council Directive of the EEC (European Economic Community) of Nov. 1973 [13,14].

The revised edition of this directive was officially implemented in 1977 [2]. This edition approved, for screening purposes, the OECD method (published June 1976) and the methods used by the member states, such as the French AFNOR test published by Association Française de Normalization in 1981 and the British "Porous pot test" (described in a technical report of the Water Research Center) [15,16]. The final European Market directives were issued in 1982 for anionics and nonionics as Directive 82/243/EEC, amending Directive 73/405/EEC, and Directive 82/242/EEC, amending Directive 73/404/EEC [17].

The ecological concerns of the 1970s are reflected in the following quotation from Council Directive 73/404/EEC [13]:

> The pollutant effects of detergents on waters, namely the formation of foam in large quantities, restricts contact between water and air, renders oxygenation difficult, causes inconvenience to navigation, impairs the photosynthesis necessary to the life of aquatic flora, exercises unfavorable influence on the various stages of processes for the purification of wastewaters, causes damage to wastewater purification plants, and constitutes an indirect microbiological risk due to the possible transference of bacteria and viruses.

The OECD screening test is based on a static shake flask method and corresponds to surface water conditions. In this procedure a mineral salt solution is inoculated and incubated with 5 mg/L MBAS or BiAS as the only source of carbon. The MBAS or BiAS losses are compared with two reference surfactants: the readily biodegradable LAS (92% loss) and poorly biodegradable DDBS (less than 35% loss). The measurements are made at fixed intervals up to 19 days. According to this screening test, primary biodegradation (given by % MBAS/BiAS removal) of several anionic surfactants, such as C14–18 α-olefinesulfonates, C16–18 fatty alcohol sulfates,

C12–13 oxo alcohol sulfates, and C16–18 α-sulfo fatty acid methyl esters, was found to be 99%, while LABS and DDBS showed a primary biodegradation of 95% and 8–25%, respectively.

Under the same test conditions, nonionic surfactants, such as C12–14 fatty alcohols 30EO and C16–18 fatty alcohols 14EO, showed 99% biodegrability, while C13–15 oxo alcohols 7EO and C8–10 *n*-alkylphenols 9EO were found to biodegrade to 93% and 84%, respectively [1].

The OECD Confirmatory Test, known also as simulation test, is a continuous procedure run under more realistic environmental conditions, simulating activated sludge plants, as shown in Figure 2 [14]. The Confirmatory Test can simulate several types of environment, such as lake, sea, and land, and can be run under aerobic or anaerobic conditions conditions [10]. In this procedure a solution of MBAS (20mg/L) or BiAS (10 mg/L) in a synthetic sewage water (containing 110 mg/L meat extract, 160 mg/L peptone, 30 mg/L urea, 7 mg/L NaCl, 4 mg/L $CaCl_2 \cdot 2H_2O$, and 2 mg/L $MgSO_4 \cdot 7H_2O$) is continuously fed into the plant model during a retention time of 3 hours. After inoculation of the test system and growth of the activated sludge, an acclimation period is run, following a predetermined procedure. After a minimum 14 days, the degradation rate reaches a plateau for readily biodegradable surfactants, while an irregular curve, with ups and downs of low biodegradation rate, is shown by "hard" surfactants. This initial period is followed by a 21-day evaluation period in which the high-biodegradation-rate plateau is maintained by the readily biodegradable substance.

Typical primary surfactant biodegradations found by the OECD Confirmatory Test as MBAS/BiAS/DSBAS for anionic/nonionic/cationic surfactants are given in Table 1 [11].

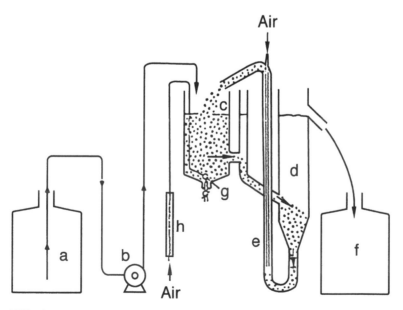

FIG. 2 OECD Confirmatory Test: (a) Storage container; (b) dosage pump; (c) activated sludge vessel (capacity, 3 L); (d) separation unit; (e) air lift; (f) collection unit; (g) aeration unit; (h) air flow meter. (From Ref. 14.)

TABLE 1 Surfactant Biodegradation in Sewage Treatment Plant Models

Surfactant	Primary biodegradation, OECD Confirmatory Test, (% MBAS/BiAS/DSABS removal)	Ultimate biodegradation, Coupled Units Test, (%C removal)
LABS	90–95	73 ± 6
DDBS	36	41 ± 9[a]
C12–14 Fatty alcohol sulfate	99	97 ± 7
C16–18 Fatty alcohol 10 EO	98	90 ± 16
C13–15 Oxo alcohols 12 EO	96	59 ± 6
C 8–10 n-Alkyl phenols 9EO	96	68 ± 3
Cetyltrimethylammonium bromide (CTAB)	98	104 ± 6
Distearyldimethylammonium chloride (DSDMAC)	94	108 ± 9

[a] Checked by COD (chemical oxygen demand).
Source: Ref. 1.

A modification of OECD Confirmatory Test, known as the *Coupled Units Test*, affords determination of ultimate biodegradation until total degradation. In this test two units are run in parallel. One is fed with the test compound at 20 mg/L concentration in the synthetic sewage, while the blank unit is fed only with synthetic sewage. The ultimate biodegradation is represented by carbon removal (C%) or chemical oxygen demand (% COD). Typical results are presented in Table 1 [1].

E. Effect of Surfactant Regulation on Sewage and Surface Water Load (1980–2000)

The implementation of the surfactant regulation solved most of the significant ecological problems in Europe, the United States, and Japan with respect to residual concentration of surfactants in sewage effluents and surface waters.

The order of magnitude of the contribution of laundry detergents and other cleansing agents to the sewage surfactant load in Germany in 1980 has been documented comprehensively [1,18,19]. In these publications, the average theoretical concentrations (mg/L) of the total detergents and different surfactant categories were calculated. This estimate was based on a 200-liter daily water consumption per capita and took into account the 1980 detergent production in Germany as 750,000 tons. These figures translate into a 33.3 g daily per capita consumption of laundry detergents and an average concentration in the municipal sewage of 167 mg/L. Based on 1980 production volumes of 151,000 tons of anionic, 91,700 tons of nonionic, and 26,100 tons of cationic surfactants, the daily per capita consumption of each group has been calculated as 6.71 g, 4.07 g, and 1.16 g, respectively. These figures lead to a calculated average concentration of 33.5 mg/L, 20.3 mg/L, and 5.8 mg/L of anionic, nonionic, and cationic surfactants, respectively, in municipal sewage. Assuming that 70% of the total anionic and nonionic surfactants comes from laundry detergents and cleansing agents while about 90% of the cationic surfactants derive from fabric softeners, their contribution to municipal sewage is 23.5 mg/L for anionics, 14.2 mg/L for nonionics, and 5.2 mg/L for cationics.

The surfactant concentration in municipal sewage has been checked analytically and found to correspond on average to the theoretical calculated values. Thus, the anionic surfactant concentration was found to be 5–35 mg/L, while that of nonionics was 2–25 mg/L.

Comprehensive and well-documented data based on extensive investigation of 140 measuring points for anionic surfactants and 20 measuring points for nonionic surfactants were summarized in the late 1980s for the main German rivers. The average concentrations were as follows [2]:

Anionic surfactants: ≤0.07 mg/L of MBAS
Anionic surfactants: ≤0.02 mg/L of LAS
Nonionic surfactants: ≤0.01 mg/L of BiAS.

In the Rhine, which prior to 1989 was collecting about 60% of all wastewater in Germany, even lower surfactant concentrations were reported [2]:

Anionic surfactants: ≤0.05 mg/L of MBAS
Anionic surfactants: <0.01 mg/L of LAS
Nonionic surfactants: <0.01 mg/L of BiAS
Cationic surfactants: <0.07 mg/L of DSBAS

These results also represent the average concentrations in the large rivers of Germany. The Rhine Water Quality Report of 1988 took note of these remarkable improvements and clearly positioned them as an interim results in the Environmental Protection Program of the German Federal Government.

The pragmatic decision of the early 1960s on the manner of solving the surfactant issue served as the model for this interdisciplinary Environmental Protection Program, which was announced in 1971 and fully achieved its defined "waters" goals in the late 1980s [2].

Several other reports measured water concentrations of LAS during the two decades, finding them in good agreement with model predictions [6,20]. Extensive references and documented reviews have been published on biodegradation of surfactant in different water media. A few selected examples are noted next. As a part of an AISE/CESIO (Association Internationale de la Savonnerie de la Detergence et des Produits d'Entretien/Comité Europeen des agents de Surface et de leurs Inter-medieres Organiques) environmental surfactant monitoring program, initiated by the European detergent and surfactant industries, a 7-day pilot study of LAS degradation was conducted on a UK sewage treatment plant. Organic load elimination of 98% as BOD and 99.9% of LAS were achieved [6,20].

Since the early 1990s, typical concentrations of anionic and nonionic surfactants in sewage influents in Europe and Israel were 9–11 mg/L and 0.8–2.5 mg/L, respectively [21–23]. Recent monitoring studies of 50 river sites just below wastewater treatment plants determined average LAS concentrations as 35 ppb; while measurements of 500,000 U.S. river miles found less than 4 ppb [21,24].

A representative survey of nonionic surfactants concentration in wastewater influents and effluents in the United States, Europe, and Israel during 1977–1995 was compiled by Zoller [21,25] and is presented in Table 2 [21]. The decrease in nonionic levels in U.S. effluent during these years is evidence for the biodegradability and efficient sewage treatment of the nonionics, mostly alcohol ethoxylates. In Israel, on

TABLE 2 Typical Concentration of Nonionic Surfactants in Wastewater Influents and Effluents in the United States, Europe, and Israel

Location	Year	Concentration (mg/L)		
		Influent	Effluent	Receiving waters
United States	1977	4.4	1.4	0.3
Switzerland	1983	1.12	0.369	n.a.
Switzerland	1983	1.11	0.040	n.a.
United States	1983	1.62	0.18	n.a.
Germany	1984	6.8	0.19	n.a.
Israel	1984	1.2	0.25	1.93
United States	1988	2.27	0.088	n.a.
Israel	1990	7.04	0.30	0.025
United States	1990	4.7–12.2	0.17–0.25	
United States	1991	1.335	0.064	
United States	1990–1991			0.002–0.015
United States	1995		<0.002–1.30	<0.002–0.014

Source: Ref. 21.

the other hand, where the use of alkylphenol ethoxylates (APEO) is permitted, the nonionic influent concentrations were even higher in 1990 (as the share of nonionics in compact detergent formulations increased) and hence also slightly higher were concentrations in effluent waters.

A study on biodegradation of anionic surfactants in the United States checked the biodegradation with the presence of petroleum pollutants. Under experimental conditions and concentrations, a synergism has been reported in the biodegradability of kerosene and surfactants [6,26]. The synergistic biodegradation behavior of a binary or a multiple-component system has been noted in another publication, recalling also the performance benefits and industrial applications of such systems [27].

F. Surfactants: Ecotoxicological Profile (1980–2000)

Water protection demands on surfactants refer mainly to biodegradability and aquatic toxicity. The latter is given by LC_{50} values—the concentration that causes 50% mortality.

The following objectives and requirements were defined by Malz [2]:

1. Achieve and stabilize water quality.
2. Protect the food chain and the variety of aquatic species.
3. Protect the drinking water supply from surface waters.
4. Protect the estuaries, tidelands, and groundwater.

According to Malz [2], attainment of these objectives required, in additon to product regulations, consideration of a number of improvements in sewage treatment, including lower sludge loading, extended residence times, multistep processes, nitrification–denitrification, phosphorus elimination, and pretreatment in the case of indirect wastewater discharge.

In 1986 the Main Committee on Detergents in Germany published the basic requirements of the ecotoxicological profile of the surfactants. The ecorelevant properties of different surfactants related to these requirements are presented in Table 3. It can be seen that the basic requirements of a minimum 80% surfactant degradation/elimination, 60% organic carbon degradation, and minimum LC_{50} values of 1 mg/L are easily surpassed by most surfactants. Based on the experience gained in the 1990s, the requirements have been revised and made more severe, meaning: a minimum of 95% surfactant degradation/elimination, a minimum 80% organic carbon degradation, and an LC_{50} value of at least 0.2 mg/L in aquatic toxicity (based on "degraded" substance) [2].

The use of a "degraded" substance simulates real-life conditions and is considered by Malz an essential criterion to avoid misinterpretations that may, and did, arise when toxicity is measured on the original material. Aquatic toxicity values in the effluents of biological sewage treatment of biodegraded material have been found to exceed those of the original materials [2].

Comprehensive reviews and studies on ecotoxicology have been reviewed by Matzner [6]. A few are recalled next.

1. Anionic Surfactants

Standard reference works on the biochemistry, toxicology, and dermatology of anionic surfactants have been published [6,28]. The U.K. department of the Environment reported on LAS biodegradability and toxicity, concluding that it poses no hazard to human health or the environment [6,29].

TABLE 3 Biodegradability and Aquatic Toxicity of Surfactants Used in Detergents and Cleaning Products (1986)

Type of surfactant	Biodegradation/elimination		Aquatic toxicity	
	Surfactant (primary degradation)	DOC (total mineralization)	Fish toxicity (LC_{50} mg/L)	Daphnia toxicity (LC_{50} mg/L)
Basic requirements	>80	>60	>1	>1
Anionic surfactants				
Alkyl benzene sulfonates	90–97	73–93	3.2–4.9	8.9–14
Alcohol ether sulfates	96	67–89	1.4–20	1–50
Alkylsulfonates	97–98	83–96	3–24	8.7–13.5
Alcoholsulphates	98–99	97–99	3–20	5–70
Sulfosuccinic acid esters	96	49	39	33
Soaps		90	6.7–150	22–72
Nonionic surfactants				
Alcohol ethoxylates	83–98	62–90	1–50	2–200
Alkyl phenol ethoxylates	85–97	50–94	1.5–1000	4–50
Cationic surfactants				
Alkyl ammonium quarternary compounds	94	100	1–6	0.1–1.0

Source: Ref. 2.

A major study on anionic surfactants along the Mississippi river showed that dissolved LAS levels were one to two orders of magnitude below toxic limits, while the LAS in the bottom-level sediment was one order of magnitude below toxic limits [6,30].

The U.S. Fish and Wildlife Service evaluated LAS toxicities in self-consistent laboratory and field experiments, showing NOEC (no-observed-effect concentrations) results. Addition of 5% sewage effluent had little effect on LAS bioavailability, while 45 days' exposure to NOEC (0.36 ppm) had no effect on survival [6,31].

2. Nonionic Surfactants

The well-known ether-type nonionic surfactants, such as fatty alcohol ethoxylates and nonylphenol ethoxylates (NPEO), as the main representatives of alkylphenol ethoxylates (APEO), were developed beginning in the 1950s. Alkylphenol ethoxylates (APEO), popular since the 1960s, are one of the two largest nonionic surfactant types because of their efficacy, cost performance, and easy handling and formulating. APEO include hexyl-, octyl-, nonyl-, decyl-, and dodecylphenol ethoxylates, of which nonylphenol ethoxylates (NPEOs) are by far the most important:

$$C_9H_{19} - C_6H_4 - O(CH_2CH_2O)_nH \text{ (NPEO)}$$

Because of the nonlinear character of the molecule, given by the branched nonyl group and the presence of the aromatic ring, NPEO biodegradation has been a controversial subject for many years. As shown in Table 1, the primary biodegradation of C_{8-10} alkylphenol 9EO is about 96% and meets the biodegradability requirements of EEC directive 82/242/EEC on nonionic surfactants [1]. The primary biodegradation was found to be influenced by temperature and by sludge retention time. However, a minimum of 90% has been found under varied conditions. Ultimate biodegradation of C_{8-10} alkylphenol 9EO, however, was found to be much lower, not over 68%. It is interesting to note also that the oxo alcohol ethoxylates with odd-numbered alkyl chain showed even lower ultimate biodegradation values (59%) as compared to natural fatty alcohol ethoxylates (90%) (Table 1). Several measurements in three standard systems on several EO lengths of NPEOs revealed average ultimate biodegradation rates of 30–60%, the highest being achieved in a continuous-flow activated sludge unit [32,33].

These findings created doubts about the environmental safety of NPEOs, and beginning in the mid-1980s Western Europe called for restrictions and bans. Detergent manufacturers in Germany and Switzerland began voluntarily to reduce NPEO consumption. In 1992 the Paris convention (PARCOM) encouraged contracting states to prohibit the use of APEO in domestic cleaning applications by 1995 and in industrial applications by 2000. In the United States there are no regulations relating to the use of APEO in any of these applications [32].

In 1995, annual world demand for APEO was about 665,000 tons. Western Europe manufactured not more than 109,800 tons in 1994, of which only 74,800 tons were consumed internally, the rest being exported. None was used in household cleaning, while 26,500 tons (35%) were used for industrial and institutional cleaning [32].

Nonylphenol (NP) and NPEO were found to be biodegradable in natural waters. Biodegradation of labeled nonylphenol in seawater was slow initially but increased rapidly [6]. A comprehensive review on biodegradability, aquatic toxicity, and environmental fate is available [32].

The APE (Alkylphenol and Ethoxylates Panel) of the Chemical Manufacturers Association has sponsored an extensive program on the toxicity of NP to aquatic life, including acute and chronic exposures [34]. Fish intervertebrats and algae vary widely in their sensitivity, as measured by acute toxicity. NP was found to be more than 10 times more toxic than NP(EO)$_9$, which has an LC$_{50}$ of greater than 1.5 mg/L, while that of NP is generally about 0.1 mg/L [35]. In most species of fish, the LC$_{50}$ value of most APEO was found to be between 4 and 14 mg/L. Studies indicated that the levels of APEO measured in the environment are well above those that would have adverse effects on aquatic life [34].

Other studies similarly show that the toxicity of APEO to aquatic organisms increases with a decrease in the number of ethylene oxide units and an increase in the hydrophobic chain length, meaning that the toxicity of the parent surfactants is less than that of the biodegradation products [32,36].

Several years ago, APEO and related compounds were reported to be estrogenic, since it has been demonstrated in laboratory studies that they mimic effects of estradiol, both in vitro and in vivo. These studies showed that the growth of cultured human breast cancer cells is affected by NP at concentrations as low as 1 μm (220 μg/L).

Estrogenic effects have been reported on rainbow trout hepatocytes, chicken embryo fibroblasts, and a mouse estrogen receptor. Moreover, studies on UK rivers receiving sewage effluents have shown that an estrogenic pollutant is present, for male rainbow trout placed in these rivers produced a female egg yolk protein [35].

Chemicals referred to as "environmental estrogens" are suspected of causing health effects in both humans and animals through disruption of the endocrine system. A comprehensive review, covering biological and physiological estrogen background, reached the conclusion that given the low reported environmental and bioconcentration factors (BCF) of APEO, the potential for these compounds to produce estrogenic effects in the environment seems to be low [34,37]. Similar conclusions were reported recently, evidencing that a worldwide research program has been set up to study the biological effects of environmental estrogens on human and animal life [32].

Zoller [21,25,38,39] reviewed the groundwater pollution by detergents and polycyclic aromatic hydrocarbons, with emphasis on nonionic surfactants, in studies that include problem presentation, environment–detergent formulation relationship, biodegradation, surfactants in surface waters, treatment, and groundwater pollution.

Groundwater differs from surface waters and is characterized by a slow environmental response. Consequently it is susceptible to a long-term contamination by surfactants like APEO or other organic intermediates with highly branched alkyl chains. These "hard" surfactants may become environmentally persistent organic pollutants (POPs). Even "soft" linear fatty alcohol ethoxylates, which are completely biodegradable under aerobic activated sewage treatment, may show lower biodegradation at lower temperatures (winter) or under anaerobic conditions. For this reason and others, like surfactant adsorption and infiltration, very low levels of surfactants are expected to be found in groundwater.

Several results in groundwater contamination were presented by Zoller [21]. A two-year (1988–89) study on anionic surfactants in Israel showed average MBAS of 0.020–0.400 mg/L, found in several observations at the wells' aquifer recharge site. The effluent concentration after mechanical-biological treatment had values of 0.3–0.6 mg/L [21].

Concern about nonionic POP contamination in Israeli groundwater is very high, because APEO comprise about 10% of the surfactant found in Israeli waste-waters. The effectiveness of surfactant removal in Israel by primary and secondary treatment processes is in the range of 85–97% for anionics and 69–81% for nonionics. Therefore nonionic surfactant concentrations of 0.3–0.5 ppm and 1.1–2.4 ppm, respectively, are found in treated and untreated representative municipal sewage influents. The results suggest that the concentration found in well water is about 23% of the initial concentration in the adjacent streams. The author's conclusion is that even though unpublished 1995 results of nonionic surfactant levels in several water wells were found to be within 0–2 ppb, the environmental concern about persistence of nonionics in Israel is an open issue [21]. Zoller's opinion is that this contamination is unavoidable and that neither the existing treatment facilities nor the naturally occuring biodegradation in the subsurface or by physical adsorption on soil is able to cope with the problem.

The persistence of APEO is considered by Zoller an environmental and health risk, since reused sewage water is a part of the Israeli national water supply system. Implementation of a preventive program has been suggested considering appropriate sewage treatment facilities and a total ban or severe restrictions of the use of "hard" APEO in detergent formulations [21].

3. Cationic Surfactants

The cationic surfactants of the 1960s were mostly amine compounds, of which the most effective were quaternary ammonium salts, with one or two long chains attached to the nitrogen nucleus or quaternary pyridine-based salts. The major representatives during more than 30 years of use were mainly the distearyldimethylammonium chloride (DSDMAC) and partly dialkyl imidazolinium chloride (DAIC), with minor quantities of amidoamine quaternaries [40]. In the 1980s, DSDMAC (also known as DHTDMAC, dihardened tallow dimethylammonium chloride) and DAIC were the two principal cationic surfactants used for fabric softeners (17,000 tons/year in 1985 in Germany) [1].

The environmental behavior of DSDMAC and DAIC could not have been predicted by early versions of detergent laws or ingredient regulations because no definitive test procedures have been adopted. However, some biodegradability data based on several investigations began to appear in Germany, indicating very poor biodegradability [1]. Only later was it understood that as the sole surfactants in biodegradation screening tests in mineral solutions, cationic surfactants act as biocides and inhibit bacterial biodegradation. Moreover, results obtained from activated sludge studies were ambiguous, since degradation could not be distinguished from elimination by sorption.

In OCED Confirmatory Test and Coupled Unit Test, high elimination rates were found for DSDMAC, most being attributed to genuine biodegradation. Nevertheless, the authorities in the Netherlands and in Germany had already concluded that DSDMAC does pose a threat to the aquatic environment and requested the detergent industry to replace it, beginning December 31, 1990. Whether the 28-day closed bottle test provides an accurate assessment of ultimate environmental fate of nitrile-based quaternaries is a question that was raised too late to affect the decision.

The impact has been huge and quick. Annual consumption of DHTDMAC dropped by over 70% practically overnight [40]. Within less than three years a new generation of cationic softeners, mainly ester quats, replaced the DSDMAC [40–42].

G. Effect of Surfactant Regulation on Surfactant Type and Consumption

The surfactant situation in Germany illustrates the impact of regulations on the consumption of surfactants and the distribution of surfactant types [2]. The 1991 surfactant consumption in detergents and cleaners is summarized in Table 4, showing a consumption of 131,000 tons of anionics, 61,000 tons of nonionics, and 4,000 tons of cationics, adding up to a total of 206,000 tons. The total surfactant consumption in 1991 in all industrial applications, including detergents and cleaners, was about 383,000 tons, of which anionics accounted for 211,000 tons, nonionics for 150,000 tons, cationics for 15,000 tons, and amphoterics for 7,000 tons. Most significant is the distribution of the various types of surfactants in 1991 compared to that in 1988 (Table 4). The ecological surfactant shift during this period is obvious:

- A substantial decrease in LAS use (52% to 36%)
- An increase in alkyl ether sulfate consumption (19% to 25%)

TABLE 4 Influence of Surfactant Regulation on Surfactant Type Distribution

	Surfactant consumption in detergents and cleaners		
	1991		1988
	Tons/year	%	%
Anionic surfactants	131,000	100%	100%
LABS	47,000	36	52
Alkane sulfonates	22,000	17	16
Fatty alcohol sulfates (FAS)	14,000	11	9
Fatty alcohol ethoxy sulfates (FAES)	33,000	25	19
Others	15,000	11	4
Nonionic surfactants	61,000	100%	100%
Fatty alcohol ethoxylates (FAEO)	2,000	69	64
Alkanolamides	4,000	6	8
Alkoxylates	6,000	10	7
Others	9,000	15	20
APEO	0	0	1
Cationic surfactants	4,000		
Total	206,000		

Source: Ref. 2.

- An increase in alcohol ethoxylates use (64% to 69%)
- A total ban on APEO

However, the surfactant categories (anionic, nonionic, cationic, amphoteric) showed a significant shift in Germany from 1988 to 1991 [2]:

- Anionics increased from 51% to 63%.
- Nonionics decreased from 37% to 29%.
- Cationics decreased from 10% to 5%.

This shift points to the increased use and/or launch of environmentally friendly anionic surfactants, seemingly at the expense of anionics. The cationics decline also reflects the beginning of ecological blame.

IV. DETERGENT REGULATION AND ENVIRONMENTAL LEGISLATION (1970–1995)

A. Decline of the Phosphate Era

An important step in the continuing development of detergents was the launch of detergents based on sodium tripolyphosphate (STP or STPP) ($Na_5P_3O_{10}$) at the beginning of the 1960s. During the '60s the detergents contained as much as 40% STPP, and even higher contents could be found in premium formulations for very hard water. These products that softened the wash water, allowing a more efficient performance, marked the beginning of the "phosphate era."

However, phosphates are not only excellent water softeners but also good fertilizers for plants in the fields and for algae in rivers and lakes. Thus, when excess phosphate was released into surface waters, it appeared that in certain lakes, slow-moving rivers, and ponds, algae started to reproduce at an unprecedented rate, depleting oxygen suplies in the lower water layers. The result was a general reduction of overall water quality that negatively affected potential drinking water and fishing. Extensive use of phosphates in detergent products has been singled out as the cause of this phenomenon.

For a more objective view, a better understanding of the phosphorus cycle in nature is necessary. Comprehensive studies of the subject are well covered in the literature [43,44]. An explicit illustration of the phosphorus cycle was published in 1992 by the Economic Commision for Europe (ECE) in a survey entitled "Substitutes for Tripolyphosphate in Detergent." It is presented in Figure 3 [44]. It can be seen that the excessive fertilization, an environmental phenomenon called *eutrophication*, is brought about by the combined effect of phosphorous from agriculture, wastewater, and detergent phosphates. An aqueous environment rich in plant nutrients is called *eutrophic* (i.e., well nourished), and eutrophication refers to nutrition by chemical means.

Sources of phosphates in the aquatic environment are divided into two categories: point sources and diffuse sources [44]. The main point sources are man-made and include municipal, industrial, and some agricultural wastes. Municipal sewage contains several sources of phosphates, such as general domestic waste, human excrement, food waste, some industrial waste, and sometimes collected rainwater. The diffuse sources are defined as those that cannot be specifically located. The major diffuse sources are of agricultural (fertilizer losses, animal excrements, soil erosion) or

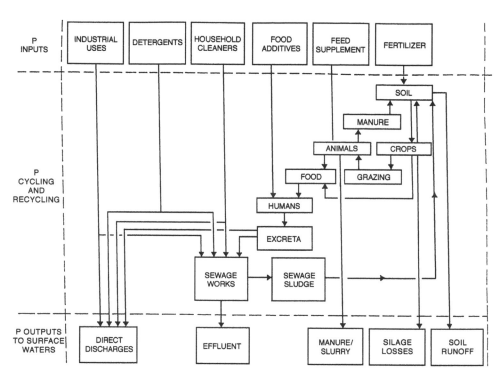

FIG. 3 The Phosphorus cycle. (From Ref. 44.)

nonagricultural origin. The latter include dust, rainwater, plant remains, and, most important for aquatic environments, lake sediments.

Since the increase in the use of detergent phosphates was accompanied by an increased use of phosphate fertilizers that also find their way into natural waters, the blame for the environmental effect of phosphorus should not have been taken solely by the detergent industry. According to the ECE, detergents seem to contribute less than 45% to the total domestic phosphate load and less than 15% to the total phosphate load in the environment, from which every living organism takes in and to which it excretes large amounts of phosphate [44]. In areas where phosphate removal is part of sewage treatment, the contribution is even smaller.

A comprehensive study on phosphates in Germany, entitled "Phosphorous: Pathways and Fate in the Federal Republic of Germany," was carried out by the Phosphate Commission within the Water Chemistry Section of the German Chemical Society [1,45]. According to this report, in 1975 only 8.6% (69,000 tons phosphorus/ year) from the total phosphate production and import in Germany (807,000 tons of phosphorus/year) was consumed by the detergent and cleaner market. The same phosphate study showed that about 60% of the phosphates encountered in municipal sewage originated from detergents and cleaners.

Partial removal of phosphate by sewage treatment plants and inflow of phosphate from various sources led to an estimated share of about 40% of detergent phosphate in surface waters. It is obvious that the complete solution of the eutrophication problem cannot be based only on the removal of phosphates from detergents but also on efficient sewage treatment.

The reduction in phosphate levels in sewage treatment plants is based on primary and secondary treatments. Primary sewage treatment includes mechanical separation, by which insoluble phosphate particles are removed, reducing the total phosphate load by about 15%. Secondary sewage involves a biological decomposition stage, by using activated sludge or percolation beds, and removes an additional 20–40% of the total phosphate [44].

Over time, even significantly higher removals of phosphate were achieved by enhanced biological treatment, which can remove approximately 55% of the phosphate, or by chemical precipitation, which can lead to over 90% removal. Today, final release phosphate levels of as low as 1 mg/L seem to be an achievable target [44].

However, even if adequate sewage treatment could provide a more satisfactory solution to the phosphate problem, the immediate reaction of lawmakers was to regulate consumer products [1,46]. From that point, phosphate regulation marked the cornerstone of the history of detergent legislation.

B. The (Second) Detergent Law on Detergents and Cleaning Products (1975)

The First Detergent Law, enacted in 1961 in Germany, specified minimum requirements for the degree of biodegradability of surfactants. However, during the following years, more ecological targets and detergent requirements began to arise, leading, on August 20, 1975, to the so-called "Second German Detergent Law," better known as the "Detergent Law," which regulated the "environmental compatibility of laundry detergents and cleaning products" [1,47].

The Detergent Law paid particular attention to the potential role of surface water in the drinking water supply and to issues related to proper operation of sewage treatment plants. The implementation of these basic requirements implied the establishment of broader demands, some of which were part of the law itself. Manufacturers, consumers, and local water authorities were involved in the law's implementation. Consumers, for instance, provided with all necessary information regarding usage recommendations, were supposed to act according to the law. The overall requirement "Detergent and cleaner usage must be consistent with the law and with the objective of water quality preservation" significantly spoke for itself.

The requirements and authorizations for implementation dealt generally with the following issues:

- Explicit product labeling, including a declaration of qualitative composition.
- Deposition of representative sample formulations with the Federal Environmental Agency.
- Incorporation only of biodegradable organic ingredients, particularly surfactants, as well as other organic compounds viewed as potential phosphate substitutes.
- Providing consumer information on the package, especially dosage instructions, including variations according to the local water hardness, where hardness was defined in terms of four ranges. The water hardness had to be made known to consumers by local water supply authorities.
- Maximum permitted phosphate levels, including some cases of a total phosphate ban, provided ecological substitutes were available.

The detergent products covered by the Detergent Law included all categories considered to be potential pollutants of surface waters, such as household laundry detergents, household cleansers, dishwashing agents, rinsing and laundry aids, industrial and institutional cleaners, detergents for commercial laundries, as well as personal care products.

C. Phosphate Regulation (1975–1991)

1. Europe

Beginning in 1975 the German Detergent Law limited the maximum allowable phosphate level in detergents and cleaners in a gradual, stepwise manner over the period from June 1980 to January 1984, when the implementation of the regulation was complete [1]. Thus, a maximum phosphate content of 2.0 g/L was permitted in home laundry detergents in areas of relatively soft water between 0 and 7 German degrees (0–125 ppm $CaCO_3$). Higher levels were allowed for harder water. In fact, the responsibility for implementation was largely left to the consumer when the manufacturers marketed a single formulation in different localities but provided the consumer with a chart showing the precise amount of detergent to be used for a laundry load in water of a given degree of hardness.

A regulatory attitude has been gradually adopted by most of the Western European countries, not as an overall European Economic Community action but on a national and/or regional basis. Water hardness level and the availability of effective phosphate substitutes played a major role in national decisions. Italy proposed phosphate limits in March 1982, including them in a more general environmental protection law. This legislation allowed a 1% maximum phosphorous content in detergents beginning June 30, 1987. This limitation covered all laundry and dishwashing detergents and cleaners for household, institutional, or industrial use. In the Netherlands the industry averted new regulations through a voluntary agreement reducing phosphate levels in two stages untill January 1983. Switzerland, which suffered most severely from eutrophication problems because of its geological characteristics, regulated phosphate content beginning in 1977. However, only in December 1980 did the federal government approve legislation requiring a two-stage reduction in allowable phosphate levels. For several years, different levels were permitted according to the geographical area, until the use of phosphates was totally banned in 1986.

The Nordic countries were more liberal on the phosphate issue, because of geographical and geological advantages. In Finland, a "gentlemen's agreement" between industry and government allowed a maximum phosphorous content of 7%, while Norway permitted a maximum level of 12% STPP in all household laundry detergents. Beginning in 1986, a special attitude was adopted by Sweden, which concluded that the best approach is the use of tertiary sewage treatment plants, especially those built by the end of 1983 to serve nearly 80% of the Swedish population [1].

A 1994 summary of the detergent phosphate regulatory situation in Western Europe and market shares (%) of phosphate detergents in each country has been reviewed [48] and quoted [6] as follows:

Greece, no restrictions (80%); Portugal, no restrictions (80%); Spain, no restrictions (70%); Ireland, no restrictions (60%); France, no restrictions, earlier ecomarketing abated (60%); United Kingdom, no restrictions (56%); Denmark, no restrictions (50%); Sweden, no legislative restrictions (40%); Belgium, no restrictions (20%); Finland, no restrictions, manufacturers agree to reduce phosphates (10%); Austria, legislative restriction to low levels depending on conditions, heavy ecomarketing (2%); Netherlands, no restriction, manufacturers agree to promote no P, ecomarketing (0%); Germany, restriction to 23% STP, heavy ecomarketing (0%); Italy, restriction to 4% (0%); Switzerland, detergent phosphate ban since 1986 (0%).

In some countries (Netherlands, Germany, Austria, Italy, Switzerland, Finland) detergents manufacturers, as a result of public environmental concern, implemented nationally even more severe phosphate restrictions than imposed by regulations. On the other hand, nine countries did not impose any restriction and based their regulation on ecological national logics, allowing phosphate-built detergent to maintain a wide range of market shares, which varied with market forces, from 20% in Belgium to 80% in Greece and Portugal. The Israeli market, in which the share of phosphate-built detergent was 100% until 2000, as required by the Israeli Standard for washing powders, recently adopted a new phosphate regulation. Beginning in 2000, a modified version of the Israeli Standard allowed marketing of phosphate- and non-phosphate-built powders as well. This issue will be further presented as a case study.

A more recent report presents the 1998 market shares of nonphosphate heavy-duty powder detergents in Europe compared to those in Israel (Figure 4) [49]. The situation in Great Britain and France, where phosphate detergents account for more than 50% market share, reflects a regulatory climate based on a justified lower

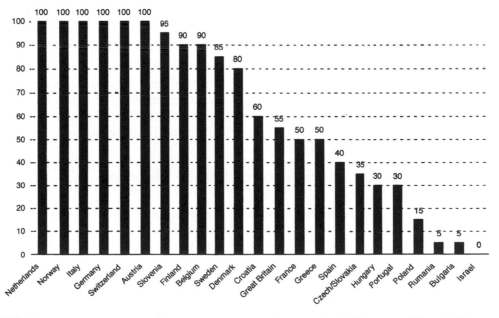

FIG. 4 1998 market shares of nonphosphate heavy-duty powder in Europe and Israel. (From Ref. 49.)

environmental concern and professional awareness of cost/performance ratio. To an even greater extent, the same criteria characterize the situation in Israel, one of the higher-water-hardness countries in the world (400 ppm $CaCO_3$).

2. The United States and Canada

A very comprehensive overview of the environmental and detergent regulatory situation in North America and Europe has been presented by Matzner, including a highly recommended list of relevant reviews on the effects of detergents on the ecology [6,46].

Beginning in 1970, a large variety of detergent phosphate bans have been enacted by individual U.S. states and Canada. Some bans totally eliminated phosphorus in detergents, whereas other permitted up to 0.5% phosphorus (equivalent to 2% phosphate) or even 2.2% phosphorus (9% phosphate).

Matzner [6] reviews in detail all the state phosphate bans on a historical and a continuing basis, showing that today total bans exist in Arkansas, the District of Columbia, Georgia, Indiana, Maine, Maryland, Massachusetts, Michigan, Minnesota, New Hampshire, New York, North Carolina, Vermont, Virginia, Washington, and Wisconsin, while only partial local bans are imposed in Florida, Idaho, Illinois (Chicago), Missouri, Montana, Ohio, and Texas. In Canada a federal limit of 2.2% phosphorus has been established. However, the situation is dynamic, and states that rejected phosphate bans have reconsidered them subsequently. Thus, total bans in Arkansas and Massachusetts were enacted in 1993 after several earlier rejections [46].

The wide range of regulations in different states illustrates the local debate on the magnitude of the numerical limits directly dependent on ecological/geophysical/geological/socioeconomic considerations. Product performance is also taken into account in establishing phosphorus limits, as in the case of automatic dishwashing powder, for which an 8.7% minimum phosphorous content has been found essential for adequate performance [46].

3. The Far East, Australia, and Africa

Japan enacted total bans on detergent phosphate by prefecture regulations. Most products were reformulated. After two years, 90% of all detergents manufactured and marketed were free of phosphates, being based mostly on zeolites [1,50]. South Korea manufactures basically only phosphate-free detergents [51]. Australia and China are still scenes of debates on the issue [1,51]. South Africa and Thailand are slowly moving to new zeolite-containing detergents [51].

4. Effect of Phosphate Regulation on Sewage Treatment and Water Quality

Phosphate regulation had a serious impact on the detergent phosphate market, which shrank drastically. STPP consumption in Western Europe dropped drastically from more than 1 million tons in 1976 to about 300,000 tons in 1999 [51]. Similar data on U.S. consumption of phosphates show figures of 423,000 tons in 1976 compared to less than 25,000 tons in 1994 [6].

This quantitative implementation of phosphate regulation was supposed to solve most of the phosphate environmental issue. The contribution of laundry detergents to the phosphate load in sewage was evaluated and checked in Germany in 1980, as it has been done for the surfactant load [1,18,19]. An estimation based on the same assumptions as already described in Section III.E for surfactants (200 liters daily water comsumption per capita and 750,000 tons detergent production in 1980) has been conducted for phosphorus consumption in detergents, for which a daily per capita consumption of 2.5 g and a calculated average concentration in municipal sewage of 12.6 mg/L were found. These data were based on a 1980 German phosphorus consumption in detergents of about 56,800 tons.

Phosphate levels in municipal sewage have been checked analytically, and the total phosphorus levels were found to vary from 2 to 40 mg/L, with 20 mg/L being the average. The total phosphorus of the mixed wastewater derived from laundering was found to have a average value of 200 mg/L [1].

Thus, phosphate regulation proved to be effective in reducing the phosphate outflow of sewage treatment plants. It also reduced the cost of the chemicals used in phosphate removal operations. Annual savings for all U.S. sewage treatment plants have been estimated up to$ 3 million [46,52].

Detergent phosphate bans had a direct impact on water quality, as measured by phosphate level, chlorophyll (algae) content, Secchi depth (a visual measure of suspended solids), and active oxygen level [1,6,46]. The total phosphorus content of drinking water as reported in 1980 in Germany was 0.2 ppm [1]. In Austria the phosphorus load into Lake Monosee decreased from 26 to 9 tons/year during 1979–1984.

Chlorophyll and phytoplankton levels of the top layers decreased, and the phosphorus concentration fell from eutrophic to mesotrophic levels. Blue-green algae populations decreased drastically [6,53]. A 1986 Swiss report found that dissolved phosphate content decreased from 1 ppm to 0.15 ppm over 10 years. This result was attributed not only to the detergent ban but also to the largest sewage treatment plant built in the area [6,54]. U.S. Water Control Board measurements in Virginia rivers found total phosporus values, due to local point sources, of 20 ppb after the phosphate ban, as compared to 110 ppb before.

Nevertheless the magnitude of the impact was not always as expected, probably because 88% of phosphorus entering lakes and streams came from nonpoint sources and because detergents contribute only a few percent of the remaining 12% [6,46]. Such evidence was found in the central basin of Lake Erie in the U.S. Total P levels decreased only by 0.21–0.28 ppm/year from 1970 to 1986, while the rate of oxygen depletion was reduced only very slightly. Moreover, similar sediment phosphorus regeneration rates were found in 1970 and 1990, despite 20 years of nutrient reductions [6,55,56].

A U.S. National Oceanic and Atmospheric Administration report stated that detergent phosphate bans could not improve the quality of the Great Lakes. Moreover an Environmental Protection Agency (EPA) study across U.S. lakes concluded that the phosphate ban would have "very little effect on water quality" [5,57,58]. In addition, an American Chemical Society monograph underlined that uncontrollable phosphorus sources contained in average domestic sewage can feed the growth of 1 g/L of blue-green algae, which is 50–100 times more than the algal concentration of most U.S. lakes and 1000 times that found in oceans [6,59].

Supporting evidence for the failure of detergent bans to achieve significant environmental improvements, apart from the importance of nondetergent phosphate sources, was provided by several studies in the United Kingdom [6,43,60,61], Canada [6,62], and Italy [6,63]. Even more intriguing reports suggested that phosphate-free detergents may contribute to unfavorable environmental effects, such as the appearance of toxic red tides in coastal waters [46,63]. Moreover, disturbing the living algae equilibrium in nature was believed to contribute to the "greenhouse effect," because the algae fix large amounts of atmospheric carbon dioxide. Global warming has become a concern today. Algae seem to be of importance in alleviating it [46,64], and phosphate bans seem to have a negative impact also.

V. NONPHOSPHATE BUILDERS—A PERPETUAL SEARCH

A. Builder Functions in "Built" Detergents

Phosphate regulation presented the detergent industry with a tremendous challenge to find suitable substitutes for STPP. The search for new complexing agents started worldwide beginning in the 1970s. R&D work to find an ideal substitute still goes on. The ulitmate target is a builder superior to STPP in terms of the challenging requirements that must be satisfied simultaneously: performance, cost, and environmental impact. Comprehensive reviews on builders in powdered detergents have been published by Rieck [51,65].

The builder has to perform several functions:

- Water softening
- Alkalinity and buffering
- Dispersant and antiredeposition activity
- Exhibiting anticorrosion properties
- Control of heavy metal ions (Fe^{2+}, Cu^{+2}, Mn^{+2}) for bleach stabilization
- Liquid (surfactant) carrying capacity.
- Lowering the CMC (critical micellar concentration) of surfactants

Three chemical elements are basic for creating nearly all builder materials: silicon, carbon, and phosphorus [51]. Carbon is present in sodium carbonate and organic carboxylates (such as soap and trisodium citrate). Silicon is present in sodium silicate, sodium disilicate, and zeolites (sodium aluminosilicates). Phosphorus is present in STPP and organic phosphonates. The structural formulae of representative builders are presented in Figure 5. The structural similarities of the ion exchanging builders are obvious.

The early builders of 1907, water glass and soda ash, were used to soften the water and to make the wash liquor alkaline. Typical European detergent formula-

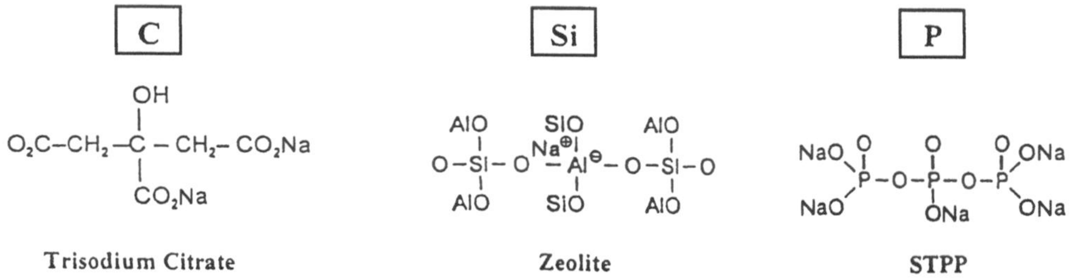

FIG. 5 Main builder elements. (From Ref. 51.)

tions from throughout this century are presented in Table 5, as compiled by Smeets [66] and quoted by Rieck [51].

B. Zeolites

Mineral zeolites were discovered in 1756 by Baron von Cronstedt, a Swedish mineralogist. The zeolites owe their name to the fact that they release large amounts of water upon heating. von Cronstedt derived the name from the greek *zeo* (boil) and *lithos* (stone).

1. Chemistry and Performance

Zeolites are crystalline aluminosilicates of natural or synthetic origin, with the following formula [67]:

$$M_{x/n}[(AlO_2)_x(SiO_2)y] \cdot zH_2O$$

M is an exchangeable cation with valence n, either an alkali metal, an alkaline earth metal, a transition metal, or a quaternary ammonium ion. The primary building unit is composed of SiO_4 and AlO_4 tetrahedra linked by so-called secondary building units to form a three-dimensional network. About 50 natural and more than 200 synthetic zeolites have been classified by structural groups and polyhedral tertiary building units [68]. On the basis of these characteristics, which lead to adsorptive capacity, zeolites were first used as "molecular sieves" by the sugar industry in 1896.

Detergent zeolites are sodium aluminosilicates of different crystalline structures. The principal zeolite used in detergents since 1976 is zeolite A ($x = y = 12, z = 27$) $Na_{12}(AlO_2)_{12}(SiO_2)_{12} \cdot 27H_2O$, known also as zeolite Na A or zeolite 4A. It possesses a well-defined crystal structure consisting of cages (diameter 1.1 μm), connected by windows (diameter 0.42 μm) that can be traversed only by small molecules or ions. Thus, calcium ions diffuse relatively easily into the porous structure, exchanging

TABLE 5 Development of European Detergent Formulations

Ingredient (%)	1907	1953	1970	1983	1987
Soap	32.0	44.0	3.5	3.0	1.5
LAS	—	—	6.5	8.0	7.0
Nonionics	—	—	2.5	3.0	5.0
Pyrophosphate	—	10.0	—	—	—
STPP	—	—	40.0	24.0	—[a]
Zeolite A	—	—	—	18.0	25.0[b]
Polycarboxylates	—	—	—	1.0	4.0
Sodium carbonate	24.0	12.0	—	—	10.0
Sodium Silicate	3.0	7.0	3.0	4.5	3.5
Sodium perborate (4H₂O)	9.0	6.0	27.0	22.0	20.0
Activator (TAED)	—	—	—	1.5	2.0
Sodium sulfate	—	3.5	4.0	4.0	10.0
Water	30	15	9	9	10

[a] A significant share of European detergent formulations continued with phosphate.
[b] In the case of phosphate-free formulations.
Source: Ref. 51.

sodium ions, while magnesium ions are impeded by a hydrate shell and are incorporated more slowly. Only at higher temperature, when the hydrate shell of the magnesium ion is gradually removed, does the rate of ion exchange increase. Therefore zeolite A requires a cobuilder to take care of the magnesium water hardness, especially at low temperatures.

The first formulations in 1976 contained combinations of STPP and zeolite A, followed in 1983 by mixtures of zeolite A, carbonate, and polycarboxylates [67]. Apart from its slow exchange rate for magnesium, zeolite A was shown to have beneficial effects in the laundering process. It promotes the inhibition of graying caused by redeposition of soil during repeated wash cycles through heterocoagulation with dirt particles. Zeolite can reduce the effect of dye transfer to noncolored textiles because it removes dyes from the wash liquor via heterocoagulation and adsorption [51,67]. A further positive feature is the high adsorptive capacity for liquid surfactants, which permits formulation of surfactant-rich compact detergents. The adsorptive capacity, measured as % linseed oil, was found to be 35% for zeolite A, compared to 10% for STPP and 15% for soda ash [67]. Zeolites are suitable builders for achieving the higher bulk densities of 800–900 g/L in supercompact powders and of 1000–1300 g/L in tablets.

For these reasons, zeolite A was the main STPP replacement in Western Europe between 1980 and 1993, increasing in consumption from about 100,000 tons in 1982 to 525,000 tons in 1993. However, stagnation in zeolite A consumption since 1994 has been reported [51]. Market data show zeolite consumption figures in Europe of about 600,000–650,000 tons/year during 1994–2000, out of a total builders market of about 1,050,000 tons/year [67]. This trend characterized the United States and East Asia as well, reflecting new environmental concerns, as will be further described.

However, it is our opinion (based on the experience gained in Israel, a country with a water hardness of about 400 ppm $CaCO_3$) that a significant reason for this stagnation in zeolite consumption may be a poorer overall washing performance in hard water areas of zeolite/soda ash/ polycarboxylate formulations compared to STPP formulations. The 50% market share of nonphosphate powders in France and Great Britain, as illustrated in Figure 4, support this judgment.

A recent development, called zeolite MAP (maximum aluminum P), is an aluminosilicate builder with the zeolite P (gismodine) structure, with a Si/Al ratio of around 1.0. The new type possesses a more flexible, adaptable layered crystal structure by which the calcium ions are bound more firmly than in the case of zeolite A. Therefore the water hardness is reduced more effectively [51,67,69,70].

A further new development on the market is a cocrystallite comprising of 80% zeolite X (similar chemical composition to zeolite A but with a larger pore diameter of 0.74 µm for better magnesium-binding capacity) and 20% zeolite A. This grade, reffered to as zeolite AX, displays calcium and magnesium exchange properties, synergistically superior to those of a blend of the pure zeolites.

2. Zeolites in the Environment

The ecological and toxicological safety of zeolite A was established in Germany between 1974 and 1978 by a comprehensive program through cooperation between industry, government, and independent research institutes and public authorities. The results of this study were published by the German Federal Environmental Agency in

1979, showing neither harmful influence on aquatic organisms nor any promoting of algae growth.

These results have since been supported by further investigation [67,71]. Phosphate-free zeolite–based detergents were found to have no harmful effect on phytoplankton, zooplankton, and fish in Swiss lakes during 10 years of use. At the same time, phosphate replacement led to a reduction in water burdens, which advanced the timing of achieving water-protection targets [67].

The water insolubility of zeolite has raised questions about potential sediment deposits that can cause clogging in domestic and municipal sewage systems. The sedimentation behavior has been investigated under laboratory and real-world use conditions as well as in large-scale studies done in Germany during 1976–1978. Neither in the domestic sewers nor in the main drains did sludge deposits accumulate or solidify over prolonged use periods.

Generally, zeolites have a positive impact on the operation of sewage treatment plants, especially with respect to sedimentation performance. It was found that about 96% of zeolite A in sewage was removed by the treatment plants and only 4% was discharged. Moreover, no detrimental effects of zeolite A were observed in the operation and performance of the biological treatment stage or on the tertiary treatment stage of phosphate removal.

Taken together, these positive findings justified environmental confidence and the expectation that the eutrophication problem would be solved by zeolite as a promising substitute for STPP. However, in 1993 several ecological and toxicological reevaluations weakened this belief.

3. Zeolites or Phosphates—A Detergent Phosphate Revival?

In 1993 *New Scientist* reported that "green detergents" may be blamed for a green bloom in the Adriatic Sea. Researchers at the University of Bologna said: "Zeolites and polycarboxylic acids (PCAs) may actually encourage the conglomeration of bacteria and microorganisms that produce globs of smelly, brown foam that since 1988/89 have driven visitors away." The researchers suspected that zeolites and PCAs, coming to the ocean from freshwater tributaries act as seeding material that allows bacteria to gather and grow in the same way as suspended clay particles behave naturally in the ocean. These findings were reported by "Happi" [72], which added that "this discovery encouraged the detergent industry, which hopes to put phosphates back into their products because of their cost performance." Even the studies conducted by the Triest Institute "Laboratorio di Biologia Marina," which found no evidence for the involvement of zeolites in this phenomenon [67], did not diminish the hope. The high aluminum molecular content in zeolites led to a risk assessment speculation that the use of sewage sludge may create a critical soluble aluminum concentration in combination with acidic soil and acid rain.

Concurrent with the ecological doubts, comparative studies of the builder systems zeolite/NTA or zeolite/NTA/copolymer vs. STPP emphasized their disadvantage in cost performance. The study by Rhone Poulenc showed that for equal performance, the cost of the builder system is about 1.4 times higher [73]. The same study also raised the incrustation problem encountered with zeolite formulations. The 1992 ECE study on "Substitutes for STPP in detergents" also emphasized this problem, still unsolved even today [44]. A British Landbank environmental research report 1994 known as "The Phosphate Report" described a comprehensive life cycle assessment (LCA) of STPP and zeolite using the Delphi technique. The environmental impact evaluated via this technique showed no significant differences between STPP (107 points) and zeolites/polycarboxylic acid (110 points) [6,60,74]. Similar reports claimed low environmental impact differences among phosphate, zeolite, polycarboxylate, and nitrilotriacetic (NTA) builder systems in European detergents [6,43]. These studies [60,74] were later criticized because they did not comply with the standardized requirements for LCAs [67].

The 20-year debate on the merits of phosphate replacement received practical answers from the 1995 decisions of the U.K. and Nordic Ecolabeling boards, which, based on life cycle analyses, accepted to use of phosphates in household laundry detergents as an environmentally acceptable builder for detergents according to the Scientific Committee on Phosphates in Europe (SCOPE) [48]. The conclusions of the executive summary of the Swedish Phosphate Report, give high priority to efficient wastewater treatment [48]:

1. The Swedish Phosphate Report completes the analysis of detergent builders which began with "The Phosphate Report," published in 1994. It assesses the fate in the environment of phosphates and zeolites-PCA after use, within the context of advanced wastewater treatment as widely practiced by Sweden and the other Nordic countries.
2. Seventeen senior scientists with expertise in wastewater treatment and sludge disposal from eight European countries took part in the Delphi study to evaluate the impact of the two builder system upon the range of receiving waters and in four disposal situations for sludge. Response curves for each scenario were constructed on the basis of the scientific consensus which emerged from the Delphi consultation process. These were used to model each of the four Nordic countries and the UK.
3. The study finds that advanced wastewater treatment is highly effective in preventing pollution, as measured by the indices of environmental damage for each of the modeled countries. Sweden and Norway, which have invested heavily in wastewater treatment, show very low indices of environmental damage by comparison with the U.K., which has not. The study also finds that phosphates are less environmentally damaging than zeolites-PCA under all modeled conditions.
4. Finally, the study discusses the necessary conditions for the sustainable management of freshwater resources in Europe. It concludes that wastewater treatment to high standards, coupled with methods to remove and recycle phosphates, are essential components of such a sustainable regime.

C. Organic Carboxylates

1. Monomeric Carboxylates [16]

The simplest monomeric carboxylate is, of course, the soap RCOONa, which reduces water hardness via precipitation. However, because of lime soap incrustation and yellowing of white clothes, the use of soap in powder detergents is limited to foam control. Softer potassium soaps of oleic and lauric acids are used as cleaning agents in liquid detergents. Soaps are biodegradable but, like other organic materials, increase biological oxygen demand (BOD).

Hydrocarboxylic acids, of which trisodium citrate is the best known, are used in combination with zeolite A to help in dispersion and solubility. Trisodium citrate is also used in liquid detergents as an STPP replacement. Citric acid and its derivatives enjoy an excellent ecotox image, being defined as GRAS (generally recognized as safe) by the FDA [21].

Good sequestration ability is also provided by gluconic and tartaric acids, but their use in alkaline detergents is limited because of their high price. However, their use in acid cleaners is worthy of mention. New candidates, unfortunately of even higher cost, have been developed, such as tartrate succinates, carboxymethoxy succinates, carboxymethoxy malonates, and lactobionic acid derivatives [51,69,75].

The *amino carboxylic acids*, of which EDTA and NTA are best known, form a group of chelants that raise environmental concerns that are still being debated. EDTA is a very strong sequestrant for calcium, magnesium, and heavy metals. Its complexing ability led to a wide range of applications in products in daily use [76]. However, it is expensive and considered nonbiodegradable [51,76]. In 1990, Germany was the leading country in EDTA use, with an estimated consumption of 6,900 ton, followed by the United Kingdom with 6,400 ton. The environmental impact of EDTA is based on its polluting of air but especially on contamination of natural aquatic systems.

As a result of improved production techniques, less than 2 kg/ton of manufactured EDTA is emitted into the air, while 10–15 kg/ton EDTA is released in production wastewater. Since wastewater treatment was poor, the effluent EDTA concentration reached about 0.5–1.0 mg/L. Awareness of the EDTA problem led to better industrial wastewater treatment in many industrial sites. However, these improvements had only a moderate environmental impact because a large percentage of the EDTA contribution to municipal wastewater comes from domestic use [77].

Since EDTA is xenobiotic, its fate in the environment began to be critically controlled. It was found that the highest mass concentration of EDTA is in surface waters, probably due to its hydrophylic character and poor biodegradability as well as to its ability to form most stable complexes with transition metals. A large-scale investigation of the chemical, photochemical and biological degradation of EDTA in recent years has led to a remarkable agreement in Germany between government and industry to reduce drastically emissions of EDTA into the aquatic environment [76].

In the 1980s, sodium nitriloacetate (NTA) seemed to be in the most promising amino carboxylic acid builder. It has been used successfully in laundry detergents in Europe and Canada because of its strong chelating ability with calcium and magnesium, in spite of a relatively poor suspending action. Though NTA was initially considered 95% biodegradable in sewage treatment plants, a November 1985 amendment to the New York State Detergent Phosphate Ban prohibited the sale or use of detergents containing NTA. In 1994, the EPA of California established a nonsignificant risk exemption proposition for NTA [51]. However, continuing opposition to NTA on a state level, based on toxicological and environmental aspects, combined with its negative image, resulted in its withdrawal from detergents [51]. In Switzerland, NTA was limited to a 5% level in detergents by a 1985 regulation. In the Netherlands, NTA is limited to 6,500 tons/year; in Germany, it is voluntarily restricted to a total use of 5,000 tons/year. NTA is banned in Turkey and Italy [6].

A comprehensive summary on NTA in environmental studies, euthrophication aspects in fresh- and seawater, as well as on the influence of NTA on heavy metal elimination in sewage plants and, especially, on remobilization of heavy metals from water sediments has been published [77]. Remobilization, which is a recomplexation of heavy metals bound to sediments or other solid surfaces, has been investigated in experimental model studies, sometimes criticized as not being sufficiently realistic [77]. However, some significant experiments did not show remobilization of heavy metal ions out of rock at an NTA concentration as low as 50 mg/L [51,78].

The development of organic chelants is growing steadily. The focus is on combining the high chelating ability of EDTA with the good biodegradability of citrate. The target is a fair chelate stability constant, reasonable cost, and biodegradability. Recently, amino acid derivatives have been reported, including MGDA (methylglycerine diacetate), GLDA (glutamic acid diacetate), SDA (serine diacetate), and ASDA (aspartic acid diacetate) [51,69].

2. Polymeric Carboxylates

Calcium sequestration capability is enhanced by increasing the number of carboxylic groups per molecule. This chemical logic led to the development of polycarboxylates, of which the sodium salts of polyacrylic and polymaleic acids are mostly widely used. Because of their high cost, the copolymers of maleic and acrylic acids are used only as dispersing and antiredeposition agents.

Since 1980, copolymers with a molecular weight of 60,000–70,000 at concentrations of not more than 4–5% have been used extensively in premium zeolite A/soda ash/polycarboxylates building systems, which reduce graying and incrustation on fabrics. However, polycarboxylates are considered to be poorly biodegradable, as has been reviewed recently by the industry [6,79].

The European Center for Ecotoxicology and Toxicology of Chemicals initiated risk assessment studies showing low acute toxicity and no evidence of mutagenic potential. The studies underlined that no toxicity is associated with these polymers, in spite of their accumulation in the environment, as a result of their poor aerobic and anaerobic biodegradation.

Nevertheless, the proposed European Union Commission Ecolabel requirements assign a relatively high negative score to polycarboxylates [6,80]. Therfore significant work is being done to develop new biodegradable polycarboxylates. For instance, insertion of polyvinyl alcohol to achieve biodegradability has been mentioned, and some acrylic maleic vinyl acetate copolymers (used in toilet-cleansing blocks) are claimed to be biodegradable [6,69,81,82].

Oxidation products of carbohydrates, polyglycosanes, starch, starch hydrolyzate, and polysaccharides have been described as potential cobuilders with good biodegradability [6,51,83,84,85]. Development of polyaspartic acid as a biodegradable replacer for polyacrylic acid has been reported and promoted in the market [6,51,64,79].

D. Silicates

1. Amorphous Silicates [3,16]

Sodium silicate has been used in detergent formulations for many decades as a source of alkalinity and buffering capacity and as a corrosion inhibitor. It is usually made by the fusion of sand containing a high proportion of silica with soda ash in an electric furnace according to the reaction

$$Na_2CO_3 + SiO_2 \rightarrow Na_2SiO_3 + CO_2$$

which yields sodium metasilicate and carbon dioxide. Adjusting the ratio of silica sand to soda ash gives rise to a variety of alkalinities. Writing the molecular formula as $Na_2O:SiO_2$ one can define the amorphous sodium silicates ("water glass") as silicates having a silica ratio above 1. $Na_2O:2SiO_2$ is known as "alkaline" glass (pH 11.3), while the ratio 1:3.3 gives a "neutral" glass (pH 10.5). A common silica ratio used in spray-dried phosphate-based detergent powder is 1:2.4. Complete water evaporation from amorphous sodium silicates at high temperatures leads to amorphous disilicates, further crystallized to sodium disilicates. The zeolite-built detergents contain less water glass added at the spray tower stage because of some incompatibility. Silicate is postadded as granular disilicate.

2. Crystalline Silicates [16]

The best-known chain silicate is sodium metasilicate $Na_2SiO_3(Na_2O \cdot SiO_2)$, widely used in laundry powders and especially in automatic dishwashing powders in its anhydrous or in the more stable pentahydrate form $(Na_2SiO_3 \cdot 5H_2O)$. Metasilicate is a chain silicate with a one-dimensional configuration.

Layered silicates with a two-dimensional configuration, such as the clay-type silicates (bentonites), are known in the nature. Their use in detergents is a relatively recent development in the field. Bentonites have only medium ion exchange properties, but have been used as softening agents in softergent formulations. However, crystalline sodium layered silicates, especially of the δ phase, showed unexpectedly good calcium ion–sequestering properties and are being claimed as effective detergent builders [51,86].

The stoichiometric composition of the newly developed sodium layered silicate, supplied under the trade name SKS-6, is $Na_2Si_2O_5$, and its crystalline layer structure is closely similar to that of the mineral natrosilite. The crystal lattice is arranged in layers, which consist of tetrahedral SiO_4 connected at the vertices in an infinite two-dimensional array [87]. The sodium ions are situated between the layers, so they are relatively mobile and can be replaced by other ions. Unlike zeolite, SKS-6 shows equal performance in both calcium and magnesium complexation. The silicate acts as a multifunctional builder because it provides alkalinity, buffers the pH value, suspends soil in the wash liquor, carries liquid surfactants, and is suitable for granulation and compact powder technology. Moreover, it absorbs moisture and binds heavy metal ions and, therefore, stabilizes bleaching agents and especially the less stable percarbonate [51,65,86,87]. The reported ecotox profile is low risk. The layered silicates do not contribute to eutrophication, nor do they cause problems of solid residue during sewage treatment. Because they will be degraded in the wastewater, it is believed that no problems of accumulation and related phenomena like remobilization should arise [86,87].

E. Phosphonates

Phosphonates are derivatives of phosphorous acid in which the hydrogen atom attached to the phosphorous atom has been replaced with another grouping, essentially forming a compound with a C–P bond. They are efficient complexing agents, with good stability in aqueous media and toward oxidizing and reducing agents. Typical examples are hydroxyethane diphosphonic acid (HEDP) and ethylenediaminetetramethylene phosphonic acid (EDTMP). They exibit a threshold-scale inhibition effect acting as crystal growth inhibitors at substoichiometric levels and retard precipitation of insoluble salts [7,51,88]. Due to their good chelating properties, they complex heavy metal ions (copper, manganese, iron) and thereby stabilize peroxobleach, which is present in some bleach-containing heavy-duty laundry detergents, liquid hypochlorite, and hydrogen peroxide bleaches.

Environmental concern about the phosphonates has always been low, especially because of the low concentration of less than 1% used in detergents [7]. Phosphonates may cause mobilization and bioavalability of heavy metal ions. However, because phosphonates are strongly adsorbed on rocks, their concentration in water is very low [51].

F. Carbonates

Sodium carbonate (Na_2CO_3) is the cheapest builder. The disadvantage of incrustations formed by precipitation of poorly soluble calcium and magnesium carbonates can be minimized by the addition of polycarboxylates. However, the increased pH level and the alkalinity supply provided by sodium carbonate, apart from a possible lowering of the surfactant CMC, lead to an overall improvement in detergency performance.

The soda ash effect at several dosages has been compared to the effect of various builders on the detergency of nonionic surfactants at a water hardness of 2 mmol calcium chloride/L. The performance of 10, 20, and 50 mmol/L of soda ash added to the surfactant was, respectively, 60, 79, 80 (expressed as percentage clay removed) and 56, 60, 68 (expressed as percentage grease removed). Under the same conditions of water hardness, the nonionic with no builder added showed a washing performance of 66% (clay removal) and 54% (grease removal); while at zero water hardness it was 92% for both clay and grease [7].

In the same report it was also shown that in hard water STPP ($Na_5P_3O_{10}$) helps to attain the best performance in clay removal (2 mmol/L STTP removed 92% clay), while sodium metasilicate (Na_2SiO_3) was the best builder for grease removal (20 mmol/L removed 80% grease) probably because of higher alkalinity [7,89].

Increased solubility at low temperatures was shown by potassium carbonate (K_2CO_3) used in some Japanese powders [7]. Such a property can be useful in solving environmental problems of sodium characteristic of Israel, a country where the reuse of wastewater for agriculture is essential.

In the past decade, cogranulates of sodium carbonates and sodium silicate have been used to improve washing efficiency at reduced detergent dosage. These cogranulates lower the zeolite buildup in the sewage sludge and thus limit the level of organic ingredients that cause excessive BOD and COD, which is harmful to fish. Several patents for cogranulates have been submitted, one of which, for instance, contains 55% soda ash, 29% disilicate, and 16% water and permits good dispersion of calcium carbonate and particulate soil [51,90].

New, interesting ion exchange resins were patented in 1996 [83,91], disclosing crystalline inorganic builders containing a carbonate anion, a calcium cation, and at least one water-soluble cation. They may be layered or not, with apparent compositions such as $Na_2Ca(CO_3)_2$, $NaKCa_2(CO_3)_3$, and $K_2Ca(CO_3)_2$. The rate of ion exchange is reported to be as good as or better than that of zeolite A.

G. The Ideal Builder—Still a Challenge

The ideal multifunctional builder has to fulfill simultaneously several requirements, which have already been listed. Basically they must achieve optimal performance and high ecological acceptance, offered at a competitive cost. Rieck's comprehensive reviews on the subject present the builder situation, suggesting a tertiary diagram ECP as assessment of a material with respect to ecology, cost, and performance [51,65].

1. Detergent Performance

The performance in detergency "building" is an overall summing up of the builder contribution on any of the functions already mentioned. The builder performance is related directly to water softening, alkalinity and buffering capacity, dispersing, and suspension properties associated with bleach stabilization and anticorrosion properties. The difficulty in chosing the most suitable builder stems from the fact that any builder acts differently in performing these various functions.

Water softening is the most decisive parameter in the builder choice, and water hardness is the key factor in the regional ecological decision. The direct connection between water hardness in a particular area and performance and, therefore, the consumer acceptability of phosphate-free formulations in the same area is striking.

Israel is an extreme case of very high hard water (about 350–400 ppm $CaCO_3$) almost all over the country. Therefore, STPP, as the most efficient sequestrant for both calcium and magnesium, is the builder of choice. Fortunately, this choice is supported by geophysical conditions, in terms of few lakes and rivers.

The different types of washing machines (top or front loading with horizontal drums in Europe compared to top loading with agitators as pulsators in the United States and Japan) and the detergent dosage related to the wash water volume impact the builder concentration per liter of water and therefore the ppm of water hardness (Table 6). The ratio of builder level per water hardness (expressed as mg/ppm $CaCO_3$),

TABLE 6 Builder Concentration per Water Hardness

	Japan	United States	Western Europe	Israel
Average water hardness (ppm $CaCO_3$)	50 ppm	100 ppm	200 ppm	350 ppm
Water consumption per washing cycle (liters)	30	60	18[a]	18
Detergent dosage per washing cycle[b]	20 g[a]	60 g	72 g[a]	40 g
Detergent (grams/liter)	0.67	1.00	4.00	2.22
Average builder content per detergent	50%	55%	55%	60%[c]
Builder (grams/liter)	0.33	0.55	2.2	1.33
Builder/water hardness (mg/ppm $CaCO_3$)	6.7	5.5	11.0	3.8

[a] New machines and new formulations, respectively.
[b] For normal soil and mean water hardness.
[c] STPP only until year 2000.
Source: Data from Ref. 51.

as presented in Table 6, is the key factor for a satisfactory washing performance [51]. The low recommended dosage in Israel seems to show the lowest ratio (3.8 mg/ppm $CaCO_3$). Without STPP, at such low concentration the washing performance should be too poor. This situation is even more acute today, when new Israeli regulations promoted by the Ministry of Environment limit the washing dosage by an official detergent standard, in order to minimize the sodium concentration per wash in recycled water for irrigation in a country fighting for this missing source (see Section VIII.E).

Alkalinity and buffering capacity, developed by the alkaline sodium silicates and soda ash, are very good, compared to only poor to moderate for STPP, zeolite, and sodium citrate [7,51].

The stabilization of the bleaching system, especially under warm and humid conditions, is important nowadays, in particular for the less stable sodium percarbonate formulations. Phosphonates and layered silicates contribute to bleach stability, while zeolite A needs a coated percarbonate to coexist.

Anticorrosion properties toward metal or porcelain enamel surfaces of sodium silicates and especially of higher silicate molar ratios are properly made use of in automatic dishwashing detergents and washing powders. The compact powders and the superconcentrates require underdosage and therefore do need builders with increased liquid loading capacity, especially for liquid nonionic surfactants. Layered silicates, zeolite MAP, and zeolite A have a higher carrying capacity than STPP and soda ash [51,67,87].

The processability requirements of the new-old techniques of dry mixing and agglomeration are different from the spray-drying needs. For instance, layered silicates are suitable for dry-blending but hydrolize under tower spray-drying conditions.

2. Cost

The total cost of a detergent formulation is influenced primarily not only by the purchasing price of the specific ingredients, but also by the manufacturing costs, work and energy as well as packaging and transportation. The price level of several builders was estimated, approximately in ascending order, as: "water < sodium sulfate < soda ash < water glass solution < zeolite A < amorphous sodium silicate < zeolite MAP < STPP < δ-disilicate < carboxylates" [51]. The cost/performance ratio is essential for builder acceptance and governs the targeted market share and profit.

3. Ecology

Today, environmental acceptance of detergent ingredients is evaluated by life cycle analysis (LCA), which attempts to quantify the environmental impact of use as well as of the manufacture of the test material.

The recent environmental dilemma regarding STPP and zeolite has been already presented. Emphasis was placed on a positive "ecobalance," i.e., an environmentally friendly life cycle of zeolites that are manufactured by combining silica from mined sand, with alumina from ore bauxite. After use, the metastable aluminosilicate redecomposes into its mineral constituents, silica and aluminum [93].

The insolubility of zeolites was criticized because it maximizes sludge in sewage and may deposit in the wastewater system. On the other hand, cogranulates of sodium silicates and soda ash were designed to keep water treatment stations sludge free [16,94].

A comprehensive summary of ideal builder requirements, including ecological concerns (consumer safety, biodegradability, eutrophication, and sludge solubility), has been compiled by Rieck [51].

An interesting approach to evaluating a builder by assessing the interrelation between ecology, cost, and performance by a ECP ternary diagram has been proposed by Rieck [51,56] and is presented here, by permission, in Figure 6. Because of changing requirements, the only materials to be used on a long-term basis are positioned in the interior region of the diagram, which is the target region for future developments [51].

H. Builder Combinations

Because there is no ideal multifunctional builder, builder combinations have been developed. Zini [95] reported on a multitude of formulations, based mainly on polymeric additives, in order to achieve an overall washing performance equivalent to that of STPP. Especially in hard water and at lower temperatures, polycarboxylated polymers improved binary and ternary zeolite–containing builder systems, but they could not equal the cost/performance of high-STPP powders (40–50%), unless about 15–20% STPP was left in the builder system [95].

1. Binary Builder Systems

Zini reviewed binary systems containing 20% STPP and showed that 2% phosphonate or 10% NTA on top of 20% STPP gave the best cleaning performance at 60°C [95]. However, at 90°C the binary mixture of 20% zeolite:20% STPP gave the best cleaning effect. Combinations of zeolite with sodium silicates (both amorphous and layered disilicate) free from carbonates, phosphonates, and polycarboxylates have been reported to show detergency comparable to ordinary commercial detergents that contain large amounts of phosphonates, carbonates, and polycarboxylates [51,96].

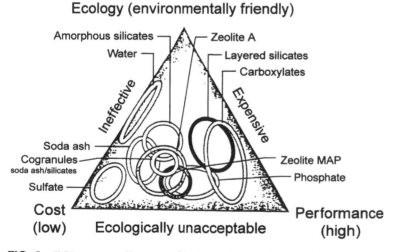

FIG. 6 ECP ternary diagram of builders (1995). (From Ref. 51.)

2. Multicomponent Systems

Ternary mixtures containing STPP showed the best cleaning performance at 60°C for STPP/zeolite/polymer (71% soil removal) compared with STPP/citrate/polymer (67% soil removal) and STPP/NTA/phosphonate (69% soil removal). At 90°C, soil removal increased to 79%, 75%, and 76%, respectively [95]. A system of zeolite, layered silicate, and citrate allows a significant reduction in the wash dosage [51,97]. An optimum ratio of zeolite:layered silicate:citrate of 3.5:1:1 to 4.5:1:1 has been disclosed [51,98].

VI. ECOLOGICAL FORMULATION TRENDS—THE COMPACT ERA (1988–2000)

Zeolite A/soda ash/polycarboxylate seemed to be the preferred combination in the late 1980s not only in Europe but also in North America and Japan [51]. This combination was suitable for conventional powders and especially for the new phosphate-free compact powders in the 1990s, as shown in Table 7 [69]. In this system, each of the separate components performs only a part of the ideal STPP function. Therefore, the combination of multiple components is essential. Zeolite A, the preferred STPP substitute, eliminates water hardness via ion exchange, but the exchange rate becomes extremely slow at low temperatures compared to the soluble STPP. Especially for cold water washing, therefore, it is necessary to combine zeolites with other water-soluble builders and dispersing cobuilders, such as soda ash. Citrates are also often used in the building system, because of their chelation properties as well as alkalinity/buffering capacity. The polycarboxylates of choice are copolymers of maleic and acrylic acids, also because of their anticorrosion properties.

TABLE 7 Typical Compositions of Phosphate-Free Compact Detergents

Final product (wt%)

Ingredient	United States	Europe	Japan
Surfactants	20–30	10–20	20–35
Anionics	15–25	10–15	5–30
Nonionics	0–5	5–15	5–20
Builders	60–70	50–70	50–60
Zeolite	25–35	25–40	20–25
Sodium citrate	0–3	0–4	0–3
Sodium carbonate	10–30	15–25	10–25
Polycarboxylates	0–5	0–5	0–5
Sodium silicates	1–3	5–10	3–15
Bleach	0–5	15–20	0–7
Perborate/Percabonate	0–5	10–20	0–5
Activator	0–5	2–7	0–5
Enzymes	0–3	0–3	0–3
FWA	0.1–0.5	0.1–0.5	0.1–0.5

Source: Ref. 69.

Ecological constraints had a direct impact on the composition and technology of detergents. Efforts to save energy and to reduce detergent consumption significantly impacted the technological development of both detergents and washing machines. The compact detergent launched in Japan by Kao in 1987 spread worldwide.

In Japan new compact brands captured a 55% share by the second half of 1988 [83]. Western Europe and the United States joined the compact trend in 1989, achieving in 1991 a share of about 20% of the powder market in both Europe and the United States. However, in 1996 compact powders attained 60% of the U.S. market and only 30% in Western Europe. In the same year, Latin America used less than 5% compact detergents and Africa about 10% [10]. By 1998, the market share of compacts in Europe accounted for about 30% in value, which corresponds to 37% of the whole powder segment. In the United States, in 1998 over 90% of the powder market was represented by compact detergents [67,99]

It is worth mentioning that the use of compact powders had started already during the early 1970s, but the concept was premature, technologically and commercially. The success in the late 1980s was driven in Japan by space efficiency and in Europe by environmental pressures, which, during the past decade, became a global concern.

The density of European and U.S. compact powders varies from about 500 to 750 g/L, much lower than the densities of up to 950 g/L in Japan.

In 1996, the first supercompact powders appeared in Europe and Japan. The latest technological advance, combined with a change in washing habits, was the launch of tablets.

The changes in detergent ingredients, accompanied by a significant reduction in detergent dosage per wash, are reflected in the leading German detergent brand presented in Table 8 [67,100]. The reduction in the recommended detergent dosage directly influenced the per capita consumption of detergents, which showed a significant decrease. Thus, consumption in Germany decreased from 10.6 kg/capita in 1988, through 8.2 kg/capita in 1993, to 8.1 kg/capita in 1998 [67]. The ecological contributions of the low-dosage instructions followed by the decreased per capita consumption are obvious.

Even at the lower values of today, the European and U.S. dosages are still much higher than the Japanese dosages, which all through history were about 25–35% of the European dosage. Japan's initial 40-g dosage in the conventional powder of the past changed to compact dosages that began at 25 g/30 L in 1987 and have been reduced to 20 g in 1995 and even to 15 g in the 1996 supercompact formulations [69].

VII. OTHER DETERGENT INGREDIENTS—
ENVIRONMENTAL IMPACT

A. Bleaching Agents

Bleaching agents are included in laundry detergents to eliminate stubborn stains. The stains are usually classified by the nature of the soil present: fatty soil (such as oil, grease, sebum), proteinaceous soil (blood, egg, meat), particulate soil (such as dust, starches), and nonfatty stain (such as tea, red wine, fruit juices, coffee). Fatty, proteinaceous, and particulate stains are efficiently removed by combinations of surfactants, builders, and enzymes. For removal of nonfatty stains, bleach is needed.

TABLE 8 Changes in Detergent Ingredients and Dosages, 1979–2000[a]

Main ingredients[b]	Heavy-duty detergent type (Germany)				
	Regular		Compact	Supercompact	Tablet
	1979	1991	1991	1996	2000
Anionic surfactants	5.9 (16.2)	7.5 (11.0)	8.0 (7.8)	14.9 (11.3)	15.2 (11.4)
Nonionic surfactant	4.0 (11.0)	5.2 (7.6)	6.0 (5.9)	4.4 (3.3)	3.7 (2.8)
Soap	3.0 (8.3)	1.5 (2.2)	0.8 (0.8)	0.9 (0.7)	0.7 (0.5)
STPP	39.0 (107.3)	—	—	—	—
Zeolite A	—	25.0 (36.5)	23.5 (23.0)	20.6 (15.7)	25.1 (18.8)
Sodium carbonate	—	12.1 (17.7)	12.6 (12.3)	9.0 (6.8)	2.6 (2.0)
Polycarboxylate	—	3.8 (5.5)	5.5 (5.4)	5.2 (4.0)	3.2 (2.4)
Silicates	4.9 (13.5)	4.3 (6.3)	3.0 (2.9)	1.6 (1.2)	1.8 (1.4)
Sodium perborate ($4H_2O$)	24.9 (68.5)	24.0 (35.0)	9.0 (8.8)	—	—
Sodium perborate ($1H_2O$)	—	10.0 (9.8)	12.2 (9.3)	—	—
Sodium percarbonate	—	—	15.0 (11.2)	—	—
TAED	—	2.0 (2.9)	5.5 (5.4)	7.0 (5.3)	
Disintegrant	—	—	—	—	6.0 (4.5)
Recommended dosage (grams) (normally soiled laundry, 14–21°d)[c]	275	146	98	76	75

[a] Values in %, values in parentheses = grams of ingredient per wash.
[b] Other ingredients (mainly water and additives) up to 100%.
[c] 14—21°d equivalent to 250–375 ppm $CaCO_3$.
Source: Ref. 67.

The bleaching action refers to removal of a colored stain from a substrate by an irreversible chemical reaction of oxidation or reduction. Typical bleachable soils consist of natural substances such as brown tannins (tea, red wine), organic polymers of humic acid type (coffee, cocoa, tea), pyrrol derivatives (chlorophyll), carotinoid dyes (carrots, tomatoes), red to blue anthocyanins (cherry, blackberry, red currant), as well as synthetic dyes and inks.

The bleaching agents are classified as

- Reducing agents (sulfites or bisulfites)
- Chlorinated products
- Oxygen-releasing compounds (peroxygen bleaches)

Reducing agents are used only in particular institutional applications. Chlorine compounds were in fact the first bleaching agents to be used as additives for the washing process. In the United States its use dates from the 1930s, and it is still used in low-temperature washing, where the peroxygen bleaches are less efficient. The oxygen-releasing compounds are typically derived from hydrogen peroxide (H_2O_2), which is used, as such, in liquid bleach laundry additives. The solid derivatives, used mostly in detergent powders, are sodium perborate tetrahydrate ($NaBO_3 \cdot 4H_2O$) and sodium percarbonate ($2Na_2CO_3 \cdot 3H_2O_2$).

1. Boron Derivatives (Sodium Borax, Perborates)

Borax decahydrate ($Na_2B_4O_7 \cdot 10H_2O$), as a popular washday additive, first appeared well over a century ago. Its appearance, together with that of Persil in 1907, is one of the cornerstones of detergent history.

Sodium perborate tetrahydrate (PBS4) (10.4% theoretical active oxygen content) was used extensively in Europe during the 20th century, especially at high temperatures (80°C). Beginning in the 1980s PBS4 gained a strong presence in the U.S., South American, and Asian detergent markets, followed by sodium perborate monohydrate (PBS1) (16% theoretical active oxygen content) in more recent formulations. In comparison to the tetrahydrate, the monohydrate has better stability, particularly at higher temperatures and relative humidities (hot and humid climates), better solubility, mainly at low temperatures, and better compatibility with other detergent ingredients.

The bleaching action caused by the perhydroxyl anion OOH^-, which is color safe and does affect animal, vegetable or synthetic fibers, supports a "fabric care" concept. Moreover, sodium perborate adds alkalinity and buffering properties to the detergent and is therefore a detergent builder on its own.

Under TAED-activated, unbuffered conditions, where pH might vary, sodium percarbonate (PCS) performance, against the tea and red wine stains, was shown to be poorer than that of both PBS4 and PBS1. This was due to the higher pH of PCS wash liquors (pH10 for PCS vs. pH 9.5 for PBS), since the stain removal of peracids is considered to decline at higher pH [101].

Multiple functions have been attributed to borates, including calcium ion sequestration, interfacial oil–water tension reduction, promotion of the dislodging of fatty soils, increasing surfactant solubility, and electrostatic stabilization of pigment salts. Moreover, borates, as derivatives of hydrogen peroxide and other oxygen donors, are highly active toward organic matter and have a remarkable disinfectant effect, based on the deactivation of proteins and enzymes [101].

In European and U.S. formulations, perborates reached a level as high as 25%, reduced to 15% in the more ecological, energy-saving, low-temperature washing formulations. Environmental concern about the release of borates into wastewater relates to H_2O_2 and boron. Released H_2O_2 is inactivated rapidly. Even at concentrations much higher than those present in laundry wastewater, H_2O_2 presents no toxicity to bacteria, nor can it affect biological sewage treatment [1].

However, boron, even though present in the environment as an essential micronutrient for plant and mammalian growth, is an element of high ecological concern. The Safe Drinking Water Act, issued by the EPA in 1993 as a Drinking Water Standard, defined a maximum boron concentration of 0.9 mg/L as a guideline. The European Parliament established 1 mg/L. The U.S. EPA's Office of Water Management issued a Drinking Water Boron Advisory, providing information on health effects and relevant treatment technology useful in dealing with water contamination, but no enforceable standards were involved [6].

A number of systematic studies in Germany of boron levels in rivers and drinking water have shown very low levels, less than 0.25 mg/L, considered toxicologically negligible. Highly polluted rivers contained 1–2 mg/L. These concentrations present no problem for fish because the boron fish toxicity is quite low, $LC_{50} > 300$ mg/L. However, boron levels as low as 1–2 mg/L possess a specific phytotoxicity with respect to certain agricultural plants, such as citrus, fruit trees, tomatoes, and vineyards. For this reason, it was recommended as early as 1970 not to recycle boron-containing wastewater for irrigation purposes [1,102]. In Israel, this ecological drawback became a problem of major detergent-environmental concern. The Israeli Standard for washing powders in 1970 set a maximum of not more than 12% PBS4, which is equal to 8.4 grams of boron per kilogram of product [103]. This standard was revised in 1999 [104], imposing a progressive reduction in boron levels, reducing it to 6% (4.2 B/kg product) by Jan. 1, 2002, and to a total ban by Jan. 1, 2008.

The boron detergent regulation in Israel is expected to solve most of the boron pollution problem, since detergents have been estimated to be responsible for about 85% of the boron contamination of wastewater. A detailed case study on Israel ecological issues be will presented later in this chapter [105].

Since most of the boron found in water is of anthropogenic origin, the level of boron in wastewater is sometimes taken as an indicator of anthropogenic pollution. For rivers, such as the Ruhr in Germany, with a very low natural boron level, it has been estimated that the additional amounts of boron introduced to the river through wastewater are largely due to detergent use [1,106]. Other German sources estimated the proportion of anthropogenic boron in the River Neckar to be 73%, while in the Rhine, at times of low water flow, to be only 50% [1].

2. Sodium Percarbonate

The perborates are considered to be true persalts (peroxidates), i.e., compounds containing an–O–O– group. Sodium percarbonate is considered not a true but a false or pseudo-persalt (hydroperoxidate) that contains oxygen in the form of a hydrogen peroxide adduct.

The percarbonate is not as stable as the perborate and has been used primarily only in bleach additive formulations. However, it is more soluble than the perborate and therefore more efficient at lower temperatures.

The solution of the progressive boron legislation in Israel led to a need to reformulate washing powders, with percarbonate as the principal sole source of active oxygen. This alternative followed an international trend imposed by the need for oxygen donors more active at lower temperatures, which are suitable for energy-saving ecological washing practices.

Percarbonates possess attractive properties such as higher levels of active oxygen (14%), good dissolution properties, a good source of alkalinity, and above all its environmental compatibility. However, greater solubility and greater activity at lower temperatures are accompanied by relatively fast degradation and hence reduced storage stability. In phosphate-based formulations, at 30°C and 80% relative humidity in protective packaging, the lowered stability is less pronounced [10].

For Israel where phosphate-based detergent powders are allowed, this is a fortunate phenomenon. However, for realistic day-to-day storage during summer, the actual temperatures and humidities are higher, and accelerated storage stability at 37°C, 80% RH do show loss in active oxygen in all formulations. Storage stability in conventional zeolite-based powders is even poorer. The instability is due to the presence of heavy metals and free water in the formula. The present compact and superconcentrated powders have a lower water level, making the challenge easier, but still the percarbonate must be stabilized.

Several patents describe coatings of percarbonate with boric acid and sodium metaborate, while others comprise sodium or potassium sulfates, chlorides, carbonates, and even polycarboxylates [6,101]. Several patent applications disclose storage-stable coated percarbonate formulations containing granular water-insoluble magnesium silicates and/or aluminosilicates and sodium silicate or containing granular carbonate and percarbonate of different particle sizes or containing anionic surfactants [6,107].

Some companies (Degussa, Solvay, EKA, Kao) tried to solve the instability by encapsulating percarbonate with organic mineral or polymeric compounds [10]. Percarbonate Q 30, developed and marketed by Degussa, is reported to display a high stability (98% active oxygen remaining after 8 weeks at 30°C and 80% RH) in a phosphate-free laundry powder [10,108]. In phosphate-based dishwashing powders, 78% of active oxygen is found after 12 weeks of storage at 35°C and 80% RH, compared to 90% remaining in a sodium perborate tetrahydrate powder stored under the same conditions [10,108].

3. Sodium Hypochlorite

The use of hypochlorite began in 1785, when a solution of potassium hypochlorite was sold to textile bleachers, under the name "Eau de Javel," after the location where it was prepared. The manufacture followed the Berthollet process, dissolving chlorine gas in a solution of "potash" lye. In 1820, potash was replaced by the cheaper soda liquor, yielding sodium hypochlorite, which has been marketed as a solution since 1869 in southern Europe and since 1918 in the United States

Hypochlorite has been used as a laundry and household bleach and disinfectant, as a sanitizer for swimming pools, and as a disinfectant for municipal drinking water and sewage. Domestic hypochlorite products, containing 3–6% active chlorine, are sold as laundry bleaches and household cleaning and disinfecting products. In France, an industrial 12.5% hypochlorite precursor is also marketed for domestic purposes. Sodium hydroxide is usually added at 0.1–1.0% concentrations to improve stability by minimizing the hypochlorite disproportionation to chloride and chlorate. In Israel, an official standard calls for a maximum 0.4% NaOH in a 3.5% maximum active chlorine product [109].

In Europe, Spain seems to be the largest consumer of hypochlorite, with a total consumption level of 464 tons/year as well as per capita consumption of 11.8 kg/year/inhabitant [110].

The human safety of the product has been well documented over the long history of its use. Apart from household acid products, it is the most regulatory labeled product. Under the Dangerous Substances Directive 67/548/EEC [111] and Dangerous Preparations Directive 88/379/EEC [112], hypochlorite solutions are classified as:

Above 10% active chlorine: "Corrosive," with the risk phrases
  R-31, contact with acids liberates toxic gas
  R-34, causes burns
and "S" phrases, S1/2}, S28, S45, S50 added to labeling.
For 5–10% active chlorine: "Irritant," with the risk phrases
  R-31 and R-36/38, irritating to eyes and skin
Below 5% active chlorine solutions: not classified.

Other R or S phrases have been or may be used according to national regulations or professional rules [110].

All hypochlorite solutions above 1% active chlorine must be labeled with "Warning! Do not use together with other products. May release dangerous gases (chlorine)." According to the European Commision Recommendation for Labeling of Detergents and Cleaning Products (89/542/EEC) [113], all hypochlorite solutions above 0.2% must be labeled as containing chlorine-based agents, with their level indicated (below 5% or 5–15%).

Under EEC directives [111,112], household hypochlorite bleaches are not classified as very toxic, toxic, or harmful by ingestion [110]. The acute oral LD50 (rats) is reported to be 13,000 mg/kg body weight for 5.25% NaOCl [110], while Clorox unpublished data report the acute dermal toxicity to be above 2,000 mg/kg body weight for the same concentration [114].

The environmental safety of hypochlorite, including ecotoxicology data on the fate and exposure of by-products and the effect of hypochlorite on sewage treatment as well as studies on organohalogens in the environment, have been reviewed comprehensively [110,114]. Sodium hypochlorite is very toxic to surface waters and aquatic life. Adverse effects for concentrations as low as 0.01 mg/L have been reported [114]. Fortunately it does not reach the sewage treatment plant in the form of free chlorine. On mixing with sewage, it is rapidly destroyed by chemical reaction with materials such as ammonia or other sewage components that are readily oxidized or chlorinated. Several studies showed that more than 96% of hypochlorite is destroyed within 2 minutes after addition to sewage and that a maximum of 0.16% of available chlorine formed chlorinated organic compounds when 75–300 mg/L of sodium hypochlorite was added to domestic wastewater. Other studies confirm that hypochlorite is converted primarily to inorganic chloride ions [114,115].

Therefore, the environmental concern about hypochlorite is not related to any aquatic threat of the product but rather to the possibility that small amounts of chlorinated by-products may form during storage and use [114,116]. Some works showed that 1–2% of the available chlorine from household hypochlorite products and as much as 3.8% of the available chlorine used in commercial laundries form chlorinated organic compounds when mixed with wastewater [110,114,117].

Several organohalogens also form during industrial processes such as acid chlorination of wood pulp, water treatment, and bleaching of cotton fibers [114]. These by-products are difficult to identify chemically, and they are measured by a group parameter called AOX, which stands for adsorbable organic halide. The AISE scientific dossier [110] summarizes the organohalogen issue:

> The levels of AOX produced are low (for example, 37 µg/L from bleach use, compared to a sewage background level of 106 µg/L), and the organohalogens produced from domestic use of hypochlorite are not believed to have an adverse effect on the environment. Available data indicate that no dioxins are produced and that the identified AOXs are typically small molecules with a low degree of chlorination and for which ecotoxicological properties are known or can be predicted.
>
> The majority of the measured AOX is unidentified, but thought to consist of high molecular weight components formed from fats, proteins, and humic acids, which are too large to bioaccumulate. In addition, studies on the whole AOX mixture in laundry wastewater indicated that the level of AOX present did not affect growth or reproduction of Ceriodaphnia and that around 70% is removed in activated sludge.

Thus household use of hypochlorite is not considered to impact the environment adversely [114].

B. Bleach Activators

A major contributor to lowering washing temperatures below 60°C was the introduction of activators in Europe in the 1970s. Beginning in 1980, most of the European formulations were based on TAED (tetra-acetyl ethylene diamine):

$$_2(_3HCOC)NCH_2CH_2N(COCH_3)_2$$

The activation is done by reaction between TAED and perborate, which leads to the formation of peracetic acid (CH_3COOOH), which subsequently acts as a bleaching agent, ionizing under the alkaline conditions of the wash liquor.

Development of activators was greatly influenced by the biodegradation requirements. TAED, the most successful activator, had a market share of 92% in 1985 out of a total European market of 25,000 tons. By 1993, its market share was 99.9% out of 70,000 tons [118]. The moderate cost and excellent biological biodegrability were the governing parameters in its high market acceptability (see Table 9) [118].

Other activators have also been developed over time. The first, TAGU (tetra-acetylglycoluril), was developed in parallel with TAED but suffered from higher manufacturing costs and, more seriously, a low biodegradation profile [7,118]. The same reasons kept DADHT (diacetyldioxo hexahydrotriazine) from market penetration [118]. However, in the United States the standard activator is SNOBS (sodium nonanoyloxybenzenesulfonate), which exhibits, under U.S. washing conditions (low washing temperatures and low detergent dosage), better washing performance than TAED. Above 40°C, the two activators show comparable performance. At 55°C the TAED system performs better [118].

C. Enzymes

Enzymes are a key ingredient in washing powders. They have been used for more than 70 years by launderers to remove blood stains, but only with a moderate success. They were not stable, coming from animal sources, and expensive. The trigger was the food industry's demand, for which enzymes from bacterial sources have been developed. Encapsulation became the preferred way to achieve good storage stability.

TABLE 9 Ecotoxicological Data on TAED

Biodegradability		
Method	Parameter	Degradation (%)
TAED		
Modified OECD Screening test	DOC	89
Zahn–Wellens test	DOC	95
Acute toxicity to aquatic organisms		
Organisms	Test duration	EC_{50} (mg/L)
Fish	96 h	$> 1,500 / > 2,500$
Daphnia magna	48 h	> 800
Algae	14 d	> 500

Source: Ref. 118.

The usage of enzymes grew rapidly during the 1980s, when second-generation proteases, lipases, amylases, and cellulases were introduced. The "bleach alternative" concept in the 1990s in the United States marked the beginning of the current period, known for the popularity of triple and multienzyme systems [83]. Modern lower-temperatures washes make soil removal more difficult, increasing the use of enzymes, which are especially efficient at wash temperatures below 60°C.

As high molecular mass proteins, enzymes are rapidly inactivated and biode-graded by sewage treatment. Environmental concerns in respect to water quality are therefore practically nil [1]. However, health hazards were always of higher concern. Since the enzymes attack proteins, it was feared that workers exposed to them might develop skin and lung irritations. In the United Kingdom, the detergent industry drew up a voluntary Code of Practice in 1968 with the aim of minimizing hazards to the respiratory system of the workers. Measurements of dust levels and regular medical examinations were instituted.

Manufacturing methods, which produce dusty products (such as spray-cooling the liquid enzyme concentrate on sodium sulfate) have been avoided. More modern methods, like spray-cooling to a granular or extruded form and especially encapsu-lation with, for instance, nonionic surfactants, have been adopted. Encapsulation overcomes the storage stability problem as well, for moisture or perborate do not come into contact with the enzyme [7].

Allergy complaints on bio-products, focusing on enzymes, were studied as early as 30 years ago on a panel of 7,000 housewives and reported by the *British Medical Journal*. A five-month panel was run on 4,116 women who used enzyme-containing products, while 2,884 used nonenzyme controls. No differences of any sort were found between the hands of the panelists. Today the human health risk assessment receives a much more methodical approach. However, for certain immunology endpoints there are as yet no standard approaches that can be used by the industry, including valid OECD models or valid in vitro methods [7,119,120].

Because enzymes were known agents of occupational asthma and allergy, a toxicology and allergy assessment was undertaken by Procter & Gamble. Since the application of enzymes in laundry products is safe for consumer use, the program proposed to control enzyme allergies only in the occupational setting [119,121].

The American Conference of Governmental and Industrial Hygienists (ACGIH) has set a threshold limit value (TLV) for protease at 60 mg protein/m^3 air. P&G has set an internal Occupational Exposure Guideline (OEG) of 15 mg protein/m^3. This guideline requirement, implemented by improved engineering and hygiene systems and followed by an employee education and training program, is intended to minimize new sensitization to enzymes and to eliminate asthma and allergies caused by exposure to them [121].

D. Fluorescent Whitening Agents (FWAs)

The fluorescent whitening agents (FWAs) give whiteness and brightness to washes. They are substantive dyestuffs that are adsorbed by the textile fibers from the liquor wash. They have the property to transform invisible ultraviolet (UV) light into visible light on the blue side of the spectrum.

Although FWAs are substantive to fabric and are adsorbed on it, repeat laundering causes part of them to be discharged into the wastewater. Contradictory results were obtained in simple laboratory tests that showed only a limited degree of biodegradability and, on the other hand, in more realistic analyses under conditions simulating a two-stage sewage treatment where elimination values of up to 96% have been found and explained on the basis of adsorption on sludge. Tertiary treatment led to total elimination. The conclusion was that proper sewage treatment ensures the absence of FWAs in surface waters.

Careful checks of concentrations of FWAs in the problematic rivers of Europe—the Thames, the Seine, and the Rhine—showed values below the limit of FWA detection (0.25 μg/L). Based on a low value of fish toxicity ($LC_{50} > 100$ mg/L), a biological safety factor in rivers of 106–108 was adopted [1].

VIII. ECOLOGY IN THE IMMEDIATE PAST AND FUTURE (1992–2005)

Environmental awareness by governmental authorities, detergent manufacturers, and consumers reached remarkable heights during the last decade of the 20th century. Environmental consciousness in Europe in 1992 was highest in Switzerland at an estimated 36%, in Germany at 32%, and in Austria at 26%, with lower estimates for France and Italy (14% and 9%, respectively) [122]. No doubt, these ratings would have much higher values today.

Eurobarometer research published by the EEC in 1993 revealed that 80% of Europeans surveyed placed the environment in the five top priority areas for action, along with unemployment, AIDS/cancer, drugs, and poverty [123]. This situation is reflected in sustainable developments, aimed to ecobalance economic, environmental, and social factors of society, with an explicit commitment to long-term consequences of legislation and regulations, in order to preserve the environment for future generations.

National legislations that follow regional consumer behavior and ecological needs show a pronounced trend toward harmonization and unity. The Earth Summit in Rio de Janeiro in June 1992 reached agreement on the Rio Declaration, replacing regional water and air pollution caused by chemical substances with global environmental issues, such as global warming, depletion of the ozone layer, desertification, and ecological destruction.

Looking back over the last few decades, we see that the history of the development of detergents seems to parallel the history of commitment to ecological issues [123,–125].

A. Environmental Safety Assessement of Detergents

1. Environmental Risk Assessment

The crucial role of biodegradability in evaluating the ecotoxic profile of surfactants has been reflected over the years in detergent legislation and, more recently, in the revised detergent directives, defining the compulsory use of surfactants for easy ultimate biodegradation. However, modern decision criteria for the evaluation of their environmental safety are based on a broader comprehensive assessment of their possible risk on ecosystems.

In the early 1990s, a European Union generally accepted concept for the environmental risk assessment of chemicals was adopted and is now widely used for safety evaluation and possible risk management measures [126–130]. The principle of this assessment [126], described in detail by the EU technical Guidance Documents [127], is based on a comparison of the predicted environmental concentration (PEC) of each individual chemical with its predicted no-effect concentration (PNEC) in relevant environmental segments. The PNEC of a specific material is defined as the highest concentration at which no adverse effects may be expected in organisms exposed to the material. PEC and PNEC values are determined experimentally or by model calculations [128]. A PEC/PENC of less than 1 is considered acceptable, and no risk is to be anticipated if PEC < PNEC [128,129]. Neither of these values can ever be predicted with certainty, and reasonable worst-case assumptions must, at times, be built into the calculations to ensure minimum uncertainty [128].

A complex and multidimensionally sound environmental risk assessment that includes basic fate and effects data, calculation models, and simulation tests for PEC determinations, as well as analytical methods for objective monitoring evaluation called for a joint scientific platform. In 1991, the Environmental Risk Assessment of Surfactants Management (ERASM) was created, combining the expertise and resources of the European detergent and surfactant manufacturers, represented by their associations AISE and CESIO, respectively [130]. The main working areas of ERASM, whose mator target was coordination among industry, regulators, and academia, were presented during the third CESIO Congress in 1996. A major effort concentrated on risk assessment of four major surfactant groups in the Netherlands [130,131].

In recent years, ERASM activities focused on the development of a project called GREAT-ER (Geography-Referenced Regional Exposure Assessment Tool for European Rivers), which is a sophisticated computer software program for calculation of realistic PEC distribution along a river. The GREAT-ER calculation model allows a visualized prediction of the concentration of chemicals along European rivers and in-depth studies into bioconcentration behavior, as well as the fate and effects of surfactants [130,132].

GREAT-ER has been validated for two chemicals, boron and LAS, in six catchment areas along European surface waters in Germany, Italy, and England. The system used an arcinfo/arcview- (ESRI-) -based geographical information system (GIS) for data storage visualization [132].

Several factors have a direct influence on PEC and must be estimated, usually based on a worst-case scenario, especially for new detergent products or ingredients [10,128]:

- Market penetration (the quantity of the product that will be sold on the market)
- Direct discharge of raw sewage into rivers or the sea
- Dilution and mixing in the receiving waters
- Interpretation of biodegradability data (with emphasis on primary biodegradation)
- Bioconcentration factors
- Useful and reliable mathematical models, validated by field observations

Factors that influence PNEC include:

- Application factors, i.e., dividing the lowest toxicity scores obtained on several aquatic species by a large safety factor that will protect all species under all extreme conditions of possible exposure

- Physicochemical properties
- Environmental risk analysis, especially of biodegradation and adsorption properties of potentially toxic by-products
- Presence of other toxic products in the environment

The accumulated ERASM expertise is expected to become a basic building block for the health and environmental risk assessment (HERA) program of the detergent industry [130].

2. Life Cycle Assessment (LCA)

The basic approach taken to evaluate the impact of a product on the environment at each stage of its manufacture, use, and disposal is the life cycle assessment (LCA). An LCA includes a life cycle inventory (LCI), which includes the determination of the resource requirements, energy, and raw materials consumed in the production of a surfactant and its feedstocks as well as in the quantification of environmental releases (such as atmospheric, water-borne, and solid waste emissions). A typical LCA will evaluate energy consumption, use of resources, air emissions, discharge into water, discharge into the ground, and amount of solid waste generated. Each parameter has to be determined at the different stages: raw material production, product manufacture, packaging, distribution, use, and disposal [10,133–135].

B. Development of Novel Environmentally Friendly Surfactants

Three major issues affect the development strategy for a new surfactant: raw material availability, cost performance, and environmental acceptability. The last has undoubtely assumed increasing importance since the 1980s. Environmental pressures strongly influenced development trends, becoming a key market driver.

The relevant properties by which the environmental acceptability of a surfactant are assessed include biodegradability, ecotoxicity, and human safety, with which mildness is associated for personal care use. Several approaches are concurrently used in tailoring a surfactant molecule to fulfill biodegradability and ecotoxicity requirements.

1. Cationic Surfactants—An Environmental Risk Assessment Development Test Case

The significant controversy over the environmental risk assessment of DHTDMAC between industry and authorities in Holland and Germany in 1990 may serve as a typical case of improper use of the risk assessment model to assess the environmental safety profile of products [40].

The industry referred to data gathered over 20 years of use of DHTDMAC showing safety to aquatic and terrestrial ecosystems, emphasizing that, under simulated field conditions, the PNEC value was greater than 10 times the PEC value. These findings were supported by subsequent data published by ECETOC in Europe on aquatic and terrestrial hazard assessment of DHTDMAC [136]. This report, compiled by a working group of nine ecotoxicologists, concluded that environmental concentrations of DHTDMAC do not pose any hazard to aquatic and terrestrial ecosystems [128,136]. On the contrary, the assessment of the German and Dutch environmental authorities was based on laboratory studies using organic solvents, under conditions of maximum bioavailability. The industry complained that the authorities ignored the field simulation tests, which showed far less toxicity of DHTDMAC in the presence of anionic surfactants and additional material, both of which reduce its bioavailability in surface waters.

Based on the laboratory findings only, the Dutch authorities concluded that in many poorly diluted Dutch surface waters, the PEC value of DHTDMAC exceeds the laboratory determined PNEC values. These conclusions were supported by the German Environmental Protection Agency and led the Dutch and German authorities to demand that detergent manufacturers replace DHTDMAC in fabric softeners by December 31, 1990.

These environmental pressures pushed the detergent manufacturers in Europe to reformulate within less than three years most fabric softener compositions with new cationics, which replaced DHTDMAC [40,42]. The development of a new generation of cationic softeners represents an illustration of environmentally based product development.

The new generation of softeners, which includes mainly ester quats and partly ester amines, are similar to DHTDMAC, for they combine two C16–C18 alkyl chains and deliver the softening effect, with a cationic nitrogen function providing substantivity to fabrics. However, they additionally contain at least one ester group between the alkyl chains and the cationic nitrogen [40–42]. Representative ester quats include the ditallow ester quat of methyl triethanolamonium methosulfate and the ditallow ester quat of dihydroxypropyl ammonium chloride, as well as ester amidoamines, mainly used in Japan.

The ester group provides the potential breaking point in biodegradation sequence (see Fig. 7) [41]. The smaller, fragmented molecules are more easily accessible to attack by microorganisms than the large molecule of the old surfactant, DHTDMAC. Some of the molecular fragments are fatty acids, known to be readily biodegradable. The main intermediate tris(hydroxyethyl)methylammonium cation (MTEA) is a water-soluble cationic, whose biodegradation, as the methylammonium sulfate salt, has been checked and found to result in more than 80% DOC removal [41].

The primary and ultimate biodegradability of the ester quats have been studied under the OECD Confirmatory Test and Coupled Unit Test, showing more than 90% degradation. Anaerobic degradation measured by the ECETOC screening test indicated a high degree of biodegradation [41].

The ecotoxicological behavior of ester quats has been checked extensively, showing a favorable range for surfactants, such as an acute aquatic toxicity (fish,

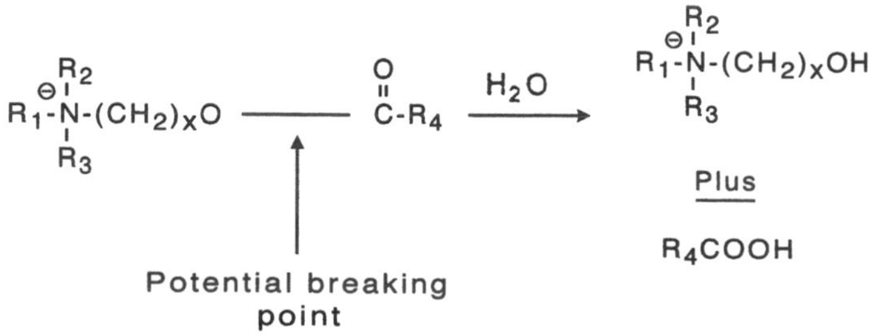

FIG. 7 New ester quat softeners: biodegradation principle. (From Ref. 41.)

daphniae, bacteria) of LC_{50} values of 3–90 mg/L, with low NOEC values for subacute/chronic toxicity (less than 3 mg/L for fish, daphniae, and bacteria and less than 0.3 mg/L for algae).

A series of case studies in Germany related the ecotoxicological test data to realistic concentrations in surface waters [41]. "Worst possible case" scenarios were assumed in the Netherlands, meaning that degradation and elimination would occur only in sewage treatment plants, without any further degradation in surface waters. PNECs were found to be greater than the PEC of ester quats in untreated wastewaters. The estimated ester quat concentrations of 0.01–0.06 mg/L in rivers were one order of magnitude less than PNEC values.

To further confirm these favorable overall conclusions, an extensive monitoring program was carried out, selecting the intermediate product MTEA as an analytically sensitive detectable indicator. Monitoring on more than 800 samples from the Netherlands, France, and Germany confirmed the results of the environmentally approved tests. Short- and long-term toxicity studies, skin irritation and sensitization tests, and mutagenicity and toxokinetic studies have been performed, proving that ester quats and MTEA present no hazard to health when used in softeners [41].

The foregoing data on ester quat development has been presented as a test case for updating present methodology to future methodology for a proven environmental development of a novel surfactant.

2. Oleochemicals vs. Petrochemicals—A Balance for the Future

Since the early 1980s, the choice of raw materials has been judged to influence environmental acceptability. Oleochemically based surfactants have been claimed to be environmentally preferable to those based on petrochemicals [122]. This approach, based on the "green" concept of using "natural," renewable raw materials is in harmony with the increasing environmental consciousness of consumers. It is also supported by risk assessment evaluations, which emphasize good biological degradability under all conditions and no negative impact on the CO_2 balance.

In 1992, Hovelman [122] evaluated the oleochemical potential for the year 2000 and predicted an increase in fats and oils usage from 12 million tons in 1991 to 15 million tons in 2000. This prediction, found to be valid, was based on the high surfactant level in superconcentrated, low-dosage, heavy-duty detergents. In 1992, the surfactant level in European formulations was 13.8% (9.3% anionics and 4.5% nonionics), while the correctly forcaste 2000 level was of 22.5% (10.0% anionics and 12.5% nonionics). Since fatty alcohol sulfates and fatty alcohol ethoxylates are the ecologically preferred species, fatty alcohols seem to be the most important surfactant intermediates. Hovelman anticipated an increasing ratio of "natural" (based on renewable raw materials) vs. synthetic fatty alcohol, following the trend since 1980 (40:60 in 1980, 47:53 in 1990), of 60:40 in 2000 [122].

The renewable resources concept triggered the development of new surfactants, the most prominent ones being the alkyl polyglycosides (APGs), characterized by a positive ecotoxicological profile and associated with superior mildness [122,137]. Strong interest in alkyl polyglycosides as an expression of environmentally conscious development is well reflected in the statistics on patent and literature publications [6,138]. More than 1800 publications between 1980 and 1993, including license agreements with major detergent manufacturers, is an impressive figure [138]. Early patent applications cover synthesis routes to create the R–O-cyclic glycoside, made in a first step with a lower alcohol (usually butanolysis of starch or glucose at 160°C and 100°C, respectively) followed by a transesterification, where one or two sugar units form the hydrophile, the hydrophobe being the fatty alcohol. Some patent applications disclose the APGs preparation by the reaction of the fatty alcohol with monosaccharides, catalyzed by quaternary alkyl sulfates and sulfonates [6].

A multitude of patents disclose applications covering mixtures of APGs with virtually all of the anionic, cationic, and nonionic surfactants (soaps, LAS, FAS, FAES, AOS, FAEO, sulfosuccinates, betaines, protein fatty acid condensate, acyl lactate, sucrose fatty acid esters) and further additives [138].

Steber [137] reviewed APG biodegradation data, including primary and ultimate biodegradation (89 ± 2%), the metabolites test (checking formation of poorly degradable intermediate metabolites), and anaerobic degradation (made via the ECETOC screening test). Scores of 101% and 98%, respectively, were obtained in the last two tests.

Because of the significant economic and political implications of the oleochemical claims, several studies on the environmental impact of the petrochemical-based surfactants were undertaken by the European surfactant industry [135,140,141].

The environmental acceptability of petrochemical and oleochemical-based surfactants has been fully evaluated in comprehensive studies that analyzed the life cycle inventories (LCIs) of all commercially important surfactants, reflecting the significant business and political implications of raw material source–based claims [135,139–141]. A summary of these comprehensive data on commercial surfactants from petrochemicals (Pc) and/or oleochemicals (Oc) such as LAS (Pc), FAS (Pc/Oc), FA(EO)$_3$ (Pc/Oc), FA (EO)$_7$ (Pc/Oc), FA (EO)$_{11}$ (Pc/Oc), APG (Oc), and soaps (Oc) has made available [135,140,141]. The overall conclusion is clearly stated in the abstract of the original study [135,141]:

> Based on the findings, no unequivocal technical rationale exists for claiming overall environmental superiority, neither for production of individual surfactants nor for the various options for sourcing from petrochemical and oleochemical/agricultural feedstocks and minerals. The value of the study lies in allowing each manufacturer to assess opportunities for improving the environmental profile of its surfactants and intermediates

An interesting remark by Vogel [139] on the lack of sustainability of the petrochemical-based surfactants is worth recalling. Since it had been suggested that these surfactants are not sustainable because of the finite nature of the basic raw material, crude oil, there is a need to make sure that the resources for the world economy required by future generations are not depleted. Vogel casts doubts on the sustainability of virtually all modern products in the years to come. Increasing pressure from the population growth rate will impact natural resources of land, energy, water, and food. The United Nations Population Fund estimates that by 2150 there will be an average of 189 people for every square kilometer of land on the surface of the earth, i.e., six times more than today. Such a dramatic scenario requires expansion of production capacities, solving sustainability problems for all raw materials, and keeping open all surfactant options, to provide the best that industry can offer for meeting the tremendous forecast demand [139] .

It seems that over the long term, both oleo- and petrochemicals will be compatible and competitive in all aspects of environmental acceptability, economics, and technology, offering a greater choice of feedstocks to detergent manufacturers.

3. Molecular Structure–Environmental Acceptability Relationship

The relation between molecular structure and surface properties of a surfactant and its biodegradability and toxicity properties has been traced by Rosen [142]. Generally, the greater the rate of biodegradation to innocuous products, the less critical is the toxicity of the undegraded surfactant and its degradation products. Surfactants with branched alkyl chains are less biodegradable, particularly if branching is adjacent to the terminal methyl groups of the chain. In general, the greater the distance between the hydrophilic group and the far end of the hydrophobic group, the more rapid is the rate of biodegradation. In contrast, the rate of degradation decreases with an increase in the number of alkyl chains in the molecule and the number of methyl groups on the alkyl chains.

For ethoxylated surfactants, biodegradability also decreases with an increase in the number of oxyethylene groups. Replacement of oxyethylene groups by oxy-propylene or oxybutylene groups further decreases biodegrability. Secondary alcohol ethoxylates degrade more slowly than primary alcohol ethoxylates, even when the alkyl chain is linear. Surfactants with lower CMC, for instance, nonionics (10–100 times lower than that of anionics), should be the more preferred surfactants for lower-dosage, environmentally friendly detergents.

Surfactant toxicity depends on both the tendency to adsorb onto organisms and the ease of penetration through their cell membranes. The tendency to adsorb onto organisms seems to be directly related to the tendency of the surfactant to adsorb on the interfaces. The ease of penetration appears to decrease with an increase in the area occupied by the surfactant at the interface.

The aqueous solubility of surfactants in the presence of other water-soluble or water-dispersible substances also has an environmental impact, since the effect of a surfactant on other chemical species depends on its concentration in the aqueous medium that contacts the species.

4. New Applications of "Green" Surfactants in Environmental Remediation

A comprehensive review of the use of surfactants in environmental remediation has been published [27]. It discusses the size and background of the environmental problem of groundwater contamination and the manner, practical and scientific, in which surfactants are being used to address the problem.

Within the past few years there have been several successful tests to demonstrate the feasibility of using surfactants for a restoration of aquifers contaminated by solvents and fuels. It was reported that the addition of surfactants appears to improve the biodegradation of refractory substances, probably due to adherence to cell membranes. This led to the challenging concept of optimum surfactant concentration in wastewater [6]. In fact, the potential for using surfactants for rapid, relatively economical remediation of contaminated aquifers has been obvious since their use in the EOR (enhanced oil recovery) field.

There had always been a general ecological reluctance to use this technology for aquifer restorations due to environmental concerns over surfactants. However, the extent of damage by the widespread release of hydrocarbon and chlorocarbon liquids into groundwater, combined with the availability of surfactants of proven biodegradability and nontoxicity, has revived interest in surfactant-enhanced aquifer remediation.

This development, as an important new application of surfactants, opens up a number of new area for research and technical optimization. A critical area is the assessment of natural biodegradability of the surfactants and their micellar forms in the aquifers. The potential environmental risk of the escape of surfactant micelles containing solubilized contaminants from the treated zone of the aquifer and migrating toward a drinking water supply, a lake, or a stream is obvious. This potential could be critical for determining if the technology of surfactant-enhanced aquifer remediation can be used. Howver, the availability of many "green" surfactants gives a much higher feasibility to this surfactant–environment-compatible technology [27].

C. Environmental Commitment of the Detergent Industry

The soap and detergent industry manages environmental and health issues by coordinated efforts of industry groups that help companies to develop and interpret studies on environmental subjects. Some leading groups in the United States are the Council for LAB/LAS Environmental Research (CLER), the Alkylphenol and Ethoxylates (APE) Panel of the Chemical Manufacturers Association (CMA), and the Technical Department of the Soap and Detergent Association (SDA) [143,144]. These groups work in coordination with sister organizations in Europe, such as the European Center of Studies on LAB/LAS (ECOSOL) and the APE Task Force of the European Chemical Industry Council. These organizations are technically focused groups, composed of scientists from major global suppliers of raw materials and detergent products, whose mission is to conduct research and to provide information on environmental, health, and safety topics, promoting a unified and consistent position on scientific, regulatory, and legislative issues [144].

In Japan, the Ministry of International Trade and Industry (MITI) is requesting each company to prepare a voluntary plan for environmental conservation. The Environment Agency has published an action guideline for an "Environment-Friendly Company." On the other hand, the Federation of Economic Organizations of Japan announced in 1991 the Keidanren global environmental charter, which includes guidelines for corporate action. According to the Environment Agency's investigation in 1993, 45% of manufacturing industries have already established management policies for environmental conservation, while 9% were planning to institute them [124,145].

The increasing cooperation between these organizations and governmental authorities is worthy of mention. The regulatory future of nonylphenol ethoxylates (NPEO) in the United States and Canada serves as a good example, as reported by the *News Bulletin* [146]. In anticipation of these forthcoming regulations, the APE research council has developed an "NPE Environmental Management Program" to help companies comply with the new water guidelines for NPs in the United States and potential future regulations in Canada. The NPE environment management program will:

- Promote the forthcoming U.S. regulations through a major outreach program to end users' NPEs and related products, including workshops with formulators and end-user associations, technical guidance materials, and a new Web site
- Develop voluntary, risk-based levels for commercial NPEs and related compounds: low-mole nonylphenol ether carboxylates, which are the transitory degradation intermediates of NPEs

- Monitor targeted waterways in conjunction with government agencies to help ensure that water quality guidelines are being met
- Work with facilities to apply appropriate product stewardship and treatment methods to ensure compliance with the guidelines, including making available technical experts to work directly with end users to determine appropriate control methods

As said by the EPA executive director: "We think that voluntarily including NPEs and the degradation intermediates in the program will make end users more confident that they are protecting the environment so they can continue using NPEs" [146].

Recently, CESIO reported [147] on an ICCA (International Council Of Chemical Associations) initiative to prepare harmonized, internationally agreed data sets, and initial hazard assessments under the Screening Information Data Set (SIDS) program of the OECD. The goal of this work is the improvement of the current database of approximately 1,000 OECD high-production-volume (HPV) chemicals based on existing information or, where necessary, by additional testing. The target date is the end of 2004. CESIO will include in this program 10 main surfactant groups: amine oxides, quarternary ammonium compounds, betaines, sorbitan oleate, fatty acid glyccrides, fatty alcohol sulphates, LAB/LAS, alpha olefin sulphonates (alkane, alkene, and hydroxyalkane sulfonic acid), hydrotropes, other surfactants (including alkanolamides).

The industry's environmental commitment is well reflected in modern management attitudes, of which the Procter & Gamble "Environmental Management Framework" is presented as a typical example that deserves to be followed. The framework is based on a "3 Es" sustainable development balance between economy, environment, and society (everyone) [134]. This responsible attitude follows the sustainable development concept [134,148] based on two widely quoted definitions: "Development that meets the needs of the present without compromising the ability of future generations to meet their own needs" [134,149] and "Improving the quality of human life within the carrying capacity of supporting ecosystems" [134,150]. The overall objective of P&G's economically and technically feasible and socially acceptable Environmental Management Framework is broken into four specific goals [134]:

1. Ensure human and environmental safety.
2. Ensure regulatory compliance.
3. Ensure efficient resource use and waste management.
4. Ensure that social concerns are addressed.

D. Code of Environmental Practice for Household Detergents

The valuable particular initiative of Procter & Gamble is fully compatible with the basic commitment of advanced European industry. The European Detergent Industry Association (AISE) developed a Code of Good Environmental Practice for Household Detergents [151]. This code (Box 1), even though it codifies some industry behavior already implemented during recent years, represents a state-of-the-art code with specific environmental goals and well-defined detergent industry responsibilities, sometimes discharged to consumers.

AISE has long been committed to protecting consumers and the environment, and AISE members have been active in establishing, publishing, and implementing environmental policies. In order to maintain and promote further this commitment to the reduction of the environmental impact of detergents, AISE developed this present code of practice. Compliance with this code of practice signifies a clear and unequivocal commitment to consumer and environmental safety as well as to environmentally sustainable development. This code requires manufacturers to commit to continued environmental innovation, in partnership with consumers and other stakeholders who also influence the environmental impact of household products. AISE is committed to reviewing this code of practice at least every five years and to report progress at least every two years.

This voluntary scheme is open to all manufacturers, importers, or other persons (subsequently referred as "the manufacturer") whether or not affiliated with AISE national associations, placing household detergents on the market:

1.   The manufacturer shall design the composition and packaging of its products, taking into account major impacts on the environment identified by acknowledged scientific criteria.

2.   The manufacturer shall provide consumer information designed to encourage the correct use of product. This information will be based on life cycle analysis considerations for the product category.

3.   The safety evaluation of products, with respect to the consumer and the environment, must comply with the principles of the Guidelines for Risk Assessment established by the EU Commission in the context of Regulation EU 1488/94, which describes the environmental and consumer parameters for safety reassurance. Any actions indicated from the evaluation shall be carried out.

4.   The manufacturer's product must fully comply with all relevant environmental and consumer protection legislation, including the European Union Directives on biodegradability of surfactants (EU 73/404, 82/242, 82/243), classification and labeling under the Dangerous Preparations Directive (EU 88/379), the limitation to marketing of some dangerous substances and preparations (EO76/769 + adaptations), and the Packaging Waste Directive (EU 94/62), in addition to complying with the EU Recommendation on Ingredients Labeling (EU 89/542) and, when appropriate, dosage instructions covering different soils, loads, and water hardness.

5.   The environmental advertising claims made by the manufacturer for its products must be truthful, supported by factual data, and designed to inform the consumer. They must meet the requirements of the specific ICC (International Chamber of Commerce) codes for environmental advertising claims or equivalent national codes, providing guidance on what kinds of claims are acceptable and how they should be supported.

6.   Any manufacturer that commits itself to this Code of Practice and its appendix shall provide a written declaration signed by its legal representative to AISE that it fully complies with all the principles of the present code of practice and gives a commitment of continued work to use life cycle analysis and risk assessment techniques to identify and collect any necessary further data to implement further reductions in environmental impact.

Box 1.   AISE Code of Good Environmental Practice for Household Detergents. (From Refs. 134,151.)

Any manufacturer, importer, or other persons (subsequently referred to as manufacturer) within the European Detergent Industry, whether or not affiliated with the AISE national association, that adopts the AISE Code of Good Environmental Practice for household laundry detergents should commit itself to striving to achieve the following targets. These are targets for the European Economic Area (EEA), starting in January 1997. They may need to be adjusted for individual countries, depending on ongoing environmental progress, washing habits, and consumer choices. They are established for five years.

ENERGY SAVING (kWh per wash cycle):
Target: 5% reduction of the energy used in the washing process.

PRODUCT CONSUMPTION
Target: 10% reduction of the detergent consumption per capita.

PACKAGING CONSUMPTION (includes primary and secondary packaging)
Target: 10% reduction of packaging consumption per capita.

BIODEGRADABILITY (measured by recognized biodegradability tests)
Target: 10% decrease of the consumption per capita of the organic ingredients of household laundry detergents that are not inherently biodegradable.

1. Manufacturers who make these commitments will also commit to provide baseline data (1996 data) from their operations on each of the environmental progress areas listed and to track and report progress made to AISE.

2. The commitment of AISE is to monitor and collect these data and to include it in one report to be published per country and on a European basis at least every two years.

---

*Covering household laundry powder and liquid, heavy- and light-duty detergents.

Box 2. AISE Code of Good Environmental Practice for Household Laundry Detergents. (From Refs. 134,141.)

Following this code, the European detergent industry proposed four targets concerning energy savings, product and packaging consumption, and biodegradability (Box 2).

E. Reducing Wastewater Salinity from Detergents—An Israeli Environmental Contribution

1. Past Detergent Regulation in Israel (1975–1999)

In Israel, detergent washing powders are regulated by Israeli Standard IS 438, issued in 1982 [103], superseding the Israeli Standard from 1975. IS 438 stipulated a set of compulsory requirements, covering several issues, such as:

- Ingredient (surfactants, phosphates, oxygen donors, silicates, chlorides) specifications and range limits.
- Packaging
- Nominal weights and permitted deviations
- Labeling
- Test methods

Specific requirements have been established according to the method of laundering (hand or machine washing) and the type of fibers for which the powder was intended (wool, cotton, polyester, combinations).

The logic of a major part of these requirements was based on the desire to guarantee a minimal level of performance to the consumer. Because Israel has very hard water (350–400 ppm $CaCO_3$), an improperly formulated detergent would give poor washing performance. The standard imposed a minimal amount of detergent (for a 4-kg wash load) that shall contain at least 20 grams of total active matter composed of anionic/soap/nonionic surfactants. Also, STPP, as the best builder, was compulsory, only 20% of the phosphates being permitted to be replaced by substitutes.

Eutrophication problems have not been reported in Israel, unfortunately, due to lack of rivers and lakes. The orginal standard allowed 0.2–2% of active oxygen, without specifying the bleaching agent. In April 1984, an amendment was added, due to environmental concern, establishing a limit on borate levels, expressed as 12% sodium perborate tetrahydrate maximum. Chlorides (as sodium chloride) were limited to 10% in machine-washing powders and to 20% in hand-washing powders.

Use of biodegradable LABS was covered by Israeli regulation beginning in the late 1960s. No restrictions were made on use of alkylphenol ethoxylates (APEO).

2. New Detergent Regulation in Israel

Several driving forces imposed regulatory reform on detergents. Officially, the Israeli government passed in November 1988 a new amendment to the Standards Law (5713) of 1953, whose objective was to ensure preservation of public health, public safety, and environmental protection. The adoption of this regulatory apparatus, associated with the elimination of performance-related requirements was expected to remove some local detergent restrictions toward market liberalization and regulatory harmonization.

Concurrently, the Technical Soaps and Detergents Committee of the Standards Institution of Israel, composed of detergent manufacturers (from the Manufacturers Association of Israel and multinational detergent companies), consumer organizations, academic institutions, and governmental authorities (Ministry of Health, Ministry of Trade and Industry, Ministry of the Environment), had began in 1997 to elaborate a new detergent standard, following previous environmental pressures of the Ministry of the Environment (MOE).

As early as 1995, the Industrial Wastewater Division of the MOE had shared new environmental concerns with manufacturers and importers. The data brought to this task force revealed severe wastewater contamination in Israel, a country that recycles its wastewater for agricultural irrigation purposes. Detergents have been found to contribute 80–90% of the total boron, about 41% of total sodium, and about 7% of total chlorides added to municipal sewage [105]. These data were used by the MOE as a basis for a proposal for a new and unique Israeli standard for detergents. While the proposal was met initially with objections and demands for risk assessment evaluations by manufacturers and importers, the severity of the problem was soon recognized. A fruitful cooperation, which included joint funding of surveys by the Manufacturers Association of Israel (MAI) and MOE began in 1996. The multinational companies involved in the Israeli market joined these efforts, supplying relevant data from EU and U.S. regulatory situations and including internal data as well.

Laboratory analyses of detergents and washing powders were undertaken in 1997 to determine the pollution levels of different types of detergents (regular, compact, and liquid). Mean boron, sodium, and chloride discharges from domestic detergents are presented in Table 10 [105]. These data were calculated for maximum

TABLE 10 Mean Boron, Sodium, and Chloride Discharges from Domestic Detergents

Detergent type	Average dosage (g/4 kg laundry)	Boron (mg/kg laundry)	Sodium (mg/kg laundry)	Chlorides (mg/kg laundry)
Regular washing powder	100	200	6,000	550
Compact washing powder	40	80	2,500	200
Heavy-duty laundry liquid	100	—	500	20

concentrations of 12% $NaBO_3 \cdot 4H_2O$, as allowed by IS 438 (1982), in regular and compact powders and for an average level of 25% sodium in washing powders and 2% sodium in laundry liquids. The wash load was based on 4 kg laundry per cycle. The calculation of sodium content of washing powders and laundry liquids is presented in Table 11. This table includes only the sodium donor ingredients. On the basis of these data, it was agreed that the boron, sodium, and chloride concentrations should be regulated in detergents in Israel because of a combination of specific factors: its high release of wastewater for irrigation, its dry climate, high background levels of these elements, and future plans to produce water through desalination.

The new Israeli standard [104] adopted and implemented these agreements, as presented in Table 12 [105] in comparison with the previous standard.

The impact on reduction of boron and sodium was anticipated by MOE as follows [105].

(a) Boron. The detergent industry was required to reduce boron concentrations by 60% in four years and 94% in eight years. The anticipated reduction in boron contamination from detergents in wastewater is shown in Figure 8 [105]. Since it is estimated that detergents are responsible for some 85% of the contamination of

TABLE 11 Typical Sodium Donor Ingredients in Israeli Laundry Detergents

Ingredient type	Powders		Liquids	
	% Ingredient	% Sodium	% Ingredient	% Sodium
LABS	4–7	0.26–0.47	10–15	0.67–1.0
FAS	1–3	0.06–0.20		
Soap (medium-chain fatty acids)	—	—	10–15	1.0–1.5
Soap (long-chain fatty acids)	3–6	0.22–0.44	3–5	0.1–0.4
STPP	25–50	7.8–15.6		
Sodium carbonate	0–30	0–13		
Sodium perborate, sodium percabonate	8–15	2.34–4.38		
Sodium citrate			0–3	0–0.5
Sodium silicate	8–12	1.5–2.25		
Sodium sulfate	10–40	3.2–12.9		
Total		24.7–27.3		1.5–3.0

TABLE 12 Boron, Chlorides, and Sodium in the Israeli Standard for Washing Powders in Comparison with the Previous Standard

Element	Previous standard content	New standard content and timetable
Boron (B) as NaBO$_3$·4H$_2$O	Equaling 8.4 g boron/kg product, 12% borates max.	8.4 g/kg product max – until 6/30/1999 7.0 g/kg product max – from 7/1/1999 5.6 g/kg product max – from 7/1/2000 4.2 g/kg product max – from 1/1/2002 3.5 g/kg product max – from 1/1/2003 0.5 g/kg product max – from 1/1/2008
Chlorides (as Cl ions)		
Powders for washing machines	60.7 g/kg product	40 g/kg product max
Powders for hand washing	121.4 g/kg product	90 g/kg product max
Sodium (Na)	Not included; current content up to 6 g/kg laundry in regular laundry powder (not compact)	5 g/kg laundry – from 7/1/1999 4 g/kg laundry – from 7/1/2001

Source: Ref. 105.

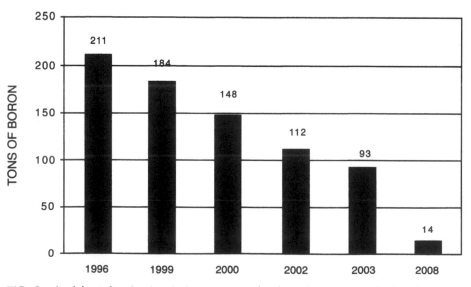

FIG. 8 Anticipated reduction in boron contamination of wastewater in Israel. (From Ref. 105.)

wastewater by boron in Israel, it is hoped that the new standard will solve the problem of boron pollution.

(b) Sodium.   The sodium concentration has not been reduced in washing powders by replacing the sodium ingredient donors, such as sodium carbonate, STPP, and sodium sulfate, with potential potassium substitutes, for it is a very costly operation for the manufacturers and consumers. Therefore the standard requirement was based on a lower dosage.

Taking into account an average of 25% sodium content in washing powders, a maximum allowable dosage of 4 g sodium/kg laundry, imposed since July 2001, translates into a maximum dosage of 56 g powder per 4 kg laundry. Compact powders comply with this requirement, the low dosage recommended by manufacturers being around 40 g. However, regular-density powders will not be able to offer the consumer a fair performance at such a dosage and will probably change to compact or perish from the market.

Since sales of regular detergents constituted about 25% of detergent sales in Israel in 1997, it is estimated by the MOE that the quantity of sodium contamination will decrease by some 700 tons. In absolute figures, the anticipated reduction in sodium contamination was 9,570 tons/year until July 1, 1999, and 8,850 tons/year after July 1, 2001. In practice, even a greater reduction is expected, due both to the decreasing share of regular detergents in the market and to the reduction of sodium chloride in all types of detergents.

The cooperation among governmental authorities, industry, and academia in achieving the pioneer regulations in Israel is one more example of outstanding environmental awareness.

F. The European Eco-Label

The "Community Legislation in Force" establishing ecological criteria for the award of the community "Eco-label" to laundry detergents may be considered the ultimate environmental directive of the European community. First proposed in 1995 [80], the Eco-label award scheme was fully revised and legislated in 1999 [152].

General requirements and specific ecological criteria established by this regulation apply to the award of an Eco-label to laundry detergents. The target of the ecological criteria is to minimize:

- Water pollution by reducing the quantity of harmful ingredients
- Waste production by reducing packaging and promoting its reuse and/or recycling
- Energy consumption by promoting the use of low-temperature detergents

In addition, the target is to maximize consumers' environmental awareness.

The ecological criteria, selected as the basis for awarding an Eco-label, cover ingredients and packaging by using a scoring system. Eight ecological parameters assess ingredients: total chemicals, critical dilution volume, phosphates, insoluble inorganics, soluble organics, nonbiodegradable organics (aerobic), nonbiodegradable organics (anaerobic), biological oxygen demand (BOD). Packaging criteria include total packaging and total virgin material in packaging. The criteria are related to grams per wash load, not percentages in the product.

FIG. 9 The Daisy, the official logo of the European Eco-label.

For awarding an Eco-label, the sum of scores related to the eight criteria must exceed a specified threshold value, defined by 99/476/EEC as 45. Weighting factors are assigned for each criterion to take into account the different impacts of the selected criteria. Thus, the maximum grams per wash amounts that disqualify a detergent at each specific criterion and the respective weighting factors (WF) are [6]:

Total chemicals   200 g/wash;   WF = 3
Phosphates   50 g/wash;   WF = 2
Insoluble inorganics 50 g/wash;   WF = 0.5
Soluble inorganics   140 g/wash;   WF = 0.5
BOD 140   g/wash;   WF = 2

A functional dosage must be recommended by the manufacturer for a water hardness of 2.5 mmol $CaCO_3$/L. This is taken as a reference dosage for calculation of ecological criteria.

Since Eco-detergents call for lower-dosage, lower-temperature washing cycles and environmentally conscious formulations, the performance of these detergents is lowered. For this reason, the products shall be compared in their washing performance to reference detergents of the same type and shall fulfill minimum requirements, according to the EU Eco-detergent performance tests. The test conditions (type of soiling and substrate) are still subject to improvement to achieve the best representative simulation of overall washing requirements.

Consumer information includes environmentally conscious use instructions, education in presorting laundry, washing with full loads, avoiding overdosage, and choosing low-temperature wash cycles.

Similar ecological criteria for the award of the Eco-label to detergents for dishwashers were established by Commission Decision 99/427/EEC [153].

The Daisy (Figure 9) is the official logo of the European Eco-label award. Any further information about the Daisy and the European Eco-labeling scheme is presented on the Eco-label Website at:

http://europa.int/comm/environment/ecolabel/index.htm

and by the Daisy News at:

http://europa.eu.int/comm/environment/ecolabel/news.htm [154].

Officially, the Daisy is the award logo given to a product that meets the minimum standard Eco-requirements. Symbolically, the Daisy is a flower given by the environment to the detergent industry with thanks.

REFERENCES

1. Jakobi, G.; Lohr, A. *Detergents and Textile Washing*; VCH Verlag:Weinheim, Germany, 1987; 167–187.
2. Malz, F. Loading surface waters with surfactants. In *Detergents in the Environment*; Schwuger, M.J., Ed.; Marcel Dekker:New York, 1997.
3. Rosen, M.J.; Goldsmith, H.A. *Systematic Analysis of Surface Active Agents,* 2nd Ed. Wiley-Interscience:New York, 1972.
4. Longman, G.F. *The Analysis of Detergents and Detergent Products*; Wiley-Interscience: New York, 1978.
5. Gesetz Uber detergentien in wasch und reinigungsmitteln, BGBI I, 1653, 5.9.1961.
6. Matzner, E.A. Overview of regulation and environmental issues affecting laundry detergents. *Powdered Detergents*; Showell, Michael S., Marcel Dekker:New York, 1998; 313–344.
7. Davidsohn, A.S.; Milwidsky, B. *Synthetic Detergents,* 7th Ed.. Longman, New York, 1987.
8. Cox, M.F. *Detergents and Cleaners: A Handbook for Formulators;* In Lange, K.R. Ed.; Hanser: Munich, 1994; 43–90.
9. Berna, J.L.; Cavalli, L. *LAS Facts & Figures*; ECOSOL Publication (European Center of Studies on LAB-LAS), 1999.
10. Ho Tan Tai, L. *Formulating Detergents and Personal Care Products*; AOCS Press: Champaign, IL, 2000.
11. LAS: The biodegradable, environmentally friendly detergent ingredient. Houston, TX, Vista Publications.
12. Heinze, J.E.; Britton, L.N. Anaerobic biodegradation: environmental relevance. In: Proceedings of the 3rd World Conference on Detergents Cahn, A. Ed.; AOCS Press:Champaign, IL, 1993; 235–239.
13. Council directive 73/404/EEC, on the approximation of the laws of the member states relating to detergents, 22.11.1973.
14. Council directive 73/405/EEC, on the approximation of the laws of the member states relating to methods of testing the biodegradability of anionic surfactants, 23.11.1973.
15. T-73-260, AFNOR, June 1981.
16. Technical report no. 70, Water Research Centre, 1978.
17. Council Directive 82/243/EEC, amending Directive 73/405/EEC, Council Directive 82/242/EEC, relating to methods of testing the biodegradability of nonionic surfactants and ammending Directive 73/404/EEC, 31.3.1982.
18. Schulze Rettmer, R. SÖFW J 1976, *107*, 427.
19. Krussman, H.; Hloch, H.G. SÖFW J 1981, *107*, 436.
20. McAvoy, D.C.; Rapaport, R.A., et al. Environ. Toxicol. Chem 1993, 977–987.

21. Zoller, U. The case of the persistent ("hard") nonionic surfactants in the environment. *Toxicological and Environmental Chemistry;* OPA, 1998; Vol. 66, 145–157.

22. Zoller, U. J. Am. Oil. Chem. Soc. 1985, *62*, 1006–1008.

23. Zoller, U. J. Environ. Sci. Health 1992, *A27*, 1521–1532.

24. Ainsworth, S.J. Chem. Eng. News January 23,1995; 30–53.

25. Zoller, U. Groundwater contamination by surfactants. ln *Groundwater Contamination and Control;* Zoller, U, Marcel Dekker:New York, 1994, 273–292.

26. Sundaram, N.S., et al. Chemosphere 1994, *29*, 1253–1261.

27. Harwell, J.H.; Knox, R.C.; Sabatini, D.A. Surfactants in environmental remediation: aquifer restoration. ln Karsa, D.R., Ed.; New Products and Applications in Surfactant Technology, Annual Surfactant Review; Sheffield Academic Press:England. 1998; 30–58.

28. Gloxhuber, G., et al. Anionic Surfactants. *Surfactant Science Series;* p. 43 Marcel Dekker:New York, 1992; 43pp.

29. 2nd Report of the Technical Committee on Detergents and the Environment, UK Dept. of the Environment, Dec. 1994.

30. Tabor, C.F. Jr, et al. Natl. Mtg. ACS Div. Environ. Chem 1992, *22*, 52–55.

31. Fairchild, J.F., et al. Environ. Toxicol. Chem 1993, *12*, 1763–1765.

32. Hager, C.D. Alkylphenol ethoxylates: biodegradability, aquatic toxicity and environmental fate. *New Products and Applications in Surfactant Technology, Annual Surfactant Review* Kasa, D.R.., Ed.;1998 Sheffield Academic Press:England, 1998; 1–29.

33. Huges, P.I.; Peterson, O.R.; Markarian, R.K. Comparative biodegradability of linear and branched alcohol ethoxylates. Proceedings of the American Oil Chemists Society Annual Meeting; AOCS Press:Champaign, IL, 1989.

34. Alkylphenol ethoxylates in the environment: an overview, AEP, Chemical Manufacturers Association, August 1994.

35. An environmental assessment of APEO and AP, published by Friends of the Earth, Scotland, Jan. 1995.

36. Talmage, S. Environmental and human safety of major surfactants; The Soap and Detergent Association:Boca Raton, FL, Lewis, 1994.

37. Benson, W.H.; Nimrod, A.C. *The Estrogenic Effects of Alkylphenols Ethoxylates;* APE, Chemical Manufacturers Association, August 1994.

38. Zoller, U. Water Sci.Technol 1993, *27*, 187–194.

39. Zoller, U. Environ. Sci. Pollut. Control Ser. 1991, *11*, 273–282.

40. Sebold, U. Fabric softeners wordlwide. ln Proceedings of the 3rd World Conference on Detergents. Cahn, A., Ed.; AOCS Press:Champaign, IL, 1993; 88–94.

41. Puchta, R.; Krings, P.; Sandkühler, P. Tenside Surf. Det. 1993, *30*, 186–191.

42. Schröder, U. Fabric softener market development worldwide. *Proceedings of the 4th World Conference and Exhibition on Detergents;* AOCS Press:Champaign, IL, 1998; 142-148.

43. Morse, G.K.; Perry, R., et al. Sci. Total Environ 1995, *166*, 179–190.

44. ECE/CHEM 80 Report on Substitues for Tripolyphosphate in Detergents, United Nations, New York, 1992.

45. Hauptausschuss Phosphate: Phosphor, Wege und Verbleibin der Bundesrepublik Deutschland, Verlag Chemie, Weinheim—New York, 1978.

46. Matzner, E.A. The detergent regulatory and environmental situation. In *Detergents and Cleaners: A Handbook for Formulators;* Lange, K.R. Ed.; Hanser:Munich, 1994; 205–233.

47. Von Lersner, H.F. Proceedings of 2nd World Conference of Detergents. Baldwin, A.R., Ed.; In AOCS Press, 1987; 199–200.

48. SCOPE, Scientific Committee on Phosphates in Europe, November 1994.

49. Mueller, T.H.; Kirschbaum, E.J. Facing future challenges—European laundry products on the threshold of the twenty-first century. Proceedings of the 4th World Conference and Exhibition on Detergents 1998; 93–106.

50. Berth, P.; Krings, P.; Verbeek, H. Tenside Deterg 1985, *22*, 169.

51. Rieck, H.P. Builders: the backbone of powdered detergents. In *Powdered Detergents*. Showell, Michael S Ed.; Marcel Dekker:New York, 1988; 43–108.

52. Clayton Associates. Potential environmental and economic benefits of discontinuing the use of phosphates in laundry detergents. Draft Report to the U.S. Environmental Protection Agency, 1992.

53. Dokulil, M.T., et al. Hydrobiologia 1992; 243–244 389-394.

54. Von Gunten, H.R.; Zobrist, J. Better drinking water quality due to lower phosphate concentrations. Neue Zurcher Zeitung, Fernausgabe No. 172, July 27, 1994; 37 pp.

55. Bertram, P.E. J. Great Lakes Res 1993, *19*, 224–236.

56. Charlton, M.N., et al. J. Great Lakes Res 1993, *19*, 290–309.

57. Lee and Jones. NOAAA Report, October 1979.

58. Lorenzen. U.S. EPA 560/11-79-011, 1979.

59. Toy, A.D.F.; Walsh, E.N. *Phosphorous Chemistry in Everyday Living*; American Chemical Society:Washington DC, 1987.

60. Wilson, B.; Jones, B. *The Phosphate Report*; Landbank Enviromental Research & Consulting:London, 1994.

61. Pollutants in cleaning agents, U.K. Department of Environment, March 1991.

62. Mayer, T., et al. Water Qual. Res.J. Can 1996, *31*, 119–151.

63. Il Sole, August 27, 1991.

64. Post, W.M., et al. Am. Sci. 1990, *78*, 310.

65. Rieck, P. Builders: ecology, cost and performance. In *Proceedings of the 3rd World Conference on Detergents*; Cahn, A., Ed. AOCS Press: Champaign, IL, 1993; 161–167.

66. Smeets, F.L.M. Natuor en Techniek 1990, *58*.

67. Zeolites for Detergents: As Nature Indeed. Zeodet (Association of Detergent Zeolite Producers) CEFIC, 2000.

68. Kurzendorfer, C.P.; Kuhm, P.; Steber, J. Zeolites in the Environment. In *Detergents in the Environment*; Schwuger, M.J., Ed.; Marcel Dekker:New York, 1997; 127 pp.

69. Saijo, H.; Tanaka, A.; Noguchi, T.; Kasai, K.; Tagata, S. Technology developments of detergent builders. Proceedings of 4th World Conference and Exhibition on Detergents; AOCS Press:Champaign IL, 1998; 183–189.

70. Adams, C.J., et al. Zeolite MAP: the new detergent zeolite in progress. *Zeolites and Microporous Materials, Studies in Surface Science and Catalysis,* Elsevier, 1997; Vol 105, 1667 pp.

71. Cristophliemk, P.; Gerike, P.; Potokar, M. Zeolites. In de Oude, N.T., *Handbook of Environmental Chemistry*; Detergents part F. Springer Verlag:Berlin, 1992; Vol 3, 205pp.

72. Happi, July 1993, 32 pp.

73. Greser, R. A Comparative Study of the New Builder Systems for Laundry Detergents. Rhone Poulenc.

74. Life cycle study heralds phosphate detergent revival. Environ. Business February 9, 1994.

75. Lactobionic Acid and Its Derivatives. Solvay Enzymes. GMBH & Co.

76. Frimmel, F.H. Physicochemical properties of EDTA and consequences for its distribution in the aquatic environment. In Schwuger, Johann, Ed.; *Detergents in the Environment, Milan*; Marcel Dekker:New York, 1997; 289–311.

77. Kiessling; Kaluza Nitrilotriacetic Acid. In *Detergents in the Environment, Milan*. Schwuger, Johann, Ed.; Marcel Dekker:New York, 1997; 270 pp.

78. Potthoff-Karl, B. SÖFW J 1994, *120*, 104–109.

79. Paik, Y.H., et al. (Rohm & Haas). Adv. Chem. Ser. 1996, *248*, 79–98.

80. Commission Decision on Ecological Criteria for the Award of the Community Ecolabel to Laundry Detergents European Commission, XI 454/94, Brussels, 2/14/1995.

81. Trolli, F., et al. J. Environ. Polym. Degrad. 1994, *2*, 89–97.

82. Beck, R. Ger. offen., DE 4, 314, 659.

83. Houston, J.H. Inform 1997, *8*, 928–938.
84. Videau, D.; Gosset, S. Eur. Pat. Appl. 511081 to Roquette Frères, 1992.
85. Santacesaria, E., et al. Eur. Pat. Appl. 472042, 1992.
86. Rieck, H.P. Eur. Pat. Appl. 164514 to Hoechst AG, 1985.
87. SKS-6, Detergent Builder of the future, Höechst AG, Frankfurt, May 1993.
88. Nijs, H.; Godecharles, V.; May, B.H. SÖFW J 1985, *111*, 149–150, 203-205.
89. Porter. Proceedings of Fourth International Congress on Surface Activity; 1980; Vol III, 187 pp.
90. Boittiaux, P.; Joubert, D.; Talvet, P. Eur. Pat. Appl, 561,656 to Rhone Poulenc Chimie, 1993.
91. Pancheri, E.J. WO 096/38524, WO 096/38525, WO 096/38526 to Procter & Gamble, 1996.
92. Bertleff, W.W. New chelating agents for detergents and cleaners. In New Horizons, an AOCS/CSMA Detergents Industry Conference. Coffey, R.T., Ed.; Champaign, IL, 1996; 97–112.
93. Coffey, R.T.; Gudowicz, T.H. Inform. 1992, *3*, 656–664.
94. Nabion, New Generation Soluble Builder, Rhone Poulenc S.A. 1994.
95. Zini, P. *Polymeric Additives for High-Performing Detergents*; Technomic:Lancaster, PA, 1995.
96. Upadek, H.; Poethkow, H.; Salz, R.; Riebe, H.J.; Seiter, W. WO Patent Appl. 92/15663 to Henkel KGaA, 1992.
97. Pretty, A.J.; Fraser, D.G.; Hardy, P.A. WO Patent 92/03525 to Procter & Gamble, 1992.
98. Pancheri, E.J. U.S. Pat 5,378,388 to Procter & Gamble, 1995.
99. Leading International Market Reserch Institute. Information by Henkel KGa A, Düsseldorf, October 1999.
100. Hauthal, H.G. SÖFW-Journal 122 1996, *13*, 899.
101. Borates as multifunction cleaning agents, Borax Europe Ltd., 1999.
102. Boron from the standpoint of environment, IVA-Report 33. Acad. Eng. Science: Stockolm, 1970.
103. IS 438, Israeli Standard for Washing Powders, May 1982.
104. IS 438, part 1: Cleaning Powders—Environmental Requirements and Labeling Requirements: Laundry Powders, October 1999.
105. Weber, B. Reducing Wastewater Salinity from Detergents. Israel Environment Bulletin 2000; 15–17.
106. Dietz, F. GWF Gas Wasserfach: Wasser/Abwasser 1975, *116*, 301–308.
107. Baillely, G.M., et al. Eur. Patent Appl. EP 634,479, EP 634,483,EP 639,637
108. German patents DE 2,651,442, DE 2,712,139, DE 2,810,379 to Degussa.
109. IS 261, Sodium Hypochlorite Solutions, 1990, revised Aug. 2000.
110. Benefits and Safety Aspects of Hypochlorite Formulated in Domestic Products, Scientific Dossier, AISE, March 1997.
111. Council Directive 67/548/EEC–relating to dangerous substances.
112. Council Directive 38/379/EEC–relating to dangerous preparations.
113. Council Directive 89/542/EEC–Labeling of detergents and cleaning products.
114. Smith, W.L. Human and environmental safety of hypochlorite. In Proceedings of 3rd World Conference on Detergents. Cahn, A. Ed.; AOCS PressChampaign, IL, 1993; 178–182.
115. Raff, J.; Hegemann, W.; Weil, L. Wasser–Abwasser 1987, *128*, 319.
116. Jolley, R.L.; Jones, G.; Pitt, W.W.; Thompson, J.E. *Water Chlorination: Environmental Impact and Health Effects;* Jolley, R.L. Ed.; Ann Arbor ScienceAnn Arbor, MI, 1978; Vol 1, 105–138.
117. Krossman, H.; Hloch, H.; Bonnen, J.; Knofe, J. Tenside Surf. Det. 1991, *28*, 487.

118. Jürges, P. Activators and peracids. In Proceedings of the 3rd World Conference on Detergents. ;Cahn, A Ed,; AOCS Press:Champaign, IL, 1993; 178–182.

119. Sarlo, K. Human health risk assessement: focus on enzymes. In Proceedings of the 3rd World Conference on Detergents. Cahn, A Ed.; AOCS Press:Champaign, IL, 1993; 54–74.

120. New Scientist & Science Journal, 3 June 1971, 556 pp.

121. Pepys, J.; Wells, I.D.; D'Souza, M.F.; Greenberg, M. Clin. Allergy 1973, *3*, 143.

122. Hovelman, P. The basis of detergents: basic oleochemicals. *Proceedings of the 3rd World Conference on Detergents* Cahn, A Ed; AOCS Press:Champaign, IL, 1993; 117–122.

123. Bircher, H.R. European environmental and regulatory trends. In *Proceedings of the 3rd World Conference on Detergents* Cahn, A. Ed.; AOCS Press:Champaign, IL, 1993;14–18.

124. Yanagawa, T. Environmenal and regulatory trends in Asia. In *Proceedings of 3rd World Conference on Detergent* Cahn, A. Ed,; AOCS Press:Champaign, IL, 1993; 25-31.

125. Klüppel, H.J. Environmental Challenges. Soaps, detergents and oleochemicals AOCS conference and exhibit. Fort Lauderdale, USA.1997.

126. Commission Directive 93/67/EEC on the assessment of risks to man and the environment of substances notified in accordance with Directive 67/584/EEC on dangerous substances. Off. J. Europ. Comm. L227., 8.9.1993.

127. Technical guidance in support of Commission Directive 93/67/EEC on risk assessment for new notified substances and the Commission Regulation (EC) 1488/94 on risk assessment for existing substances ECB. Ispra, 19.4.1996.

128. Gilbert, P.A. Environmental safety assessment of detergents. In *Proceedings of the 3rd World Conference on Detergents*; Cahn, A., Ed.; AOCS Press:Champaign, IL, 1993; 50-53.

129. Report of the Second AIS Workshop on Practical Aspects of Environmental Hazard Assessment of Detergent chemicals in Europe, AIS, 1992.

130. Steber, J. ERASM–Joint industry research programs for the environmental risk assessment of surfactants. Proceedings of the 5th World Surfactants Congress, CESIO 2000; 4-10.

131. Van Der Plassche, E.; DeBruijn, J.; JFeijtel, T.C. Risk assessment of the major four surfactant groups in the Netherlands. Tenside Surfactants Detergents 1997, *34*, 242–248.

132. Feitfel, TCJ, et al. Pan-European development of GREAT-ER, a Geography-Reference Regional Exposure Assessment Tool for European Rivers. SÖFW-Journal 11/99, *125*, 46–50.

133. Fussler, C. Life cycle assessment: a new business tool? In Proceedings of the 3rd World Conference on Detergents. Cahn, A., Ed.; AOCS Press:Champaign, IL, 1993; 58-63.

134. Hindle, P.; White, P.R. Managing toward sustainability: on environmental management framework. In Proceedings of the 4th World Conference on Detergents: Strategies for the 21st century Cahn, A. Ed.; AOCS Press:Champaign IL, 1998; 71-79.

135. Nielsen, A.M. Current environmental issues for surfactants. Inform Jan. 1997, *8*, 28–38.

136. DHTDMAC-Aquatic and Terrestrial Hazard Assessment, ECETOC, Technical Report No. 53, 1993.

137. Steber, J. The biodegradability of mild cosmetic surfactants based on renewable raw materials, Skin Care Forum, Henkel KGaA, No. 12; July 1995; 8–10.

138. Fabry, B. Alkyl polyglycosides: an overview of the patent situation. Skin Care Forum, Henkel KGaA, No. 12; July 1995; 10–12.

139. Vogel, W.J.B. Trends in surfactant raw materials: petrochemicals. Proceedings of the 3rd World Conference on Detergents; Cahn, A., Ed,; AOCS Press:Champaign, IL, 1993; 123–126.

140. Tenside Surfactant Detergents 1995, *32*, 82–193.

141. Stalmans, H., et al. European life-cycle inventory for detergent surfactant production. Tenside Surfactant Detergents 1995, *32*, 84–109.

142. Rosen, M.J.V.; Dahanayake, M. *Industrial Utilization of Surfactants: Principles and Practice*; AOCS Press:Champaign IL, 2000; 8–9.

143. Wright, D.R. An overview of environmental and regulatory trends in the Americas. In Proceedings of the 3rd World Conference on Detergents. Cahn, A Ed.; AOCS Press Champaign, IL, 1993; 19–24.

144. Kirschner, E.M. Soaps and Detergents. C&EN; Jan 26,1998.

145. Environment Agency of Japan, Environmental White Paper (general remarks), Ministry of Finance of Japan, 1993.

146. New U.S.-Canada guidelines to lift regulatory "cloud" over nonylphenol ethoxylates, Washington, March 21, 2000.

147. CESIO News, 3: 1-2, Mat 2000.

148. Towards shared responsibility. *EPE Workbook 1.1, European Partners for the Environment*; Westmalle:Belgium, 1994.

149. *Our Common Future, World Commission on Environment and Development, WCED*; Oxford University Press:New York, 1987.

150. International Union for the Conservation of Nature;1991.

151. European Commission Recommendation for Good Environmental Practice for Household Laundry Detergents, EC Press Release, 23rd July 1998.

152. Directive 1999/476/EEC: Commission decision estabilishing the ecological criteria for the award of the Community Eco-Label to laundry detergents, Community Legislation in force, 10 June 1999.

153. Directive 1999/427/EEC: Commision decision establishing the ecological criteria for the award of the Community Eco-label to detergents for dishwashers, Community Legislation in force, 28 May, 1999.

154. Market News. SÖFW Journal June 2001, *127*, 66.

3

Distribution, Behavior, Fate, and Effects of Surfactants and Their Degradation Products in the Environment

GUANG-GUO YING CSIRO Land and Water, Glen Osmond, Australia

I. INTRODUCTION

Surfactants are a diverse group of chemicals designed to have cleaning or solubilization properties. They generally consist of a polar headgroup (either charged or uncharged), which is well solvated in water, and a nonpolar hydrocarbon tail, which is not easily dissolved in water. Hence, surfactants combine hydrophobic and hydrophilic properties in one molecule. Synthetic surfactants are economically important chemicals. They are widely used in household cleaning detergents, personal care products, textiles, paints, polymers, pesticide formulations, pharmaceuticals, mining, oil recovery, and the pulp and paper industries. Worldwide production of synthetic surfactants amounts to 7.2 million tons annually [1].

Surfactants consist mainly of three types: anionic, nonionic, and cationic (Table 1). Commonly used commercial surfactants are linear alkylbenzene sulfonates (LAS), alkyl ethoxy sulfates (AES), alkyl sulfates (AS), alkylphenol ethoxylates (APE), alkyl ethoxylates (AE), and quaternary ammonium-based compounds (QAC). LAS, APE, and QAC are the most extensively studied surfactants. In the following, we use abbreviations for each class of surfactants, for example, C12EO9 (EO = ethylene oxide unit) having 9 EO units and an alkyl chain of 12 carbon atoms, C14LAS having an alkyl chain of 14 carbon atoms, NPE9 or NPEO9 for nonylphenol ethoxylates with 9 EO units.

Linear alkylbenzene sulfonates (LAS) are the most popularly used synthetic anionic surfactants. They have been extensively used for over 30 years, with an estimated global consumption of 2.8 million tons in 1998 [2]. Commercially available products are very complex mixtures containing homologues with alkyl chains ranging from 10 to 14 carbon units (C10–C14 LAS). Furthermore, since the phenyl group may

TABLE 1 Acronyms of the Most Widely Used Surfactants

Class	Common name	Acronym
Anionic surfactants	Linear alkyl benzene sulfonates	LAS
	Secondary alkane sulfonates	SAS
	Alcohol ether sulfates (Alkyl ethoxy sulfates)	AES
	Alcohol sulfates (Alkyl sulfates)	AS
Nonionic surfactants	Alkylphenol ethoxylates	APE (or APEO)
	Nonyl phenol ethoxylates	NPE (or NPEO)
	Octyl phenol ethoxyales	OPE (or OPEO)
	Alcohol ethoxyaltes (Alkyl ethoxyaltes)	AE (or AEO)
Cationic surfactants	Quaternary ammonium-based compounds	QAC
	Alkyl trimethyl ammonium halides	TMAC
	Alkyl dimethyl ammonium halides	DMAC
	Alkyl benzyl dimethyl ammonium halides	BDMAC
	Dialkyl dimethyl ammonium halides	DADMAC
	Dihydrogenated tallow dimethyl ammonium chloride	DHTDMAC or DTDMAC
	Ditallow trimethyl ammonium chloride	DTTMAC
	Diethyl ester dimethyl ammonium chloride	DEEDMAC

be attached to any internal carbon atom of the alkyl chain, each homologue contains five to seven positional isomers.

Alkylphenol ethoxylates (APE) constitute a large portion of the nonionic surfactant market. The worldwide production of APEs was estimated at 500,000 tons in 1997, with 80% of nonylphenol ethoxylates (NPE) and 20% of octylphenol ethoxylates (OPE) [3]. Concern has increased recently about the wide usage of APE because of their relatively stable biodegradation products, nonylphenol (NP) and octylphenol (OP). NP and OP have been demonstrated to be toxic to both marine and freshwater species [4,5] and to induce estrogenic responses in fish [6,7].

Quaternary ammonium-based surfactants (QAC) are molecules with at least one hydrophobic hydrocarbon chain linked to a positively charged nitrogen atom, the other alkyl groups being mostly short-chain substituents, such as methyl or benzyl groups. The major uses of this group of cationic surfactants are as fabric softeners and antiseptic agents in laundry detergents as well as other industrial uses. Until recently, the most widely used active ingredient in fabric softeners has been dihydrogenated tallow dimethyl ammonium chloride (DTDMAC). However, the replacement of DTDMAC by ester cationic surfactants such as diethyl ester dimethyl ammonium chloride (DEEDMAC) has recently begun in Europe [8].

After use, residual surfactants and their degradation products are discharged to sewage treatment plants or directly to surface waters and then dispersed into different environmental compartments. Due to their widespread use and high consumption, surfactants and their degradation products have been detected at various concentrations in surface waters, sediments, and sludge-amended soils. In order to assess their environmental risks, we need to understand the distribution, behavior, fate, and biological effects of these surfactants in the environment.

II. DISTRIBUTION OF SURFACTANTS AND THEIR DEGRADATION PRODUCTS

Surfactants are widely used in households and industry in large volumes and then disposed of after use into the environment. The majority of this waste stream is treated in wastewater treatment plants (WWTPs), with some proportion directly entering the environment. Although WWTPs can remove these surfactants at high rates, some of the surfactants still remain in the sewage effluents and sludges, which are normally discharged into surface waters or disposed of on lands. These surfactants in the receiving environment may further be distributed into different environmental compartments (air, water, sediment, biota), thus impacting the ecosystems.

A. Sewage Effluents and Sludges

Surfactants in raw sewage or wastewater can easily be treated by modern treatment technology at high rates [9–12]. Based on the influent and effluent concentrations, the removal rates for four anionic surfactants (LAS, AS, AES, and SAS) in the Ratingen wastewater treatment plant in Germany are in the range of 99.7% (LAS) and 99.99% (AES) [13]. In a Greek municipal wastewater treatment plant, NPE concentrations in influents ranged from 1180 to 1620 µg/L, whereas NPE concentrations in effluents ranged from 35 to 130 µg/L [10]. Therefore, removal of NPE ranged from 92% to 97%. In six UK sewage treatment plants, LAS concentrations in raw sewage ranged from 1.73 to 5.58 mg/L, while in final effluent its concentrations were reduced to 40–1090 µg/L [12]. The removal rates for LAS in the six sewage treatment plants varied between 55% and 99%. The removal of surfactants in the treatment plants is due mainly to their adsorption on solids and to physiochemical and biological degradation processes. The removal efficiency of a surfactant depends on its physiochemical properties, treatment plant design, and waste load (Table 2).

In addition to the surfactants, their degradation products (or metabolites) are also widely detected in sewage effluent. Of greatest concern are alkylphenols, which are the degradation products of nonionic surfactant APE. During sewage treatment, APE are biodegraded through a mechanism involving stepwise loss of ethoxy groups to form shorter APE homologues, carboxylated products (alkylphenol ethoxycarboxylates, i.e., APECs), and finally alkylphenols such as nonylphenol (NP) and octylphenol (OP) [21–24]. NP and OP are known to be more toxic than their ethoxylate precursors and to mimic the effect of the hormone estrogens [3]. The concentrations of NP and OP in sewage final effluents vary widely among various sewage treatment plants from less than the limit of detection (LOD) to 343 µg/L (Table 3). For example, NP was detected in effluent samples collected from municipal sewage treatment plants (STPs) in Michigan, with concentrations ranging from 0.017 to 37 µg/L [18]. OP and NPE were also detected, with concentrations ranging from less than the LOD to 0.673 µg/L and from less than the LOD to 332 µg/L, respectively [18].

In wastewater treatment, a proportion of the surfactants is removed by adsorption on sewage solids during primary settlement of sewage. Holt et al. [12] found that approximately 11–45% of LAS were associated with suspended solids during primary treatment. Concentrations of surfactants in sludges from Switzerland and Germany were between 1.15 and 11.8 g/kg for anionic surfactants, between 2.0 and 15.0 g/kg for cationic surfactants, and between 0.2 and 1.2 g/kg for nonionic

TABLE 2 Concentrations of Surfactants in Sewage Influents and Effluents

Surfactant	Location	Influent (mg/L)[a]	Effluent (μg/L)[a]	Ref.
LAS	Germany	0.5–3.5	7–16	13
LAS	The Netherlands	3.1–7.3	<8.1–491	14
LAS	UK	1.73–5.58	40–1090	12
LAS	Italy	3.4–10.7	21–290	15
LAS	The Netherlands	3.4–8.9 (5.2)	19–71 (39)	16
AES (C12–C15)		1.2–6.0 (3.2)	3.0–12 (6.5)	16
AS (C12–C15)		0.1–1.3 (0.6)	1.2–12 (5.7)	16
Soap		14–45 (28)	91–365 (174)	16
AE (C12–C15)		1.6–4.7 (3.0)	2.2–13 (6.2)	16
AE (C12–C15)	United States	0.68–3.67	11–114	17
NPE	United States		<LOD–332 (9.3)	18
NPE	Italy		2–27 (10)	19
NPE	Greece	1.18–1.62 (1.4)	35–130 (62)	10
DTDMAC	United States	0.36–2.62	20–62	20
DTDMAC	The Netherlands	0.563	37	20
DTDMAC	Germany	1.05–1.40	30–50	20

[a] Concentration range and median in parentheses. LOD = limit of detection.

surfactants [33]. These sludges are generally digested under anaerobic conditions; but in some plants sludges are aerobically treated. Bruno et al. [34] determined the surfactants and their metabolites in untreated and anaerobically digested sludges and found that the removal rates for anionic surfactants LAS, AS, and AES (7%, 28%, and 8%) were lower than nonionic surfactants AE and NPE (54% and 63%). However, OP and NP concentrations in the sludges increased after treatment from 14 and 242 mg/kg to 17 and 308 mg/kg, respectively. This is because OPE and NPE in the sludges were degraded into OP and NP during the anaerobic digestion.

Dried sludge samples from European sewage treatment plants have yielded concentrations of 0.47–4000 μg/g of 4-NP, 0.66–680 μg/g of NPE1, and 0.04–280 μg/g of NPE2 [28,35–40].

TABLE 3 Concentration of Alkylphenols in Effluents of Sewage Treatment Plants

Location	Sample no.	Concentration (μg/L)[a]		Ref.
		NP	OP	
Canada	8	0.8–15.1 (1.9)	0.12–1.7 (0.69)	26
United Kingdom	16	<0.2–5.4 (0.5)		27
Switzerland	2	5–11		28
Spain	3	6–343		29
Japan	10	0.08–1.24	0.02–0.48	30
United States	1	16	0.15	31
United States	6	0.171–37 (1.02)	<LOD–0.673 (0.072)	18
Germany	16	<LOD–0.77 (0.111)	<LOD–0.073 (0.014)	32
Italy	12	0.7–4 (1.8)		19

[a] Concentration range and median in parentheses. LOD = limit of detection.
Source: Adapted from Ref. 25.

In many countries, the treated sludges (biosolids) may be applied onto agricultural lands as fertilizers for plants. High concentrations of surfactants were found in treated sludges and ranged from 47 mg/kg dry weight for AS to 30,200 mg/kg dry weight for LAS (Table 4). Surfactants in aerobically treated sludges are found in much lower concentrations than in anaerobically digested sludges because of their quicker aerobic biodegradation.

B. Surface Waters, Sediments, and Soils

Surfactant concentrations up to 416 µg/L have been reported in surface waters (Table 5). Waters and Feijtel [9] summarized the LAS monitoring data in European rivers with concentrations ranging from under 2.1 to 130 µg/L in water and from 0.49 to 5.3 mg/kg in sediment. Eichhorn et al. [46] monitored LAS and its main degradation product, sulfophenyl carboxylates (SPC), in Rio Macacu, Brazil. The LAS concentrations ranged between 14 and 155 µg/L and the levels of SPC were found to be from 1.2 to 1.4 µg/L. SPC was also found in the drinking water samples from Niteroi, Sao Goncalo, and Rio de Janeiro, with its concentration ranging from 1.4 ± 0.2 µg/L to 3.7 ± 0.7 µg/L [46].

Surfactants were also detected in sediments and soils (Tables 5 and 6). DTDMAC was detected in the marine sediments near sewage outfalls from Barcelona, Spain, at a concentration of 0.88 ± 0.02 g/kg [45]. High levels of DTDMAC have been determined in sediments (<3–67 mg/kg) from Rapid Creek (Pennington County, SD) [52] and in Japanese sediments (6–69 mg/kg) (cited in Ref. 45).

Carlsen et al. [58] studied the occurrence of LAS in a series of soil samples with or without sludge amendment in Denmark. LAS levels in the soils with no history of sludge application ranged from less than the detection limit to 1.19 mg/kg, whereas the concentrations in soils with a history of sludge application were found to range from 0.36 to 19.25 mg/kg. LAS were also detected in the soil layers 50 cm deep [58].

TABLE 4 Concentrations of Surfactants in Treated Sludges

Surfactant	Location	Treatment	Concentration (mg/kg dry weight)	Ref.
LAS	Switzerland	Anaerobically digested sludge	2900–11,900	41
LAS	Germany	Aerobically treated sludge	182–432	42
		Anaerobically digested sludge	1327–9927	
LAS	Spain	Aerobically treated sludge	100–500	43
		Anaerobically treated sludge	7000–30200	
LAS	United States	Aerobically treated sludge	152 ± 119	44
		Anaerobically digested sludge	10,462 ± 5,170	
LAS	Italy	Anaerobically digested sludge	4342	34
AS			47	
AES			69	
AE			143	
NPE			81	
OP			17	
NP			308	
DTDMAC	Switzerland	Anaerobically digested sludge	150–5870	45

TABLE 5   Concentrations of Surfactants in Surface Waters and Sediments

Surfactant	Location	River water ($\mu g/L$)[a]	Sediment ($\mu g/kg$)	Ref.
LAS	Brazil	14–155		46
LAS	Philippines	1.2–102		47
LAS	Switzerland		190–3400	48
LAS	Taiwan	11.7–135		49
LAS	The Netherlands	<2.1–168		14
LAS	United Kingdom	22–130		50
LAS	United Kingdom	5–416		51
DTDMAC	The Netherlands	2–60		20
DTDMAC	United States	<2–51		20
DTDMAC	Spain (Barcelona)		880 ± 20	45
DTDMAC	United States	1–92	<3000–67000	52
AE	United States	2–37		18
NPE	United States	<LOD–17.8 (6.97)		19
NPE	Taiwan	2.8–25.7 (21.3)		49

[a] Concentration range and median in parentheses. LOD = limit of detection.

The occurrence of APE degradation products (NP, OP) has been widely reported in surface waters (rivers, lakes, and coastal waters as well as aquatic biota) around the world (Table 7). Concentrations in surface waters were found to be up to 644 $\mu g/L$ for NP and up to 0.47 $\mu g/L$ for OP, respectively. Owing to their hydrophobic nature, the reported alkylphenol levels in sediments were much higher than in the corresponding surface waters. Their concentrations varied between less than 0.1 and 13,700 $\mu g/kg$ for NP and up to 670 $\mu g/kg$ for OP in the sediments.

III. BEHAVIOR OF SURFACTANTS IN THE ENVIRONMENT

A. Chemistry of Surfactants

A fundamental property of surfactants is their ability to form micelles in solution. This property is due to the presence of both hydrophobic and hydrophilic groups in each surfactant molecule. It is the formation of micelles in solution that gives surfactants their detergency and solubilization properties. When dissolved in water

TABLE 6   Concentrations of LAS in Sludge-Amended Soils

Location	Soil concentration postapplication (mg/kg)	Monitoring period	Final soil concentration (mg/kg)	Half-life (days)	Ref.
Germany	16	76 days	0.19	13	53
Spain	16	90 days	0.3	26	54
Spain	53	170 days	Not reported	33	54
Switzerland	45	12 months	5	9	55
United Kingdom	2.6–66.4	5–6 months	<1	7–22	56,57

TABLE 7 Concentration of Alkylphenols in Surface Waters and Sediments

Location	Concentration (µg/L) in water[a]		Concentration (µg/kg) in sediment		Ref.
	NP	OP	NP	OP	
Canada	<LOD–0.92	<LOD–0.084	0.1–72	<LOD–1.8	59
United Kingdom	(<LOD)	(<LOD)	(10.6)[b]	(0.41)[b]	27
	<0.03–53 (1.3)				
	<0.2–22 (<0.2)		<0.1–15 (<0.1)		60
Switzerland	0.7–26 (2.7)[b]				61
	(1.8 ± 0.52)[b]				61
	<LOD–0.48				62
Spain	<LOD–644 (51)				29
Japan	0.05–1.08	0.01–0.18	30–13000	3–670	30
	0.11–3.08	<LOD–0.09			63
	<LOD–1.9 (0.25)[b,c]				64
	<LOD–3.0 (0.15)[b,d]				64
United States	<LOD–1.19 (1.52)	<LOD–0.081 (0.017)			18
	12–95 (48)[b]				65
	0.077–0.416 (0.2)	0.00156–0.007 (0.002)	6.99–13700 (2107)[b]	<LOD–45 (30)[b]	66
	<0.11–0.64 (0.12)[b]		<2.9–2960 (162)[b]		67
Germany	0.0067–0.134 (0.023)	0.0008–0.054 (0.0038)			32
Taiwan	1.8–10 (3)				49

[a] Concentration range and median in parentheses. LOD = limit of detection.
[b] Arithmetic mean (± standard deviation) in parentheses for the data in this row.
[c] Summer sampling.
[d] Autumn sampling.
Source: Adapted from Ref. 25.

at low concentration, surfactant molecules exist as monomers. At higher concentrations the system's free energy can be reduced by the aggregation of the surfactant molecules into clusters (micelles), with the hydrophobic groups located at the center of the cluster and the hydrophilic headgroups toward the solvent. The concentration at which this occurs is known as the *critical micelle concentration* (CMC) [68]. Nonionic surfactants have lower CMC levels than anionic and cationic surfactants (Table 8).

At concentrations above the CMC level, surfactants have the ability to solubilize more hydrophobic organic compounds than would be dissolved in water alone (Fig. 1). The effectiveness of surfactants in solubilizing water-insoluble or poorly soluble compounds is dependent on the sorbed compounds, the environmental media, and the surfactant [73]. Surfactants may affect the mobility and degradation of hydrophobic organic compounds in soil or sediment [74,75]. Aronstein et al. [73]

TABLE 8 Octanol/Water Partition Coefficients (K_{ow}) and Critical Micelle Concentrations (CMC) of Surfactants

Compound	Log K_{ow}	CMC (mM) (distilled water)
C12LAS	1.96	1.1
C13LAS	2.54	0.46
C12SO4	1.60	8.2
C14SO4	na[a]	2.1
C16SO4	na	0.52
C12EO3SO4	na	2.8
C12EO5SO4	na	1.9
NPEO10	na	0.094[d]
NPEO12	na	0.057[b]
NPEO15	na	0.114[d]
NPEO30	na	0.206[d]
OPEO9-10	na	0.24[c]
C14EO7	2.47	0.0095
C12EO4	na	0.064
C12EO8	na	0.11
C12EO16	na	0.25
C18DMAC	2.69	0.0046
C18TMAC	na	0.4
C16TMAC	1.81	1.6
C12TMAC	na	20
C8TMAC	na	220

[a] na = not available.
[b] Ref. 70.
[c] Ref. 71.
[d] Ref. 72.
Source: Adapted from Ref. 69.

found that the extent of phenanthrene biodegradation was markedly increased at nonionic surfactant concentrations of 10 μg/kg soil in both a mineral and an organic soil, despite lack of desorption enhancement in the organic soil.

In sewage sludge–amended soils, there are many other hydrophobic organic compounds in addition to surfactants at high concentrations. These surfactants may interact with those hydrophobic compounds. Kile and Chiou [76] studied the effect of anionic, cationic, and nonionic surfactants on the water solubility of DDT and trichlorobenzene. As would be expected, the solubility was enhanced when the surfactant was present at concentrations greater than the critical micelle concentration. There was also a solubility enhancement at surfactant concentrations less than the CMC levels. However, the studies by Klumpp et al. [77] and Edwards et al. [74] found that surfactants below the CMC enhanced the sorption uptake of hydrophobic organic pollutants due to the formation of hemimicelles. At higher concentrations the same surfactants in micellar form remobilized those hydrophobic compounds already adsorbed by solubilization. The concentration of surfactants required to mobilize contaminants is significantly above those normally found in sewage sludge–amended soils [78].

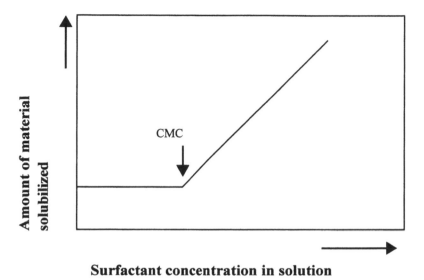

Surfactant concentration in solution

FIG. 1 Solubilization of an insoluble or poorly soluble material as a function of surfactant concentration. CMC is the critical micelle concentration of a surfactant.

B. Sorption of Surfactants

Once surfactants enter the environment through sewage discharge into surface water, pesticide application, or sludge disposal on land, they undergo many processes, such as sorption and degradation. Knowledge of the processes involved in distributing these surfactants among ecosystem compartments is essential to an understanding of their behavior in the environment. Sorption of a surfactant onto sediment/soil depends on many factors, including its physiochemical properties, sediment nature, and environmental parameters. The information from sorption process of a surfactant can be used to estimate the distribution of the surfactant in different environmental compartments (sediment/soil and water). Sorption data can also be used to estimate the bioavailability of the surfactant. Furthermore, sorption has a significant influence on the degradation of the surfactant in the environment.

Sorption can be described by using sorption isotherms. The commonly used Freundlich equation defines a nonlinear relationship between the amount sorbed and the equilibrium solution concentration:

$$S = K_f C^n$$

where S is the concentration of a surfactant sorbed by the solid phase (mg/kg), K_f is the Freundlich sorption coefficient (L/kg), C is the equilibrium solution concentration (mg/L), and n is a power function related to the sorption mechanism. When the value of n is unity, we have the simplest linear isotherm:

$$S = K_d C$$

where K_d is the sorption coefficient (L/kg). The parameter K_d is frequently used to characterize the sorption of a chemical in sediment/soil and is an important parameter governing the partitioning and mobility of the chemical in the environment. Sorption of some chemicals, especially those nonpolar compounds, closely depends on organic matter in the sediment/soil. Therefore, the organic carbon sorption coefficient (K_{oc}) is often used to describe the sorption of those compounds on sediment/soil.

Due to their chemical features, surfactant molecules may sorb directly onto solid surfaces or may interact with sorbed surfactant molecules. The sorption mechanism is dependent on the nature of the sorbent and the surfactant concentration [71,79–81]. At low concentrations, the surfactant molecules may be sorbed to a mineral surface or clean sediment that has very few sorbed surfactant molecules, and sorption may occur mainly due to van der Waals interactions between the hydrophobic and hydrophilic moieties of the surfactant and the surface. There are no significant sorbate–sorbate interactions at the low concentrations. As the surfactant concentration increases, active sorption sites on the solid surface become less and less available, and more and more hemimicelles form. At higher concentrations, such sorption may entail the formation of more structured arrangements, including the formation of monomer surfactant clusters on the surface or a second layer, for which these arrangements may be governed mainly by interactions between hydrophobic moieties of the surfactant molecules. Therefore, two-stage sorption isotherms (Fig. 2) have been reported for nonionic surfactants NPE and AE and anionic LAS, although the sorption behavior is different for nonionic and anionic surfactants [71,79–81].

The sorption of LAS on natural soils has two stages: linear and exponentially increasing isotherms [81]. At low LAS concentration (<90 μg/mL), the sorption isotherms were linear and K_d ranged from 1.2 to 2.0. At high levels (>90 μg/mL), cooperative sorption was observed and the sorption amount of LAS increased exponentially with the increasing of LAS concentration in solution [81]. This enhanced sorption of LAS on soils was also observed by Fytianos et al. [80]. In a real soil environment or aquatic environment, where LAS levels are rather low, the LAS sorption ability of a soil or sediment is very weak.

In contrast, the sorption of a nonionic surfactant reached a maximum on the solid surface when the solution is near or just at the CMC level of the surfactant. The decreased sorption of nonionic surfactants (APE and AE) on sediment at higher

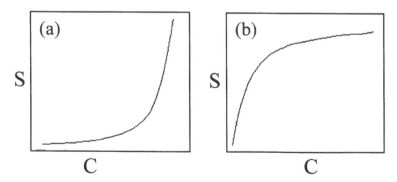

FIG. 2 Sorption isotherms for anionic and nonionic surfactants (a. LAS; b. APE and AE). S is the sorbed surfactant concentration and C is the surfactant concentration in the solution.

concentrations was observed [71,72]. A Langmuir isotherm, as described by the following equation, provides a reasonable fit to the sorption data:

$$S = S_{max}K_1 C/(1 + K_1 C)$$

Here, S is the sorbed concentration of the surfactant on the solid surface (mg/kg), S_{max} is the maximum sorbed concentration (mg/kg), C is the aqueous-phase surfactant concentration (mg/L), and K_1 is the Langmuir constant (L/kg).

Surfactant concentrations in the environment are normally in a low concentration range, below the CMC of the surfactant. Surfactant sorption onto environmental sorbents (sediment or soil) is therefore mostly Freundlich type [79,82]. Table 9 lists reported sorption coefficients for certain surfactants. Anionic surfactant LAS have much lower K_d values than nonionic surfactants APE and AE. However, cationic surfactants tend to adsorb strongly onto sediment/soil [68]. For example, the calculated K_d value was 5,000 L/kg for DEEDMAC on sludge [8]. Topping and Waters [88] and Games and King [84] reported that less than 95% of the cationic surfactants were adsorbed on the surface of particulate matter in activated sludge. Sorption coefficients of AE suspended sediment increased with increasing alkyl and ethoxylate chain lengths [87]. The dominant influence of the alkyl chain suggests a hydrophobic sorption mechanism.

Ferguson et al. [66] investigated the partitioning of APE degradation products to suspended solids in Jamaica Bay, New York, and found that log K_{OC} values did not vary greatly among the APE degradation products and were 5.39 for NP, 5.18 for OP, 5.46 for NPE1, 5.18 for NPE2, and 4.87 for NPE3, respectively. John et al.

TABLE 9 Sorption Coefficients (K_d) of Surfactants

Compound	Sorbent	K_d (L/kg)	K_{oc} (L/kg)	Ref.
C12LAS	Sable soil (TOC 2.9%)	3.3–45	114–1552	83
	Spinks soil (TOC 1.2%)	3.6–9	300–750	83
	Sarpy soil (TOC 0.6%)	2–3.5	333–583	83
C12LAS	Soil (TOC 0.46–1.08%)	1.2–2.0	185–261	81
C18TMAC	Activated sludge solids	1.8×10^4 to 4.9×10^4		84
NPEO3	River sediment	1460		85
NPEO10	River sediment	450		85
C15EO9	Sediment (TOC 0.2–2.8%)	350–2100		82
C13EO6	Sediment (TOC 0.76–3.04%)	40–62		79
C13EO3	Sediment (TOC 0.2–2.8%)	110–500		86
C13EO9		110–590		86
C10EO3	Suspended sediment (TOC 11%)	41		87
C10EO5		48		
C10EO8		126		
C12EO3		257		
C12EO5		724		
C12EO8		1230		
C14EO3		2951		
C14EO5		3467		
C14EO8		3548		

[85] measured sorption coefficients (K_d) of NPE3-13 homologues onto native sediment, organic-free sediment, kaolinite, silica, and sewage sludge and found that K_d values for native sediment decreased progressively from 1,460 L/kg for NPE3 to 450 L/kg for NPE10 and then increased again slightly for higher homologues. In contrast, K_d values for organic-free sediment (230–590 L/kg) or kaolinite (190–490 L/kg) increased steadily from NPE3 to NPE13. Adsorption to sewage sludge was very strong with K_d values ranging from 12,000 to 33,000 L/kg. These data indicated that interactions with organic matter were important in controlling sorption of AP and short ethoxylate APE. However, as the level of APE ethoxylation increased, association with mineral surfaces became the dominant contributor to APE sorption.

C. Bioconcentration of Surfactants and Their Degradation Products

When a chemical in sewage effluent is discharged into the environment, it distributes into the different phases, such as water, air, sediment, and biota, and equilibrium is formed, depending on the properties of the chemical and the phases. Therefore, the water-to-biota transfer is of critical importance because we are principally concerned with adverse effects on biota. The process involving the direct transfer of a chemical from water to biota is known as *bioconcentration* [89]. At equilibrium, bioconcentration is characterized by the bioconcentration factor (BCF), the ratio between the concentration in biota, C_B, and the concentration in water, C_w; i.e., BCF $= C_B/C_w$.

Since a surfactant has to be taken up into an organism before it can elicit an effect, the processes and factors influencing uptake are relevant when assessing the environmental risk. Lipophilic compounds are the organics most likely to bioaccumulate. Mackay [90] has demonstrated that the lipid phase in biota is the dominant phase for their accumulation. Lipophilicity, or hydrophobicity, measured as the octanol-to-water partition coefficient (K_{ow}) has been identified as the driving force behind bioconcentration. Bioconcentration increases with increasing K_{ow} value.

1. Linear Alkylbenzenesulfonates

It has been found that longer LAS homologues have higher K_{ow} values (Table 8). LAS are taken up from water via the fish gills rather than skin [91]. The concentrations of the selected LAS homologues (C10LAS to C13LAS) in the liver and the internal organs of juvenile rainbow trout increased rapidly, demonstrating fast uptake into systemic circulation. The relatively slow increase of LAS concentrations in the less well-perfused tissue pointed to internal redistribution being controlled by perfusion. The BCFs in rainbow trouts ranged between 1.4 and 372 L/kg. The BCFs in fathead minnows were higher, ranging from 6 to 990 L/kg [91]. Water hardness was found to influence the aqueous-phase behavior of LAS [91]. Increased water hardness can bring higher fluxes of LAS from water into fish.

In the terrestrial environment, BCFs are significantly lower than in the aquatic environment and bioaccumulation of LAS in terrestrial biota is mostly unlikely [92]. Figge and Schöberl [53] used radiolabeled LAS to estimate uptake by plants in two mesocosm studies. The concentrations of LAS were estimated to be up to 210 mg/kg in roots of grass, between 106 and 134 mg/kg in roots of radishes and garden beans, and up to 66 mg/kg in potatoes. BCFs were estimated to be between 2 and 7 for the four plant species, based on the initial LAS concentrations of 16 and 27 mg/kg in soil.

TABLE 10 Bioconcentration Factor (BCF, Wet Weight) Data for Alkylphenols

Species	4-Nonylphenol	4-t-Octylphenol	Ref.
Ayu fish (field)	21 ± 15	297 ± 194	95
Killifish	167 ± 23	267 ± 62	95
Sticklebacks (field)	1300		96
Salmon	282		5
Fathead minnow	270–350		97
Rainbow trout		471	98

2. Alkylphenols

Alkylphenols (nonylphenols and octylphenols) are the degradation products of a widely used class of nonionic surfactant APE during wastewater treatment [34]. Nonylphenol (NP) and octylphenol (OP) have attracted a lot of scientific attention because of their estrogenic effects and ability to bioaccumulate in aquatic organisms. These two chemicals have been widely detected in the environment due to the discharge of sewage effluents into surface waters [25]. NP has a log K_{ow} of 4.48 and a water solubility of 5.4 mg/L, whereas OP has a log K_{ow} of 4.12 and a water solubility of 12.6 mg/L [93,94]. This physiochemical profile indicates that NP and OP may bioaccumulate in aquatic organisms. This has been documented in some species of fish from natural waters and from controlled laboratory exposure (Table 10). The reported bioconcentration factors (BCF values) in whole fish range from 21 to 1,300 for 4-NP and 267 to 471 for 4-t-OP. The differences in the BCF values of NP and OP among fish species are probably due to their different metabolic abilities, functioning of their gills, etc. [95].

Alkylphenols can be rapidly metabolized by phase I and II enzymes in fish [98–100]. Arukwe et al. [99] studied the in vivo metabolism and organ distribution of 4-n-NP in juvenile salmon and found that 4-n-NP was metabolized mainly to its corresponding glucuronide conjugate and to a lesser extent to various hydroxylated and oxidated compounds. The half-life of residues in carcass and muscle was between 24 and 48 hours after exposure. Similar results were found by Ferreira-Leach and Hill [98] in a study on bioconcentration and distribution of 4-t-OP in juvenile rainbow trout. The concentrations of 4-t-OP residues were higher in bile, followed by feces, pyloric ceca, liver, and intestine. In these tissues the majority of alkylphenol was in the form of two metabolites, which were identified by GC-MS as the glucuronide conjugates of 4-t-OP and t-octylcatechol. 4-t-OP accumulated as the parent compound in fat with a BCF of 1,190, and in brain, muscle, skin, bone, gills, and eye with BCF values of between 100 and 260. This suggests that exposure to waterborne alkylphenols results in rapid conjugation and elimination of the chemical via the liver/bile route but that high amounts of the parent compound can accumulate in a variety of other fish tissue [98].

IV. BIODEGRADATION OF SURFACTANTS IN THE ENVIRONMENT

Degradation of surfactants through microbial activity is the primary transformation occurring in the environment. Biodegradation is an important process for treating surfactants in raw sewage in sewage treatment plants, and it also enhances the removal of these surfactants in the environment, thus reducing their impact on biota. During biodegradation, microorganisms can either utilize surfactants as substrates for energy and nutrients or cometabolize the surfactants by microbial metabolic reactions. There are many chemical and environmental factors that affect biodegradation of a surfactant in the environment. The most important influencing factors are chemical structure and the physiochemical conditions of the environmental media. Different classes of surfactants exhibit different degradation behavior in the environment (Table 11). Most of the surfactants can be degraded by microbes in the environment, although some surfactants, such as LAS, may be persistent under anaerobic conditions [24].

A. Anionic Surfactants

1. Linear Alkylbenzenesulfonates

LAS can be degraded by consortia of aerobic microorganisms and attached biofilms in the environment [102,103,105]. LAS biodegradation intermediates are mono- and dicarboxylic sulfophenyl acids (SPC) that are formed by ω-oxidation of the alkyl chain terminal carbon followed by successive β-oxidation [19,105,126]. A variety of SPCs have been identified having an alkyl chain length of 4–13 [105,127]. Then the SPCs are further desulphonated. The ω-oxidation of the alkyl chain and the cleavage of the benzene ring require molecular oxygen; therefore, under anaerobic conditions degradation via these pathway is unlikely. There is no evidence that LAS can be easily degraded anaerobically [101,104].

Due to the incomplete removal of LAS in sewage treatment plants, some residues of the surfactant, together with its aerobic breakdown intermediates, the sulfophenyl carboxylates (SPC) enter the receiving waters via the discharge of the sewage effluents. Aerobic degradation of LAS in river water is well documented, with half-lives less than 3 days [128]. LAS could be biodegraded primarily to more than 99% by natural microbial flora of river water even at 7°C [129]. However, in the marine environment, the degradation of LAS and its intermediate SPCs is slower,

TABLE 11   Biodegradability of Surfactants in the Environment

Surfactant	Aerobic condition	Anaerobic condition
LAS	Degradable (46,101–105)	Persistent (24,104)
SAS	Readily degradable (24,106)	Persistent (24)
Soap	Readily degradable (24,107)	Readily degradable (24,107)
Fatty acid esters (FES)	Readily degradable (108,109)	Persistent (110)
AS	Readily degradable (111,112)	Degradable (113,114)
AES	Readily degradable (24)	Degradable (24)
Cationic surfactants (e.g., TMAC, DTDMAC)	Degradable (8,84,115)	Persistent (116)
APE	Degradable (22,117,118)	Partially degradable (119)
AE	Readily degradable (120–123)	Degradable (23,124,125)

which is due mainly to the lower microbial activity and their association with Ca^{2+} and Mg^{2+} [127]. In sewage-contaminated groundwater, the rate of LAS biodegradation increased with increasing dissolved oxygen concentrations; but under low oxygen conditions (<1 mg/L), only a fraction of the LAS mixture biodegraded [104].

Concentrations of LAS in raw sewage sludges are very high due to its widespread use and strong sorption on sludge during the treatment. Sewage sludge that had been aerobically treated had LAS concentrations of 100–500 mg/kg dry weight, while anaerobically treated sludge had much higher LAS concentrations, ranging from 5,000 to 15,000 mg/kg dry weight [92]. Of course, the LAS level in sludge also depends on the individual wastewater treatment plant, because the input of LAS into a sewage treatment plant and its treatment method and efficiency are different. However, McEvoy and Giger [41] measured LAS concentrations in sludge before and after anaerobic digestion and found that no degradation of LAS occurred during anaerobic treatment. This further substantiates the conclusion that the degradation of LAS under anaerobic conditions is not favored.

LAS can be easily degraded in aerobic soil, with a half-life of 7–33 days (Table 6). Once sludge is applied on land, LAS are rapidly metabolized by aerobic bacteria in sludge-amended soil and will not accumulate in soil, as demonstrated by field experiments [53,54,56,57]. Holt et al. [57] concluded that degradation of LAS in soil was primarily microbially driven and that soil type, agricultural land use, application method, and whether a soil had been plowed or not had no effect on degradation rates. LAS homologue distribution showedno significant changes postapplication, suggesting no differential degradation.

2. Alkyl Sulfates

Alkyl sulfates (AS) are among the most rapidly biodegradable surfactants. Both primary and ultimate biodegradations are fast and complete in a wide range of test designs [24,130]. The biodegradation is found to involve the enzymatic cleavage of the sulfate ester bonds to give inorganic sulfate and a fatty alcohol. The alcohol is oxidized to an aldehyde and subsequently to a fatty acid, with further oxidation via the β-oxidation pathway, thus achieving ultimate biodegradation [131]. This pathway is further confirmed by the identification of alkylsulfatase enzymes, which catalyze the initial desulfation step, and long-chain alcohol dehydrogenases that follow them [132]. Alkylsulfatase-producing strains, such as *Pseudomonas* sp., are widely distributed in the environment [132]. Lee et al. [133] found faster degradation of sodium dodecyl sulfate (SDS) by riverine biofilms. SDS biodegradation was reported in Antarctic coastal waters with half-lives of 160–460 hours [111]. AS was readily degraded under anaerobic conditions using municipal digester solids as a source of anaerobic bacteria [23]. Therefore, AS can be readily bioavailable under aerobic and anaerobic conditions and easily degradable both primarily and ultimately. Treatment in a sewage treatment plant can sufficiently remove AS, with little possibility to reach the environment by effluent discharge and sludge disposal.

B. Cationic Surfactants

Quaternary ammonium compounds (QACs) are cationic surfactants used increasingly as fabric softeners or disinfectants. Most uses of QACs lead to their release to wastewater treatment plants. Cationic surfactants sorb strongly onto suspended particulates and sludge, which are predominantly negatively charged. Cationic surfactants are considered biologically degradable under aerobic conditions, although the biodegradation for individual surfactants varies. The degradation pathway for alkyl trimethyl ammonium and alkyl dimethyl ammonium halides (TMAC and DMAC) is believed to begin initially by N-dealkylation, followed by N-demethylation [134]. Trimethylamine, dimethylamine, and methylamine were identified as the intermediates of alkyl trimethyl ammonium salts in activated sludge obtained from a municipal sewage treatment plant [134]. In this pathway, alkyl trimethyl ammonium salts are initially degraded to trimethylamine by N-dealkylation. The trimethylamine is then degraded to dimethylamine, and this intermediate is further degraded to methylamine, which is rarely detected. Long-chain alkyl trimethyl ammonium salts are ultimately biodegradable in activated sludge. Games and King [84] reported a half-life of 2.5 hours for octadecyl trimethyl ammonium chloride (C18TMAC) primary biodegradation in a laboratory-based activated sludge system.

Cationic surfactants containing a quaternary ammonium (e.g., R_4N^+, where R = alkyl chain and N = quaternary nitrogen) often have a strong biocidal nature [135]. The alkyl chain length not only determines the physical-chemical properties of a surfactant, but also may have a decisive role in the fate and effects of these compounds in the environment. Under aerobic conditions, the biodegradability of QACs generally decreases with the number of nonmethyl alkyl groups (i.e., R_4N^+ < R_3MeN^+ < $R_2Me_2N^+$ < RMe_3N^+ < Me_4N^+, where Me = methyl radical) [130]. Substitution of a methyl group in a QAC with a benzyl group can decrease biodegradability further [115,130]. Garcia et al. [115] reported time to achieve 50% of primary biodegradation for a series of QAC homologues ranged from 3 to 8 days in the modified OECD screening test and seawater, except for hexadecyl dimethyl ammonium chloride (C16DMAC) (>15 days). The degradation of these compounds in coastal waters was associated with an increase in bacterioplankton density, suggesting that the degradation takes place because the compound is used as a growth substrate.

In contrast, under anaerobic conditions, QACs showed no or very poor primary biodegradation, and no evidence of any extent of ultimate biodegradation was found [116,136]. Primary biodegradation in sludge under anaerobic conditions was found to range from 19 to 38 for mono-alkyl quaternary ammonium–based surfactants and the toxicity to methanogenesis decreased with increasing alkyl chain length [116]. No degradation was observed for ditallow dimethyl ammonium chloride (DTDMAC) in anaerobic screening tests [136]. Due to its poor biodegradation kinetics, diethyl ester dimethyl ammonium chloride (DEEDMAC) was introduced to replace DTDMAC, the major cationic surfactant used in fabric softener formulations worldwide for over 30 years. DEEDMAC differs structurally from DTDMAC by the inclusion of two ester linkages between the ethyl and tallow chains. These ester linkages allow DEEDMAC to be rapidly and completely degraded in standard laboratory screening tests and a range of environmental media such as sludge, soil, and river water, with half-lives ranging from 0.8 days to 18 days [8]. DEEDMAC can be completely degraded under aerobic and anaerobic conditions, and it has a half-life of around 24 hours in raw sewage. Therefore, removal of DEEDMAC during sewage treatment is greater than 99% [8].

C. Nonionic Surfactants

1. Alkylphenol and Ethoxylates

The biodegradation of APE in conventional sewage treatment plants is generally believed to start with a shortening of the ethoxylate chain, leading to short-chain APE containing one or two ethoxylate units. Complete deethoxylation with formation of alkylphenols (AP) has been observed only under anaerobic conditions [35]. Further transformation proceeds via oxidation of the ethoxylate chain, producing mainly alkylphenoxy ethoxy acetic acid and alkylphenoxy acetic acid [137]. The three most common groups of intermediates reported were: (a) alkylphenols (e.g., NP and OP); (b) short-chain alkylphenol ethoxylates having 1–4 ethoxylate units, with APE2 predominating; (c) a series of ether carboxylates, including alkylphenoxy acetic acid and alkylphenoxy ethoxy acetic acid. Recalcitrant decarboxylated NPE biotransformation products with the alkyl chain carboxylated (CAPEs) were also detected in a sewage treatment plant effluent [138]. Previous investigations showed that APE metabolites degraded more easily under aerobic than under anaerobic conditions [38].

The measured removal rates of NPE through sewage treatment plants varied from 93% to 99% in the United States [97], from 66% to 99% in Japan [139], from 74% to 98% in Italy [19,140], and from 47% to 89% in Switzerland [141]. This suggests that only partial degradation occurs.

Due to the amphiphilic nature, APE and their metabolites show an affinity for particulate surfaces; a significant proportion is observed in sludge. Concentrations of APE ranged from 900 to 1,100 mg/kg in anaerobically digested sludge, which are much higher than in aerobically digested sludge (0.3 mg/kg) [24]. APE degradation appears restricted under anaerobic conditions. However, under aerobic conditions, APE undergo almost complete primary degradation. Jones and Westmoorland [142] reported 98% reduction of NPE in composted Australian wool, scouring sludge within 100 days.

The primary degradation of NPE9 showed half-lives of 4 days in water and <10 days in sediments in a river die-away test [143]. Manzano et al. [118] conducted a river die-away test on the biodegradation of a nonylphenol polyethoxylate in river water and found that temperature had a strong influence on the period of acclimation of the microorganisms and on the rate of biodegradation. The percentages of primary biodegradation vary from 68% at 7°C to 96% at 25°C, at all the temperatures studied, metabolites (NPE2, NPE1, NPEC1, and NPEC2) were generated during the biodegradation process that do not totally disappear at the end of the assay (30 days). The mineralization rates reached in the various assays, ranging from 30% at 7°C to 70% at 25°C. Similar results were generated from a static die-away test of NPE in estuarine water in the dark at 28°C for 183 days [22]. Primary degradation was complete in 4–24 days, with a lag period of between 0 and 12 days. The intermediates detected include NPE2 and NPEC2, with much smaller amounts of NPE1 and NPEC1. But NP was not detected. In the primary biodegradation, light was found to be a retarding factor for biodegradation [117].

2. Alkyl Ethoxylates

Alkyl ethoxylates (AE) are easily degradable under aerobic and anaerobic conditions. High primary biodegradation (96 ± 0.5%) was found for AEs in the continuous-flow activated sludge test with a high concentration of metabolites, free fatty alcohol (FFA) and poly(ethylene glycols) (PEG) [123]. However, in a static test, a primary degradability of 75%–98% in an aqueous environment was achieved in 10 days, without significant accumulation of metabolites PEG [121]. Knaebel et al. [144] showed AE to be readily biodegraded in a variety of different soil types, suggesting AE will not accumulate in aerobic sludge-amended soils.

The mechanism for aerobic biodegradation of AE was believed to be initiated by the central cleavage of the molecule, leading to the formation of PEG and FFA, followed by ω- or β-oxidation of the terminal carbon of the alkyl chain, and the hydrolytic shortening of the terminal carbon of the polyethoxylic chain [120,121]. In contrast to aerobic biodegradation, where central cleavage prevails, the first step of anaerobic microbial attack on the AE molecule is the cleavage of the terminal ethoxy unit, releasing acetaldehyde stepwise and shortening the ethoxy chain until the lipophilic moiety is reached [124].

V. BIOLOGICAL EFFECTS OF SURFACTANTS AND THEIR DEGRADATION PRODUCTS

Surfactants entering the environment through the discharge of sewage effluents into surface waters and application of sewage sludge on land have the potential to impact

TABLE 12 Aquatic Toxicity Data for Anionic Surfactants

Chemical	Species	Endpoint	Ref.
C10LAS	*Daphnia magna*	LC50—48 h, 13.9 mg/L	2
C12LAS		LC50—48 h, 8.1 mg/L	
C14LAS		LC50—48 h, 1.22 mg/L	
C12LAS	*Dunaliella* sp. (green alga)	EC50—24 h, 3.5 mg/L	146
C11–12LAS	*Oncorhynchus mykiss* (rainbow trout fry)	NOEC—54 d, 0.2 mg/L	147
C12LAS (SDBS)	*Salmo gairdneri* (rainbow trout)	Immobilization, EC50—48 h, 3.63 mg/L 148	148
	Gammbusia affinis (mosquito fish)	Immobilization, EC50—48 h, 8.81 mg/L	
	Carassius auratus (goldfish)	Immobilization, EC50—48 h, 5.1 mg/L	
C12AS (SDS)	*Salmo gairdneri* (rainbow trout)	Immobilization, EC50—48 h, 33.61 mg/L	148
	Gammbusia affinis (mosquito fish)	Immobilization, EC50—48 h, 40.15 mg/L	
	Carassius auratus (goldfish)	Immobilization, EC50—48 h, 38.04 mg/L	
Sodium dodecyl ethoxy sulfate (SDES)	*Salmo gairdneri* (rainbow trout)	Immobilization, EC50—48 h, 10.84 mg/L	148
	Gammbusia affinis (mosquito fish)	Immobilization, EC50—48 h, 13.64 mg/L	
	Carassius auratus (goldfish)	Immobilization, EC50—48 h, 12.35 mg/L	

the ecosystem, owing to their toxicity on organisms in the environment. The toxicity data from laboratory and field studies are essential for us to assess the possible environmental risks from the surfactants.

A. Aquatic Toxicity

Aquatic toxicity data are widely available for anionic, cationic, and nonionic surfactants. Lewis [145] has summarized the chronic and sublethal toxicities of surfactants to aquatic animals and found that chronic toxicity of anionic and nonionic surfactants occurs at concentrations usually greater than 0.1 mg/L. Tables 12–14 list some recently published toxicity data for the three classes of surfactants on several test organisms (algae, invertebrates, fish) from the literature. Singh et al. [148] tested seven surfactants for toxicity (immobility, EC50—48 h) on six freshwater macrobes and found that cationic surfactants were more toxic than anionic surfactants and anionic surfactants more toxic than nonionic surfactants. Utsunomiya et al. [146] studied the toxic effects of C12LAS and three quaternary alkylammonium chlorides on unicellar green alga *Dunaliella* sp. by measuring ^{13}C glycerol. The 24-hour median

TABLE 13 Aquatic Toxicity Data for Cationic Surfactants

Chemical	Species	Endpoint	Ref.
TMAC	*Dunaliella* sp.	EC50—24 h, 0.79 mg/L	146
DADMAC	(green algae)	EC50—24 h, 18 mg/L	
BDMAC		EC50—24 h, 1.3 mg/L	
C16TMAC	*Salmo gairdneri* (rainbow trout)	Immobilization, EC50—48 h, 1.21 mg/L	148
	Gammbusia affinis (mosquito fish)	Immobilization, EC50—48 h, 8.24 mg/L	
	Carassius auratus (goldfish)	Immobilization, EC50—48 h, 3.58 mg/L	
DTDMAC	*Salmo gairdneri* (rainbow trout)	Immobilization, EC50—48 h, 0.74 mg/L	148
	Gammbusia affinis (mosquito fish)	Immobilization, EC50—48 h, 7.91 mg/L	
	Carrassius auratus (goldfish)	Immobilization, EC50—48 h, 2.37 mg/L	
DTDMAC	*Daphnia magna*	LC50—48 h, 0.49 mg/L NOEC—21 d, 0.38 mg/L	52
DEEDMAC	*Daphnia magna*	Immobilization, LC50—24 h, 14.8 mg/L Growth, NOEC—21 d, 1 mg/L	8
	Pimephales promelas (fathead minnow)	Growth, NOEC—35 d, 0.68 mg/L	
	Selenastrum capricornutum (algae)	Growth inhibition, EC50—96 h, 2.9 mg/L	
TMAC	*Daphnia magna*	Immobilization, IC50—24 h, 0.13–0.38 mg/L	115
BDMAC	*Daphnia magna*	Immobilization, IC50—24 h, 0.13–0.22 mg/L	115

TABLE 14 Aquatic Toxicity Data for Nonionic Surfactants

Chemical	Species	Endpoint	Ref.
C12EO6	*Salmo gairdneri* (rainbow trout)	Immobilization, EC50—48 h, 22.38 mg/L	148
	Gammbusia affinis (mosquito fish)	Immobilization, EC50—48 h, 29.26 mg/L	
	Carassius auratus (goldfish)	Immobilization, EC50—48 h, 28.02 mg/L	
C9–11EO6	*Pimephales promelas* (fathead minnow)	LC50—10 d, 2.7 mg/L	149
OPEO6	*Salmo gairdneri* (rainbow trout)	Immobilization, EC50—48 h, 6.44 mg/L	148
	Gammbusia affinis (mosquito fish)	Immobilization EC50—48 h, 9.65 mg/L	
	Carassius auratus (goldfish)	Immobilization, EC50—48 h, 9.24 mg/L	
NPEO8	Australian native frogs	Full narcosis, EC50—48 h, 2.8–3.8 mg/L	150
NPEO9	Fathead minnow	LC50—96 h, 4.6 mg/L	97
	Daphnia magna	LC50—48 h, 14 mg/L	
NP	Fathead minnow	LC50—96 h, 0.3 mg/L	97
	Daphnia magna	LC50—48 h, 0.19 mg/L	

effective concentrations (EC50—24 h) were 3.5 mg/L for LAS, 0.70 mg/L for alkyl trimethyl ammonium chloride (TMAC), 18 mg/L for dialkyl dimethyl ammonium chloride (DADMAC), and 1.3 mg/L for alkyl benzyl dimethyl ammonium chloride (BDMAC); the toxic potencies were in the order of TMAC > BDMAC > LAS > DADMAC.

LAS acute toxicity to *Daphnia magna* increases with the alkyl chain or homologue molecular weight, probably due to higher interaction of heavier homologues with cell membranes [2]. It is also found that a very high water hardness (>2,000 mg/L as $CaCO_3$) may be a stress factor, giving a much lower LC50—48 h than at lower water hardness and the same LAS concentrations. Although 0.2 mg/L is considered as the no-observed-effect concentration (NOEC), lamellar gill epithelia of rainbow trout fry hypertrophied, and its swimming capacity was reduced after 54 days' exposure [147]. Temara et al. [151] conducted risk assessment of LAS in the North Sea. The LAS concentration range in the estuaries around the North Sea ranged from 1 to 9 μg/L, while in the offshore sites it is below the detection limit (0.5 μg/L). The predicted no-effect concentrations (PNEC) were 360 and 31 μg/L for freshwater and marine pelagic communities, respectively. Given that the maximum expected estuarine and marine concentrations are 3 to over 30 times lower than the PNEC, the risk of LAS to pelagic organisms in these environments is judged to be low.

Kimberle and Swisher [152] found that SPC, biodegradation intermediates of LAS, give LC50 values that are 120%–240% higher than that of LAS. No estrogenic effects were observed by Navas et al. [126] for LAS and SPC by two in vitro assays: the yeast estrogen receptor assay and the vitellogenin assay with cultured trout hepatocytes. Garcia et al. [115] carried out acute toxicity tests on *Daphnia magna* and *Photobacterium phosphoreum* for two families of monoalkyl quaternary ammonium surfactants: alkyl trimethyl ammonium and alkyl benzyl dimethyl ammonium halides. The 24-h immobilization EC50 on *D. magna* ranged from 0.13 to 0.38 mg/L for the six cationic surfactants, whereas EC50 on *P. phosphorem* ranged from 0.15 to 0.63 mg/L. Although the substitution of a benzyl group for a methyl group increases the toxicity, an incremental difference in toxicity between homologues of different chain length were not observed. This could be attributed to a lower bioavailability of the longest-chain homologues due to their decreasing solubility. This assumption is enhanced by the results on *D. magna* for DTDMAC, which is a less soluble compound than QACs. QACs exhibited a lower toxicity [153]. DTDMAC is being replaced by less toxic and easily degradable DEEDMAC. The 21-day-growth NOEC values on *D. magna* are 0.38 mg/L for DTDMAC [52] and 1.0 mg/L for DEEDMAC [8].

Mann and Bidwell [150] studied the acute toxicity of NPE and AE to the tadpoles of four Australian and two exotic frogs. All species exhibited nonspecific narcosis following exposure to both these surfactants. The 48-h EC50 values for NPE ranged between 1.1 mg/L (mild narcosis) and 12.1 mg/L (full narcosis). The 48-h EC50 values for AE ranged between 5.3 mg/L (mild narcosis) and 25.4 mg/L (full narcosis).

A stream mesocosm study by Dorn et al. [149] demonstrated that fish and invertebrates were most responsive to the effects of AE. For C9 11EO6, the 10-day lab LC50 value for fathead minnows was found to be 2.7 mg/L, compared to a 10-day mesocosm LC50 of 6.4 mg/L. The 30-day mesocosm LC50 value was 5.5 mg/L, which indicates that there is little change after 10 days' exposure to this surfactant. Fathead minnows were particularly sensitive to AE, with an NOEC of 0.73 mg/L for egg production and larval survival. Bluegill were less sensitive than fathead minnows, with an NOEC for survival and growth of 5.7 mg/L. The stream mesocosm results for fish and invertebrates were similar to those obtained using laboratory single-species tests.

APE are found to be much less acutely toxic than their degradation products (NP and OP) to aquatic organisms [97]. Yoshimura [143] found that the LC50—48-h value for NPE increased with EO unit chain length, therefore becoming less toxic. NP showed much more toxicity than NPEO9 to aquatic organisms [97,143].

B. Terrestrial Toxicity

Sewage sludges are increasingly being applied on agricultural lands as fertilizers for plants. These sludges have been found to contain high concentrations of surfactants as well as other contaminants. The terrestrial environment has become a significant sink for the surfactants. In order to sustainably use sewage sludge, it is therefore necessary to assess the toxicity of those surfactants to the soil-dwelling organisms, especially plants.

The terrestrial toxicity data are quite limited and they are mainly measured for LAS on plants (Table 15), but limited toxicity data are also available on soil fauna [92,158]. Unilever [155], as cited in Mieure et al. [156], studied the effect of LAS on sorghum (*Sorghum bicolor*), sunflower (*Helianthus annuus*), and mung bean (*Phaseolus aureus*) via the OECD Terrestrial Plant Growth Test (OECD 208). Using test

TABLE 15 Terrestrial Toxicity Data for Anionic Surfactants (LAS)

Species	Test condition	Endpoint	Ref.
Bush beans, radish, and grasses	Field tests in sludge-amended clay soil	Yield and growth, NOEC—76 d, 27 mg/kg	53
Potatoes	Field, sandy soil	Yield and growth, NOEC—106 d, 16 mg/kg	53
Ryegrass	Field, two soils	Yield, NOEC 500 kg/ha, with necrosis and chlorosis observed	154
Sorghum	Lab, potting compost	Growth, EC50—21 d, 167 mg/kg	155,156
Sunflower		Growth, EC50—21 d, 289 mg/kg	
Mung bean		Growth, EC50—21 d, 316 mg/kg	
Oats	Lab, sandy loam	Growth, EC50—14 d, 50 mg/kg	92,157,158
		Growth, EC50—14 d, 300 mg/kg	
Turnip		Growth, EC50—14 d, 90 mg/kg	
		Growth, EC50—14 d, 200 mg/kg	
Mustard		Growth, EC50—14 d, 200 mg/kg	
		Growth, EC50—14 d, 300 mg/kg	

concentrations of 1, 10, 100, and 1,000 mg/kg LAS in a potting soil, they determined the 21-day-growth EC50 of 167, 289, and 316 mg/kg for sorghum, sunflower, and mung bean, respectively. The highest reported NOEC was 100 mg/kg for the three species. Gunther and Pestemer [157], as cited in Jensen [92], performed a series of toxicity tests with LAS on oat (*Avena sativa*), turnip (*Brassica rapa*), and mustard (*Sinapis alba*) in a sandy loam at different concentrations and measured the fresh weight of shoots after 14 day's exposure. The lowest 14-day EC50 value was determined for oats (50 mg/kg soil). But its EC50 value was similar to that of turnip or mustard.

Litz et al. [154] observed considerable short-term acute physiological damage on ryegrass from LAS in a field experiment using an application rate of 500 kg/ha, but no reduction in yield was found after harvest. Figge and Schöberl [53] conducted an extensive study of LAS effects on plants (and potato) using a plant metabolism box. They estimated the field NOEC values to be 16 mg/kg for bush beans, grass, and radish and 27 mg/kg for potatoes. From the limited terrestrial toxicity data available, LAS can be considered to be of relatively low toxicity to terrestrial organisms.

C. Endocrine Disruption

Some chemicals in the environment can disrupt the normal functioning of the endocrine system in wildlife as well as human beings; these chemicals are called *endocrine-disrupting chemicals* (EDCs) [159]. The estrogenic properties of alkylphenols were recognized as early as 1938 [160]. The ability of 4-alkylphenols to displace estradiol from the estrogen receptors was reported in 1978 [161]. More recently, the estrogenic activities of alkylphenols have been demonstrated both in vitro [162] and in vivo [6]. NP and OP have been shown to be capable of inducing the production of vitellogenin in male fish, a protein usually found only in sexually mature females under the influence of estrogens [163]. The relative potencies of OP and NP to β-estradiol were measured to be 1×10^{-4} and 1.3×10^{-5} [164]. Therefore, alkylphenols (OP and NP) are weak estrogen-mimic compounds.

The study by Jobling et al. [165] showed widespread sexual disruption in wild fish in UK rivers. The lowest-observable-effect level (LOEL) values were reported to be 5 and 20 µg/L for 4-t-OP and 4-NP on rainbow trout, respectively [6]. This suggests that in some highly contaminated rivers and estuaries in the United Kingdom, levels of alkylphenols are high enough to affect the reproductive health of fish. Hence, alkylphenols may play a significant role in the feminization of fish in UK rivers.

VI. CASE STUDY: ENVIRONMENTAL FATE OF NONYLPHENOL AND OCTYLPHENOL

A. Introduction

In Australia, sewage effluents from sewage treatment plants in big cities around the coast are discharged mainly into the marine environment through sewage outfalls. The discharge of sewage treatment plant effluents into the marine environment creates the potential for coastal contamination due to the existence of various classes of contaminants in the effluents. Water quality degradation and seagrass loss have been recorded in the Australian coastal environment. In recent years, treated effluents have been increasingly reused for irrigation. Effluents are stored in underground aquifers through artificial injection in rainy seasons and recovered in dry seasons for irrigation [166]. Nonylphenol and octylphenols as the degradation products of APE have been found in sewage effluents at µg/L levels [25]. These two chemicals have been demonstrated to be weak endocrine-disrupting chemicals [167]. Therefore, it is necessary to understand the behavior and fate of NP and OP in the marine and coastal environments as well as during aquifer storage and recovery (ASR). This case study determined the levels of NP and OP in effluents from the Bolivar sewage treatment plant and investigated the sorption of NP and OP on an aquifer material and their degradation in marine sediment and aquifer material under aerobic and anaerobic conditions [168].

B. Materials and Methods

1. Chemicals

4-tert-Octyl phenol (4-t-OP) was obtained from Chem Service, while 4-n-nonyl phenol (4-n-NP) was purchased from Fluka (Riedel-de Haën). HPLC-grade methanol and acetonitrile were obtained from BDH (England). Stock solutions (100 mg/L) of each standard as well as mixtures were prepared in methanol.

2. Sewage Effluents and Sediments

Three sewage effluent samples were collected at different times in 2001 from the Bolivar sewage treatment plant in South Australia. Most of the effluent from this plant is discharged into the sea, with a proportion reused for horticulture nearby. The sediment samples were collected from St. Kilda beach near the sewage treatment plant, while seawater was from a jetty nearby. The aquifer materials (sediment and groundwater) used in this study were collected from the ASR well at Bolivar in South Australia, corresponding to a relatively permeable horizon. Sediment samples were from a depth of 153–154 m, while groundwater was from the well 300 m from the injection well. This is representative of native groundwater at this site. The physiochemical properties of the sediment and water samples were analyzed and are presented in Table 16.

3. Sorption Tests

Sorption of OP and NP on the aquifer material was measured at room temperature (about 25°C) by a batch equilibration method. Sediment (2 g) was weighed into each 250-mL bottle; 100 mL of groundwater was added into each bottle. The concentrations used were as follows: 40, 60, 80, and 100 µg/L for 4-*n*-NP and 5, 10, 20, and 40 µg/L for 4-*t*-OP. The sediment solutions were equilibrated by shaking in a mechanical shaker for 16 hours. After equilibration, the tubes were centrifuged at 3,000 rpm for 30 min, and then the supernatants were further filtered through Whatman glass fiber filters (GF/C). The filtrates were then analyzed by a Varian HPLC with a fluorescence detector. All tests were done in duplicate.

Blanks without sediment but with different concentrations of each chemical were prepared at the same time during the sorption test. No significant loss (<3%) was found in the blanks for 4-*t*-OP during the sorption process, but some loss (about 38%) was found for 4-*n*-NP after overnight shaking. For 4-*n*-NP, shaking time was reduced to 2 hours, giving a recovery of 90%. Blanks with different concentrations of 4-*n*-NP and 4-*t*-OP in groundwater were prepared during the sorption and analysis processes as quantitation controls.

4. Degradation Experiments

Biodegradation of NP and OP in marine sediment and aquifer material was determined under aerobic and anaerobic conditions. In the experiments, 5 g of marine sediment and 5 mL of seawater were used for the study of biodegradation in

TABLE 16 Physiochemical Properties of the Sediment and Water Samples

	pH	NO_3-N (mg/L)	PO_4-P (mg/L)	TC (mg/L)	IC (mg/L)	DOC (mg/L)
Seawater	8.4	<0.01	0.04	46	26	20
Groundwater	7.9	0.02	0.02	56	49	7

	pH	TC (%)	OC (%)	CEC(NH4) cmol(+)/kg	CO_3 as $CaCO_3$ (%)	Clay (%)	Silt (%)	Sand (%)
Marine sediment	9.5	8.8	0.1	2.4	72	3	0.8	22.1
Aquifer material	8.9	1.9	0.5	2.4	12	3.1	1.1	83

TC = total carbon, IC = inorganic carbon, DOC = dissolved organic carbon, OC = organic carbon, CEC = cation exchange capacity.

marine sediment, whereas 5 g of aquifer sediment and 5 mL of groundwater were used for the study of biodegradation in aquifer material. The concentration applied for each chemical was 1 µg/g in the sediment. The incubation temperature used in the studies was 20°C. The concentrations of the five compounds were monitored at certain intervals following treatment. All experiments were performed in duplicate, and duplicate sterile controls were monitored at the same times.

(a) Aerobic Study.   Sediment and water were weighed into 100-mL Schott bottles. Half of the bottles were sterilized and used as controls. Chemicals were spiked at a required concentration (1 µg/g). All bottles were incubated in a temperature-controlled chamber. Those bottles were shaken for 1 minute at each sampling time.

(b) Anaerobic Study.   Sediment and water were weighed into Hungate anaerobic culture tubes (16 × 125 mm). Half of them were sterilized and used as controls. These Hungate tubes were placed into an anaerobic incubation chamber filled with nitrogen gas. The lids of these tubes were loosened to facilitate gas exchange. Resazurin was added at a concentration of 0.0002% into two tubes as a redox indicator. Once reducing conditions were reached in the tubes, those with resazurin turned from a red color to colorless. All tubes were left in the anaerobic incubation chamber for nearly a month, until the tubes having the redox indicator resazurin turned colorless. Chemicals were spiked into each tube at a required concentration (1 µg/g), and the lids of all tubes were tightened after spiking, which was performed inside the anaerobic incubation chamber. Then all the tubes were incubated in the same chamber as in the aerobic study.

5. Extraction and Analysis

Sediment samples were extracted with 20 mL of ethyl acetate. The extracts were dried under a gentle nitrogen stream and redissolved in methanol. The recoveries were 105 ± 2% for 4-*t*-OP and 91 ± 3% for 4-*n*-NP. Samples (sediment and water) were then analyzed by a Varian HPLC with a fluorescence detector [169].

C. Results and Discussion

1. OP and NP Levels in the Effluents

OP and NP were detected in the three effluent samples. The concentrations in the effluents were found to vary from 0.30 to 1.85 µg/L for 4-*t*-OP, from 5.59 to 11.29 µg/L for 4-NP, and from 4.09 to 41.53 µg/L for NPE. These concentrations are comparable to the data reported in Table 3.

2. Sorption on an Aquifer Material

Sorption of 4-*t*-OP and 4-*n*-NP was very high on the aquifer material. The sorption coefficients (K_f) values on the aquifer material were 90.9 L/kg for 4-*t*-OP and 195 L/kg for 4-*n*-NP, respectively.

The organic carbon normalized sorption coefficients (K_{oc}) for the two chemicals were 18,200 L/kg for 4-*t*-OP and 38,900 L/kg for 4-*n*-NP, which are comparable to the reported values [66,170,171]. Ferguson et al. [66] investigated the partitioning of APE metabolites to suspended solids in Jamaica Bay, New York. The log K_{oc} values were 5.39 for NP and 5.8 for OP. Sekela et al. [171] measured similar log K_{oc} values for NP on five samples (log K_{oc} = 4.7–5.6). Johnson et al. [170] used laboratory batch techniques to study the sorption of OP on sediments from three

English rivers of contrasting water quality. The results showed that given either sufficient time or mixing, a large proportion of the OP in solution will sorb to the bed sediments, with distribution coefficients (K_d) of 6–700 L/kg and organic carbon normalized partition coefficients (K_{oc}) of 3,500–18,000 L/kg (cf. current results: 18,200 L/kg). The sediments that sorbed the highest quantities of OP had higher total organic carbon and a greater proportion of clay and silt particles. From the reported K_d values and current study results, OP and NP tend to sorb strongly onto sediments.

3. Degradation in the Aquifer Media

Biodegradation experiments showed different behavior for the two chemicals (4-*t*-OP and 4-*n*-NP) under aerobic and anaerobic conditions in the aquifer material (Figs. 3 and 4). Under the aerobic conditions, 4-*n*-NP was degraded very quickly, while 4-*t*-OP remained unchanged over 70 days. Within 10 days, nearly 50% of 4-*n*-NP was degraded in the aquifer media, with an estimated half-life of 7 days based on the first-order reaction equation. However, under anaerobic conditions, the two chemicals remained almost unchanged over 70 days. Clearly the two compounds were persistent in the anoxic aquifer material.

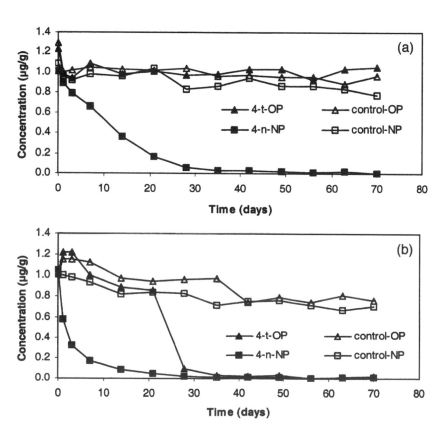

FIG. 3 Degradation of 4-*t*-OP and 4-*n*-NP under aerobic conditions in (a) aquifer material and (b) marine sediment.

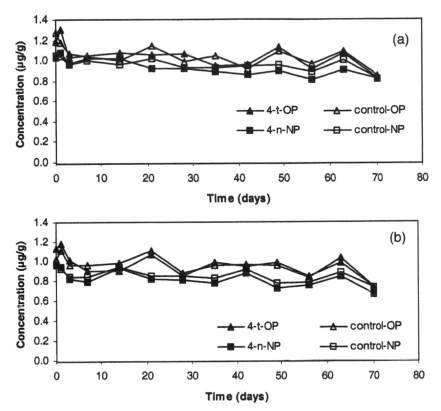

FIG. 4 Degradation of 4-*t*-OP and 4-*n*-NP under anaerobic conditions in (a) aquifer material and (b) marine sediment.

4. Degradation in the Marine Sediment

Similar degradation results were obtained for 4-*t*-OP and 4-*n*-NP in the marine sediment, except for 4-*t*-OP under aerobic conditions (Figs. 3 and 4). After 70 days of experiment, 4-*t*-OP and 4-*n*-NP in the marine sediment were degraded under aerobic conditions. 4-*n*-NP was degraded very quickly, with a half-life of 5.8 days. The concentration of 4-*n*-NP decreased to 0.58 µg/g at 1 DAT (days after treatment) and to 0.05 µg/g at 21 DAT. But 4-*t*-OP was degraded much more slowly, with a dissipation half-life of less than 20 days. For 4-*t*-OP, there was an acclimation period, with a concentration of 0.84 µg/g at 21 DAT, followed by fast degradation, with its concentration decreasing to only 0.09 µg/g within a week.

5. Discussion

The present study showed that 4-*n*-NP can be degraded more easily than 4-*t*-OP under aerobic conditions. This is believed to be related to the branched alkyl group in 4-*t*-OP. Biodegradation has been reported to play a major role in the removal of these two chemicals from aquatic environments. Laboratory and field experiments all demonstrated that alkylphenols could be degraded under aerobic conditions but were persistent under anaerobic conditions (Ref. 25 and references therein). Persistence of NP and OP has been reported in the marine sediments from the Canadian coast [172], the Venice lagoon [173], and Tokyo Bay [30]. Owing to the hydrophobic nature of these chemicals, they tend to adsorb on to the suspended and bed sediments. Therefore, biodegradation in sediment is detrimental for the fate of these two EDCs discharged into the marine environment through sewage effluent outfalls. They will be persistent and accumulate in anoxic marine sediment.

It would appear that in the anaerobic natural environment for the Bolivar aquifer material tested, it is unlikely that OP and NP will be removed through biodegradation processes because they are persistent under anaerobic conditions. However, it is possible that significant removal of OP and NP may occur during aquifer storage and recovery due to their strong retention in the aquifer material. But without biodegradation, breakthrough of these two compounds at the recovery well would be expected ultimately.

VII. SUMMARY

Surfactants have been found in the aquatic environment at µg/L levels mainly due to their widespread use and discharge of sewage effluents into surface waters. They are also measured in sludge-amended soils because of the high residual concentrations in sludges. Surfactants have relatively high sorption on sediment and soil, and their sorption is in the order of: cationic > nonionic > anionic. They are strongly associated with particulates or sediment; therefore biodegradation of surfactants in sediment is detrimental in determining their fate in the environment.

Surfactants can be degraded under aerobic conditions; however, some of them are persistent under anaerobic conditions, such as LAS and DTDMAC. APE are partially degraded in the anaerobic environment to form alkylphenols (NP and OP), which are also persistent and have estrogenic activities to organisms such as fish.

Elevated concentrations of surfactants and their degradation products may affect organisms in the environment. The environmental risks posed by surfactants and their degradation products can be assessed based on the comparison of the predicted environmental concentration and the predicted no-effect concentration. But more toxicity data are needed for terrestrial risk assessment of surfactants and their degradation products.

ACKNOWLEDGMENTS

The authors would like to thank internal reviewers Drs. D. Steve and E. Smith (CSIRO Land and Water) for their useful comments.

REFERENCES

1. Di Corcia, A. J. Chromatogr. A. 1998, *794*, 165–185.
2. Verge, C.; Moreno, A.; Bravo, J.; Berna, J.L. Chemosphere 2001, *44*, 1749–1757.
3. Renner, R. Environ. Sci. Technol. 1997, *31*, 316A–320A.
4. Comber, M.H.I.; Williams, T.D.; Stewart, K.M. Water Res. 1993, *27*, 273–276.
5. McLeese, D.W.; Zitko, V.; Sergeant, D.B.; Burridge, L.; Metcalfe, C.D. Chemosphere 1981, *10*, 723–730.
6. Jobling, S.; Sheahan, D.; Osborne, J.A.; Matthiessen, P.; Sumpter, J.P. Environ. Toxicol. Chem. 1996, *15*, 194–202.

7. Purdom, C.E.; Hardiman, P.A.; Bye, V.J.; Eno, N.C.; Tyler, C.R.; Sumpter, J.P. Chem. Ecol. 1994, *8*, 275–285.
8. Giolando, S.T.; Rapaport, R.A.; Larson, R.J.; Federle, T.W.; Stalmans, M.; Masscheleyn, P. Chemosphere 1995, *30*, 1067–1083.
9. Waters, J.; Feijtel, T.C.J. Chemosphere 1995, *30*, 1939–1956.
10. Fytianos, K.; Pegiadou, S.; Raikos, N.; Eleftheriadis, I.; Tsoukali, H. Chemosphere 1997, *35*, 1423–1429.
11. Prats, D.; Ruiz, F.; Vazquez, B.; Rodriguez-Pastor, M. Water Res. 1997, *31*, 1925–1930.
12. Holt, M.S.; Fox, K.K.; Burford, M.; Daniel, M.; Buckland, H. Sci. Total Environ. 1998, *210/211*, 255–269.
13. Schroder, F.R.; Schmitt, M.; Reichensperger, U. Waste Management 1999, *19*, 125–131.
14. Feijtel, T.C.J.; Matthijs, E.; Rottiers, A.; Rijs, G.B.J.; Kiewiet, A.; de Nijs, A. Chemosphere 1995, *30*, 1053–1066.
15. Crescenzi, C.; Di Corcia, A.; Marchiori, E.; Samperi, R.; Marcomini, A. Water Res. 1996, *30*, 722–730.
16. Matthijs, E.; Holt, M.S.; Kiewiet, A.; Rijs, G.B.J. Environ. Toxicol. Chem. 1999, *18*, 2634–2644.
17. Fendinger, N.J.; Begley, W.M.; Mcavoy, D.C.; Eckhoff, W.S. Environ. Sci. Technol. 1995, *29*, 856–863.
18. Snyder, S.A.; Keith, T.L.; Verbrugge, D.A.; Snyder, E.M.; Gross, T.S.; Kannan, K.; Giesy, J.P. Environ. Sci. Technol. 1999, *33*, 2814–2820.
19. Di Corcia, A.; Samperi, R. Environ. Sci. Technol. 1994, *28*, 850–858.
20. Versteeg, D.J.; Feijtel, T.C.J.; Cowan, C.E.; Ward, T.E.; Rapaport, R.A. Chemosphere 1992, *24*, 641–662.
21. Jonkers, N.; Knepper, T.P.; de Voogt, P. Environ. Sci. Technol. 2001, *35*, 335–340.
22. Potter, T.L.; Simmons, K.; Wu, J.; Sanchez-Olvera, M.; Kostecki, P.; Calabrese, E. Environ. Sci. Technol. 1999, *33*, 113–118.
23. Salanitro, J.P.; Diaz, L.A. Chemosphere 1995, *30*, 813–830.
24. Scott, M.J.; Jones, M.N. Biochim. Biophys. Acta. 2000, *1508*, 235–251.
25. Ying, G.G.; William, B.; Kookana, R. Environment Int. 2002, *28*, 215–226.
26. Lee, H.B.; Peart, T.E. Anal. Chem. 1995, *67* (13), 1976–1980.
27. Blackburn, M.A.; Waldock, M.J. Water Res. 1995, *29*, 1623–1629.
28. Ahel, M.; Giger, W. Anal. Chem. 1985, *57*, 1577–1583.
29. Sole, M.; Lopez de Alda; Castillo, M.; Porte, C.; Ladegaard-Pedersen, K.; Barcelo, D. Environ. Sci. Technol. 2000, *34*, 5076–5083.
30. Isobe, T.; Nishiyama, H.; Nakashima, A.; Takada, H. Environ. Sci. Technol. 2001, *35*, 1041–1049.
31. Rudel, R.A.; Melly, S.J.; Geno, P.W.; Sun, G.; Brody, J.G. Environ. Sci. Technol. 1998, *32*, 861–869.
32. Kuch, H.M.; Ballschmiter, K. Environ. Sci. Technol. 2001, *35*, 3201–3206.
33. Kuhnt, G. Environ. Toxicol. Chem. 1993, *12*, 1813–1820.
34. Bruno, F.; Curini, R.; Di Corci, A.; Fochi, I.; Nazzari, M.; Samperi, R. Environ. Sci. Technol. 2002, *36*, 4156–4161.
35. Giger, W.; Brunner, P.H.; Schaffner, C. Science 1984, *225*, 623–625.
36. Waldock, M.J., Thain, J.E. Environmental considerations of 4-nonylphenol following dumping of anaerobically digested sewage sludges: a preliminary study of occurrence and acute toxicity. International Council for the Exploration of the Sea Marine Environmental Quality Committee Report CM 1986/E: 16, 1986.
37. Marcomini, A.; Giger, W. Anal. Chem. 1987, *59*, 1709–1715.
38. Brunner, P.H.; Capri, A.; Marcomini, A.; Giger, W. Water Res. 1988, *22*, 1465–1472.
39. Wahlberg, C.; Renberg, L.; Wideqvist, U. Chemosphere 1990, *20*, 179–195.
40. Chalaux, N.; Bayona, J.M.; Albaige, J. J. Chromatogr. A. 1994, *686*, 275–281.

41. McEvoy, J.; Giger, W. Naturwissenschafen 1985, 72, 429–431.
42. Matthijs, E.; DeHenau, H. Tenside Surfact. Deterg. 1987, 24 (4), 193–199.
43. Berna, J.L.; Ferrer, J.; Moreno, A.; Prats, D.; Bevia, F.R. Tenside Surfact. Deterg. 1989, 26, 101–107.
44. McAvoy, D.C.; Eckhoff, W.S.; Rapaport, R.A. Environ. Toxicol. Chem. 1993, 12, 977–987.
45. Fernandez, P.; Alder, A.C.; Suter, M.J.F.; Giger, W. Anal. Chem. 1996, 68, 921–929.
46. Eichhorn, P.; Rodrigues, S.V.; Baumann, W.; Knepper, T.P. Sci. Total Environ. 2002, 284, 123–134.
47. Eichhorn, P.; Flavier, M.E.; Pae, M.L.; Knepper, T. Sci. Total Environment 2001, 269, 75–85.
48. Reiser, R.; Toljander, H.O.; Giger, W. Anal. Chem. 1997, 69, 4923–4930.
49. Ding, W.H.; Tzing, S.H.; Lo, J.H. Chemosphere 1999, 38, 2597–2606.
50. Holt, M.S.; Waters, J.; Comber, M.H.I.; Armitage, R.; Morris, G.; Newbery, C. Water Res. 1995, 29, 2063–2070.
51. Fox, K.; Holt, M.; Daniel, M.; Buckland, H.; Guymer, I. Sci. Total Environ. 2000, 251/252, 265–275.
52. Lewis, M.; Wee, V. Environ. Toxicol. Chem. 1983, 2, 105–108.
53. Figge, K.; Schöberl, P. Tenside Surfact. Deterg. 1989, 26, 122–128.
54. Berna, J.L.; Ferrer, J.; Moreno, A.; Prats, D.; Bevia, F.R. Tenside Surfact. Deterg. 1989, 26, 101–107.
55. Marcomini, A.; Capel, P.D.; Lichtenseiger, T.H.; Brunner, P.H.; Giger, W. J. Environ. Qual. 1989, 18, 523–528.
56. Waters, J.; Holt, M.S.; Matthijs, E. Tenside Surfact. Deterg. 1989, 26, 129–135.
57. Holt, M.S.; Matthijs, E.; Waters, J. Water Res. 1989, 23, 749–759.
58. Carlsen, L.; Metzon, M.B.; Kjelsmark, J. Sci. Total Environ. 2002, 290, 225–230.
59. Bennie, D.T.; Sullivan, C.A.; Lee, H.B.; Peart, T.E.; Maguire, R.J. Sci. Total Environ. 1997, 193, 263–275.
60. Blackburn, M.A.; Kirby, S.J.; Waldock, M.J. Marine Pollut. Bull. 1999, 38 (2), 109–118.
61. Ahel, M.; Schaffner, C.; Giger, W. Water Res. 1996, 30, 37–46.
62. Ahel, M.; Molnar, E.; Ibric, S.; Giger, W. Water Sci. Technol. 2000, 42 (7–8), 15–22.
63. Tsuda, T.; Takino, A.; Kojima, M.; Harada, H.; Muraki, K.; Tsuji, M. Chemosphere 2000, 41, 757–762.
64. Tabata, A.; Kashiwa, S.; Ohnishi, Y.; Ishikawa, H.; Miyamoto, N.; Itoh, M.; Magara, Y. Water Sci. Technol. 2001, 43 (2), 109–116.
65. Dachs, J.; Van Ry, D.A.; Eisenreich, S.J. Environ. Sci. Technol. 1999, 33, 2676–2679.
66. Ferguson, P.L.; Iden, C.R.; Brownwell, B.J. Environ. Sci. Technol. 2001, 35, 2428–2435.
67. Naylor, C.G.; Mieure, J.P.; Adams, W.J.; Weeks, J.A.; Castaldi, F.J.; Ogle, L.D.; Romano, R.R. JAOCS 1992, 69 (7), 695–703.
68. Haigh, S.D. Sci. Total Environ. 1996, 185, 161–170.
69. Tolls, J.; Sijm, T.H.M. Environ. Toxicol. Chem. 1995, 14, 1675–1685.
70. Brix, R.; Hvidt, S.; Carlsen, L. Chemosphere 2001, 44, 759–763.
71. Adeel, Z.; Luthy, R.G. Environ. Sci. Technol. 1995, 29, 1032–1042.
72. Kibbey, T.C.G.; Hayes, K.F. J. Contam. Hydrol. 2000, 41, 1–22.
73. Aronstein, B.N.; Calvillo, Y.M.; Alexander, M. Environ. Sci. Technol. 1991, 25, 1728–1731.
74. Edwards, D.A.; Adeel, Z.; Luthy, R.G. Environ. Sci. Technol. 1994, 28, 1550–1560.
75. Tiehm, A. Appl. Environ. Microbiol. 1994, 60, 258–263.
76. Kile, D.E.; Chiou, C.T. Environ. Sci. Technol. 1989, 23, 832–838.
77. Klumpp, E.; Heitman, H.; Schwuger, M.J. Tenside Surfact. Deterg. 1991, 28, 441–446.

78. Sweetman, A., Rogers, H.R., Watts, C.D., Alcock, R., Jones, K.C. Organic contaminants in sewage sludge - Phase III (Env 9031). Final Report to the Department of the Environment, UK, Rep. No. DOE 3625/1, 1994.

79. Brownawell, B.J.; Chen, H.; Zhang, W.; Westall, J.C. Environ. Sci. Technol. 1997, *31*, 1735–1741.

80. Fytianos, K.; Voudrias, E.; Papamichali, A. Chemosphere 1998, *36*, 2741–2746.

81. Ou, Z.; Yediler, A.; He, Y.; Jia, L.; Kettrup, A.; Sun, T. Chemosphere 1996, *32*, 827–839.

82. Cano, M.L.; Dorn, P.B. Environ. Toxicol. Chem. 1996, *15*, 684–690.

83. Orth, R.G.; Powell, R.L.; Kute, G.; Kimerle, R.A. Environ. Toxicol. Chem. 1995, *14*, 337–343.

84. Games, L.M.; King, J.E. Environ. Sci. Technol. 1982, *16*, 483–488.

85. John, D.M.; House, W.A.; White, G.F. Environ. Toxicol. Chem. 2000, *19*, 293–300.

86. Cano, M.L.; Dorn, P.B. Chemosphere 1996, *33*, 981–994.

87. Kiewiet, A.T.; de Beer, K.G.M.; Parsons, J.R.; Govers, H.A.J. Chemosphere 1996, *32*, 675–680.

88. Topping, B.W.; Waters, J. Tenside Surfact. Deterg. 1982, *19*, 164–169.

89. Connell, D.W. Rev. Environ. Contam. Toxicol. 1988, *101*, 117–154.

90. Mackay, D. Environ. Sci. Technol. 1982, *16*, 274–276.

91. Tolls, J.; Haller, M.; Seinen, W.; Sijm, D.T.H.M. Environ. Sci. Technol. 2000, *34*, 304–310.

92. Jensen, J. Sci. Total Environ. 1999, *226*, 93–111.

93. Ahel, M; Giger, W. Chemosphere 1993, *26*, 1461–1470.

94. Ahel, M.; Giger, W. Chemosphere 1993, *26*, 1471–1478.

95. Tsuda, T.; Takino, A.; Muraki, K.; Harada, H.; Kojima, M. Water Res. 2001, *35*, 1786–1792.

96. Ekelund, R.; Bergman, A.; Granmo, A.; Berggren, M. Environ. Pollut. 1990, *64*, 107–120.

97. Naylor, C.G. Text Chem. Color 1995, *27*, 29–33.

98. Ferreira-Leach, A.M.R.; Hill, E.M. Marine Environ. Res. 2001, *51*, 75–89.

99. Arukwe, A.; Goksøyr, A.; Thibaut, R.; Cravedi, J.P. Marine Environ. Res. 2000, *50*, 141–145.

100. Thibaut, R.; Debrauwer, L.; Rao, D.; Cravedi, J.P. Sci. Total Environ. 1999, *233*, 193–200.

101. De Wolf, W.; Feijtel, T. Chemosphere 1998, *36*, 1319–1343.

102. Takada, H.; Mutoh, K.; Tomita, N.; Miyadzu, T.; Ogur, N. Water Res. 1994, *28*, 1953–1960.

103. VanGinkel, C.G. Biodegradation 1996, *7* (2), 151–164.

104. Krueger, C.J.; Radakovich, K.M.; Sawyer, T.E.; Barber, L.B.; Smith, R.L.; Field, J.A. Environ. Sci. Technol. 1998, *32*, 3954–3961.

105. Yadav, J.S.; Lawrence, D.L.; Nuck, B.A.; Federle, T.W.; Reddy, C.A. Biodegradation 2001, *12*, 443–453.

106. Field, J.A.; Field, M.A.; Poiger, T.; Siegrist, H.; Giger, W. Water Res. 1995, *29*, 1301–1307.

107. Prats, D.; Rodriguez, M.; Varo, P.; Moreno, A.; Ferrer, J.; Berna, J.L. Water Res. 1999, *33*, 105–108.

108. Gode, P.; Guhl, W.; Steber, J. Fat Sci. Technol. 1987, *89*, 548–552.

109. Steber, J.; Wierich, P. Tenside Surfact. Deterg. 1989, *2*, 406–411.

110. Maurer, E.W.; Weil, J.K.; Linfield, W.M. J. Am. Oil Chem. Soc. 1965, *54*, 582–584.

111. George, A.L. Marine Environ. Res. 2002, *53*, 403–415.

112. Margesin, R.; Schinner, F. Int. Biodeterioration Biodegradation 1998, *41*, 139–143.

113. Feitkenhauer, H.; Meyer, U. Bioresource Technol. 2002, *82*, 115–121.

114. Feitkenhauer, H.; Meyer, U. Bioresource Technol. 2002, *82*, 123–129.

115. Garcia, M.T.; Ribosa, I.; Guindulain, T.; Sanchez-Leal, J.; Vives-Rego, J. Environ. Pollut. 2001, *111*, 169–175.

116. Garcia, M.T.; Campos, E.; Sanchez-Leal, J.; Ribosa, I. Chemosphere 1999, *38*, 3473–3483.

117. Mann, R.M.; Boddy, M.R. Chemosphere 2000, *41*, 1361–1369.

118. Manzano, M.A.; Perales, J.A.; Sales, D.; Quiroga, J.M. Water Res. 1999, *33*, 2593–2600.

119. Charles, W.; Ho, G.; Cord-Ruwisch, R. Water Sci. Technol. 1996, *34* (11), 1–8.

120. Marcomini, A.; Pojana, G. Analusis 1997, *25* (7), M35–M37.

121. Reznickova, I.; Hoffmann, J.; Komarek, K. Chemosphere 2002, *48*, 83–87.

122. Salanitro, J.P.; Diaz, L.A.; Kravetz, L. Chemosphere 1995, *31*, 2827–2837.

123. Szymanski, A.; Wyrwas, B.; Swit, Z.; Jaroszynski, T.; Lukaszewski, Z. Water Res. 2000, *34*, 4101–4109.

124. Huber, M.; Meer, U.; Rys, P. Environ. Sci. Technol. 2000, *34*, 1737–1741.

125. Mezzanotte, V.; Bolzacchini, E.; Orlandi, M.; Rozzi, A.; Rullo, S. Bioresource Technol. 2002, *82*, 151–156.

126. Navas, J.M.; Gonzalez-Mazo, E.; Wenzel, A.; Gomez-Parra, A.; Segner, H. Marine Pollut. Bull. 1999, *38* (10), 880–884.

127. Gonzalez-Mazo, E.; Honing, M.; Barcelo, D.; Gomez-Parra, A. Environ. Sci. Technol. 1997, *31*, 504–510.

128. Larson, R.; Payne, A. Appl. Environ. Microbiol. 1981, *41*, 621–627.

129. Perales, J.A.; Manzano, M.A.; Sales, D.; Quiroga, J.A. Int. Biodeterioration Biodegradation 1999, *43*, 155–160.

130. Swisher, R.D. *Surfactant biodegradation*; Surfactant Science Series. Marcel Dekker: New York, 1987; Vol. 18.

131. Thomas, O.R.T.; White, G.F. Biotechnol. Appl. Biochem. 1989, *11*, 318–327.

132. White, G.F. Water Sci. Tech. 1995, *31* (1), 61–70.

133. Lee, C.; Russell, N.L.; White, G.F. Water Res. 1998, *32*, 2291–2298.

134. Nishiyama, N.; Toshima, Y.; Ikeda, Y. Chemosphere 1995, *30*, 593–603.

135. Baleux, B.A.; Caumette, P. Water Res. 1977, *11*, 833–841.

136. Garcia, M.T.; Campos, E.; Sanchez-Leal, J.; Ribosa, I. Chemosphere 2000, *41*, 705–710.

137. Talmage, S.S. Environmental and human safety of major surfactants: alcohol ethoxylates and alkylphenol ethoxylates. *A report to the Soap and Detergent Association*; Lewis: Boca Raton, FL, 1994.

138. Di Corcia, A.; Costantino, A.; Cresenzi, C.; Marinoni, E.; Samperi, R. Environ. Sci. Technol. 1998, *32*, 2401–2409.

139. Nasu, M.; Goto, M.; Kato, H.; Oshima, Y.; Tanaka, H. Water Sci. Technol. 2001, *43* (2), 101–108.

140. Crescenzi, C.; Di Corcia, A.; Samperi, R. Anal. Chem. 1995, *67*, 1797–1804.

141. Ahel, M.; Giger, W.; Koch, M. Water Res. 1994, *28*, 1131–1142.

142. Jones, F.W.; Westmoreland, D.J. Environ. Sci. Technol. 1998, *32*, 2623–2627.

143. Yoshimura, K. JAOCS 1986, *63*, 1590–1596.

144. Knaebel, D.B.; Federle, T.W.; Vestal, J.R. Environ. Toxicol. Chem. 1990, *9*, 981–988.

145. Lewis, M.A. Water Res. 1991, *25*, 101–113.

146. Utsunomiya, A.; Watanuki, T.; Matsushita, K.; Nishina, M.; Tomita, I. Chemosphere 1997, *35*, 2479–2490.

147. Hofer, R.; Jeney, Z.; Bucher, F. Water Res. 1995, *29*, 2725–2729.

148. Singh, R.P.; Gupta, N.; Singh, S.; Singh, A.; Suman, R.; Annie, K. Bull. Environ. Contam. Toxicol. 2002, *69*, 265–270.

149. Dorn, P.B.; Rodgers, J.H.; Dubey, S.T.; Gillespie, W.B.; Lizotte, R.E. Ecotoxicology 1997, *6*, 275–292.

150. Mann, R.M.; Bidwell, J.R. Environ. Pollut. 2001, *114*, 195–205.

151. Temara, A.; Carr, G.; Webb, S.; Versteeg, D.; Feijtel, T. Marine Pollut. Bull. 2001, *42*, 635–642.

152. Kimerle, R.A.; Swisher, R.D. Wat Res 1977, *2*, 31–37.

153. Roghair, C.J.; Buijze, A.; Schoon, H.N.P. Chemosphere 1992, *24*, 599–609.

154. Litz, N.; Doering, H.W.; Thiele, M.; Blume, H.P. Ecotoxicol. Environ. Saf. 1987, *14*, 103–116.

155. Unilever. Effects on the growth of *Sorghum bicolour, Helianthus annuus* and *Phaselous aureus*. Contract study conducted for Unilever Research Port Sunlight laboratory, UK, by ICI, Brixham laboratory, UK. Study report no. BL/B/3078, 1987.

156. Mieure, J.P.; Waters, J.; Holt, M.S.; Matthijs, E. Chemosphere 1990, *21*, 251–262.

157. Gunther, P.; Pestemer, W. Hall, J.E. Sauerbeck, D.E., In *Effects of organic contaminants in sewage sludge on soil fertility, plants and animals*; Hermit, P.L., Eds.; As cited in Jensen [Ref. 92] and Kloepper-Sams et al. [Ref. 158].

158. Kloepper-Sams, P.; Torfs, F.; Feijtel, T.; Gooch, J. Sci. Total Environ. 1996, *185*, 171–185.

159. Ying, G.G.; Kookana, R.S. AWA J. Water 2002, *29* (6), 53–57.

160. Dodds, E.C.; Goldberg, L.; Lawson, W.; Robinson, R. Nature 1938, *1*, 247–248.

161. Mueller, G.C.; Kim, U.H. Endocrinology 1978, *102*, 1429–1435.

162. Soto, A.M.; Justicia, H.; Wray, J.W.; Sonnenschein, C. Environ Health Perspect 1991, *92*, 167–173.

163. Pedersen, S.N.; Christiansen, L.B.; Pedersen, K.L.; Korsgaard, B.; Bjerregaard, P. Sci. Total Environ. 1999, *233*, 89–96.

164. Gutendorf, B.; Westendorf, J. Toxicology 2001, *166*, 79–89.

165. Jobling, S.; Nolan, M.; Tyler, C.R.; Brighty, G.; Sumpter, J.P. Environ. Sci. Technol. 1998, *32*, 2498–2506.

166. Dillon, P.; Toze, S.; Pavelic, P.; Ragusa, S.; Wright, M.; Peter, P.; Martin, R.; Gerges, N.; Rinck-Pfeiffer, S. AWA J. Water 1999, *26* (5), 21–29.

167. Jobling, S.J.; Sumpter, J.P. Aquat. Toxicol. 1993, *27*, 361–372.

168. Ying, G.G.; Kookana, R.S.; Dillon, P. In *Water Quality Improvements During Aquifer Storage and Recovery (AWWARF Project 2618 report)*; Dillon, P., Ed.; CISRO Land and Water: Adelaide, 2002; Vol. 1, 105–118.

169. Ying, G.G.; Kookana, R.S.; Chen, Z. J. Environ. Sci. Health 2002, *B37* (3), 225–234.

170. Johnson, A.C.; White, C.; Besien, T.J.; Jurgens, M.D. Sci. Total Environ. 1998, *210/211*, 271–282.

171. Sekela, M.; Brewer, R.; Moyle, G.; Tuominen, T. Water Sci. Technol. 1999, *39* (10–11), 217–220.

172. Shang, D.Y.; MacDonald, R.W.; Ikonomou, M.G. Environ. Sci. Technol. 1999, *33*, 1366–1372.

173. Marcomini, A.; Pojana, G.; Sfriso, A.; Alonoso, J.M.Q. Environ. Toxicol. Chem. 2000, *19*, 2000–2007.

4

Relevance of Biodegradation Assessments of Detergents/Surfactants

M. RODRIGUEZ and D. PRATS University of Alicante, Alicante, Spain

I. INTRODUCTION

Detergents constitute the main source of chemical substances of human origin that are discharged into the environment, due to their high volume of consumption and widespread use. The main compounds that normally appear in the formulation of detergents are surfactants, adjuvents, and other additives (bleaches, perfumes, softeners, builders, etc.). The most important from an environmental point of view are surfactants (anionic, cationic, nonionic, and amphoteric) and adjuvents of these compounds. Some studies exist that provide information about the environmental effect of the compounds used as cobuilders. It should be pointed out that the vast majority of the methods mentioned in this chapter are applicable to these compounds and are subject to the same risk assessments as surfactants.

The measurement of the biodegradation percentages of detergents (and in particular of their most relevant compounds, surfactants) has been a widely analyzed subject, and these percentages have been used to determine their environmental acceptability by means of the formulation of laws at practically all possible levels.

The introduction of all the terminology relative to the biodegradation of detergents and surfactants dates from the 1960s, when the first laws were introduced to reduce the effect that these substances were having on the environment (formation of foams in rivers and lakes, accumulation in the soil and sediments, etc). The detergent industry, which has always had a high respect for all aspects regarding the environment and public acceptability, introduced modifications in its detergent formulations by modifying its hard surfactants to others that were softer on the environment and easily biodegradable. Painter [1] sums up the legal requirements and the test strategy for analyzing the biodegradation of surfactants.

Surfactants and certain adjuvents have an effect on flora and fauna and are therefore subjected to control by various methods. Mukherjee et al. [2] present data regarding a discharge into a contaminated lagoon of around 400–1000 mg/L of detergent, the phosphate content of which eliminates the restriction of the phosphate as a limiting nutrient and causes an unrestrained growth of the algae. Freeman and Bender [3] analyzed the elimination of polycarboxylates by observing a polyacrylate with low molecular weight in a wastewater treatment plant (WWTP) and finally calculating a discharge into surface waters of less than 0.02 mg/L, considering a 1:10 dilution of the effluent from the WWTP.

The biodegradation tests established during the 1960s, and since modified based on later experience, are today complemented by studies of risk assessment. At the European Community (EC) level, The Technical Guidance Documents [4] lay out the principles for the risk assessment of new substances (Commission Directive 93/67/EEC) and for existing substances (Commission Regulation No.1488/94). Data on these analyses already exist for the main surfactants used in the formulation of detergents: LAS [5–9], LAB [10], ethoxylated alcohols [9,11], and ethoxysulfates and alcohol sulfates [12]. Certain authors indicate the possibility that the results of the biodegradation of surfactants must not be extrapolated for studying risk assessment. Schöberl [13] points out that the use of data obtained from biodegradation experiments that simulate the environment should not be used for the evaluation of risk assessment because it is the surfactant metabolites (normally less toxic than the original products) that reach the various environmental compartments. Kloepper-Sams et al. [14] carried out risk assessment of soils amended with sludge from the WWTP, because this matrix represents an important discharge of surfactants from WWTPs, as a consequence of the adsorption/precipitation of these compounds.

The formulation of a methodology for the determination of the biodegradation is a difficult task, and the following general aspects (among others) must be taken into account: concentration of the substance to be tested, cultivation medium, concentration of microorganisms (pure or mixed strains), strains previously acclimatized to these substances, origin of the inoculum. At the same time, some aspects that correspond to operation variables should be considered, such as the duration of the experiment (or the residence time in the case of dynamic tests), temperature, oxygen concentration, among many other variables that can have an influence on the final results. One final parameter must also be considered, since this has considerable influence on the final result and on the analytical method used to quantify the disappearance of the compound studied. A very significant parameter and one with great influence on the determination of the percentage of degradation of a surfactant is the adsorption/precipitation that the organic material may undergo. The presence of Ca and Mg salts in the cultivation medium may represent differences in the results, but these ions must be considered when studies for environmental analysis are being carried out. Surfactants such as anionics (principally soaps and LAS) have low solubility constants and precipitate very easily in hard water. This fact results in two opposite effects. On the one hand, the degradation constants that may be calculated will be low and will not reflect their degradation potential. On the other hand, the indexes of toxicity that are shown for the biosphere will be low.

It must never be forgotten that these tests attempt to simulate the behavior of these substances in the environment, and so the previously mentioned variables must be adapted to achieve this objective.

II. DEFINITIONS

Many terms exist in the literature regarding the biodegradation of detergents that must be known before going into more detail. All these terms make reference to aspects related to the subject, but these must be listed before continuing. With reference to biodegradation in its most general sense, the following definitions may be found.

Initial or primary biodegradation: This corresponds mainly to the loss of the characteristics that define the compound. In the case of surfactants, this can be assimilated to the loss of the foam properties of surfactants, which shows the wide extent of this definition and, when using specific analytical techniques, to the loss of identification of the initial molecule. The quantification of this parameter may therefore be carried out by means of nonspecific or specific techniques.

Final or last biodegradation, or mineralization: This corresponds to the reduction of the initial molecule into simple chemical substances, such as carbon dioxide, methane, water, sulfates, sulfites, and nitrates.

Period of acclimatization or adaption: This is the time necessary for a microbial community to adapt itself to a particular compound, undergoing changes both in number and in predominant species to generate the degradation of the compound in question. This time of adaptation is variable and depends on the inoculum used, on the concentration used, and, logically, on the compound analyzed.

Ready or inherent biodegradation: These terms were introduced by the Group of Experts of the OECD [15] and differentiate between the compounds that start their degradation immediately on being exposed to a microbial community of natural origin (i.e., rapidly biodegradable) and the compounds that make these communities need a certain period of adaptation before commencing their degradation.

Biodegradation and elimination: These two terms must be clarified because they cause many misunderstandings in the results in studies on elimination. *Biodegradation* is considered the destruction of the molecules tested (these may be primary or final). The term *elimination* has been widely used in studies in WWTPs and corresponds to the elimination of surfactants in a water stream by biodegradation or by adsorption/precipitation on solids.

Inoculum: This represents the source of microorganisms that are used to carry out the degradation of the substance. This has a great influence on the final result, because on selecting the source of the inoculum, the microbial families that must carry out the degradation are being selected. At the same time, one must take into account that this inoculum source, as commented previously, may or may not show an adaptation to the compound to be analyzed. In this way, the cycle is completed and, if an adapted inoculum is not selected or one with a highly diverse microbial population, the degradation of these compounds may not take place. The source of the inoculum must be the most representative possible of the environmental compartment being simulated. This is very difficult to achieve because it is unlikely that there are identical microbial populations in the different geographical zones where tests are carried out.

Cultivation medium: This represents the medium in which the process of biodegradation develops. Thus two points must be borne in mind. On the one hand, the inorganic nutrients must be present that are necessary to allow the microbial development throughout the biodegradation experiment. If the source of the inoculum and the doses of the substance test do not provide sufficient organic nutrients, these must also be included in the cultivation medium. The cultivation medium must represent as adequately as possible the environmental behavior that is being simulated. Another group of important definitions refers to the oxygen content in the medium: *Aerobic* refers to a compartment in which there is sufficient oxygen to degrade any kind of material present in the medium; *anoxic* refers to the compartment in which the oxygen is present as a limiting reagent, and the so-called *anaerobic* refers to the compartment in which there is an absence of oxygen.

III. ENVIRONMENTAL COMPARTMENTS

The main classification that may be made of the various existing environmental compartments is that of the oxygen concentration that is present. As commented previously, these compartments may be classified as aerobic, anoxic, and anaerobic. The classical aerobic compartments that these substances can encounter during their life span are practically all those that exist in the biosphere, which is predominantly aerobic: circulation in well-designed sewage systems, water treatment lines in WWTPs, aerobic treatment of the stabilization of sludges, surface streams in rivers, the upper parts of lakes, reservoirs, seas and oceans, freshwater and marine sediments that are not very deep (at their upper layers), and soils.

The most relevant anoxic compartments to be found are in sewage systems and in areas of the WWTP with an insufficient supply of oxygen, lakes and eutrophized reservoirs, or with an insufficient supply of oxygen in the middle layers of soil and sediment layers. These types of compartments are widely represented and have changeable characteristics. In accordance with the chemical definition, in these compartments oxygen is present as a limiting reagent. The limits of these compartments in nature are not clear and can vary with time, with temperature (difference in the solubility of the oxygen), and with many other parameters. The presence of high quantities of nutrients (the example of eutrophication) causes oxygen to be consumed and compartments that are clearly aerobic to be transformed into ones that are anoxic or anaerobic.

The most relevant anaerobic compartments are in the stabilization of sludge via anaerobic digestion and in deep layers of soil and also in freshwater and marine sediments, although the depth necessary depends on many factors and oscillates from 2–3 mm up to higher values. Other, less relevant anaerobic compartments (and in many cases due to special reasons) can be insufficiently drained soils and underground water with oxygen consumption by contamination of organic material. The influence of these different compartments must be taken into account when defining the concentrations in the test, the residence time in the biodegradation study system, and the oxygen concentration present. This influence must also be considered when defining the concentration of biomass that is to be used to study the percentage of biodegradation that the substances may reach.

Another point that must be considered is the origin of the inoculum, mentioned previously. This parameter is of great importance when quantifying the biodegradation values, in any case the source of the inoculum must be taken from the compartment where the simulation takes place in the biodegradation test and that is representative of this.

The oxygen content also has an influence, and three types of common microbial populations may be defined—aerobic, facultative, and anaerobic—corresponding to bacteria that need oxygen to live, bacteria that can survive with or without oxygen, and bacteria that cannot survive in the presence of oxygen, respectively.

Directly related to the previous, there is another parameter that characterizes a compartment: its redox potential. This parameter principally governs the use of the substance by the bacteria as an energy source (Table 1) and may produce great errors in the determination of the percentages of final biodegradation when manometric techniques are employed (oxygen consumption or CO_2 and CH_4 formation), for we may find ourselves in areas that do not produce decreases or increases in pressure (reduction of nitrates, Fe^{+3}, sulfates).

It is possible to establish the path of the surfactants and of the substances adjacent to them as they pass through the environment. A scheme of this path is shown in (Figure 1).

Surfactants are used mainly in homes and in waste deposits through sewage systems. This is when surfactants start their degradation process (primary, for the most part) and the first products of this degradation start to appear. During this process, most of the adsorption/precipitation occurs that can cause differences in the relationships between adsorbed and dissolved surfactants. (Since the wide application of wastewater treatment, a very small proportionate percentage is discharged into septic tanks, and an even smaller percentage is discharged directly into freshwater or seawater without any prior treatment.)

When the residual water stream reaches the WWTP it will already have reached (depending on the length of the sewage system) the adsorption/ precipitation equilibria and in the first treatment stages is extracted from the water stream and directed to the sludge stream. In well-designed and correctly operated WWTPs, the percentage of the elimination of solids in suspension is around 60%; on considering the surfactant adsorption factors in the solids in suspension, the percentages of the elimination of sludge in the primary sedimentation processes are between 25% and 40%, depending on the surfactant, the content of solids in the stream, and the system of operation.

IV. BIODEGRADATION ANALYSIS METHODS

Special emphasis must be made on the analytical methods used to calculate the percentage of biodegradation. Two well-known types of analytical methods can be used to determine the primary biodegradation of surfactants: nonspecific methods (normally based on the formation of colored complexes, MBAS, CTAS, BIAS, etc.)

TABLE 1 Redox Potentials

Redox potential (mV)	Reaction	Denomination
1229	$O_2 \rightarrow 2H_2O$	Aerobic respiration
750	$NO_3^- \rightarrow N_2$	Denitrification
360	$NO_3^- \rightarrow NH_4^+$	Nitrate reduction
50	$Fe^{3+} \rightarrow Fe^{2+}$	Iron reduction
−220	$SO_4^{2-} \rightarrow H_2S$	Sulfate reduction
−250	$CO_2 \rightarrow CH_4$	Methanogenesis

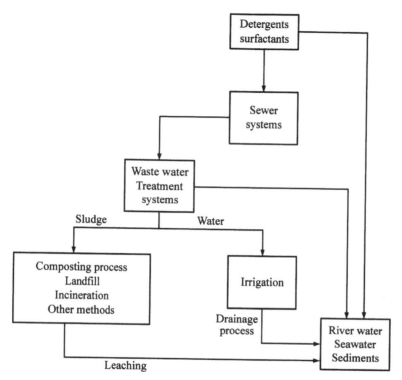

FIG. 1 Pathways of surfactants in the environment.

and specific methods, based on modern analytical techniques and that are developed for only a few surfactants.

A. Nonspecific Analysis Methods

The nonspecific methods of analysis are those that have been and are still widely used to carry out degradation studies. Many classifications may be established within this category. It is not necessary to detail them all, so only the most used methods will be mentioned. More details on analysis techniques can be found elsewhere [16]. The colorimetric methods based on the formation of colored compounds and their quantification by means of the use of spectophotometers have for a long time been the classical methods for the determination of surfactants. In this group of methods may be mentioned MBAS (methylene blue–active substances), BIAS (bismuth-iodine-active substances), and CTAS (cobalt thiocyanate–active substances). All these methods analyze the primary biodegradation, because small modifications in the structure of the surfactants impedes the formation of the colored compound. These methods do not allow us to differentiate between surfactants, but they do permit the difference between the initial and the final quantity to be quantified. Then, when the biodegradation of detergent formulations (normally more than one surfactant) are analyzed, the results give information about the mixture, and they cannot be extrapolated to the surfactants belonging to the formulation. The main disadvantages are the great number of possible interferences in the results.

The other large group of methods for quantifying the biodegradation of surfactants is based on metabolic parameters, and these are not specific.

We can find the measurement of the organic material, in all its possible forms. Among the methods that can be used to quantify the organic material are the chemical oxygen demand (COD) and the determination of total organic carbon (TOC). It must be remembered that only in those cases in which the organic content of the cultivation medium (biodegradation test conditions) and the amount of test substance used are in similar proportions may correct results be obtained. If a large amount of organic material exists in the medium (besides the test substance), the results obtained for organic material will be extremely doubtful. Therefore, the possibility must be considered of interferences in the measurement of the COD, while the measurement of TOC will not contain interferences. In these cases it is essential to use systems in which the addition of the test substance is not carried out, in order to know the amount of organic material that belongs to the inoculum or to the cultivation medium that has been degraded and that must be compared with that corresponding to the system that contains the test substance.

Another series of methods is based on the measurement of the biological oxygen demand (BOD), by measuring the oxygen necessary to carry out the degradation of the substances. The same considerations must be borne in mind as with the measurements regarding the content of organic material.

The main studies developed that include specific analytical determination have been carried out with LAS [17], soaps [18], and AE [19]. So in order to study the general biodegradation of the substances, other analytical techniques can be used, such as the biological oxygen demand (BOD_5), total organic carbon (TOC), absorption/consumption of O_2, and production of CO_2.

B. Specific Analysis Methods

Specific analysis methods are those that provide further information on the elimination of a specific substance. These methods are based on techniques such as gas chromatography, liquid chromatography, and their associations with the technique of mass spectrometry. Prior purification steps are normally always necessary [16].

The main technique that can be included here is the use of substances that contain radioactive elements. This has many advantages, such as being able to employ low concentration levels (similar to those to be found in environmental matrices), to perfectly determine the final distribution of the carbon (formation of biogas, dissolving in an aqueous medium, adsorption/precipitation on solids, incorporation into biomass, etc.). On the other hand, there is the question (already commented on) of the necessity of costly equipment and the formation of the molecules necessary to develop the process.

C. Kinetic Analysis

It must be pointed out that the majority of biodegradation measurement methods have not been designed to obtain exact data on biodegradation velocities, and they are limited to reporting on the biodegradation characteristics that different compounds possess. The possibilities proposed to obtain data necessary for a kinetic study to be made involve modifications in the existing methods for studying biodegradation.

The possible modifications to the static tests are of two kinds: extraction of aliquots to carry out the quantification of the biodegradation by means of methods that analyze the content of the surfactant and, or in the case of the manometric methods, noting the pressure versus time. First of all it must be borne in mind that this extraction does not significantly disturb the test and that after the extraction the study continues as normal. When the extraction affects the process (as a consequence of the volumes needed), it is necessary to consider studies that develop at different contact times and to analyze the degradation obtained for each period of time and afterwards carry out an adequate kinetic study. Normally the use of first-order kinetics is accepted to quantify the degradation of the compounds studied. In this type of kinetics, the velocity of degradation depends on only two parameters, one constant characteristic of the compounds and in chemical terms dependent on the temperature in accordance with the Arrhenius expression and on the concentration of the compound. The expression corresponding to this first-order kinetics and to a perfectly mixed discontinuous reactor

$$r_A = \frac{dC_A}{dt} = -kC_A \tag{1}$$

The value of the constant of degradation (k) of the different compounds can easily be obtained from fitting the experimental data by means of techniques as a minimum squares of the neperian logarithm of the concentration versus to time:

$$\ln(C_t) = -kt \tag{2}$$

where C_t represents the concentration of the compound at the moment t and t represents time, normally measured in days.

The use of a first-order kinetics assumes a very useful simplification, but it must be considered that many data are removed due to other factors that have great influence on the degradation of a surfactant and that are reflected in other kinds of kinetics.

Contrary to chemical systems, whose tests can be described by means of thermodynamic and kinetic data, biological systems need additional information, such as the strain and the phenotypes of the microorganisms used (or of the strains if mixed strains are involved), the morphology, and the cultivation medium. The models that contain this information are known as *structured models*, those that do not contain this information are considered *nonstructured models*.

The multiple stages of cellular growth can be adequately described in a discontinuous cultivation. Figure 2 shows the cellular concentration in function of the cultivation time. Seven phases of growth can be observed:

1. *Initiation (or adaption) phase.* When the microorganisms that grow in a cultivation unit are inoculated into another, adjacent cultivation unit, they undergo an environmental impact. The substrate can be different, the concentration of the cells and of the metabolites can be very low, etc. The microorganisms need a certain period of time to adapt themselves to these new conditions. In this way, when a new cultivation starts, this period of adaption appears, it may be longer or shorter, depending on the extent of the changes undergone in the cultivation medium. During this period the concentration of cells remains constant.

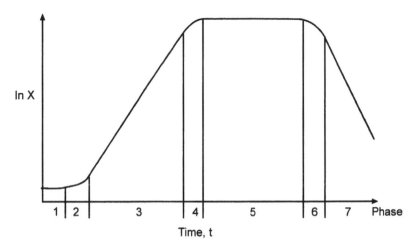

FIG. 2 Curve of microbial growth.

2. *Acceleration phase.* Some of the microorganisms will have adapted themselves to the new medium and will start to propagate. The next (exponential growth) phase starts when all the microorganisms have completed this phase.
3. *Exponential growth phase.* In this stage, except for a limitation in some of the necessary nutrients, the bacteria duplicate their concentration for each growth cycle completed. In this phase,

$$R_x = \frac{dX}{dT} = \mu X \tag{3}$$

where X represents the microorganism concentration and μ is the specific growth velocity. Given that in this phase the growth velocity is at maximum, the following can be applied:

$$\mu = \mu_m \tag{4}$$

Integrating Eq. (3) with the limit conditions $X = X_0$ when $t = 0$ and taking into account Eq. (4), the cellular population at a moment t of this phase:

$$X = X_0 \exp(-\mu_m t) \tag{5}$$

Using this equation, the time necessary to double the cellular mass could be determined in the same way as when determining the average time of the radioactive lifespan. Thus,

$$t_{1/2} = \frac{\ln 2}{\mu_m} \tag{6}$$

4. During this period, the growth velocity decreases slowly as a consequence of the depletion of some necessary nutrient or because metabolism products that have toxic effects accumulate in the cultivation medium. The duration and velocity of transition of this phase vary greatly.
5. In the stationary phase, the velocity of propagation and that of the death of the microorganisms are equal. The concentration of the cellular mass reaches its

maximum value, and its duration may vary within a very wide range, the same as in all the other phases.

6. As soon as the velocity of death overtakes the velocity of generation, the concentration of the cellular mass starts to decrease.
7. In the exponential reduction phase, the number of living cells decreases rapidly, due to the insufficient amount of nutrients or to the presence of toxins.

Given that the growth of microorganisms is a highly complex process, simplifications must be carried out, in some cases drastic ones, before a model can be developed. The biological populations consist of the union of individuals that all differ in age, size, biological activity, morphology, and possibly genetically. The activities of these individuals combine and the effect of the different distributions of groups cannot be assessed (we can only assess the overall effects), so the distribution states must be replaced by mixed microbial populations that represent as well as possible the biological system. This constitutes one of the drastic simplifications that must be done in order to conveniently analyze the system.

Since the study of the differentiation of strains and other sources of micro-organisms is not the objective of this chapter, only nonstructured models with normal growth will be analyzed.

The kinetic models that are most used in biodegradation processes are various. The most simple are those based on first-order kinetics, although those normally used are based on the Monod growth equation [20]:

$$\mu = \mu_m \frac{S}{S + K_s} \tag{7}$$

where μ is the specific growth velocity, μ_m is the maximum specific growth velocity, S represents the concentration of the substrate, and K_s is the semisaturation constant of the substrate (equivalent to a concentration where the specific velocity is half that of the maximum growth velocity).

Thus, the growth velocity is expressed in the model by

$$R_x = \frac{dX}{dt} = \mu X = \mu_m \frac{SX}{S + K_s} \tag{8}$$

On occasion it is not possible to describe the growth using Eq. (8). Numerous modifications can be found to the Monod model to reduce the differences between the measurements taken and the data of the model.

Some authors [21,22] introduce a coefficient of maintenance b related to the consumption of substrate necessary for the survival of the microorganisms, although other authors use this as a coefficient of extinction. In this case, it is considered that the death of the bacteria follows first-order kinetics:

$$\mu = \mu_m \frac{S}{S + K_s} - b \tag{9}$$

The concentration of cellular mass can also inhibit the growth of the micro-organisms when high cellular concentrations are used. This deviation from the Monod model was described by Contois [40] and involves the addition of a term that includes the concentration of cells in the denominator:

$$\mu = \mu_m \frac{S}{S + K_s X} \tag{10}$$

As in the case of inhibition by microorganisms, inhibition may be produced by the substrate itself. Andrews [23] applies this modification to the Monod model:

$$\mu = \mu_m \frac{S}{S + K_s + \frac{S^2}{K_I}} \tag{11}$$

where K_I represents an inhibition constant.

Chen and Hashimoto [24] propose a variation to the Contois model that has been widely used:

$$\mu = \mu_m \frac{(S/S_0)}{K + (1 + K)(S/S_0)} \tag{12}$$

where K is a nondimensional constant, S_0 is the initial substrate concentration, S, y, and μ are as already described, and μ_m is the maximum growth velocity that is produced when $S = S_0$.

It must be pointed out that the majority of the standard methods that exist do not consider this evolution with time, and the final result is a report on the biodegradation characteristics that these compounds may possess. This limitation normally impedes the use of many data to carry out the corresponding risk assessments and involves the use of values of the constants established arbitrarily. In the technical document of the European Commission [4] regarding risk assessment of new and existing substances, the following constants are suggested for chemical products in general:

$k = 0 \text{ h}^{-1}$ for substances that are not biodegradable.

$k = 0.1 \text{ h}^{-1}$ for chemical substances inherently degradable.

$k = 0.3 \text{ h}^{-1}$ for chemical products that are degraded in a rapid test in less than 28 days but that do not pass the test of the window of 10 days (after the necessary acclimatization period, the substance degrades but needs more than 10 days to accomplish this completely).

$k = 1.0 \text{ h}^{-1}$ for chemical products that degrade in a rapid test passing the window of 10 days in less than 28 days.

V. BIODEGRADATION STUDY METHODS

In this section, aerobic and anaerobic analysis methods for the biodegradation of existing surfactants (and other substances) are examined. First of all, methods for the determination of aerobic biodegradation will be dealt with. The existing methods are used mainly to obtain the biodegradability quantification of commercial surfactants. The European Commission requires that a static test be carried out for anionic surfactants; if the limit of 80% of elimination of MBAS is not reached, then a dynamic test must be carried out (Fig. 3). There is a great number of exact methods for carrying this out. The most significant experiments are listed in Table 2.

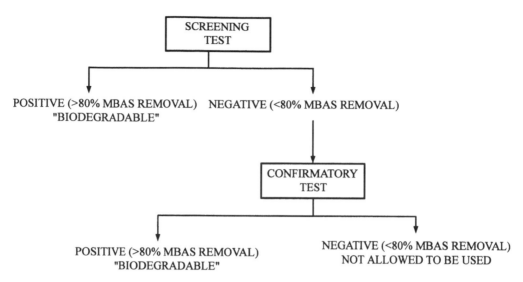

FIG. 3 Sequence of the biodegradation test.

The study of anaerobic biodegradation is very recent, but even so the number of tests existing is already significant. The most used are based on the use of static systems and only quite recently have dynamic studies been employed. Static studies differentiate clearly between methods that consider the concentration of CO_2 in the liquid medium by means of the analysis of dissolved inorganic carbon present and those methods that estimate the concentration of CO_2.

As commented previously, the parameters that must be defined and clearly established when describing a method for the analysis of biodegradation are test substance, concentration, time necessary to carry out the test, temperature, nutrients, inoculum, analytic control, and the necessary equipment. These parameters will influence the reproducibility, the validity, and its relevance to the environment.

A. Static Biodegradation Tests

These tests are used to rapidly analyze the biodegradation of substances by means of simple systems. In the case when a surfactant shows characteristics of biodegradability

TABLE 2 Main Screening Test Methods

---

1. OECD screening test
2. Warburg test
3. Sturm CO_2 production test
4. Closed bottled test
5. Closed bottle test (O_2 saturated)
6. Zahn and Wellens test
7. River die-away test
8. British STCSD test
9. Treccani enrichment test
10. Laboureur test

---

in this type of test, further study will not be necessary (known as dynamic test or screening test).

As commented previously, there exists a multitude of analysis methods with these general characteristics, the only difference being the method of measuring the biodegradation, cultivation medium, concentration of the test substance, and the inoculum source.

There are several OECD guidelines for the study of the biodegradation of surfactants. The current OECD guidelines are enumerated in Table 3. OECD Test Guideline 301 D [25] measures oxygen consumption using sealed bottles, a surfactant concentration of between 2 and 10 mg/L (as the active material), and an inoculum source taken from the final effluent of WWTPs that treat predominantly domestic water. OECD Test Guideline 301 F [26] uses a manometric respirometry and concentrations of the test substance of 50 mg/L, where the inoculum is 30 mg of dry solid per liter and is taken from the aeration tank of this type of plant.

B. Dynamic or Confirmation Biodegradation Tests

The dynamic study methods are various, based on systems that simulate the biological treatment of a WWTP. The most classical methods are listed in Table 4. All these cases consist of a unit that simulates a biological reactor and a secondary settling tank that carries out the sedimentation of the sludge formed and permits its recirculation. The modified OECD test uses Hussman units, a hydraulic retention time of 6 hours, and a solids retention time of 10 hours. The concentration of the test substance is 10 mg/L as DOC, and the control is carried out according to this parameter.

These units have recently been modified to simulate the changes experimented within many WWTPs, which include denitrification stages in their biological process. This modification has been transferred to the biodegradation study units.

TABLE 3 Adopted Test Guidelines from OECD

301 Ready biodegradability
    A. DOC die-away test
    B. CO_2 evolution test
    C. Modified MITI test
    D. Closed bottle test
    E. Modified OECD screening test
    F. Manometric respirometry test
302A Inherent biodegradability: modified SCAS test
302B Inherent biodegradability: Zahn–Wellens/EMPA test
302C Inherent biodegradability: modified MITI test (II)
303 Simulation test—aerobic sewage treatment
303A Activated sludge units
303B Biofilms
304A Inherent biodegradability in soil
305 Bioconcentration: flow-through fish test
306 Biodegradability in seawater
307 Aerobic and anaerobic transformation in soil
308 Aerobic and anaerobic transformation in aquatic sediment systems

TABLE 4   Activated Sludge
Tests

1. OECD confirmatory test
2. SDA test
3. Swisher miniature units
4. Fischer activated sludge test
5. WRC porous pot method

C. Other Aerobic Tests

There exists a multitude of studies outside the field of application of surfactant
detergents that apply the same methods of analysis of biodegradation and that must be
taken into here. Nyholm et al. [27] suggest a methodology for the estimation of kinetic
constants of biodegradation using short batch experiments and concentrations at the
level of µg/L, using adapted sludge and nonadapted to the substances studied,
obtaining data in accordance with the existing values for a few substances (acetate,
aniline, 4-cloroaniline, and pentachlorophenol) in accordance with existing data.
Nevertheless, the analytical technique used by this author implies a temperature of
14°C, which necessitates specific expensive instruments.

D. Methods for Studying Anaerobic Biodegradation

Contrary to the case of anaerobic biodegradation, few firmly established methods
exist to quantify the percentages of degradation of chemical products. The main
reason is that public attention has only recently been centered on this type of
compartment. Additionally, the difficulties involved with these methods must be
pointed out, together with the great number of variables that can influence the
results.

The points that must be established in a methodology for the analysis of the
anaerobic degradation of chemical compounds are various:

Matrix: The anaerobic matrix to be analyzed must be considered. As commented
previously, there exist various anaerobic compartments: marine sediments,
lagoon water, contaminated reservoirs, and the anaerobic digesters of WWTPs.
Of these, the anaerobic digesters of WWTPs are the most important because of
the attention they have been receiving of late.

Concentration of the test substance to be used: There are principally two ranges, one at
the level of designs representative of the environmental matrices and the other at
higher levels representative of the anaerobic digesters of the WWTP.

Cultivation medium: Several different aspects exist that must be considered. The
cultivation medium used should supply sufficient nutrients to carry out the
degradation of the substance and maintain the bacterial populations at adequate
levels and not be so high as to avoid a large amount of production of biogas. In
accordance with this, some authors [28] indicate that the sludge of the WWTP
used as the source of the inoculum should be washed with the inorganic

cultivation medium to reduce the content of organic material within the system and at the same time to reduce the amount of inorganic carbon that exists.

Quantification of the test substance mineralized. There exist two tendencies for quantifying the amount of test substance mineralized that can be considered as equivalent. The measurement of the mineralized substance is carried out measuring the volume of biogas produced or the increase in pressure produced in the system. In both cases the dead volume of the system used must be as small as possible or be perfectly determined. At the same time, some authors deduce the CO_2 that can remain dissolved at the end of the test by using correlations with respect to the concentration existing in the biogas and the temperature; others carry out the quantification of this carbon by measuring the inorganic carbon dissolved.

In the references at the end of this chapter many methods based on anaerobic respirometric techniques are described. In these methods the production of methane and carbon dioxide is measured, these are the final products of anaerobic degradation. Of the existing methods, that proposed by Shelton and Tiedje [29] is the most extensive. It is based on previous studies by Owen et al. [30], Healy and Young [31], and Gledhill [32], consisting of the use of a syringe or a pressure transducer to monitor the production of biogas.

In all these studies, a sample of anaerobic sludge is diluted in a medium of mineral salts, and an adequate amount of the compound to be studied is added. The mixture is subjected to digestion in a sealed bottle and the net production of gas (test − control) is obtained by measuring the pressure in the space at the top of the bottle. The amount of biogas produced is corrected to take into account the CO_2 and CH_4 dissolved in the liquid phase. In this way the carbon liberated from the added test substance is determined. The correction factors for the solubilities need to be experimentally determined previously.

Although the method proposed by Shelton and Tiedje is accurate enough to be used as a screening test (biodegradation test), a series of problems exist that are still unsolved because of their methodology. The principal difficulty in the exact quantification of the biogas production lies in the solubility of the CO_2 in the digestion liquor. This solubility depends on numerous factors, such as pressure, pH, the relationship between the space at the top of the bottle and the volume of the liquid, temperature, and the complex thermodynamic equilibrium between the CO_2 and the carbonates/bicarbonates of calcium and magnesium. Thus, to evaluate the anaerobic biodegradation of a compound and correct these quantities with the data corresponding to the solubilities of both gases and in this way obtain the theoretical volume of biogas, the Tarwin and Buswell equation [33] can be used. But to do this, it is necessary to know the empirical formula of the substance, which in many commercial formulations is extremely difficult. In accordance with Tarwin and Buswell, the substances that contain C, H, and O are converted to carbon dioxide and methane, as shown in the following empirical expression:

$$C_nH_aO_b + \left(n - \frac{a}{4} - \frac{b}{2}\right)H_2O \rightarrow \left(\frac{n}{2} - \frac{a}{8} - \frac{b}{4}\right)CO_2 + \left(\frac{n}{2} - \frac{a}{8} - \frac{b}{4}\right)CH_4$$

The problems with using these methods can increase when large amounts of carbon dioxide and methane are produced due to the sludge used as an inoculum of the system. In this case, the net production of biogas can be affected by large errors. This problem can be solved by using large quantities of test compounds that provide significant differences between test and control. The application of these solutions is impossible when inhibiting effects are present as a consequence of the high concentrations necessary. Among this generic type of methods are those used in the United States (ASTM, 1987) and EPA (1988) and the UK procedure (HMSO, 1989).

The Shelton and Tiedje method was later modified by the ECETOC (European Chemical Industry and Technology Center) with a series of proposals to avoid the problems mentioned previously [28]. The main details of this proposal are as follows.

- Measurement at the end of the DIC test (dissolved inorganic carbon), thus avoiding corrections to the volume of theoretical biogas using solubilities of CO_2.
- Reduction in the content of organic material (reduction in the volume of the biogas produced by the inoculum) and DIC (reduction of the final DIC in the liquid) of the sludge used in the inoculum by washing this with a mineral medium of the cultivation.
- The amounts of the test substance used is situated in the range between 20 and 50 mg C/L.
- This method has been adopted and estandarized as ISO Method 11734. A very simple scheme for the method is shown in Figure 4.

There exist relatively few tests developed in the dynamic mode, the most recent and most developed being that of Baumann and Mueller [34], which proposes a method for determining the anaerobic biodegradation using a continuous reactor with a fixed bed. The quantification of the biodegradation is carried out measuring the methane produced (the carbon dioxide produced is absorbed when it passes through the biogas through a bed of sodium hydroxide) and the maximum amount of theoretical methane, calculated from the COD of the test substance.

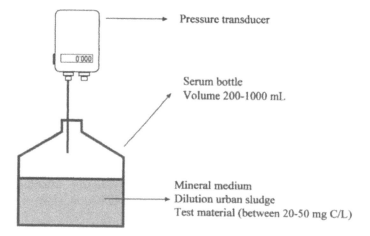

Test material $\longrightarrow$ Dioxide Carbon (gas) + Methane (gas) + DIC (liquid)

FIG. 4 Scheme of measurement of the anaerobic degradation. Metodo Ecetoc (European Centre for Ecotoxicology and Toxicology of Chemicals).

VI. REPRESENTATIVENESS OF THE BIODEGRADATION ANALYSIS METHODS

The extrapolation of the data obtained in the laboratory on the various environmental compartments has always been a controversial subject. The reasons for this controversy were mentioned previously: different concentrations, different microbial populations, and different concentrations. The differences in the results that can be obtained are due mainly to the degradation velocity and to the upper limit of the same. In studies with environmental samples it is normal that highly varied periods of acclimatization of the microorganisms exist. Madsen et al. [35] used a test of the production of gas to determine the biodegradation of 22 chemical compounds. The results obtained were in good agreement with the results of the potential of biodegradation obtained for sludge in WWTPs, marsh water, and marine sediments, but they showed differences in the final levels of biodegradation and in the period of acclimatization necessary (this could extend to several weeks before a net production of biogas was observed). Sulfates also exist in the sediments, and this could lead to processes of sulfate reduction.

Although regulations do not exist in European legislation or the OECD for testing biodegradation in anaerobic environmental compartments, some researchers suggest different methodologies for carrying out this task. Schöberl et al. [36] suggest a simulation model for natural surface waters. Gotvajn and Zagore-Koncan [37] compare the results obtained in classical biodegradation tests (with a sealed bottle and modifications) and the results obtained after carrying out modifications (using various natural waters, different conditions with nutrients, and different microbial species).

It must be taken into account that the addition of a natural substrate or any of the analysis methods mentioned (aerobic and anaerobic, static or dynamic) can cause changes in the overall results. Thus, the addition of a sediment (marine or freshwater) as a matrix in a static biodegradation test may occasion the adsorption of the surfactant on this matrix and cause it to be nonbioavailable so that it will not undergo biodegradation processes. Marchesi et al. [38], on the other hand, found that the addition of sediment to the biodegradation tests accelerated the elimination of SDS because there existed a union of the species used to carry out the degradation of the sediments. Some studies exist that try to extrapolate the results obtained in these standardized tests to natural environments [27,39].

REFERENCES

1. Painter, H.A. In *Biodegradation Surfactants*; Karsa, D.R., Porter, M.R.B.; Eds.; Glasgow: U.K., 1995; 118133.
2. Mukherjee, B.; Pankajakshi, G.V.N.; Bose, P.; Kumar, R.; Kumar, P.; Kumar, D.; Kumar, A.; Kumari, R.; Singh, S.; Kanogia, S. J. Environ. Biol. 1994, *15* (1), 27–39.
3. Freeman, M.B.; Bender, T.M. Environ. Technol. (Letters) 1993, *14* (2), 101–112.
4. EEC. Luxembourg, Office for Official publications of the European Communities, 1996. Part II.
5. Waters, J.; Feitjel, T.C.J. Chemosphere 1995, *30*, 1939–1956.
6. Schöberl, P. Tens. Surf. Deterg. 1995, *32*, 25–35.
7. Feitjel, T.C.J.; Matthijs, E.; Rottiers, A.; Rijs, G.B.J.; Kiewiet, A.; Nijs, A.; Chemosphere 1995, *30*, 1053–1066.
8. Holt, M.S.; Comber, M.H.I. Principles involved in the organization of well-designed

environmental monitoring programs for surfactants. Proc. 4th CESIO World Surfactants Congress, Barcelona, June 3–7, 1996; *3*, 69–80.

9. Cavalli, L.; Cassani, G.; Lazzarin, M. Tens. Surf. Deterg. 1996, *33* (2), 158–160.

10. Gledhill, W.E.; Saeger, V.W.; Trehy, M.L. Environ. Toxicol. Chem. 1991, *10* (2), 169–178.

11. Matthijs, E.; Holt, M.S.; Kiewiet, A.; Rijs, G.B.J. Env. Toxicol. Chem. 1999, *18* (11), 2634–2644.

12. Fendinger, N.J.; Begley, W.M.; Begley, D.C. Environ. Sci. Technol. 1992, *26*, 2493–2498.

13. Schöberl, P. Tens. Surf. Deterg. 1997, *34* (1), 28–36.

14. Kloepper-Sams, P.; Torfs, F.; Feitjel, T.; Gooch, J. International Symposium on Organic Contaminants in Sewage Sludges; Lancaster, U.K. May 16–17, 1995.

15. Gerike, P.; Fisher, W.K. Ecotoxicol. Env. Safety. 1981, *5*, 45–55.

16. Schmitt, T.M. *Analysis of Surfactants*; Marcel Dekker: New York, 1992.

17. Swisher, R.D. *Surfactant Biodegradation*; Marcel Dekker: New York, 1987.

18. Mix-Spagl, K. Muench. Beitr. Abwasser-,Fisch.-Flussbiol (Umweltvertraegglichkeit WaschReinigungsm) 1990, *44*, 153–171.

19. Marcomini, A.; Pojana, G.; Carrer, C.; Giacometti, A.; Cavalli, L.; Cassani, G. Biodegradation behavior of alcohol polyethoxylates (AE): an update. Proceedings of the 5th CESIO World Surfactants Congress, Florence, May 29–June 2, 2000; *2*, 1380–1386.

20. Monod, J. *Recherches sur la Croissance des Cultures bacteriennes*, 2nd ed.; Hermann: Paris, 1942.

21. Mosey, F.E. Trib Cebedeau 1981, *453–454* (34), 289–400.

22. Kennedy, K.J.; Hamoda, M.F.; Droste, R.L. J. Water Pol. Control Fed. 1987, *59* (4), 212–221.

23. Andrews, J.F. Anaerobic Digestion Process 1969, *2*, 95–116.

24. Chen, Y.R.; Hashimoto, A.G. Biotechnol. Bioeng. 1978, *22*, 2081–2095.

25. OECD Test Guideline 301 D, 1992.

26. OECD Test Guideline 301 F, 1992.

27. Nyholm, N.; Ingerslev, F.; Berg, U.T.; Pedersen, J.P.; Frimer–Larsen, H. Chemosphere 1996, *33*, 851–864.

28. Birch, R.R.; Biver, C.; Campagna, R.; Gledhill, W.E.; Pagga, U.; Steber, J.; Reust, H.; Bontik, W.J. Chemosphere 1989, *19* (10–11), 1527–1550.

29. Shelton, G.M.; Tiedje, J.M. Appl. Env. Microl. 1984, *47*, 850–857.

30. Owen, W.F.; Stuckley, D.C.; Healy, J.B.; Young, J.C.; McCarthy, P.L. Water Res. 1979, *13*, 485–492.

31. Healy, J.B.; Young, J.C. Appl. Env. Microbiol. 1979, *38*, 84–89.

32. Gledhill, W.E.; Appl. Microbiol. 1974, *17*, 265–293.

33. Tarvin, D.; Buswell, A.M. J Am. Chem. Soc. 1934, *56*, 1751–1755.

34. Baumann, U.; Mueller, M.T. Water Res. 1997, *31* (6), 1513–1517.

35. Madsen, T.; Rasmussen, H.B.; Nilsson, L. Chemophere. 1995, *31* (10), 4243–4258.

36. Schöberl, P.; Guhl, W.; Scholz, N.; Taeger, K. Tens. Surf. Deterg. 1998, *35* (4), 279–285.

37. Gotvajn, A.Z.; Zagore-Koncan, J. Biodegradation studies as an important way to estimate the environmental fate of chemicals. International Specialized Conference on Chemical Process Industries and Environmental Management; Cape Town, South Africa, Sept. 1997; 8–10.

38. Marchesi, J.R.; Graham, G.F.; Russell, N.J.; House, W.A. FEMS Microbiol. Ecol. 1997, *23* (1), 55–64.

39. Struijs, J.; VanderBerg, R. Water Res. 1991, *29*, 255–262.

40. Contois, D.E. J. Gen. Microbiol. 1959, *21*, 40–50.

5

Toxicology and Ecotoxicology of Detergent Chemicals

YUTAKA TAKAGI, SHINYA EBATA, and TOSHIHARU TAKEI
LION Corporation, Kanagawa, Japan

I. INTRODUCTION

The main chemicals used in detergent products are surfactants. Therefore, surfactants are treated as the representative example for detergent chemicals. In this chapter, we outline how safety evaluations of detergent chemicals are performed.

In recent years, the effect of chemical products on human health and the environment has become a major concern. Such concern can have a serious economic impact on suppliers (companies produce the ingredients), formulators (companies incorporate the ingredients) and importers (companies import the ingredients or the products).

Various detergents are used on a daily basis over a long period of time. While they are convenient and effective, their chemical components might be harmful to the environment and to humans, possibly even to the next generation.

For that reason and from the viewpoint of product liability as well, risk assessments of detergent chemicals must be carried out according to the newest and most exacting scientific methods.

II. TOXICOLOGY OF DETERGENT CHEMICALS

A. Introduction

We are exposed to innumerable chemical substances every day. Detergent products, such as laundry and dishwashing products, household cleaners, body soaps, and shampoos, are among the most familiar products that contain chemical substances. Detergent products are used by vast numbers of consumers over long periods of time, so toxicological studies both on ingredients and finished products must be conducted to guarantee their safety before they are marketed. In this chapter, we discuss the several components of a safety evaluation, using a surfactant as a model ingredient of detergent products.

B. Components of Safety Evaluation

1. Risk/Benefit Balance

All chemicals are toxic at some level of exposure. Therefore, the essence of a safety evaluation is to determine potential adverse effects and the exposure level at which they occur. Because there is no way to guarantee total absence of risk, some kind of approach to reaching decisions on risk acceptability is needed. The concept of risk/ benefit balance has been developed as one of the approaches to that decision. In the case of life-saving drugs, the concept can be accepted easily. On the other hand, household products are thought to have rarely been associated with serious hazard to health and to be safe (Fig. 1). In the case of a detergent intended for daily use, *foolproof* (anyone can use it safely no matter how they use it) or *fail-safe* (no adverse effect will be caused even with misuse or accidental ingestion) is ideal, but it is impossible to be 100% certain that a product will cause no harm. Thus, *safety in use* is a more practical goal.

2. Guarantee of Safety in Use

For household detergents, safety in use means that (1) usual use causes no adverse or side effects and (2) effects of misuse or ingestion are transient and not serious if appropriate first aid measures are taken. *Usual use* means use in foreseeable conditions. That includes (1) intended use and (2) unintended but reasonably foreseeable use, such as using too much or using incorrectly. Toxicological studies for safety evaluations must consider misuse as well as usual use conditions (Fig. 2).

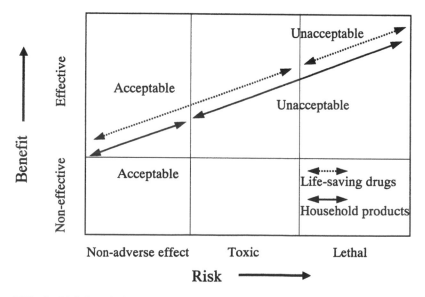

FIG. 1 Risk/benefit balance.

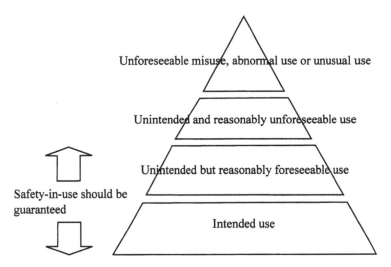

Unforeseeable misuse, abnormal use or unusual use

Unintended and reasonably unforeseeable use

Unintended but reasonably foreseeable use

Intended use

Safety-in-use should be guaranteed

FIG. 2 Relation between use conditions and guarantee of safety in use.

3. Identification of Potential Risk and Risk Assessment

(a) Exposure Assessment. To identify the potential risk from a substance, it is necessary to consider how the human body contacts the substance through the life cycle of the product (manufacture, use, and disposal) under both usual and misuse conditions. In the case of detergents, possible consumer exposure is by skin, ingestion, and inhalation. Consumer exposure conditions can be studied by checking investigation items, as shown in Table 1 [1].

(b) Hazard Identification. To identify the potential risk of a substance, its potential harmful effects have to be investigated. Existing data can be used, but in cases where very little or no information is available, some additional toxicological tests may need to be conducted in vitro (e.g., testing methods using cultured cells) and/or in vivo (such as in animal and/or human).

TABLE 1 Investigation Items of Consumer Exposure

1. Class of products where the ingredient may be used
2. Method of usage
3. Concentration of ingredients in product
4. Quantity of product at each use
5. Frequency of use
6. Total area of skin contact
7. Site of contact
8. Duration of contact
9. Foreseeable misuse that may increase exposure
10. Nature of consumers
11. Quantity likely to enter the body, resulting from bioavailability studies
12. Projected number of consumers

(c) Risk Assessment. We evaluate whether a substance can be used safely based on exposure conditions and potential harmful effects. Following from toxicological studies, a no-adverse-effect level (NOAEL) or threshold for each endpoint is established. By comparing these levels with the amount of exposure, considering species variation and individual differences, we evaluate whether it can be safely included in a finished product. For some endpoints, such as skin and eye irritation, NOAEL is not always determined. Instead, the data are assessed in a comparative approach, using existing similar commercial products as a benchmark.

C. Toxicological Requirements and Studies for Safety Evaluation

Considering consumer exposure to detergents, toxicological test items, which are shown in Table 2, are commonly required to evaluate safety in use.

1. Safety in Usual Use

(a) Local Effects. In usual use, consumers make contact with detergent chemicals via skin directly and as a residue on clothes. Therefore, dermal effects must be evaluated first.

   Skin Irritation. Some chemicals can permeate the skin and might irritate it. Skin irritation is the product of a reversible inflammatory reaction. Albino rabbits or guinea pigs, which have more sensitive skin than humans, are generally used for skin testing. The test substance is applied topically with or without an occlusive dressing, and the skin reaction, such as erythema, edema, and desquamation, is observed. If the skin reaction is positive, the substance is not automatically rejected; rather, its propriety use is judged by comparing the reaction to that of a similar product that has already been safely used.

   In human studies, the test substance is applied to the back or the arm of volunteers by means of an occlusive dressing, and the skin reaction is observed [2,3].

   Skin Sensitization. Skin sensitization is an immunologically mediated reaction. Some absorbed chemicals are recognized as antigens after binding to epidermal membrane proteins. The formed antigen is recognized by T cells and induces a hyper-

TABLE 2 Toxicological Test Items for Safety Evaluation of Detergents

In usual use	Local effects
	Skin irritation
	Skin sensitization
	Photosensitivity (phototoxicity, photoallergy)
	Systemic effects
	Chronic toxicity
	Reproductive toxicity
	Mutagenicity
	Carcinogenicity
In the case of misuse or accidental exposure	Local effects: eye irritation
	Systemic effects: acute toxicity

sensitive state. Once a hypersensitive state develops, a slight quantity of the substance can cause allergic reactions and sensitized consumers must avoid contact permanently. Therefore, allergenicity is an important issue in the safety evaluation of detergent chemicals, and the following step-by-step assessment is done in some cases: (1) determination of skin sensitization potential in an animal test and/or by alternative studies (see later); (2) confirmation of safety in a human study.

As animal test models, adjuvant tests (e.g., guinea pig maximization test) and nonadjuvant tests (e.g., Buehler test with guinea pigs) can be used [4–6]. Both of these tests consist of the following two steps. First, the test substance is applied to the skin repeatedly (*induction exposure*); a hypersensitive state may develop during a certain period of time. After the induction period, the test substance is applied again (*challenge exposure*), and the skin reaction is observed.

The repeated insult patch test (RIPT) and human maximization test are representative testing methods to confirm nonsensitization potential in human. In these tests, induction exposure is made by repeated occlusive patch with a test material [7–9]. After a certain resting period, challenge is made by occlusive patch, and the skin reactions are evaluated.

In vitro models that use immunocompetent cells or that measure binding activity of the chemicals to the protein are not yet sufficiently developed to play a significant role in the assessment of skin sensitization potential. Some alternative methods, however, such as the local lymph node assay, which includes an ex vivo process, have been shown to reliably detect moderate to strong sensitizers and are useful for assessing sensitization potential [10].

Photosensitivity. Photosensitivity includes toxicity and allergicity mediated by UV absorption. The former causes primary irritation and the latter causes an immunologic contact dermatitis. Photosensitivity studies are not necessary for substances that do not absorb UV light.

Phototoxicity. Some testing methods, such as the Morikawa method, have been developed to evaluate phototoxicity [11]. The tested substance is applied to the skin of animals, followed by UV irradiation. Phototoxicity is evaluated based on the difference in skin reactions between irradiated and nonirradiated sites.

Photosensitization. There are some testing methods, such as the adjuvant and strip method [12]. Topical application and UV irradiation are repeated as photo-induction exposure. After challenge exposure and UV irradiation (photochallenge), photosensitization potential is examined based on the difference in skin reactions between irradiated and nonirradiated sites.

(b) Systemic Effects. In addition to exposure via skin, detergents can be ingested from residues left on dishes. Although such exposure is likely to be very low, it could occur repeatedly over long periods, and systemic effects must also be considered in the toxicological evaluation of detergent chemicals in usual use.

Chronic Toxicity. Studies in experimental animals (e.g., mice or rats) reveal the effects of repeated dermal or oral exposure to a substance. The Organization for Economic Cooperation and Development (OECD) guidelines for testing of chemicals provide the methodology. The ratio of the no-observed-adverse-effect level (NOAEL) obtained in the study to the actual human exposure is the safety factor. A safety factor of 100 calculated from interspecies and intraspecies variability is usually used as a standard to assess the safety of a substance.

Reproductive Toxicity. In a reproductive toxicity study, the effects of test substances on male and female animals (includeing dams, fetuses, and pups) during the reproductive cycle (spermatogenesis, ovogenesis, fertilization, gestation, parturition, lactation) are examined. The OECD guidelines for testing of chemicals provide the methodology. Although exposure may be low, assurance of the safety to a pregnant woman is important.

Mutagenicity. Mutagenicity tests are conducted to research the effects of a substance on DNA. These tests may also be used to screen for carcinogenicity. In vitro tests (e.g., reverse mutation test in bacteria, chromosomal aberration test with cultured mammalian cells, and the mouse lymphoma tk assay) and in vivo tests (e.g., chromosomal damage test in rodent hematopoietic cells) have been used in the standard genotoxicity test battery [13–18]. Taking the toxicological profile of the substance and its chemical structure into account, further carcinogenicity studies may be performed.

2. Safety in the Case of Misuse or Accidental Exposure

(a) Local Effects. Because it is possible for detergent chemicals to contact the eyes accidentally, the effect of such exposure must be evaluated.

The Draize test for eye irritation in the rabbit has been used for many years [19]. A test substance is placed in the eye, and reactions at the cornea, iris, and conjunctiva are observed. In order to evaluate the effect of rinsing, eyewashing treatment can be added.

Because the Draize test has inherent scientific and ethical problems, various alternative methods have been developed. For example, Griffith and colleagues have developed the low-volume eye irritation test, which is more realistic than the Draize test [20]. Moreover, some testing methods that do not use animals have been reported. These altenative methods use cultivated cells, cultivated organs, cholioallantoic membranes (CAM), etc., and will probably become standard in the future [21].

3. Systemic Effects

Acute toxicity tests provide information on the effects of relatively large exposures produced by accidental ingestion. Acute toxicity tests reveal the effects that occur within a short time following a single exposure to a substance or multiple exposures within 24 hours. These tests indicate the inherent toxicity of a substance. In acute toxicity tests, the test substance is administrated to animals (e.g., rats or mice) and the toxicological effects and LD50 (dose that is lethal to 50% of the animals treated) are determined.

D. Alternatives to Animal Testing

Much valuable information has been acquired from animal tests, but more predictive and more humane alternatives to animal testing are necessary for both scientific and ethical reasons.

The development of alternative tests has advanced according to the 3R concept (replacement, reduction, and refinement) proposed in 1959 by Russell and Burch [22]. In the OECD guidelines for testing of chemicals, for example, the limit test has been brought into an acute toxicity test, thus reducing the number of animals required [23].

Also, the mouse ear swelling test and the local lymph node assay, which was developed for assessing sensitization potential, are introduced as alternatives that offer the advantages of objective endpoint, short duration, and minimal animal treatment [10].

Examples of refinement are the use of anesthetics and low dosages in eye irritation tests on rabbits [20,24].

For replacement, alternatives are not yet sufficiently developed for realistic assessments. As already mentioned, various approaches are being studied to replace the animal eye irritation test, but none has yet been validated. In the evaluation of surfactants, however, some in vitro tests (the neutral red uptake assay, the isolated bovine cornea test, the isolated, enucleated chicken eyes) show high correlation with in vivo tests [25]. Application of those tests to the evaluation of detergent chemicals could be investigated.

For phototoxicity testing, in vitro neutral red uptake by Balb/c 3T3 cells has been validated under the leadership of the European Union/the European Cosmetic, Toiletry and Perfumery Association (EU/COLIPA), and incorporation into the OECD guidelines for testing of chemicals has been proposed [26–29].

E. Conclusion

We have discussed the methods used in the safety assessment of detergent chemical products. Toxicological studies and safety evaluation techniques, however, are changing rapidly, and product safety should be evaluated according to the latest scientific standards. Because new potential risks may appear as a result of new scientific discoveries, it must be understood that a need to reassess currently marketed detergent chemicals may arise.

III. ECOTOXICOLOGY OF DETERGENT CHEMICALS

A. Introduction

Detergents products, including laundry and dishwashing detergents, household cleaners, and body soaps and shampoos are used in the home regularly and in large amounts, contributing to the health and quality of life of the consumer. Surfactants are a major ingredient of detergents products, and they are used as well in emulsifiers and dispersants. They are manufactured in increasingly large quantities every year, and regardless of their use they are ultimately discharged into the aquatic environmental. Thus, it is important to evaluate their environmental effects. In this section, we discuss the following two points about the ecological effects of surfactants: the water area chosen as a place for evaluation, and the surfactants to be examined. The water area is of most concern as a shift system for the outflow of detergent, and surfactants should be the main ingredients for securing the basic performance of detergents.

B. Environmental Risk Assessment of Detergent Chemicals

In this section, we describe the introduction of the environmental risk assessment. The details will be given in Sections III.C, III.D, and III.E concerning assessment approach.

1. Approach for Environmental Risk Assessment

The environmental risk assessment of chemicals is based on a comparison of their predicted environmental concentration (PEC) and their predicted no-effect concentration (PNEC). An overview of the risk assessment approach for the ecosystem is shown in Figure 3. Environmental risk assessment is usually verified by a multiple-step or tiered process. In the early phase of risk assessment, biological effect levels and environmental concentrations are estimated from standardized ecotoxicity tests, structure–activity relations, and simple exposure models. Since there is relatively high uncertainty on the accuracy of these first stages, a wider margin between the PEC and PNEC is needed to conclude that the chemical is safe to use. PEC and PNEC are determined with sufficient accuracy to demonstrate that a reliable margin of safety exists.

For hazard identification, test organisms for which the trophic levels are fundamentally different are selected and examined. When evaluating a specific water area, the method of carrying out an examination with the species of aquatic organisms that inhabit the water area is adopted. After the completion of testing, the PNEC value is determined from the relevant EC_{50} (effect concentration for 50% of the population), LC_{50} (lethal concentration for 50% of the population), or NOEC (no-observed-effect concentration) value in the most sensitive species observed in the laboratory, divided by UF (uncertainty factors). The need for high-quality testing and reliable, accurate data is therefore very important because the numerical values obtained directly affect the risk assessment model.

2. Estimation of Predicted Environmental Concentration (PEC)

We can calculate the PEC either by measuring the actual concentrations in the affected waters or by estimating it on the basis of the amount discharged per consumer, the rate of removal by sewage processing, and the rate of dilution. Because actual measurement is costly, time-consuming, and difficult, the PEC is usually estimated.

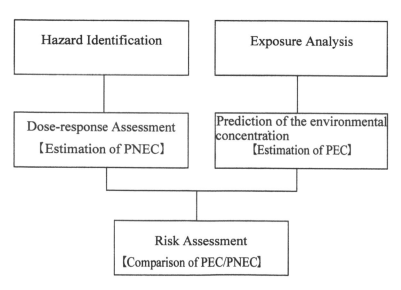

FIG. 3 Conceptual figure for environmental risk assessment.

Estimation presumption using a guess formula that combines the factors affecting environmental concentration becomes effective in exposure analysis. Primary environmental risk assessment can be carried out by presuming the environmental concentration using the estimate equation proposed based on the scientific foundation.

3. Risk Assessment

If the PEC-to-PNEC ratio is sufficiently less than 1 (i.e., PEC/PNEC < 1), the use of the chemical is considered safe, and further testing is unnecessary and wasteful of resources. On the other hand, if PEC/PNEC is close to 1 or greater than 1, a more detailed program of work will usually be necessary to enable the estimates of PEC and PNEC to be refined and the environmental hazard to be quantified with greater confidence.

C. Hazard Identification

Chemical substances present unique hazards, and toxicological effects may vary between test species. Therefore, in order to calculate the PNEC, it is necessary to carry out a toxicity test for various species. Various aquatic-organism testing methods have been released for the purpose of evaluating the ecological effects of chemical substances. An international examination method is carried out in many cases in accordance with OECD guidelines for the testing of chemicals [30]. The toxicity tests specified by these test guidelines are set up in consideration of the food chain, because the chain cycle runs from energy production to consumption in an ecosystem. In addition to this, regarding the aquatic organisms used for the toxicity test, one must take into consideration the biological characteristics of the test species and the method of breeding in the laboratories that is established. Although there are various ecotoxicity tests, for technical and economic reasons the ecological effects on algae, the cladoceran crustaceans (*Daphnia*), and fish are widely used as the minimum data set. These ecotoxicity tests are the most fundamental data sets also prescribed by SIDS (Screening Information Data Set) when the initial risk assessment of HPV (high production volume) is carried out. In these tests, the concentration of a test substance is changed in a geometric series, test solutions are prepared, and the aquatic organisms are exposed to them. The concentration that begins to influence the degree of toxicity, the EC_{50}, and the LC_{50} is then calculated on the basis of the results. Dechlorinated drinking water, good natural water, or distilled water to which is added calcium, magnesium, etc., is used as the examination water for fish and *Daphnia*. In the case of the Alga Growth Inhibition Test, a culture solution containing a mineral salt required for the growth of alga is prepared and is considered as a test liquid. The significance and features of each ecotoxicity test are described next.

1. Alga Growth Inhibition Test

Alga is grown and increased via photosynthesis by taking in nutritive salts. Therefore, alga occupies an important position as a primary producer in a water system food chain. When used as a test aquatic organism, because its life cycle is short, single-cell green algae has the feature of needing only a comparatively small number of days to estimate the influence over several generations. As an aquatic test species, it is comparatively easy to use, and results with sufficient reproducibility are obtained; thus, *Selenastrum* sp. and *Chlorella* sp. are used. In OECD method 201, algae is exposed to a chemical substance for 72 hours. EC_{50} and NOEC are calculated according to their growth and breeding.

2.   *Daphnia* sp., Acute Immobilization Test

Because *Daphnia* sp. is an animal plankton with vegetable feeding habits, it is important as an aquatic organism located in the middle of the food chain that connects higher-order predators to a primary producer. Moreover, because the life cycle is comparatively short and the first brood is born in about 10 days, there is an advantage by which we can evaluate the influence on reproduction classified into long-term toxicity. In the ecotoxicological evaluation using an aquatic invertebrata, *Daphnia* is used as a typical species for the following. They have high sensitivity to toxic substances, such as pesticides, their handling is easy, and large-scale equipment is unnecessary because of their very small body size. In OECD method 202, *Daphnia* is exposed to a chemical substance for 48 hours. EC_{50} and NOEC are then calculated by immobilization. In 1998, a reproduction test was separated as OECD method 211. *Daphnia* is exposed to a chemical substance for 21 days and the influence (NOEC, EC_{50}, etc.) is investigated for batches of young and the parental generation.

3.   Fish Toxicity

Fish occupy an important position in the food chain and are located in the best grade according to the aquatic ecosystem. They not only become food for humans, but they are also an aquatic organism that brings grace to our life. Moreover, because many ecotoxicity data have been reported for various fish species, they are important as an index organism for environmental pollution. In OECD method 203, fish are exposed to a chemical substance for 96 hours. LC_{50} is then calculated from their mortality.

   Moreover, since it involves evaluation when a chemical substance is exposed for a long time, there is an early-life-stage toxicity test (ELST) in OECD method 210. The early life stages consist of the egg, fry, and early juvenile stages. Early-life-stage toxicity tests are often preferred to tests with later developmental stages because the embryolarval and early juvenile stage are frequently the most sensitive. The egg, fry, and juvenile are exposed to a chemical substance at least until all the control fish have been free-feeding. LOEC and NOEC are then calculated on the basis of the results.

4.   Relevance of Structure and Ecotoxicological Effects
of Surfactants

The degree of ecotoxicity changes with the chemical molecular structure. Isomers and homologues show a fixed tendency for the kind of surfactants. Generally, ecotoxicity increases when the hydrophobic force becomes strong and decreases when the hydrophilic force becomes strong. That is, ecotoxicity increases with increasing alkyl chain length (Fig. 4) and decreases with increasing ethylene oxide chain length [31–35]. In the case of LAS (linear alkylbenezene sulfonates), its effect is influenced by the phenyl position relative to the alkyl chain. That is, ecotoxicity increases as the position of the hydrophilic group approaches the end of the chain [36,37]. The acute toxicity value (LC_{50}) of LAS, due to the difference in the binding site of the phenyl group relative to the alkyl chain length, for Japanese killifish (*Oryzias latipes*) is shown in Figure 5. For LAS, a similar result is reported for

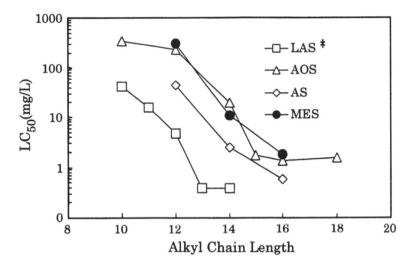

FIG. 4 Relationship between acute toxicity and alkyl chain length, 48-h LC_{50} values for Japanese killifish (*Oryzias latipes*). [*: fathead minnow (*Pimephsles promelas*).]

Daphnia [38]. With nonionic surfactants, fish toxicity decreases with increasing ethylene oxide chain length (Fig. 6) [39]. It is necessary to fully consider the relevance of structure and ecotoxicological effects of surfactants. That is, the concentration of a surfactant in an aquatic environment should be analyzed with consideration of the alkyl chain or EO chain lengths. The ecotoxicological effects of the surfactants are then presumed to exist in rivers [40].

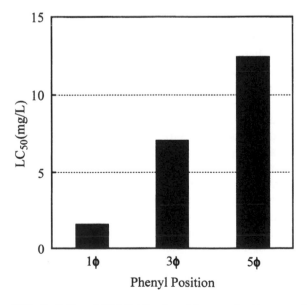

FIG. 5 LC_{50} of C12-LAS; phenyl isomer for Japanese killifish (*Oryzias latipes*).

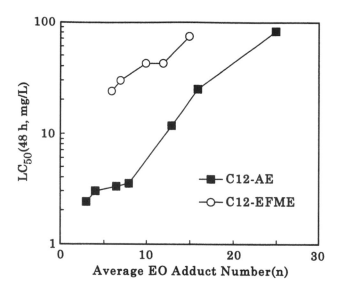

FIG. 6 Relationship between acute toxicity and ethylene oxide chain length, 48-h LC_{50} values for Japanese killifish (*Oryzias latipes*). (EFME: ethoxylated fatty methyl esters.)

5. Structure–Activity Relationships

It is understood that the structure of a surfactant has a close relevance to its eco-toxicity. From this viewpoint, the so-called QSARs (quantitative structure–activity relationships) are also considered for ecologiccal effects. Specifically, QSAR is "a mathematical model corresponding to the chemical structures and their biological physiology activity shown by chemical or physical character." Various models have been constructed, so the sacrifice of a valuable test organism is avoided as much as possible, and the approach is simple and does not exert an economic burden. In the U.S. EPA (U.S. Environmental Protection Agency), the prediction technique via QSAR has been used in the toxicity evaluation of new chemical substances for aquatic organisms since 1981. Since 1990, predicting acute and chronic toxicity for fish, *Daphnia*, and green algae by QSAR has been carried out routinely for all industrial chemicals regardless of the existence of data in a notification document. The U.S. EPA has continued to examine and improve the QSAR prediction technique. They introduced ECOSARs (Ecological Structure–Activity Relationships) in 1994. QSAR can be used in the initial environmental risk assessment or the priority determination for implementation of toxicity tests and application to identifying a highly toxic substance from many candidate substances.

   In the hazard identification relative to aquatic organisms for surfactants, there are two mainstream approaches for construction of QSAR equations that predict the toxicity. One is the method of presuming based on the octanol–water partition coefficient (P_{ow}) that Roberts proposed [41]. Another is the method of presuming based on the physical-properties values of a surfactant, e.g., critical micellar concentration and surface tension. They use the presumed P_{ow} (the so-called $C \log P$ or calculated P_{ow}) value acquired by calculation based on chemical structure according to the technique of Leo and Hansch, not as a physical-properties value, which could acquire the characteristics of a target surfactant experimentally, but as an alternative index [42]. This $C \log P$ value serves as an indispensable input item to ECOSAR of the U.S. EPA. Roberts has established QSAR equations for anionic surfactants and alcohol ethoxylates (AE) on the basis of $C \log P$ [43]. AE are common, nonionic surfactants widely used in consumer product applications. In commercial form, AE are a complex mixture of varying alkyl chain and ethoxymer chain lengths. Each homologue of the mixture has unique physical and chemical properties that alter the sorption affinities, biodegradation rates, and ecotoxicity. As opposed to ecotoxicity evaluation of such a mixture, after computing $C \log P$ for the composition of a homologue, the environmental risk assessment by the toxicity of prediction using QSAR equations is then introduced [44].

D. Ecotoxicity Evaluation Under Actual Environmental Conditions

The ecotoxicity tests used for hazard identification mentioned earlier involved investigating the direct effects of various surfactants on each test species. When the ecological effects are evaluated, as well as the potential effects of the surfactant itself, evaluating the surfactants in an actual environmental state has an important meaning: A detergent chemical enters the environment as part of a complex mixture of wastewater sludge or effluent. This can change the detergent's chemical form and distribution in the water as well as its bioavailability and toxicity. Moreover, in practice these substances and natural substances with a different toxicological profile form complex mixtures. For example, it is also believed that dissolved organic matter and suspended solids in natural waters may reduce the bioavilability as a result of complexation or adsorption.

1. Relevance of the State of Existence and the Ecotoxicological Effects of Surfactants

If the water temperature is high, higher ecotoxicological effects may be found [45]. The ecotoxicity of nonionic surfactants is not affected by an increase in the hardness of water in many cases. However, higher ecotoxicological effects are found in the case of anionic surfactants [45–47]. This is considered to be based on the increase in the affinity for the cell membrane of the surfactant, increasing hydrophobicity via salt exchange, i.e., Na to Ca or Mg [46]. The highest ecotoxicological effects on the growth of algae are found in cationic surfactants. The sorbability of a cationic surfactant on pectin, which plays an important role as a vegetable growth factor, is presumed to be the main factor exerting a strong influence [48]. Because pectin is negatively charged, a cationic surfactant is considered to be sorbed and to control the action of the growth factor [49]. However, when the ecotoxicological effects of cationic surfactants are evaluated, it is necessary to consider interaction with anionic surfactants. Analysis of river water clearly shows that cationic surfactants coexist with LAS or suspended matter. As a result, it was determined that many DAD-MACs (dialkyl dimethyl ammonium chloride) form a complex with coexisting substances and exist in a state where they can be adsorbed with the insoluble ingredient. If the ecological effects of DADMAC are actually examined, when LAS in an amount 10 times that of DADMAC is present, EC_{50} on algae growth reduced from 26 μg/L to 10,500 μg/L [50].

2. Biodegradation and Ecological Effects

A surfactant is affected simultaneously with microorganisms by the physiochemical action of light and temperature in the environment. That is, a surfactant is photo-degraded, hydrolyzed, or biodegraded. In ecotoxicological safety evaluation of chemical substances, the biodegradation of microorganisms is the most important. For this reason, biodegradation has contributed greatly to the degradation of chemical substances. Biodegradation is one of the most significant factors in consid-ering the persistence of chemical substances in the environment. For example, if the substance is readily biodegradable, then when it flows into a sewage treatment plant, it is assumed that decomposition removal is carried out by activated sludge processing. Even when chemical substances are released into environmental groundwater, they are biodegraded by microorganisms and do not remain in the environment over a long period. If a chemical is not persistent, accumulation in the environment can be avoided by continuous retention and elevated concentration in a specific place. Therefore, the adverse effects on the ecosystem or humans are decreased. Further-more, complete mineralization via degradation of microorganisms is a very important endpoint from the viewpoint of environmental exposure. As a result, chemical substances can disappear completely, and they are degraded to organic and inorganic components. Moreover, they are reused by the organism, i.e., utilization, and are incorporated in natural components. Therefore, especially in the case of a substance consumed in large quantities, such as the surfactants used in detergents, perfect mineralization by microorganisms is the most important ecotoxicological item from the standpoint of removal from the environment. In the process of biodegradation, an intermediate product may be generated temporarily [51]. Therefore, intermediate products from biodegradation are being investigated for their ecological effects. The biodegraded products of surfactants have shown significant and rapid reduction in ecotoxicity, except for the biodegradation of a branched alkyl chain type of poly-oxyethlene alkylphenolethers, because ecotoxicity increases with increasing hydro-phobicity due to shortening of the ethylene oxide chain [52–54]. For example, although the LC_{50} value is 4.6 mg/L for 96 hours' exposure of the fathead minnow to LAS, in the early stages of biodegradation it is 1,200 mg/L in sulfophenyl undecanoic acid sodium salt ($HOOCC_{10}H_{20}C_6H_5SO_3Na$), the alkyl chain terminal of which is generated in response to oxidization (β-oxidization) [55]. Fish toxicity of LAS decreased with a shorter alkyl chain length (β-oxidization), which indicates the progression of biodegradation of LAS [55]. Moreover, the influence of alpha-sulfonated fatty acid methyl ester (MES) on the mortality of fish during the biodegradation was investigated using a mixed alkyl chain length (C_{14}/C_{16}) as a test MES material. As the result, the toxicity to fish was markedly and rapidly reduced by the degradation (Table 3) [56].

E. Exposure Analysis

Although the environmental effect of detergent chemicals (i.e., surfactants) has been shown, in this section the exposure assessment is explained. Exposure assessment is an important matter, like hazard identification, in order to evaluate the environ-mental risk assessment of a surfactant. That is, it is necessary to evaluate quanti-tatively the amount of exposure of environmental organisms to the surfactant. The purpose of the exposure analysis is to estimate or measure the concentration of chemical substances to which organisms are exposed. PECs of detergent chemicals

TABLE 3 Change in Acute Toxicity of Mixed C_{14}/C_{16}-MES to Japanese Killifish (*Oryzias latipes*) During Biodegradation

Biodegradation (%) (BOD/TOD)	96-h LC_{50} (mg/L)	
	As intact MES[a]	MBAS
0	2.4	2.4
5	3.2	2.7
7	7.1	4.2
18	> 100	—
44	> 100	—

[a] Calculated based on dilution factor.
BOD: biochemical oxygen demand; TOD: theoretical oxygen demand; MBAS: methylene blue–active substances.

are generally derived from volume of discharge, removal/biodegradation efficiency during sewage treatment, and effluent dilution by receiving waters. Much labor and expense are needed to measure the detergent chemicals in the environment. In the first screening evaluation, a simple dilution model is applied that is drawn from the amount used and river flux. When the outcome of the environmental safety assessment of a detergent chemical indicates that the safety margin between PEC and PNEC may be small, additional information may be sought by more complex models. These models take a great number of parameters (e.g., adsorption and biodegradation) into account.

The surfactants currently used widely, such as LAS, AS, AE, AES, AOS, and soap, are satisfactory for collecting monitoring data. However, obtaining actual concentrations in an aquatic environment is difficult for the newly developed surfactants. Therefore, exposure assessment of new surfactants is required to presume environmental concentrations based on a certain model. According to the environmental behavior and predictive exposure model, information on the behavior of the substances in various situations can be acquired from the environmental conditions and the released conformation of the substances. Various mathematical models have also been developed in the water area that is the discharge place of a surfactant. The model design is carried out as follows. In the case of behavior evaluation for a stationary state in a river, U.S. EPA ReachScan is used; QWASI (Quantitative Water Air Sediment Interaction model) is the compartmental model used for behavior evaluation in lakes and marshes. Furthermore, EXAMS (Exposure Analysis Modeling System) of the U.S. EPA. is a model that performs detailed behavior evaluation. Recently, after being based on the characteristic geographic and weather conditions, e.g., precipitation and the change in the amount of water in a river, developmental research on an environmental concentration prediction system (GREAT-ER) for a surfactant processed in sewage was also performed. The exact presumption of PEC is attained by enriching these environmental predictive models.

F. Conclusion

Detergent products are manufactured in relatively large quantities, used by almost all consumers, and disposed of broadly into the environment and therefore warrant thorough environmental hazard evaluation and risk assessment. Hence, much scientific knowledge has been accumulated and an effective framework has been established for risk assessment and measurement of the various interactions of detergent products with the environment. If an environmental risk assessment of surfactants is performed, the results of two research studies must be extensively. The most thorough aquatic risk assessment of currently used surfactants was undertaken by the Dutch government in cooperation with the European detergent and surfactant industries. This investigation was also conducted for over five years [57]. Another is the environmental risk assessment for LAS by the U.S. SDA (Soap and Detergent Association) [58]. Based on these results, the overall conclusion is that, despite the large quantities of surfactants entering the environment, they do not represent a serious threat to it. Similarly, for the majority of other detergent chemicals, the aquatic risk assessment approach predicted a relatively large margin between the estimated exposure and toxicity threshold concentrations.

IV. CONCLUSION

This chapter outlined how the health and the environment effects of detergent chemicals are evaluated. This is an important subject because detergent chemicals are used on a daily basis for a long period of time. Safety evaluation must always be based on the newest and most exacting scientific methods.

REFERENCES

1. Loprieno, N. Food Chem. Toxicol. 1992, *30* (9), 809–815.
2. Curry, A.S.; Gettings, S.D.; McEwen, G.N., Jr. *The Cosmetic, Toiletry, and Fragrance Association Safety Testing Guidelines: Evaluation of Primary Skin Irritation Potential.* CTFA, Washington DC, 1991; 1–5.
3. Walker, A.P.; Basketter, D.A.; Baverel, M.; Diembeck, W.; Matthies, W.; Mougin, D.; Paye, M.; Röthlisberger, R.; Dupuis, J. Food Chem. Toxicol. 1996, *34*, 651–660.
4. Magnusson, B.; Kligman, A.M. J. Invest Dermatol. 1969, *52* (3), 268–276.
5. Magunusson, B.; Kligman, A.M. *Allergic Contact Dermatitis in the Guinea Pig*; Charles C Thomas: Springfield, IL, 1970.
6. Ritz, H.L.; Buehler, E.V. *Current Concepts in Cutaneaus Toxicity, Planning, Conduct, and Interpretation of Guinea Pig Sensitization Tests*; Academic Press: 1980; 25–40.
7. Shelanski, H.A.; Shelanski, M.V. Proceedings of Scientific Section, The Toilet Goods Association, 1953, *19*, 46–49.
8. Marzulli, F.N.; Maibach, H.I. J. Soc. Cosmet Chem. 1973, *24*, 399–421.
9. Kligman, A.M. J. Invest. Dermatol. 1966, *47* (5), 393–409.
10. Organization for Economic Cooperation and Development Guideline for Testing of Chemicals: Test Guideline 406 "Skin Sensitization," OECD, Paris, 1992
11. Morikawa, F., et al. *Sunlight and Man*; University of Tokyo Press: Tokyo, 1974; 529–557.
12. Ichikawa, H.; Armstrong, R.B.; Harber, L.C. J. Invest. Dermatol. 1981, *76*, 498–501.
13. Organization for Economic Cooperation and Development Guideline for Testing of Chemicals: Test Guideline 471 "Bacterial Reverse Mutation Test," OECD, Paris, 1997.
14. Organization for Economic Cooperation and Development Guideline for Testing of Chemicals: Test Guideline 472 "Genetic Toxicology: *Escherichia coli*, Reverse Mutation Assay," OECD, Paris, 1983.
15. Organization for Economic Cooperation and Development Guideline for Testing of

Chemicals: Test Guideline 473 "In Vitro Mammalian Chromosome Aberration Test," OECD, Paris, 1997.

16. Organization for Economic Cooperation and Development Guideline for Testing of Chemicals: Test Guideline 474 "Mammalian Erythrocyte Micronucleus Test," OECD, Paris, 1997.

17. Organization for Economic Cooperation and Development Guideline for Testing of Chemicals: Test Guideline 475 "Mammalian Bone Marrow Chromosome Aberration Test," OECD, Paris, 1997.

18. Organization for Economic Cooperation and Development Guideline for Testing of Chemicals: Test Guideline 476 "In Vitro Mammalian Cell Gene Mutation Test," OECD, Paris, 1997.

19. Draize, J.H. Appraisal of the Safety of Chemicals in Foods, Drugs, and Cosmetics. Association of Food and Drug Officials of the United States, 1959; 49–51.

20. Griffith, J.F.; Nixon, G.A.; Bruce, R.D.; Reer, P.J.; Bannan, E.A. Toxicol. Appl. Pharmacol. 1980, 55, 501–513.

21. Luepke, N.P. Food Chem. Toxicol. 1985, 23, 287–291.

22. Russell, W.M.S.; Burch, R.L. *The Principles of Humane Experimental Technique*; Methuen: London, 1959.

23. Organization for Economic Cooperation and Development Guideline for Testing of Chemicals: Test Guideline 423 "Acute Oral Toxicity—Acute Toxic Class Method," OECD, Paris, 1996.

24. Arthur, B.H.; Kenney, G.L.; Pennisi, S.C.; North-Root, H.; Dipasquale, L.C.; Penney, D.A.; Re, T.; Sekerke, H.J.; Dinardo, J. J. Toxicol. Cut Ocular Toxicol. 1986, 5 (3), 215–227.

25. Spielmann, H.; Liebsch, M.; Kalweit, S.; Moldenhauer, F.; Wirnsberger, T.; Holzhutter, H.-G.; Schneider, B.; Glaser, S.; Gerner, I.; Pape, W.J.W.; Kreiling, R.; Krauser, K.; Miltenburger, H.G.; Steiling, W.; Luepke, N.P.; Muller, N.; Kreuzer, H.; Murmann, P.; Spengler, J.; Bertram-Neis, E.; Siegemund, B.; Wiebel, F.J. ATLA 1996, 24, 741–858.

26. Spielmann, H.; Balls, M.; Brand, M.; Doring, B.; Holzhutter, H.G.; Kalweit, S.; Klecak, G.; Eplattenier, H.L.; Liebsch, M.; Lovell, W.W.; Maurer, T.; Moldenhauer, F.; Moore, L.; Pape, W.J.W.; Pfanenbecker, U.; Potthast, J.; de Silva, O.; Steiling, W.; Willshaw, A. Toxicol in Vitro 1994, 8, 793–796.

27. Spielmann, H.; Balls, M.; Dupuis, J.; Pape, W.J.; Pechovitch, G.; de Silva, O.; Holzhutter, H.-G.; Clothier, R.; Desolle, P.; Gerberick, F.; Liebsch, M.; Lovell, W.W.; Maurer, T.; Pfannenbecker, U.; Potthast, J.M.; Csato, M.; Sladowski, D.; Steiling, W.; Brantom, P. Toxicol in Vitro 1998, 12, 305–327.

28. Spielmann, H.; Balls, M.; Dupuis, J.; Pape, W.J.; de Silva, O.; Holzhutter, H.-G.; Gerberick, F.; Liebsch, M.; Lovell, W.W.; Pfannenbecker, U. ATLA 1998, 26, 679–708.

29. de Silva, O.; Basketter, D.A.; Barratt, M.D.; Corsini, E.; Cronin, M.T.D.; Das, P.K.; Degwert, J.; Enk, A.; Garrigue, J.L.; Hauser, C.; Kimber, I.; Lepoittevin, J.-P.; Peguet, J.; Ponec, M. ATLA 1996, 24, 683–705.

30. Organization for Economic Cooperation and Development, Guidelines for Testing of Chemicals: Effects on Biotic Systems, OECD, Paris, 1984.

31. Environmental and Human Safety of Major Surfactants, Volume I. Anionic Surfactants Part 1. Linear Alkylbenezene Sulfonates, Final Report To: The Soap and Detergent Association, New York, A.D. Little, Cambridge, MA, 1991

32. Kikuchi, M.; Wakabayashi, M. Bull. Japan Soc. Sci. Fish 1984, 50 (7), 1235–1240.

33. Lion Corp. (1990). *unpublished report*.

34. Lion Corp. (1999). *unpublished report*.

35. Lion Corp. (1972). *unpublished report*.

36. Oba, K.; Sugiyama, T.; Miura, K.; Ishimatsu, T.; Nishino, T.; Morisaki, Y. 7th International Congress on Surface-Active Substances, Section D, Volume 4, Moscow, 1976.

37. Oba, K.; Sugiyama, T.; Miura, K.; Morisaki, Y. Bull. Japan Soc. Sci. Fish 1977, *43* (8), 1001–1008.
38. Verge, C.; Moreno, A. Tenside Surfact. Deterg. 2000, *37* (3), 172–175.
39. Hama, I.; Sasamoto, H.; Tamura, T.; Nakamura, T.; Miura, K. J. Surfact. Deterg. 1998, *1* (1), 93–97.
40. Environmental Annual Report; Japan Soap and Detergent Association: Tokyo, 2000; 35–40.
41. Roberts, D.W. J. Surfac. Deterg 2000, *3* (3), 309–315.
42. Leo, A.J.; Hansch, C. *Substituent Constants for Correlation Analysis In Chemistry and Biology*; Wiley: New York, 1979.
43. Roberts, D.W.; Marshall, S.J.; Hodges, J. *Quantitative Structure–Activity Relationships for Acute Aquatic Toxicity of Surfactants*, Proceedings 4th World Surfactants Congress, A.E.P.S.A.T., Barcelona, Spain, 1996; Vol. 4, 340–351.
44. Belanger, S.E.; Guckert, J.B.; Bowling, J.W.; Begley, W.M.; Davidson, D.H.; LeBlanc, E.M.; Lee, D.M. Aquat Toxicol. 2000, *48*, 135–150.
45. Wakabayashi, M. Jpn. J. Environ. Toxicol. 1998, *1*(2), 27–40.
46. Tovell, P.W.A.; Newsome, C.; Howes, D. Water Res. 1974, *8*, 291–296.
47. Tovell, P.W.A.; Newsome, C.; Howes, D. Water Res. 1975, *9*, 31–36.
48. Darvill, A.; Augur, C.; Bergmann, C.; Carlson, R.W.; Cheong, J.J.; Eberhard, S.; Hahn, M.G.; Lo, V.M.; Marfa, V.; Meyer, B. Glycobiology 1992, *2* (3), 181–198.
49. *Molecular Biology of the Cell*; 2nd Ed.; Kyoikusya: Tokyo, 1990:1140 pp.
50. Environmental Annual Report; Japan Soap and Detergent Association: Tokyo, 1998; 5–10.
51. Utsunomiya, A. J. Health Sci. 1999, *45* (2), 70–86.
52. Kimerle, R.A.; Swisher, R.D. Water Res. 1977, *11*, 31–37.
53. Yoshimura, Y. J. Am. Oil Chem. Soc. 1986, *63* (12), 1590–1596.
54. Kurata, N.; Koshida, K.; Fujii, F. Yukagaku 1977, *26* (2), 115–118.
55. Swisher, R.D., Gledhill, W.E., Kimerle, R.A., Taulli, T.A. 7th International Congress of Surface-Active Agents, Paper No. 2, Moscow, 1976.
56. Masuda, M.; Odake, H.; Miura, K.; Oba, K. J. Jpn. Oil Chem. Soc. (Yukagaku) 1994, *43* (7), 551–555.
57. Environmental Risk Assessment of Detergent Chemicals. Proceedings of the A.I.S.E./CESIO Limelette III Workshop, Brussels, November 28–29, 1995.
58. Linear Alkylbenzene Sulfonate. SDA Monograph, 1996.

6

Generation and Use of Data to Predict Environmental Concentrations for Use in Detergent Risk Assessments

M. S. HOLT European Center for Ecotoxicology and Toxicology of Chemicals, Brussels, Belgium

K. K. FOX University of Sheffield, Sheffield, United Kingdom

I. INTRODUCTION

Risk-based decision making or risk management of a substance includes the identification of an acceptable level of risk and, if needed, the actions that should be taken to reduce an unacceptable level of risk to an acceptable level. Part of the overall risk management process is risk assessment, which is based on the recognition that risk requires two elements:

- The inherent ability of a chemical or material to cause adverse effects
- The exposure or interaction of the chemical or material with an ecological component or with a human population at sufficient intensity and duration to elicit the adverse effect(s)

The risk assessment process can be a stepwise process in which assessment of potential adverse effects and exposure is integrated and compared with increasing realism. The ratio of the predicted environmental concentration (PEC) to the predicted no-effect concentration (PNEC) is used as a measure of this risk.

One of the essential requirements for promoting improved understanding of the environmental fate and distribution of detergent ingredients is an accurate determination of their concentration in the various environmental compartments. These exposure assessments may require an in-depth assessment of the intrinsic physico-chemical properties and abiotic and biotic degradation processes and can be carried out directly, by measuring exposure, or indirectly, by estimating exposure based on models. In order to carry out meaningful environmental risk assessments for detergent ingredients, one of the fundamental requirements for promoting improved understanding of their exposure is for accurate determination of their environmental fate.

Measurements of their environmental concentration during the various stages of sewage treatment and in the receiving water are a vital part of the exposure data collection. The need to establish the validity of the data from both the analytical and the sampling standpoints is essential if we are to avoid the significant consequences that can result from a poorly planned monitoring program, which may generate unrepresentative samples and hence inaccurate exposure estimates. Furthermore, sampling instrumentation and protocols and analytical procedures must be appropriate to the needs of the program and reflect the use to which the data will be put. To be applied appropriately, exposure assessment results need to be representative, reproducible, and at a sufficient level of accuracy to support their end use. This chapter addresses the important issues relating to sampling practices and design protocols, makes some recommendations for future monitoring programs and illustrates how measured data can be used in exposure assessments.

II. ROUTES OF ENTRY INTO THE ENVIRONMENT

The release scenarios for detergent ingredients assumes 100% release of the chemicals at the use stage. After use, the disposal route involves discharge into the environment via wastewater sewer systems, which may or may not be connected to a sewage treatment works (STW). As a consequence their release is primarily into the aquatic compartment, as either treated, partially treated, or untreated wastewater.

Some surfactants will adsorb to solids in sewers and during sewage treatment and under certain conditions may not undergo biodegradation, with the result that varying amounts of some classes of surfactant may be released to the terrestrial environment during sewage sludge disposal. The pathways of detergent ingredients into the environment are shown in Figure 1. As the figure shows, risk assessments may be required in fresh and saline waters and in sediments and soils. A thorough assessment of the release pathway, environmental fate, and distribution of the chemical is required based on exposure measurements and/or mathematical models. In environmental exposure assessment the concentration of a substance in the different environmental compartments is estimated from the amount of the substance released in the production and formulation processes, the use and disposal patterns and the releases resulting from them, the physicochemical properties of the substance, and the properties of the environmental compartments into which the substance is released. The PEC can therefore be calculated from knowledge of the quantity of the substance that will enter the environment and the distribution and degradation processes occurring in the environment. This may be done using generic, representative model environments, as, for example the generic regional environment specified in the Technical Guidance Document [1] for Risk Assessment of New and Existing Chemicals in the European Union.

An alternative to such calculations is to measure the concentrations in the relevant environmental compartments according to a preplanned sampling strategy. This is possible only for substances released in quantities large enough to be detectable by appropriate analytical methods after dilution in the environment. Measured data should take precedence over the predicted PECs in cases where reliable high-quality monitoring data are available that have satisfied the requirements of statistical and temporal evaluation and adequate spatial representation. It is not the purpose of this chapter to review the analytical techniques and methods available for measuring the

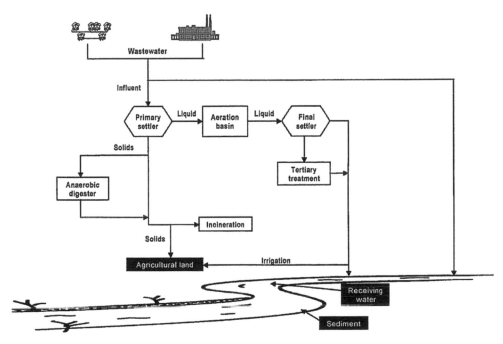

FIG. 1 Routes of entry of detergent ingredients into the environment.

concentration of detergent chemicals in the various environmental compartment. Instead, the intention is to give guidance on how to carry out monitoring programs and to discuss the general approaches currently used to derive an exposure assessment. In this respect, a distinction will be made between data specifically generated to address risk assessments and the use of data from routine monitoring programs that may have been designed originally for other purposes (termed "existing" data in the text). Some of the points will be illustrated with specific experiences gained during AISE/CESIO monitoring programs and the ERASM- (and UK Environment Agency)-sponsored Geographically Referenced Regional Exposure Assessment Tool for European Rivers (GREAT-ER) project [2] (see also http://www.great-er.org).

III. GENERATION OF MEASURED DATA

Information on measured concentrations is the product of a series of activities, consisting of program design, execution, and reporting.

There are a number of practical considerations when designing a monitoring program that are generic, regardless of the compartment being monitored. In Europe, the TGD [1] uses two different concepts for the predicted environmental concentration, PEC_{local} and $PEC_{regional}$. Most of the following discussion relates to the design of monitoring studies to determine PEC_{local} since the criteria for the derivation of $PEC_{regional}$ are still poorly defined. Many of the principles apply to both types of study, however, and the case study addresses the issue of $PEC_{regional}$ in some detail. The topics considered are practical, and they concern the specifics of study design and implementation.

The resources needed to carry out a monitoring program can be significant. To ensure that these resources are effectively and efficiently deployed it is important that the relevant parties, whose endorsements are necessary for the acceptance of the study, be identified and involved at all stages from project conception. Following initial discussions with interested parties, which should establish a commitment to the study, a team should be set up whose role is to agree to a project protocol. This protocol should clearly state the aims, objectives, and deliverables, identify individual responsibilities, and allocate resources and a strategy to achieve those goals within a specified time plan.

Some of the key stages in designing a monitoring program are shown in Figure 2. Defining the objectives of the study sets the degrees of precision and accuracy required. These in turn will determine the complexity of the study, determine the site selection, the sampling strategy (duration, type, and numbers of samples), the data requirements, how those data will be collected, and the limitations of available resources and manpower and therefore the cost of the study. The following practical aspects of conducting a monitoring program should be considered:

- *Representativeness*: Will the results be valid for the specific exposure scenario?
- *Data quality*: What sampling and analysis are appropriate, and how should these be documented?
- *Design practicalities*: What are the generic issues related to planning and coordination?

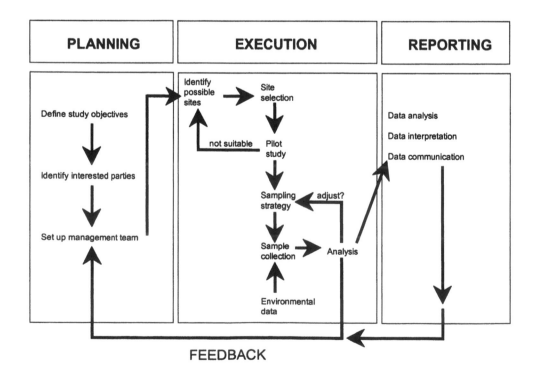

FIG. 2 Stages in planning and execution of a monitoring program.

A. Representativeness

It is important that the design criteria of a monitoring programme be selected so that the results of the monitoring represent a well-defined environmental concentration. There are two distinct types of representativeness.

- *Sample representativeness* relates to the confidence in the result, i.e., the number of samples taken, how far apart they are, and how frequently they are taken. The sampling frequency and pattern should be sufficient adequately to represent the concentration at the selected site.
- *Scenario representativeness* relates to the extent to which the site at which the sampling is done is representative of the local or regional setting and scenario in question.

The idea of measuring a representative concentration is a complex statistical issue [3]. If the distribution of an environmental concentration of a substance were homogeneous and constant, the problems involved in assuring representativeness could be handled with simple sampling statistics and an evaluation of the random errors (bias) inherent in the equipment and the process of sampling and analysis. A statistical understanding of the resultant data variability leads directly to a calculation of the number of samples needed to report a mean concentration at a level of confidence that meets the objectives of the study [4,5].

The environment, however, is both heterogeneous and dynamic, and thus the concentration of any particular chemical will vary in space and time as a result of variations in its rates of emission, transport, and degradation. Transport is greatly influenced by the state of the particular medium. Some examples of environmental properties that affect transport in different compartments are:

- *Air*: wind speed and direction, turbulence, precipitation
- *Water*: flow rate, turbulence, stratification, depth, dilution volume
- *Soil*: percolation rate (porosity, rainfall, surface features), organic matter, moisture

Environmental factors, such as temperature, UV exposure, and redox potential, affect degradation rates, as will interaction with other chemicals and biota.

Consideration should also be given to the way in which concentrations of detergent ingredients vary with time and space. Certainly they are influenced by daily and seasonal environmental conditions. Temporal variability is highly influenced by the emission characteristics for the chemical. If the chemical is entering the environment from a point source, the frequency and duration of the emission are directly correlated with the environmental concentrations. Rather than random sampling, a knowledge of the times and duration of the emission will allow sampling intervals to reflect temporal variation in emissions. This approach is far more cost effective than random sampling over an extended period of a year or more.

The most useful measured data will be those that capture all of these sources of spatial and temporal variability. The data may then, with appropriate statistical analysis, be used to estimate probability distributions of the chemical in space and time. Such probability distributions may be used to choose locations for subsequent sampling programs and, in risk assessment, to estimate the likelihood of reaching a particular concentration at specific locations and times.

The following criteria for representativeness are considered important:

- *Monitoring location.* Details of the monitoring location should be given (i.e., geographical coordinates, grid references). The monitoring site should be shown to be representative of the location and scenario chosen. The type of site will determine the homogeneity of the samples and therefore the need for replication and frequency of sampling. It is also important to know if any changes (product regulation, risk reduction measures, new production plants, increased capacity, etc.) have occurred at the site since the sample was collected. The type of location will also determine the homogeneity of the samples and therefore the need for replication or frequency of sampling.
- *Proximity of discharges.* For the aqueous environment, detailed information on the distance from and location of other sources, flow, and dilution are needed. For soils the application rate is important.
- *Discharge emission pattern.* Is there a constant and continuous discharge, or is the chemical under study released as a discontinuous emission showing variations in both volume and/or concentration with time? Peak emissions are characterized by relatively large discharges over short periods, and the time between peaks can vary greatly. Block emissions are characterized by reasonably constant flow over a certain time followed by regular intervals with low or zero emission.
- *Sampling frequency and pattern.* The temporal and spatial representativeness needs to address seasonal variation and whether or not the samples are time averaged or extremes.
- *Date of sampling.* The time, day, month, and year may all be important, depending upon the release pattern of the chemical. For some modeling and trend analysis, the year of sampling will be the minimum requirement.

B. Site Selection

The criteria imposed for site selection will vary with the aims of the program. The general principle is that the site should be very well characterized and that the data generated can be readily interpreted and positioned. Inevitably there will be a number of features about the site that will be essential to achieve study aims, and these will differ according to the aims. It is of particular importance to have a good historical database. Information should be available to establish baseline data for the proposed study and to identify the normal inputs and characteristics of the site. The history of the site is also important. Prior activities on the site may lead to background chemical interference in the analyses or unexpected physical obstructions to sampling. Site history should also include some idea of the stability of the local environment surrounding the site. Have there been changes in surrounding land or water due to human activities or catastrophic weather or geological events (e.g., changes in river flow, flooding, mud slides, new construction)? If past history indicates changes in either the site use or local environment, consideration needs to be given to the possibility that the site may not remain as intended over the future course of the monitoring program. In some cases, site geology should be evaluated to assess the background concentration of naturally occurring substances and the possible influence of the geology on water chemistry and thus on bioavailability.

The state of the environment surrounding the site may also be impacted by other nearby activities. If neighboring facilities or communities contribute to emissions of the chemical being studied, their influence must be taken into account when considering site representativeness. Furthermore, the influence of these neighboring emission sources must be accounted for by collection of adequate background samples (i.e., upstream).

1. Sewage Treatment Works

For studies at sewage works, the characteristics and operating conditions of the plant should be established to confirm that the site is representative for the study in question [6]. The key factors that need to be understood to establish the operating conditions are:

- *Sewage works treatment type* (e.g., activated sludge, trickling filter, oxidation ditch, carousel, rotating disc, anaerobic digester)
- *Sewage works size* (e.g., plant capacity, population served, population equivalents, tons BOD_5 or COD/day)
- *Percentages of trade and domestic waste* in the raw sewage
- *Sewage works performance* (BOD_5 removal, nitrified effluent, etc.)
- *Geographical location* (as a crude surrogate for temperature and other climatic effects)
- *The hydraulics* [i.e., flow pattern, sludge retention time (SRT), hydraulic retention time (HRT), sludge wastage, solids recycling, liquid recycling, etc.]
- *Type, if any, of tertiary treatment* (e.g., reed beds, sand filters)

A schematic of the works can be useful in confirming the hydraulics and characteristics of the works, that the works are operating normally, that the details match the project protocol, and that no development work is scheduled over the proposed study period.

Information should be collected to establish baseline data for the proposed study and to identify the normal inputs, including types of trade effluent treated. It is important that there be no compounds present in the samples that will interfere with the specific analytical method to be applied. Furthermore, certain trade effluents can have a significant, although probably temporary, affect on the works performance.

In addition, accurate historical removal efficiency and flow data are essential to confirm that the plant is performing normally and under typical conditions during the study. Data interpretation will be simpler and the data validity greater if it is possible to avoid or identify abnormal loads to the system. For example, septic tank waste is often delivered by tankers to STW and discharged into the influent channels. This may or may not be a significant load in terms of volume or biochemical oxygen demand (BOD_5) to the works, may or may not be added to the influent before the flow detectors, and may or may not be a daily event. One of the key parameters essential to understanding works performance is biochemical oxygen demand (BOD_5). There are, however, no suitable methods available to guarantee preservation of samples prior to determination, and this should be considered when BOD_5 data for 24-hour, flow-related composite samples for sewage influents, effluents, and river waters are analyzed. Grab (spot) samples should also preferably be collected in cooled containers (4°C) and analyzed as soon as possible.

It is important, in order to be able to correctly interpret and position any data, to include in the study protocol a supporting program designed to measure works performance and water quality, unless this is the objective of the monitoring study.

The timing, duration, and location of the study should be defined by the objectives and practical feasibility of the study. It is advisable to avoid sustained periods of heavy rainfall if at all possible, since storm events generally result in very different operational conditions at the works. In addition to the obvious problems associated with flow and dilution, heavy rainfall can result in premature operation of combined sewer overflows or storm overflows discharging directly to the receiving water. Heavy rainfall can affect the primary settlers, leading to an increase in solids in the aeration tank and final effluent. Furthermore the increased flow to the works may, when the storm overflow tanks are full, result in the discharge of untreated sewage into the receiving water. The operational procedures for discharge from the storm tanks (i.e., direct release to river or after treatment via, for example, aeration tanks) need to be understood. Storm sewage overflows may be triggered into operation for a variety of reasons, and it is essential to know the location of the storm discharge point(s). Results from a Dutch study [7] investigating the removal of linear alkylbenzene sulphonate (LAS) in an activated sludge type of STW were complicated by heavy rainfall that caused a malfunctioning of the primary settler and a concomitant increase in solids in the aeration tank and final effluent. This resulted in lower LAS removal figures and higher effluent concentrations. The specific considerations to be taken into account when designing monitoring programs in STWs are listed in Table 1.

Matthijs et al. [8,9] have recommended that boron be used as an effective tracer for understanding the hydraulics of STWs since it is not removed at any stage of the treatment process. Boron originates mainly from detergent use and is not removed in the sewers or treatment works. Thus it is possible to predict the raw sewage boron concentration from measured sewage flows, the combined number of domestic inhabitants connected to the works, and information on per capita surfactant consumption. The ability to generate real-time flow data is essential for the preparation of flow-related composite samples, which are necessary if the overall load of boron delivered to the STW is to be determined. Comparison of the predicted boron concentration with the measured concentration can therefore be used to provide a good basis for the removal of chemicals during treatment [10–12].

2. Receiving Environment

(a) Aquatic. In a similar way it is essential to understand fully the nature of the receiving water. Sampling programs may be complex where wide variations caused by extreme changes of temperature, flow patterns, or intermittent discharges occur. Knowledge of the extent of the mixing zone, including streaming and stratification, needs to be available. The most important factors to be considered when choosing a river site are the flow characteristics. Flow characteristics that will influence the degree of dilution and mixing include the total flow rate (volume/time), the current velocity, the depth, and the degree of turbulence caused by irregular channel or bottom features. The direction of flow may be changeable. In particular, details of tidal limits, locks, navigational use or recreational use, and major confluences should be known. The flow rate may vary considerably and be more dependent on the amount of navigational use

TABLE 1 Considerations for the Design of Monitoring Programs

1. Technical aspects of sampling
   a. Select appropriate sample size (i.e., volume, weight, etc.).
   b. Select appropriate sampling techniques: apparatus (type and materials), duration, transfer methods.
   c. Select appropriate containers (size, composition, type, labeling, cleaning procedures, closures), replication, methods to avoid (cross-) contamination.
   d. Select appropriate sample preservation, cooling, transport-storage times, and conditions.
   e. Select subsampling techniques: methods to avoid volatile losses, methods to remove sorbed or precipitated chemical from container surfaces, separation of suspended particulates, mixing, homogenization, etc.
2. Technical aspects of analysis
   a. Describe techniques for sample preparation, concentration, or extraction.
   b. Describe approach used to correct for losses during the analytical scheme (recoveries of laboratory and field spikes at appropriate concentrations).
   c. Describe techniques to evaluate and account for background interference or contamination.
   d. Select or develop the selective chemical analysis with appropriate sensitivity and reproducibility.
   e. Quote method detection limit and ensure that it is low enough for the intended purpose.
   f. Specify the method and frequency of calibration/standardization of the analysis.
3. Quality control and quality assurance
   a. Establish procedures for tracking and documenting sample integrity, transfers, etc. (e.g., inventories, chain of custody).
   b. Establish procedures to ensure consistent sampling and analysis methods (protocol, standard operating procedures, personnel training, etc.).
   c. Establish procedures to associate the final analytical result with the original sample (sample coding and unique sample IDs data sheets, transfer forms, etc.).
   d. Establish procedures to track and document analytical reliability (e.g., QC charting, sample reanalysis, blind performance samples).
   e. Establish criteria for reanalysis or omission of "outliers."
   f. Define monitoring or auditing responsibilities for supervisory or independent QA checks on conformity of sampling and analysis with the protocol requirements.
4. Sample location in time and space
   a. Select sample locations: frequency and time (of day, month, year)—is there a sampling window? distance up- and downstream, control sites, transects, distance apart, distance from river bank, proximity to obstructions or structures causing water flow changes, depth, etc. For bottom sediment, select the depth from which the sediment sample will be taken.
   b. Select sample sizes and decide on whether to composit or not.
   c. For composite samples, select locations or sample intervals to be combined — for aqueous samples, flow or time composites?
   d. Consider the influence of the season and the weather on sample representativeness.
   e. Sample location (macro): Select catchment, country, geographic location (e.g., latitude/longitude), distance from source(s), etc.
   f. Sample location (micro): Select sampling locations relative to the source and surface of the environmental compartment and relative to other samples taken (i.e., distance, direction, time interval). Positively identify a site for future sampling.

TABLE 1 Continued

5. Sample characteristics
   a. For water, decide on measurement of the following:
      i. Temperature
      ii. Suspended solids content and dissolved organic carbon content
      iii. Water chemistry (hardness, salinity, alkalinity, pH, other characteristics, depending on concerns)
      iv. Indicators of pollution (dissolved oxygen, oxygen demand (BOD/COD), H_2S, etc.)
      v. Inorganic contaminants (nitrate, nitrite, ammonia, sulphate, phosphate, metals, etc.)
      vi. Organic indicator of pollution (specific contaminants)
   b. For sediment, decide on measurements of the following:
      i. Temperature and pH
      ii. Total organic carbon content
      iii. Physical properties (granularity, density, ion exchange properties)
      iv. Characteristics of overlying water (as in point 5a)
6. Environment characteristics
   a. Select the water body that meets the study objectives with respect to:
      i. Type (lake, canal, river, estuary, coastal or open sea)
      ii. Flow at time of sampling (current speed, flow volume, tidal range, phase of the lunar cycle, spring tide, neap tide, high water, low water, none)
      iii. Size (width, depth, volume)
      iv. Ancillary information (abundance and diversity of aquatic and benthic biota, bottom characteristics, shoreline biota, etc.)
   b. Take the following characteristics into account when selecting a site:
      i. Distance downstream from STWs and industrial effluents of the same chemical
      ii. Land usage (residential, industrial, mixed use, agricultural, natural)
      iii. Representative water traffic (major port, barge transport, pleasure boats, none)
      iv. Water quality (polluted, unimpacted, eutrophic, oligotrophic, etc.)
   c. For the specific purpose (scenario) for which the data are intended, the following are appropriate:
      i. Distance from industrial effluents of the same chemical
      ii. Population (urban, suburban, rural)
      iii. Climate and weather (temperature variations, rainfall, etc.)
      iv. Physical characteristics (topography, distance inland, etc.)
   d. Establish the presence and location of other sources of pollution in the region of the sampling site.
   e. Provide data on the site history regarding physical characteristics and pollution.
7. Local source or site characteristics
   a. Select a site representative of the emission source (point source or diffuse source).
   b. Select a site representative of the intended life stages of the chemical (production, processing, storage, etc.)
   c. Select a site typical of the industry in annual production or throughput of the chemical.
   d. Select a site that uses typical processes for the industrial step.
   e. Select a site that uses waste disposal and wastewater treatment practices similar to those that are representative of the industry.
   f. Select a site that has a stable operational schedule: The time frame for releases of the chemical are known and not likely to change; no changes in schedule, new equipment, or process changes are likely to occur over the time frame of the study.

TABLE 1 Continued

g. Does all aqueous effluent undergo wastewater treatment (storm water runoff, cooling water, etc.)? What is the type, flow rate, etc. of the STW, or is it discharged untreated?

8. Reporting
   a. Give details of the sampling scheme, sample locations and time intervals, sample sizes and numbers, local environment and sample characteristics, site coordinates.
   b. Provide available data on site representativeness, including the direction and magnitude of source and site characteristics regarding industrial activities, tonnage throughput, etc.
   c. Provide details of the sample-handling and tracking procedures. Give results of the background or procedure blank analyses.
   d. Provide details of the sample preparation procedures, including mixing, subsampling, extractions or concentration, dilution of extracts.
   e. Give details of the analytical methods and data on selectivity and background interferences, as well as blank/background analyses and sample analytical results.
   f. Provide data on the sensitivity and reproducibility of the sample handling and analysis, including results for spikes, QC samples, and QC charting.
   g. Provide a statistical summary of the data: number, distribution, mode, mean, std. dev., 90th percentile, etc.
   h. Describe the value used to incorporate nondetects in the mean and the actual number of nondetects.
   i. Discuss the relevance of the data, how concentrations compare with background (upstream) data, historical data, data from other, similar monitoring studies, etc.

(number of locking operations, for example,) than upon prevailing weather conditions. Similarly, any abstraction or discharge points can affect significantly river flows, and all such features should be identified and the relevant abstraction or discharge volumes obtained, plus any potential variability due to batch discharges etc. Whenever possible, information on the volume and location of other point and diffuse sources discharging to the area should be obtained, and to help interpretation of the data there should be reasonable distance between discharges. Confluences and abstraction zones should also be located and avoided if possible. When the purpose of the monitoring is to determine the effects of a particular point discharge, then sampling both upstream and downstream should be carried out. Identifying suitable sampling locations that can be easily identified (by reference to fixed features on the river bank) means that comparable samples can be taken over a period of time.

A discussion about mixing zones is a more general topic related to the risk assessment approach, but it warrants comment here since it has distinct implications for monitoring at a local level. In the past there has been much confusion and debate over the definition of a *mixing zone* and what it really represents, particularly for larger water bodies. However, it is an important concept for local risk assessment, and there is a need to address its meaning in the context of risk assessment, particularly with respect to larger rivers, estuaries, and open seas.

In practice, each river is different, and the distance to complete mixing can be quite variable. If a monitoring program is intended to represent a scenario of complete mixing, it will be necessary to establish whether this has occurred, which may require monitoring at various depths and across the width of the watercourse at different distances downstream and under various conditions of flow. The PEC_{local} in risk assessment relates primarily to the end of the mixing zone rather than to a fixed, arbitrary distance from the discharge.

In an estuary, the effective dilution is often determined more by the tidal regime and local mixing characteristics than by the freshwater flow. When a substance is released it will travel both upstream and down and may be characterized either by a long, thin effluent plume or perhaps by one that is shorter and wider. Often, this may also affect only the side of the estuary where the discharge is located. Monitoring in estuaries, coastal waters, and open seas requires some specific considerations that are not always appropriate to address in river systems. There are many factors that distinguish marine systems from freshwater. But in terms of monitoring the environment and the implications for risk assessment, two, in particular, demand consideration.

- Much larger volumes of water result in higher dilution.
- Coastal systems are normally influenced by the tide.

The first point is particularly important with respect to detection levels. Quite simply, if concentrations are very low, it may be difficult to measure them, even with advanced analytical techniques, and the volumes of water available for dilution are often such that levels may fall below detection limits within a short distance from a discharge point. Because the concentrations to be measured are generally very low, it is essential to employ the right sampling technique. There are, for example, water samplers that can concentrate several hundred liters of seawater on a filter and adsorption resin. This, however, makes sampling relatively time consuming and expensive. The implications of this are that it may be extremely difficult to measure anything at a remote distance from any discharge location (at least in the water phase), and for most substances it will be necessary to undertake monitoring close to outfall locations in order to detect and quantify the amount entering the environment. Thus local monitoring may be the only realistic option for some substances.

An exception to this might be for high-volume discharges of substances that are persistent in the environment. Such substances, released over time, may build up to detectable concentrations at more remote locations. The adsorption characteristics, however, and potential for bioaccumulation will also provide important indicators in such cases on which compartment (water, sediment, or biota) should be monitored to look for elevated concentrations on a regional scale.

The second point concerns tidal effects. Many tidal phenomena are extremely complex, and the potential list of specific considerations for risk assessment of the marine environment is large. In terms of monitoring, however, the main factor is that the tide goes in and out. Thus a substance discharged at one moment may, for example, be traveling north with the tidal current, whereas in six hours it may be traveling south. Over the space of a few hours, therefore, given the dilution characteristics outlined earlier, the concentration at any one location close to the discharge point may vary over several orders of magnitude, depending on which direction the tide happens to be flowing. Hence any local monitoring of the tidal environment will need to be targeted to take into account the direction of the tide, in order to be able find detectable levels.

There is another important consideration for tidal and other time-varying exposure scenarios, in that different organisms will "experience" different exposure times. Monitoring should therefore be designed to give time-variable concentration fields or exposure doses in the actual area. Alternatively, or additionally, monitoring of body burdens of the discharged compounds in biota can be performed. Such data can be used alongside models that predict the dilution, exposure, effect, and environmental risk on a local and a regional scale.

Much the same sort of considerations must be given to estuaries and the open sea. Both are characterized by tidal regimes and much larger volumes of water available for dilution. Estuaries tend to be much wider toward the sea, and discharges on one side of an estuary may not have any impact on the opposite bank, depending on wind speed and direction etc. Thus any "local" monitoring must often be targeted in the same way as in the open sea, also taking into account the turning of the tide, in order to detect measurable concentrations.

Though risk assessment in the marine environments is in its infancy in Europe, both the Oslo–Paris Commission (OSPAR) and the European Chemicals Bureau (ECB) are currently drafting guidance documents. It is worth noting that:

- The tidal effect is such that substances may remain in the estuary and potentially accumulate for several tidal cycles (perhaps hundreds), depending on the degradation characteristics of the substance and the location of the discharge in terms of distance from the sea.
- Many industrial sites are located on estuaries.
- There are often several discharges from different industrial and domestic wastewater sites in one estuary, each discharging different types of effluent.

All these points recognize that estuaries are potentially more vulnerable environments than the open sea, but they are also naturally robust and capable of withstanding extremes, such as wetting/drying, fresh/saline waters, and freezing/baking. Monitoring activities should be tailored to take this into account, and they will vary from estuary to estuary, depending on the local geomorphology and ecology.

(b)   Terrestrial.   A significant difference between soil and the other environmental compartments is that soil is generally much less mobile. Mixing processes occur more slowly than for other compartments, and spatially variable deposition, compounded by variability in soil parameters such as texture and organic content, results in heterogeneous patterns of contamination. Consequently, the concentration in a single soil grab sample may differ greatly from a sample taken close by. As before, a monitoring program must take account both spatial heterogeneity and temporal variability.

In the context of risk assessment of detergent ingredients, the $PEC_{regional}$ for soil will be determined primarily by the application of sewage sludges or effluents. Given limited resources and an expectation that the PEC_{local} will be greater than the $PEC_{regional}$, the objective of most monitoring programs will be to determine the PEC_{local}. Thus the focus of the rest of this section will be on the requirements of such a program.

International Standard ISO 10381, Parts 1–6 [13], provides detailed information and guidance on soil sampling. The British Standards Institution has issued a draft for development covering the same subject [14].

Samples (usually not less than 500 g) may be taken on grid systems, along transects, or at single points [13,15]. Grid systems may be nonsystematic (irregular) or systematic. The advantages and disadvantages of the different systems are discussed in Ref. 13. Clearly the choice of system will depend on whether the objective is a spot check, a qualitative prestudy, or an extensive quantitative study, including statistical analysis. The number of samples will also depend on the objective of the sampling strategy. A grid system is the most appropriate for monitoring soil concentrations after sludge application. In such a scenario, the choice of site would be made after discussions with the authorities responsible for sludge application to determine sites where the history of sludge application (e.g., number of applications and/or time since the last application [16]) will produce a worst-case PEC_{local}. A regular grid pattern of sampling would then be appropriate, with repeated sampling between applications to establish the potential for losses by degradation, leaching (c.f. groundwater monitoring), and uptake by plants.

If possible, samples should be taken in open fields without vegetation or disturbance of air movements/deposition by surrounding buildings. When this is not possible, surface litter should be removed. Contaminants may or may not be most concentrated at the surface, depending on the site history and physicochemical properties. However, the recommended sample depth is under 10 cm for aerial deposition and plow depth (i.e., 0–30 cm) for sludge-amended soil sampling.

There are essentially two approaches to soil sampling [17,18]:

- The excavation method, using a shovel
- The probe method, using an auger

The advantage of the probe method is that it causes less disturbance to the site, the samples sizes can be small (important when preparing a composite sample), and it is easy to sample down to 30 cm.

Other issues, in addition to representativeness, need to be addressed with regard to site selection.

- The site must be accessible. If the study objectives require numerous samples over a range of distances from the emission source or over an extended time period, then the intended locations for sampling must continue to be readily accessible. Evaluation of the site for these sorts of details is most readily accomplished through site visits.
- Weather conditions (ice cover, storms, floods, etc.) that would affect sampling or change the characteristics of the environment need to be anticipated.
- The safety of sampling personnel must be ensured. Water and sediment samples may need to be taken by boat, so dangerous currents or obstructive conditions such as dams, weirs, and low bridges need to be considered.
- The site should be secure from interference in sampling, from vandalism, or from theft of sampling equipment.

C. Site Reconnaissance and Prestudy

Many of the points just described can be addressed by a site visit. The suitability of the site and sampling locations that can easily be identified should be confirmed in the prestudy. The feasibility of protocol details such as sample preservation, transport, and analytical methods should be evaluated. Another useful purpose of a prestudy is to gather data for final selection of representative sample locations. For river water samples, this might entail describing flow patterns in order to establish the distance downstream of the mixing zone, movements of plumes, etc. For soil monitoring, samples may need to be collected to determine soil chemistry, soil type, and prior history. Execution of the plan at an STW, for example, benefits from involvement and "ownership" of the project by site operatives, who may require specific training. The nature, reasons for, and value of the program should be explained to all concerned. It is a practical way to get local knowledge and a history of the site, which may not necessarily be officially documented. Another important function of a site visit is to confirm that the local facility or facilities are stable and operating normally and that no development work or change in facility design, operating conditions, or equipment is scheduled over the proposed study period. Site records of all events that may have affected facility operation should be consulted. Reliance solely on the memory (or honesty) of the operator can be unwise, and it is advisable to confirm which records will be kept during the period of the monitoring study. Finally, during the prestudy the nature of, reasons for, and value of the program may be explained to all concerned.

D. Sampling Program

After establishing a suitable site, a program for sampling and recording environmental conditions should be developed. The spatial distribution and sampling frequency will depend directly on the objectives of the monitoring program and should represent the anticipated spatial and temporal variability in river flows and discharge volumes or concentrations.

   In surface waters, a targeted approach is usually required, which takes account of prior information about where and when a chemical might be found. This is normally the most cost-effective approach. The discussion of mixing zones in the previous section is also relevant, and it may be necessary to target effluent plumes to facilitate a proper interpretation of the results. Sampling locations outside mixing zones, particularly in large water bodies such as estuaries and open seas, may yield concentrations that are simply too low to measure or require expensive "accumulative" sampling techniques.

   For rivers, a number of features regarding sampling location and frequency are common to many monitoring programs. A background (upstream) concentration should be measured at a site shown not to be contaminated with the discharge. Effluent samples at the discharge point are normally required and at a number of points downstream, depending on the scale and objectives of the monitoring program. As well as the chemical monitoring data, it is normally essential to record the discharge and river flows during the sampling period.

   The spatial distribution of sampling points depends largely on the homogeneity of the actual distribution of a chemical, which can be affected by many different factors. For example, stratification in the receiving water will affect the concentration at different depths, so the depth of sampling is an important variable. Differences in temperature or density between the effluent and receiving water will also influence the distance the effluent plume will travel downstream before it is well mixed. The use of mathematical models, from simple flow models to complex computer models, can be extremely useful for estimating the flow-mixing properties and choosing appropriate sampling locations. For more detailed information, advice and guidance on water monitoring see Refs. 19–28.

   The number of samples collected for analysis must be sufficient to meet the intended use of the data. There are many types of sampling situations, some of which can be satisfied by a simple sampling regime, whereas others may require sophisticated sampling techniques. If the objective is to have a "spot check" on compliance, a single sample may do. Should the objective be to investigate trends, for example, of water quality for a region or a catchment over a number of years, then obviously a large number of samples will need to be taken. Protocols for the sampling regime (time, place, type, and frequency) and the training of the sampling staff should be such as to guarantee that these criteria are fulfilled and the same locations are sampled.

The two most common sample types are:

1. *Grab or spot samples*: a discreet sample taken randomly with respect to time. The random nature of these samples may result in high or low values.
2. *Composite samples*: two or more samples or subsamples mixed together in appropriate proportions (based on either time-related or flow-related considerations).

These samples are representative of larger sampling periods from which an average result of the desired determinand may be obtained.

There are a number of issues that need to be considered when deciding whether to use grab or composite samples. Both have their limitations, and the appropriate technique should be used. Due regard should be given to the resource implications of using either of these two methods. The manpower and equipment needed to carry out a composite sampling program are considerably more than those required for grab sampling. The statistical validity of spot versus composite samples is an area of continuing debate and study [29]. If the concentration varies but the flow is constant, then a time proportional composite sample should be collected. If both the flow rate and concentration of determinand vary significantly, then flow proportional composite samples would currently be recommended. Flow proportional samples, however, are of no value in identifying transient peaks in the concentration of a determinand. For effluents it is advisable always to analyze flow proportional composites, since this would reflect the averaging implicit in the steady-state risk assessment concept. For large rivers there may be less need to collect flow proportional samples, since there may have been multiple sewage effluents discharged along the length of the river. In these cases the time of travel from the discharge points to the sampling points will make the data difficult to interpret.

Samples may be taken over a long duration, or numerous individual samples may be composited in order to smooth out short-term variation in concentrations. This compositing of samples also reduces the number (and cost) of sample analyses. For rivers and effluents, flow- or time-related samples should be considered. For soil, chemical concentrations tend to vary less with time; but owing to localized differences in soil properties, concentrations of chemicals in soil often show more variability with sample location. Sampling of bed sediment is usually limited to the surface layer of the bottom of the river or other water body. Normally it is the horizontal spatial distribution that is of interest. But for some longer-term studies the concentration profile of a chemical with depth in the sediment can provide useful information on the history of contaminant deposition. However, the accuracy of such determination depends very much on the sampling method deployed and the conditions of formation of the sediment and its stability. Again, this variability may be averaged by taking more samples or by compositing a number of samples taken over a larger area. If the objective of the study is to obtain a representative concentration for a specific sample site, then such compositing of samples may be more cost effective than analyzing numerous samples and then compositing the data through calculating an average. Compositing samples does not, however, provide data on the extent of variability.

For risk assessment in the EU there are implicit expectations regarding the degree to which temporal and spatial variability are quantified within a monitoring study. It is expected that seasonal variations in chemical production, dilution water flows, etc. are measured. Thus a measuring program for establishing data would need to have sampling intervals planned to correspond to those temporal changes. Good records should be kept, so unreasonable concentrations caused by abnormal situations such as spills and atypical weather conditions should not be included. Sufficient flexibility needs to be built into the sampling strategy to avoid these situations.

A number of other practical considerations need to be kept in mind. Sample size may be limited by the sampling equipment used and sample containers available. Sample number may be limited by sample size, storage capacity, analytical capacity, etc. At times, sampling may be affected by practical feasibility. The skill and training of the sampling staff are important to obtaining the samples planned in the protocol; so is the extent of available resources. Sampling times or locations need to be planned to allow the staff sufficient time to store the sample properly and to take the next sample at the expected time. Unforeseen on-site conditions such as accidents and weather may cause deviation from planned activities.

Many of the points discussed in this chapter are a direct consequence of five national pilot studies that were carried out in Europe in the 1990s [7,30–33], the results of which are summarized by Waters and Feijtel [34], to establish a protocol for monitoring detergent ingredients at STWs. These studies addressed three main issues of importance to risk assessment: what the day to day variation is in load to a STW (whether there is a washday), whether there is diel variationsin influent and effluent LAS concentrations, and what the percentage removal is at the different types of works. To study the first issue, flow-related composite samples were collected daily for one week. There was no evidence of any one day contributing a significantly higher load to the plant than any other day. For example, in the United Kingdom [30], the mean concentrations of LAS in raw sewage ranged from 11.8 to 18.2 mg L^{-1}, and the daily loads varied from 102 to 139 kg d^{-1}. By analyzing the individual grab samples (from which the composite samples were prepared), it was possible to demonstrate distinct diel variations in the concentration of LAS in the influent to the plant and, although less obvious, in the effluent. The findings from these pilot studies were then used as a basis for a study monitoring the fate of alcohol ethoxylates (AE), alcohol ethoxy sulphates (AES), alcohol sulphates (AS), and soap [10] and for the GREAT-ER project [2]. The nature of the consumer use pattern of detergent products and the characteristics of wastewater treatment make the routine use of grab samples inadvisable for STW studies carried out for risk assessment purposes. Hazelton [35] has also concluded that the results from water quality monitoring programs reliant on grab samples can be misleading and that diel variations need to be taken into account when designing a sampling program.

There are further problems associated with the use of grab-sample data when looking at the effect of discharges on receiving water concentrations because of the "time of travel" effect. The issue of sampling the same body of water as it travels down a river will be discussed in the case study (Section V).

E. Data Quality

Before representative measured data are included in a risk assessment, their quality must be assessed. The quality issues relate to the technical aspects of both sampling (sample integrity) and analysis. Quality control objectives [36] ought to be defined in the planning phase of the study.

1. Sample Integrity

Since it is impossible to measure concentrations in the total environment, the objective of sampling is to collect a sample that is small enough to be handled, transported, and analyzed conveniently but that still accurately reflects that part of the environment sampled; i.e., the sample is truly representative. Not only should the sample be representative of the environment, but no change should occur during sample handling.

2. Sample Handling and Storage

It is imperative to try to avoid contamination, physical loss, degradation, and mistakes over identification of samples [37]. Sampling equipment needs to be appropriately cleaned between samples to avoid cross-contamination. If possible, sampling times or locations should be planned so that sample progression is from higher to lower expected concentrations. Sample volume should be kept to a minimum relative to the needs of the analytical procedure but be sufficient to allow for replicate or repeat analyses if necessary. Careful packing for transport is necessary to avoid breakage, leakage, or loss of samples during shipment.

3. Sample Preservation

The use of an effective preservative to eliminate or minimize degradation of surfactants between collection and analysis of the sample is of the utmost importance. The samples should be extracted or otherwise prepared for analysis as soon as practicable. Storage of samples under the wrong conditions can result in considerable changes in the analyte concentration. The main approaches used to preserve samples are pH adjustment, addition of chemicals, and refrigeration. Refrigeration or cooling from the point of collection until analysis is usually advisable, even when other preservation techniques are employed. In the case of the very readily biodegradable chemicals, it is essential to ensure that preservative is added to the sample container prior to the sample's being collected. Failure to do so will result in significant losses of parent material. The choice of preservative is also important. Sodium azide (0.01 M) was used by Kiewiet et al. [38] to preserve samples for AES analysis but was reported to have a limited effect, even when stored at 7°C. In another study, by Syzmanski et al. [39], chloroform, copper(II) at 50 mg L^{-1}, and mercury(II) at 25 mg L^{-1} proved to be ineffective preservatives for nonionic surfactants in river waters. Dubey (personal communication, 1994) found that the addition of 8% formaldehyde was necessary to prevent losses of AE, particularly with sewage influent and effluent samples. Samples collected during a monitoring exercise at a sewage treatment works are subject to rather different problems than samples of freshwater or seawater:

- High biomass (microbial) activity
- A complex chemical cocktail

The high biomass in influents and samples of sludges in particular makes preservation a key issue, and therefore it is essential to confirm the efficacy of the preservative. It is worth reemphasising that a high level of quality assurance for both sampling and analysis is required during any environmental monitoring program at wastewater treatment works because of problems associated with sample stability (primarily biodegradation), contamination of equipment, and the very sensitive extraction and analytical methodologies involved. It is important in order to correctly position the data to measure in some way the general STW performance. Although BOD_5 is often used for this purpose, caveats regarding BOD_5 preservation should be borne in mind, particularly when 24-h composite samples are used.

4. Sample Identification

Sample containers should be prepared and labeled in advance. Each sample should be identified by its specific location (grid reference) and date (and time, if needed) and should be initialed, at the time of collection, by the individual who collected it. Labels should be affixed in such a manner that weather or storage conditions will not remove the label or deface the writing. On arrival at the analytical laboratory, the samples should be inspected and an inventory made to ensure that the correct sample number and identification, as specified by the sampling protocol, have been delivered. Any peculiarities, such as broken containers, leakage, and unexpected visual appearance, should be noted on the inventory.

Sample containers need to be composed of materials that will not influence subsequent analyses through leakage, absorption, or contamination of the sample. Whenever possible, containers ought to be unbreakable.

Each of the steps in the sampling regime is a potential source of error that can influence the concentration finally determined. Since the aim is to measure the actual concentration as accurately as possible, all of these sources of error must be minimized (see Fig. 3). These sampling issues are best addressed through a quality control program.

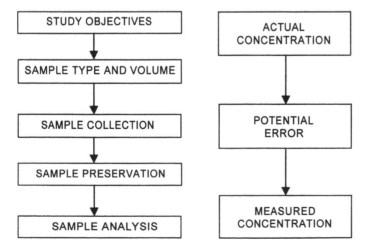

FIG. 3 Potential sources of error.

5. Sample Analysis

The quality of the data is defined by two key factors: the integrity of the sample and the limitations of the analytical methodology. The three major quality features of any analytical method are selectivity, sensitivity, and reproducibility. The *selectivity* of the analysis should be such that all of the target analyte and only the target analyte is measured and that other related chemicals or contaminants do not interfere with the response. Thus, a chromatographic method is preferred, one that has sufficient *sensitivity* to measure the chemical concentrations expected to be found. The method must have sufficient *reproducibility* to limit the number of replicate samples but still provide information on the actual variability in chemical concentration between the samples. The appropriate validated methods for environmental analysis should be used. Currently the techniques for surfactant analysis (with the exception of LAS) to enable trace analysis in the various environmental compartments are still being developed.

Irrespective of the technique, it is important to provide convincing evidence that the analytical scheme is accurate and precise. Methods should be used and reported that standardize the assay and correct for background and losses. The detection limit of the analytical scheme must be appropriate to the expected concentrations encountered in the field, since the detection limit may contribute significantly to the final calculated concentration if a large proportion of the samples analyzed is at or below the detection limit of the assay. At concentrations approaching the limit of detection of an analytical method, percentage errors will be greater than at higher concentrations.

The measures used to assess the quality of the data should include:

- Laboratory blanks (instrument, reagent, matrix, etc.) and field blanks (matrix-match, equipment, etc.) to assess laboratory and field contamination
- Laboratory-standard additions (at concentrations applicable to the sample matrix) to assess the recovery or efficiency of the analytical methodology
- Field-standard additions (spiked blanks or samples) to assess the preservation technique, stability of the samples, etc.
- Replicate field samples to assess overall variability in sampling and analysis
- Replicate analysis of the same sample to assess variation in analytical procedure

Interlaboratory analysis (or repeat, "blind" analysis in the same laboratory) of spiked and unspiked samples (i.e., sample subdivided in the field) may also serve as a check on reproducibility. For all types of locations it is important to collect samples that can be checked in the laboratory for the presence of any materials that cause interference with the analytical determination, recovery of standard additions, and minimum levels of detection. For frequent analyses, it is useful to have a continuing check on reproducibility through the use of quality control (QC) charts [40,41]. In QC charting, the mean analytical response to the specific chemical and its associated variability is calculated and charted. If the standards fall outside the expected range of normal variability for a particular day's analysis, this fact is immediately apparent and may be investigated.

6. Quality Control and Quality Assurance

Many of the problems associated with bias and systematic errors encountered during sampling and analysis may be assessed by a variety of internal quality control measures. The study plan should include the quality control and quality assurance procedures to be used. All methods and steps undertaken in the program require trained, skilled personnel. Minimum requirements for the test laboratory are listed in EN 45001 or ISO Guide 25 [42] standards. Standard operating procedures, conformance with the protocol, and clear and timely record keeping are necessary. These issues may be addressed with a quality assurance inspection or audit of the written records of the work in progress and of the final report.

For use in risk assessment, the most important factors to be addressed are the analytical quality control (AQC) and the representativeness of the sample. AQC activities are the basis for establishing some checks on the analytical performance, and though they are now widely used, they were not employed in the majority of environmental monitoring programs until very recently.

The criteria relate to either quality or representativeness. Information on the following are very important.

- Precisely what has been analyzed should be clear. Details of the sample preparation, including, for example, whether the analysis was of the dissolved fraction, the suspended matter (i.e., adsorbed fraction), or the total (aqueous and adsorbed), should be given. Also important is how the sample was collected, preserved, and stored, including the material that sample collection and storage containers were made of and what measures were taken to avoid contamination. The chemical speciation in the environment under investigation must be easily determined from the information given. Therefore more information will be required for some chemicals than for others.
- The analytical method, including sample extraction, cleanup, and end analysis (e.g., specific, such as GC/MS or HPLC/MS, or nonspecific, such as colourimetric) should be given in detail, or the scientific publication, ISO/DIN method, or standard operating procedure should be referenced.
- The minimum level of detection and details of possible interfering substances should be quoted.
- Concentrations in systems blanks should be given to support the minimum level of detection.
- Recovery of laboratory- and field-standard additions (spikes) should be quoted.
- Accuracy should be detailed. The relationship between the measured concentration and the minimum level of detection should be given.
- Reproducibility should be stated; that is, the degree of confidence and standard deviation in the results from repeat analyses should be given.

7. Reporting

It is essential that the data be used and reported correctly and transparently. It is important to include in the study design a supporting program aimed at obtaining the necessary data and/or information to allow the correct interpretation and positioning of the data set obtained and how the data are to be recorded.

Procedures used for sampling, sample handling, and analysis all need to be reported so that the data may be adequately evaluated. The limitations of the method as regards background interference, recovery, and detection limits should be clearly stated. The details of the procedures employed to check the selectivity, sensitivity, and reproducibility of the analytical method need to be reported. To allow an adequate evaluation of the results for use in risk assessment, additional information on the characteristics of the sample and the site need to be described.

Anomalous high or low sample results are often due to errors in sample handling, labeling, etc. Where outliers have been identified, their inclusion/exclusion should be justified and discussed. The data should be critically examined to establish whether high values reflect an increased or new release or a recent change in emission pattern. The data should also be examined to check that the analytical procedure was the cause of the discrepancy. It may be appropriate to reanalyze the samples or to disregard the result. The criteria for reanalysis or data deletion ought to be established prior to sample collection, either in the protocol or by written standard operating procedure. The reported results should indicate the number of samples analyzed for each particular time and location and whether and why data were omitted. Any statistical procedures employed to summarize the data need to be clearly described, and the individual data points ought to be available.

The need for the use of appropriate statistics to summarize the data is clear. When a detailed monitoring program is properly conducted, large numbers of data points are produced encompassing differing locations and times of sampling and blanks, spikes, standards, and quality control samples. The limits of detection and confidence in the result must be assessed and reported. The detection limit (DL) of the analytical method, which is normally defined by the analytical technique being used, should be suitable for the risk assessment. All measurements under the detection limit constitute a special problem and should be considered on a case-by-case basis. As a minimum, sufficient statistics ought to be conducted to quantify the sensitivity and reproducibility of the assay and to represent the statistical distribution of the data. For samples from multiple locations or different times, a simple mean value omits much of the valuable quantitative information on the variability of the concentration in space and time. Such data should be inspected or assessed statistically for the possible presence of time trends, either monotonic or cyclical. For use in risk assessment, a mean value is insufficient for a PEC, since the TGDs define the reasonable worst case as the 90th-percentile concentration.

The major generic areas to consider when designing a monitoring program to collect detergent chemical concentration data to be used for risk assessment are summarized in Table 1.

The information considered important in characterizing measured chemical concentrations is subdivided into six categories:

1. Technical details of how the sample was collected, preserved, and analyzed
2. Degree of quality assurance and quality control employed in arriving at a final environmental concentration
3. Location in space and time of the sample(s)
4. Condition and characteristics of the sample and the environment from which it came
5. Local manufacturing facilities, the characteristics of the facility
6. Data handling, how many analyses a concentration represents, and how the numbers were composited

Categories 1, 2, and 6 pertain to the data quality issues in sampling and analysis, while categories 3–5 relate to issues necessary or useful in establishing the representativeness of the samples and of the site (geographically and within a particular risk assessment scenario).

Close adherence to the principles specified in Table 1 should ensure that the resulting monitoring program generates results that will be both suitable and acceptable for use in environmental risk assessments.

F. Useful Supplementary Data

Numerous aspects of the sample, beyond its size and the concentration of the chemical, provide useful data in determining sample representativeness or in interpretation of the chemical concentrations in the context of risk assessment.

One of the most important aspects of water sampling is the flow rate. A number of devices are available that can be used at the time of water sampling to measure concurrently the flow.

For water, useful support analyses include suspended sediment concentration, dissolved organic carbon concentration, alkalinity, and hardness, depending upon the specific chemical interactions of the chemical of interest. Analogous data in sediment are particle size distribution, organic carbon content, and ion exchange capacity.

Other sample characteristics may provide information on the type of environment that was sampled and how much it may have been impacted by pollution. Data on salinity, temperature, hardness, buffering capacity, etc. help define the general water quality. Data on the dissolved organic carbon and biochemical oxygen demand may be used to evaluate the extent of pollution. Direct measures of contaminants such as ammonia, heavy metals, cyanides, and specific pollutant chemicals are also useful in putting the data into the risk assessment context. Finally, ancillary data on the presence of microorganisms, higher organisms, colored matter, surface films, etc. may also be valuable indicators of the type of water or sediment sampled.

Mass flows in wastewater treatment plants, rivers, soil, and other compartments allow the risk assessor to interpret monitoring data and to refine exposure models, increasing confidence in their use in risk assessment. The identification of mass flows in rivers and wastewater discharges is of particular importance at the boundaries between countries, districts, or water systems. Mass flows are the subject of international negotiations and are an input for mass balances for specific substances. Guidelines on water quality monitoring and assessment of transboundary rivers were recently published by a UN/ECE Task Force [43].

IV. USE OF EXISTING DATA

Criteria on how to decide whether the data are adequate for use in exposure assessment and how much importance should be attached to them have been proposed by OECD [44] and ECETOC [26]. To address the problem, three quality levels for existing data have been proposed (see Table 2). In recommending these criteria, the OECD stressed that

these criteria should be applied in a flexible manner. For example, the data should not always be discounted because they do not meet all the criteria. Risk assessors should make a decision on whether or not to use the data on a case-by-case basis, according to their experience and expertise and the needs of the risk assessment.

TABLE 2 Quality Criteria for Use of Existing Data

Criterion	Study category: use		
	1: Valid without restriction—may be used for measured PEC	2: Valid with restrictions—may be used to support exposure assessment (data interpretation difficult)	3: Valid only at screening level—should not be used in exposure assessment (but may be useful for other purposes)
What has been analyzed[a]	✔	✔	✔
Analytical method[b]	✔	✔	
Minimum detection limit[c]	✔	✔	
Blank concentration[d]	✔		
Recovery[e]	✔		
Accuracy[f]	✔		
Reproducibility[g]	✔		
Sample collection[h]	✔		
One shot or mean[i]		✔	
Location[j]	✔	✔	✔
Date (dd/mm/yy)[k]	✔	Minimum is knowledge of year	✔
Compartment characteristics	✔		
Sampling frequency and pattern	✔	✔	✔
Proximity of discharge points[l]	✔	✔ (for local scale)	
Discharge emission pattern and volume[m]	✔		
Flow and dilution or application rate	✔	✔ (for local scale)	
Explanation of value assigned to nondetects if used in a mean	✔	✔	

[a] Precisely what has been analyzed should be clear. Details of the sample preparation, including, for example, whether the analysis was of the dissolved fraction, the suspended matter (i.e., adsorbed fraction), or the total (aqueous and adsorbed) should be given.
[b] The analytical method should be given in detail, or a scientific publication (e.g., the relevant ISO/DIN method or standard operating procedure) should be referenced.
[c] The detection limit and details of possible interfering substances should be quoted.
[d] Concentrations in system blanks should be given to support the minimum detection limit.
[e] Recovery of laboratory- and field-standard additions (spikes) should be quoted.

The most important factors that should be addressed, however, are QC and representativeness. Govaerts et al. [45] studied the different alternatives available to summarize data collected over time and suggest a practical methodology on how to use such measured data to predict regional-exposure distributions.

V. CASE STUDY: GENERATION AND USE OF DATA TO PREDICT AQUATIC CONCENTRATIONS FOR USE IN DETERGENT INGREDIENT RISK ASSESSMENTS

As discussed in the introduction, the concept of environmental risk assessment has been designed to protect the environment from the adverse effects of chemical substances. Most risk assessment protocols compare a predicted environmental concentration, or PEC, with a predicted no-effect concentration, or PNEC. PNEC is generally based upon measured laboratory ecotoxicity test data. The lowest of the endpoints of the laboratory tests will be multiplied by an application factor whose magnitude depends both upon the nature of the test (i.e., acute, chronic, or mesocosm) and upon the protective regulatory framework of the country concerned. Thus the lowest of the acute tests from three different trophic levels will receive an application factor of 100 in the United States but 1000 in Europe. In contrast, the PEC is often based upon concentrations predicted by multimedia or other exposure models, though it is preferable for this to be supported by measured environmental concentration data. If the PEC/PNEC ratio is less than 1, the substance is considered to pose no danger for the environment. Risk assessment protocols are designed to be conservative so that they will err on the side of giving enhanced environmental protection.

The case study addresses the environmental risk assessment approach used in Europe. The EU TGD [1] suggests that ideally both monitoring data and modeling

---

[f] Results of analysis of standard "reference samples," containing a known quantity of the substance, should be included. Accuracy is connected to the analytical method and the matrix.

[g] The degree of confidence and standard deviation in the result from repeat analysis should be given. Reproducibility is also connected to the analytical method and the matrix.

[h] Whether the sampling frequency and pattern relate to the emission pattern or whether they allow for effects such as seasonal variations needs to be considered.

[i] The assessor needs to know that the data have been treated, e.g., whether the values reported are single values, means, or 90th percentiles.

[j] The monitoring site should be representative of the location and scenario chosen. If data represent temporal means, the time over which concentrations were averaged should be given too.

[k] The time, day, month, and year may all be important, depending upon the release pattern of the chemicals. Time of sampling may be essential for certain discharge/emission patterns and locations. For some modeling and trends analysis, the year of sampling will be the minimum requirement.

[l] For the local aqueous environment, detailed information on the distance of other sources, in addition to qualitative information on flow and dilution, is needed.

[m] Whether there is a constant and continuous discharge or whether the chemical under study is released as a discontinuous emission showing variations in both volume and concentration with time needs to be considered.

Source: Ref. 44.

results will be available for complementary use as part of the overall exposure assessment process. European environmental exposure assessment uses the concepts PEC_{local} and $PEC_{regional}$ to define two different concepts for predicted environmental concentrations. These are defined according to a "Standard EU Environment," using a steady-state model. PEC_{local} and $PEC_{regional}$ conceptually represent the predicted environmental concentrations at the point of complete mixing downstream of a sewage treatment works and a "background concentration" in which the substance concentration has reached a steady state, distributed over the standard EU region, respectively. Relating this steady-state risk assessment world to the real world, with its dynamic processes and explicit spatial dependence of environmental processes, is not straightforward. Of course, environmental monitoring data is obtainable only from the real, dynamic, spatially explicit world. To use measured data in environmental exposure assessment, data obtained from any specific place and at any specific time must somehow be related to the generalized concepts of PEC_{local} and $PEC_{regional}$.

If measured data are selected to represent PEC_{local}, the sites chosen should be close to their local source. The representativeness of the sites chosen should be considered, because the data are to be used in a generalized, not a site-specific, context. The sampling methodology should, if possible, consider the averaging of time-dependent concentration magnitudes at any specific site, either through the use of time or flow proportional sampling or by an appropriate use of a randomized spot sample collection routine. It is possible to generate a set of measured data, for example, from the points of complete mixing downstream of a number of "representative" sewage treatment works, which could reasonably represent PEC_{local}. The 90th percentile of this set of measured data would be considered to give a "reasonable worst case" for the environmental exposure, for PEC_{local}. Often, all that is available is a set of data collected at sites near a local source, but with no attention paid to the representativeness of the sampling sites or to the averaging of time-dependent factors. The 90th percentile for this type of data set could also be used in the environmental exposure assessment, although the results would clearly be inferior.

The definition of *representativeness* and the completeness of coverage for measured data sets whose 90th percentile could represent $PEC_{regional}$ are much more difficult. Although the European TGD suggests that, as a starting point, sites far from any local source, and thus representing a "regional background," could be selected for $PEC_{regional}$ derivation, it would be a great improvement to be able to define and use "representative" sites in a region. As with the PEC_{local} scenario, ideally the time dependence of the data obtained at any specific, representative site should be suitably averaged by the monitoring protocol chosen. However, the practical difficulties of representative site definition limit the feasibility of this approach. An alternative to the definition of *representative* sites is the consideration of all sites in a region. Although it is not be possible to obtain measured data from all sites in a region, it is possible, using GIS-based models, to model the concentrations of chemicals in all river stretches in a region and to validate the predicted concentrations at selected sites. This concept is currently being developed, on a catchment rather than a regional basis, to provide a better basis for the calculation of $PEC_{regional}$.

The GIS-based model that is being developed contains the rivers, with appropriate flow distribution curves for each river stretch, and the sewage treatment works, with associated populations served and type of treatment offered, for specific catchments. The model also includes processes for removal in the sewage treatment works and in the receiving waters. With this model, actual effluent release locations, with the appropriate release levels, dilution ratios, river flows, and river residence times, can be used to predict chemical concentrations in all river stretches in the catchment. A distribution of these concentrations can then be used as the basis for regional environmental exposure assessment.

This model, named GREAT-ER (Geography-referenced Regional Exposure Assessment Tool for European Rivers) is a deterministic steady-state model, with stochastic treatment of selected parameters. In principle, the local concentrations it predicts in specific sewage treatment works effluents and in rivers at specific locations above and below sewage treatment works can be evaluated by comparison with measured data. However, measured data, of necessity, relate to the specific place and time at which they were collected and to the environmental conditions that were then prevalent. Thus to evaluate a steady-state model, the monitoring program must collect sufficient, representative data covering all seasons, giving a representative distribution of high, medium, and low flow conditions.

As well as predicting the local concentrations of a substance found in specific river stretches, GREAT-ER provides the ability to generate selected distributions of the concentrations found throughout a catchment. For example, a distribution of the means, or perhaps the 90th percentiles, of concentrations found in stretches immediately below sewage treatment works can be generated. The mean, or perhaps 90th percentile, of this distribution can then be calculated and chosen to represent the PEC_{local} for European risk assessment. Note that while the TGD concept of a PEC_{local} cannot be validated by environmental measurements, it is possible to evaluate the site-specific predictive capability of a model such as GREAT-ER. Transparent description of the calculations the model can carry out with the validated site-specific predicted concentrations can then give a defined PEC_{local} for a selected set of sites. For example, PEC_{local} for a selected catchment could be defined as the mean of the 90th percentiles of the concentrations calculated for all stretches immediately below sewage treatment works, where each stretch is weighted by the volume of water it contained (if, indeed, such weighting is included in the calculation).

Calculations can similarly be performed with selected descriptors (e.g., the 90th percentiles) of each concentration distribution of all the stretches in a catchment, to generate background or perhaps whole catchment distribution functions. Mean, median, or 90th percentile values calculated from these background or catchment distributions can be generated for regional risk assessment or for screening assessments for the protective purposes of the EU Water Framework Directive [46]. Weighting river stretches by, e.g., length or volume may be considered appropriate [12,47]. The calculated distributions of concentrations at each specific site can be compared with the distribution of measured concentrations found at that site, but only for these sites can the model be validated. It is assumed that the model will work for other, nonvalidated sites, and thus the choice of representative monitoring sites is important.

A comparison of the use of the validated model to calculate an overall catchment concentration with a simple distribution calculated from the monitoring data used to evaluate this same model has been made for the Aire catchment, in Yorkshire, United Kingdom [12]. The result for boron are shown in Table 3. In these calculations, a reasonable worst-case concentration has been determined using the 10th percentile of the flow distribution curve, i.e., the flow that is exceeded 90% of the time. The means of

TABLE 3 Summary of Regional PEC Values for Boron in the Aire Catchment

Mean of 90th percentiles of monitored data	Calculated 90th percentile data, normalized by site flow volume, excluding upstream sites	Calculated 90th percentile data, normalized by site flow volume, including upstream sites	Calculated 90th percentile data, normalized by river stretch length, including upstream sites
242 μg L^{-1}	238 μg L^{-1}	186 μg L^{-1}	28 μg L^{-1}

the calculated concentrations in the selected stretches at these low-flow conditions are then calculated by the model, using different methods for site selection and for the weighting of the different stretches.

In this specific example, the results show that the mean of the site-specific 90th percentile boron concentrations, as obtained from monitoring data, provides a conservative estimate of the 90th percentile of the volume-weighted catchment averaged boron concentration. However, if stretch-length-dependent averaging is appropriate, then the 90th percentile boron concentration will be overpredicted by more than half an order of magnitude. The overall approach used here demonstrates the conceptual advantages of combining modeling and monitoring in environmental exposure assessment [12].

The GREAT-ER model has been evaluated with monitoring studies in the United Kingdom, Italy, and Germany. A major part of the initial evaluation of the GREAT-ER model was carried out in the Yorkshire region of the United Kingdom, from 1996 to 1998. The design of this model evaluation program will be used as an illustrative case study for the generation and use of data to predict environmental concentrations for use in detergent risk assessments.

A. Planning the Validation Study

The environmental exposure assessment of substances in detergent products contains several distinct parts, each of which should, if possible, be supported by environmental measurements or predicted by a validated model. As Figure 1 shows, substances that, due to their physical chemical properties, will reside predominantly in the aqueous compartment will be released from households, proceed through sewers, receive sewage treatment, and then enter receiving waters, where dilution and natural removal processes will occur. Information on the following points will be necessary to predict the concentration of the substance at any point in space and time along this pathway from the household to the receiving environment.

- The amount of substance released (per person per day and ultimately per treatment works per day)
- The removal rate of the substance in the sewer
- The removal rate of the substance during sewage treatment, which may depend upon sewage treatment type
- The removal rate of the substance in the receiving water

These points have been considered in this case study, the development and evaluation of the GREAT-ER model.

1. Defining the Objectives

The GREAT-ER model validation study should ideally consist of validation of the four component parts of the model described earlier, plus an evaluation of the overall model performance by site-specific comparison of the concentrations predicted in river stretches with the concentrations found in a monitoring exercise. However, the objectives of the validation study were essentially set, within the overall financial constraints of the project, by the overall GREAT-ER project team. This team comprised representatives of European detergent ingredient producers and detergent product formulators who wished to provide a modeling tool that would yield an improved technical basis for realistic environmental exposure assessment. Evaluation of the model by comparison with measured data was considered essential, both to ensure that a realistic model had been developed and to gain public acceptance of the model as a useful tool. Ideally, validation with several chemical substances in several European countries would be attempted, but financial constraints precluded this. Previous experience [34] suggested that measured data for a minimum of two substances present in household detergent products, boron (present in a bleaching agent) and the surfactant LAS, would enable the evaluation of many of the processes included in the model. Boron, which is not readily removed from the environmental system, would be used as a conservative marker to validate the calculated loads that reached the sewage treatment works from consumer use and the concentrations found in effluents and in rivers. LAS would be used to evaluate the predicted removal due to the biodegradation and adsorption processes. The choice of countries in which to carry out the monitoring study was heavily influenced by the availability of an existing monitoring data set generated by the LOIS project [48] in the United Kingdom, which contained the concentrations of many substances in rivers, many of which were located in Yorkshire. The availability of experienced and enthusiastic industry volunteers to carry out the project in the United Kingdom was also taken into account.

The detailed UK monitoring program was focused on providing the specific environmental measurements needed for model evaluation, consisting of river and effluent concentrations throughout the catchments to be modeled. The model also required, as input information, data on the removal of LAS in European trickling-filter type of sewage treatment works. Representative data on the removal of LAS in activated-sludge plants, the most common European sewage treatment facility, had been obtained in a previous exercise [7,30–33]. Although some information on the removal rate of LAS in surface waters was available, data on the removal rate of LAS in European rivers was felt to be necessary. To meet these needs, the UK GREAT-ER team carried out two further campaigns, in addition to the main model evaluation study. The first determined LAS removal in six trickling-filter-type sewage treatment works [49] (point 3 from earlier), and also allowed (point 1 earlier) the evaluation of the use of consumer sales data to calculate the load reaching the sewage treatment plant. The second measured LAS removal in specific rivers in the Yorkshire area [50] (point 4 earlier). Of the four components that needed evaluation, the point least addressed was the second point, that of overall removal of LAS in sewers. However, comparison of LAS and boron reaching the sewage works in the trickling-filter removal study [49] and of the measured LAS STW removal percentages with the measured effluent measurements indicated that LAS removals approaching or even exceeding 50% were probable.

2. The Identification of Interested Parties, and the Setting Up of the Project Management Team

The UK GREAT-ER team was extended from its initial industrial membership following discussions with the Environment Agency for England and Wales, who were simultaneously initiating an Environment Agency (EA) project to monitor surfactants in the environment. Both Industry and the EA agreed to join forces, and the project, under overall industrial leadership but with several EA members, met regularly at the EA headquarters in Leeds. The EA theme leader was able to interest the major sewage treatment provider in the area, Yorkshire Water, in the project, and Yorkshire Water provided access to the sewage treatment facilities and personnel who helped to design and carry out the monitoring program. The EA permitted the use of standard EA sampling sites and sample collectors to collect the samples. Also, analysis of the samples was provided, on a cost-plus-overhead basis only, at the EA analytical facilities in Leeds. In addition, modelers provided by the EA and hydrologists from CEH Wallingford, who were funded by the overall GREAT-ER project, completed the team, ensuring that the modelers were modeling processes that the monitoring team was measuring. The roles of the team members are shown in Table 4.

B. GREAT-ER Model Validation Study

1. Execution

Although the broad objectives of the GREAT-ER model validation study were as just described, the input of local knowledge and local collection and analytical capability that the detergent industry/EA/water industry collaboration made possible meant that the detailed objectives were able to be much more extensive than the initial

TABLE 4   Roles of UK GREAT-ER Monitoring Team Members

Role	Provider
Pilot study	Industry/EA
Analysis of all samples	EA analytical laboratory
Sample collection–rivers	EA
Collection of complementary water quality data	EA
Sample collection–effluents	Yorkshire Water
Project design–monitoring expertise	All
Project design–modeling interaction	All
Knowledge of local area for sampling point selection	EA/Yorkshire Water
Hydrological information	CEH Wallingford
STW characterization	Yorkshire Water/EA
Overall UK project leadership	Industry
Report and publication writing	All
Interaction with regulators and communication about GREAT-ER	EA/industry

funding would have allowed. However, using the existing EA sample collection program did limit the model evaluation program to a collection of spot samples of river waters and effluents, as used by the EA in building their water quality models. This was carried out over a two-year period, with site selection and sampling frequency that the EA had previously determined to be satisfactory for their own catchment monitoring and model-building program. The project team was satisfied that the EA sampling regime, based upon a statistical evaluation of significant differences between sites previously identified from knowledge of the local area, would be satisfactory for generating a representative sample set for evaluating the performance of the GREAT-ER model. The monitoring program covered river sites above and below significant inputs from both sewage treatment works and tributaries. A schematic of the selected sites for the river Aire is given in Figure 4. The details of the execution of the main validation study are given in Table 5.

During the GREAT-ER model evaluation program carried out in the United Kingdom, almost 2000 samples from sewage treatment plant effluents and from EA river-monitoring sites in the four Yorkshire catchments (Aire, Calder, Went, and Rother) were collected. The sampling period spanned almost two years. The number of samples collected in each catchment is shown in Table 6.

2. Reporting

The data analysis for LAS, boron, and water quality parameters was carried out at the EA laboratory in Leeds. The data interpretation was carried out primarily by the detergent industry participants, with support from all. The correlation of the water quality results and the LAS and boron results for each site and time of sampling was provided by the EA. A "lesson learned" here was that giving separate sampling codes to the water quality and LAS/boron data made the later link between samples collected at the same site and the same time a major computational task. This will be avoided in the future by devising a sample information retrieval methodology before sample collection begins.

Statistical analysis of the information collected was provided by the detergent industry, with excellent and essential statistical advice provided by WRc, who also advised the EA in this respect. The results were means and standard deviations of the LAS, boron, and water quality parameters at each site. These were compared with the model predictions. It was found necessary to adapt the model to allow it to provide for a naturally present background boron in two of the four UK catchments. Also, one sewage treatment works had to be remodeled to allow for the provision of tertiary sewage treatment in the model, because this had been provided in reality. When this was carried out, the model predictions and the monitoring results agreed within one standard deviation of the measured values, which allowed the model to be evaluated as satisfactory. A comparison of the modeled and predicted results is shown in Figure 5.

C. Trickling-Filter Removal Survey

1. Execution

The trickling-filter removal survey was carried out to determine a representative removal of LAS from trickling-filter works. Details of the study have been reported previously [49]. Although the measurement of daily composite samples of influent

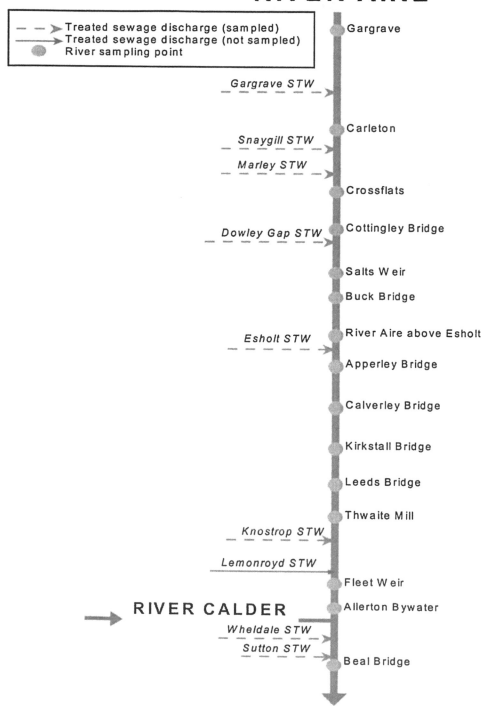

FIG. 4 Schematic of the River Aire showing sewage inputs and sampling points.

TABLE 5 Execution of Monitoring Studies

	Main validation study	TF removal study	In-stream removal study
Identify possible sites	Set by EA monitoring strategy. Number and location of essential sites determined with the aid of the WRc Aardwark program [51].	Yorkshire Water staff identified representative trickling-filter works receiving waste from a range of population sizes, covering a range of industrial influent content.	Three sites were considered initially. The STW of the first site was upgraded before the survey could take place, and the second was rejected due to heavy vegetation and lack of site access at critical points along the stream.
Site selection	Set by EA using Aardwark [51]	Six sites, ranging from small to large works, and with input ranging from purely domestic to 34% trade effluent, were selected. Three sites in the Aire catchment were covered in one week and three sites in the Calder catchment in the next week.	A section of Red Beck below an inadequately performing STW, scheduled for replacement, was selected. The LAS concentration in effluent from this STW was sufficiently high to allow analysis in the lower reaches of the stream. This site had good access at several points along the stream, which allowed adequate time of travel, without input from major tributaries.
Pilot study	Carried out to verify the working of the sample collection, analytical methods, and preservation protocol. Standard addition and recovery studies were carried out as part of the prestudy.	Site previsits established site access procedures and located safe sites to set up the composite samplers. Grab samples taken from each site established the validity of the analytical procedures and established the concentration ranges of the influents' and effluents. Yorkshire Water arranged for detailed data on the flow volume of input sewage to be available for each site and provided works diagrams establishing the treatment flow pathways, volumes of the various treatment components, and operational conditions.	Pilot studies established the high LAS level in the effluent and satisfactorily detectable LAS levels at several points along the stream. Another pilot study determined the site logistics—easy access, siting of sampling points, locations of joining streams. A fluorescent dye-tracer pilot study carried out by University of Sheffield researchers experienced in measuring river flows established approximate travel times and appropriate dye dilution levels. Also, EA staff discussed the survey with residents of the area, to answer any questions they might have.

TABLE 5   Continued

	Main validation study	TF removal study	In-stream removal study
Sampling strategy	Monthly grab samples at each selected site, to fit with normal EA sampling strategy. The number of samples was limited to 20–22 at most sampling sites, although 36 data points would ideally have been obtained to meet EA monitoring quality guidelines (Julian Ellis, personal communication). Occasional sample breakages and financial limitation to a 2-year project determined the final number of samples available and determined the overall statistical quality of the GREAT-ER monitoring program.	Influent raw sewage and works effluent were sampled at all sites. In addition, primary settled sewage effluent was measured at one site. At least 3 days of flow proportional composite sampling were planned for each site. Midweek sampling was preferred, and 12 2-hourly time-proportional samples prepared from four half-hourly shots were collected each day. Appropriate proportions, determined from the works influent flow volume data, were mixed in the laboratory to prepare 24-h flow proportional samples.	At each of five sites, manual collection of fluorescent samples detected the approaching dye (Rhodamine WT) concentration maximum. Samples for analysis for LAS, boron, and water quality parameters were collected throughout the peak, with more frequent sample collection near the peak maximum. Approximately 50 samples were collected for each site, though only the first, last, and a composite of several samples taken about the centroid of the fluorescence peak were analyzed. In addition, two automatic samplers captured solution for analysis at sites that could not be covered by the available manpower. The EA provided a team to measure river flow volumes at three sites, using a transept and flow meters. Sewage flow volumes were provided by Yorkshire Water.
Sample collection	Washed and labeled sample bottles containing preservative for the LAS samples sent by EA analyst with other EA sample bottles to normal EA sampling teams. Filled bottles returned to EA analyst with other bottles containing samples for water quality. Regular auditing of the samples arriving at the laboratory was used to	The large number of clean bottles required were prepared before the campaign by the EA analyst. Special attention was paid to the labeling of the sample bottles, which referred to time and date of collection as well as to the material contained and to the inclusion of suitable preservative. The	Several hundred clean glass bottles containing the appropriate preservative were prepared by the EA analyst before the start of the study. In addition, clean polythene bottles were prepared for collection of samples for water quality analysis. The samples were returned to the EA laboratory for storage at appropriate temperatures on the evening of the survey.

TABLE 5 Continued

	Main validation study	TF removal study	In-stream removal study
	identify any problems with the sampling collection program. Occasional difficulty was found with the collection service in one area, which was remedied by the EA analyst collecting the samples in his own time.	automatic samplers required for composite sample preparation were provided by the industry and by Yorkshire Water. Sewage flow volume information was provided by Yorkshire Water.	
Sample analysis and quality control	Carried out by the EA analyst under Good Laboratory Practice and AQC. Attention paid to storage conditions, including temperature, and time allowed to elapse before analysis. Standard addition and recovery studies were carried out routinely using state-of-the-art analytical methods and instrumentation.	Carried out by the EA analyst under Good Laboratory Practice and AQC. Attention paid to field and laboratory spike recovery determinations, storage conditions, including temperature, and time allowed to elapse before analysis.	Fluorescent samples were analyzed at the stream side, to establish the passage of the dye maximum and to trigger the collection of the other samples. All other samples were returned to the EA laboratory in Leeds for analysis as described for the main validation study.
Environmental data	Samples linked to water quality samples collected at the same time. Difficulties in correlating sampling codes caused some of this information to be lost.	Water quality parameters measured: BOD_5, COD, and ammonia. Special attention was applied to the collection and analysis of the BOD_5 samples, to ensure minimum deterioration of samples	Water temperature, dissolved oxygen, BOD_5, total organic carbon, ammoniacal nitrogen, total organic nitrogen, nitrate, nitrite, suspended solids, boron, and calcium were recorded or measured for each manual sampling site.

and effluent were required for this work, specific analysis of the diel variation of LAS in influent and effluent was also carried out for some works' on some days. Comparison of measured boron found in raw sewage with boron predicted from consumer sales data was also carried out. Table 4 emphasizes the attention paid to various aspects of the study during its design and execution.

Extensive rainfall during some of the study meant that some work had to be repeated, because days when any influent sewage was diverted either to storm tanks or directly to the river due to heavy rainfall were not suitable for relating the market load to the chemical found in influent sewage. Vandalism at one site caused that site to be abandoned and an alternate site selected. These problems effectively doubled the initially planned experimental time to obtain a sufficient number of samples.

TABLE 6 Number of Samples Collected During the GREAT-ER Model Evaluation
Program

	Catchment			
	Aire	Calder	Rother/Don	Went
River sites	15	18	15	8
River samples	273	418	301	179
Effluent sites	8	14	6	7
Effluent samples	205	298	156	139

2. Reporting

The data analysis for LAS, boron, and water quality parameters was carried out at
the EA laboratory in Leeds. The data interpretation was carried out primarily by the
detergent industry participants, with support from all.

For the trickling-filter removal survey, the data interpretation consisted of the
determination of the percentage removal found at each of the six treatment works
studied, which were chosen to cover a range of size and domestic-to-industrial
effluent ratios. The results indicated that, during the study, over 90% of the LAS that
reached the trickling-filter plants was removed during treatment. An analysis of LAS
versus BOD_5 removal showed that, under similar sample preservation regimes, LAS
was removed somewhat more extensively than BOD_5. This part of the survey

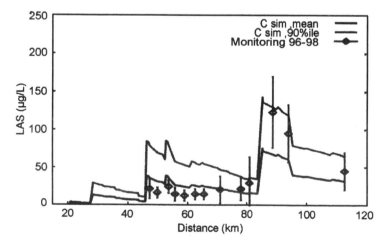

FIG. 5 Comparison of predicted and measured concentrations for LAS in the River Aire,
determined as part of the GREAT-ER project. "Csim, mean" refers to concentrations
simulated using the mean of the flow distribution curve for each site. "Csim, 90% ile" refers to
concentrations simulated using the 10th percentile of the flow distribution curve for each site.
The lines through the points representing the mean of the measured data represent 1 standard
deviation. At the two most up stream measuring points, the mean LAS concentration was
below the 5 $\mu g l^{-1}$ detection limit.

indicated the inadequate nature of current BOD_5 preservation techniques. A part of the survey involving filtered and unfiltered effluent samples determined the percentage of LAS associated with solids in the effluent. Analysis of the diel variation of LAS in influent and effluent was carried out at two sites, and this confirmed the smaller diel variation in the effluent, which was predicted well by a simple model. Finally, the boron data were used to show that the average boron load per person per day that reached the six treatment works during the duration of the survey agreed, within 95% confidence limits, with the amount of boron sold in consumer products during the same time period. This confirmed that boron was an excellent tracer for ingredients in detergent products. Based upon this analysis, it was possible to conclude that during this study, carried out during a warm summer period, over 60% of the LAS sold in washing products was removed in the sewer.

D. LAS In-Stream Removal Survey

1. Execution

The LAS in-stream removal survey was carried out to provide information on the removal of LAS in rivers, for input to the GREAT ER model. Details of the first study, carried out in Red Beck, have been published [50]. Two subsequent surveys were carried out in the river Maun, at different seasons and different river flow conditions. The site requirements were for a sewage treatment works that would provide LAS in effluent to the receiving water, which would then have a flow time of several hours before either receiving further LAS input or experiencing a major dilution event—e.g., from tributaries joining the receiving water. A high initial LAS input from a less efficient sewage treatment facility was required, to keep the amount of LAS in the stream above the minimum level needed for analytical detection, especially in the lower reaches of the stream. Identification of suitable effluents came from the routine effluents monitoring program outlined earlier. A low level of vegetation in the stream was preferred, to minimize removal by adsorption on the surface of aquatic vegetation. The initial site selection proved to be the most difficult part of this survey. Also, considerable attention was given to determining the river flow volume so that biodegradation and other removal processes could be distinguished from a simple dilution effect. Most importantly, another participant, a dye-tracer expert from the University of Sheffield, was persuaded to join the project, to provide time-of-travel information on a "packet" of water moving downstream. The opportunity to carry out later work in the river Maun arose from this collaboration. The highlights of the Red Beck study are summarized in Table 4.

2. Reporting

The data analysis for LAS, boron, and water quality parameters was carried out at the EA laboratory in Leeds. The interpretation of the fluorescent dye data to give a time of travel and to indicate the dilution caused by additional water entering the stream as it progressed down the catchment was carried out by Sheffield University. The data interpretation for other aspects of this survey was carried out primarily by the detergent industry participants, with support from all.

The in-stream removal survey also required several different types of data interpretation. The dye-tracer data were interpreted to determine the time at which the centroid of the dye packet passed each sampling site and thus gave the time of travel down the stream. The dye trace obtained from the Red Beck survey is shown in Figure 6, which also illustrates the results of dispersion on the shape of the dye pulse with passage downstream. The LAS, boron, and water quality samples taken near the centroid were analyzed, and the results were used to determine the trends in the concentrations of the respective determinands.

A major effort was devoted to determining the effect of dilution upon the LAS concentrations measured. It was found that adsorption of dye meant that the dye-tracer method indicated greater dilution than that found by the boron dilution and current flow-meter measurements. The greater number of boron flow measurements made this the recommended method for future work. After correction for dilution, a LAS removal half-life of 2.2 hours was found in the Red Beck study.

Comparison with water quality parameters showed that LAS removal was greater than ammonia removal and was approximately twice the BOD_5 removal rate.

In-stream removal studies were also carried out in a larger river, the Maun, in a stretch downstream of an activated-sludge sewage treatment works. Because of the superior performance of activated-sludge treatment, the Maun receives a lower LAS concentration in the effluent from the sewage works than that received by the Red

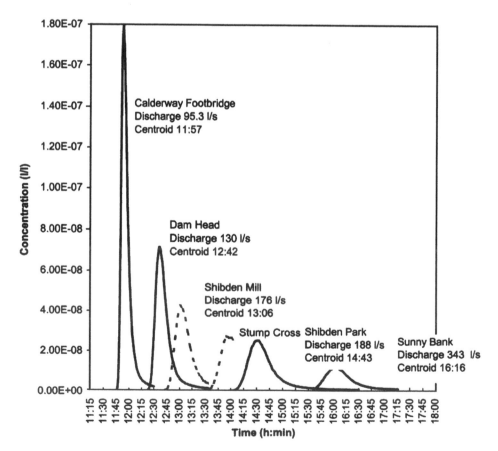

FIG. 6   Dye trace curves obtained at the four manned stations (solid lines) and two automatic stations (dashed lines) in the Red Beck in-stream removal study.

TABLE 7 Half-Lives from "In-Stream" Removal Studies

River	Month	Temperature (°C)	Water depth (m)	LAS concentration (μg L^{-1})	Half-life (h)
Red Beck	March	9	0.21	155	2.66
Maun	February	11	0.3	20	10.4
Maun	June	14	0.6	7	9.6

Beck. The LAS half-lives in the Maun and in the Red beck are compared in Table 7, along with some parameters that may influence the LAS removal rate. Based upon these measurements, a half-life for LAS of 10 hours was taken as the default value for the GREAT-ER model calculations.

The results of the in-stream and trickling-filter removal studies were used to determine the LAS removals used as inputs for the GREAT-ER model. The output for the model in the Aire catchment is shown in Figure 7. This shows the sewage treatment works (black dots) that are the source of the chemical to the river and the color-coded results of dilution, additional chemical input, and in-stream removal on the concentration of the chemical in the river. The removal from household to treatment plant effluent stream has been determined from stochastic treatment of the experimentally determined removal ranges from trickling-filter plants and the previously observed range for activated-sludge plants. The output from this model

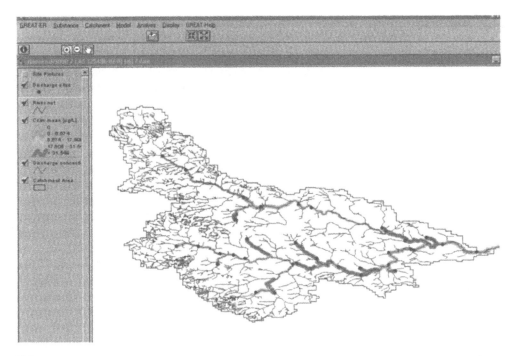

FIG. 7 GREAT-ER output for the Aire catchment, showing the color-coded concentrations of LAS in the rivers that result from inputs from sewage treatment facilities and in-stream dilution and removal processes. The display scale is 1/50,000.

was used in the comparison between predicted and measured river concentration data shown in Figure 5.

E. Data Communication

Several types of data communication have been utilized, with both the detergent industry and the EA encouraging the dissemination of information. The initial communications were internal, with both the detergent industry and the EA informing their own members of the results via internal reports and verbal presentations. As a second stage, several poster and oral presentations were made at external conferences, which described the growing body of information that was emerging from this work. The GREAT-ER organization also held a workshop with participants from the competent authorities and other interested stakeholders, which featured the model validation campaigns as part of the overall model development strategy.

Publication in peer-reviewed journals and reviews is the preferred form of communication for scientific reporting. The overall UK model validation effort has resulted in 11 official GREAT-ER peer-reviewed publications, and two other publications [11,12] have used the monitoring results obtained in the United Kingdom to derive associated risk assessment methodologies. In addition, at least two other publications presenting or otherwise using the UK data are in preparation. The methodology has also been used for other detergent ingredients [52].

F. Conclusion

This case study demonstrates the range of scientific disciplines needed to successfully address the monitoring and modeling requirements for the assessment of environmental exposure to detergent ingredients. It has also illustrated several ways in which the combined use of measured data and a modeling framework can be used to improve environmental exposure assessments of detergents and our understanding of the environmental exposure assessment process. The need for a more representative way to carry out regional exposure assessment in Europe was the original driver for the GREAT-ER project. The model required input information on the removal of the surfactant both from rivers and by trickling-filter treatment works, which was obtained from monitoring surveys. After this input information was obtained, environmental monitoring data from a series of statistically distinct river and effluent locations was required, in order to evaluate the performance of the model for two representative chemicals. Once these monitoring data were obtained and showed that the model performance was satisfactory, then the ability to perform mathematical calculations within the modeling framework enabled concepts such as $PEC_{catchment}$ to be calculated, as a surrogate for $PEC_{regional}$. Comparison of the 90th percentile of the monitoring data used to validate the model with $PEC_{catchment}$ calculated in various ways by using the GREAT-ER model then allowed the evaluation of the concept of using various 90th percentiles to define a reasonable worst-case exposure for detergent ingredients in the catchment. The expansion of models such as GREAT-ER to a wider range of catchments will allow further evaluation of the dependence of the various 90th percentiles on catchment characteristics and will allow further improvements in the determination of environmental concentrations of detergent ingredients for environmental risk assessment.

VI. GLOSSARY

Definitions with an asterisk are from Ref. 53.

Advection	Physical transport or movement of a substance with its medium (air, water, sediment).
Assessment (as distinct from analysis)	The combination of analysis with policy-related activities such as identification of issues and comparison of risks and benefits (as in risk assessment and impact assessment).*
Background concentration	The concentration of a chemical in a medium prior to the action under consideration or the concentration that would have occurred in the absence of the prior action.* For some chemicals, geogenic sources will contribute to the background concentration.
Bioavailability	The ability of a substance to interact with the biosystem of an organism. Systemic bioavailability will depend on the chemical or physical reactivity of the substance and its ability to be absorbed through the gastrointestinal tract, respiratory surface, or skin. It may be locally bioavailable at all these sites.*
Blank	Term used interchangeably with *control.**
Catchment	The area from which rainfall flows into a river.
COD	Chemical oxygen demand: the amount of oxygen consumed by a specified oxidizing agent during chemical oxidation of the matter present in a sample.
Composite sample	Two or more samples or subsamples mixed together in appropriate proportions (based on either time-related or flow-related measurements).
Control	A treatment in a study that duplicates all the conditions of the exposure treatments but contains no test material. The control is used to determine the absence of substance under basic study conditions.
DOC	Dissolved organic carbon.
DOM	Dissolved organic matter.
Emission	Release of a substance from a source, including discharges into the wider environment.*
Environmental compartments	Subdivisions of the environment that may be considered as separate boxes and that are in contact with each other. A simple model would separate the environment into air, water, and soil, with biota, sediment (bottom and suspended), layering of water bodies, and many other refinements being allowed if data to support their inclusion are available.
Exposure	(1) Concentration, amount, or intensity of a particular physical or chemical agent or environmental agent that reaches the target population, organism, organ, tissue, or cell, usually expressed in (numerical) terms of substance concentration, duration, and frequency (for chemical

agents and microorganisms) or intensity (for physical agents such as radiation); (2) process by which a substance becomes available for absorption by the target population, organism, organ, tissue, or cell by any given route.*

Exposure scenarios
Descriptive pathways that portray the specific release pattern of a chemical into the environment and the specific properties of that environment that will influence the fate of the chemical within it.

Fate
Disposition of a material in various environmental compart-ments (e.g., soil or sediment, water, air, biota) as a result of transport, partitioning, transformation, and degradation.*

GIS
A geography-referenced information system

Hazard
The set of inherent properties of a substance or mixture that makes it capable of causing adverse effects in man or to the environment when a particular level of exposure occurs. C.f. *Risk*.*

Local scale
A specific concept in EU Environmental Risk Assessment, which defines a specific or local release site. Further details may be found in the TGDs.

Measurement error
Error that results from an inaccuracy in the measurement of parameter values.*

Mixing zones
Sections of water downstream of a discharge within which the discharge and the receiving water have not yet been fully mixed.

Model
A formal representation of some component of the world or a mathematical function with parameters that can be adjusted so that the function closely describes a set of empirical data. A *mathematical* or *mechanistic* model is usually based on biological, chemical, or physical mechanisms, and its parameters have real-world interpretations. By contrast, *statistical* or *empirical* models are curve-fitted to data where the mathematical function used is selected for its numerical properties. Extrapolation from mechanistic models (e.g., pharmacokinetic equations) usually carries higher confidence than extrapolation using empirical models (e.g., the logistic extrapolation models). A model that can describe the temporal change of a system variable under the influence of an arbitrary "external force" is called a *dynamic* model. To turn a *mass balance* model into a dynamic model, theories are needed to relate the internal processes to the state of the system, expressed, e.g., in terms of concentrations. The elements required to build dynamic models are called *process* models.*

Model error
The element of uncertainty associated with the discrepancy between the model and the real world.*

Monitoring	Long-term, standardized measurement, evaluation, and reporting of specified properties of the environment, in order to define the current state of the environment and to establish environmental trends. Surveys and surveillance are both used to achieve this objective.
Monitoring strategies	Approaches used to achieve the objectives of a monitoring exercise within the constraints of available resource and manpower.
Parameterize	The allocation of values to variables.
Parameter uncertainty	The element of uncertainty associated with estimating model parameters. It may arise from measurement or extrapolation.*
PEC	Predicted environmental concentration: the concentration of a chemical in the environment, predicted on the basis of available information on certain of its properties, its use and discharge patterns, and the quantities involved.*
PEC_{local}	In the EU TGDs, the PEC predicted for the vicinity of a point source, e.g., a production or formulation site, or for a wastewater treatment plant.
$PEC_{regional}$	In the EU TGDs, the PEC averaged over a standard European region of 200 km × 200 km, with almost five times the average European population density and 10% of the total European production capacity.
PNEC	Predicted no-effect concentration: environmental concentration that is regarded as a level below which the balance of probability is that an unacceptable effect will not occur.
Probablistic	The characterization of a property by a distribution function (incorporating distribution shape, standard deviation, mean, median, and other statistical descriptors) rather than by a single value.
Reasonable worst case	Reasonably unfavorable but not unrealistic situation; combining the most adverse environmental circumstances and worst-case release parameters necessarily results in an unrealistic overall worst-case estimation, which is extremely unlikely to occur.*
Receiving water	Surface water (e.g., in a stream, river, or lake) that has received a discharged waste or is about to receive such a waste (e.g., just upstream or up-current from the discharge point).*
Reproducibility	Measure of the extent to which different laboratories obtain the same result with the same reference test compound.*
Risk	The probability of an adverse effect on man or the environment resulting from a given exposure to a chemical or mixture. It is the likelihood that a harmful effect or effects will occur due to exposure to a risk

factor (usually some chemical, physical, or biological agent). Risk is usually expressed as the probability that an adverse effect will occur, i.e., the expected ratio between the number of individuals that would experience an adverse effect in a given time and the total number of individuals exposed to the risk factor.*

Risk management
A decision-making process that entails the consideration of political, social, economic, and engineering information together with risk-related information in order to develop, analyze, and compare the regulatory options and to select the appropriate regulatory response to a potential health or environmental hazard.*

Sampling strategy
A plan, consistent with manpower and analytical constraints, for collecting a sufficient number of discrete samples from a site and for combining these in such a way that the combined sample will be representative of the property of interest at the site.

Spot or grab sample
A discrete sample.

Steady state
The nonequilibrium state of a system in which matter flows in and out at equal rates so that all of the components remain at constant concentrations (dynamic equilibrium). In a chemical reaction, a component is in a steady state if the rate at which the component is being synthesized (produced) is equal to the rate at which it is being degraded (used). In multimedia exposure models and bioaccumulation models it is the state at which the competing rates of input/uptake and output/elimination are equal. An apparent steady state is reached when the concentration of a chemical remains essentially constant over time. Bioconcentration factors are usually measured at steady state.*

Stochastic
Due to, pertaining to, or arising from chance and, hence, involving probability and obeying the laws of probability. The term indicates that the occurrence of effects so named would be random.*

Surveillance
Measurement of environmental characteristics over an extended period of time to determine status or trends in some aspects of environmental quality.*

Survey
A sampling program of a finite duration and for a specific purpose.

TOC
Total organic carbon, often expressed as mg/L in water or kg OC/kg solid. The organic matter content of soil and sediment is often determined by measurement of organic carbon. Typically, about half of all natural organic matter consists of carbon (OC $\approx 0.6 \times$ OM).*

Validation (of a physically based model)
The process of establishing that the predictions of a model agree with the results of an experiment and that the agreement is not fortuitous but is the result of the

correctness and the applicability of the theories that are intended to capture the natural processes the model intends to predict.

Validity (of a physically based model)	A property of a model that requires that it be (1) useful, (2) able to capture natural processes, (3) able to reproduce natural patterns.
Verification	Comparing predicted with measured values and test assumptions and internal logic of the model. This includes (1) scientific verification that the model includes all major and salient process; (2) the processes are formulated correctly; and (3) the model suitably describes observed phenomena for the use intended.
Worst-case assumptions	The most adverse environmental circumstances or the highest possible release parameters. Combining these necessarily results in an unrealistic overall worst-case estimation, which is extremely unlikely to occur.

REFERENCES

1. EC. 2nd edition of the Technical Guidance Document in Support of the Commission Directive 93/67/EEC on Risk Assessment for New Notified Substances and Commission Regulation (EC) No. 1488/94 on Risk Assessment for Existing Substances, Luxembourg, 2003.
2. Feijtel, T.C.J.; Boeije, G.; Matthies, M.; Young, A.; Morris, G.; Gandolfi, C.; Hansen, B.; Fox, K.; Holt, M.; Koch, V.; Schröder, R.; Cassani, G.; Schowanek, D.; Rosenblom, J.; Niessen, H. Development of a geography-referenced regional exposure assessment tool for european rivers—GREAT-ER. Contribution to GREAT-ER #1. Chemosphere 1997, *34*, 2351–2374.
3. Barcelona, M.J. In *Principles of Environmental Sampling*; Keith, L.H., Ed.; American Chemical Society: Washington, DC, 1988; 3–20.
4. Taylor, J.K. In *Principles of Environmental Sampling*; Keith, L.H., Ed.; American Chemical Society: Washington, DC, 1988; 101–107.
5. Natrella, M.G. Experimental Statistics. National Bureau of Standards: U.S. Dept. of Commerce, Washington DC, Handbook 91, 1966.
6. Holt, M.S.; Comber, M.H.I. Principles involved in the organization of well-designed environmental monitoring programs for surfactants. Proc. 4th CESIO World Surfactants Congress, June 3–7, Barcelona 1996; 69–80.
7. Feijtel, T.C.J.; Matthijs, E.; Rottiers, A.; Rijs, G.B.J.; Kiewiet, A.; de Nijs, A. Chemosphere 1995, *30*, 1053–1066.
8. Matthijs, E.; Debaere, G.; Itrich, N.; Masscheleyn, P.; Rottiers, A.; Stalmans, M.; Federle, T. Water Sci. Technol. 1995, *31*, 321–328.
9. Matthijs, E.; Holt, M.S.; Kiewiet, A.; Rijs, G.B.J. Tenside Surf. Deter. 1997, *34*, 238–241.
10. Matthijs, E.; Holt, M.S.; Kiewiet, A.; Rijs, G.B.J. Environ. Toxicol. Chem. 1999, *18*, 2634–2644.
11. Fox, K.K.; Cassani, G.; Facchi, A.; Schröder, F.R.; Poelloth, C.; Holt, M.S. Measured variation in boron loads reaching European sewage treatment works. 2001. Accepted by Chemosphere.
12. Fox, K.K.; Daniel, M.; Morris, G.; Holt, M.S. Sci. Total Environ. 2000, *251/252*, 305–316.
13. ISO 10381-1 Soil quality—Sampling—Part 1: guidance on the design of sampling programs (draft), International Organization for Standardization, Geneva, Switzerland, 1995.

14. British Standards Institution. Code of practice for the identification of potentially contaminated land and its investigation—draft for development, DD 175, 1988.

15. HMSO. The sampling and initial preparation of sewage and waterworks sludges, soils, sediments, plant materials and contaminated wildlife prior to analysis. In *Methods for the Examination of Waters and Associated Materials,* 2nd ed; Her Majesty's Stationary Office: London, 1986.

16. Holt, M.S.; Matthijs, E.; Waters, J. Water Res. 1989, *23,* 749–759.

17. ISO 10381-2 Soil quality—Sampling—Part 2: guidance on sampling techniques (draft), International Organization for Standardization, Geneva, Switzerland, 1995.

18. Barnard, T.E. In The Handbook of Environmental Chemistry—volume 2, Part G, Chemomometrics in Environmental Chemistry—Statistical Methods; Hutzinger, O., Ed.; Springer-Verlag: Berlin, 1995; 1–40.

19. Berg, E.L. Handbook for Sampling and Sample Preservation of Water and Wastewater. U.S. EPA No. 600/4-82-029, Cincinnati, OH, 1982.

20. Plumb, R.H. *Procedure for Handling and Chemical Analysis of Sediment and Water Samples.* U.S. EPA/U.S. Corp. Army Engineer Waterways Experiment Station: Vicksburg, MS, 1981.

21. Greenburg, A.E., Trussel, R.R., Clesceri, L.S., Ed.; Standard Methods for the Examination of Water and Wastewater, 16th Ed. American Public Health Association, American Water Works Association, and Water Pollution Control Federation: Washington, DC, 1985.

22. WRC. *Handbook on the design and interpretation of monitoring programs.* Water Research Center: Medmenham, UK, 1989.

23. Kristensen, P. *Water quality of large rivers. Topic Report 4—Inland Waters.* European Environment Agency: Copenhagen, 1996.

24. Kristensen, P.; Bogestrand, J. *Surface Water Quality Monitoring. Topic Report 2—Inland Waters*; European Environment Agency: Copenhagen, 1996.

25. Nixon, S.C.; Rees, Y.J.; Gendebien, A.; Ashley, S.J. *Requirements for Water Monitoring. Topic Report 1—Inland Waters*; European Environment Agency: Copenhagen, 1996.

26. ECETOC. *Monitoring and modelling of industrial organic chemicals, with particular reference to aquatic risk assessment. Technical Report No 76*; European Center for Ecotoxicology and Toxicology of Chemicals: Brussels, 1999.

27. EPA-600/4-82-029. *Handbook for Sampling and Sample Preservation of Water and Wastewater*; Environmental Protection Agency, Environmental Monitoring and Support Laboratory Office of Research and Development: Cincinnati, OH, 1982.

28. Holt, M.S.; Fox, K.; Griessbach, E.; Johnsen, S.; Kinnunen, J.; Lecloux, A.; Murray-Smith, R.; Peterson, D.R.; Schröder, R.; Silvani, M.; ten Berge, W.F.J.; Toy, R.J.; Feijtel, T.C.M. Chemosphere 2000, *41,* 1799–1808.

29. National Consent Translation Project FR/CL 003. Foundation for Water Research, Marlow, UK, 1994.

30. Holt, M.S.; Waters, J.; Comber, M.H.I.; Armitage, R.; Morris, G.; Newbery, C. Water Res. 1995, *29,* 2063–2070.

31. Di Corcia, A.; Samperi, R.; Bellioni, A.; Marcomini, A.; Zanette, M.; Lemnr, K.; Cavalli, L. Riv. Ital. Delle Sostanze Grasse 1994, *71,* 467–471.

32. Sanchez Leal, J.; Garcia, M.T.; Tomas, R.; Ferrer, J.; Bengoechea, C. Tenside Surf. Deterg. 1994, *31,* 253–256.

33. Schoberl, P.; Klotz, H.; Spilker, R.; Nitschke, L. Tenside Surf. Deterg. 1994, *31,* 243–252.

34. Waters, J.; Feijtel, T.C.J. Chemosphere 1995, *30,* 1939–1956.

35. Hazelton, C. J. CIWEM 1998, *12,* 124–129.

36. Keith, L.H. *Environmental Sampling and Analysis: A Practical Guide*; Lewis: Chelsea MI, 1992.

37. Mascarinec, M.P.; Moody, R.L. In *Principles of Environmental Sampling*; Keith, L.H., Ed.; American Chemical Society: Washington, DC, 1988; 145–155.
38. Kiewiet, A.T.; van der Steer, J.M.D.; Parsons, J.R. Anal. Chem. 67:4409–4415.
39. Syzmanski, A.; Swit, Z. Lukaszewski. Anal. Chim. Acta, 1995; 31–36.
40. Lewis, D.L. In *Principles of Environmental Sampling*; Keith, L.H., Ed.; American Chemical Society: Washington, DC, 1988; 145–155.
41. Black, S.C. In *Principles of Environmental Sampling*; Keith, L.H., Ed.; American Chemical Society: Washington, DC, 1982; 109–118.
42. ISO Guide 25. *General Requirements for the Competence of Calibration and Testing Laboratories*; International Organization for Standardization: Geneva, Switzerland, 1990.
43. UN/ECE. *Guidelines on Water Quality Monitoring and Assessment of Transboundary Rivers*; RIZA: Lelystad, The Netherlands, 1996.
44. OECD. *Report of the 1998 OECD Workshop on Improving the Use of Monitoring Data in the Exposure Assessment of Industrial Chemicals. OECD Series on Testing and Assessment, No 18*; OECD: Paris, 2000.
45. Govaerts, B.; Beck, B.; Lecoutre, E.; le Bally, C.; Vanden Eeckaut P. From monitoring data to regional distributions: a practical methodology applied to water risk assessment. 2001. Submitted to Envirometrics.
46. Proposal for a Council Directive Establishing a Framework for Community Action in the Field of Water Policy. COM (97) 49 final, 1997.
47. Boeije, G.; Wagner, J.-O.; Koormann, F.; Vanrolleghem, P.; Schowanek, D.; Feijtel, T.C.J. Chemosphere 2000, *40*, 255–265.
48. LOIS. Vols 194/195 (1997), 210/211 (1998), and 251/252 (2000). Special issues containing the results of the UK portion of the Land Ocean Interaction Study (LOIS). Sci. Total Environ. 1997, *194/195*. 1998, *210/211*. 2000, *251/252*.
49. Holt, M.S.; Fox, K.K.; Burford, M.; Daniel, M.; Buckland, H. Sci. Total Environ. 1998, *210/211*, 255–269.
50. Fox, K.K.; Holt, M.S.; Daniel, M.; Buckland, H.; Guymer, I. Sci. Total Environ. 2000, *251/252*, 265–275.
51. Aardwark. http://www.wrcplc.co.uk/.
52. Sabaliunas, D.; Webb, S.F.; Hauk, A.; Eckhoff, W.S. Environmental fate of triclosan in the River Aire basin. . 11th Annual SETAC Europe Meeting, May, 6–10, 2001; Madrid.
53. Van Leeuwen, C.J.; Hermens, J. In *Risk Assessment of Chemicals. Research Institute Toxicology*; University Utrecht., Ed.; Kluwer: Dordrecht, Netherlands, 1996.

7

Life Cycle Assessment

A Novel Approach to the Environmental Profile of Detergent Consumer Products

ERWAN SAOUTER   The Procter & Gamble Company, Petit Lancy, Switzerland

GERT VAN HOOF   The Procter & Gamble Company, Strombeek-Bever, Belgium

PETER WHITE   The Procter & Gamble Company, Newcastle upon Tyne, United Kingdom

I. INTRODUCTION

As we move into the early years of the 21st century, we see an increased public, political, and business awareness of the need for sustainable development. As defined by the government of the United Kingdom, "Sustainable development is about ensuring a better quality of life for everyone, now and for generations to come" [1]. Consumer goods companies can contribute to sustainable development by providing products and services that help improve quality of life while ensuring environmental protection, social responsibility, and economic development. Today, it is increasingly accepted that products and services need to be environmentally, socially, and economically sustainable. Environmental sustainability requires that products improve lives, with minimum environmental impact in terms of consumption of both energy and materials and production of emissions and solid wastes. Social sustainability has been less well defined, but it requires corporate social responsibility from the companies concerned. Last but by no means least, economic sustainability means that products must deliver more value to consumers in terms of both performance and price. This part is vital, since only if a product provides high value and is sold will any environmental or social benefits in the product be realized. A more sustainable product needs to be sold as replacement for a less sustainable competitor for there to be any net benefit to the consumer or to society as a whole.

It is also becoming increasingly accepted that any consideration of product sustainability, and in particular the environmental aspects, must be done on a "cradle-to-grave," or life cycle, basis. There are several good reasons for this. First and foremost, looking at the whole life cycle helps prevent "problem shifting." This will help ensure that any environmental improvements made in one part of a life cycle do not produce greater deteriorations at another stage, time, or place in the life cycle. Second, the life cycle approach attempts to allocate all environmental burdens to the particular service provided. This allows the burdens to be compared with the value of the service provided [2]. Third, a life cycle approach allows for optimization of the life cycle. It can show where the largest environmental burdens occur in the life cycle so that these can be addressed and where possible reduced.

Thus a life cycle approach is essential, but there is no single tool that can cover all environmental aspects of a product, across all stages of its life cycle. Tools such as the life cycle inventory (LCI) or broader life cycle assessment (LCA) attempt to assess *some* environmental aspects on a life cycle basis, but there are many areas that these tools fundamentally cannot address. For example, since current LCI and LCA methodology aggregates emissions over time and over all sites in the life cycle, it is not possible to predict concentrations or whether thresholds are likely to be exceeded and thus whether any actual impacts will occur. For this reason neither LCI nor LCA can be used to assess human and environmental safety.

The interactions of a complete product life cycle with the environment are complex, so it is not surprising that they cannot be managed using a single tool. What is needed is an overall environmental management framework that includes a range of different tools. Together these can ensure that all dimensions of a product's life cycle are covered and that no important issues are overlooked.

This chapter looks at the use of life cycle assessment as a tool to assess the environmental profile of detergent consumer products. It starts by looking at how the LCA tool fits into an overall framework for environmental management. It then goes on to look at both how and where it has been used.

II. A FRAMEWORK FOR ENVIRONMENTAL MANAGEMENT

The environmental management framework (EMF) that P & G has developed is shown in Table 1 [3]. The framework shows how the many tools used to manage product life cycles can be integrated, along with experience and judgement, for overall environmental management within a company. Some of these tools have been derived from the environmental sciences; others are more traditional business management tools.

The overall objective of the framework is sustainable environmental management. To achieve this there are four separate elements.

1. *Ensuring human and environmental safety.* For a consumer goods company such as P & G, it is of prime importance to ensure that products, packaging, and operations are safe for consumers, production workers, and the environment. Safety must be ensured through all stages of a product's manufacture, use, and disposal. This requires use of well-established tools such as human and ecological risk assessment [4].
2. *Ensuring regulatory compliance.* There is a need to ensure compliance with all health and environmental regulations and legislation. Specific tools can help in collection of the necessary data.

TABLE 1 P & G Framework for Environmental Management

Goal	Elements of Goal	Available tools
Sustainable development	1. Human and environmental safety	Human health risk assessment (occupational and domestic exposure) Ecological risk assessment (plant-site and consumer releases)
	2. Regulatory compliance	Manufacturing site management system auditing Manufacturing site wastes reporting (e.g., SARA, TRI) Material consumption reporting (e.g., Dutch packaging covenant) New-chemicals testing and registration Product and packaging classification and labeling
	3. Efficient resource use and waste management	Material consumption monitoring and reduction Manufacturing site management system auditing Manufacturing site environmental auditing Auditing major and new suppliers Disposal company auditing Product LCI Eco-design Economic analysis
	4. Addressing societal concerns (i.e., understand/ anticipate and interact)	Understand/anticipate: Opinion surveys Consumer and market research Networking (antenna function) Interact: Information through presentation and publications to key audiences (consumers, employees, retirees, scientists, etc.) Academic, policy, and industry work groups (e.g., think tanks, professional bodies, consultants) Lobbying to influence future policy and regulations Corporate reporting Specific problem solving with others

3. *Ensuring efficient resource use and management of emissions and wastes.* Having established human and environmental safety and regulatory, compliance, there are further needs for sustainability: efficient use of resources and management of wastes. Several tools are of use here, based on materials and energy accounting.
4. *Addressing societal concerns.* It is first necessary to understand and then respond to issues of concern to society at large.

For each element, several different tools can be of use; conversely, some tools serve more than one element. The first two elements can be considered essential for conducting business today. The second two elements are emerging as clear business needs for the future.

Table 1 also shows clearly where LCA fits into the overall environmental management framework. It is particularly useful in the third element, i.e., in ensuring efficient resource use and waste management. Note that this is done after other tools have been used to ensure that products are safe for both humans and the environment.

III. LIFE CYCLE ASSESSMENT

Life cycle assessment (LCA) is a methodology developed to evaluate the mass balance of inputs and outputs of systems and to organize and convert those inputs and outputs into environmental themes or categories relative to resource use, human health, and ecological areas. The quantification of inputs and outputs of a system is called a *life cycle inventory*. At this stage, all emissions are reported on a volume or mass basis (i.e., x kg CO_2, y kg cadmium, z m^3 solid waste). The conversion of these emissions into environmental themes is called *life cycle impact assessment* (LCIA). The themes usually covered in an LCIA include the greenhouse effect (or potential global warming), natural resource depletion, ozone depletion, acidification, eutrophication, human toxicity, and aquatic toxicity. The methods to convert emissions into themes are, for some of them, based on internationally and scientifically accepted approaches; others are still under development (i.e., human toxicity, aquatic toxicity) and need careful evaluation when used.

The mass balance of these inputs and outputs, or life cycle inventory (LCI), spans the entire life cycle of a product: raw material extraction, production of energy and energy feedstocks, manufacture of ingredients (raw materials), processing of the final product (in this case the laundry detergent), transport, packaging, use, and disposal (Fig. 1).

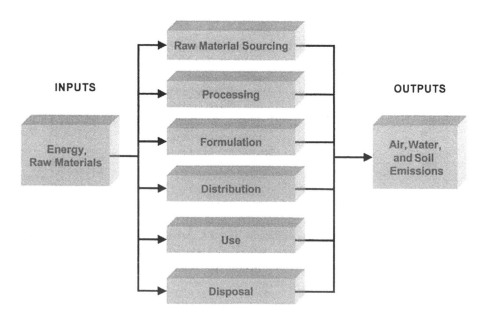

FIG. 1 Life cycle assessment is an accounting of inputs (energy and raw material demand) and outputs (air, water, and soil emissions) through the life cycle of a product, from raw material extraction, production of energy and energy feedstocks, manufacturing of ingredients (raw materials), processing of the final product (formulation), transport, and packaging to use and disposal.

In recent years, LCA methodology has evolved considerably. Since the publication of the Society of Environmental Toxicology and Chemistry code of conduct [5], its standardization has taken place within ISO with the 14040 series [6–9]. These guidelines, by clearly describing the methodology and the domain of application of LCA, are essential documents to be consulted before considering doing an LCA on a product or service. Because LCA can involve hundreds of different processes, the results can often be influenced by the quality of the data collected and the way the boundaries have been defined. It is therefore critical that all assumptions be made very transparent when a decision is based on an LCA study.

Procter and Gamble has implemented life cycle assessment practices in its business activities since the early 1990s. Recently, an important effort has been undertaken to adapt our work process to the recent ISO guidelines so that managers can (1) analyze detergents from a systemwide, functional-unit point of view in a consistent, transparent, and reproducible manner; (2) analyze energy and resource use in the detergent system; (3) analyze various emissions, wastes, and resources using environmental themes; (4) identify what parameters are most likely to be significant to monitor and control; (5) identify opportunities for improving overall system performance; and (6) benchmarking the product over time and reporting progress. A detailed description of our LCA system can be found in Ref. 10.

A. Goal and Scope Definition

The objective of an LCA is to assist scientists, industries, regulators, associations, etc. in (1) identifying opportunities to improve a service or product throughout its life cycle (i.e., cradle to gate); (2) helping the decision-making process (e.g., choose between two production processes); and (3) selecting the relevant parameters to monitor or improve. Depending on the goal of the study, the level of detail of an LCA may vary considerably. If it is for internal and screening purposes, the quality of the data may be less scrutinized (or less important) than if the work is going to be used for external claims (e.g., "Product A is superior to product B"). In the latter, full compliance with ISO guidelines is a must.

B. Functional Unit

The definition of the functional unit is an important step, one related to the definition of the function that a product or service will deliver. When conducting an LCA on laundry detergents, we often report the results on the basis of 1000 wash cycles. All energy and raw materials consumption as well as associated environmental emissions are calculated on the basis of this functional unit. No indications on when and where these 1000 wash cycles occurred are reported in the study. These two elements, however, are essential to understanding and quantifying the impact of the emissions on the receiving environment.

For a cradle-to-gate (up to manufacturing's gates), the results can be reported on a mass basis, e.g., 1 kg of finished product, but it is important to keep in mind that consumers will use the products differently. For example, product A may be used at 70 g per wash and product B at 100 g per wash. For the functional unit of 1000 wash cycles, the totally quantity of detergent needed to fulfill the same function is higher with product B and will likely generate more emissions.

C. Laundry Detergent Database

To construct a full life cycle of laundry detergent, which involves many different processes, the requirement for data is very important. From the making of the raw materials, which can take place in different parts of the world, to the making of the detergent product, which takes place in a few well-identified locations (P & G has only a few plants to manufacture its laundry detergent products for all of Europe). Data on use and disposal are critical to collect to analyze and understand the life cycle impact of a "product." People from the north of Europe do not wash at the same temperature as do people living in the south; the dosage of detergent in the washing machine is also different from north to south due to differences in water hardness. All these will affect the result of the LCA. Comparing washing systems between countries involves more than just comparing two boxes of detergents.

1. Detergent Chemicals

Several inventories related to the production of detergent ingredients have been published since early 1990s. An overview of the inventories incorporated in our database is presented in Ref. 10. Several of these were either compiled or provided by Franklin Associates, Ltd., for the purpose of constructing LCIs for commonly used surfactants [11] or to support LCA research previously conducted for laundry detergents by P & G. Another important source of inventories for detergent ingredients is the work performed by the Swiss Federal Laboratories for Materials Testing and Research, or EMPA [12–15]. Various groups within the European Chemical Industry Council (CEFIC) have commissioned LCI work from EMPA. These LCIs are considered representative of European production processes. An effort to harmonize the data related to energy sources and emission parameters was recently made by EMPA for a number of chemicals based on Buwal 250 [16]. The fraction of these updated inventories needed to construct an LCI of our products was entered into our database.

All energy and raw material consumption and environmental emissions listed in an individual inventory are allocated to a product LCI on a mass basis, according to the specified functional unit.

In practice, this means that emissions associated with the production of oleochemical ingredients (i.e., surfactants) in Malaysia or the Philippines are added to the emissions associated with the making of the detergent in Europe, to the emissions generated at the use phase that may occur in another country, etc. In a comparative LCA (product A versus product B), especially when using impact assessment categories, it is therefore important to understand the boundaries of both systems. Two products may have the same quantity of chemicals emitted to the water compartment, but the real impact on the environment can be very different if the boundary of one product is spatially very limited as compared to that of another product where the boundaries include different locations in the world.

2. Detergent Manufacturing (or Formulation)

The life cycle inventories of the manufacture of traditional, compact powder detergents and compact liquid detergents were published in the *Tenside* journal in 1995 [17]. These data represent an average of different manufacturing sites from different companies located in Germany. The manufacture of a traditional detergent starts from a hot slurry (~100°C), which is spray-dried in a tower (up to 300°C). The first generation of compact detergents, around the 1990s, were produced from a mixed technology involving a spray-drying tower and dry mix. This LCI was compiled by the German Detergent Industry Association (IKW) and follows SETAC guidelines (1993) and the draft ISO guidelines [6] on the LCI. The companies participating in this study represented more than 90% of the German detergent market volume.

3. Packaging Materials

Inventories of packaging raw materials have been published [16,18] and are used as such to construct our life cycle inventories of detergent boxes and refills of bottles. Environmental emissions from the production of packaging are listed in the inventories of the corresponding raw materials (i.e., cardboard, plastic, etc.). Energy consumption and environmental emissions from the disposal of the packaging are not currently included in our system. The quantity of packaging needed for a particular life cycle phase is considered to become a solid waste following its use and is therefore included in the total solid waste, along with others, such as the ashes from energy generation and the sludge from wastewater treatment plants. The energy consumption and emissions associated with the disposal of our packaging will be soon implemented in our system.

4. Use Phase

Once the product is made at P & G manufacturing sites, it is distributed to consumers via a distribution network and retailers. The LCI input data for the transportation between manufacturing sites and retailers can be estimated from the transportation inventories (train, truck, etc.). These inventory data have been published by the Swiss Agency for the Environment [16]. The emissions are reported on the basis of tons of merchandise transported per kilometer.

The data for the consumer use phase required a reasonable understanding of how consumers are using laundry detergents all over Europe: the amount of chemicals used per wash load, what fraction of loads is prewashed, what wash program parameters are used (wash and rinse temperatures, water level, fabric softener use, bleach use, number of rinses, etc.). All of our data are country specific. The amounts of chemicals per wash load are based on the recommended dosage indicated on the package, which is based on local water hardness.

The distribution of wash temperatures (average between liquid and powder detergent) per country was published in the Annual Report of the AISE [19] (Fig. 2).

The energy consumption of the washing process for each wash temperature selection, as reported by the European Washing Machine Manufacturer Association (internal report), varies from country to country, depending on the efficiency of washing used in each country. It varies from 0.6 to 1.14 kWh per wash cycle (country average).

For the use phase, energy and energy feedstock consumption and environmental emissions are calculated by the system based on the percentage of wash loads performed at 30, 40, 60 and >60°C.

5. Disposal Phase (Discharge of the Wash Liquor into the Sewer to Wastewater Treatment Plants)

Once the wash is completed, the wash water is discharged to the sewer. Depending on the country, this wastewater is treated in primary (settler) and/or secondary (activated sludge, trickling filters, etc.), possibly followed by tertiary treatment (sand filtration,

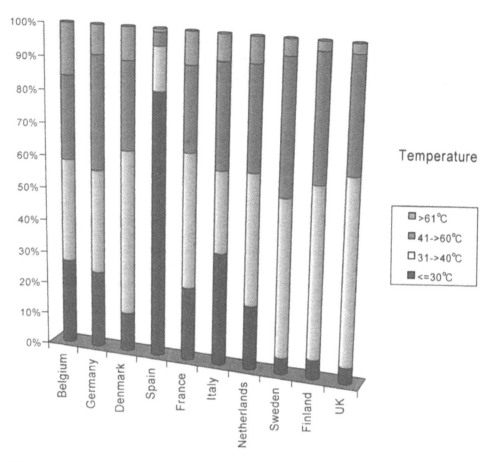

FIG. 2 Wash temperature distribution across European countries.

nutrient removal, etc.), or it is discharged directly into the environment without any type of treatment.

The removal of each ingredient of a detergent by wastewater treatment is taken into account to calculate the amount potentially discharged into the environment. If the ingredient is eliminated through sorption and hence contributes to sludge generation, this is also taken into account in calculating the amount of sludge produced. Two types of elimination are therefore considered: (1) total removal, due to biodegradation and sorption, which is used to calculate the amount of chemical discharge with the effluent, and (2) removal through sorption on solids only, which is used to calculate sludge production.

Emissions due to the waste water treatment plant (WWTP) operation were also accounted for. All emissions are attributable to the generation of energy by on-site burning of gas or oil. The data were derived from emission factors for Dutch WWTPs [20–22].

During the wastewater treatment plant stage, the CO_2 and CH_4 from biodegradation of the detergent are added to the CO_2 emissions from the operation of the plant. The fraction of each ingredient that is not removed by sewage treatment is reported as a waterborne emission. Any counterions that are not removed by sewage treatment are also considered as waterborne emissions.

IV. LIFE CYCLE INVENTORY OF LAUNDRY DETERGENT

When all these data are assembled together to represent the life cycle of the product system (i.e., 1000 wash cycles), taking into account the specific level of each ingredient in the product, the formulation of the detergent, its use in the consumer's house (specific dosage per wash and specific temperature), and its disposal (see Table 2), it is possible to quantify the overall energy and raw material requirements and associated environmental emissions.

Such assembly has been constructed under the Dutch condition to illustrate a typical LCA of a hypothetical laundry detergent. The Netherlands electricity grid is, therefore, used as the basis for the energy calculations. For a few selected inventory endpoints (primary energy, CO_2, biological oxygen demand, and total solid waste) the distribution between the different stages is presented in Figure 3.

Several observations can be made on the results of the LCA.

1. The majority of the energy demand occurs during consumer use (~76%), followed by the manufacturing of the ingredients (~18%). The formulation process, the disposal of spent wash water, and the manufacture of packaging raw materials each constitute only a minor fraction (1.6, 3.5, and 0.7%, respectively) of the total energy use. This distribution reflect the electricity grid of the country, the consumer habits, and to a lesser extent the composition of the detergent. Consumer habits refer here to both the quantity of detergent the consumer uses per wash and the wash temperature selected. The latter has actually the greater influence on the overall energy demand of this type of system. If the same analysis is conducted for two countries where the wash temperature selection is very different (e.g., Germany and

TABLE 2 Chemical Inventories and Other Data Used as Inputs for an LCI of a Hypothetical Granular Laundry Detergent Used in The Netherlands

Selected ingredient inventory		Other data	
AE11-PO (CEFIC/ECOSOL)	2%	Dosage of product per load	100 g
AE7-pc (CEFIC/ECOSOL)	4%	Distribution of wash temperatures in The Netherlands	
LAS-pc (CEFIC/ECOSOL)	7.8%	30°C	20%
Citric acid (FAL 94)	5.2%	40°C	41%
Na-silicate powder (EMPA 1999)	3%	60°C	32%
Zeolite (EMPA 1999)	20.1%	>60°C	8%
Sodium carbonate (FAL 94)	17%	Packaging materials	
Perborate mono (EMPA 1999)	8.7%	Paper woody U B250 (1998)	21.7 g
Perborate tetra (EMPA 1999)	11.5%	Corrugated cardboard	108.2 g
Antifoam S1.2-3522 (Dow 1999)	0.5%	HDPE B250 (barrier)	8.1 g
FWA DAS-1 (EMPA 1999)	0.2%		
Polyacrylate (EMPA 1999)	4%		
Protease (Schmidt 1997)	1.4%		
Sodium sulfate (Buwal 250)	0.4%		
Water	14.2%		

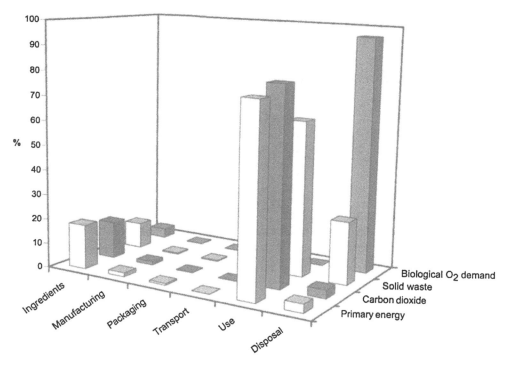

FIG. 3 Life cycle inventory results of selected endpoints (FU: 1000 wash cycles): primary energy, carbon dioxide, solid waste, and biological oxygen demand.

Spain; see Fig. 2), the energy demand as well as CO_2 and solid waste is up to 40% less in Spain than in Germany (Fig. 4). This difference is due solely to a lower wash temperature in Spain (most of the washes are done at 40°C), given that the electricity grids of these two countries are quite similar [16]. It should be noted, however, that Spanish consumers compensate the low-temperature usage by the use of an additional bleach treatment. If these two countries were going to be compared for their laundry practices, this will need to be factored in.

When analyzing energy figures from two LCAs, it is important to keep in mind the meaning of "primary energy" in the LCA. For example, at the use stage, the primary energy reported is the energy consumption of the washing machine corrected for the electricity production efficiency of the specific country. Consequently, an electricity grid based on fossil fuels will result in a much higher total energy consumption at the consumer use stage than a grid based on hydroelectric power, even if the consumption of electricity by the washing machine is the same. This is illustrated in Figure 5 from a recent work aimed at comparing P & G laundry detergents in The Netherlands and Sweden from 1988 to 1998 [23]. While the energy demand of the washing machines in each country is very similar, the total primary energy is 1.29 times higher in the Netherlands than in Sweden. This difference is due to lower energy efficiency in the Netherlands (~31%), where the production is based mainly on coal and gas, whereas in Sweden the efficiency is higher (~41%), with a production based mainly on hydroelectric power and nuclear energy.

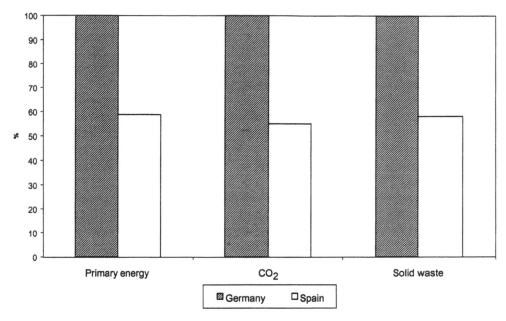

FIG. 4 Effect of lower wash temperature on primary energy, CO_2, and solid waste emissions (average wash temperature: 33°C in Spain versus 49°C in Germany).

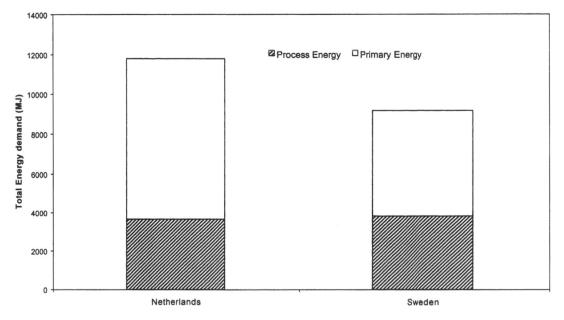

FIG. 5 Primary energy and process energy for 1000 wash cycles performed in Dutch and Swedish consumers' homes.

An important observation is that an evident way to decrease the total energy consumption by laundry detergents is by promoting the use of lower wash temperatures. This would require the design of laundry detergents than can deliver high cleaning performance at lower temperatures. The ideal laundry detergent would deliver good performance at ambient temperatures.

2.   About 63% of total solid waste is generated at the consumer stage and is correlated directly with the production of ashes from energy generation. The two next largest contributors are the disposal (due to sludge from WWTP) and ingredient supply phases (25% and 10%, respectively). Disposal of packaging and the solid waste associated with the manufacturing of the packaging and its raw materials represent merely 2% of the total solid waste production.

3.   Air emissions occur primarily during the supplier and consumer use stages, are proportionally higher during consumer use, and are correlated directly with energy generation from fossil fuels. In countries that derive most of their energy from nuclear or hydroelectric power, such as France and Sweden, air emissions such as CO_2, SO_x, NO_x are expected to be much lower. For other air pollutants, such as dust particles and volatile organic carbons (VOCs), the highest emissions are reported at the supplier stage. For example, in the case of citric acid, 50% of dust particle emissions is associated with the production of the material (data not shown).

4.   Emissions to water have a totally different profile (Fig. 3). Their distribution among the different stages is highly dependent on the chemical considered. More than 96% of biological oxygen demand (BOD) emissions to water occurs during the disposal stage. This is not surprising, since almost 100% of the chemicals used during the wash is discharged to the sewer. These discharges represent a very low percentage of the total BOD originally present in the detergent, because a large fraction (90% on average for BOD) is removed during wastewater treatment. More than 96% of metal emissions occurs during the disposal stage. Most of the discharged metal (99.95%) is sodium (data not shown), which is the counterion for many detergent ingredients and is discharged as a salt (sodium carbonate, sodium silicate, sodium perborate, etc.).

V.   IMPACT ASSESSMENT

Impact assessment methods were developed as tools to broaden the information and context of LCI data, which are largely on mass and energy. In addition to the amounts of resources used and pollutants released, the environmental context can be obtained, e.g., the conversion and aggregation of carbon dioxide, methane, etc., into an overall greenhouse gas release or burden by the system. The fact that an LCIA indicates that certain emissions are associated with certain environmental themes or impact categories does not imply that the detergent actually causes effects. It means, however, that in the course of the detergent's life cycle, emissions are generated that contribute to a pool of similar emissions known to be associated with these environmental themes or impact categories. Used this way, the LCA is the appropriate tool to help determine the extent to which particular product, process, or ingredient's emissions may be associated with a particular impact category.

A variety of impact assessment methods can also be applied to the full inventory table. It is beyond the scope of this chapter to discuss the merits and limitations of impact assessment within an LCA. In this work, the CML92 method is used to

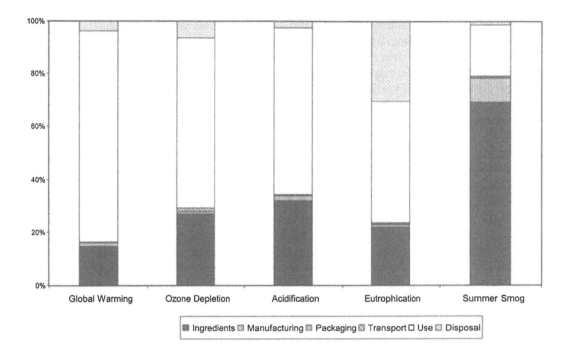

FIG. 6 Results of an impact assessment per 1000 wash cycles for a granular laundry detergent used under Dutch conditions.

analyze the inventory tables [24]. If desired, other impact assessment methods can be used to address specific needs in line with the goal and scope of each individual study.

The result of the LCA identifies the use stage as the largest contributor for the following related impact assessment categories: global warming, acidification, ozone depletion, and eutrophication (Fig. 6). In each case, the major contributing emissions are energy related due to the heating of the water in the washing machine and not to the use of a specific detergent or ingredient.

As can be expected, the disposal stage is also a large contributor to eutrophication impact assessment categories. This is due to the discharge of the fraction of chemicals assumed not to be removed in wastewater treatment plant effluents. Since the detergent analyzed in this study was phosphate free, any eutrophication impact potential would be attributable to nutrients other than phosphate, in addition to any organic matter (BOD) released.

Having 69% of the total score associated with it, the supplier stage was identified as the largest contributor to the summer smog impact assessment category. These emissions consisted almost entirely (98.8%) of volatile organic carbons from process fuel emissions.

VI. CONCLUSION

The database was customized specifically to conduct life cycle inventories and impact assessments of P & G laundry detergents. An LCI can take into account the manufacture of the ingredients and the formulation of the end product, transportation, the packaging operation, consumer use of the product, the disposal of the product by the consumer, and the wastewater treatment plant operation. For the ingredients supply, the manufacturing process, and packaging, the life cycle inventories used are representative of the average European situation. For the consumer use and disposal stages, the LCI relies on country-specific data (wash habits, disposal practices).

The construction of the database allows a rapid, consistent, and transparent execution of an LCI for P & G laundry detergents. It enables the ranking of the life cycle stages in terms of their contributions to certain emissions or impact category. The analysis presented here clearly demonstrates the qualitative conclusion that, from an LCA point of view, the product use stage is the most important one; most of the emissions, and therefore most of the environmental impact scores, are driven by how the consumer uses the detergent. Most of these emissions are generated during the production of energy to heat the water. Quantitatively, the impact of the consumer use stage is very sensitive to variability in consumer habits as well as to the characteristics of the local electricity grid.

The validity of a comparison between two isolated stages of the life cycle of a detergent (i.e., comparing two products only) is more limited with the current database. The number of inventories available is rather small when compared to the multitude of ingredients used in laundry detergents. In some cases more robust links need to be established between the product's ingredients and the available inventories. Some inventories are outdated; some are of limited use because they are representative of only one manufacturing site. Different energy databases have also been used to calculate the ingredient inventories, and the consistency between fuel values, reduction to elementary flow, inclusion and exclusion criteria, allocation, etc. has not been assessed.

REFERENCES

1. DETR. A Better Quality of Life. A strategy for sustainable development in the United Kingdom. Department of the Environment, Transport and the Regions, London, 1999.
2. Hindle, P.; White, P.R.; Minion, K. Long-Range Planning 1993, 2, 36–48.
3. White, P.R.; De Smet, B.; Owens, J.W.; Hindle, P. Resources, Conservation Recycling 1995, 14, 171–184.
4. Beck, L.W.; Maki, A.W.; Artman, N.R.; Wilson, E.R. Regul. Toxicol. Pharmacol. 1981, 1, 19–58.
5. Guidelines for Life Cycle Assessment: A code of Practice; Society of Environmental Toxicology and Chemistry, Pensacola, FL, 1993.
6. ISO 14040. ISO/FDIS/TC207SC514040/1997(E), 1997.
7. ISO 14041. ISO/TC207/SC5/DIS 14041, 1998.
8. ISO 14042. ISO/TC207/SC 5N 97, 1999.
9. ISO 14043. ISO/TC207/SC 5N 104, 1999.
10. Saouter, E.; Van Hoof, G.; Int J Life Cycle Assessment 6 (http://dx.doi.org/10.1065/lca2001.09.065), 2001.
11. FAL. Resource and environmental profile analysis of product and packaging of four granular detergent formulations. Report prepared for The Procter and Gamble Company. Franklin Associates, Ltd. 1994.
12. Fawer, M. Life Cycle Inventory for the Production of Zeolite A for Detergents, Report

No. 234, Swiss Federal Laboratories for Materials Testing and Research (EMPA). St. Gallen, Switzerland, 1996.

13. Fawer, M. Life Cycle Inventory for the Production of Sodium Silicates for Detergents, Report No. 241. Swiss Federal Laboratories for Materials Testing and Research (EMPA). St Gallen, Switzerland, 1997.

14. Boustead, I.; Fawer, M. Ecoprofile of Perborates. CEFIC, av E. Van Nieuwenhuyse 4, Box 1. B. 1160. Brussels, 1998.

15. Boustead, I.; Fawer, M. Ecoprofile of Hydrogen Peroxyde. CEFIC, av E. Van Nieuwenhuyse 4, Box 1. B. 1160. Brussels, 1998.

16. Buwal 250. Okoinventare fur Verpackungen, Schriftenreihe Umwelt 250 Bern, 1996.

17. Franke, M.; Klüppel, H.; Kirchert, K.; Und Olschewski, P. Tenside Surfactant and Deter. 1995, *32*, 508–514.

18. Boustead, I. Ecoprofile of European Plastic Industry, Report 2: Olefin Feedstock Sources. Association of Plastic Manufacturers in Europe (APME). Brussels, 1993.

19. AISE. Annual Review. Association Internationale de la Savonnerie, de la Détergence et des Produits d'Entretien. Brussels, 1998.

20. Ecolabel. Official Journal of the European Communities 95/365/EC. L217: 0014–0030, 1995.

21. Peek, C.J. Rioolwaterzuiverringsinrichtingen. Werkgroep Emissies Servicebedrijven en Produktgebruik. RIVM (Report 773003003), Riza (notanr. 93.046/H1), DGM en GBS, 1993.

22. Peek, C.J.; Mulschlegel, J.H.C.; Versteegh, J.F.M. Waterleidengbedrijven. Werkgroep Emissies Servicebedrijven en Produktgebruik. RIVM (Report 773003003), Riza (notanr. 93.046/H4), DGM en CBS. Bilthoven, 1995, 22.

23. Saouter, E.; Van Hoof, G.; FTCJ, Owens, JW. Int J Life Cycle Assessment 6 (http://dx.doi.org/10.1065/lca2001.06.057.2), 2001.

24. Heijungs, R.; Guinée, J.B.; Huppes, G.; Lankreijer, R.M.; a Udo De Haes, H.; Wegener Sleeswijk, A.; Ansems, A.M.M.; Eggels, P.G.; Van Duin, R.; De Goede, H.P. Environmental life cycle assessment of products. Guide LCA. CML Leiden, Netherlands, 1992.

8

Biodegradability and Toxicity of Surfactants

TORBEN MADSEN   DHI Water and Environment, Hørsholm, Denmark

I. INTRODUCTION

The high volume of surfactants consumed with household detergents and industrial and institutional cleaning products and the numerous applications where surfactants are used to enhance product and process performance imply that these substances are widespread in the environment. The complex chemistry of many surfactants and their possible interactions with humic materials and minerals further makes it necessary to obtain an adequate understanding of the fate and effects of surfactants in relevant environmental compartments. This chapter presents a compilation of data describing the biodegradability and aquatic toxicity of the four main groups of surface-active substances, i.e., anionic, nonionic, cationic, and amphoteric surfactants. Both high-production-volume (HPV) and non-HPV surfactants are included. The consumption of some of the major surfactants on the market, i.e., linear alkylbenzene sulfonates (LAS) and alcohol derivatives constituted by alkyl sulfates (AS), alkyl ether sulfates (AES), and alcohol ethoxylates (AE), is presented in Chapter 14 of this volume.

Most of the major surfactants produced today are readily biodegradable, which implies that these surfactants are assumed to degrade rapidly in aerobic environments, provided they are bioavailable. The elimination of readily biodegradable surfactants during wastewater treatment normally leads to concentrations in the effluent below 200 µg/L and resulting predicted environmental concentrations (PEC_{water}) in the aquatic recipient below 20 µg/L (Table 1). The measured environmental concentrations of surfactants in Table 1 were taken from the extensive monitoring program carried out in the Netherlands [1]. The estimated values in Table 1 were based on the assumptions that sewage influent concentrations are correlated with the volumes used of the surfactants (this was supported by values for surfactant consumption [2]), and that the removal efficiencies during sewage treatment are the same for readily degradable surfactants.

Because of the sorption of surfactants to organic solids, a minor part of the surfactants is conveyed to aquatic sediments, which are normally dominated by

TABLE 1 Estimated Surfactant Concentrations in Wastewater and Aquatic Recipients in Europe

Surfactant	Sewage influent (mg/L)	Treated sewage effluent (µg/L)	PEC_{water} (µg/L)[e]
C_{12-15} AS	0.6 (0.1–1.3)[a]	5.7 (1.2–12)[a]	0.6
C_{12-15} AES	3.2 (1.2–6.0)[a]	6.5 (3.0–12)[a]	0.7
LAS	5.2 (3.4–8.9)[a]	39 (19–71)[a]	4
SAS	0.8[b]	5.9[b]	0.6
Soap	28 (14–45)[a]	174 (91–365)[a]	17
C_{12-15} AE	3.0 (1.6–4.7)[a]	6.2 (2.2–13)[a]	0.6
APG/glucose amides	0.4[c]	0.8[c]	0.1
Esterquats	0.9[d]	1.9[d]	0.2

[a] Average measured concentrations in effluents from seven municipal sewage treatment plants in the Netherlands. Values in parentheses indicate measured concentration ranges. *Source*: Ref. 1.

[b] Estimate based on the assumptions that total consumption of SAS corresponds to 15% of that of LAS and that the removal percentages of SAS and LAS during wastewater treatment are the same.

[c] Estimate based on the assumptions that total consumption of APG and glucose amides corresponds to 10–15% of that of C_{12-15} AE and that the removal percentages of these surfactants during wastewater treatment are the same.

[d] Estimate based on the assumptions that total consumption of esterquats corresponds to 30% of that of C_{12-15} AE and that the removal percentages of these surfactants during wastewater treatment are the same.

[e] Predicted environmental concentrations (PEC_{water}) for emissions of treated effluents were estimated from average effluent concentrations by assuming an initial 10× dilution in the aquatic recipient.

anoxic conditions. Surfactants that degrade anaerobically have achieved attention among manufacturers due to authority and customer requests for detergents with low environmental impact. The inclusion of anaerobic biodegradability in the criteria for the European "Flower" Eco-label is one example of the new requirements' application to detergents. The information on the anaerobic biodegradability of surfactants is relatively scarce, and a compilation of results of biodegradation studies under anoxic conditions was therefore given priority in the present chapter. Surfactants that are not degraded in the absence of molecular oxygen accumulate in sewage sludge digesters, which implies that these surfactants enter the terrestrial environment when the sludge is used as a fertilizer on agricultural soil. The final part of the chapter reviews the current data on the fate and effects of linear alkylbenzene sulfonates in sludge-amended soils, which have been extensively studied by researchers in Denmark.

II. ANIONIC SURFACTANTS

The largest volume of anionic surfactants is used in laundry detergents, cleaning and dishwashing agents, and personal care products. By volume, the most important groups of anionic surfactants are alkyl sulfates (AS), alkyl ether sulfates (AES), linear alkylbenzene sulfonates (LAS), and fatty acid soaps. Anionic surfactants are negatively charged in aqueous solutions due to the presence of a sulfonate, sulfate, phosphate, or carboxylate group. Commercial anionic surfactants contain mixtures of homologues with different alkyl chain lengths. For some types of surfactants, the existence of different isomers also adds to the complex nature of commercial anionic surfactants. Representative structures of anionic surfactants are shown in Figure 1.

A. Alkyl Sulfates and Alkyl Ether Sulfates

1. Aerobic Biodegradability

The biodegradation of AS is initiated by a hydrolytic cleavage of the sulfate ester bond, resulting in inorganic sulfate and fatty alcohol. The fatty alcohol is oxidized to fatty acids that are degraded by β-oxidation [3]. Primary biodegradation of essentially linear AS is usually rapid and occurs within a few days, but branching considerably reduces the rate of primary degradation. For example, a level of 95% removal of methylene blue–active substances (MBAS) was attained within 1, 3, and 12 days, respectively, for different AS containing 99%, 50%, and less than 3% linear material [4]. The ultimate biodegradability of linear and monobranched AS complies with the OECD criteria for ready biodegradability, but some multibranched AS do

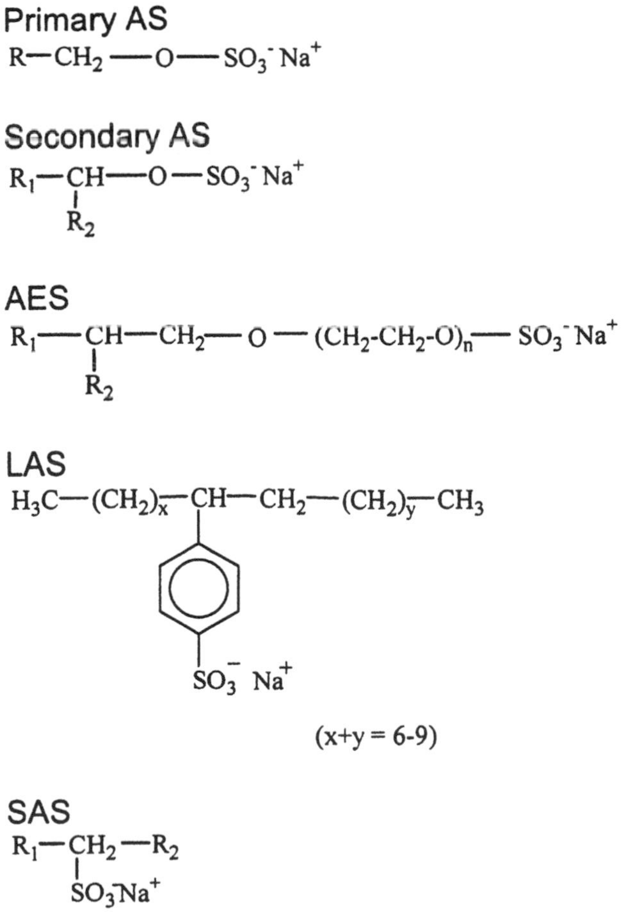

Primary AS

$R-CH_2-O-SO_3^- Na^+$

Secondary AS

$R_1-CH-O-SO_3^- Na^+$
$\quad\;\; |$
$\quad\;\; R_2$

AES

$R_1-CH-CH_2-O-(CH_2\text{-}CH_2\text{-}O)_n-SO_3^- Na^+$
$\quad\;\;\; |$
$\quad\;\;\; R_2$

LAS

$H_3C-(CH_2)_x-CH-CH_2-(CH_2)_y-CH_3$

$SO_3^- Na^+$

$(x+y = 6\text{-}9)$

SAS

$R_1-CH_2-R_2$
$\quad\;\; |$
$\quad\;\; SO_3 Na^+$

FIG. 1 Structures of anionic surfactants.

not reach the specified pass levels and cannot be characterized as readily biodegradable. The effect of branching was examined in a study in which the ready biodegradability of sulfate derivatives of commercially available alcohols was determined in the closed-bottle or manometric respirometry test [5]. Three mono-branched C_{12} AS (2-methyl-, 2-butyl-, and 2-hexyl-) reached 85–100% of the theoretical oxygen demand (ThOD) in the closed-bottle test [5]. The same authors showed that the biodegradability in the manometric respirometry test attained 50% ThOD for a C_{13} AS butylene trimer and 36% ThOD for a C_{13} propylene tetramer with 10% quaternary carbons (Table 2).

The first step in the biodegradation of AES is usually the cleavage of an ether bond, leaving a fatty alcohol or an alcohol ethoxylate and ethylene glycol sulfates. The alcohol is degraded by ω,β-oxidation, whereas the ethylene glycol sulfates are eliminated by cleavage of C_2 units and desulfation [3]. Primary degradation of linear and moderately branched AES may reach a level of 90–100% within a few days, but a less extensive removal of MBAS may be seen for highly branched structures like, e.g., propylene tetramer AES [4]. Apparently, the ultimate biodegradability of AES does not depend on the length of the alkyl chain or the number of EO, and both linear and moderately branched surfactants are normally readily biodegradable in the standardized OECD tests (Table 2).

2. Anaerobic Biodegradability

Linear AS and moderately branched AS are ultimately biodegradable in screening tests conducted in the absence of molecular oxygen. For example, the ultimate biodegradation of a linear C_{12-14} AS reached 20% of the theoretical gas production (ThGP) after 28 days and increased to 85% of ThGP after 56 days in the ISO 11734 screening test with digested sludge [7]. In another study, using anaerobically digested sludge [8], a mixture of 80% linear and 20% monobranched C_{14-15} AS approached 100% biodegradation, which confirms that both the linear and branched AS were ultimately degraded (Table 3). Anaerobic biodegradability screening tests with AES

TABLE 2 Ultimate Aerobic, Ready Biodegradability of AS and AES

Surfactant	Method	Result	Ref.
C_{12-14} AS	Closed bottle, 28 d	90–94% ThOD	3
C_{12} AS [b][a,b]	Closed bottle, 28 d	85–100% ThOD	5
C_{13} AS [b][c]	Manometric respirometry, 28 d	50% ThOD	5
C_{13} AS [b][d]	Manometric respirometry, 28 d	36% ThOD	5
C_{15} AS [b][e]	Closed bottle, 28 d	0% ThOD	5
C_{16-18} AS	Closed bottle, 28 d	77% ThOD	3
AES/oxo-AES[f]	Closed bottle, 28 d	58–100% ThOD	6
AES/oxo-AES[f]	CO_2 evolution, 28 d	65–83% $ThCO_2$	6

[a] [b], branched alcohol.
[b] Various monobranced AS (2-methyl-, 2-butyl-, and 2-hexyl-).
[c] Butylene trimer, three internal CH_3 groups.
[d] Propylene tetramer, four internal CH_3 groups, 10% quaternary carbons.
[e] 2-Butyl-5,7,7-trimethyl octanol sulfate; three internal CH_3 groups, quaternary carbon.
[f] Mixture of C_{12-14} AE$_2$S and oxo-C_{12-15} AE$_3$S.

TABLE 3 Ultimate Anaerobic Biodegradability of AS and AES in Digested Sludge

Surfactant	Test conditions and incubation[a]	Result	Ref.
C_{12-14} AS	1 g SS/L, 35°C, 56 d[e]	85% ThGP	7
C_{14-15} AS [b][b]	1 g SS/L, 35°C, 40–50 d	65; 78% $ThCH_4$	8
$[^{14}C]C_{14}$ AS[c]	24–29 g SS/L, 35°C, 15 d	80%; $^{14}CO_2 + {}^{14}CH_4$	9
C_{18} AS	35°C, 56 d[e]	88% ThGP	10
$[^{14}C]C_{14}$ AE_3S[d]	24–29 g SS/L, 35°C, 17 d	88%; $^{14}CO_2 + {}^{14}CH_4$	9

[a] SS, suspended solids dry weight.
[b] [b], branched alcohol; 80% linear and 20% 2-alkyl-branched material.
[c] $[1-{}^{14}C]C_{14}$ AS.
[d] ^{14}C-[1,3-ethoxylate]C_{14} AE_3S.
[e] ECETOC/ISO 11734 test.

have indicated that these surfactants are inhibitory to anaerobic bacteria at the high initial concentration, which is required to determine ultimate biodegradability by measurements of gas production [11]. Alkyl ether sulfates may, however, degrade ultimately under anoxic conditions in experiments applying a higher inoculum to test substrate ratio [9]. By using 24–29 g of digested sludge (dry weight) per liter of medium, the recovery of $^{14}CO_2$ and $^{14}CH_4$ corresponded to 88% of the added ^{14}C-labeled C_{14} AE_3S after 17 days of incubation at 35°C (Table 3).

3. Toxicity to Aquatic Organisms

The acute toxicity of AS to crustaceans and fish increases with increasing alkyl chain length. This may be illustrated by the median lethal concentrations (LC50) obtained in experiments with rice fish (*Oryzias latipes*) as the LC50 was 51 mg/L for C_{12}, 5.9 mg/L for C_{14}, and 0.50 mg/L for C_{16} [12]. Typical median effect concentrations (EC50) or LC50 of commercial AS (with alkyl chain lengths from C_{12} to C_{16}) range between 1 and 30 mg/L for algae, between 0.5 and 80 mg/L for crustaceans, and between 2.5 and 51 mg/L for fish (Table 4). The marine crustacean *Acartia tonsa* is relatively sensitive to surfactants, as an LC50 value of 0.55 mg/L (96 h) was reported for lauryl sulfate [13]. Experiments in which *Ceriodaphnia dubia* was exposed to different AS under continuous-flow conditions showed that the chronic 7-d NOECs on reproduction were 0.88 mg/L for C_{12} AS, <0.06 mg/L for C_{14} AS, 0.23 mg/L for C_{15} AS, 0.20 mg/L for C_{16} AS, and 0.60 mg/L for C_{18} AS [18].

The relations between structure and aquatic toxicity of AES are complex, but in general changes in the number of ethylene oxides (ethoxylates, EO) appear to have a larger effect on the toxicity than changes in the alkyl chain length. The toxicity tends to decrease with increasing numbers of EO for AES with alkyl chains of less than C_{16}, which includes the most frequently used chain lengths in commercial products. Lethal concentrations (LC50) of 0.9 and 0.8 mg/L, respectively, have been reported for C_{16} AE_4S and C_{16} AE_6S in studies with fathead minnow, *Pimephales promelas* [4]. Representative EC50/LC50 values of C_{12-15} $AE_{1-3}S$ in short-term tests are between 4 and 65 mg/L for algae, between 1 and 50 mg/L for crustaceans, and between 1 and 28 mg/L for fish. The chronic toxicity of a C_{13-15} $AE_{2.25}S$ (average alkyl chain length, $C_{13.67}$) was examined during 21 days of continuous-flow exposure of *Daphnia magna* and during 30 days of exposure of *P. promelas* [16]. The no-observed-effect concentrations (NOECs) for the reproduction of *D. magna*

TABLE 4 Aquatic Toxicity of AS and AES

Organism	Surfactant	EC/LC50 (mg/L)[a]	Time	Ref.
Algae	C_{12-14} AS	1–30	—	7
Prorocentrum minimum	Lauryl sulfate	1.32 (0.22–3.42)[b]	—	13
Skeletonema costatum	Lauryl sulfate	2.33 (1.57–3.54)[b]	—	13
Pseudokirchneriella subcapitata	C_{12} AS	27	72 h	14
Algae	C_{12-15} $AE_{1-3}S$	4–65	—	7
Pseudokirchneriella subcapitata	C_{12-14} AES	32	72 h	14
Crustaceans	C_{12-14} AS	0.5–80	—	7
Daphnia magna	C_{12} AS	1.8	48 h	15
Acartia tonsa	Lauryl sulfate	0.55	96 h	13
Crustaceans	C_{12-15} $AE_{1-3}S$	1–50	—	7
Daphnia magna	$C_{13-15}AE_{2.25}S$	1.17 (0.82–1.66)[b]	96 h	16
Daphnia magna	$C_{13-15}AE_{2.25}S$	NOEC: 0.27	21 d	16
Fish	C_{12-14} AS	2.5–51	—	7
Cyprinodon variegatus	C_{12} AS	4.1 (3.83–4.47)[b]	96 h	13
Menida menida	C_{12} AS	2.8 (2.55–2.98)[b]	96 h	13
Fish	C_{12-15} $AE_{1-3}S$	1–28	—	7
Oncorhynchus mykiss	C_{12-13} AE_2S	28	96 h	4
Salmo trutta	C_{12-15} AE_3S	1.0–2.5	96 h	17
Pimephales promelas	C_{16} AE_4S	0.9	24 h	4
Pimephales promelas	C_{16} AE_6S	0.8	24 h	4
Pimephales promelas	C_{13-15} $AE_{2.25}S$	NOEC: 0.1	30 d	16

[a] Median effect concentration (EC50), median lethal concentration (LC50), or no observed effect concentration (NOEC; when indicated).
[b] Parentheses indicate 95% confidence limits.

and *P. promelas* were 0.27 and 0.10 mg/L, respectively (Table 4). The NOEC values were defined as the highest measured test concentrations causing no observable effects relative to control data. The median effect concentrations (21-d EC50) to the survival and reproduction of *D. magna* in this study were 0.74 mg/L (95% confidence limits, 0.56–0.94 mg/L) and 0.37 mg/L (95% confidence limits, 0.22–0.54 mg/L), respectively [16].

B. Alkylbenzene-, Alkane-, and Olefine Sulfonates

1. Aerobic Biodegradability

The initial step in the aerobic biodegradation of LAS is an ω-oxidation of the terminal methyl group of the alkyl chain, whereupon further degradation proceeds by β-oxidation and results in the formation of short-chained sulfophenyl carboxylic acids. The cleavage of the aromatic ring apparently requires molecular oxygen and occurs before desulfonation [3]. The primary biodegradation of LAS has been confirmed in standardized tests, in which 93–97% removal of MBAS was observed [6]. The ultimate aerobic biodegradability of LAS complies with the OECD criteria for ready biodegradability as between 73% and 84% removal of dissolved organic carbon (DOC) was reached in the modified OECD screening test [6]. Also secondary alkane sulfonates (SAS) and α-olefine sulfonates (AOS) are rapidly transformed by aerobic bacteria as the MBAS removal may exceed 90% within a few days [4]. In a continuous activated sludge test (CAS test), the ultimate biodegradability of a ^{14}C-labeled C_{17} SAS resulted in the formation of 47% of the added ^{14}C as ^{14}CO$_2$, whereas 25% was associated with the sludge biomass [3]. In the closed-bottle test, the ultimate biodegradability of a C_{13-17} SAS reached 99% of ThOD [19], whereas a C_{14-18} AOS attained 85% of ThOD [6], which implies that both of these surfactants may be characterized as readily biodegradable.

2. Anaerobic Biodegradability

Sulfonates are generally believed to resist ultimate biodegradation in the absence of molecular oxygen. Until now, only primary anaerobic biodegradation of LAS has been confirmed in some microbial communities. Denger and Cook [20] showed that commercial LAS and a pure C_{12}-3 LAS were desulfonated under sulfur-limited anoxic conditions. Anaerobic transformation of LAS was also observed in experiments conducted with continuously stirred tank reactors with digested sludge, operated at 37°C, in which the removal of a pure C_{12}-2 LAS corresponded to between 20% and 25% of the initial concentration [21]. In a study by Wagener and Schink [22], no biodegradation of a sodium-dodecylsulfonate (SAS) was noted in experiments with either digested sludge or creek sludge incubated under anoxic conditions.

3. Toxicity to Aquatic Organisms

Due to the large number of aquatic species that have been used in toxicity tests with LAS, the reported effect concentrations cover a wide range. The effect concentrations below originate from studies of LAS with alkyl chain lengths between C_{10} and C_{13} as for they are representative of the so-called European LAS cut, which has an average alkyl carbon number of 11.6–11.8. The EC50 or LC50 of LAS in short-term tests with a duration of 96 h or less may be within the ranges of 0.9–270 mg/L for algae, 2–30 mg/L for crustaceans, and 0.4–58 mg/L for fish (Table 5). The lowest effect concentrations in these tests include a 96-h EC50 of 0.9 mg/L ($C_{11.8}$ LAS) for the marine alga *Microcystis aeruginosa* [23] and a 48-h LC50 of 0.4 mg/L (C_{13} homologues = 99%) for fathead minnow, *P. promelas* [26]. Extension of the duration of the exposure may lead to lower effect concentrations, as exemplified by an 8-d LC50 of 0.54 mg/L (C_{10-13} LAS; average alkyl chain length, $C_{11.6}$) for the marine crustacean *A. tonsa* [25] and a 7-d LC50 of 0.71 mg/L ($C_{11.9}$ LAS) for *P. promelas* [27]. The chronic NOEC values found in long-term tests with *D. magna* were determined to be 1.18 and 0.57 mg/L for two C_{10-14} LAS mixtures with average alkyl chain lengths of $C_{11.8}$ and $C_{13.3}$, respectively (Ref. 16; Table 5). The NOEC for *A. tonsa* in the study by Kusk and Petersen [25] can be estimated as being equivalent to the 8-d EC10, which was found at 0.27 mg/L (inhibition of development rate) and 0.28 mg/L (survival). Studies with the aforementioned C_{10-14} LAS resulted in fish long-term NOECs at 0.90 mg/L (average alkyl chain length, $C_{11.8}$) and 0.15 mg/L (average alkyl chain length, $C_{13.3}$) during 30 days of exposure of *P. promelas* (Ref. 16; Table 5). The acute aquatic toxicity of narrow cuts of SAS to *D. magna* is illustrated by 24-h EC50 values of 319 mg/L for $C_{10.3}$, 133 mg/L for $C_{11.2}$, 111 mg/L for C_{14}, 34.2 mg/L for C_{15}, 30.1 mg/L for C_{16}, 12.3 mg/L for C_{17}, 3.31 mg/L for $C_{18.9}$, and 6.30 mg/L for $C_{20.7}$ [28]. The lowest short-term-effect concentrations of SAS include a 96-h LC50 of 1.3 mg/L for C_{18} SAS in tests with bluegill sunfish,

TABLE 5 Aquatic Toxicity of LAS

Organism	LAS homologues	EC/LC50 (mg/L)[a]	Time	Ref.
Algae	C_{10-13}	0.9–270	—	7
Scenedesmus subspicatus	C_{12}	48	72 h	14
Scenedesmus subspicatus	C_{12}	NOEC: 18	72 h	14
Microcystis aeruginosa	$C_{11.8}$	0.9	96 h	23
Microcystis aeruginosa	C_{13}	5.0	96 h	23
Navicula pelliculosa	C_{13}	1.4	96 h	23
Dunaliella sp.	C_{12}	3.3 (3.0–3.7)[b]	24 h	24
Chlorella pyrenoidosa	C_{12}	29 (28–31)[b]	96 h	24
Crustaceans	C_{10-13}	2–30	—	7
Daphnia magna	C_{10-14} (avg 11.8)	3.94 (2.87–6.83)[b]	96 h	16
Daphnia magna	C_{10-14} (avg 11.8)	NOEC: 1.18	21 d	16
Daphnia magna	C_{10-14} (avg 13.3)	2.19 (1.85–2.82)[b]	96 h	16
Daphnia magna	C_{10-14} (avg 13.3)	NOEC: 0.57	21 d	16
Acartia tonsa	C_{10-13} (avg 11.6)	2.1 (1.2–3.1)[b]	48 h	25
Acartia tonsa	C_{10-13} (avg 11.6)	NOEC: 0.27[c]	8 d	25
Fish	C_{10-13}	0.4–58	—	7
Pimephales promelas	C_{10-14} (avg 11.8)	NOEC: 0.90	30 d	16
Pimephales promelas	C_{10-14} (avg 13.3)	NOEC: 0.15	30 d	16
Pimephales promelas	C_{12}	4.7	48 h	26
Pimephales promelas	C_{13}	0.4	48 h	26

[a] Median effect concentration (EC50), median lethal concentration (LC50), or no-observed-effect concentration (NOEC, when indicated).
[b] Parentheses indicate 95% confidence limits.
[c] Reference states 0.27 mg/L as the EC10, which, however, could hardly be distinguished from NOEC.

Lepomis macrochirus, whereas slightly lower 96-h LC50 values of 0.5 mg/L were observed for a C_{16-18} AOS in tests with harlequin fish, *Rasbora heteromorpha*, and brown trout, *Salmo trutta* (Table 6).

C. Fatty Acid Soaps

1. Aerobic Biodegradability

Fatty acid soaps are considered rapidly primarily degraded, although the primary biodegradability of soaps cannot be determined by measurements of MBAS removal. The ultimate biodegradation of fatty acid soaps generally exceeds the pass criteria in tests for ready biodegradability. For example, the ultimate biodegradability attained 100% for sodium laurate (C_{12}), ≥90% for sodium palm kernel soap (C_{8-14}), 100% for sodium oleate (C_{18}), 100% for sodium tallow soap (C_{16-18}), >85% for sodium stearate (C_{18}), and >75% for sodium behenate (C_{22}) during 10 days in a BOD test [3].

2. Anaerobic Biodegradability

The anaerobic biodegradability of fatty acid soaps has been examined in laboratory screening tests with digested sludge and aquatic sediments. The ultimate anaerobic

TABLE 6 Aquatic Toxicity of SAS and AOS

Organism	Surfactant	EC/LC50 (mg/L)[a]	Time	Ref.
Algae				
Chlamydomonas variabilis	C_{14} SAS	32.4[b]	4 h	28
Chlamydomonas variabilis	C_{17} SAS	3.93[b]	4 h	28
Pseudokirchneriella subcapitata	C_{14-18} AOS	45	48 h	29
Crustaceans				
Daphnia magna	C_{13-18} SAS	8.7–13.5	—	6
Daphnia magna	C_{15-18} SAS	0.7–6	—	4
Daphnia magna	C_{14-16} AOS	16.6	—	4
Daphnia magna	C_{16-18} AOS	7.7	—	4
Fish	C_{13-18} SAS	1.3–144	—	7
Phoxinus phoxinus	C_{14} SAS	34.5	24 h	28
Phoxinus phoxinus	C_{16} SAS	3.11	24 h	28
Lepomis macrochirus	C_{16} SAS	4.6	96 h	4
Lepomis macrochirus	C_{18} SAS	1.3	96 h	4
Idus idus melanotus	C_{14-16} AOS	3.4	96 h	17
Idus idus melanotus	C_{16-18} AOS	0.9	96 h	17
Rasbora heteromorpha	C_{14-16} AOS	3.3	96 h	17
Rasbora heteromorpha	C_{16-18} AOS	0.5	96 h	17
Salmo trutta	C_{14-16} AOS	2.5–5	96 h	17
Salmo trutta	C_{16-18} AOS	0.5	96 h	17
Oncorhynchus mykiss	C_{14-16} AOS	5.1	24 h	4
Oncorhynchus mykiss	C_{16-18} AOS	0.8	24 h	4

[a] Median effect concentration (EC50), median lethal concentration (LC50).
[b] The effect on the motility of the cells was estimated.

TABLE 7 Ultimate Anaerobic Biodegradability of Fatty Acid Soaps

Surfactant	Test conditions and incubation[a]	Result	Ref.
Digested sludge			
Na-cocoate (C_{8-18})	0.15 g SS/L, 35°C, 56 d	93% ThGP	30
K-cocoate (C_{12-16})	1 g SS/L, 35°C, 56 d[c]	99% ThGP	7
Na-palmitate, C_{16}	35°C, 28 d[c]	79–94% ThGP	10
Palmitic acid, C_{16}[b]	Approx. 25 g SS/L, 35°C, 28 d	97%; $^{14}CO_2 + {}^{14}CH_4$	31
Laurate	Semicontinuous fermentor	95% ThGP	3
Oleate and palm kernel soap	Semicontinuous fermentor	70% ThGP	3
Tallow soap	Semicontinuous fermentor	60% ThGP	3
Behenate	Semicontinuous fermentor	14% ThGP	3
Sediment			
Na-cocoate (C_{8-18})	Freshwater swamp, 35°C, 56 d	84% ThGP	30
Na-cocoate (C_{8-18})	Marine sediment, 35°C, 56 d	96% ThGP	30

[a] SS, suspended solids dry weight.
[b] [U-^{14}C]Palmitic acid.
[c] ECETOC/ISO 11734 test.

biodegradability of sodium cocoate (C_{8-18}) was examined in screening tests by using digested sludge, freshwater swamp material, or marine sediment as inoculum. Following 28 and 56 days of incubation at 35°C the biodegradability attained was 70% and 93% of ThGP, respectively, for the digested sludge, 60% and 84% of ThGP for the freshwater swamp material, and 63% and 96% of ThGP for the marine sediment [30]. Steber and Berger [3] report the results of experiments that were conducted by use of a fermentor simulating an anaerobic digester to which fatty

TABLE 8   Aquatic Toxicity of Fatty Acid Soaps

Organism	Soap/fatty acid	EC/LC50 (mg/L)[a]	Time	Ref.
Algae	Soaps	10–320	—	7
P. subcapitata[b]	C_{8-18} soap	10–50	48 h	29
Nitzschia fonticola	C_{8-18} soap	20–50	48 h	29
Microcystis aeruginosa	C_{8-18} soap	10–20	72 h	29
Scenedesmus subspicatus	Na-laurate	53	72 h	32
Scenedesmus subspicatus	Na-oleate	58	72 h	32
Scenedesmus subspicatus	Na-palm kernel.	140	72 h	32
Scenedesmus subspicatus	Na-tallowate	190	72 h	32
Scenedesmus subspicatus	Na-behenate	230	72 h	32
Crustaceans	Soaps/fatty acids	2–88	—	7
Daphnia magna	Na-laurate	48	24 h	32
Daphnia magna	Palm kernel soap	25	24 h	32
Daphnia magna	Tallow soap	40	24 h	32
Daphnia magna	Na-laurate	32 (w.h. 215)[c]	—	32
Daphnia magna	Soap	NOEC: 10	21 d	33
Daphnia magna	Na-oleate	4.2	24 h	32
Daphnia magna	Lauric acid	2.0; 5.4 (w.h. 54)[c]	48 h	32
Gammarus pulex	Hard. tallow soap	88 (w.h.: 25)[c]	72 h	32
Fish	Soaps/fatty acids	0.1– >1000	—	7
Idus idus melanotus	Ca-soap	6.7 (w.h. 0)[c]	—	6
Idus idus melanotus	Na-soap	54 (w.h. 269)[c]	48 h	32
Oncorhynchus mykiss	C_{8-18} soap	42 (w.h. 120)[c]	96 h	32
Oryzias latipes	Na-soap	5.9 (w.h. 0)[c]	48 h	12
Oryzias latipes	Na-laurate	11	—	32
Oryzias latipes	Na-myristate	118	—	32
Oryzias latipes	Na-stearate	125	—	32
Oryzias latipes	Na-palmitate	150	—	32
Oryzias latipes	Na-oleate	217	—	32
Lepomis macrochirus	Oleic acid	66.6 (w.h. 40)[c]	96 h	32
Oncorhynchus kisuth	Oleic acid	12	33 h	32
Oncorhynchus mykiss	Oleic acid	0.1; 0.5; 2.1 (w.h. 32)[c]	96 h	32
Pimephales promelas	Oleic acid	205	96 h	32

[a] Median effect concentration (EC50), median lethal concentration (LC50), or no-observed-effect concentration (NOEC, when indicated).
[b] *Pseudokirchneriella subcapitata*.
[c] Parentheses indicate water hardness in mg $CaCO_3$/L.

acid soaps were added in a semicontinuous mode. The ultimate anaerobic biodegradability in the fermentor reached 95% for laurate, 70% for oleate and palm kernel–based soap, 60% for tallow-based soap, and only 14% for behenate (Table 7). A single addition of laurate, stearate, or behenate to the fermentor resulted in ≥90% degradation after 10–13 days of incubation.

3. Toxicity to Aquatic Organisms

The aquatic toxicity of soaps is highly variable and depends on the species tested, the specific fatty acid soap, and the water hardness. The single most important factor affecting the toxicity of soaps is probably water hardness. The effect of water hardness is illustrated in a study with rice fish (*O. latipes*) [12]. The 48-h LC50 of sodium soap was 5.9 mg/L in soft water (0 mg $CaCO_3$/L), whereas no effects were seen at 84 mg/L in water containing 25 mg $CaCO_3$/L [12]. Short-term EC50 or LC50 values of soaps and fatty acids are found within the intervals of 10–320 mg/L for algae, 2–88 mg/L for crustaceans, and 0.1–217 mg/L for fish. The lowest EC50 or LC50 values range between 1 and 10 mg/L, with the exception of the high acute toxicity of oleic acid in tests with rainbow trout, *Oncorhynchus mykiss* (Table 8). The chronic toxicity of fatty acid soaps is scarcely investigated. A chronic NOEC is reported for an unspecified soap in a 21-d test with *D. magna* (ref. 33; Table 8).

III. NONIONIC SURFACTANTS

Nonionic surfactants are widely used in laundry detergents, cleaning and dishwashing agents, and personal care products. In addition, nonionic surfactants are used as emulsifying agents in a large number of industrial applications. Nonionic surfactants are substances that do not ionize in aqueous solutions. The commercially most important nonionic surfactants are included in the very versatile group of alcohol ethoxylates (AE) and alcohol alkoxylates (AA). Alcohol ethoxylates are composed of a hydrophobic alkyl chain, which is combined with a number of ethoxylate (EO) units. Special AE surfactants can be end-capped with *n*-butyl, which provides foam-reducing properties, which are also known for alcohol alkoxylates (AA). The term *alcohol alkoxylates* is frequently used for fatty alcohol-based nonionic surfactants containing propylene oxide (propoxylate, PO) or butoxylate (BO) in their hydrophilic structure, normally in combination with EO. Commercial nonionic surfactants consist of a mixture of homologous structures with linear or branched alkyl chains that differ in the number of carbons. Also, the hydrophilic moieties of commercial surfactants exhibit a homologue distribution, and, e.g., the homologues in technical AE contain different numbers of EO. The indicated range of alkyl chain lengths in this text and these tables normally reflects the dominating alcohol cut, whereas the average numbers of EO or PO are normally given for the different surfactants. The distribution of EO in commercial AE usually covers a range from 0 to more than 15 EO units. For example, a C_{12-13} AE with an average number of 5 EO (NEODOL 23-5) contains structures with 0 through 14 EO (and a minor fraction ≥ 15 EO), whereas a C_{14-15} AE with an average number of 13 EO (NEODOL 45-13) contains structures with 4 through 18 EO (and a minor fraction ≤ 3 EO) [34]. Nonionic surfactants belonging to the group of alkyl phenol ethoxylates (APE) are still being used in some detergent formulations. The consumption of APE has been reduced because of their persistence in the environment [35] and their endocrine-disrupting activity [36,37]. Alkyl phenol ethoxylates have not been included in this chapter because the environmental properties of these surfactants have been reviewed previously [38]. Representative structures of nonionic surfactants are shown in Figure 2.

A. Alcohol Ethoxylates and Alcohol Alkoxylates

1. Aerobic Biodegradability

Three different mechanisms have been proposed for the biological degradation of AE under aerobic conditions [39,40].

1. The first mechanism is a central scission, or ether cleavage, which leads to the formation of fatty alcohols and polyethylene glycols (PEG). The fatty alcohols are

Lineary primary AE, C_{13} EO7

$CH_3\text{-}(CH_2)_{12}\text{-}O\text{-}(CH_2\text{-}CH_2\text{-}O\text{-})_7 H$

Iso-C_{13} branched primary AE, EO7

$H\text{—}(CH_2\text{—}\underset{\underset{CH_3}{|}}{CH})_4\text{—}CH_2\text{—}O\text{—}(CH_2\text{—}CH_2\text{—}O)_7 H$

Linear primary AE, C_{13} EO10, end-capped with n-butyl

$CH_3\text{-}(CH_2)_{12}\text{-}O\text{-}(CH_2\text{-}CH_2\text{-}O\text{-})_{10}\text{-}CH_2\text{-}CH_2\text{-}CH_2\text{-}CH_3$

Linear primary AA, EO5, PO4

$R\text{—}O\text{——}(CH_2\text{—}CH_2\text{—}O)_5\text{——}(CH_2\text{—}\underset{\underset{CH_3}{|}}{CH}\text{—}O)_4 H$

APG

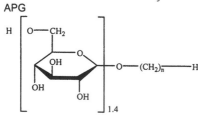

Glucose amides

$R\text{—}CH_2\text{—}\overset{\overset{O}{\|}}{C}\text{—}N\overset{\diagup CH_3}{\diagdown CH_2\text{—}(CHOH)_4\text{—}CH_2\text{—}OH}$

Fatty acid monoethanol amides

$R\text{—}\overset{\overset{O}{\|}}{C}\text{—}NH\text{—}CH_2\text{—}CH_2\text{—}OH$

Fatty acid diethanol amides

$R\text{—}\overset{\overset{O}{\|}}{C}\text{—}N\overset{\diagup CH_2\text{—}CH_2\text{—}OH}{\diagdown CH_2\text{—}CH_2\text{—}OH}$

FIG. 2 Structures of nonionic surfactants.

transformed to fatty acids by ω-oxidation of the terminal carbon, after which the fatty acids are degraded by β-oxidation. The β-oxidation of the fatty acid releases pairs of C atoms from the carbon chain, which are mineralized to CO_2. The PEG are degraded via a nonoxidative shortening, which releases one glycol unit at a time from the terminus of the PEG and/or via an oxidative hydrolysis forming monocarboxylated PEG.

2. The second mechanism is a microbial attack on the terminal carbon of the alkyl chain, via an ω-oxidation, followed by series of β-oxidations. By this mechanism, the AE is transformed to a carboxylated AE (with the carboxylic group at the alkyl chain), which is further degraded via the formation of monocarboxylated and dicarboxylated PEG.

3. The third mechanism is an ω-oxidation of the terminal carbon of the polyethoxylic chain. This mechanism proceeds via the formation of a carboxylated AE (with the carboxylic group at the polyethoxylic chain), which is further degraded via dicarboxylated AE (with carboxylic groups at both alkyl and polyethoxylic ends) and dicarboxylated PEG.

The relations between the biodegradation mechanisms and the structure of the AE have been elucidated in experimental studies [39,40]. The formation of PEG was observed only for a linear AE and an oxo-AE (composed of linear AE and monobranched AE with 2-methyl-, 2-ethyl-, 2-propyl-, and 2-butyl side chains), whereas only carboxylated AE (with the carboxylic group at the polyethoxylic chain) were detected during biodegradation of a multibranched AE. The absence of carboxylated AE in the experiments with the linear and the monobranched oxo-AE indicates that the central scission (mechanism 1) was the primary mechanism for the biodegradation of linear and most monobranched AE in the examined commercial mixtures, whereas the multibranched AE was degraded via ω-oxidation of the polyethoxylic chain (mechanism 3). Biodegradation of an oxo-2-butyl-substituted AE resulted in only carboxylated AE (mainly metabolites with the carboxylic group at the alkyl chain), suggesting that ω-oxidation of the alkyl chain was the primary mechanism (mechanism 2). The results obtained with the 2-butyl-substituted AE show that a shift from the central scission to the ω,β-oxidation is introduced when the length of the 2-alkyl branch exceeds three carbon atoms [40].

Far less is known about the biodegradation mechanisms for AA and end-capped AE. Alcohol alkoxylates are degraded via cleavages of the hydrophilic chain, which may be either nonoxidative or oxidative, like the degradation of PEG. A secondary carbon atom in the hydrophilic moiety, e.g., in PO groups, inhibits the oxidative route [41]. End-capped AE are degraded by a combination of ω-oxidation of the hydrophilic chain and central hydrophobe–hydrophile scission. The ω-oxidation is inhibited by the presence of PO in the hydrophilic chain, whereas the extent of central scission is determined by the degree of 2-alkyl branching [41]. The findings in the aforementioned studies with 2-butyl-substituted AE [40] further illustrate the effect of the length of the 2-alkyl substituent.

Linear AE are normally readily degraded under aerobic conditions, and only small differences are seen in the time needed to reach the 60% or 70% pass level in standardized tests for ready biodegradability (Table 9). Alcohol ethoxylates with long polyethoxylic chains (e.g., more than 20 EO) are usually degraded more slowly because the large molecular size reduces the bioavailability and limits the transport of the AE through the cell wall. Branching of the alkyl chain may also reduce the rate

TABLE 9 Ultimate Aerobic, Ready Biodegradability of AE and AA

Surfactant	Method	Result	Ref.
C_{9-11}, 8EO	Closed bottle, 28 d	80% ThOD	19
C_{12-15}, 7EO	CO_2 evolution, 28 d	82% $ThCO_2$	42
C_{16-18}, 30EO	Closed bottle, 28 d	27% ThOD	6
Iso-C_{10}, 7–8EO [b][a,b]	Die-away screening, 28 d	90% DOC	43
C_{13}, 7–8EO [b][c]	Die-away screening, 28 d	95% DOC	43
C_{13}, 7–8EO [b][d]	Die-away screening, 28 d	50% DOC	43
C_{11-15}, 7EO [b][e]	CO_2 evolution, 28 d	40% $ThCO_2$	44
Oxo-C_{12}, 5EO [b][f]	CO_2 evolution, 28 d	65% $ThCO_2$	40
Oxo-C_{14-15}, 9–20EO [b]	CO_2 evolution, 28 d	65–75% $ThCO_2$	6
C_{12-18}, 10EO, butyl[g]	Manometric respirometry, 28 d	98% ThOD	7
C_{13-15}, 6EO, 3PO	Ready biodegradability, 28 d	> pass level	45
Iso-C_{13}, 6EO, 3PO [b][b]	Ready biodegradability, 28 d	> pass level	45
C_{12-18}, 6EO, 2PO	Closed bottle, 28 d	83% ThOD	6
C_{12-18}, 2.5EO, 6PO	Closed bottle, 28 d	36% ThOD	6

[a] [b], branched alcohol.
[b] Propylene trimer, three internal CH_3 groups.
[c] 25% branched, 2-methyl-C_{12}, one internal CH_3 group.
[d] 46% branched, 2-propyl-C_{10} + 2-pentyl-C_8, two internal CH_3 groups.
[e] Four internal CH_3 groups, quaternary C-atom.
[f] Monobranched, 2-butyl-C_8.
[g] End-capped with *n*-butyl.

and extent of ultimate biodegradation. The rate of biodegradation of monobranched AE usually decreases with increasing length of the side chain. Although the 60% pass level was fulfilled in the CO_2 evolution test (but not the 10-day window), the degradation of an oxo-2-butyl-substituted AE (oxo-C_{12}, 5EO) occurred more slowly than the degradation of an oxo-AE blend containing 2-methyl-, 2-ethyl-, 2-propyl- and 2-butyl side chains [40]. The presence of more than one internal CH_3 group may also reduce the extent of biodegradation of AE, especially if the methyl groups are located at the same carbon atom, resulting in a quaternary structure (Table 9). The ultimate biodegradability of AA generally decreases with increasing number of PO or BO units. For example, a C_{12-18} AA with 2.5 EO and 6 PO attained only 36% ThOD in the closed-bottle test, whereas 83% ThOD was observed when the hydrophilic part was composed of 6 EO and 2 PO [6]. Generally, linear AA should contain less than 6–7 PO units in order to comply with the requirements for primary biodegradability [41]. Considering the requirements of at least 90% primary biodegradability in the European legislation, fewer PO units can be tolerated in branched AA. This implies that the extent of branching of the alcohol chain determines the construction of the hydrophilic part of commercial AA on the market in the European Union. This relation between the alcohol branching and the number of PO may be illustrated by studies conducted by Naylor et al. [46]. These authors showed that the primary biodegradation exceeded 80% for three different AA: (1) a slightly branched AA (20% branching) with up to 3.5 PO, (2) an AA with one internal CH_3 group containing up to 2.0 PO, and (3) an AA with two internal CH_3 groups containing up to 0.4 PO.

2. Anaerobic Biodegradability

The anaerobic biodegradation of linear AE is apparently initiated at the end of the polyethoxylic chain by a stepwise release of C_2 units as acetaldehyde, which leads to an AE with a shortened ethoxylate chain and, eventually, a fatty acid [47]. This pathway was recently confirmed in anaerobic biodegradation experiments with a linear pure C_{12} AE with 8 EO and a linear technical C_{12} AE with an average of 9 EO [48]. The first identifiable metabolites were AE with a shortened ethoxylate chain, and subsequent metabolites included dodecanoic acid and acetic acid. No PEG was observed during the degradation of the linear AE, which indicates that central scission of the AE was not the mechanism of the anaerobic biodegradation.

The anaerobic biodegradability of AE has been examined primarily in studies with surfactants composed of essentially linear alkyl chains. Divergent results of AE biodegradability have been obtained in anaerobic screening tests with diluted digested sludge, in which a high substrate concentration is required for measurements of anaerobic gas production (Table 10). The initial concentration of AE in these tests (>20 mg C/L) normally inhibits the anaerobic bacteria for several weeks, which may explain why extensive gas production, exceeding the gas production in control vessels, is frequently not observed. Compared to the screening test, environmentally more realistic biodegradation potentials may be obtained in mineralization experiments using ^{14}C-labeled surfactants and a higher inoculum to test substrate ratio. For example, the $^{14}CO_2$ and $^{14}CH_4$ recovered during anaerobic mineralization of a ^{14}C-labeled linear C_{18} AE with 7 EO corresponded to 84% of the added ^{14}C, when the AE was applied at 10 mg/L in a test system containing approx. 10 g of activated sludge (dry weight) and 13–20 g (dry weight) of anaerobic sludge per liter of medium [31]. Ultimate anaerobic biodegradation of linear AE has also been confirmed in sediments (Table 10).

TABLE 10 Ultimate Anaerobic Biodegradability of AE

Surfactant	Test conditions and incubation[a]	Result	Ref.
Digested sludge			
C_{9-11}, 8EO	0.15 g SS/L, 35°C, 56 d	79% ThGP	30
C_{12-15}, 7EO	0.15 g SS/L, 35°C, 56 d	38% ThGP	42
$[^{14}C]C_{18}$, 7EO[b]	Approx. 25 g SS/L, 35°C, 28 d	84%; $^{14}CO_2 + {}^{14}CH_4$	31
C_{12-18}, 10EO, butyl[c]	1.0 g SS/L, 35°C, 56 d[d]	54% ThGP	7
Sediment			
C_{9-11}, 8EO	Freshwater swamp, 35°C, 56 d	77% ThGP	30
C_{9-11}, 8EO	Marine sediment, 35°C, 56 d	66% ThGP	30
C_{10-12}, 7.5EO	Polluted creek mud, 28°C, 37 d	70% ThCH$_4$	22
C_{12}, 23EO	Polluted creek mud, 28°C, 37 d	80% ThCH$_4$	22
C_{11-12}, 8–9EO	Polluted pond sediment, 22°C, 87 d	24–40% ThGP	49
C_{11-12}, 8–9EO	Pond sediment, 22°C, 87 d	13% ThGP	49

[a] SS, suspended solids dry weight.
[b] Stearyl alcohol-[U-^{14}C]7EO.
[c] End-capped with *n*-butyl.
[d] ECETOC/ISO 11734 test.

3. Toxicity to Aquatic Organisms

The aquatic toxicity of AE and AA varies markedly, dependent on the molecular structure and homologue distribution of technical surfactants. Generally, the toxicity of linear AE increases with increasing alkyl chain length and decreasing ethoxylate chain length. Increasing branching of the alkyl chain normally leads to a decrease in the toxicity to aquatic organisms as it is seen particularly for multibranched AE [7]. Typical EC50 or LC50 values of commercial AE with an essentially linear C_{12-15} alkyl chain and 4–9 EO range between 0.05 and 50 mg/L for algae, between 0.2 and 10 mg/L for crustaceans, and between 0.96 and 7.5 mg/L for fish (Table 11). Apparently, the toxicity of linear AE either does not increase or increases slightly when the time of the exposure is increased from days to weeks. For example, the EC50 in tests with *D. magna* were 1.14 mg/L (96 h) and 0.93 mg/L (21 d) for a C_{12-13} AE with 6.5 EO [16]. No-observed-effect concentrations (NOECs) of linear AE in long-term tests range between 0.18 and 1.0 mg/L (Table 11). The effects of a linear C_{14-15} AE with 7 EO on aquatic invertebrates were examined in outdoor stream mesocosms [52]. In this study, the nominal AE concentrations in the artificial streams varied between 0 and 0.6 mg/L, with exposure periods of 28 or 30 days. No significant effects ($p > 0.05$) on population densities of Copepoda, Cladocera, Chironomidae, Nematoda, or Annelida were observed, and no effects were seen on the numbers of any of these organisms collected in drift nets. However, there was a significant decrease ($p \leq 0.05$) in the densities of Simuliidae (Diptera) at surfactant concentrations of 0.16 or more mg/L. The lowest-observed-effect concentration (LOEC) for aquatic invertebrates was 0.16 mg/L, and the NOEC was 0.08 mg/L [52]. Multibranched AE containing more than one internal CH_3 group in the lipophilic moiety have a lower aquatic toxicity than the linear AE, as indicated by typical EC50 and LC50 values between 5 and 10 mg/L (Table 12). The few available data describing the toxicity of end-capped AE and AA indicate that these surfactants follow the general trends that the toxicity increases with increasing alkyl chain length and that branching of the alkyl chain reduces the toxicity. For example, the EC50 of a C_{10-13} AA with 6 EO and 3 PO units in tests with algae ranged between 1 and 10 mg/L, whereas a C_{13-15} alkyl chain with the same hydrophilic moiety produced an EC50 between 0.1 and 1 mg/L [45]. A branched iso-C_{13} AA with 6 EO and 3 PO showed an EC50 above 10 mg/L in tests with algae (Table 13).

B. Alkyl Glycosides and Glucose Amides

1. Aerobic Biodegradability

The biodegradation pathways have not been elucidated for alkyl glycosides or glucose amides. The ultimate biodegradability of a C_{12-14} alkyl polyglycoside (APG) was investigated in a modified coupled units test, which allowed the detection of possible stable metabolites by a circulation of the effluent. In this system, the biodegradation of the C_{12-14} APG corresponded to 100% removal of the DOC, which indicates that the surfactant approached complete mineralization without accumulation of metabolites [54]. Both linear alkyl and monobranched alkylglycosides are usually rapidly degraded under aerobic conditions and comply with the pass criteria in tests for ready biodegradability (Table 14). For example, in the closed-bottle test, the period required to reach the 60% pass level was 5 days for a

TABLE 11 Aquatic Toxicity of Linear AE

Organism	Surfactant	EC/LC50 (mg/L)[a]	Time	Ref.
Algae	C_{12-15}, 4–9EO	0.05–50	—	7
Scenedesmus subspicatus	C_{12-14}, 7–8EO	0.5	72 h	43
Pseudokirchneriella subcapitata	C_{12-15}, 7EO	0.85 (0.84–0.86)[b]	72 h	42
Pseudokirchneriella subcapitata	C_{12-15}, 7EO	NOEC: 0.50	72 h	42
Microcystis aeruginosa	C_{14-15}, 6EO	0.60	96 h	23
Scenedesmus subspicatus	C_{14-15}, 6EO	0.09	96 h	23
Crustaceans	C_{12-15}, 4–9EO	0.2–10	—	7
Daphnia magna	C_{9-11}, 6EO	5.3 (2.9–8.5)[b]	48 h	34
Daphnia magna	C_{9-11}, 8EO	12 (9.0–18)[b]	48 h	34
Daphnia magna	C_{12-13}, 5EO	0.46 (0.39–0.56)[b]	48 h	34
Daphnia magna	C_{12-13}, 6.5EO	1.14 (0.96–1.31)[b]	96 h	16
Daphnia magna	C_{12-13}, 6.5EO	NOEC: 0.24	21 d	16
Daphnia magna	C_{12-15}, 12EO	1.4 (1.1–1.6)[b]	48 h	34
Daphnia magna	C_{14-15}, 13EO	1.2 (0.65–1.9)[b]	48 h	34
Daphnia magna	C_{14}, 4EO	0.24 (0.20–0.27)[b]	48 h	50
Daphnia magna	C_{14}, 4EO	1.76 (1.43–2.03)[b]	48 h	50
Daphnia pulex	C_{14}, 4EO	0.21 (0.18–0.23)[b]	48 h	50
Daphnia magna	C_{14-15}, 7EO	0.43 (0.37–0.51)[b]	96 h	16
Daphnia magna	C_{14-15}, 7EO	NOEC: 0.24	21 d	16
Daphnia magna	C_{16-18}, 18EO	20	48 h	38
Daphnia magna	C_{16-18}, 30EO	18	48 h	38
Fish	C_{12-15}, 4–9EO	0.96–7.5	—	7
Pimephales promelas	C_{9-11}, 6EO	8.5 (6.0–12)[b]	96 h	34
Pimephales promelas	C_{9-11}, 8EO	11 (8.5–17)[b]	96 h	34
Danio rerio	C_{12-15}, 7EO	1.0–2.0	96 h	42
Pimephales promelas	C_{12-13}, 5EO	1.0 (0.84–1.3)[b]	48 h	34
Pimephales promelas	C_{12-13}, 6.5EO	NOEC: 0.32	30 d	16
Pimephales promelas	C_{14-15}, 7EO	NOEC: 0.18	30 d	16
Pimephales promelas	C_{9-11}, 6EO	NOEC: 1.01[c]	28 d	51
Pimephales promelas	C_{12-13}, 6.5EO	NOEC: 0.82[c]	28 d	51
Pimephales promelas	C_{12-15}, 12EO	1.4 (1.2–1.5)[b]	96 h	34
Pimephales promelas	C_{14-15}, 13EO	1.0 (0.62–1.9)[b]	96 h	34
Pimephales promelas	C_{14-15}, 7EO	NOEC: 0.37[c]	28 d	51
Oncorhynchus mykiss	C_{14-15}, 18EO	5.0–6.3	96 h	38

[a] Median effect concentration (EC50), median lethal concentration (LC50), or no-observed-effect concentration (NOEC, when indicated).
[b] Parentheses indicate 95% confidence limits.
[c] Early life stage test (endpoint: fry growth).

C_{12} ethyl glycoside 6-O monoester (EGE), 10 days for a monobranched C_8 APG, and 14 days for a C_{12-14} APG [42]. The aerobic biodegradability of a C_{12-14} glucose amide in the CO_2 evolution test attained 89% and 86% of the theoretical CO_2 production (ThCO_2), respectively, for substrate concentrations of 10 and 20 mg/L (Table 14). In an experiment with activated sludge, the accumulated $^{14}CO_2$ from mineralization of a ^{14}C-labeled C_{12} glucose amide reached 89% of the added ^{14}C during 28 days, and the mineralization half-life was calculated to be 1.26 days [56].

TABLE 12 Aquatic Toxicity of Branched AE

Organisms	Surfactant	EC/LC50 (mg/L)[a]	Time	Ref.
Algae				
P. subcapitata[b]	C_{12-15}, 7EO,	10.0	96 h	53
	4 int. CH_3 groups,	NOEC: 4.0	96 h	53
	quaternary C-atom			
Scenedesmus subspicatus	Iso-C_{10}, 7–8EO,	50	72 h	43
	3 int. CH_3 groups			
Scenedesmus subspicatus	Iso-C_{13}, 7–8EO,	5	72 h	43
	3 int. CH_3 groups			
Crustaceans				
Daphnia magna	Iso-C_{10}, 7–8EO,	50	48 h	43
	3 int. CH_3 groups			
Daphnia magna	C_{13}, 7EO,	9.8 (9.0–10.7)[c]	48 h	53
	3.6 int. CH_3 groups,	NOEC: 2.0	7 d	53
	quaternary C atom			
Daphnia magna	C_{12-15}, 7EO,	11.6 (11.0–12.2)[c]	48 h	53
	4 int. CH_3 groups,	NOEC: 4.0	7 d	53
	quaternary C atom			
Daphnia magna	Iso-C_{13}, 7–8EO	5	48 h	43
	3 int. CH_3 groups			
Fish				
Pimephales promelas	C_{13}, 7EO,	4.5 (3.0–5.3)[c]	96 h	53
	3.6 int. CH_3 groups,	NOEC: 1.0	7 d	53
	quaternary C atom			
Pimephales promelas	C_{12-15}, 7EO,	6.1 (5.8–6.3)[c]	96 h	53
	4 int. CH_3 groups,	NOEC: 1.0	7 d	53
	quaternary C atom			

[a] Median effect concentration (EC50), median lethal concentration (LC50), or no-observed-effect concentration (NOEC, when indicated).
[b] *Pseudokirchneriella subcapitata*.
[c] Parentheses indicate 95% confidence limits.

2. Anaerobic Biodegradability

Anaerobic screening tests have confirmed the ultimate biodegradation of alkyl glycosides with linear alkyl chains by use of various inocula (Table 15). For example, in the ECETOC test using anaerobically digested sludge as inoculum, the anaerobic biodegradation of linear C_{8-10} and C_{12-14} APG reached 95% and 84% of ThGP, respectively, during 56 days of incubation at 35°C [54]. Branching of the alkyl chain may reduce the biodegradation of glycoside surfactants under anoxic conditions. The gas production in test vessels containing diluted digested sludge and a mono-branched C_8 APG attained a level of approx. 20% of ThGP within 14 days, but only 22% of ThGP was recorded after 56 days (Table 15). The recalcitrant nature of the monobranched C_8 APG under anoxic conditions was confirmed by measurements of the accumulated CH_4 in the test vessels, which did not exceed the CH_4 formation in controls without addition of surfactant [42].

TABLE 13 Aquatic Toxicity of Alcohol Alkoxylates and End-Capped Alcohol Ethoxylates

Organisms	Surfactant	EC/LC50 (mg/L)[a]	Ref.
Algae	C_{8-10}, 6EO, 3PO	10–100	45
Algae	C_{9-11}, 6EO, 3PO	1–10	45
Algae	C_{10-13}, 6EO, 3PO	1–10	45
Algae	Iso-C_{13}, 6EO, 3PO [b][b]	10–100	45
Algae	C_{13-15}, 6EO, 3PO	0.1–1	45
Algae	C_{12-14}, 9EO, butyl[c]	0.3	6
Daphnia magna	C_{12-14} AE, 9EO, butyl[c]	1.0–2.0	6
Daphnia magna	C_x, 2–5EO, 4PO[d]	2.4–6.0	6
Fish	C12–14, 9EO, butyl[c]	0.5–4.6	6
Fish	C_x, 2–5EO, 4PO[d]	0.7–5.7	6

[a] Median effect concentration (EC50), median lethal concentration (LC50).
[b] [b], branched alcohol.
[c] End-capped with *n*-butyl
[d] Type of alcohol not indicated.

3. Toxicity to Aquatic Organisms

The aquatic toxicity of alkyl glycosides and glucose amides is characterized by EC50 or LC50 between 3.0 and 101 mg/L for surfactants with linear alkyl chains, whereas the data on a single monobranched C_8 APG indicate that branching of the alkyl chain may decrease the toxicity markedly (Table 16). The aquatic toxicity increases with increasing alkyl chain length for linear types of glucose-based surfactants. The effect of alkyl chain length may be illustrated by the effect concentrations (EC50) of the commercially important linear APG and glucose amides that were obtained in tests with freshwater green algae. The 72-h EC50 in tests with *Scenedesmus subspicatus* were 21 mg/L for a linear C_{8-10} APG and 6.0 mg/L for a linear C_{12-14} APG [54]. The glucose amides showed 96-h EC50 of 57 mg/L for C_{12} and 3.9 mg/L for C_{14} in tests with *Pseudokirchneriella subcapitata* [56]. The lowest chronic NOECs include a value of 1.0 mg/L for C_{12-14} APG (*D. magna*, 21 days) and 2.9 mg/L for C_{14} glucose amide (*P. subcapitata*, 96 hours) (Table 16).

TABLE 14 Ultimate Aerobic, Ready Biodegradability of Alkyl Glycosides and Glucose Amides

Surfactant[a]	Method	Result	Ref.
C_{8-10} APG	Closed bottle, 28 d	81–82% ThOD	54
C_{12-14} APG	Closed bottle, 28 d	73–88% ThOD	54
C_{8-16} APG	Closed bottle, 30 d	80% ThOD	55
C_{12-16} APG	Closed bottle, 30 d	77% ThOD	55
C_8 APG [b]	CO_2 evolution, 28 d	78% $ThCO_2$	42
C_{12} EGE	CO_2 evolution, 28 d	78% $ThCO_2$	42
C_{12-14} Glucose amide	CO_2 evolution, 35 d	86%; 89% $ThCO_2$	56

[a] APG, alkyl polyglycoside; APG [b], monobranched APG; EGE, ethyl glycoside monoester.

TABLE 15 Ultimate Anaerobic Biodegradability of Alkyl Glycosides

Surfactant[a]	Test conditions and incubation[b]	Result	Ref.
Digested sludge			
C_8 APG [b]	0.15 g SS/L, 35°C, 56 d	22% ThGP	42
C_{8-10} APG	3.0 g SS/L, 35°C, 56 d[c]	95% ThGP	54
C_{12-14} APG	3.0 g SS/L, 35°C, 56 d[c]	84% ThGP	54
C_{10} EGE	0.15 g SS/L, 35°C, 56 d	96% ThGP	30
C_{12} EGE	0.15 g SS/L, 35°C, 56 d	84% ThGP	30
Sediments			
C_{12-14} APG	Freshwater swamp, 35°C, 56 d	76% ThGP	30
C_{10} EGE	Freshwater swamp, 35°C, 56 d	83% ThGP	30
C_{12} EGE	Freshwater swamp, 35°C, 56 d	89% ThGP	30
C_{10} EGE	Marine sediment, 35°C, 56 d	79% ThGP	30

[a] APG, alkyl polyglycoside; APG [b], monobranched APG; EGE, ethyl glycoside monoester.
[b] SS, suspended solids dry weight.
[c] ECETOC/ISO 11734 test.

C. Fatty Acid Amides

1. Aerobic and Anaerobic Biodegradability

Most fatty acid amides are ultimately biodegraded under aerobic conditions in standardized tests for ready biodegradability. For example, the widely used coco-monoethanolamide and cocodiethanolamide attained 82% and 71% of ThOD, respectively, in the closed-bottle test [57], whereas a C_{12-18} diethanolamide reached 74% removal of DOC in the modified OECD screening test [6]. The available data describing the biodegradability of ethoxylated fatty acid amides are contradictory. Schöberl et al. [6] report that only 47% and 35% of ThOD were obtained in the closed-bottle test for C_{12-14} monoethanolamide with 4 EO and 10 EO, respectively. However, according to data from manufacturers, more than 60% $ThCO_2$ was reached for C_{12-14} monoethanolamides with 5 EO and 12 EO in the CO_2 evolution test [58].

The biodegradation potential in the absence of molecular oxygen has been examined for cocomonoethanolamide by use of the ECETOC test and the ISO 11734 screening test. The ultimate biodegradability of cocomonoethanolamide attained 79% of ThGP during 42 days at 35°C [57]. In the ISO 11734 test, the biodegradability of cocomonoethanolamide corresponded to 66% of ThGP after 28 days and 81% of ThGP after 56 days of anaerobic incubation at 35°C [7].

2. Toxicity to Aquatic Organisms

The available results of aquatic toxicity tests indicate that fatty acid diethanolamides are more toxic than the corresponding monoethanolamides. The typical EC50 or LC50 values for cocodiethanolamides were between 1 and 10 mg/L in tests with algae, daphnia, and fish, whereas EC50/LC50 values above 10 mg/L are generally found for cocomonoethanolamides. The ethoxylated fatty acid amides show the same level of aquatic toxicity as the corresponding nonethoxylated fatty acid amides (Table 17).

TABLE 16 Aquatic Toxicity of Alkyl Glycosides and Glucose Amides

Organisms	Surfactant[a]	EC/LC50 (mg/L)[b]	Time	Ref.
Algae				
P. subcapitata[c]	C_8 APG [b]	1543 (1474–1621)[d]	72 h	42
		NOEC: 100		
Scenedesmus subspicatus	C_{8-10} APG	21	72 h	54
		NOEC: 5.7		
P. subcapitata[c]	C_{12-14} APG	11 (10–13)[d]	72 h	42
		NOEC: 3.1		
Scenedesmus subspicatus	C_{12-14} APG	6.0	72 h	54
		NOEC: 2.0		
P. subcapitata[c]	C_{12} EGE	38 (37–38)[d]	72 h	42
		NOEC: 11		
P. subcapitata[c]	C_{12} Glu. amide	56.8 (50.4–63.9)[d]	96 h	56
		NOEC: 21.3		
P. subcapitata[c]	C_{12-14} Glu. amide	12.6 (11.6–13.6)[d]	96 h	56
		NOEC: 5.6		
P. subcapitata[c]	C_{14} Glu. amide	3.9 (2.5–6.4)[d]	96 h	56
		NOEC: 2.9		
Crustaceans				
Daphnia magna	C_8 APG [b]	557 (465–717)[d]	48 h	42
Daphnia magna	C_{8-10} APG	20	48 h	54
Daphnia magna	C_{12-14} APG	7.0	48 h	54
Daphnia magna	C_{12-14} APG	NOEC: 1.0	21 d	54
Daphnia magna	C_{12} EGE	23 (21–25)[d]	48 h	42
Daphnia magna	C_{12} Glu. amide	44 (38–53)[d]	48 h	56
Daphnia magna	C_{12-14} Glu. amide	18 (16–21)[d]	48 h	56
Daphnia magna	C_{12-14} Glu. amide	NOEC: 4.3	21 d	56
Daphnia magna	C_{14} Glu. amide	5.0 (3.3–9.2)[d]	48 h	56
Fish				
Danio rerio	C_8 APG [b]	558 (447–718)[d]	96 h	42
Danio rerio	C_{8-10} APG	101	96 h	54
Danio rerio	C_{12-14} APG	3.0	96 h	54
Danio rerio	C_{12-14} APG	NOEC: 1.8	28 d	54
Danio rerio	C_{12} EGE	11–17	96 h	42
Danio rerio	C_{12-14} Glu. amide	7.5	96 h	56
Pimephales promelas	C_{12} Glu. amide	39 (31–51)[d]	96 h	56
Pimephales promelas	C_{14} Glu. amide	2.9 (2.4–3.7)[d]	96 h	56
Pimephales promelas	C_{12-14} Glu. amide	NOEC: 4.8	21 d	56

[a] APG, alkyl polyglycoside; APG [b], monobranched APG; EGE, ethyl glycoside monoester; Glu. amide, glucose amide.
[b] Median effect concentration (EC50), median lethal concentration (LC50), or no-observed-effect concentration (NOEC, when indicated).
[c] *Pseudokirchneriella subcapitata*.
[d] Parentheses indicate 95% confidence limits.

TABLE 17 Aquatic Toxicity of Fatty Acid Amides

Organisms	Surfactant[a]	EC/LC50 (mg/L)[b]	Time	Ref.
Algae				
Pseudokirchneriella subcapitata	Coco MEA	17.8 (16.2–19.2)[c,d]	72 h	7
		26.2 (25.6–26.8)[c,d]		
		NOEC: 10.0		
Scenedesmus subspicatus	Coco MEA	16.6 (15.2–18.4)[c,d]	72 h	7
		36.4 (34.4–38.8)[c,d]		
		NOEC: 1.0		
Scenedesmus subspicatus	Coco MEA	1.0; 1.1	96 h	57
Scenedesmus subspicatus	Coco DEA	2.2	72 h	57
		NOEC: 0.32		
Scenedesmus subspicatus	C_{12-14} MEA, 4EO	14	72 h	58
Scenedesmus subspicatus	C_{12-14} MEA, 5EO	20	96 h	58
Crustaceans				
Ceriodaphnia dubia	Coco DEA	2.25	48 h	57
Daphnia magna	Coco MEA	10; 24.8; 37.5	24 h	57
Daphnia magna	Coco DEA	4.2; 5.4	24 h	57
Daphnia pulex	Coco DEA	2.39	48 h	57
Daphnia magna	C_{12-14} MEA, 4EO	10–100	—	6
Daphnia magna	C_{12-14} DEA, 4EO	2–3	—	6
Fish				
Danio rerio	Coco MEA	28.5; 31	96 h	57
Danio rerio	Coco DEA	3.6; 4.0	96 h	57
Leucistus idus	Coco MEA	13.5; 20.7	48 h	57
Oryzias latipes	C_{12} DEA	10.8–13.8	24 h	57

[a] MEA, monoethanolamide; DEA, diethanolamide.

[b] Median effect concentration (EC50), median lethal concentration (LC50), or no-observed-effect concentration (NOEC, when indicated).

[c] Parentheses indicate 95% confidence limits.

[d] Upper value: EC50, biomass; lower value: EC50, growth rate.

IV. CATIONIC SURFACTANTS

Cationic surfactants serve multiple functions as biocides and disinfectants, emulsifiers, wetting agents, and antistatic agents. Typical products containing cationic surfactants include cleaning agents, hair care preparations, and fabric softeners. Cationic surfactants are positively charged in aqueous solutions. Of the cationic surfactants the quaternary ammonium salts especially are widely used. These surfactants are characterized by a positively charged quaternary nitrogen atom. Two main groups of cationic surfactants are discussed in this section. The first group includes the quaternary alkyl ammonium salts that are characterized by a quaternized nitrogen, to which one or two alkyl chains or an alkyl chain and a benzene group are linked. The second group contains the quaternary alkyl ester ammonium salts, in which one or two ester linkages are inserted in the molecular structure. The alkyl ester ammonium salts include different types of esters, e.g.: (1) *N*-methyl-*N*,*N*-bis[2-(C_{16-18}-acyloxy)ethyl]-*N*-(2-hydroxyethyl) ammonium methosulfate (esterquat); (2)

N,N,N-trimethyl-N-[1,2-di-(C_{16-18}-acyloxy)propyl] ammonium (diesterquat); (3) the diethyl ester dimethylammonium chloride (DEEDMAC); and (4) 3-behenoyloxy-2-hydroxypropyl trimethylammonium chloride (BTMAC). Representative structures of quaternary ammonium compounds are shown in Figure 3.

A. Alkyltrimethyl, Dialkyldimethyl, and Alkyldimethylbenzyl Ammonium Salts

1. Aerobic Biodegradability

The biodegradation pathways of quaternary alkyl ammonium salts are scarcely investigated. It has been proposed that aerobic biodegradation of quaternary alkyl

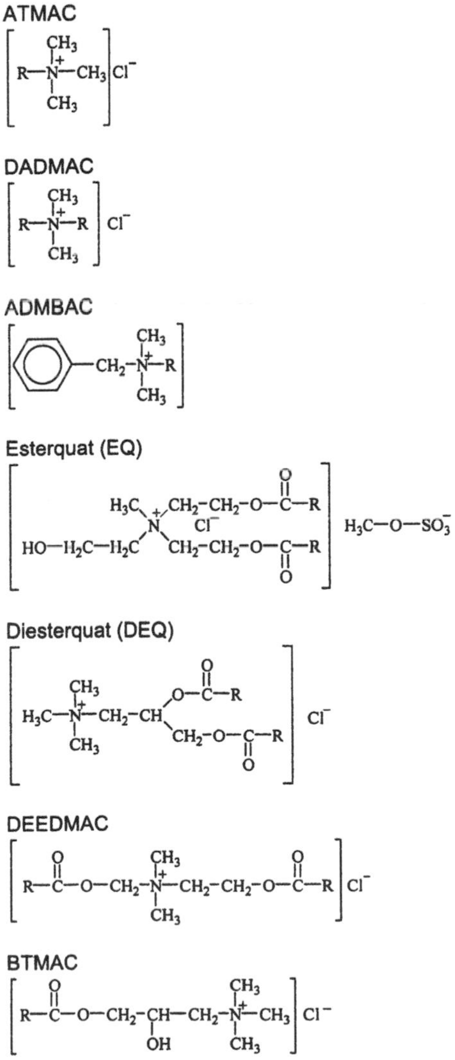

FIG. 3 Structures of cationic surfactants.

ammonium salts proceeds via a fission of a C–N bond, by which the alkyl chain or a methyl group is cleaved from a tertiary amine [59]. Qualitative analyses of the metabolites formed during degradation of C_{14} alkyldimethylbenzylammonium chloride (ADMBAC) in pilot activated-sludge plants showed an accumulation of tetradecyldimethylamine [60], which indicates that the initial degradation step was a cleavage of the bond linking the benzene group to the alkyldimethylammonium.

The ultimate aerobic biodegradability of quaternary ammonium compounds has been examined in standardized biodegradation tests by measuring the O_2 uptake or the evolution of CO_2. Respirometric parameters are more reliable in biodegradability tests with cationic surfactants than, e.g., removal of DOC, because of the strong sorption of these substances. However, the principal problem related to the use of standardized biodegradability tests is the inhibitory effects of quaternary alkyl ammonium compounds at the required substrate concentration (mg/L). The studies of Masuda et al. (cited by Ginkel [59]) indicate that the biodegradability of quaternary alkyl ammonium compounds in screening tests decreases with increasing alkyl chain length. The biodegradability of alkyltrimethylammonium chlorides (ATMAC) attained 73% of ThOD for C_8, 63% for C_{10}, 59% for C_{12}, 35% for C_{14}, and 0% for C_{16} and C_{18} during 10 days in the MITI test. The same tendency was seen for ADMBAC as the results obtained after 10 days in the MITI tests were reported to be 79% of ThOD for C_8, 95% for C_{10}, 89% for C_{12}, 83% for C_{14}, 5% for C_{16}, and 0% for C_{18} [59]. Information related to the inoculum used in the MITI tests is lacking, therefore, it is difficult to verify whether or not the OECD criteria for ready biodegradability were fulfilled. However, the results of Masuda et al. show that extensive ultimate biodegradation of ATMAC and ADMBAC may occur under aerobic conditions. The toxic effects of quaternary alkyl ammonium salts may imply that results from screening tests underestimate the biodegradation potential of these substances in the aquatic environment. The bacterial toxicity of especially the longer-chained quaternary alkyl ammonium salts can be mitigated by the addition of equimolar amounts of anionic surfactants. Several studies have shown that ATMAC may be mineralized when complexated with the anionic surfactant LAS. For example, Games et al. [61] showed that a C_{18} ATMAC at 20 mg/L inhibited the endogenous CO_2 production in a semicontinuous activated-sludge test (SCAS test), whereas a mineralization corresponding to 81% of $ThCO_2$ was attained during 25 days in a mixture of C_{18} ATMAC and LAS (Table 18). Experiments conducted by use of silica gel or radiolabeled model chemicals represent alternative approaches allowing a low exposure concentration of cationic surfactants. The primary biodegradability attained 100% removal of $C_{(10)2}$ and $C_{(18)2}$ dialkyldimethylammonium chloride (DADMAC) during 4 and 8 days, respectively, in silica gel columns inoculated with an axenic bacterial culture [68]. Rapid mineralization was observed when a ^{14}C-labeled C_{18} ATMAC was added to a SCAS test system at initial concentrations of 0.1 and 1.0 mg/L as between 63% and 88% of the radiolabeled carbon was recovered as $^{14}CO_2$ after approx. 7 days [61]. Compared to the accumulation of $^{14}CO_2$ in the SCAS test with ^{14}C-labeled ATMAC [61], the accumulated $^{14}CO_2$ during mineralization of a ^{14}C-labeled C_{16-18} DADMAC attained between 22% and 53% of the added ^{14}C after 39 days in a semibatch reactor with activated sludge [62]. This confirms that DADMAC are more resistant to aerobic biodegradation than ATMAC and ADMBAC (see also data in Table 18).

TABLE 18 Ultimate Aerobic Biodegradability of Quaternary Ammonium Salts

Surfactant	Method[a]	Result	Ref.
Alkyl ammonium salts			
C_{16} ATMAC	Manometric respirometry, 28 d	40% ThOD	7
C_{18} ATMAC	CO_2 evolution, ATMAC + LAS, 25 d	81% $ThCO_2$	61
$[^{14}CH_3]C_{18}$ ATMAC	Unacclimated SCAS, 1 g SS/L, 0.1 mg/L, 7 d	88%; $^{14}CO_2$	61
$[1-^{14}C]C_{18}$ ATMAC	Unacclimated SCAS, 1 g SS/L, 0.1 mg/L, 7 d	67%; $^{14}CO_2$	61
$C_{(16-18)2}$ DADMAC	Closed bottle, 283 d	68% ThOD	59
$[^{14}CH_3]C_{(16-18)2}$ DADMAC	Activated sludge reactor, 39 d	40%, 53%; $^{14}CO_2$	62
$[1-^{14}C]C_{(16-18)2}$ DADMAC	Activated sludge reactor, 39 d	31%; $^{14}CO_2$	62
$[U-^{14}C]C_{(16-18)2}$ ADMAC	Activated sludge reactor, 39 d	22%, 31%; $^{14}CO_2$	62
C_{12} ADMBAC	CAS test	83% DOC	63
Alkyl ester ammonium salts			
MTEA[b]	CO_2 evolution, 28 d	76–94% $ThCO_2$	64
Diesterquat	CO_2 evolution, 28 d	87% $ThCO_2$	65
DEEDMAC	CO_2 evolution, 28 d	80% $ThCO_2$	66
BTMAC	Closed bottle, 28 d	74% ThOD	67

[a] SS, suspended solids dry weight.
[b] Main metabolite of esterquat.

2. Anaerobic Biodegradability

Although cationic surfactants sorb to sludge particles and eventually reach the digester during the treatment of wastewater sludge, there is very limited information about the biodegradability of these compounds under anoxic conditions. It has been demonstrated, however, that the concentration of quaternary alkyl ammonium salts did not decrease or decreased only slightly in an anaerobic digester [59].

3. Toxicity to Aquatic Organisms

Quaternary alkyl ammonium compounds are acutely toxic to a number of aquatic organisms, including algae, crustaceans, annelids, snails, molluscs, and fish [7]. The lowest EC50 or LC50 values for ATMAC, DADMAC, and ADMBAC are below 1 mg/L in short-term tests. Algae are very sensitive to DADMAC, as indicated by EC50 values below 0.1 mg/L (Table 19). Although it is evident that these cationic surfactants have inherent ecotoxic properties, it is likely that their toxicity in aquatic environments is reduced by the sorption of the surfactants to suspended solids or by the formation of complexes with anionic surfactants. Studies conducted with DADMAC showed a considerably lower acute and chronic toxicity in tests with river waters as compared with the toxicity observed in corresponding tests with

TABLE 19 Aquatic Toxicity of Quaternary Alkyl Ammonium Salts

Organisms	Surfactant	EC/LC50 (mg/L)[a]	Time	Ref.
Algae				
Chlorella pyrenidosa	C_{16-18} ATMAC	0.28 (0.22–0.36)[b,c]	96 h	24
Dunaliella sp.	C_{16-18} ATMAC	0.38 (0.33–0.45)[b,c]	24 h	24
Microcystis aeruginosa	C_{12} ATMAC	0.12	96 h	23
Navicula pelliculosa	C_{12} ATMAC	0.20	96 h	23
Pseudokirchneriella subcapitata	C_{12} ATMAC	0.19	96 h	23
Chlorella pyrenidosa	$C_{(16-18)2}$ DADMAC	6.0 (5.5–6.5)[b,c]	96 h	24
Dunaliella sp.	$C_{(16-18)2}$ DADMAC	18 (13–24)[b,c]	24 h	24
Microcystis aeruginosa	$C_{(18)2}$ DADMAC	0.05	96 h	23
Navicula pelliculosa	$C_{(18)2}$ DADMAC	0.07	96 h	23
Pseudokirchneriella subcapitata	$C_{(18)2}$ DADMAC	0.06	96 h	23
Chlorella pyrenidosa	C_{12-14} ADMBAC	0.67 (0.62–0.73)[b,c]	96 h	24
Dunaliella sp.	C_{12-14} ADMBAC	1.8 (1.6–2.1)[b,c]	24 h	24
Crustaceans				
Daphnia magna	ATMAC	1.2–5.8[d]	—	69
Gammarus sp.	C_{16} ATMAC	0.1 (0.08–0.14)[b]	48 h	70
Daphnia magna	$C_{(16-18)2}$ DADMAC	0.19 (0.15–0.24)[b]	48 h	71
Daphnia magna	$C_{(18)2}$ DADMAC	0.16	48 h	71
Fish				
Idus melanotus	ATMAC	0.36–8.6[d]	—	69
Gasterosteus aculeatus	$C_{(16-18)2}$ DADMAC	4.5 (4.1–4.9)[b]	96 h	72
Gasterosteus aculeatus	$C_{(16-18)2}$ DADMAC	NOEC: 0.75	28 d	72
Lepomis macrochirus	$C_{(16-18)2}$ DADMAC	0.62 (0.45–0.85)[b]	96 h	71
Lepomis macrochirus	$C_{(18)2}$ DADMAC	1.04 (0.74–1.45)[b]	96 h	71
Carrasius auratus	ADMBAC	2.0	—	69
Lepomis macrochirus	ADMBAC	0.5	—	69

[a] Median effect concentration (EC50), median lethal concentration (LC50), or no-observed-effect concentration (NOEC, when indicated).
[b] Parentheses indicate 95% confidence limits.
[c] The purities of the surfactants were greater than 97.5%.
[d] The specified ranges include tests with C_{12}, C_{14}, C_{16}, C_{18}, and C_{20-22}.

filtered laboratory water. For example, the 48-h EC50 values for $C_{(18)2}$ DADMAC in tests with *Daphnia magna* were 0.16 mg/L in filtered laboratory water and 3.1 mg/L in river water containing 3–5 mg of suspended solids per liter [71]. The toxicity of C_{16-18} ATMAC and C_{12-14} ADMBAC complexated with LAS was lower than the toxicity of the cationic surfactants alone in growth inhibition tests with the algae *Dunaliella* sp. and *Chlorella pyrenoidosa*. This tendency was less clear for $C_{(16-18)2}$ DADMAC because a reduction of the toxicity of the LAS complex compared to that of the DADMAC alone was noted only for *C. pyrenoidosa* [24]. The acute and chronic toxicity of a technical surfactant ditallowdimethyl ammonium chloride ($C_{(16-18)2}$ DADMAC) was determined in short-term and long-term tests with the three-spined stickleback *Gasterosteus aculeatus*, the midge larva *Chironomus riparius*, and the pond snail *Lymnaea stagnalis* [72]. The acute 96-h LC50 values were 4.5 mg/L for *G. aculeatus* (95% confidence limits, 4.1–4.9 mg/L), 9.2 mg/L for *C. riparius* (95% confidence limits, 8.1–11 mg/L), and 18 mg/L for *L. stagnalis* (95% confidence limits, 15–21 mg/L). The lowest NOECs that can be derived from long-term exposures (28 days) of these organisms were 0.75 mg/L for *G. aculeatus* (behavior and growth), 1.34 mg/L for *C. riparius* (survival and development), and 0.32 mg/L for *L. stagnalis* (behavior) [72].

B. Alkyl Ester Ammonium Salts

1. Aerobic Biodegradability

The ester linkages in the alkyl ester ammonium salts are readily attacked by microorganisms, and the subsequent cleavage of the linkages results in smaller molecules that are readily biodegraded. The aerobic biodegradability of the poorly water-soluble N-methyl-N,N-bis[2-(C_{16-18}-acyloxy)ethyl]-N-(2-hydroxyethyl) ammonium methosulfate (esterquat) has been examined under simulated sewage treatment plant conditions in the coupled units test, in which more than 90% degradation was found [64]. The main metabolite formed from the degradation of esterquat was a tris-(hydroxyethyl) methylammonium methosulfate (MTEA). The higher water solubility of MTEA compared to that of esterquat made it possible to confirm the ready biodegradability of MTEA in the CO_2 evolution test as between 76% and 94% $ThCO_2$ was reached during 28 days [64]. N,N,N-Trimethyl-N-[1,2-di-(C_{16-18}-acyloxy) propyl] ammonium (diesterquat) is hydrolyzed via a 3-monoester quaternary to a diol quaternary ammonium compound, which is very water soluble [65]. In spite of poor aqueous solubility, the parent molecules of diesterquat, diethyl ester dimethyl-ammonium chloride (DEEDMAC), and 3-behenoyloxy-2-hydroxypropyl trimethyl-ammonium chloride (BTMAC) all have proven to be readily biodegradable under screening test conditions (Table 18).

Studies simulating the fate of diesterquat in aerobic surface waters have demonstrated a rapid biodegradation of the entire molecule under river die-away conditions. The accumulated $^{14}CO_2$ from the mineralization of diesterquat reached 94%, 88%, and 95% of the added ^{14}C for ^{14}C-stearyl-, ^{14}C-methyl-, and ^{14}C-dihydroxypropyl- labeled isotopes, respectively. The associated mineralization half-lives of diesterquat were 0.65–0.70 days (^{14}C-stearyl), 7.1–7.7 days (^{14}C-methyl), and 6.1–6.7 days (^{14}C-dihydroxypropyl) [65]. The mineralization of ^{14}C-labeled DEEDMAC was examined in activated sludge and river waters containing 0.1% and 1.0% acclimated activated sludge. The total recovery of $^{14}CO_2$ from the mineralization of DEEDMAC attained 76% and 82% of the added ^{14}C for the batch activated sludge and the river die-away test, respectively. The mineralization half-lives of DEEDMAC were determined to be 1.0 days in the activated sludge and 1.1 days in the river water die-away test [66].

2. Anaerobic Biodegradability

The ultimate biodegradability of DEEDMAC has been examined under methano-genic conditions in the ECETOC screening test. The total gas production from mineralization of DEEDMAC reached 90% of ThGP during 60 days [66], which indicates that DEEDMAC has the potential for complete biodegradation under anoxic conditions. The anaerobic biodegradability of another alkyl ester ammonium salt BTMAC was confirmed in the ISO 11734 screening test as 65% of ThGP was attained after 56 days of incubation at 35°C (T. Madsen, unpublished data). Finally, esterquat is reported to be degradable under anaerobic screening test conditions, but no specific data were provided by the authors [64].

3. Toxicity to Aquatic Organisms

The aquatic toxicity of alkyl ester ammonium salts is generally lower compared to the toxicity of the quaternary alkyl ammonium compounds discussed in the previous section. The lowest reported EC50 or LC50 values of alkyl ester ammonium salts vary between 2 and 10 mg/L, which is similar to typical ranges for many anionic and nonionic surfactants. The NOEC values found in laboratory tests (0.3–4.0 mg/L) are three orders of magnitude higher than the predicted environmental concentrations, which were estimated for the emission of treated sewage effluents to aquatic recipients (0.2 µg/L; see Table 1). The main metabolite produced during degradation of esterquat, i.e., MTEA, is characterized by a lower aquatic toxicity compared to that of the intact surfactant. The transient hydrolysis product of diesterquat, the 3-monoester quaternary, is characterized by a higher bioavailability than the parent substance, which explains the high acute toxicity of the 3-monoester to daphnia (Table 20).

TABLE 20   Aquatic Toxicity of Alkyl Ester Ammonium Salts

Organisms	Surfactant	EC/LC50 (mg/L)[a]	NOEC (mg/L)[b]	Ref.
Algae	Esterquat		0.3	64
Algae	MTEA[c]		300	64
Scenedesmus subspicatus	Diesterquat		1.8 (72 h)	65
Scenedesmus subspicatus	3-monoester[d]	3.7 (72 h)	1.8 (72 h)	65
Pseudokirchneriella subcapitata	DEEDMAC	2.9 (96 h)		66
Crustaceans				
Daphnia magna	Esterquat	78	3.0 (21 d)	64
Daphnia magna	MTEA[c]		3.0 (21 d)	64
Daphnia magna	Diesterquat	7.7 (48 h)	1.0 (21 d)	65
Daphnia magna	3-monoester[d]	0.6; 0.7 (48 h)	0.26 (48 h)	65
Daphnia magna	DEEDMAC	14.8 (24 h)	1.0 (21 d)	66
Fish	Esterquat	3.0	4.0 (14 d)	64
Fish	MTEA[c]		1,000 (14 d)	64
Oncorhynchus mykiss	Diesterquat	7.0 (96 h)	≥3.5 (28 d)	65
Danio rerio	DEEDMAC	5.2 (96 h)		66
Pimephales promelas	DEEDMAC		0.68 (35 d)	66

[a] Median effect concentration (EC50), median lethal concentration (LC50).
[b] No-observed-effect concentration.
[c] Metabolite of esterquat.
[d] Metabolite of diesterquat.

V. AMPHOTERIC SURFACTANTS

Amphoteric surfactants are used in personal care products, such as hair shampoos and conditioners, liquid soaps, and cleansing lotions. Other applications include cleaning agents and dishwashing detergents for household or industrial use. Amphoteric surface-active compounds comprise betaines and surfactants based on fatty alkyl imidazolines. Betaines are characterized by a fully quaternized nitrogen atom and do not exhibit anionic properties in alkaline solutions, which means that betaines are present only as "zwitterions," i.e., molecules with two ionic groups with equivalent charges. The other group of amphoterics is designated as imidazoline derivatives because of the formation of an intermediate imidazoline structure during the synthesis of some of these surfactants. This group contains the real amphoteric surfactants, such as alkylamphoacetates, alkylamphopropionates, and alkyliminopropionates, which form cations in acidic solutions, anions in alkaline solutions, and "zwitterions" in mid-pH-range solutions. Representative structures of amphoteric surfactants are shown in Figure 4.

A. Betaines and Imidazoline Derivatives

1. Aerobic Biodegradability

The primary biodegradation of betaines approaches 100%, as the loss of surface activity was reported to be 100% for C_{12} alkyl betaine, 98% for cocoamidopropyl betaine, and 96–100% for C_{14-15} hydroxysulfobetaine [73]. The results from ultimate biodegradability tests of alkyl betaines include degradation percentages below and above the pass level for ready biodegradability, especially, if older data are taken into account. However, several studies confirm the aerobic ready biodegradability of both

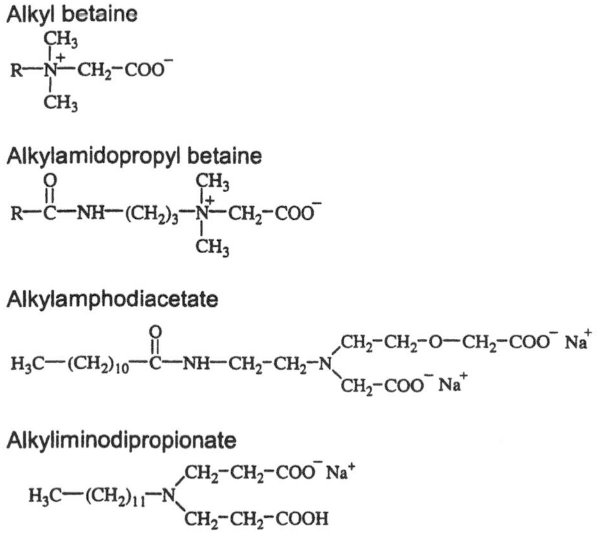

Alkyl betaine

$$R-\overset{\overset{CH_3}{|}}{\underset{\underset{CH_3}{|}}{N^+}}-CH_2-COO^-$$

Alkylamidopropyl betaine

$$R-\overset{\overset{O}{\|}}{C}-NH-(CH_2)_3-\overset{\overset{CH_3}{|}}{\underset{\underset{CH_3}{|}}{N^+}}-CH_2-COO^-$$

Alkylamphodiacetate

$$H_3C-(CH_2)_{10}-\overset{\overset{O}{\|}}{C}-NH-CH_2-CH_2-N\overset{CH_2-CH_2-O-CH_2-COO^-\ Na^+}{\underset{CH_2-COO^-\ Na^+}{}}$$

Alkyliminodipropionate

$$H_3C-(CH_2)_{11}-N\overset{CH_2-CH_2-COO^-\ Na^+}{\underset{CH_2-CH_2-COOH}{}}$$

FIG. 4 Structures of amphoteric surfactants.

TABLE 21 Ultimate Aerobic, Ready Biodegradability of Amphoteric Surfactants

Surfactant	Method	Result	Ref.
C_{12-14} Alkyl betaine	Closed bottle, 28 d	63% ThOD	19
Cocoalkyl betaine	Closed bottle, 30 d	>60% ThOD	73
Cocoamidopropyl betaine	Closed bottle, 30 d	84% ThOD	57
Cocohydroxysulfo betaine	CO_2 evolution	33% $ThCO_2$	73
C_{14-15} hydroxysulfo betaine	Closed bottle	40% ThOD	73
Cocoamphodiacetate	Closed bottle, 30 d	66% ThOD	73
Cocoamphodiacetate	Modified OECD screening	>70% DOC	73
C_{12-18} Alkyl amphopropionate	Modified OECD screening	79% DOC	73
C_{12} Alkyl iminodipropionate	Manometric respirometry, 28 d	99% ThOD	7

alkyl betaines and cocoamidopropyl betaines (Table 21). A special type of betaines is the hydroxysulfobetaines, in which the carboxylic group of alkyl betaine is replaced by sulfonate and an OH group is inserted in the hydrophilic moiety. These amphoteric surfactants cannot be regarded as readily biodegradable in standard screening tests on the basis of the available data. According to results reported by Domsch [73], the biodegradability of two hydroxysulfobetaines attained 33% and 40% in respirometric tests. The few available data describing the ultimate aerobic biodegradability of alkylamphodiacetate, alkylamphopropionate, and alkyliminodipropionate indicate that these substances are readily biodegradable (Table 21).

2. Anaerobic Biodegradability

Studies describing the anaerobic biodegradability of amphoteric surfactants are scarce. The biodegradability of cocoamidopropyl betaine and disodium cocoamphodiacetate has been examined in the ISO 11734 screening test. Under the methanogenic test conditions, the ultimate biodegradability of cocoamidopropyl betaine

TABLE 22 Aquatic Toxicity of Alkyl and Amidopropyl Betaines

Organisms	Surfactant	EC/LC50 (mg/L)[a]	Time	Ref.
Algae	C_{12-14} Alkyl	2.5	72 h	74
Scenedesmus subspicatus	Cocoamidopropyl	0.55; 1.84	96 h	57
Scenedesmus subspicatus	Cocoamidopropyl	NOEC: 0.09; 0.3	96 h	57
Scenedesmus subspicatus	Cocoamidopropyl	30; 48[b]	72 h	75
		NOEC: 3.2		
Crustaceans				
Daphnia magna	Cocoamidopropyl	6.5; 21.7	48 h	57
Fish				
Danio rerio	C_{12-14} Alkyl	21.9	96 h	74
Danio rerio	Cocoamidopropyl	2.0; 6.7	96 h	57

[a] Median effect concentration (EC50), median lethal concentration (LC50), or no-observed-effect concentration (NOEC, when indicated).
[b] First value: EC50, biomass; second value: EC50, growth rate.

attained 45% and 75% of ThGP after 28 and 56 days of incubation at 35°C, respectively [7]. Also, disodium cocoamphodiacetate was rapidly degraded in the anaerobic ISO 11734 test, as 78% ThGP was achieved after 56 days of incubation at 35°C (T. Madsen, unpublished data).

3. Toxicity to Aquatic Organisms

The aquatic toxicity of betaines varies considerably, and a large variation may be seen between tests conducted with the same species. This becomes evident when evaluating the effect concentrations determined for the green alga *S. subspicatus* because EC50 values have been found in the range from 0.55 to 48 mg/L (Table 22). No data describing the aquatic toxicity of alkylamphoacetates, alkylamphopropionates, and alkyliminopropionates were found in the literature.

VI. SURFACTANTS IN SLUDGE-AMENDED SOILS

Almost every surfactant will sorb to sludge solids during wastewater treatment; hence, some of the major surfactants may reach high concentrations in sludge intended for use on agricultural soil. The use of sludge as a nutrient source in agriculture is the primary route for the introduction of surfactants into the soil environment, although also other routes exist, such as leaking of sewer lines and septic tanks, use of sewage effluents as irrigation water, and use of surfactants in pesticide formulations. The application of sludge to agricultural soil is regulated in Denmark by Statutory Order No. 49 of January 20, 2000. This statutory order establishes certain cutoff values for organic substances in sludge to be used in agriculture, building upon the philosophy that low levels of the specified contaminants ensure that the sludge constitutes no risk to a sustainable land use. Maximum acceptable concentrations in sludge used for agricultural purposes have been defined for four groups of organic substances: (1) LAS, 1,300 mg/kg; (2) Σ PAH, 3 mg/kg; (3) Σ nonylphenol and nonylphenol ethoxylates with 1–2 EO, 10 mg/kg; and (4) di(2-ethylhexyl)phthalate, 50 mg/kg.

The present knowledge of the fate and effects of surfactants in the terrestrial environment is scarce, and the only surfactant for which an adequate set of data exists is LAS. Concentrations of LAS in aerobic sludge are usually lower than 500 mg/kg (dry weight), whereas levels between <1,000 and 30,000 mg/kg (dry weight) may be found in anaerobic sludge [76]. The concentrations of other surfactants in sludge have not been monitored as frequently as in the case of LAS. Cavalli and Valtorta [76] report surfactant concentrations in anaerobic sludges of 270–800 mg/kg for SAS, 18,800–51,900 mg/kg for soap, 0.3–1,360 mg/kg for alkylphenol ethoxylates, and <700 mg/kg for AE (all values in mg/kg dry weight). This indicates that surfactants that precipitate as poorly water-soluble Ca salts (LAS and soap) and/or resist degradation in the absence of molecular oxygen may accumulate in sludge, which has been treated by anaerobic digestion.

A. Biodegradation and Fate

When the sludge is applied as a fertilizer in agriculture, the sludge-bound organic contaminants enter a soil environment dominated by aerobic conditions. Aerobic biodegradation is considered to be the most important removal mechanism for surfactants in soil. The typical half-lives of LAS in the surface soil of sludge-amended fields range between 10 and 33 days [77,78]. The degradation of LAS in the field was confirmed in laboratory studies conducted with mixtures of sandy soils and sludge under defined conditions [79]. The effects of available oxygen on the mineralization of LAS were examined in sludge–soil mixtures with sludge:soil ratios of either 1:20 or 1:100 (dry weight), which were moistened to either 40% or 80% of the water-holding capacity (WHC). The mixtures were transferred to glass tubes and gently pressed into a core with a length of approx. 25 mm. During incubation at 15°C with a gas phase of atmospheric air, the mineralization of a pure ^{14}C-labeled C_{12}-2 LAS reached a level of 69–81% of the added ^{14}C after two months when aerobic conditions prevailed (C_{12}-2 LAS was randomly ^{14}C-labeled in the benzene ring). Sludge–soil mixtures moistened to 80% of WHC were only aerobic at the surface of the cores. In these mixtures, the LAS mineralization after two months was 39% at the 1:100 sludge:soil ratio (31% of core was aerobic) and 19% at the 1:20 sludge:soil ratio (19% of core was aerobic) [79]. The mineralization of ^{14}C-labeled C_{12}-2 LAS and a pure ^{14}C-labeled C_{13} AE with 4 EO was examined at 15°C in sludge–soil mixtures, in which either aerobic, denitrifying, or methanogenic conditions were maintained [80]. The C_{13} AE was ^{14}C-labeled at the alkyl carbon linked to the hydrophilic group, and the sludge:soil ratio applied in these experiments was 1:100. Mineralization of LAS occurred only under aerobic conditions, and, following a lag period of 16 days, the ^{14}CO$_2$ recovered after 75 days corresponded to 70% of the added ^{14}C. The aerobic mineralization of C_{13} AE occurred without a visible lag period and attained 65% of the added ^{14}C after 75 days. The C_{13} AE was more slowly mineralized under denitrifying conditions as a lag period of 12 days preceded the 41% mineralization, which was reached after 75 days. The mineralization of AE in actively denitrifying sludge–soil mixtures also occurred when the temperature was reduced to 4°C, although with a lower rate, as indicated by the ^{14}CO$_2$ recovery corresponding to 17% of the added ^{14}C after 75 days [81]. Strictly methanogenic conditions resulted in a very slow mineralization of the C_{13} AE, which corresponded to only 3.9% of the added ^{14}C after 75 days at 15°C [80,81]. The poor mineralization of the AE in the methanogenic sludge–soil mixtures contrasts to the findings in aqueous systems with anaerobic digested sludge (Table 10) and may be due to several factors, such as low adaptation of the biomass, low concentration of AE, which may have prevented growth of specific degraders, and stronger sorption of AE in soil systems. One of the concerns related to the use of sludge on agricultural fields is the possible leaching of contaminants to the groundwater, which is the predominant drinking water source in Denmark. The leaching of sludge-bound LAS in sludge-amended soils has been simulated for two Danish catchment areas, i.e., the Karup stream catchment, which consists mainly of sandy soil, and the Langvad stream catchment, which consists of loamy moraine clay soils with a topsoil of sandy loam [82]. The basic calculations for the two catchment areas included the following parameters: (1) soil characteristics and climate data, (2) sludge application rate of 6,000 kg/ha every three years, (3) sludge harrowing depth of 15 cm, (4) concentration of LAS in the sludge of 16,100 mg/kg, and (5) half-lives of LAS of 25 days in the sludge-containing surface layer and 15 days in the soil below the soil harrowing depth. The calculations showed that 98–99% of the sludge-bound LAS is expected to be degraded in the upper soil layer within one year after the application of sludge.

Furthermore, the simulations predicted that a high initial concentration of LAS occurs in the upper 0–15 cm soil profile following application of sludge containing LAS at 16,100 mg/kg. The initial concentration in the 0–15 cm is predicted to decrease to between 1 and 10 mg/kg within 3–4 months (Fig. 5). Leaching of LAS to depths below 1 m was predicted to be less than 1.3% of the LAS applied with sludge, and the concentrations of LAS in the percolate were estimated to be 1.4 μg/L for the Karup catchment and 2.3–8.5 μg/L for the Langvad catchment (all estimates were based on 16,100 mg LAS/kg sludge). The higher leaching potential for LAS predicted for loamy moraine clay soils in Langvad was due to transport of LAS in macropores [82].

B. Effects on Soil-Living Organisms

The adverse effects of sludge-bound LAS have been investigated in the field [83]. Briefly, two sludge samples, to which LAS was added (at 10 and 50 g/kg), and a control sludge (69 mg LAS/kg) were placed in well-defined strings in a sandy soil in southern Jutland (Lundgaard, Denmark) at a depth of 6–8 cm and were covered with soil. The oxidation of ammonium is usually a sensitive microbial parameter, and this process was completely inhibited in the two LAS-spiked sludge fractions during the first 3–4 weeks in the field, after which the activity returned to the level observed in the control sludge. However, the ammonium oxidation was also inhibited in some of the control sludge samples, which indicates that other factors than LAS affected the ammonium-oxidizing bacteria in the sludge [83]. Also, enchytreids living in the

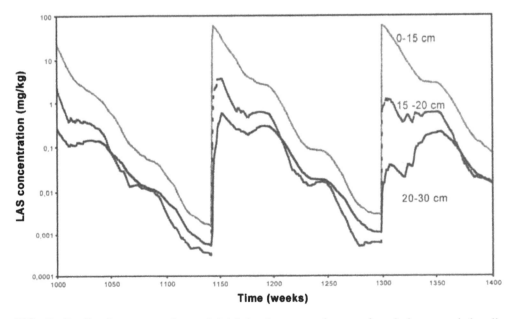

FIG. 5 Predicted concentrations of LAS in the upper layers of a sludge-amended soil (Langvad catchment; loamy clay soil). Sludge dosage was 6,000 kg/ha. The concentration of LAS in the sludge was 16,100 mg/kg (dry weight).

sludge were affected by LAS as these organisms were 15 and 9 times less abundant in the sludge samples at LAS concentrations of 10 and 50 g/kg, respectively, during the first three months as compared to the number of enchytreids in the control sludge. The enchytreid population in the LAS-contaminated sludge started to recover after three months, but the normal population size was not regained after five months [83].

The effects of LAS on microbial processes have also been examined in laboratory experiments conducted with sludge–soil mixtures. Elsgaard et al. examined the toxicity of LAS to ammonium oxidation, iron reduction, dehydrogenase activity, and microbial biomass in a sandy soil [84] and in sludge–soil mixtures [85]. The study using sludge–soil mixtures [85] showed that particularly the iron reduction was sensitive to LAS, as indicated by the effect concentrations: EC10, 24 mg/kg (95% confidence limits, 0–32 mg/kg); EC50, 47 mg/kg (95% confidence limits, 40–52 mg/kg); and NOEC, 22 mg/kg (all values are mg/kg dry weight for 10 d of incubation). The ammonium oxidation in the sludge–soil mixtures was also inhibited by LAS, as indicated by the effect concentrations: EC10, 102 mg/kg (95% confidence limits, 91–118 mg/kg); EC50, 272 mg/kg (95% confidence limits, 244–294 mg/kg); and NOEC, 62 mg/kg (all values are mg/kg dry weight for 14 d of incubation). Gejlsbjerg et al. [86] found a slightly lower toxicity of LAS to the ammonium oxidation in sludge–soil mixtures as the EC10 was 157 mg/kg (95% confidence limits, 98–207 mg/kg), the EC50 was 431 mg/kg (95% confidence limits, 373–483 mg/kg), and the NOEC was 250 mg/kg (all values are mg/kg dry weight). Anaerobic CH_4 formation was also affected by LAS, as the EC10, EC50, and NOEC values were 57 mg/kg (95% confidence limits, 23–90 mg/kg), 277 mg/kg (95% confidence limits, 222–335 mg/kg), and 125 mg/kg, respectively (all values are mg/kg dry weight). No effects of LAS were observed on aerobic respiration and denitrification, as the NOEC were $\geq$2,500 mg/kg and $\geq$5,000 mg/kg (dry weight), respectively [86].

Studies of the effects of sludge-bound LAS on soil invertebrates were conducted with the enchytreid *Enchytraeus albidus* and the springtail *Folsomia candida* (all values are expressed in mg/kg dry weight; Ref. 86). By using the reproduction of *E. albidus* as the endpoint, it was found that the EC10 was 447 mg/kg (95% confidence limits, 177–648 mg/kg), the EC50 was 1,143 mg/kg (95% confidence limits, 877–1,400 mg/kg), and the NOEC was 750 mg/kg. The reproduction-effect concentrations for *F. candida* were: EC10, 480 mg/kg (95% confidence limits, 352–567 mg/kg); EC50, 1,437 mg/kg (95% confidence limits, 1,235–1,894 mg/kg); and NOEC, 500 mg/kg. Lower effect concentrations of LAS are obtained in laboratory studies conducted with soil without addition of sludge. The following effect concentrations toward the reproduction of soil invertebrates were found in dose–response experiments with LAS by using an agricultural, sandy soil (all values are expressed in mg/kg dry weight; Ref. 87): *E. albidus* (EC10, 6 mg/kg; EC50, 41 mg/kg), *Aporrectodea caliginosa* (earthworm; EC10, 14 mg/kg; EC50, 129 mg/kg), *Aporrectodea longa* (earthworm; EC10, 27 mg/kg; EC50, 137 mg/kg), *Folsomia fimetaria* (springtail; EC10, 85 mg/kg; EC50, 424 mg/kg), *Hypogastrura assimilis* (springtail; EC10, 99 mg/kg; EC50, 421 mg/kg), and *Hypoaspis aculeifer* (mite; EC10, 82 mg/kg; EC50, 236 mg/kg). The higher toxicity of LAS to *E. albidus* and springtails observed by Holmstrup and Krogh [87] as compared to the results of Gejlbjerg et al. [86] may at least partly be due to a higher bioavailabilty of LAS in soil without sludge organics. Because of the high bioavailability of LAS in sandy soil, the results of Holmstrup and Krogh [87] represent a conservative estimate of the effects of LAS in sludge-amended soils.

In a monitoring program executed in Denmark, the median concentration of LAS attained 300 mg/kg (dry weight) in dewatered sludge samples collected from 20 wastewater treatment plants, whereas the LAS concentration exceeded 900 mg/kg (dry weight) in 25% of the samples [88]. This implies that the predicted environmental concentration (PEC $_{soil}$) of LAS in sludge-amended soil can be estimated to be 1.2 mg/kg (LAS sludge concentration, 300 mg/kg) and 3.6 mg/kg (LAS sludge concentration, 900 mg/kg), assuming a 1:250 dilution of the sludge in the soil. The lowest NOEC or EC10 values of LAS in soil were observed for microbial iron reduction (sludge–soil mixtures; NOEC, <8 and 22 mg/kg dry weight for 5 d and 10 d of incubation, respectively; Ref. 85) and for the enchytreid *E. albidus* (soil; EC10, 6 mg/kg dry weight; Ref. 87). The predicted no-effect concentration (PNEC $_{soil}$) can be estimated to be 0.6 mg/kg by using the EC10 for *E. albidus* and an assessment factor of 10 [89]. Risk quotients (= PEC $_{soil}$/PNEC $_{soil}$) in sludge-amended soil can then be calculated to be 2 and 6 for LAS sludge concentrations of 300 and 900 mg/kg, respectively. This indicates a risk of adverse effects of LAS immediately after the application of the sludge on agricultural soil. Linear alkylbenzene sulfonates (and many other major surfactants) are rapidly degraded in the aerobic soil environment, as illustrated by model simulations predicting a degradation of 98–99% of the initial LAS added with sludge during one year [82]. The concentration of LAS in sludge-amended soil is expected to decline rapidly with time (Fig. 5), and the risk of long-term toxic effects is therefore considered to be low.

REFERENCES

1. Matthijs, E.; Holt, M.S.; Kiewiet, A.; Rijs, G.B.J. Environ. Toxicol. Chem. 1999, *18*, 2634–2644.
2. European Council on Studies on LAB/LAS (ECOSOL). Surfactant consumption in household detergents. European market, 1996.
3. Steber, J.; Berger, H. *Biodegradability of Surfactants*; Karsa, D.R., Porter, M.R., Eds.; Blackie Academic & Professional: Glasgow, 1995; 134–182.
4. Painter, H.A. *Detergents*. de Oude, N.T., Ed.; Springer-Verlag: Berlin, 1992; 1–88.
5. Battersby, N.S.; Kravetz, L.; Salanitro, J.P. Effect of branching on the biodegradability of alcohol-based surfactants. Proceedings of 5th World Surfactant Congress, Florence, 2000; 13971–407.
6. Schöberl, P.; Bock, K.J.; Huber, L. Tenside Surfact. Det. 1988, *25*, 86–98.
7. Madsen, T.; Boyd, H.B.; Nylén, D.; Pedersen, A.R.; Petersen, G.I.; Simonsen, F. Environmental and health assessment of substances in household detergents and cosmetic detergent products. Environmental Project No. 615, http://www.mst.dk/udgiv/publications/2001/87-7944-596-9/html/. Danish Environmental Protection Agency, Ministry of Environment and Energy, Copenhagen, 2001.
8. Salanitro, J.P.; Diaz, L.A. Chemosphere 1995, *30*, 813–830.
9. Nuck, B.A.; Federle, T.W. Environ. Sci. Technol. 1996, *30*, 3597–3603.
10. Birch, R.R.; Biver, C.; Campagna, R.; Gledhill, W.E.; Pagga, U.; Steber, J.; Reust, H.; Bontinck, W.J. Chemosphere 1989, *19*, 1527–1550.
11. Madsen, T.; Rasmussen, H.B.; Nilsson, L. Chemosphere 1995, *31*, 4243–4258.
12. Kikuchi, M.; Wakabayashi, M.; Nakamura, T.; Inoune, W.; Takahashi, K.; Kawana, T.; Kawahara, H.; Koido, Y. Ann. Rep. Tokyo. Metrop. Res. Inst. Environ. Prot. 1976, 57–69.

13. Roberts, M.H.; Warinner, J.E.; Tsai, C.F.; Wright, D.; Cronin, L.E. Arch. Environ. Contam. Toxicol. 1982, *11*, 681–692.
14. Verge, C.; Moreno, A.; Roque, S. Tenside Surf. Det. 1996, *33*, 166–169.
15. Bishop, W.E.; Perry, R.I. *Aquatic Toxicology and Hazard Assessment: 4th Conference, STP 737*; Branson, D.R., Dickson, K.L., Eds.; ASTM: Philadelphia, 1979; 421–435.
16. Maki, A.W. J. Fish. Res. Board. Can. 1979, *36*, 411–421.
17. Reiff, B.; Lloyd, R.; How, M.J.; Brown, D.; Alabaster, J.S. Water Res. 1979, *13*, 207–210.
18. Dyer, S.D.; Stanton, D.T.; Lauth, J.R.; Cherry, D.S. Environ. Toxicol. Chem. 2000, *19*, 608–616.
19. Madsen, T.; Damborg, A.; Rasmussen, H.B.; Seierø, C. Evaluation of methods for screening surfactants. Ultimate aerobic and anaerobic biodegradability. Working Report No. 38, Danish Environmental Protection Agency, Ministry of Environment, Copenhagen, 1994.
20. Denger, K.; Cook, A.M. J. Appl. Microbiol. 1999, *86*, 165–168.
21. Angelidaki, I.; Haagensen, F.; Ahring, B.K. Anaerobic transformation of LAS in continuous stirred tank reactors treating sewage sludge. Proceedings of 5th World Surfactant Congress, Florence, 2000; 1551–1557.
22. Wagener, S.; Schink, B. Water Res. 1987, *21*, 615–622.
23. Lewis, M.A.; Hamm, B.G. Water Res. 1986, *20*, 1575–1582.
24. Utsunomiya, A.; Watanuki, T.; Matsushita, K.; Tomita, I. Environ. Toxicol. Chem. 1997, *16*, 1247–1254.
25. Kusk, S. Petersen, Environ. Toxicol. Chem. 1997, *16*, 1629–1633.
26. Kimerle, R.A.; Swisher, R.D. Water Res. 1977, *11*, 31–37.
27. Fairchild, J.F.; Dwyer, F.J.; La Point, T.W.; Burch, S.A.; Ingersoll, C.G. Environ. Toxicol. Chem. 1993, *12*, 1763–1775.
28. Lundahl, P.; Cabridenc, R. Water Res. 1978, *12*, 25–30.
29. Yamane, A.N.; Okada, M.; Sudo, R. Water Res. 1984, *18*, 1101–1105.
30. Madsen, T.; Rasmussen, H.B.; Nilsson, L. Methods for screening anaerobic biodegradability and toxicity of organic chemicals. Environmental Project No. 336, Danish Environmental Protection Agency, Ministry of Environment and Energy, Copenhagen, 1996.
31. Steber, J.; Wierich, P. Water Res. 1987, *21*, 661–667.
32. BKH Consulting Engineers. Environmental Data Review of Soaps. Delft, The Netherlands, 1994.
33. Canton, J.H.; Slooff, W. Chemosphere 1982, *11*, 891–907.
34. Wong, D.C.L.; Dorn, P.B.; Chai, E.Y. Environ Toxicol Chem 1997, *16*, 1970–1976.
35. Ahel, M.; McEvoy, J.; Giger, W. Environ. Pollut. 1993, *79*, 243–248.
36. Laws, S.C.; Carey, S.A.; Ferrell, J.M.; Bodman, G.J.; Cooper, R.L. Toxicol. Sci. 2000, *54*, 154–167.
37. Hemmer, M.J.; Hemmer, B.L.; Bowman, C.J.; Kroll, K.J.; Folmar, L.C.; Marcovich, D.; Hoglund, M.D.; Denslow, N.D. Environ. Toxicol. Chem. 2001, *20*, 336–343.
38. Talmage, S.S. *Environmental and Human Safety of Major Surfactants*; Lewis: Boca Raton, FL, 1994.
39. Marcomini, A.; Zanette, M.; Pojana, G.; Suter, M.J.-F. Environ Toxicol Chem 2000, *19*, 549–554.
40. Marcomini, A.; Pojana, G.; Carrer, C.; Cavalli, L.; Cassani, G.; Lazzarin, M. Environ. Toxicol. Chem. 2000, *19*, 555–560.
41. Balson, T.; Felix, M.S.B. *Biodegradability of Surfactants*; Karsa, D.R., Porter, M.R., Eds.; Blackie Academic & Professional: Glasgow, 1995; 183–203.
42. Madsen, T.; Petersen, G.; Seierø, C.; Tørsløv, J. J. Am. Oil. Chem. Soc. 1996, *73*, 929–933.
43. Kaluza, U.; Taeger, K. Tenside Surf. Det. 1996, *33*, 46–51.
44. Kravetz, L.; Salanitro, J.P.; Dorn, P.B.; Guin, K.F. J. Am. Oil. Chem. Soc. 1991, *68*, 610–618.

45. Bertleff, W.; Baur, R.; Gümbel, H.; Welch, M. SÖFW-J. 1997, *123*, 222–233.
46. Naylor, C.G.; Castaldi, F.J.; Hayes, B.J. J. Am Oil. Chem. Soc. 1988, *65*, 1669–1676.
47. Wagener, S.; Schink, B. Appl. Environ. Microbiol. 1988, *54*, 561–565.
48. Huber, M.; Meyer, U.; Rys, P. Environ. Sci. Technol. 2000, *34*, 1737–1741.
49. Federle, T.W.; Schwab, B.S. Water Res. 1992, *26*, 123–127.
50. Maki, A.W.; Bishop, W.E. Arch. Environ. Contam. Toxicol. 1979, *8*, 599–612.
51. Lizotte, R.E., Jr.; Wong, D.C.L.; Dorn, P.B.; Rodgers, J.H., Jr. Arch. Environ. Contam. Toxicol. 1999, *37*, 536–541.
52. Gillespie, W.B., Jr.; Rodgers, J.H., Jr.; Crossland, N.O. Environ. Toxicol. Chem. 1996, *15*, 1418–1422.
53. Dorn, P.B.; Salanitro, J.P.; Evans, S.H.; Kravetz, L. Environ. Toxicol. Chem. 1993, *12*, 1751–1762.
54. Steber, J.; Guhl, W.; Stelter, N.; Schröder, F.R. Tenside Surf. Det. 1995, *32*, 515–521.
55. Garcia, M.T.; Ribosa, I.; Campos, E.; Sanchez Leal, J. Chemosphere 1997, *35*, 545–556.
56. Stalmans, M.; Matthijs, E.; Weeg, E.; Morris, S. SÖFW-J. 1993, *119*, 795–806.
57. IUCLID. CD-ROM. Public data on high-volume chemicals, Joint Research Centre, European Chemicals Bureau, Ispra, Italy, 2000.
58. Akzo Nobel Surface Chemistry, Stenungsund, Sweden. Product information. Bermodol SPS 2525, SPS 2532 and SPS 2543, 1999.
59. van Ginkel, C.G. *Biodegradability of Surfactants*; Karsa, D.R., Porter, M.R., Eds.; Blackie Academic & Professional: Glasgow, 1995; 183–203.
60. Fenger, B.H.; Mandrup, M.; Rohde, G.; Sørensen, J.C.K. Water. Res. 1973, *7*, 1195–1208.
61. Games, L.M.; King, J.E.; Larson, R.J. Environ. Sci. Technol. 1982, *16*, 483–488.
62. Sullivan, D.E. Water Res. 1983, *17*, 1145–1151.
63. Gerike, P.; Gode, P. Chemosphere 1990, *21*, 799 812.
64. Puchta, R.; Krings, P.; Sandkühler, P. Tenside Surf. Det. 1993, *30*, 186–191.
65. Waters, J.; Kleiser, H.H.; How, M.J.; Barratt, M.D.; Birch, R.R.; Fletcher, R.J.; Haigh, S.D.; Hales, S.G.; Marshall, S.J.; Pestell, T.C. Tenside Surf. Det. 1991, *28*, 460–468.
66. Giolando, S.T.; Rapaport, R.A.; Larson, R.J.; Federle, T.W.; Stalmans, M.; Masscheleyn, P. Chemosphere 1995, *30*, 1067–1083.
67. Chemische Fabrik CHEM-Y GmbH, Emmerich, Germany. Closed bottle test of Akypoquat 131, 1991.
68. van Ginkel, C.G.; Venema, M.A.; Geurts, M.G.J. Rapid biodegradation of dissolved long-chain dialkyldimethylammonium salts. Proceedings of 5th World Surfactant Congress, Florence, 2000; 1408–1413.
69. Boethling, R.S.; Lynch, D.G. *Detergents*; de Oude, N.T., Ed.; Springer-Verlag: Berlin, 1992; 145–177.
70. Lewis, M.A.; Suprenant, D. Ecotoxicol. Environ. Saf. 1983, *7*, 313–322.
71. Lewis, M.A.; Wee, V.T. Environ. Toxicol. Chem. 1983, *2*, 105–118.
72. Roghair, C.J.; Buijze, A.; Schoon, H.N.P. Chemosphere 1992, *24*, 599–609.
73. Domsch, A. *Biodegradability of Surfactants*; Karsa, D.R., Porter, M.R., Eds.; Blackie Academic & Professional: Glasgow, 1995; 231–254.
74. Berol Nobel. Amphoteen 24. Produktinformation, 1993.
75. Th. Goldschmidt AG. Prüfung auf Hemmung der Algenzellvermehrung von Cocamidopropyl Betaine–F 3006, 1993.
76. Cavalli, L.; Valtorta, L. Tenside Surf. Det. 1999, *36*, 22–28.
77. Berna, J.L.; Ferrer, J.; Moreno, A. Tenside Surf. Det. 1989, *26*, 101–107.
78. Holt, M.S.; Matthijs, E.; Waters, J. Water Res. 1989, *23*, 749–759.
79. Gejlsbjerg, B.; Klinge, C.; Madsen, T. Environ. Toxicol. Chem. 2001, *20*, 698–705.
80. Gejlsbjerg, B.; Andersen, T.T.; Madsen, T. Mineralization of organic contaminants under aerobic and anaerobic conditions in sludge–soil mixtures, J. of Soils and Sediments, *in press*.

81. Andersen, T.T. Biodegradation and bioaccumulation of organic contaminants in sludge-amended soil. MSc thesis (available via contact with DHI Water & Environment, Hørsholm, Denmark). University of Copenhagen, Denmark, 2001.
82. Madsen, T.; Winther-Nielsen, M.; Rasmussen, D. CLER Rev. 1999, *5*, 14–19.
83. Brandt, K.K.; Krogh, P.H.; Cassani, G.; Sørensen, J. Does LAS affect the soil ecosystem in sludge-amended soil? Results from a field trial with well-defined strings of LAS-amended sludge in soil. Proceedings of 5th World Surfactant Congress, Florence, 2000; 1590–1597.
84. Elsgaard, L.; Petersen, S.O.; Debosz, K. Environ. Toxicol. Chem. 2001, *20*, 1656–1663.
85. Elsgaard, L.; Petersen, S.O.; Debosz, K. Environ. Toxicol. Chem. 2001, *20*, 1664–1672.
86. Gejlsbjerg, B.; Klinge, C.; Samsøe-Petersen, L.; Madsen, T. Environ. Toxicol. Chem. 2001, *20*, 2709–2716.
87. Holmstrup, M.; Krogh, P.H. Environ. Toxicol. Chem. 2001, *20*, 1673–1679.
88. Tørsløv, J.; Samsøe-Petersen, L.; Rasmussen, J.O.; Kristensen, P. Use of waste products in agriculture. Environmental Project No. 366, Danish Environmental Protection Agency, Ministry of Environment and Energy, Copenhagen, 1997.
89. European Commission. Technical guidance document in support of Commission Directive 93/67/EEC on risk assessment for new notified substances and Commission Regulation (EC) No. 1488/94 on risk assessment for existing substances. Chapter 3, Environmental risk assessment, revised guidance document, 2002.

9

The Biodegradability of Detergent Ingredients in an Environmental Context

A Practical Approach

JOHN SOLBÉ Consultant, Denbighshire, North Wales, United Kingdom

I. INTRODUCTION

The reader interested in definitive lists of the biodegradability properties of detergent ingredients is referred to Swisher (1987) or Hennes-Morgan and de Oude (1998). Useful texts on biodegradation test methods may be found in Hales et al. (1997), de Henau (1998), and the publications of the European Union (EEC, 1996; ECB, 2003), the protocols of the International Organization for Standardization (ISO), and the guidelines of the Organization for Economic Cooperation and Development (OECD). The ISO and OECD methods have been usefully summarized and compared in Hales et al. (1997). Instead of repeating this information in detail, the chapter places biodegradability in a context of pathways to the environment following use and in interpreting biodegradability as a contribution to environmental risk assessment. Examples are provided of the effects of detergent ingredients on processes that may themselves be important in the breakdown of chemicals following use.

Biodegradability, the potential of a chemical to be broken down by living organisms, is recognized as an intrinsic property of an organic substance. But in the same way that toxicity will vary according to the conditions in which the substance occurs, biodegradability is achieved in different ways and to different extents, depending on the characteristics of the environment. Examples of environmental variables controlling the rate and extent of biodegradability include oxygen concentration, water content of soils, temperature, and the availability of organisms capable of degrading the given substance. Some substances, as a result of other properties of the molecule (steric hindrance etc.) will not degrade under any conditions; other substances require a very special environment, and yet others can be freely utilized (degraded) by living organisms. There are abiotic mechanisms by which a substance can degrade in the environment, such as photolysis (OECD, 1997) and hydrolysis. Ingredients may, in or after use, form different salts from those in which they were formulated (thus a sodium salt in use may become a calcium salt after use), modifying the chemical and physical behavior of the substance, for example, in terms of solubility. All of these factors should be considered when examining the degree of persistence of substance (see Section V.C) in the environment. Note that it is the *substance*, not the *product* or formulation, that should be the subject of discussions of degradation. To the environment the concept of a product has no relevance. The product has lost its integrity the moment it is used or accidentally reaches the environment. The ingredients of a product each go their own way, according to their physical properties (solubility, volatility, degradability, etc.). Statements about the biodegradability of a product, tested as a whole, are generally meaningless and can be seriously misleading. A claim such as "99% biodegraded" should be regarded with little comfort without knowing the identity and properties of the remaining 1% of the mixture.

II. USES OF BIODEGRADABILITY DATA

Degradability of man-made substances is generally a property that is undesirable, at least until they have performed the functions for which they were designed. For example, we do not want a milk carton or the milk it contains to degrade until a reasonable period has elapsed. Once the intended period of use is over, however, we generally look at degradability as a positive attribute. But the period is dependent on the intended use: A detergent ingredient should remain undegraded for months before use, with the packaging retaining the product correctly. Perhaps fresh milk should remain intact for a few days, but mature cheese is expected to retain its integrity for months. Degradability of naturally occurring and persistent substances, e.g., wood, humic acids, DNA, oils, is not normally a matter of concern or debate. For man-made

TABLE 1 Impact of Failure to Pass a "Ready Test" of Biodegradability on Classification under European Directive 67/548

Toxicity (acute EC50: mg/L)	Substance readily biodegradable?	Classification
<1.0	Yes	Very toxic to aquatic organisms
<1.0	No	Very toxic to aquatic organisms; may cause long-term adverse effects in the aquatic environment
1–10	Yes	Not classified
1–10	No	Toxic to aquatic organisms; may cause long-term adverse effects in the aquatic environment
10–100	Yes	Not classified
10–100	No	Harmful to aquatic organisms; may cause long-term adverse effects in the aquatic environment

chemicals, however, judgments on the rate and extent of biodegradability are required for two purposes in chemicals management: classification and risk assessment.

A. Classification and Labeling

In the European Union, general chemicals are classified according to Directive 67/548, as later amended (EEC, 1992). The criterion used is "ready biodegradability" (see upcoming Section V.B). *Lack of ready biodegradability on its own is not a harmful property*, nor does it constitute a hazard of sufficient importance for the chemical to be classified. But this attribute when combined with other features of the chemical, especially toxicity, influences (increases the severity of) classifications made on the basis of toxicity. Table 1 demonstrates the impact of the property "poor degradability" on classification in Europe.

In the OECD program of work on high-production-volume chemicals, the preparation of Screening Initial Data Sheets (SIDS) contains the following list of information under the heading "Environmental Fate and Pathways," of which the items in **bold** are of particular relevance to this chapter. The information largely concerns intrinsic properties, but items such as "Additional information: sewage treatment" give a broader scope.

OECD SIDS Endpoints

Stability:	*Bioaccumulation*
Photodegradation	*Transport and distribution*:
Stability in water	Transport between environmental compartments
Stability in soil	Distribution.
Aerobic biodegradation	*Additional information*:
BOD-5, COD, or ratio BOD-5/COD	**Sewage treatment**
Monitoring data (environment)	Other information

B. Risk Assessment

The assessment of risk is made by considering the intrinsic properties of a chemical and its exposure to the situations potentially at risk. Guidelines have been provided in Europe for completing a chemicals risk assessment (EEC, 1996; ECB, 2003). To briefly demonstrate risk: Cyanide has an intrinsic toxicity, but its risk to human beings or to, say, fish, will depend on how toxic it is to the sensitive systems of the species under consideration, how it is stored, how accessible and appropriate the container is, etc., as well as some other physicochemical features of the situation in which possible exposure might become a reality, such as the pH of water into which the cyanide may be spilled (the lower the pH, the higher the proportion of undissociated molecule and the more toxic the cyanide). The same understanding of intrinsic properties and exposure to these properties is required when considering detergent ingredients.

Since 1999 an A.I.S.E*/Cefic† initiative (Project HERA—Human and Environmental Risk Assessment) has been running in Europe, with collaborative links to the United States and Japan. This project has developed risk assessment methods for evaluating the risk to human health and the environment of the ingredients of household detergent cleaning chemicals. The methods and results are posted on the HERA Web site (www.heraproject.com). The intention is to publish assessments within the next few years of all the major ingredients used by the A.I.S.E. companies in these types of products. Biodegradation is obviously an important component of the HERA assessments, which follow to a considerable extent the EU Technical Guidance Document (EEC, 1996; ECB, 2003), although modified to take advantage of specific knowledge of the use of detergent ingredients in Europe (Fox, 2001).

The role of biodegradability in risk assessment is not due to the "persistence" property alone, because this is not, in itself, a harmful feature. It is much more to do with the possibility that low rates of degradation may give other features of the chemical *more time* to express their intrinsic harm. Thus toxicity is influenced by *potency* (the inherent capacity of the chemical to poison), *dose* (the amount of chemical to which the subject is exposed at any time), and *period of exposure* (more or less the longer the exposure the greater the chance of toxicity, even at levels below those that can cause harm in short exposures: acute toxicity). The lower the rate of breakdown of a chemical, the greater the chance that harm may be seen, but only if the substance has any reasonable chance of being toxic in the first place. For example, many substances are inert under normal environmental conditions, due either to special features of their structure or to the sheer size of the molecule (molecules with a relative molecular mass greater than 700 are generally considered unable to penetrate living membranes).

III. ROUTES OF WASTE PRODUCTS TO THE ENVIRONMENT

The role and extent of biodegradability and the consequent environmental profiles of detergent chemicals will vary according to the route by which the substance moves toward the environment. A number of possibilities may be considered.

a. Direct entry of a detergent chemical, due to use in an environmental compartment, e.g., as part of the formulation of an agricultural pesticide.
b. Disposal of used washing water to the local area, e.g., to irrigate a vegetable plot.
c. Transport in a sewer to rudimentary treatment facility, e.g., sedimentation only.
d. Transport to a higher-technology wastewater treatment plant, including, e.g., a biological oxidation stage, or to a plant incorporating tertiary treatment, such as sand filters, lagoons, or grass plots, to *polish* the effluent. (Polishing of effluents includes final removal of harmful bacteria, of suspended solids, and of nutrients. In these processes there is an opportunity for further biodegradation.)

Described next is the effect of the various stages and types of treatment where treatment is provided. The consequences where there is no treatment but *direct*

---

*Association Internationale de la Savonnerie, de la Détergence et des Produits d'Éntretien (AISE).
†European Chemical Industry Council.

discharge of wastes containing detergent chemicals is described in Section IV: Responses of Receiving Waters and Soils.

A. Role of Infrastructure: Sewers, Sewage Works, Storm Sewers, Septic Tanks

1. Sewers

Detergent products are typically disposed of to some form of sewer, which may be connected to a sewage treatment works. The sewer itself is not biologically inert, and both abiotic and biological processes may act on the ingredients of the detergent. Insoluble materials may settle out, particularly if the flow of sewage is low or the gradient of the sewer is inadequate to allow continual scouring of deposits or lack of settlement. Aerobic and anaerobic conditions may both exist, allowing aerobes and anaerobes to utilize the ingredients as an energy source and to build their tissues. When less detergent ingredient emerges from a sewer than went into it, the conclusion should not be drawn that it has necessarily degraded in the sewer; it may simply be accumulating until the next high-flow event sweeps it out. Bearing in mind these provisos, the data for surfactants in Table 2 have been reported by A.I.S.E. and CESIO (1995).

2. Sewage Works

Within the sewage works a number of processes may be applied that have a direct bearing on the nature and extent of biodegradation. These are considered next in a sequence typical of works treating the wastes of a small urban community (say, 10,000 population equivalents). A synopsis of the ability of sewage treatment plants to degrade and remove detergent ingredients is given later and in Table 3.

(a) Screening and Grit Removal. This removal of large debris at the start of the process is unlikely to influence degradation of detergent chemicals. It exists to protect the works from problems such as clogging of pipework and damage to machinery.

(b) Primary Settlement/Sedimentation. The *raw sewage* is given a period (around six hours) of quiescent settling, often with the addition of chemicals such as lime to aid flocculation. Detergent ingredients vary in their solubility and tendency to adsorb onto solids. High adsorption and low solubility may dictate that some ingredients settle out in this preliminary stage of sewage treatment to form, with other settleable constituents of the sewage, the *primary sludge*, or *raw sludge*. In this way *removal* of the detergent ingredient from the waste stream has begun. The primary sludge is then

TABLE 2 Observed In-Sewer Removal of Surfactants

Surfactant	In-sewer removal (%), mean (range)
Linear alkyl benzene sulfonate	50 (10–68)
Alcohol ethoxylate (C_{12-15})	42 (28–58)
Alcohol ethoxy sulfate (C_{12-15})	11 (0–40)
Alcohol sulfate (C_{12-15})	55 (18–85)
Soap	Increased, probably from hydrolysis of fats and oils in sewage

Source: A.I.S.E. and CESIO (1995).

TABLE 3 Observed Treatment Efficiency in Dutch Sewage Treatment Plants

Surfactant or general biochemical oxygen demand (BOD)	Removal in sewage treatment (%), average (range)
BOD	98.1 (96.4–99.2)
Linear alkyl benzene sulfonate	99.2 (98.0–99.6)
Alcohol ethoxylate ($C_{12–15}$)	99.8 (99.6–99.9)
Alcohol ethoxy sulfate ($C_{12–15}$)	99.6 (99.3–99.9)
Alcohol sulfate ($C_{12–15}$)	99.2 (99.0–99.6)
Soap	99.1 (97.7–99.6)

Source: A.I.S.E. and CESIO (1995).

treated in various ways. It can be disposed of directly to land or *digested* under hot or cold, aerobic or anaerobic conditions. If a sludge digester is used, it will normally also receive *secondary sludge*, as described later.

(c) Biological Oxidation. The supernatant from the settling tanks (*settled sewage*) is passed to one of several types of biological treatment stage, which are typically classed as *biological oxidation*. Whether the technology is relatively advanced, as in the *activated-sludge process*, or relatively simple, as in *percolating filters* (*trickling filters/ bacteria beds*), the principle is the same: to bring the constituents of the sewage into contact with organisms in the presence of air so that biodegradation can occur. This stage of sewage treatment is of crucial importance. The oxidation must include that of ammonia to nitrate (we talk of a *nitrifying plant*) to prevent the exceedingly toxic undissociated ammonia from reaching the environment. (Ammonia is approximately 150 times more toxic than typical surfactants.)

Although the principles of biological oxidation remain the same, the conditions for micro- and macroorganisms vary considerably. In activated-sludge processes, the oxidation occurs in a tank or tanks filled with wastewater. On its own this would quickly become anoxic, but various methods are used to bring air into contact with the full volume of the tank, for example, by spinning agitators at the surface or by diffusers delivering air at the base of the tank, which can thus be called an *aeration tank*. There is an influence of the presence of surface-active agents at the air/water interface reducing the reaeration coefficient of water (see Section IV), but in the sewage works this is of no great significance to operations.

The microflora is dominated by bacteria, but there is also a rich community of protozoa and other organisms. There are many texts on the subject. The reader is referred to the two volumes of Curds and Hawkes (1975) for details on the organisms and the processes. In essence the bacteria are utilizing the constituents of the sewage, including detergent ingredients, for growth,[*] some of which is harvested by the protozoa. An excess of bacteria is formed, and this is allowed to settle in the final settlement tank, from which most is returned to the primary settlement tank, with the rest returned to the input region of the aeration tank as an inoculum. This inoculum gives its name to the process—*activated sludge*. This proportion of the total mass in the

---

[*] This fact is of great importance when interpreting the results of biodegradability tests.

aeration tank, the *mixed liquor suspended solids*, can be considered to comprise a community that has already responded to the conditions of the aeration tank, including the presence of everyday chemicals, such as those found as detergent ingredients. It is likely therefore that all, or almost all, the degradation pathways are in place instantly to utilize the materials in the incoming settled sewage. (Half-lives of chemicals of a few minutes are not unknown in a properly working aeration tank.) Birch (1984) calculated that the degradative ability of an activated-sludge plant is such that at steady state the effluent concentration would be independent of influent concentration. In other words the daily or hourly variation in concentration of detergent ingredients in the sewage would not affect the effluent quality, the plant coping with sudden changes.

The manager of the activated-sludge plant can manipulate the proportion of sludge "wasted" to the digester or returned to the aeration tank. The more that is wasted, the lower the average age of the remaining sludge. The term used to describe sludge age is *sludge retention time* (SRT). This too can have an influence on the degradation of detergent chemicals in the plant because the bacteria have a relatively slow rate of growth. Those that oxidize ammonia, for instance, require around six days of SRT or they will be swept out of the system faster than they can reproduce. The same is true for the degradation of nitrilo-tri-acetic acid (NTA). In fact for every detergent chemical there will be a critical SRT, but we need only concern ourselves with ingredients needing particularly slow-growing groups of bacteria to achieve effective degradation. All the others will be accommodated by the normal SRTs. A typical SRT will be around seven days. In an overloaded plant it may be far less, and the consequence will be that ammonia and complex residues of the constituents of the sewage will appear in the final effluent, contaminating or polluting (damaging) the receiving waters.

Activated-sludge plants can generally be considered powerful mechanisms for degrading chemicals, but they do have one drawback: Being fully mixed systems they are vulnerable to shock loads of toxic chemicals. No detergent chemical in its normal use is toxic enough to place an activated-sludge plant at risk, but an accidental discharge to sewer of some industrial chemicals could cause this problem.

Percolating filters are not as sensitive to such shock loads because they are not fully mixed systems. The microflora in a filter (known as the *film*) coats a solid medium such as blast furnace slag or plastic shapes. The medium is held in a tank, but this is not full of wastewater. The bottom of the tank is open, and there is a natural upward draught of air through the medium, providing oxygen and carrying away any gases produced during this type of sewage treatment. Settled sewage is trickled or sprayed onto the top of the filter and percolates its way downward to leave as effluent through the underdrains. The percolating filter organisms are very diverse, ranging from the bacteria and fungi (the latter typically in cold weather) through grazing organisms of many types, from protozoa to earthworms. The organisms are maintained in the filter, so there is no need for an inoculum. Detergent ingredients will be open to degradative processes, and the excess microflora will be consumed by the macrofauna. This, and the arrangement of filters, some receiving settled sewage, others downstream receiving the effluent from the upstream filters, allows the system to be controlled, particularly with regard to the mass of film. Percolating filters have the disadvantage that in cold weather the ability of the grazing organisms is reduced and fungal growths may limit the upward movement of air and the downward movement of sewage. As a result, the filter may occasionally lose its normal predominantly aerobic state and perform-ance will decline (Solbé et al., 1974). Generally speaking the percolating filter is not as efficient as an activated-sludge plant, but it is a robust process and requires little energy input.

Detergent ingredients may be expected to have the opportunity for almost total biodegradation in biological oxidation plants.

(d) Secondary Settlement.   The effluent from the percolating filters and the acti-vated-sludge plants is passed to settlement tanks, as described earlier. This allows the sludge to be recycled to the start of the activated-sludge process and the excess solids from both types of treatment to be passed back to the sludge digester, leaving a clear effluent to be discharged to the environment.

Taking all the foregoing into consideration, the following synopsis of removal of surfactants was observed in a number of sewage treatment works in the Netherlands (Table 3) (A.I.S.E. and CESIO, 1995). In addition, the removal of BOD from the total sewage input is given. Some of this BOD will be due to the surfactants, whose excellent biodegradability therefore improves the overall figure for BOD reduction. Among the surfactants, soap showed marginally the poorest removal and alcohol ethoxylate marginally the best.

(e) Sludge Digestion and Subsequent Treatment.   For those detergent chemicals that are adsorbed onto solids and settle out in the primary settler or that are returned as excess solids from the activated-sludge plant or with "humus solids" from the percolating filter, sludge digestion may offer a different environment for degradation, because it typically occurs under anaerobic conditions at around 37°C. Many detergent ingredients will degrade to methane and other gases under these circum-stances. The methane is burned to provide the heat needed to keep the plant at 37°C. It is a characteristic of sulphonates such as linear alkyl benzene sulphonate that they do not degrade under anaerobic conditions. This in itself is not a problem, since LAS has been shown not to affect the performance of sewage treatment plants, even at the relevant high concentration of about 3% in sludge (Berna et al., 1989). To make a full assessment of the risks to the environment from the properties of surfactants it can be useful to follow through the postdigestion fate of the sludge, as discussed later.

Digested sludge has a slightly lower water content, less odor, fewer pathogens, and far lower organic content than raw sludge. It is still very bulky, due to its water content, and the next stage in treatment is usually dewatering. This can be achieved by low-technology methods, such as spreading the sludge on beds of sand in the open and allowing drainage and evaporation to dry the sludge to the state of a moist solid (during which period there is biological activity and further degradation), or high-tech methods, such as centrifugation or pressing the water from the sludge, again to give a moist solid. Sludges can also be burned before or after drying. After drying or even without it, sludges can be disposed of to landfill or utilized as a soil conditioner in agriculture. There is increasing use of composting too, where sludges are mixed with other solid organic wastes to generate friable compost. The composting process produces very high temperatures (ca. 55°C), at which considerable degrees of degradation can be achieved.

In several studies LAS has been found to be 100% degraded in composting systems. See, for example, SPT (1999). In this case (Table 4) the process took from December to June and the compost consisted of sewage sludge from three wastewater

TABLE 4 Degradation and Final Concentrations of Contaminants of Municipal
Compost in a Danish Study

Substance	LAS	NPE	DEHP	PAH
Degradation (%)	100	78–95	63–82	56–72
Content following composting (mg/kg dry matter)	0	0.63–2.3	1.5–4.3	0.33–0.92

Source: Hausted Petersen, in SPT (1999).

treatment plus shredded wastes from gardens and parks and shredded straw. Di-ethyl hexyl phthalate (DEHP), nonyl phenol ethoxylates (NPE), and polycyclic aromatic hydrocarbons (PAH) were also substantially degraded.

The fate of detergent ingredients therefore depends on the sludge treatment process. It is usually the case that sludges from an anaerobic digester will be taken into aerobic environments, where the degradation of organics will continue, whether or not they degraded in the absence of free oxygen.

3. Storm Sewers

Sewerage systems are designed to carry the full dry-weather flow as well as volumes that exceed this basic flow by factors chosen at the design stage. At the end of the sewer, the sewage works can only carry a flow that is normally smaller than the carrying capacity of the sewer. Some of the surplus flow can be stored in *storm tanks*, but when these are full (and while waiting for treatment through the main works) the rest of the flow has to be discharged to surface waters to protect the sewage works from flooding and becoming inoperable. The overflow system works by including (for example) lateral weirs in the sewer that carry the excess to the storm sewer and thus to the river, etc. It is therefore possible that at times of high rainfall, detergent ingredients will be discharged untreated to surface waters, albeit diluted to a considerable extent as dictated by the chosen design factor (e.g., sixfold). Biodegradation of ingredients at such times will then depend on the ability of the surface water body to assimilate the wastes, as described later.

4. Septic Tanks

Depending on the country, small or large percentages of domestic sewage will be treated at source rather than being carried to a sewage works. Examples are the United Kingdom, where approximately 4% of sewage may be treated in septic tank, and the United States, with ca. 25% of sewage treated in this way.

Over the years there have been recurrent questions concerning the ability of septic tanks to handle detergent ingredients and the issue of whether surfactants can poison a septic tank system. In the 1950s this concern was addressed by the Robert A. Taft Sanitary Engineering Center in Cincinnati and found to be groundless. Similarly, in the United Kingdom (Ministry of Technology, 1966), the Water Pollution Research Laboratory undertook a long-term study, eventually adding 20 and then 50 mg/L Dobane JNX as measured against Monoxol OT to one set of septic tanks, leaving another set detergent free (apart from the small contribution from human bile in the sewage). The septic tanks reduced the concentration of surfactant during the 1–3 days of retention by about 60%. In the effluent the ammonia content and suspended solids content did not differ from those in the control septic tanks, nor did the formation of the important capping crust in the tanks. BOD was also similar in the experimental and treated tanks until the addition of 50 mg/L surfactant, when there was an increase in BOD, probably due to some of the extra surfactant added.

There appears to be no cause for concern from the disposal of surfactants to septic tanks, and they are capable of limited degradation.

The fate of detergent ingredients in the effluents from septic tanks will depend on the disposal system used. If shallow tile fields* are employed, the effluent will emerge near the surface of the ground and typically, being rich in plant nutrients, will encourage the growth of plants. Provided the tile drains do not collapse (which might cause all the effluent to emerge in one small area, with insufficient dispersal to allow adequate degradation), this should be a satisfactory method of dispersing the effluent and exposing it to oxygen. (The septic tank has minimal free oxygen available for aerobic degradation.) The alternative route for the effluent is a much deeper discharge point, and the dispersion an fate of the detergent ingredients or their residues then depend on the lateral movement and oxygen content of the groundwater.

IV. RESPONSES OF RECEIVING WATERS AND SOILS

From the earlier sections it can be seen that effluents or raw sewage containing detergent ingredients and their breakdown products are going to reach the environment in various ways. The environmental compartments involved include

Streams, rivers, and lakes
Estuaries and coastal waters
Soil, subsoil, and the saturated zone.

The responses of the receiving environment to the entry of detergent ingredients may be considered under three headings, *toxicity, oxygen demand*, and *effects on physicochemical processes*, the first one of which is outside the scope of this chapter.

The atmospheric compartment should not be forgotten, since some detergent ingredients will be sufficiently volatile to reach the air, but the transformations that may then occur (photolysis, etc.) are also outside the focus of this chapter. However, the release to the atmosphere of the gases that result from the breakdown of detergent ingredients is a subject worth a brief mention later, under the heading "Anaerobic Biodegradability."

A. Streams, Rivers, and Lakes

Confining attention to oxygen demand and physical processes, the situation observed in these surface waters will of course depend on their own properties and on the extent of pollution or contamination that accompanies the entry of the detergent chemicals and their breakdown products.

---

*A *tile field* is an arrangement of (typically earthenware) pipes spreading and branching from the discharge point of the septic tank. The pipes are installed at shallow depth, typically 0.5—2 m, often loosely jointed or perforated so that they diffuse the effluent in an area of soil.

1. Oxygen Demand

Most detergent ingredients are organic substances. Their molecular structures include features common in the natural world, and they generally have accessible value as a source of energy and nutrient for microorganisms. In aerobic environments it is therefore inevitable that microorganisms will utilize this value and consequently place an oxygen demand on the immediate environment. This demand will be met by diffusion of oxygen into the depleted zone, but there may be a shortfall where the rate of demand exceeds the rate of supply. Such could be the case in watercourses overloaded with organic wastes. Of course in the sewage works, as described earlier, provision is made for an abundant supply of oxygen to be available to meet the demand instantly.

The term *biochemical oxygen demand* (BOD) has been coined to describe a measurement of this demand under standard conditions: demand exerted over 5 days at 20°C in the dark and in the presence of an inoculum of mixed microorganisms (typically a drop of effluent from a sewage works). Often one particular aspect of oxygen demand is that imposed by the oxidation of ammonia (nitrification). At times it is important to distinguish this form of demand from carbonaceous oxidation. To observe this, the nitrification is deliberately suppressed by the addition of allyl thiourea, which inhibits the nitrifying organisms. One of the simplest measures of biodegradability is to compare the oxygen demand of a substance in the standard test with a calculated demand derived from an empirical formula. The ratio, observed to theoretical BOD, is expressed as a percentage. The higher the percentage, the more easily degraded is the substance.

In the water column, a reduced concentration of dissolved oxygen will have serious consequences for aerobic organisms, depending on how low the concentration falls compared with the saturation value for water of the salinity and temperature under discussion. (The difference between the observed oxygen concentration and the concentration at equilibrium with the air is known as the *saturation deficit*.)

Degradation occurs everywhere throughout a water body where bacteria and other microorganisms live. For a river that means in the water column, on the surfaces of plants, and on the sides and bottom of the river regardless of whether this is muddy or stony. Whereas the degrading organisms in the water column have their effect as the water moves to the sea (and afterwards if they can tolerate the salinity), those on the solid surfaces are retained in the system and may represent a long-term competency in degradation. Once able to utilize a chemical substrate, they will retain this ability even though the occurrence of the substrate is intermittent.

2. Physicochemical Processes

The rate of increase in the concentration of dissolved oxygen in water that is deficient in oxygen and in which (to obtain the absolute definition) there is no production or utilization of oxygen is known as the *oxygen exchange coefficient* (loosely, the *reaeration coefficient*). Surfactants can affect this coefficient, impairing the movement of oxygen into oxygen-depleted water. In river basins where there is inadequate wastewater treatment this can be an additional problem to the existing problems of high biochemical oxygen demand. An example of the extent of the problem is given by the following example.

Work at the Water Pollution Research Laboratory, United Kingdom, during the preparation of planning to clean up the Thames Estuary, showed that both soap and synthetic anionic surfactants had an equally inhibitory effect on oxygen uptake in estuaries. Over a range of exchange-rate conditions the effect of 1 ppm soap or synthetic anionic surfactant was to suppress the exchange coefficient by as much as 60–80% (Department of Scientific and Industrial Research, 1964). In critical cases this partial impairment of reaeration of water could delay the assimilation of organic pollutants and recovery of the system.

A second issue with detergent ingredients is concerned with chelation or sequestration of toxic metals. Chelating agents (NTA; ethylene diamine tetra-acetic acid) are used in some detergent formulations to sequester cations, including traces of metals. The latter might otherwise catalyze the breakdown of certain ingredients before they can perform their designed function. The chelating capacity in the pack of product is likely to be greater than needed, and so unused chelating power will enter the wastewater pathway. There, any heavy metals will be taken up by the chelator until its capacity is exhausted. The question is: Does this mobilization of metals matter in terms of risk assessment? The answer to this question was found several decades ago. The toxicity of sequestered metals is virtually eliminated by sequestration. The metals may be mobilized, and eventually the sequestering bond may sever, but by then the metals will be far removed from their origins, dispersed and hugely diluted in the sea.

B. Estuarine and Coastal Waters

These waters, now known as *transitional waters* in the Water Framework Directive (European Council, 2000), receive detergent ingredients from a number of types of discharge:

- Rivers carrying treated or untreated wastes
- Discharges directly to transitional waters from treated or untreated sources
- Discharges from ships or from the direct application of detergents in an attempt to disperse spilled oil at sea

In grossly polluted estuaries the concentrations of surfactant may be sufficient to place an extra demand on oxygen supply and inhibit reaeration, as described earlier. In coastal waters where there has been inadequate treatment of sewage, cases have been reported of surfactant in spray reducing the protective waxy covering of the needles of certain species of coniferous trees and allowing the airborne salt to damage the tree. This is a matter of indirect toxicity, however, and will not be considered further here. Biodegradation occurs in estuaries (and causes some of the oxygen demand seen there). Degradation also occurs further from the coast. Rates will depend on the water temperature and on the abundance of degrading organisms, which may in turn depend on a suitable supply of essential elements to drive the community of microorganisms. (It should not be assumed that organisms adapted to cold or saline environments will necessarily degrade detergent chemicals more slowly than organisms in temperate or tropical freshwaters.)

Among the direct discharges of materials that contained detergent ingredients, the dumping of sewage sludge at sea used to be important. Bans on sea dumping in the EU now restrict this problem. If the dump site was carefully chosen, the tidal scour was sufficient to disperse the sludge in the aerobic zone. If the site was outside the influence of tidal scour or wave action, there could be accumulations of materials, possibly under anaerobic conditions (see Section VI.C.2).

C. Soils, Subsoils, and the Saturated Zone

Detergent ingredients or their breakdown products may reach the soil, and thus the underlying zones from

- Contaminated irrigation water
- Direct application in pesticide formulations
- Direct application of sewage sludge to land, as a liquid slurry or a semidried cake
- Direct application of compost (although, as explained already, it is most unlikely that the majority of organic chemicals used in detergents could survive the intense degradation to be found in composting systems)

Where sludges are disposed of to landfill, the likelihood that the chemicals will reach the groundwater depends on the integrity of the landfill liner and is not considered specifically here.

1. Irrigation Water

Croplands in arid zones may be irrigated with undiluted sewage effluent or with river water that contains detergent ingredients or their breakdown products. Domestic-scale vegetable plots may also be irrigated with used washing water. Provided that the chemicals are not toxic to the crop or the soil organisms, this will be a beneficial technique, bringing with the irrigation water valuable nutrients, such as potassium, nitrate, and phosphate, some of which will have been derived from detergents. The irrigation process is typically intermittent, allowing the soil to dampen and drain repeatedly. These are excellent conditions for biodegradation; it has been suggested that for degradation activity in soils, a varying environment may have advantages over a constant environment.

There are, of course, problems if the volume of irrigation water supplied is not sufficient to perform two vital functions: (1) compensation for losses from evapotranspiration and (2) leaching out excess salts from the soils to the subsoils or underdrains. Excessive irrigation too would be undesirable, not only waterlogging the roots of the plants but also influencing the oxygen content of the soil pore water and thus preventing aerobic degradation.

2. Detergent Ingredients in Pesticide Formulations

For such substances the same general principles apply as in sludge applied to land, so these subjects are dealt with together.

3. Sewage Sludge Application to Farmland and Forests

The biodegradability of detergent ingredients in sludge (normally digested sludge) will depend on the physical form of the sludge (liquid or solid), the means of adding it to the land (sprayed, spread, or injected), and the nature of the soil (chemistry, including organic content, and drainage characteristics). Climate will also have a part to play. Surfactants in pesticide formulations may also reach soils by a variety of means, depending on the mode of application. Both aerobic and anaerobic zones will occur in soil, so a variety of metabolic pathways are available. The principal agents in the degradation are assumed to be soil bacteria and fungi, although there will be cases where the activities of larger organisms will aid the process by disturbing the soil promoting the growth of microorganisms.

A growing number of field observations of surfactants in soils is now available. Due to its failure to biodegrade under anaerobic conditions, LAS has attracted the most attention in this respect. One example is typical of them all: the observed half-lives of LAS in sludge-amended soils of 51 fields in the United Kingdom ranged from 7 to 22 days (Waters et al., 1989). The application of sewage sludge to agricultural land is subject to local, national, and international regulations. These limit the amounts and frequency of application, as does normal farming practice. Considering a typical farming practice of applying sludge once per year, a half-life of only 22 days suggests an adequate chance for extensive degradation (~15 half-lives) between applications, so there should be no buildup of detergent ingredients with degradation properties similar to LAS.

Other mechanisms of degradation and removal of detergent ingredients from ecosystems may be considered, although they are likely to be of less importance than microbial activity. Where chemicals are taken up by higher organisms, the digestive and metabolic systems of these species may have a small part to play in achieving the half-lives shown earlier.

V. DEFINITIONS

Having discussed the receiving environment and its properties that may influence biodegradation of detergent ingredients, definitions can be placed in context. It is convenient to discuss the language used in biodegradability discussions in pairs of contrasting terms: *primary* contrasting with *ultimate*, and *ready* contrasting with *inherent*. There is an additional term, persistent, which implies lack of any form of degradation in the normal environment.

A. Primary and Ultimate

By common usage the term *primary biodegradation* is generally confined to the degree of breakdown of surfactants, but there is no reason why it should not be used more widely for other types of chemical. In primary degradation the molecule undergoes one or a series of structural changes by which some specified property of the original molecule is lost. For example, a surfactant that has broken down to the extent that it can no longer induce foaming in mixture with water can be said to have undergone primary degradation. (This lack of sufficient surface activity to produce the physico-chemical effect is also accompanied by a loss of acute toxicity.) The nonspecific analytical methods (see later) correspond exactly to the physical effect too. In contrast, ultimate degradation implies what it sounds like—that the molecule has been broken down as far as it can without creating submolecular fragments. Thus an organic detergent ingredient containing carbon, hydrogen, oxygen, sodium, and sulfur would ultimately form water, carbon dioxide, and sodium sulfate under aerobic conditions.

B. Ready and Inherent

These two terms describe the speed of degradation under specified test conditions. In the case of ready biodegradability, the molecule is converted by the bacterial test inoculum to at least 60% of its theoretical CO_2 within 28 days, with the rest of the carbon either being taken into the bacterial cells for use in growth or remaining in the test environment as unchanged or only partly changed parent material. The conditions are made stringent by excluding the presence of organic matter other than the test substance, except for the very small amount that accompanies the drop of inoculum. This reduces the chances of cometabolism of the test substance with other substances and means that the measurements made (typically the carbon dioxide produced or the oxygen consumed by the inoculum as it grows) refer to the test substance and do not have to be separated from other organics, e.g., by costly studies involving radio-labeled material.

Inherently biodegradable substances are those that fail the Ready Test but pass a given threshold of degradation in a less stringent study. In an Inherent Test the period of exposure of the inoculum to the test substance is far longer (e.g., 100 days), the inoculum is far larger, and the organic substrates, including test substance, are partially replaced every day. This allows the possibility of cometabolism as well as the growth of strains of bacteria competent to take advantage of the substrate/s offered by the test chemical.

It should be noted that all the foregoing terms are used in standardized testing that is not intended to represent the real world. In contrast "simulation tests"are designed specifically with this in mind and may be considered as supplying definitive data concerning the particular situation they are intended to mimic, such as sewage treatment.

C. Persistent

When a detergent ingredient shows no sign of any kind of degradation in any test under environmental conditions (extending the range of conditions to reasonable values in terms of, e.g., temperature, salinity, pH, presence of other organic substrates, wavelengths of ambient light) it may be considered persistent. Thus *persistent* is not a term confined to *bio*degradability but is extended to all potentially destructive forces to which a chemical may be exposed in the natural environment and the parts of the man-made environment that most closely resemble natural situations, i.e., farmland, canals, reservoirs, etc.

There is a move to classify some substances as *very persistent, very bioaccumulative.* The latter part is scientifically (and grammatically) acceptable, and we could consider an arbitrary cutoff, such as "measured bioconcentration factor exceeds 100,000." The former is not grammatically correct: *persistent* is an absolute; it should not be qualified. However, it is easy to see the problem—it is not possible to measure something defined as a non-event. Van Leeuwen and Hermens (1995) define *persistence* as an "Attribute of a substance that describes the length of time that the substance remains in a particular environment before it is physically removed or chemically or biologically transformed." This allows a focus on one environmental compartment of concern at a time, so a substance could be not persistent in a river but persistent in the sea.

Persistence certainly gives rise to problems of definition, but, provided the term is never used alone (in which state there is no harmful condition attached to it), a pragmatic definition such as "half-life exceeds x years" may be derived.

VI. MEASUREMENT OF BIODEGRADABILITY

Biodegradability of detergent ingredients is measured using two types of approach, surrogate and actual. *Surrogate* approaches include the measurement of the creation of breakdown products such as carbon dioxide and methane or the requirement for dissolved oxygen dictated by the activity of the organisms carrying out the degradation. Actual (direct) measurements include specific analysis for the decreasing concentrations of the parent *compound* and nonspecific analysis for the decreasing concentrations of the chemical *family* (e.g., anionic surfactant). The measurement system (the conditions of the test) varies according to the requirements of the objective of the measurement and may be standardized (locally or globally) or individually designed to address a specific issue.

A. Role of Standardization

International standards, for example, those developed by the International Organization for Standardization (ISO 1994, 1995, 1997, 1998, 1999a, 1999b), have a particular place in the provision of data on biodegradability. They are designed as rigid protocols and are taken up by national (e.g., AFNOR, BSI, DIN) and international (e.g., CEN) bodies as part of items of legislation, including legislation on surfactants.

The rigidity of a standard provides a "level playing field" so that companies or others submitting data on a particular substance are placed at no advantage or disadvantage and so that comparisons within and between substances may be made. That is not to say that the result of a biodegradability study can be perfectly replicated. Biodegradability is not a physical process, and allowance must be made for inherent biological variability. Nevertheless, the results of replicates delivered by several laboratories according to a standardised protocol should have a close resemblance. The disadvantage of standardized systems is that they do not permit adjustment to meet individual situations. The guidelines approach of the Organization for Economic Cooperation and Development (OECD) is a basis for designing individual studies to meet particular needs, such as degradation in a particular type of soil or water body.

B. Methods

The methods used in assessing the biodegradability of organic chemicals may be set in the framework of a testing strategy, which in simplified structure resembles the following:

SCREENING TEST
(Ultimate, Ready Test with attention to possible toxic effects)

|

INHERENT BIODEGRADABILITY TEST
(If substance is not readily degradable)

|

CONFIRMATORY TEST
(If substance shows some degradability in either of preceding tests)

Within each testing tier, the choice of method is made with regard to the properties of the test chemical, particularly its solubility and tendency to adsorb onto solids.*

Only the confirmatory test can be used directly to indicate environmental concentrations. Any of the tests can be used to support the classification of the substance, as indicated in Section II.A. A brief summary follows.

1. Screening Tests

The Ready Test is useful as a screening tool because it is stringent. The following factors give the necessary stringency:

Relatively small inoculum of organisms at start of test
Artificial medium of mineral salts
Test substance the sole source of organic carbon
Much higher concentration than expected in environment (even more than 1,000-fold)
Tests are on single batches (i.e., no replenishment of test substance or microorganisms)

OECD test guidelines 301A-F (OECD, 1998b) are applicable, relating to the following ISO standard test protocols: 7827, 9439, nil, 10707, 7827, and 9408, respectively. A test period of 28 days is used.

The measurements made in these tests involve the removal of dissolved organic carbon (DOC) (OECD 301A and 301E), the evolution of CO_2 (301B), and the uptake of oxygen (301C, D, and F). The significance of these determinations is given in Section VI.C. The operator must be aware of the possibility that the substance, at the high concentrations required by the system of measurement, may be toxic to the inoculum. If this is the case, lower test concentrations may be possible or more sophisticated methods, such as the use of radiolabeled material may be needed. The pass levels for these tests are as follows:

Removal of DOC: 70%
Evolution of CO_2: 60% of theoretical
Uptake of oxygen: 60% of theoretical

A carbon analyzer, manometry, and electrolytic chemistry are used as analytical measurement techniques. An interpretation of the results of the different tests is given in Section VI.C.

2. Inherent Biodegradability Tests

When a substance fails to reach the pass levels just given, a different type of test is employed to check if there is a possibility of degradation under less stringent conditions:

Large initial inoculum of bacteria (typically the mixed liquor suspended solids from an activated-sludge plant)
Daily replenishment of test substance and organic substrate (settled sewage)
Retention of the inoculum throughout the test

---

*An excellent resumé of the various tests and their abilities to cope with different types of substance may be found on pages 74–80 of the SETAC publication edited by Hales et al. (1997).

OECD tests 302A–C apply. The endpoints are removal of DOC and the uptake of oxygen, depending on the method. 302A, the Semicontinuous Activated-Sludge Test, can continue for 26 weeks if needed. The pass level is 70% removal of DOC and 70% uptake of theoretical oxygen demand.

3. Confirmatory Tests

Ready and Inherent Tests of biodegradability have a role to play in the classification of substances. Arbitrary rate constants for biodegradation have even been assigned to the results of such tests so that they can be utilized in risk assessments for estimating the predicted environmental concentration (PEC). However, to obtain the closest approximation to the real world, as far as sewage treatment is concerned, requires the use of a Simulation Test (OECD 303). There are two forms of this test, although they both use DOC removal as the criterion for measurement. The Continuous Activated-Sludge Test includes an aeration stage and a settler for recycling sludge, as in a real works. The porous pot dispenses with the settler and uses the porous walls of the aeration vessel to permit the effluent to pass through but retain the bacteria. The bacteria are then wasted under a controlled pumping regime. This allows the system to mimic any real-life process (and the whole apparatus can be assembled in a constant-temperature room of any required temperature).

4. Anaerobic Biodegradability Tests

All the foregoing studies are made in the presence of free oxygen. But to make an estimate of the performance of the chemical in anoxic zones (sludge digesters, deep sediments, microhabitats in soils and in percolating filters, where more than 10% of the nitrogen in the sewage can be returned to the atmosphere as N_2), anaerobic tests are needed. The tests (ISO 11734 is relevant) are equivalent to the aerobic screening test, such as OECD 301D and ISO 10707. The inoculum is taken from an anaerobic digester, and the test chemical is the sole source of organic carbon. A sealed vessel is used, and the air is purged from it using nitrogen. The gases resulting from the degradation are trapped within the vessel, and analysis is made with a carbon analyzer, with and without the use of a furnace to convert any methane to CO_2 before infrared detection.

5. Biodegradability in Soils

A small-volume soil test is available, but it is regarded as a test of inherent biodegradability. The problems of "background noise" of CO_2 evolution from the living soil itself are typically overcome by use of radiolabeled test substance. A useful modification of the apparatus is given by Haigh (1993), which facilitates the absorption of labeled $^{14}CO_2$.

C. Practical Interpretation of Biodegradability Data

In this section a number of issues concerning interpretation are discussed. To those not well versed in biodegradation science and its microbiological and biochemical basis, it can at first be confusing to be informed that a 60% pass level is highly satisfactory or that the rapid die-away of DOC may be misleading. There is a particular issue with the apparently negative finding that a substance will not degrade anaerobically. These points are considered next.

1. Aerobic

In an aerobic screening test, several things may happen at once to the test material.

1. Absorption by the organisms (causes loss of DOC)
2. Degradation by the organisms (causes a requirement for oxygen and a release of CO_2)
3. Anabolic use of the test substance or, more likely, breakdown products of the test substance (causes growth of the organisms, e.g., cell division)

It may be seen from item 1 that if this is all that is happening, a percentage removal of DOC does not guarantee any degradation of the substance. Care is therefore needed in interpreting DOC results. Raising the pass level from 60% to 70% for DOC removal is an acknowledgment of the equivocal nature of this endpoint. The combination of items 2 and 3 suggests that a complete primary degradation of the substance may occur without evolution of the stoichiometric amount of CO_2. Indeed, the process of biodegradation may require 50–90% of the carbon content for energy production, to support the energy-requiring processes that maintain and allow the growth of living cells. One minor point is that the respiratory quotient (RQ)[*] associated with the utilization of detergent ingredients by bacteria is assumed to be close to unity (which it would be if all ingredients were carbohydrates). The fact that the RQ may not be exactly unity will not make a major difference to conclusions that assume CO_2 and O_2 to balance.

2. Anaerobic

In anaerobic biodegradation, some of the organic carbon of the detergent ingredient is broken down to yield methane. Not all detergent ingredients are amenable to such breakdown. On the one hand this is a good thing, because methane is over 25 times more effective as a greenhouse gas than is CO_2. On the other hand it means that a surfactant of the sulfonate family not only will pass through an anaerobic digester undegraded but will possibly double in concentration in the degraded sludge compared with the raw sludge, due to the concentrating effect of the degradative losses of other organic substances in the sludge. This fact has so concerned some authorities that linear alkyl benzene sulfonate cannot be used in their countries. But is this concern well founded? The answer depends on the subsequent route and uses of the digested sludge containing LAS. (It is assumed that there is, as there should be in developed countries, adequate aerobic treatment of sewage in the relevant location.)

*   If the sludge is to be incinerated, there is no problem of environmental contamination by LAS.
*   If the sludge is to be composted (see Section III.A), LAS will be totally destroyed in the hot, aerobic process.
*   If the sludge is to be used as a soil conditioner and fertilizer, it will reenter aerobic zones in the soil, with or without the additional aid of the cultivation techniques subsequently employed. The half-life figures (or similar values) given in Section IV will apply, and with annual applications of sludge there should be no accumulation of LAS in soil.

---

[*] The ratio of moles of CO_2 evolved to moles of O_2 absorbed in respiration.

There is another possibility—that through insufficient infrastructure the LAS will be discharged directly to surface waters, where banks of sediment may trap it. This could be in still freshwaters, estuaries with a low scour, or coastal waters not subject to major influences of the tides or wind. If this occurs, is toxicity a significant risk? The scenarios described suggest a permanent entrapment of LAS in sediment. If this is the case, the only toxicity problem could be to organisms that live in anaerobic muds. Once those muds have been disturbed and exposed to oxygen, the degradation of LAS should continue.

REFERENCES AND FURTHER READING

References shown in italics are not cited in the main text but may provide useful additional reading on the subject.

A.I.S.E. and CESIO. Environmental risk assessment of detergent chemicals. Proceedings of the A.I.S.E./CESIO Limelette III Workshop on 28–29 November 1995.

Berna, J.L.; Ferrer, J.; Moreno, A.; Prat, D.; Ruiz Bevia, F. The fate of LAS in the environment. Tenside Surf. Det. 1989, *26* (2), 101–107.

Birch, R. Biodegradation of nonionic surfactants. J. A. O. C. S. 1984, *61*, 340–343.

de Henau, H. Biodegradation. In *Handbook of Ecotoxicology*; Calow, Peter, Ed.; Blackwell Science: Oxford, 1998; 353–377.

Curds, C.R., Hawkes, H.A., Eds. Ecological Aspects of Used Water Treatment. Academic Press: New York, 1975.

Department of Scientific and Industrial Research. Effects of Polluting Discharges on the Thames Estuary. Water Pollution Research Technical Paper No. 11. Her Majesty's Stationery Office: London, 1964.

ECB. Technical Guidance Document on Risk Assessment of Chemical Substances following European Regulations and Directives. 2nd ed. European Chemicals Bureau, JRC-Ispra, (VA) Italy, April 2003. Website: http://ecb.jrc.it/tgdoc, 2003.

EEC. EU Directive 92/32/EEC (1992) 7th Amendment of Directive 67/548/EEC on the approximation of the laws, regulations and administrative provisions relating to the classification, packaging and labelling of dangerous substances, 1992.

EEC. Technical Guidance Document in support of Commission Directive 93/67/EEC on risk assessment for new notified substances and Commission Regulation (EC) 1488/94 on risk assessment for existing substances. European Chemicals Bureau, Ispra, 19 April 1996.

European Commission. Commission Directive 93/67/EEC on the assessment of risks to man and the environment of substances notified in accordance with Directive 67/584/EEC on dangerous substances. Off. J. Europ. Comm. L 227 of 8/9/1993.

European Council. Directive 2000/60/EC of the European Parliament and of the Council of 23 October 2000 establishing a framework for community action in the field of water policy. Official Journal of the European Communities, N.L 327, 22 December 2000; 1–72.

Federle, T.W.; Itrich, N.R. Comprehensive approach for assessing the kinetics of primary and ultimate biodegradation of chemicals in activated sludge: Application to linear alkylbenzene sulfonate. Environ. Sc. Technol. 1997, 31 (4), 1178–1184.

Fox, K. Environmental risk assessment under HERA: Challenges and solutions. J. Com. Esp. Deterg. 2001, *31*, 213–223.

Gerike, P.; Fischer, W.K. A correlation study of biodegradability determinations with various chemicals in various tests. Ecotox. Environ. Safety 1979, 3, 159–173.

Haigh, S.D. A modified flask for the measurement of mineralisation of ^{14}C-labelled compounds in soil. J. Soil Sci. 1993, *44*, 479–483.

Hales, S.G., Feijtel, T., King, H., Fox, K., Verstraete, W., Eds.; Biodegradation kinetics: Generation and use of data for regulatory decision making. In Proceedings of a SETAC-Europe Workshop, Port Sunlight, UK, 4–6 Sept. 1996. SETAC-Europe, Brussels, 1997.

Hennes-Morgan, E.C.; de Oude, N.T. Detergents. In *Handbook of Ecotoxicology*; Calow, Peter, Ed.; Blackwell Science: Oxford, 1998; 594–618.

ISO. ISO 10707. Water Quality—Evaluation in an aqueous medium of the "ultimate" aerobic biodegradability of organic compounds—Method by analysis of biochemical oxygen demand (closed-bottle test), 1994.

ISO. ISO 11734. Water Quality—Evaluation of the "ultimate" anaerobic biodegradability of organic compounds in digested sludge—Method by measurement of the biogas production, 1995.

ISO. ISO 14239. Soil Quality—Laboratory incubation systems for measuring the mineralisation of organic chemicals in soil under aerobic conditions, 1997.

ISO. ISO 14593. Water quality—Evaluation in an aqueous medium of the ultimate aerobic biodegradability of organic compounds—Method by analysis of inorganic carbon in sealed vessels (CO_2 headspace test), 1998.

ISO. ISO/DIS 14592 Part 1. Water quality—Evaluation of the aerobic biodegradability of organic compounds at low concentrations in water. Part 1: Shake flask method, 1999a.

ISO. ISO/DIS 14592 Part 2. Water quality—Evaluation of the aerobic biodegradability of organic compounds at low concentrations in water. Part 2: River simulation test, 1999b.

Karsa, D.R., Porter, M.R., Eds. Biodegradation of surfactants; Blackie Academic and Professional: London, 1995.

Ministry of Technology. Water Pollution Research 1965. Her Majesty's Stationery Office: London, 1966.

OECD. OECD No. 7 Series on testing and assessment. Guidance document on direct photo-transformation of chemicals in water. Environment Directorate. Organization for Economic Cooperation and Development, Paris, 1997.

OECD. OECD 100 Series. Tenth addendum to the OECD guidelines for the testing of chemicals. Section 1 Degradation and Accumulation. Organization for EconomicCooperation and Development, Paris, 1998a.

OECD. OECD 300 Series. Tenth addendum to the OECD guidelines for the testing of chemicals. Section 3 Degradation and Accumulation. Organization for Economic Cooperation and Development, Paris, 1998b.

OECD. OECD Draft Proposal for a new guideline. Simulation test—Aerobic transformation in surface water. Organization for Economic Cooperation and Development, Paris, 2000a.

OECD. OECD Draft 307. Aerobic and anaerobic transformation in soil. Draft August 2000. Organization for Economic Cooperation and Development, Paris, 2000b.

Painter, H.A. OECD Test Guidelines Program. Periodical Review. Detailed Review Paper on Biodegradability Testing, 1992.

Solbé, J.F.; de, L.G.; Ripley, P.G.; Tomlinson, T.G. The effects of temperature on the performance of experimental percolating filters with and without mixed macro-invertebrate populations. Water Res. 1974, *8*, 557–573.

SPT; Hausted Petersen, P. Degradation of xenobiotics in composing. In *Report from Workshop Organized by the Association of Danish Cosmetics, Toiletries, Soap and Detergent Industries (SPT), in Coordination with the Danish EPA, 19–20 April 1999, Copenhagen.* (This report is available in The CLER Review, vol. 5, no. 1, December 1999).

Stanier, R.Y.; Adelberg, E.A.; Ingraham, J.L. The Microbial World, 4th ed.; Prentice Hall: Englewood Cliffs, NJ, 1976; 283–284.

Swisher, R.D. Surfactant biodegradation. Surfactant Science Series, Marcel Dekker: New York, 1987; Vol 18.

Van Leeuwen, C.J.; Hermens, J.L.M., Eds.; *Risk Assessment of Chemicals: An Introduction*; Kluwer Academic: Dordrecht, 1995.

Waters, J.; Holt, M.S.; Matthijs, E. In *The Levels and Fate of Linear Alkylbenzene Sulphonate in Sludge-Amended Soils. Organic Contaminants in Wastewater, Sludge, and Sediment. Occurrence, Fate, and Disposal*; Quaghebeur, D., Temmerman, I., Angeletti, G., Eds.; Elsevier Applied Science: London, 1989; 161–178.

10

Environmental Risk Assessment of Surfactants

Quantitative Structure–Activity Relationships for Aquatic Toxicity

DAVID W. ROBERTS Safety and Environmental Assurance Center (SEAC), Unilever R&D, Sharnbrook, Bedford, United Kingdom

I. INTRODUCTION

The subject area of quantitative structure–activity relationships (QSARs) developed from the need to be able to predict the biological activity of chemicals, for molecular design purposes, as in the development of pharmaceuticals, and for environmental impact, for example, aquatic toxicity. The QSAR approach is based on the concept that the biological activity of a compound is determined by its chemical structure. From this it follows that if appropriate structural descriptors are chosen, it should be possible to find a mathematical relationship between these structural descriptors and the magnitude of a particular biological effect. Rather than applying a rigorous mathematical modeling approach, which would usually require a more detailed understanding than is available of the biological mechanism, the QSAR approach typically applies statistical methods to find a correlation between a simple measure of biological activity and structural descriptors. QSAR models can range from purely statistical, being derived in the absence of any knowledge or preconceptions of the mechanism of action, to mechanism based, where the descriptors are selected on the basis of knowledge or hypothesis of the mechanism of action. In aquatic toxicity QSAR, the latter tend to predominate, mechanistic insights having been available for many years. In the field of surfactant technology, QSAR has potential application to risk assessment purposes, investigation of mechanism of action, and product development.

The parameter EC50, defined as the concentration causing a specified effect in 50% of the population of test organisms, is normally used as the measure of biological activity. In many of the publications referred to here, particularly those relating to fish toxicity, the specified effect was lethality, and the original authors often use LC50 to refer to toxicity. For tests on *Daphnia*, EC50 is related to immobilization.

A wide range of structural descriptors has been used, ranging from "pure" structural descriptors that can be derived directly from the structural formula (e.g., connectivity indices, calculated molecular orbital parameters) to experimentally determined physicochemical parameters, such as membrane–water partition coefficients. For some physicochemical parameters, methods have been developed for estimation from chemical structure. An important example is the octanol–water partition coefficient, widely used in QSAR as its logarithm, log P (also frequently written as log $K_{o/w}$). Throughout this chapter, P is used, unless otherwise indicated, to denote the octanol–water partition coefficient. The statistical approaches used in QSAR range from complex methods, such as principle components analysis, to simple linear regression analysis. For the surfactants that are the subject of this chapter, the simpler QSAR approaches, using linear regression analysis, have proved to be adequate.

The remainder of this chapter is structured as follows. Firstly, mechanisms of aquatic toxicity for chemicals in general are discussed. For compounds of low chemical reactivity, partitioning properties, which can be modeled by log P, govern the toxicity. The next section discusses how log P values can be calculated for surfactants. The next two sections review published aquatic toxicity QSAR studies on surfactants and discuss the implications for the mechanisms of action of the various surfactant types. These are followed by sections on toxicity QSARs based on surfactant properties, on solubility cutoffs, and on other environmental endpoints.

II. AQUATIC TOXICITY MECHANISMS

Surfactants used in commercial products are chemically rather inert. In particular, they do not react as electrophiles. In general, chemicals of low reactivity exert their toxic effects toward aquatic organisms by narcosis mechanisms.

There is evidence for two mechanisms of narcosis, general narcosis and polar narcosis, which are distinguished by different QSARs based on log P.

The "classic" general narcosis QSAR was developed by Könemann [1] for toxicity to *Poecilia reticulata* (guppy). With EC50 in mol.L^{-1} (these units are used throughout this chapter) the QSAR is

$$pEC50 = 0.87 \log P + 1.13 \tag{1}$$

$$n = 50, \qquad R^2 = 0.976, \qquad s = 0.24$$

where p denotes the negative logarithm. This QSAR was developed from a set of acute toxicity data for a range of chemicals including inter alia hydrocarbons, halogenated hydrocarbons, alcohols, ethers, and ketones. Subsequently it has been found to be applicable to prediction of toxicity to a range of aquatic species, not only other fish species but other organisms, such as *Daphnia magna* [2]. There are very few chemicals that have been found to be less toxic than predicted by this equation (the few that are less toxic can be rationalized in terms of diminished availability due to causes such as low solubility or high volatility), and so it is often referred to as the *baseline toxicity equation*. Compounds that fit this equation are usually referred to as *general narcotics.* As discussed in more detail later, nonionic surfactants act as general narcotics.

For polar narcotics, the "classic" QSAR was developed by Saarikoski and Viluksela based on phenols and their toxicity to *Poecilia reticulata* [3]:

$$pEC50 = 0.63 \ \log P + 2.52 \tag{2}$$

$$n = 17, \qquad R^2 = 0.964, \qquad s = 0.16$$

This polar narcotic equation is also applicable to other aquatic species. Besides phenols, several other groups of organic compounds act as polar narcotics, including nitroaromatic compounds and aromatic amines [4]. As discussed in more detail later, anionic surfactants act as polar narcotics.

There has been some controversy as to whether there is a real distinction between the general and polar narcosis mechanisms, or whether the apparent distinction has arisen purely because log P is an inadequate descriptor of partitioning into a lipid membrane. In support of the latter viewpoint, it has been found that using experimentally derived membrane–water partition coefficients $K_{m/w}$ or liposome–water partition coefficients in place of the octanol–water partition coefficient, a single QSAR covering both general narcotics and polar narcotics can be derived [5,6]. In 2003 the evidence was analyzed in detail, with the conclusion that although practically useful QSARs covering both general and polar narcotics can be developed based on log $K_{m/w}$, there is nevertheless a real mechanistic difference between general and polar narcosis, which manifests itself by significantly different QSARs, even when based on log $K_{m/w}$, for general and polar narcotics treated separately; differences in symptoms preceding the onset of narcosis (basically lethargy for general narcosis, hyperactivity for polar narcosis); and nonadditivity between general narcotics and polar narcotics in mixture toxicity studies [4].

It has been proposed [4] that the mechanistic difference between general and polar narcosis reflects two different types of membrane partitioning. Phospholipids are the major structural component of biological membranes, the most common being phosphatidylcholine, which is an amphiphilic compound (Fig. 1). Figure 2 represents the structure of a typical membrane at the molecular level. Although membranes are often depicted as ordered arrays of amphiphilic molecules, they are far from static, as noted in Figure 2, and to a large extent the membrane interior is similar to a liquid hydrocarbon [7].

There are two ways in which solutes can partition between water and a membrane (Fig. 3), analogous to the different ways in which organic chemicals can partition between water and ionic surfactant micelles [8]. In three-dimensional (3-D) partitioning, the solute molecules can dissolve in the hydrocarbon-like interior of the membrane, where they are free to move in any direction. Although movement may be somewhat restricted relative to the situation in a liquid hydrocarbon solvent, the solute molecules are more mobile in this environment than they are at the interface in two-dimensional (2-D) partitioning. In 2-D partitioning, the solute molecules can partition into the membrane in such a way that part of the molecule is associated to the polar headgroups of the phospholipid while the rest of the molecule is in the hydrocarbon-like environment of the membrane interior. A solute molecule partitioning in this way, like the phospholipid molecules of which the membrane is

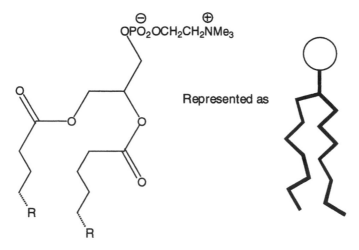

FIG. 1 Phosphatidyl choline.

composed, is able to move freely in two dimensions only, the xy plane, as shown in Figure 3.

The differences between general narcosis and polar narcosis are simply rationalized in terms of general narcotics acting via 3-D partitioning and polar narcotics acting via 2-D partitioning, these two mechanisms of action being independent of each other.

Whether a given chemical will act as a general or a polar narcotic will depend on the relative extent to which it partitions into the membrane as a 3-D or a 2-D solute. The fundamental toxicity equations may be written as follows.

For general narcosis:

$$pEC50_{(3-D)} = \log P_{3-D} - \log[N50]_{M,3-D} \tag{3}$$

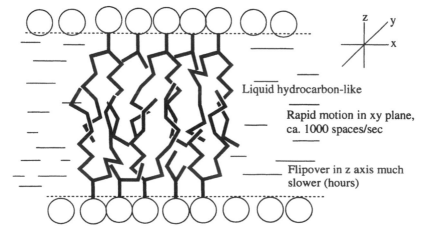

FIG. 2 Membrane structure.

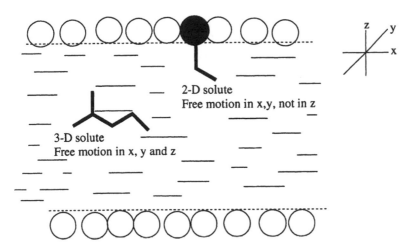

FIG. 3 Solute partitioning in a membrane.

where $\log P_{3\text{-D}}$ is the membrane 3-D partition coefficient and $[\text{N}50]_{\text{M},3\text{-D}}$, expressed in molar or molecular units (e.g., number of narcotic molecules per 1000 membrane phospholipid molecules), is the concentration of narcotic N in 3-D solution in the membrane that produces the toxic response in 50% of the test subjects.

Similarly for 2D narcosis:

$$\text{pEC}50_{(2\text{-D})} = \log P_{2\text{-D}} - \log[\text{N}50]_{\text{M},2\text{-D}} \tag{4}$$

where $\log P_{2\text{-D}}$ and $[\text{N}50]_{\text{M},2\text{-D}}$ are defined analogously to their 3-D counterparts.

Any chemical will have an intrinsic general narcotic toxicity, given by Eq. (3), and an intrinsic polar narcotic toxicity, given by Eq. (4). Chemicals for which $\text{pEC}50_{2\text{-D}} > \text{pEC}50_{3\text{-D}}$ will act as polar narcotics, and chemicals for which $\text{pEC}50_{3\text{-D}} > \text{pEC}50_{2\text{-D}}$ will act as general narcotics, apart from those chemicals that, having more specific modes of action available to them (e.g., electrophiles), are more toxic than either Eq. (3) or Eq. (4) would predict.

Three-dimensional membrane/water partition should be similar to octanol/water partition, both being dominated by deaquation energy, with a positive contribution in favor of the organic phase from the hydrophobic parts of the molecule and a negative contribution from the hydrophilic parts of the molecule. Thus $\log P_{3\text{-D}}$ should be related to $\log P$ by a Collander relationship [9], in which the coefficient a is positive:

$$\log P_{3\text{-D}} = a \log P + b \tag{5}$$

Thus Eq. (3) for general narcosis becomes

$$\text{pEC}50 = a \log P + b - \log[\text{N}50]_{\text{M},3\text{-D}} \tag{6}$$

According to the concept of narcotic action $[\text{N}50]_{\text{M},3\text{-D}}$, if expressed in molar or molecular units (e.g., number of narcotic molecules per 1000 membrane phospholipid molecules), should be independent of the nature of N, so good-quality general narcosis QSARs based on $\log P$ can therefore be expected. There is ample evidence that this is the case.

In the case of 2-D membrane/water partitioning, deaquation free energies again play a part, but there will also be a contribution from headgroup interaction energies HG, i.e., the binding forces between the phospholipid headgroups and the parts of the solute molecules associated with them.

$$\log P_{2\text{-D}} = a^* \log P + b^* + \text{HG} \tag{7}$$

Thus Eq. (4) for polar narcosis becomes

$$\text{pEC50} = a^* \log P + b^* + \text{HG} - \log[\text{N50}]_{\text{M},2\text{-D}} \tag{8}$$

The headgroup interaction term, HG, should be constant for a set of compounds all with the same headgroup binding substituent (e.g., aromatic OH of phenols), but it is unlikely to be constant across all such substituents, since, as discussed in some depth by Escher and Schwarzenbach [10] and by Scherer and Seelig [11], they will not all complex with the phospholipid headgroups in the same way. Aromatic OH groups, being strong hydrogen bond donors, will complex differently from aromatic NO_2 groups, which are strongly polar and good hydrogen bond acceptors, and differently again from sulphonate groups, which can complex by electrostatic attraction between oppositely charged groups, in anionic surfactants. Thus while a good-quality $\log P$–based QSAR for a set of polar narcotics all with the same headgroup binding substituent might be expected, the statistical quality for a set of polar narcotics covering a diversity of such substituents should be inferior. This agrees with what is found in practice.

The differences between general and polar narcosis mean that the mechanistic pathways to the narcotic effect from 2-D solution and 3-D solution in the membrane must be different and mutually independent. Although there is no direct evidence as to the nature of these pathways, the following has been proposed [4]. Narcosis results from perturbation of the activity of functional membrane-bound proteins. A 3-D solute will decrease the thickness of the membrane by allowing the long chains of the phospholipid to interpenetrate better. The membrane-bound protein will have to adjust its conformation to maintain its hydrophobic region in the hydrocarbon-like interior of the membrane, and this will perturb its function (e.g., regulation of Na/K cation balance). Two-dimensional solutes can affect the function of the same protein, but in a different way, by perturbing the headgroup interactions at the interface.

III. ESTIMATION OF Log *P* FOR SURFACTANTS

Although it is experimentally difficult to measure $\log P$ for surfactants, because of their tendencies to reside on the water/octanol interface and to solubilize octanol in water and water in octanol, there is no conceptual problem in defining a partition coefficient between octanol and water for a surfactant. *P* is simply the ratio, at equilibrium, of the concentrations of the compound in true solution in each of the two solvent phases. The fact that other equilibria exist (with concentrations at the interface and in the micellar phase if present) does not affect the definition. Surfactant in micelles is not in true solution, and the "micellar concentration" in the aqueous phase is not included in the *P* value. Although the presence of surfactant affects the solubilities of octanol and water in each other, the same phenomenon is encountered with nonsurfactant solutes and can in principle be accounted for by measuring log P at various concentrations below the CMC and extrapolating to infinite dilution.

Although measured log P values are difficult to obtain for surfactants, it is not difficult to calculate them using the Leo and Hansch method [12], with some modifications that have been introduced to deal with certain structural features often encountered in surfactants [13–16]. The Leo and Hansch method, which can be done either manually or by use of commercial software, is based on the concept that the various fragments of a molecule contribute additively to its log P value.

Log P is calculated by conceptually breaking the molecule down into its component fragments, summing the partial log P values (referred to as *hydrophobic fragment values, f*), and applying factors F to allow for variation in how the fragments are combined together in the whole molecule. The f and F values were originally derived from experimental log P values for a large number of chemicals (very few of which were surfactants). The rules for applying them are based on mechanistic considerations of the solvation free-energy changes when the solute partitions. The dominant contributor to these solvation free-energy changes is taken to be the deaquation free energy when the solute leaves the aqueous phase.

Although commercial software for the Leo and Hansch calculations is now widely available, the structure of most surfactants is sufficiently simple for the log P values to be calculated manually. The following guidelines for calculation of log P can be given.

It is important to be aware that the computerized version does not contain fragment values for anionic groupings, does not deal adequately with compounds containing several ethyleneoxy (EO) groups, and does not distinguish between different branching patterns. If using the computerized version, the best approach is the following.

If the compound contains anionic groups, enter the structure, with the anionic groups replaced by H. Adjust the answer manually by subtracting the fragment values for these H groups (0.23) and adding the bond factor (-0.12) plus the fragment values for the anionic groups as originally given by Leo and Hansch [12]. Some fragment values relevant to surfactants are:

Aliphatic–SO_3^-	−5.87
Aromatic –SO_3^-	−4.53
Aliphatic –OSO_3^-	−5.23
Aliphatic –CO_2^-	−5.19
Aliphatic ether–O–	−1.82
Aliphatic–OH	−1.64
Aliphatic ester–CO_2^-	−1.49

The first four of these are not in the computerized version of the method. Note that all of these values are negative, indicating a hydrophilic contribution from these groups.

If the compound contains a chain of ethyleneoxy (EO) units, enter the structure with a single EO unit and adjust the answer manually by applying a fragment value f_{EO}, equal to −0.10 or −0.25, for each additional EO unit in nonionic surfactants or anionic surfactants, respectively [16].

The original Leo and Hansch method contains branch factors, but these do not adequately account for the variation in properties between different isomers with different branching patterns. Based on consideration of how the number of water molecules required to solvate an alkyl group changes, for a given total carbon number, as the length of a branch increases, a position-dependent branch factor (PDBF) has been derived and validated (vide infra). The rule for application is as follows. If the compound contains branched alkyl chains, first calculate log P of the structure with these replaced by linear chains with the same carbon number. Adjust the answer by applying the PDBF, which can be calculated as

$$PDBF = -1.44 \log(S + 1)$$

where S is the carbon number of the shorter chain from the branching position. Some PDBF values are:

Methyl branch	−0.43
Ethyl branch	−0.69
Propyl branch	−0.87
Butyl branch	−1.01
Pentyl branch	−1.12
Hexyl branch	−1.22

A further recent refinement is in the way proximity factors are calculated for situations where two hydrophilic groups are close together in a molecule. The standard Leo and Hansch method, including the computerized version, has proximity factors calculated as:

Separation by one carbon	$-0.42(f_1 + f_2)$
Separation by two carbons	$-0.26(f_1 + f_2)$
Separation by three carbons	$-0.10(f_1 + f_2)$

where f_1 and f_2 are the fragment values of the two hydrophilic groups; since these are negative, the proximity factors are positive. However, this method does not deal adequately with the situation where one of the two groups is much more hydrophilic than the other—a situation that arises in several common surfactants, such as ether sulfates and ester sulfonates For such situations an approach to the calculation of proximity values based on consideration of the overlap between hydration sheaths of neighboring hydrophilic groups has been developed [17]. For example, in ester sulfonates the original method gives a proximity factor value of 3.09, which is unrealistically large (being twice as positive as the fragment value for the ester group is negative). The new approach gives a proximity factor value of 1.49. In alcohol ethoxy sulfates the anionic sulfate group is separated by two carbons from an ether oxygen group. The original method gives a proximity factor value of 1.83, while the new method gives a value of 1.48. Table 1 shows proximity values, calculated by the new method, for various pairings and separations of hydrophilic groups frequently encountered in surfactants.

To calculate log P de novo for a given surfactant structure, a simple way to proceed is first to calculate the log P value for the simplest homologue, i.e., where aliphatic R groups are taken as methyl, then to adjust by 0.54 (sum of the CH_2 fragment value and the bond factor) for each additional carbon atom in the alkyl chain, and then as appropriate to adjust for branching as detailed earlier and for any unsaturation by −0.55 per double bond.

TABLE 1 Some Proximity Factors F_p Relevant to Surfactants

Fragments		Fragment value		F_p for separation by		
1	2	f_1	f_2	1C	2C	3C
SO_3^-	CO_2	−5.87	−1.49	1.49	1.34	0.58
SO_3^-	CONH	−5.87	−2.71	2.71	1.99	0.79
SO_3^-	CO_2^-	−5.87	−5.19	4.37	2.71	1.04
SO_3^-	O	−5.87	−1.82		1.54	0.64
SO_3^-	CON	−5.87	−3.04	3.04	2.13	0.84
SO_3^-	OH	−5.87	−1.64		1.43	0.61
CO_2^-	CO_2	−5.19	−1.49	1.49	1.29	0.55
O	CO_2	−1.82	−1.49	1.37	0.86	0.33
SO_4^-	O	−5.23	−1.82		1.48	0.61

Once the log P value for one homologue of a given surfactant type has been calculated, this can be treated as the "parent structure" and the "parent log P" value can be adjusted to give log P for any homologue. Parent structures and parent log P values for some common surfactant types are listed next. In most cases it is convenient to take a homologue with a dodecyl chain as the parent structure, so that for the surfactant under consideration, if, for example, the alkyl chain has C carbon atoms, 12 can be subtracted from C and the remainder $(C − 12)$ can be multiplied by 0.54 to give the adjustment for alkyl group size.

A. Alcohol Sulfates (Linear or Branched, Primary or Secondary)

Parent structure n-$C_{12}H_{25}OSO_3Na$, log P = 1.60
For $ROSO_3Na$, log P = 1.60 + 0.54(C − 12), where C is the carbon number of R.
If R is branched, of the form $R^1R^2(CH_2)_n$, subtract 1.44 log(C_s + 1), where C_s is the carbon number of R^1 or R^2, whichever is the shorter.
Example: n-$C_{10}H_{21}CHMe.CH_2OSO_3Na$

$$\log P = 1.60 + 0.54 - 0.43 = 1.71$$

B. Alcohol Ethoxy Sulfates

Parent structure n-$C_{12}H_{25}OCH_2CH_2OSO_3Na$, log P = 2.22
For $R(OCH_2CH_2)_nOSO_3Na$, log P = 2.22 + 0.54(C − 12) − 0.25(n − 1), where C is the carbon number of R. Adjust for branching as for alcohol sulfates

C. Alcohol Propoxy Sulfates

Although these are not commonly used surfactants, the calculation method is shown to illustrate how the effect of branching in the alkyleneoxy chain can be quantified.
Parent structure n-$C_{12}H_{25}OCH_2CHMeOSO_3Na$, log P = 2.33
For $R(OCH_2CHMe)_n OSO_3Na$, log P = 2.33 + 0.54(C − 12) − 0.14(n − 1). Adjust for branching in R as for alcohol sulfates

D. Ester Sulfonates

Parent structure MeCH(SO$_3$Na)CO$_2$Me, log P = −4.02
For R^{1}CH(SO$_3$Na)CO$_2$R^2, log P = −4.02 + 0.54(C − 2), where C is the combined
   carbon number of R^1 and R^2. Adjust for branching in R^1 or R^2 as for alcohol
   sulfates.

E. Linear Alkylbenzene Sulfonates

Parent structure n-C$_{12}$H$_{25}$C$_6$H$_4$SO$_3$Na (para), log P = 3.97
For R^{1}R^{2}CHC$_6$H$_4$OSO$_3$Na, log P = 3.97 + 0.54(C − 11) − 1.44 log(C_s + 1), where
   C is the combined carbon number of R^1 and R^2 and C_s is the carbon number of
   the shorter of R^1 and R^2.

F. Isethionates

Parent structure n-C$_{11}$H$_{23}$CO$_2$CH$_2$CH$_2$SO$_3$Na, log P = 1.23
For RCO$_2$CH$_2$CH$_2$SO$_3$Na, log P = 1.23 + 0.54(C − 11), where C is the carbon
   number of R. Adjust for branching as for alcohol sulfates

G. Alcohol Ethoxylates

Parent structure n-C$_{12}$H$_{25}$OCH$_2$CH$_2$OH, log P=5.23
Adjust for alkyl group carbon number and branching as for alcohol sulfates. For
   higher degrees of ethoxylation, log P is adjusted by −0.10 for each additional
   EO unit.
Example: n-C$_{10}$H$_{21}$(OCH$_2$CH$_2$)$_3$OH

   Log P = 5.23 − 2 × 0.54 + 2 × (−0.10) = 3.95

H. Alcohols (Often Present in Commercial Alcohol Ethoxylates)

Parent structure n-C$_{12}$H$_{25}$OH, log P = 5.19
Adjust for alkyl group carbon number and branching as for alcohol sulfates.

IV. AQUATIC TOXICITY QSARs BASED ON Log P FOR SURFACTANTS

Although some studies have been carried out on pure single-component surfactants,
in many cases surfactant toxicity data relate to mixtures of several isomers and
homologues. There are several approaches to dealing with this problem [16]. The
simplest approach, although the least accurate, is to calculate a log P value (or any
alternative parameter) for the "average structure" or to measure the parameter
experimentally for the mixture. This may be the only approach possible if the levels
of the individual components are not known. If the levels of the individual
components are known, then parameters such as log(weighted average P) can be
used. P is calculated for each component, each P value is multiplied by the mole
fraction of the corresponding component, the products are summed, and the
logarithm is taken to give log WAP (WA = weighted average).

The most rigorous method is to calculate log P for each component i, apply the classical QSAR equation [Eq. (1) or Eq. (2), depending whether the surfactant is expected to act as a general or a polar narcotic] to calculate EC50$_i$, and to calculate EC50 for the mixture from the additive mixture toxicity equation:

$$1/\text{EC50 (calc., mixture)} = \sum \text{fr}_i/\text{EC50}_I \tag{9}$$

where fr$_i$ is the mole fraction of component i if EC50 is in mole concentration units or the weight fraction if EC50 is in weight concentration units. The value fr$_i$/EC50$_I$ is often referred to as the toxic contribution of component i. Observed pEC50 values can then be compared against pEC50 (calc., mixture) values by regression analysis to develop a QSAR.

The additive mixture toxicity equation [Eq. (9)] applies only if all of the components of the mixture act by the same toxic mechanism. For a mixture of substances with different mechanisms of action, the mixture shows independent joint action, and the EC50 of the mixture is the concentration at which one of the components reaches its own EC50 value. This provides a useful test. For example, if a surfactant is suspected of acting by polar narcosis, then if it is tested in a mixture with a known polar narcotic such as a phenol, then the mixture should obey Eq. (9); if it is tested in a mixture with a known general narcotic, then the mixture should not obey Eq. (9) but should show independent joint action.

A. Anionic Surfactants

The first QSAR to be developed for aquatic toxicity of surfactants was for 6-h acute toxicity to goldfish of a series of 20 pure isomers and homologues of LAS. Since the original references are not easily retrievable [13,18] the QSAR is described in some detail here. The toxicity data, published by Divo in 1976 [19], are shown in Table 2, together with log P data. At the time this analysis was carried out, the position-dependent branching factor had not been developed. To model the effect of branching, a water-sharing function (WSF) was defined as log(S + 1), where S is the carbon number of the shorter branch. This function was selected as having a value of zero for $S = 0$ (no branching) and increasing as S increases but tending to level off at high values of S.

The QSAR equation resulting from regression analysis is

$$\text{pEC50} = 0.78 \log P \text{ (no branch factor)} - 1.13 \log(S+1) + 2.06 \tag{10}$$

$$N = 20, \quad R^2 = 0.994, \quad s = 0.04, \quad F = 1212$$

Dividing the first two terms of the right-hand side of this equation by 0.78 gives, assuming the role of the second term to be solely that of modeling the effect of branching on log P,

$$\log P = \log P \text{ (no branch factor)} - 1.44 \log(S+1) \tag{11}$$

That is, the position-dependent branch factor, PDBF, is $-1.44 \log(S + 1)$. On this basis, Eq. (10) can be rewritten

$$\text{pEC50} = 0.78 \log P + 2.06 \tag{12}$$

This is how the PDBF was originally derived. It has subsequently been validated by its successful application in developing log P–based QSARs for other

TABLE 2   Goldfish Toxicity Data and Hydrophobicity Data for LAS

Compound[a]	pEC50[b]	log P (no branch factor)	WSF ($=\log(S + 1)$)
5-C10	3.62		0.70
4-C10	3.69		0.60
3-C10	3.76	2.89	0.48
2-C10	3.95		0.30
6-C11	3.86		0.78
5-C11	3.94		0.70
4-C11	4.04	3.43	0.60
3-C11	4.17		0.48
2-C11	4.36		0.30
6-C12	4.23		0.78
5-C12	4.32		0.70
4-C12	4.44	3.97	0.60
3-C12	4.61		0.48
2-C12	4.89		0.30
7-C13	4.65		0.85
6-C13	4.72		0.78
5-C13	4.81	4.51	0.70
4-C13	4.91		0.60
3-C13	5.06		0.48
2-C13	5.27		0.30

[a] Nomenclature: 5-C10 = 5-(4-sulphophenyl)decane, etc.
[b] EC50 = lethal concentration, $mol.L^{-1}$.

surfactants, including nonionic surfactants (see later), and log P–based QSPRs (QSPR = quantitative structure–property relationship) for critical micelle concentration (CMC) [20]. Aquatic toxicity QSARs based on log P (calculated with the PDBF) have also been developed for LAS isomers and homologues tested on *Daphnia magna* and *Gammarus pulex*, in hard water and in soft water [18]. These QSARs, Eqs. (13–16), are summarized in Table 3.

Equations (13–16) are similar to each other and to Eq. (12) in terms of their log P coefficients and their intercepts. Equations (12–16) are all more similar to the

TABLE 3   QSARs for LAS Toxicity to *Daphnia* and *Gammarus*

	Daphnia, H	*Daphnia*, S	*Gammarus*, H	*Gammarus*, S
Equation	**13**	**14**	**15**	**16**
Value of a	0.70	0.64	0.76	0.71
Value of b	2.23	2.44	2.46	2.27
n	9	12	9	11
R^2	0.974	0.912	0.933	0.902
s	0.07	0.15	0.15	0.16
F	263	103	98	83

Hard water (H) 250 $mg.L^{-1}$ $CaCO_3$; soft water (S) 25 $mg.L^{-1}$ $CaCO_3$.
pEC50 = $a \log P + b$.

classic polar narcosis equation (2) than to the classic general narcosis equation (1). Equations (13–16) give good agreement between predicted and observed toxicity to *Daphnia* and *Gammarus* for a range of aliphatic anionic surfactants (alkene and hydroxysulphonates, primary alcohol sulfate, and an alcohol ethoxy sulfate) in the same test systems [21]. This is illustrated in Table 4, which compares the observed and predicted toxicities for an alcohol ethoxy sulfate. The predicted figures shown here are slightly different from those in the original reference, since the current proximity factors and fragment values for EO units had not been developed at the time of the original work.

More recently [17], *Daphnia magna* studies in hard water (240 mg.L^{-1} CaCO$_3$) have been reported for six LAS homologues, each consisting of several positional isomers, and 21 single-component ester sulfonates of general formula R^{1}CH(SO$_3$Na) CO$_2$R^2. Using weighted average log P values for the LAS homologues, the QSAR equations obtained by regression analysis were:

$$LAS: \quad pEC50 = 0.77 \log P + 2.47 \tag{17}$$

$$n = 6, \quad R^2 = 0.991, \quad s = 0.08$$

$$\text{Ester sulfonates:} \quad pEC50 = 0.78 \log P + 2.45 \tag{18}$$

$$n = 21, \quad R^2 = 0.896, \quad s = 0.26$$

$$\text{LAS and ester sulfonates combined:} \quad pEC50 = 0.77 \log P + 2.45 \tag{19}$$

$$n = 27, \quad R^2 = 0.926, \quad s = 0.23$$

These three equations, like Eqs. (12–16), are very similar to the classic polar narcosis equation. Confirmation that both LAS and ester sulfonates act by the polar narcosis mechanism came from mixture toxicity studies in which the mixture toxicity equation [Eq. (9)] was found to be obeyed for mixtures of LAS and ester sulfonates with each other and with known polar narcotics but not for mixtures of the surfactants with general narcotics.

B. Nonionic Surfactants

The major nonionic surfactants used commercially are ethoxylated primary alcohols and, to a lesser extent, ethoxylated secondary alcohols. Ethoxylated alkylphenols, although not generally used in detergent products, are of environmental importance due to their use in products such as pesticide sprays. Nonionic surfactants of these types are simply aliphatic alcohols that also contain ether linkages and, in the case of the ethoxylated alkylphenols, an aromatic group remote from the hydroxide group.

TABLE 4 Observed and Calculated pEC50 Values for an Alcohol Ethoxy Sulfate
R(OCH$_2$CH$_2$)$_n$OSO$_3$Na, R $=$ n-C12/C14 (2:1 mole ratio), n (average) $=$ 3

	Daphnia, H	*Daphnia*, S	*Gammarus*, H	*Gammarus* S
Equation	13	14	15	16
pEC50 calc.	3.90	3.96	4.28	3.96
pEC50 obs.	4.41	3.99	4.06	4.29

log WAP (calculated as logarithm of weighted average P of C$_{12}$H$_{25}$(OCH$_2$CH$_2$)$_3$OSO$_3$Na and C$_{14}$H$_{29}$(OCH$_2$CH$_2$)$_3$OSO$_3$Na) = 2.39.

All of these functional groups, in isolation, are represented in the set of compounds that were used to derive the classic general narcosis QSAR shown in Eq. (1), and therefore these surfactants would be expected to act as general narcotics. A comparison of published fish EC50 data for nonionics of the types listed earlier against toxicity estimated from log P (calculated for the average structures) using Eq. (1) showed good agreement between calculated and observed values [21]. The only anomalies were three nonionics found to be significantly less toxic than predicted by the general narcosis equation. For one of these, the predicted EC50 was close to the CMC; for the other two, some of the components were insufficiently soluble to produce their full toxic effect. Similar findings were reported [22] in a study of *Daphnia* toxicity data for a series of 18 ethoxylated primary and secondary alcohols with average degrees of ethoxylation ranging from 2 to 7 and alkyl group branching ranging from 0 to 100%. In this study the "average structure" principle was applied for the degree of ethoxylation, and the mixture toxicity approach was applied for the parent alcohol distribution. In other words the mixture was treated as though each of the parent alcohols ROH had been converted to a hypothetical single ethoxamer having the same number (which may be a noninteger) of EO units as the overall average. Thus, for example, linear C12/C14 alcohol ethoxylated to an average degree of 2.78 was treated as a two-component mixture of the hypothetical compounds

$$C_{12}H_{25}(OCH_2CH_2)_{2.78}OH \quad \text{and} \quad C_{14}H_{29}(OCH_2CH_2)_{2.78}OH$$

log P was calculated for each of these hypothetical compounds, and from log P their EC50 values were calculated. The mixture toxicity Eq. (9) was then used to estimate the toxicity of the mixture. The correlation between observed *Daphnia* toxicity and toxicity calculated from the general narcosis equation 1 was

$$pEC50(obs.) = 0.91pEC50(calc.) - 0.005 \tag{20}$$

$$n = 18, \qquad R^2 = 0.953$$

A more rigorous mixture toxicity approach was carried out for a *Daphnia* acute toxicity data set [23], including materials with substantially higher degrees of ethoxylation [14]. The data set was based on three different alcohols, each ethoxylated to various degrees ranging from 0.4 EO (i.e., an average of 0.4 EO units) to 19.8 EO, in both narrow-range distributions and broad-range distributions. Two of the parent alcohols were linear (one being dodecanol and the other being a commercial mixture of dodecanol and tetradecanol), and the third was a C12–15 OXO alcohol mixture with 60% branched and 40% linear. Analytical data on the parent alcohol compositions and the ethoxamer distributions were used to estimate pEC50 (calc) from the general narcosis equation (1) and the mixture toxicity equation (9) by summing the toxic contributions of each component. For cases where the parent alcohol has only a small number of components, this calculation can be done simply using a spreadsheet. Ethoxamers above E20 can be neglected as being too hydrophilic to contribute significantly to the overall toxicity, so the number of components to be considered is manageable.

For the OXO alcohol ethoxylates, the total number of components runs into several hundreds. In such cases the calculation is conveniently done using two spreadsheets. In the first, using the known homologue/isomer distribution in the parent alcohol, the EC50 values are calculated for the parent alcohol mixture, for a hypothetical mixture of E1 ethoxamers and for hypothetical E2, E3 ... up to E20 single-ethoxamer mixtures. In a second spreadsheet, the ethoxamer distributions and the EC50 values for the hypothetical single-ethoxamer mixtures are used to calculate the EC50 value for each substance tested. The QSAR equation obtained by regression analysis was

$$pEC50 \text{ (obs)} = 0.88(\pm0.08) \text{ pEC50 (calc)} + 0.20(\pm0.06) \tag{21}$$

$$n = 35, \qquad R^2 = 0.929, \qquad s = 0.13, \qquad F = 429$$

The slope and intercept are very similar to those of equation 20.

All of these QSAR findings for nonionics suggest strongly that they act as general narcotics.

Other QSAR studies on nonionic surfactants are consistent with the preceding findings. Thus Uppgård et al. [24], in a multivariate QSAR study with 36 commercial ethoxylated alcohol mixtures tested on fairy shrimps (*Thamnocephalus platyuris*) and rotifers (*Brachionius calyciflorus*), observed that toxicity was positively correlated with log P (calculated by a different method to that described earlier), the carbon number of the hydrophobe (C), and the number of carbon atoms in the longest chain of the hydrophobe (redC). All of these parameters contribute positively to the overall hydrophobicity of the compound. For linear alkyl chains, redC is equal to C; but for branched chains, redC is smaller than C. The greater the difference between these two parameters, the more branched the surfactant, so in effect redC plays a role similar to the position-dependent branch factor used in the log P calculation. Negatively correlated with toxicity were the Davis hydrophilic–lipophilic balance (HLB), the critical packing parameter (RedCPP, a parameter associated with branching), and the critical micelle concentration (CMC). All of these parameters contribute negatively to the overall hydrophobicity of the compound. Hence this study confirms that nonionic surfactant toxicity is positively correlated with hydrophobicity, which is well modeled by log P calculated by the method described earlier.

In a study with commercial ethoxylated alcohol mixtures tested on fathead minnows (*Pimephales promelas*) and *Daphnia magna*, Wong et al. [25] carried out multiple linear regression analyses to correlate toxicity with the average carbon number C of the alkyl group and the average number of ethyleneoxy units (EO). They found toxicity to be positively correlated with C and negatively correlated with EO, which is consistent with the positive and negative contributions of methylene groups and ethyleneoxy units respectively to log P.

C. Cationic Surfactants

Until recently there has been little work reported on the QSAR modeling of aquatic toxicity for cationics. However, a recent paper by Singh et al. [26] has stimulated significant progress. Singh et al. report investigations into the toxicity of quaternary ammonium salts to the fathead minnow and give a QSAR equation for these cationics. Eight cationics were studied and the log P values were calculated

RNMe₃ Br   R = n-C14, n-C16, n-C18

PhCH₂NR₃ Cl   R = ethyl, n-butyl [N.B. These are nonsurfactants]

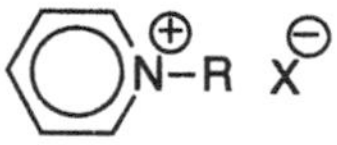

N–R X⁻ n-C12, Cl; n-C12, Br; n-C16, Br

FIG. 4 Cationics studied by Singh et al. (From Ref. 26.)

according to the Leo and Hansch approach. The structures are shown in Figure 4, and the data are presented in Table 5. The equation derived from these data is

$$\log EC50 = 1.08 \log P + 4.19 \tag{22}$$

$$n = 8, \quad R^2 = 0.967, \quad s = 0.36, \quad F = 178.5$$

The log P coefficient and the intercept are both larger than in the general narcosis and polar narcosis equations. This would seem to imply that cationics are intrinsically more toxic than general and polar narcotics and have a different, although still hydrophobicity-dependent, mode of action. However, it has been argued [27] that the apparent difference arises from a systematic deviation between the calculated and real log P values for the cationics. The calculation of log P for cationics by the Leo and Hansch method is quite complex. Leo and Hansch argue that the electronic influence of the N^+ fragment appears to extend quite far down the hydrocarbon chain, making the chain carbon atoms more polar and therefore less hydrophobic. To account for this effect, they apply enhanced bond factors whose values vary depending on the position of the carbon atom in the hydrocarbon chain. The principle is shown in Figure 5. Roberts and Costello [27], noting that in its influence on the CMC of surfactants a given alkyl chain seems to have the same effect irrespective of whether it is attached to a cationic (e.g., trimethylammonium or pyridinium) headgroup or an anionic headgroup (e.g., sulfate), argued as follows. A polar group, such as an anionic or cationic fragment, may be considered to be surrounded, in aqueous solution, by a "hydration sphere" of water molecules whose

TABLE 5 Aquatic Toxicity Data and log P Data for Cationics

Compound	pEC50[a]	log P (L&H method)[a]	log P (modified method)[b]
C₁₄H₂₉NMe₃Br	5.53	1.20	3.31
C₁₆H₃₃NMe₃Br	6.37	2.04	4.44
C₁₈H₃₇NMe₃Br	7.18	3.03	5.73
PhCH₂NEt₃Cl	1.82	−2.19	−2.19
PhCH₂NBu₃Cl	2.99	−0.63	0.36
C₁₂H₂₅PyCl	4.97	0.18	1.98
C₁₂H₂₅PyBr	4.54	0.38	2.18
C₁₆H₃₃PyBr	7.55	1.94	4.34

EC50 values in mol/L. Py = pyridinium.
[a] *Source*: Ref. 26.
[b] *Source*: Ref. 27.

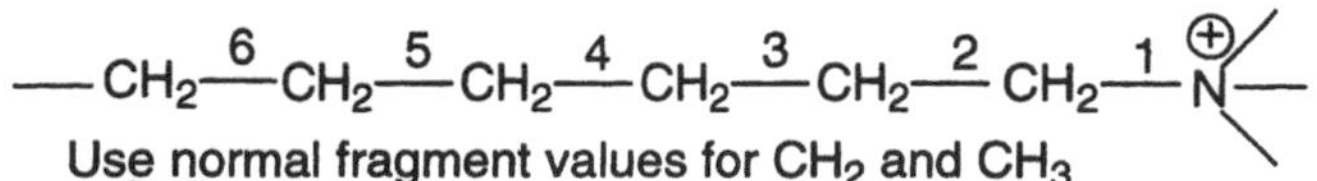

$$-CH_2 \overset{6}{—} CH_2 \overset{5}{—} CH_2 \overset{4}{—} CH_2 \overset{3}{—} CH_2 \overset{2}{—} CH_2 \overset{1}{—} \overset{\oplus}{N} —$$

Use normal fragment values for CH_2 and CH_3

$f(\overset{\oplus}{N}\overset{\ominus}{Br}) = -3.20$

Apply different bond factors depending on position

Bond 1 $F_b = -0.90$
Bond 2 $F_b = -0.60$
Bond 3 $F_b = -0.45$
Bond 4 $F_b = -0.35$
Bond 5 $F_b = -0.31$
Bonds >6 -0.27

Example $n\text{-}C_{12}H_{25}NMe_3Br$

$\log P = f(\overset{\oplus}{N}\overset{\ominus}{Br}) + 4\ f(CH_3) + 11\ f(CH_2) + 4\ F_b\ (1)$
$+ F_b\ (2) \ldots + F_b\ (5) + 7\ F_b\ (>5) = 0.66$

FIG. 5 Standard Leo and Hansch calculation for a cationic.

free energy is lowered by their interaction with the polar fragment. The hydrocarbon fragments within this hydration sphere will lose their hydrophobicity, being surrounded by water molecules whose free energy is lowered by their interaction with the polar fragment. In effect the hydrocarbon units are shielded from the bulk water by the polar group hydration sphere. Thus the log P contribution of a polar fragment may be considered to be made up of two components: the intrinsic hydrophilicity (negative hydrophobicity) of the polar fragment—related to the volume of the hydration sphere—and the degree to which attached hydrocarbon fragments are shielded by the polar group hydration sphere.

Although this argument applies equally to anionic and cationic fragments, there is an important difference (Figure 6). Suppose the polar group has a hydration sheath whose effective shielding radius is n C–C bond lengths (n will vary depending on the polar group). The common anionic fragments, such as sulfate or sulfonate, are monovalent, so for any compound likely to be of interest the shielding effect of a given anionic group will be constant and equal to $-nf(CH_2)$. Thus even though n may not be known, the shielding effect, being invariable for a given anionic polar fragment, is incorporated into the experimentally derived fragment value. For cationic fragments, usually based on quaternary nitrogen, there may be 1, 2, 3, or 4 hydrocarbon units extending beyond the effective shielding radius, and the shielding effect for a given cationic fragment is in consequence variable. Thus for calculation of log P for cationics, it is necessary to use a fragment value reflecting simply the intrinsic hydrophilicity of the cationic fragment and to allow for the shielding effect separately. From analysis of CMC data for cationics, a modified approach was derived for dealing with the hydrophobe shielding effect when alkyl groups are attached to positive nitrogen [26]: Methyl and ethyl groups and the first two methylene groups of longer chains should be treated as in the normal Leo and Hansch method, i.e., applying the special bond factors, but the rest of the alkyl chain should be treated in the same way as for uncharged compounds. Figure 7 shows an

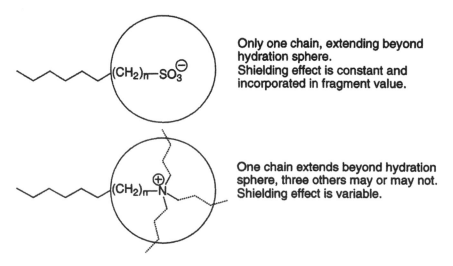

Only one chain, extending beyond hydration sphere. Shielding effect is constant and incorporated in fragment value.

One chain extends beyond hydration sphere, three others may or may not. Shielding effect is variable.

FIG. 6   Hydrophobe shielding by ionic fragments.

example of the modified log P calculation, applied to the compound n-$C_{12}H_{25}$ NMe_3Br. It can be seen from Figure 7 and Table 5 that the modified log P values are substantially different from those calculated by the normal Leo and Hansch method.

When the modified log P calculation was applied to all the cationics used by Singh et al., regression analysis gave a new fish toxicity QSAR with substantially different slope and intercept from Eq. (22):

$$pEC50 = 0.72 \ (\pm 0.09) \ \log P + 3.19 \ (\pm 0.30) \tag{23}$$

This QSAR for cationics is now not greatly different from the polar narcosis equation, suggesting that cationics may act by polar narcosis. Preliminary results from mixture toxicity studies [25] indicate that cationics obey the mixture toxicity Eq. (9) in mixtures with known polar narcotics but not in mixtures with known general narcotics. Thus the evidence points strongly to polar narcosis as the mechanism of action of cationic surfactants.

Use the Leo and Hansch special bond factor (F_b) values for the first and second bonds from N.

Normal F_b for all others

Example n-$C_{12}H_{25}NMe_3Br$

$$\log P = f(\overset{\oplus}{N}\ \overset{\ominus}{Br}) + 4\ f(CH_3) + 11\ f(CH_2) + 4\ F_b\ (1)$$
$$+ F_b\ (2) + 9\ F_b = 2.23\ (cf\ 0.66\ before)$$

FIG. 7   Modified log P calculation for cationics.

V. TOXICITY QSARs BASED ON SURFACTANT PROPERTIES

A publication by Rosen et al. [28] describes a QSAR study for aquatic toxicity in river water of pure single-component anionic surfactants to the rotifer *Brachionus calciflorus*. Unlike the studies referred to earlier, these were chronic toxicity studies. The toxicity was correlated with a combination of two surfactant specific parameters, pC20 and A_{min}. The pC20 value is the negative log of the concentration of surfactant when the surface tension of the solution is 20 mN/m less than that of the river water, and A_{min} is the minimum cross-sectional area of the surfactant at the water surface, which is calculated from the slope of a surface tension/concentration plot. The toxicity values are correlated with the ratios pC20/A_{min}. Table 6 shows the surfactants, their toxicities, their pC20, and A_{min} values, and the log P values calculated by the method summarized here.

Rosen et al. argue that the pC20 value represents the ability of the surfactant to adsorb onto the external tissues of the organism and that A_{min}, being a function of the work required to remove the headgroup hydration sheath so as to allow the molecule to pass through a membrane, is an inverse measure of the ability of the surfactant to penetrate to the interior of a cell. However, this argument seems inconsistent with the view that the site of action is the cell membrane [16]. Nevertheless the correlation between toxicity and the pC20/A_{min} parameter is good:

$$pEC50 = 0.25 \ (\pm 0.02) \ pC20/A_{min} + 3.45 \ (\pm 0.17) \tag{24}$$

$$n = 8, \quad R^2 = 0.991, \quad s = 0.06, \quad F = 676$$

Using log P in place of pC_{20}/A_{min} gives a correlation of similar statistical quality [16]:

$$pEC50 = 0.71(\pm 0.08) \ \log P + 4.03 \ (\pm 0.18) \tag{25}$$

$$n = 8, \quad R^2 = 0.982, \quad s = 0.08, \quad F = 330$$

Comparing Eqs. (24) and (25) it seems likely that both log P and the combination of pC20/A_{min} model the in vivo membrane/water 2-D partition coefficient, the latter parameter performing this role slightly better than the former.

TABLE 6 Rotifer Toxicity of Anionic Surfactants

Surfactant	log (1/EC50)	pC20	A_{min}	log P
C12S	5.30	3.68	0.504	1.60
C12E2S	5.47	4.32	0.512	1.97
C14S	5.88	4.68	0.485	2.68
C14E2S	6.28	5.24	0.454	3.05
C12SO3	4.62	3.46	0.709	0.96
C12E4S	5.09	4.35	0.688	1.47
C14E4S	5.85	5.51	0.569	2.55
C15E4S	6.17	5.96	0.557	3.09

EC50 in mol/L. Surfactant nomenclature: S = sulfate; SO3 = sulfonate.

In a later paper [29], Rosen et al. modify their surfactant parameter pC_{20}/A_{min} by substituting a standard free energy of adsorption, ΔG_{ad}^0 for pC_{20}. ΔG_{ad}^0 is a function of both pC_{20} and A_{min}:

$$\Delta G_{ad}^0 = -5.708pC_{20} - 0.1205A_{min} - 9.932 \tag{26}$$

and therefore $\Delta G_{ad}^0/A_{min}$ and pC_{20}/A_{min} are not perfectly correlated (R^2 for a plot of $\Delta G_{ad}^0/A_{min}$ against $pC20/A_{min}$ is 0.985). They apply the new parameter $\Delta G_{ad}^0/A_{min}$ to correlation of aquatic toxicity data presented in their earlier paper, for eight anionic surfactants, plus two further compounds, the nonionic surfactants

$$n\text{-}C_{12}H_{25}(OCH_2CH_2)_6OH \qquad \text{and} \qquad n\text{-}C_{14}H_{29}(OCH_2CH_2)_6OH$$

(abbreviated nomenclature C12E6 and C14E6). They show a plot of log EC50 against $\Delta G_{ad}^0/A_{min}$ with an R^2 of 0.95.

This finding raises the issues of whether $\Delta G_{ad}^0/A_{min}$ is a better parameter than pC_{20}/A_{min} for QSAR generation, and whether, in view of the nonionic and anionic surfactants both being covered by the correlation, nonionic and ionic surfactants act by the same mechanism in chronic toxicity. The analysis presented next has been carried out during the writing of the present work.

For all 10 surfactants, i.e., the two nonionics plus the eight anionics, the regression equations for $\log(1/EC50)$ against pC_{20}/A_{min} and $\Delta G_{ad}^0/A_{min}$ are:

$$\log(1/EC50) = 0.23(\pm 0.03)(pC_{20}/A_{min}) + 3.58(\pm 0.29) \tag{27}$$

$$n = 10, \quad R^2 = 0.964, \quad s = 0.11, \quad F = 214$$

$$\log(1/EC50) = 0.021\ (\pm 0.003)\ (\Delta G_{ad}^0/A_{min}) + 2.75\ (\pm 0.45) \tag{28}$$

$$n = 10, \quad R^2 = 0.957, \quad s = 0.12, \quad F = 177$$

The quality of the pC_{20}/A_{min} Eq. (27) is slightly better than that of the $\Delta G_{ad}^0/A_{min}$ Eq. (28), but the difference is marginal. A clearer distinction can be seen when the two nonionic surfactants are left out:

$$\log(1/EC50) = 0.25\ (\pm 0.02)\ (pC_{20}/A_{min}) + 3.45(\pm 0.17) \tag{29}$$

$$n = 8, \quad R^2 = 0.991, \quad s = 0.06, \quad F = 676$$

C12E6 calc. 5.98; obs. 5.69 : C14E6 calc. 6.56; obs. 6.34

$$\log(1/EC50) = 0.022\ (\pm 0.003)\ (\Delta G_{ad}^0/A_{min}) + 2.67\ (\pm 0.42) \tag{30}$$

$$n = 8, \quad R^2 = 0.970, \quad s = 0.11, \quad F = 195.5$$

C12E6 calc. 5.91; obs. 5.69 : C14E6 calc. 6.37; obs. 6.34

The better quality of the pC_{20}/A_m Eq. (29) is clearly apparent.

It is instructive to consider how well the toxicities of C12E6 and C14E6 are predicted by the QSAR equations derived from the anionic surfactant data. For the statistically better Eq. (29), based on pC_{20}/A_{min}, the residuals for the nonionics are 0.29 (C12E6) and 0.22 (C14E6). These are more than three times the standard deviation, s, of residuals for the QSAR (0.06). On that basis C12E6 and C14 E6 are best considered as outliers to the QSAR. This is reflected by the deterioration in statistical quality when C12E6 and C14E6 are included in the QSAR [compare R^2, s, and F values for Eqs. (27) and (29)]. For the statistically inferior Eq. (30), based on $\Delta G_{ad}^0/A_{min}$, the residuals for the nonionics are smaller (0.23 and 0.03 for C12E6 and C14E6, respectively) as compared with a higher s value (0.11) for the QSAR (although the residual for C12E6 is still more than twice the s value for the QSAR). Accordingly there is less of a deterioration in statistical quality when these compounds are included in the QSAR [compare Eqs. (28) and (30)]. However, C12E6 remains the least well-fitted point. The conclusions to be drawn from this analysis are therefore: For the anionic surfactants, pC_{20}/A_{min} is a better QSAR descriptor than $\Delta G_{ad}^{0}/A_{min}$, but the latter enables the two nonionics C12E6 and C14E6 (particularly the latter) to be fitted better. However, the evidence is best interpreted in terms of the two types of surfactant being modeled by different QSARs corresponding to different modes of action, as has been shown to be the case for acute aquatic toxicity, where anionics surfactants act as polar narcotics and nonionics as general narcotics.

The pC_{20}/A_{min} parameter is potentially very useful for surfactant QSAR studies, particularly for screening new surfactants in product development. For example, if a surfactant is synthesized with a novel structure such that a log P value cannot be calculated with confidence, experimental determination of pC_{20}/A_{min} can be done quickly and easily to enable a QSAR prediction to be made—provided a QSAR for the biological endpoint of interest has been established.

VI. SOLUBILITY CUTOFFS

Many commercial surfactants contain relatively insoluble components. In the use situation the major components are usually in nominal solution at above the CMC, and the less soluble components can be solubilized by incorporation into the micelles. Almost invariably, however, EC50 values are substantially below the corresponding CMC (some nonionic surfactants based on higher-chain-length alcohols, such as C16–18 with a high degree of ethoxylation, may be exceptions [22]). Thus in the toxicity testing situation the less soluble components may not be soluble enough to contribute significantly to the toxic effect. No firm rules can be given, since whether a particular compound will make its full toxic contribution can depend on water hardness and other specific features of the toxicity test, but the following guidelines are useful.

For LAS, higher homologues and isomers with the 4-sulphophenyl group near the end of the chain tend to be relatively insoluble. Thus in the development of the *Daphnia* and *Gammarus* acute toxicity QSAR Eqs. (13–16) for LAS [13,18] (Table 3): 2-C14, 3-C14, 3-C15, and 6-C15 (no other C15 materials were tested) were omitted as negative outliers for Eq. (13) (*Daphnia*, hard water); 3-C15 was omitted for Eq. (14) (*Daphnia*, soft water); 2-C12, 3-C14, and 3-C15 were omitted for Eq. (15) (*Gammarus*, hard water); 2-C14 and 3-C15 were omitted for Eq. (16) (*Gammarus*, soft water).

For nonionics, linear C14 and higher alcohols and their ethoxamers tend to be insufficiently soluble. In the development of the *Daphnia* acute toxicity QSAR of Eq. (21), all linear C14 and higher derivatives were omitted from the mixture toxicity calculations. However, as indicated earlier, these guidelines are not fixed. In this context it is instructive to consider in more detail the QSAR study of Wong et al. [25] on commercial alcohol ethoxylate surfactants. They report toxicity tests on fathead minnows (*Pimephales promelas*) and *Daphnia magna* for nine ethoxylated OXO-alcohols. The parent alcohol mixtures contain linear and β-branched alcohols in molar proportions of ca. 80:20. Previous toxicity data to fish (not *P. promelas* in all cases) and *Daphnia* on a further set of six ethoxylated OXO-alcohols of the same type are presented and compared against the predictions of the QSARs (based on average carbon number C of the alkyl group and the average number EO of ethyleneoxy units) derived from the first nine surfactants. Table 7 shows the toxicity data and relevant physicochemical parameters for the nine nonionics used to develop the QSARs (the training set) and the six further nonionics (the training set). The QSARs reported by Wong et al. are, after converting EC50 units to mol/L:

Pimephales promelas: pEC50 = 0.34C − 0.05EO + 1.65   (31)

$n = 9$, $R^2 = 0.99$, $s = 0.05$, $F = 324.7$

Daphnia magna: pEC50 = 0.38C − 0.10EO + 1.77   (32)

$n = 9$, $R^2 = 0.96$, $s = 0.10$, $F = 100.7$

The log P increments for a unit increase in alkyl carbon number and unit increase in EO are 0.54 and −0.10, respectively. Thus, if the parameters C and EO in Eqs. (31) and (32) are together acting as surrogates for log P values of the average structures and nothing more, then the ratio of the C and −EO coefficients should be 0.54:0.10. For Eq. (31) the ratio is 0.54:0.08, quite close to the expected value. However, for Eq. (32) the ratio is 0.54:0.14, implying that for the *Daphnia* toxicity QSAR the EO parameter plays an extra role besides modeling log P. As can be seen from Table 7, the two nonionics Neodol 25-12 and Neodol 45-13, which have the highest EO values (12 and 13, as compared with values in the range 5–8 for the other surfactants in the training set) are the two that contain C14 and C15 alkyl homologues. If some of these higher homologues were not completely soluble under the test conditions used by Wong et al., then the toxicities of Neodol 25-12 and Neodol 45-13 would be lower than their log P values indicate, and this effect would be modeled by a larger negative coefficient for the EO term in Eq. (32).

In agreement with the foregoing reasoning, the *Pimephales promelas* toxicities for the training set shown in Table 7 are well correlated with log (weighted average P) values (log WAP), whereas for *Daphnia magna* the correlation is poorer, with Neodol 25-12 and Neodol 45-13 underpredicted. When the log WAP values for Neodol 25-12 and Neodol 45-13 are recalculated so as to exclude any contribution from the linear C15 components, the *Daphnia magna* QSAR based on log WAP becomes similar in statistical quality to Eq. (31) based on C and EO. The log WAP QSAR equations are:

Pimephales promelas: pEC50 = 0.61 (± 0.05) log WAP + 2.52 (± 0.12)   (32a)

$n = 9$, $R^2 = 0.988$, $s = 0.06$ $F = 554.8$

Daphnia magna (*based on all components*):

pEC50 = 0.63 (±0.14)log WAP + 2.57 (±0.66)   (33)

$n = 9$, $R^2 = 0.917$, $s = 0.16$, $F = 77$

Daphnia magna (assuming no contribution from linear C15 components):

pEC50 = 0.77 (±0.11)log WAP + 2.04 (±0.52)   (34)

$n = 9$, $R^2 = 0.963$, $s = 0.11$, $F = 180.4$

TABLE 7 Toxicity and Hydrophobicity Data for Ethoxylated OXO-Alcohols

| Surfactant | Alkyl C number | | EO | log WAP[a] | pEC50[b] | |
	Range	Mean			Fish	_Daphnia_[c]
Training set						
Neodol 23-5	12–13	12.5	5	5.11	5.62	5.96
Dobanol 23-4.5/6	12–13	12.5	5.25	5.09	5.64	5.85
Neodol 23-6.5	12–13	12.5	6.5	4.96	5.57	5.82
Dobanol 91-6	9–11	10	6	3.68	4.70	4.89
Dobanol 91-8	9–11	10	8	3.48	4.68	4.64
Neodol 1-7	11	11	7	4.02	5.09	5.36
Neodol 1-9	11	11	9	3.82	4.89	4.92
Neodol 25-12	12–15	13.5	12	5.22 (4.84)	5.72	5.72 (5.82)
Neodol 45-13	12–15	14.5	13	5.39 (5.05)	5.89	5.82 (6.05)
Test set						
Neodol 91-2.5	9–11	10	2.5	4.03	4.67	5.05
Neodol 1-5	11	11	5	4.22	5.32	5.00
Neodol 25-3	12–15	13.5	3	6.13 (5.74, 5.65)	5.35 (5.58)	6.38 (6.48)
Neodol 25-7	12–15	13.5	7	5.73 (5.34, 5.25)	6.03 (6.25)	5.95 (6.05)
Neodol 25-9	12–15	13.5	9	5.53 (5.14, 5.05)	5.58 (5.80)	5.70 (5.80)
Neodol 45-7	14–15	14.5	7	5.99 (5.66, 5.52)	5.61 (6.31)	6.19 (6.41)

[a] log WAP: The calculation is illustrated by Neodol 23-5. This is treated as a mixture of the single ethoxamers n-$C_{12}H_{25}(OCH_2CH_2)_5OH$ (n-C12E5) (40 mol%), branched-C12E5 (10 mol%), n-C13E5 (40 mol%), and branched-C13E5 (10 mol%). The branched components of OXO alcohols contain C1, C2, C3 ... branches, usually in decreasing proportions, and for present purposes it is a good approximation to treat the branched C12 components together as having the same log P value as the linear C11 homologue and likewise to treat the branched C13 components together as having the same log P value as the linear C12 homologue. Log WAP is then calculated as:

$$\log[0.4 \times P (\text{n-C12E5}) + 0.1 \times P (\text{n-C11E5}) + 0.4 \times P (\text{n-C13E5}) + 0.1 \times P (\text{C12E5})]$$

The same principle is applied for the other surfactants. Where one other log WAP value is shown in brackets, this is a modified value based on exclusion of linear C15 homologues. Thus, for example, Neodol 45-13 with linear C15 homologues excluded is treated as a mixture of the single ethoxamers n-C14E13, branched-C14E13, and branched-C15E13 in molar proportions 40:10:10.

Where two other log WAP values are shown in brackets, the first of these is based on exclusion of linear C15 homologues and the second is based on exclusion of both linear C14 and linear C15 homologues.

[b] pEC50 values: These are based on EC50 in units of mol/L. Where a bracketed figure is shown in the Fish pEC50 column, the pEC50 value has been adjusted on the basis that the linear C14 and C15 homologues do not contribute. On this basis, with Neodols 25-3, 25-7, and 25-9, only 60% of the total surfactant expresses its toxicity, and accordingly the experimental pEC50 value is adjusted by subtraction of log(0.6). Likewise for Neodol 45-7, on this basis only 20% of the total surfactant expresses its toxicity and the experimental pEC50 value is adjusted by subtraction of log(0.2).

[c] Where a bracketed figure is shown in the _Daphnia_ pEC50 column, the pEC50 value has been adjusted on the basis that the linear C15 homologue does not contribute.

Source: Based on data from Ref. 25.

For the *Daphnia magna* test data on the six surfactants in the test set, regression analysis gives a good fit between pEC50 and log WAP, again assuming no contribution from the linear C15 components in the four surfactants that contain C15 alkyl groups:

$$pEC50 = 0.88 \; (\pm 0.13) \; \log WAP + 1.38 \; (\pm 0.65) \tag{35}$$

$$n = 6, \qquad R^2 = 0.979, \qquad s = 0.10, \qquad F = 187.5$$

Superficially, Eqs. (34) and (35) seem rather different, Eq. (34) appearing like a polar narcosis QSAR [cf. Eq. (2)], whereas Eq. (35) is more like a general narcosis QSAR [cf. Eq. (1)]. However, in the log P range covered by the surfactants in Table 7 the general narcosis QSAR and the polar narcosis QSAR give similar predictions, and the difference between Eqs. (34) and (35) can be attributed to statistical noise. This becomes clear when the *Daphnia* data for all 15 nonionics are plotted together (Fig. 8). The regression equation for the combined *Daphnia* toxicity data is

$$pEC50 = 0.78 \; (\pm 0.09) \; \log WAP + 1.98 \; (\pm 0.44) \tag{36}$$

$$n = 15, \qquad R^2 = 0.955, \qquad s = 0.13, \qquad F = 274.8$$

The test set fish toxicity data are not well correlated with log WAP when all components are assumed to contribute. Four of the six surfactants in this data set contain C14 and C15 alkyl homologues, all with a lower average degree of ethoxylation (ranging from 3 to 7) than the C14- and C15-containing surfactants in the *P. promelas* training set. When the test set data are adjusted on the basis of no contribution from the linear C14 and C15 components, a reasonable correlation is obtained, although the surfactant with the lowest degree of ethoxylation (Neodol 25-3) is a negative outlier. Presumably for this surfactant, other components

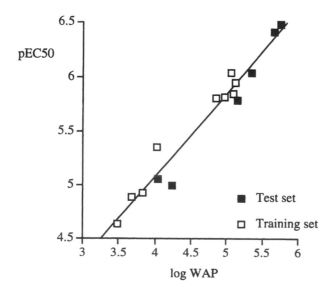

FIG. 8 *Daphnia* toxicity vs. log WAP for ethoxylated OXO alcohols. The line shown is the regression line for the combined test set and training set data, corresponding to Eq. (36). For both the training set and the test set, linear C15 alkyl homologues are assumed not to contribute to toxicity.

(possibly some of the nonethoxylated alcohols or some of the C15 branched components) were not soluble under the test conditions. The regression equation for the fish toxicity test set is

$$pEC50 = 1.01 \ (\pm 0.33) \log WAP + 0.81 \ (\pm 1.62) \tag{37}$$

$$n = 5, \quad R^2 = 0.924, \quad s = 0.22, \quad F = 36.5$$

Outlier: Neodol 25-3. pEC50: calc. 6.30, obs. 5.36

Similar to the case for the *Daphnia* QSARs, the regression equations [Eqs. (32) and (37)] for the fish toxicity training set and the test set, respectively, appear different, but the differences in the gradients and the intercepts largely cancel each other out and can be attributed to statistical noise. Figure 9 shows the data plotted all together. The composite regression equation is

$$pEC50 = 0.74 \ (\pm 0.15) \log WAP + 2.00 \ (\pm 0.69) \tag{38}$$

$$n = 14, \quad R^2 = 0.893, \quad s = 0.19, \quad F = 100.5$$

Outlier: Neodol 25-3. pEC50: calc. 5.96, obs. 5.36

The foregoing analysis of the Wong et al. [25] findings illustrates the effects of solubility cutoffs and demonstrates how, although they cannot currently be predicted confidently, they can be allowed for by retrospective recalculation.

It is important to bear in mind that although a surfactant component may not be sufficiently soluble to display its intrinsic toxicity in an acute toxicity test, chronic EC50 values can be up to two orders of magnitude lower than acute EC50 values. For this reason, the following principle is recommended. In developing a QSAR, higher homologues suspected of being insoluble or that manifest themselves as

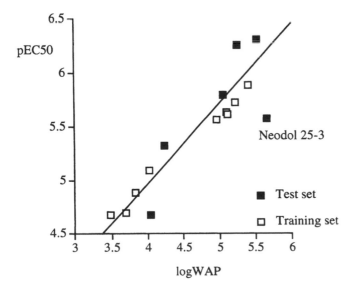

FIG. 9 Fish toxicity vs. log WAP for ethoxylated OXO alcohols. The line shown is the regression line for the combined test set and training set data, corresponding to Eq. (38). For the training set, all homologues are assumed to contribute to toxicity. For the test set, linear C14 and C15 homologues are assumed not to contribute to toxicity.

negative outliers should be omitted from the regression. If they are included, not only will the resulting QSAR be of poorer statistical quality, but more importantly it will tend to underestimate toxicity if used predictively. But when a QSAR is used predictively for risk assessment purposes, all components should be treated as though they are soluble, in order to ensure that chronic toxicity is not underestimated.

VII. OTHER ENVIRONMENTAL ENDPOINTS

The focus in this chapter has been entirely on aquatic toxicity. Surfactant QSAR for other environmental endpoints is much less well developed, although a good start has been made in bioconcentration. Thus the uptake rate constants and the bioconcentration factors (BCF) for LAS isomers and homologues and for linear ethoxylated alcohols have been found to be well modeled by log P (calculated by the methods described here) and by the $\Delta G^0_{ad}/A_{min}$ parameter, with separate QSARs applying for the two types of surfactants [30]. More work needs to be done to establish how general these QSARs are across wider ranges of surfactant structures. Among other endpoints, river sediment sorption partition coefficients for LAS have been found to be well correlated with log P [13,18].

VIII. CONCLUSIONS

Much progress has been made since the early 1980s in aquatic toxicity QSAR for surfactants. Although log P cannot easily be measured for surfactants, the log P calculation method has been extended to apply to surfactant structures, and calculated log P values have proved very useful in developing surfactant toxicity QSARs. The use of experimentally derived surfactant property parameters, in particular pC_{20}/A_{min}, seems very promising, and further work in this area, for example, application to acute toxicity QSAR, would be valuable.

QSAR and mixture toxicity studies indicate that anionic and cationic surfactants (although for these further confirmation would be useful) act as polar narcotics and that nonionic surfactants act as general narcotics. Preliminary indications are that zwitterionics, as would be expected, act as polar narcotics [31].

Statements can often be found in the literature to the effect that the toxicity of surfactants is caused by their ability to disrupt membrane function in the respiratory organs of aquatic species. Although true, such statements can mislead by implying that surfactants have special toxic properties. It is important to bear in mind that disruption of membrane function is also the basis of toxicity of nonsurfactants. The main conclusion from the QSAR studies reviewed here is that overall the evidence is very strong that surfactants in general are narcotic toxicants and do not differ from unreactive nonsurfactants in their toxic action.

REFERENCES

1. Könemann, H. Quantitative structure–activity relationships in fish toxicity studies. Part 1: Relationship for 50 industrial pollutants. Toxicology 1981, *19*, 209–221.
2. Sloof, W.; Canton, J.H.; Hermens, J. Comparison of the susceptibility of 22 freshwater species to 15 chemical compounds: I. (Sub)acute toxicity tests. Aquat. Toxicol. 1983, *4*, 112–128.

3. Saarikoski, J.; Viluksela, M. Relation between physicochemical properties of phenols and their toxicity and accumulation in fish. Ecotoxicol. Environ. Saf. 1982, 6, 501–512.

4. Roberts, D.W.; Costello, J.F. Mechanisms of action for general and polar narcosis: a difference in dimension. QSAR and Combinat. Sci. 2003, 22, 226–233.

5. Vaes, W.H.J.; Ramos, E.U.; Verhaar, H.J.M.; Hermens, J.L.M. Acute toxicity of nonpolar versus polar narcosis: Is there a difference? Environ. Toxicol. Chem. 1998, 17, 1380–1384.

6. Escher, B.I.; Eggen, R.; Vye, E.; Schreiber, U.; Wisner, B.; Schwarzenbach, R.P. Baseline toxicity (narcosis) of organic chemicals determined by membrane potential measurements in energy-transducing membranes. Environ. Sci. Technol. 2002, 36, 1971–1979.

7. Israelachvili, J.N. *Intermolecular and Surface Forces*, 2nd Ed.; Academic Press: London, 1991, Chapter 17.

8. Roberts, D.W. Use of octanol/water partition coefficients as hydrophobicity parameters in surfactant science. Proceedings of 5th CESIO World Surfactants Conference, Florence, Italy, May 29– June 2, 2000; 1517–1524.

9. Collander, R. The partition of organic compounds between higher alcohols and water. Acta. Chem. Scand. 1951, 5, 774–780.

10. Escher, B.I.; Schwarzenbach, R.P. Evaluation of liposome–water partitioning of organic acids and bases. 1. Development of a sorption model. Environ. Sci. Technol. 2000, 34, 3954–3961.

11. Scherer, P.G.; Seelig, J. Electric charge effects on phospholipid headgroups. Phosphatidylcholine in mixtures with cationic and anionic amphiphiles. Biochemistry 1989, 28, 7720–7728.

12. Leo, A.J.; Hansch, C. *Substituent Constants for Correlation Analysis in Chemistry and Biology*; Wiley: New York, 1979.

13. Roberts, D.W. Aquatic toxicity of linear alkyl benzene sulphonates (LAS)—QSAR analysis. J. Com. Esp. Deter. 1989, 20, 35–43.

14. Roberts, D.W.; García, M.T.; Ribosa, I.; Hreczuch, W. QSAR analysis of aquatic toxicity of ethoxylated alcohols. J. Com. Esp. Deter. 1997, 27, 53–63.

15. Roberts, D.W.; Marshall, S.J.; Hodges, G. Quantitative structure–activity relationships for acute aquatic toxicity of surfactants. 4th World Surfactants Congress 1996, 4, 340–351.

16. Roberts, D.W. Aquatic toxicity—Are surfactant properties relevant? J. Surfact. Deter. 2000, 3, 309–315.

17. Hodges, G. QSAR studies of surfactant toxicity to *Daphnia Magna*. PhD dissertation; John Moores University: Liverpool, 1998.

18. Roberts, D.W. Aquatic toxicity of linear alkyl benzene sulphonates (LAS)—a QSAR analysis. In Knoxville, T.N., Turner, J.E., England, M.W., Schultz, T.W., Kwaak, N.J., Eds.; QSAR 88. Proceedings of the Third International Workshop on Quantitative Structure-Activity Relationships in Environmental Toxicology; (Available from the National Technical Information Service, U.S. Dept. of Commerce, Springfield, VA, 1988; 22–26.

19. Divo, C. Investigation of fish toxicity and biodegradability of linear alkylbenzene sulphonates. Riv. Ital. Sostanze Grasse 1976, 53, 88–93.

20. Roberts, D.W. Application of octanol/water partition coefficients in surfactant science: a quantitative structure–property relationship for micellization of anionic surfactants. Langmuir 2002, 18, 345–352.

21. Roberts, D.W. QSAR issues in aquatic toxicity of surfactants. Sci. Total Environ. 1991, 109–110, 557–568.

22. Roberts, D.W.; Marshall, S.J. Application of hydrophobicity parameters to prediction of the acute aquatic toxicity of commercial surfactant mixtures. SAR QSAR. Environ. Res. 1995, 4, 167–176.

23. García, M. T.; Ribosa, I.; Campos, E.; Salvía, R. Toxicidad de los alcoholes grasos

etoxilados en función del tipo de distribución de los homólogos oxietilenados. J. Com. Esp. Deterg. 1995, *26*, 43–52.

24. Uppgård, L.-L.; Lindgren, Å.; Sjöström, M.; Wold, S. Multivariate quantitative structure–activity relationships for the aquatic toxicity of technical nonionic surfactants. J. Surf. Deterg 2000, *3*, 33–41.

25. Wong, D.C.L.; Dorn, P.B.; Chai, E.Y. Acute toxicity and structure–activity relationships of nine alcohol ethoxylate surfactants to fathead minnow and *Daphnia magna*. Env. Tox. Chem. 1997, *16*, 1970–1976.

26. Singh, W.P.; Lin, G.H.; Bockris, J O'M. QSAR modelling of toxicity of quaternary ammonium salts to the Fathead Minnow. Submitted for publication.

27. Roberts, D.W.; Costello, J.F. QSAR and mechanism of action for aquatic toxicity of cationic surfactants. QSAR Combinat. Sci. 2003, *22*, 220–225.

28. Rosen, M.J.; Zhu, Y.P.; Morrall, S.W.; Versteeg, D.J.; Dyer, S.D. Estimation of surfactant environmental behavior from physical chemical parameters. 4th World Surfactants Congress 1996, *3*, 304–313.

29. Rosen, M.J.; Fei, L.; Zhu, Y.-P.; Morrall, S.W. The relationship of the environmental effect of surfactants to their interfacial properties. J. Surf. Deterg. 1999, *2*, 343–347.

30. Tolls, J. Bioconcentration of Surfactants. PhD dissertation, University of Utrecht, 1998.

31. Davies, J.; Ward, R.S.; Hodges, G.; Roberts, D.W. QSAR modeling of acute toxicity of quaternary alkylammonium sulfobetaines to *Daphnia magna*. Env. Tox. Chem. *In press*.

11

Environmental Impacts of Detergent Packaging

The Case of Plastic Bottles

SUSAN E. SELKE   Michigan State University, East Lansing, Michigan, U.S.A.

I. DETERGENT BOTTLE STRUCTURE

A. Resin

Liquid laundry detergent is usually packaged in high-density polyethylene (HDPE) bottles. These bottles are made from a copolymer HDPE, polymerized primarily from ethylene but containing a few percent of added comonomer, usually butene or hexene. The effect of the randomly incorporated butene units is to reduce somewhat the crystallinity of the HDPE. This small reduction in crystallinity reduces the strength and stiffness of the bottle, but greatly increases its resistance to environmental stress cracking. Since detergents tend to be powerful stress-crack agents, this is essential in order for the bottle to provide adequate performance. If a homopolymer HDPE were used, the bottle would be prone to catastrophic failure.

B. Bottle Design and Manufacture

Bottles are manufactured using a resin with a relatively low melt flow index and, consequently, good melt strength. This allows the use of extrusion blow molding, the most economical processing method for making bottles. In the extrusion blow molding process, the resin is melted in an extruder and released through an annular die in the shape of a hollow tube. The hollow tube is captured inside a mold, and air pressure is used to expand the plastic into the shape of the bottle. The plastic solidifies in the water-cooled mold and is ejected, and the excess material (*flash*) at the bottom and top of the bottle is then trimmed off. The bottles usually have handles, to facilitate handling by the consumer, which results in additional generation of flash. Usually, the flash is ground up and fed back into the process, so this regrind becomes part of another bottle. The bottles are usually pigmented to give them a bright, attractive appearance.

In the United States, as in some other countries, detergent bottles are usually manufactured with a three-layer structure. The inside layer is a thin layer of virgin unpigmented copolymer HDPE. The outside layer is a thin layer of virgin pigmented HDPE. The middle layer, considerably thicker, consists of a blend of postconsumer recycled HDPE and regrind. Production of a multilayer bottle requires the use of coextrusion blow molding, a more demanding and expensive process than ordinary extrusion blow molding. However, the cost savings in using recycled resin combined with the savings in pigment cost due to the ability to confine the color to the thin outer layer make this an economically sound choice. Some detergent bottles have used recycled HDPE in single-layer bottles, in which the virgin and recycled materials are blended together.

The standard method for decorating the bottle is by in-mold labeling. In this process, the preprinted label is placed in the empty mold before the bottle is blown, and held in place using a light vacuum. When the hot plastic contacts the back of the label, it adheres, locking the label in place. The label becomes, in essence, part of the container wall. While originally most labels were paper, plastic, usually polypropylene, is now the predominant label material.

In-mold labeling has several advantages. It avoids the requirement for a subsequent labeling step and consequently for investment in equipment, time, and labor for the labeling. Also significant, the label is much less exposed to scuffing and scratching, because it does not protrude as much from the bottle wall as a subsequently applied label. The bottle is also somewhat less susceptible to bulging. According to some experts, the bottle can be lightweighted slightly, due to the contribution of the label to the strength of the bottle. However, other experts do not feel this is a significant factor.

Bottles have polypropylene (PP) closures (caps), which are usually designed to incorporate a measuring device to make it easy for the consumer to pour out the desired amount of detergent. Bottles also usually incorporate a fitment, which provides a pouring spout. This fitment may be either polypropylene, like the cap, or polyethylene, like the bottle. It may be pigmented to match the bottle or the cap or may be natural in color. The pouring fitment is usually designed with a drain feature so that when the cap is inverted on the bottle after use, the excess detergent flows back inside the bottle. The bottles and fitments are usually produced by injection molding, in which the melted plastic is forced through a small opening into a mold, where it solidifies into the desired shape.

Some detergent is packaged in less elaborate bottles, consisting of an unpigmented HDPE bottle with no dispensing features. These no-frills containers are intended to be used for refilling previously used dispensing bottles.

II. ENVIRONMENTAL IMPACTS: OVERVIEW

In examining the impacts associated with plastic detergent bottles, we need to consider impacts associated with production of the raw materials for the bottles, production of the bottles themselves, filling and distributing the bottles, their use, and finally their disposal. In keeping with the life cycle assessment approach, the first step is to set the goals and boundary for the analysis, and the next is to quantify the inputs and outputs associated with all these steps, analyze the effects of these inputs and outputs, and examine the process for areas where improvements can be made.

Environmental impacts associated with the production of plastic resins derive from the material inputs, the energy used, emissions to air and water associated with various steps in the process, and the solid wastes generated. Impacts are also dependent on the characteristics of the ecosystems from which these process inputs are derived and to which the outputs (emissions) go. For example, a requirement for a given amount of cooling water will have different environmental impact in a water-rich region than in one where water supplies are scarce.

This chapter will not attempt to provide a full life cycle analysis, but rather to provide a qualitative discussion, supplemented with some publicly available quantitative information about the material and energy requirements and process emissions for HDPE and PP resins and the bottles and caps.

III. IMPACTS ASSOCIATED WITH BOTTLE MANUFACTURE

A. Resin Production

Production of plastics, including HDPE and PP, begins with production of the starting raw materials, natural gas and crude oil. Manufacture of products, including plastics, consumes only a very small fraction of the amount of natural gas and oil used annually for fuel. The environmental impacts associated with the production of natural gas and oil and with their processing and refining include those associated with the exploration for the materials, the drilling, the operation of the wells (production), and the eventual closure as well as releases associated with accidents such as spills.

Next the raw materials must be refined or processed and then chemically treated (cracked) to yield the starting monomers for polymer production. The relative proportions of oil and natural gas used for the production of HDPE and PP vary from country to country as well as from company to company. The Tellus packaging study [1] reported polyethylene to be produced from roughly 33% natural gas and 67% oil. The Association of Plastics Manufacturers in Europe (APME) bases its Eco-profiles for polyethylene on about 25% natural gas and 75% oil as feedstocks [2]. Tellus reported polypropylene manufacture to use roughly 11% natural gas and 89% oil. APME estimates polypropylene to use 35% natural gas and 66% oil. In addition to feedstocks, various other materials are required for production of monomers: fuels, cooling water, catalysts, lubricants, water treatment chemicals, etc.

The next step in polymer production is polymerization, in which the starting monomers are combined into large polymer molecules. As mentioned, HDPE for detergent bottles is composed primarily of ethylene units but does contain some comonomer units, as well. Polypropylene for the caps, labels, etc. is composed mostly of propylene units but may contain some ethylene for added impact resistance. When polymerization is complete, the polymer is typically melted in an extruder and formed into small pellets. The pelletized plastic resins will contain, in addition to the base polymer, various deliberate or incidental additives, such as catalyst residues, solvent residues, antioxidants, and pigments, in small amounts. The precise amounts and nature of these will be a function of both the polymerization process and the intended end use for the resin. As the plastic resin is processed, it will also undergo some chemical change as a result of degradation processes and will pick up additional

TABLE 1 Raw Material Requirements for Production of 1000 kg HDPE and PP Resin

Raw material	HDPE (kg)	PP (kg)
Oil (feedstock only)	770	660
Natural gas (feedstock only)	250	350
Air	120	170
Nitrogen	65	67
Bauxite	36	2.3
Sodium chloride	33	2.7
Limestone	0.96	0.56
Fluorspar	0.64	0.032
Granite	0.62	0.027
Sulfur (elemental)	0.33	0.059
Sulfur (bonded)	0.16	0.025
Sand	0.15	0.13
Oxygen	0.039	0.032
Bentonite	0.025	0.026
Clay	0.011	0.014
Water, for processing	3,200	2,300
Water, for cooling	52,000	59,000

Source: Data from Ref. 2.

contaminants, usually in extremely small amounts, such as metal residues from the processing equipment.

Table 1 lists some material requirements for production of high-density polyethylene and polypropylene. A number of other materials are also required, but in extremely small amounts. Energy requirements associated with production of high-density polyethylene and polypropylene resins are summarized in Table 2, by fuel type. These values do not include the feedstock energy of the oil and natural gas listed in Table 1. Feedstock energy totals 48.52 MJ/kg for HDPE, 60.7% of the total energy required. For PP, it totals 49.03 MJ/kg, 63.5% of the total energy required [2].

TABLE 2 Energy Requirements for Production of HDPE and PP Resins (not Including Feedstock Energy), MJ/kg

Energy source	HDPE	PP
Oil	14.71	12.43
Natural gas	10.18	12.17
Coal	2.46	1.74
Nuclear	3.33	1.94
Hydroelectric	1.00	0.53
Other	0.36	0.28
Recovered energy	−0.64	−0.93
Total	31.40	28.16

Source: Data from Ref. 2.

Both air and water effluents are produced during the manufacture of plastic resins. Table 3 shows the major emissions released from the sequence of operations, beginning with acquiring raw materials and ending with resin production, for HDPE and PP. Avoided emissions, those that are not released due to the use of various types of emission control systems, are not included in these totals. Table 4 presents data for water emissions, on the same basis. Very small quantities of other substances are released as well.

A variety of types of solid wastes are generated during the production of plastic resins. These arise from the mining and processing operations, from the emission control systems, etc. Major solid wastes generated are listed in Table 5 for both HDPE and PP.

B. Container Production and Filling

The next step is production and filling of the container. In this stage, the polymer is molded into the desired shape, as was described earlier, and the bottles are filled with the detergent, capped, packaged, and distributed. At various points, the raw materials, polymer resin, bottles and caps, etc. are packaged and shipped from one location to another. Therefore, analysis of the impacts of detergent bottles should take into account these ancillary packaging materials and their manufacture, the equipment and energy required for handling and distributing the resins and containers, the equipment and energy required for filling the bottles, etc. Again, the relative magnitudes of these quantities and their precise nature will vary, depending on the processes and distances involved, the characteristics of the industries, the design and efficiency of the equipment, and other such factors.

Studies directly examining the material and energy requirements and emissions for HDPE detergent bottles are not readily available. However, APME has pro-

TABLE 3 Major Air Emissions from Production of 1000 kg HDPE and PP, from Raw Material Production Through Resin Production, After Reductions Achieved Through Use of On-Site Emission Controls

Emission	HDPE Production (g)	PP Production (g)
Carbon dioxide	1,700,000	1,900,000
Sulfur oxides	14,000	13,000
Nitrogen oxides	9,900	9,600
Dust	2,900	1,500
Carbon monoxide	820	720
Methane	5,700	6,100
Aromatic hydrocarbons	140	3
Other hydrocarbons	5,900	2,300
Other organics	5	1
Hydrogen	100	77
HCl	48	33
Metals	8	7

Source: Data from Ref. 2.

TABLE 4 Major Water Emissions from production of 1000 kg HDPE and PP, from Raw Material Production Through Resin Production, After Reductions Achieved Through Use of On-Site Emission Controls

Emission	HDPE Production (g)	PP Production (g)
Chemical oxygen demand (COD)	200	180
Biological oxygen demand (BOD)	150	34
Hydrocarbons	51	51
Phenol	4	4
Acid (H^+)	47	56
Ammonia (NH_4)	11	10
Na^+	370	250
Mg^{2+}	3	<1
Ca^{2+}	21	1
Al^{3+}	<1	18
Metals, unspecified	48	57
NO_3^-	6	18
Other nitrogen	8	5
Cl^-	340	1,300
F^-	18	1
SO_4^{2-}	49	56
CO_3^{2-}	25	31
Phosphate	1	3
Sulfur/sulfide	5	1
Dissolved organics	27	65
Other suspended solids	2,100	340
Other dissolved solids	350	100
Detergent/oil	68	69

Source: Data from Ref. 2.

duced an analysis for extrusion-blow-molded HDPE bottles in general, based on the average of seven facilities in the United Kingdom [2]. The analysis covers production of the polymer resin, transport of the resin to the converter, the conversion process, and packaging of the finished bottles, but it does not include delivery to the bottle filler or any subsequent steps in the bottle life cycle. The material and energy requirements for bottle production are shown in Tables 6 and 7. Figure 1

TABLE 5 Solid Waste Resulting from Production of 1000 kg HDPE and PP Resin

Waste type	HDPE Production (kg)	PP Production (kg)
Mining waste	74	17
Slag and ash	5.8	3.9
Chemical process waste	8.3	11.5
Mixed industrial waste	2.9	2.1
Construction waste	0.28	0.19
Other and unspecified	1.26	0.45

Source: Data from Ref. 2.

TABLE 6 Material Requirements for Production of 1000 kg Extrusion-Blow-Molded HDPE Bottles and Injection-Molded PP Components from Resin

Input	HDPE (kg)	PP (kg)
Resin	1000.3	1005
LDPE film for packaging	30	50
Corrugated board for packaging	96	
PP strapping for packaging		1
Pallets for packaging	8	50
Water	3000	11,112.5

Source: Data from Ref. 2.

shows the breakdown of total energy, including feedstock energy, for the various parts of the system. As can be seen, feedstock energy is the largest portion, with the remaining energy requirement fairly evenly divided between resin production and bottle production.

For PP caps and fixtures for detergent bottles, no specific life cycle information is readily available. APME has produced a study examining injection-molded PP in general, similar to that for blow-molded HDPE bottles, that provides some information [2]. The analysis includes transport of the resin to the converter, the conversion process, and packaging of the finished product, but it does not include delivery to the user or any subsequent parts of the life cycle. Further, the data should be regarded with caution, because they are based on a single facility in the United Kingdom. The material and energy requirements for injection-molded PP are shown in Tables 6 and 7. Figure 2 shows the breakdown of total energy, including feedstock energy, for the various parts of the system. As was the case for HDPE bottles, feedstock energy is the largest portion. The energy requirement for molding is about the same as for resin production, but space heating requires additional energy totaling nearly 11% of the total.

TABLE 7 Energy Requirements for Production of Blow-Molded HDPE Bottles and Injection-Molded PP Resins (Not Including Feedstock Energy), MJ/kg

Energy source	HDPE Bottles	PP Components
Oil	15.79	15.76
Natural gas	13.55	29.96
Coal	11.28	12.73
Nuclear	9.69	9.85
Hydroelectric	1.22	0.82
Other	0.43	0.41
Recovered energy	−0.64	−0.93
Total	51.34	68.58

Source: Data from Ref. 2.

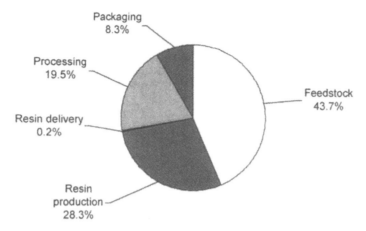

FIG. 1 Energy requirements for production of extrusion-blow-molded HDPE bottles. (From Ref. 2.)

Production of bottles, closures, and fitments also results in additional air and water emissions and solid waste. Information on these factors separate from resin production is not readily available. Tables 8–10 summarize major air and water emissions and solid waste for production of blow-molded HDPE bottles and injection-molded PP components, including those originating in resin production. While these data can be compared qualitatively with the data for resin production, subtracting resin production figures to arrive at emissions and waste associated with bottle or closure production cannot be done quantitatively with any precision, because the base data sets differ.

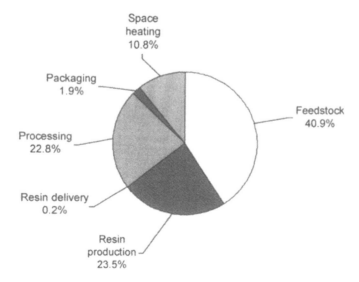

FIG. 2 Energy requirements for production of injection-molded PP components. (From Ref. 2.)

TABLE 8 Major Air Emissions from Production of 1000 kg Extrusion-Blow-Molded HDPE Bottles and Injection-Molded PP Components, from Raw Material Production Through Resin Production, After Reductions Achieved Through Use of On-Site Emission Controls

Emission	HDPE Bottles (g)	PP Components (g)
Carbon dioxide	3,000,000	4,000,000
Sulfur oxides	24,000	27,000
Nitrogen oxides	15,000	27,000
Dust	7,100	7,900
Carbon monoxide	1,500	2,300
Methane	8,300	20,000
Aromatic hydrocarbons	140	3
Other hydrocarbons	6,200	3,200
Other organics	5	1
Hydrogen	100	78
HCl	230	250
Metals	9	13
HF	13	15

Source: Data from Ref. 2.

IV. IMPACTS ASSOCIATED WITH BOTTLE DISPOSAL

Consumer concerns about the environmental impacts of plastic detergent bottles often focus on what happens to the bottle at the end of its useful life. Several options exist. The bottle may be reused by being refilled with detergent. It may be reused for some other purpose (this is rare). It may be recycled into new bottles or into some other product. It may be incinerated, with or without the recovery of energy. Finally, it may be disposed of by landfilling.

A. Reuse

Refill of the bottle with detergent presents opportunities for source reduction. While it is conceivable that the process could involve return of the bottle to the detergent manufacturer for refill at the production facility, this does not generally occur. If the bottle is reused, the refill will occur in the household, using detergent purchased in a no-frills refill container, often in an unpigmented HDPE bottle with no dispensing features. Since the refill containers are lighter in weight than the "regular" dispensing bottles, their use provides for a reduction in environmental impact roughly proportional to the weight reduction, since less HDPE and PP will be manufactured, processed, transported, and disposed to make these bottles.

Reuse of the bottle for some purpose other than dispensing of detergent is of limited value. While some people have created innovative products, ranging from decorations to trash baskets, out of HDPE bottles, such uses cannot accommodate a significant fraction of the HDPE detergent bottles manufactured.

B. Recycling

Recycling of the bottle has benefits both in terms of avoided disposal impacts and in terms of avoided production of HDPE resin. The energy used and emissions

TABLE 9 Major Air Emissions from Production of 1000 kg Extrusion-Blow-Molded HDPE Bottles and Injection-Molded PP Components, from Raw Material Production Through Resin Production, After Reductions Achieved Through Use of On-Site Emission Controls

Emission	HDPE Bottles (g)	PP Components (g)
Chemical oxygen demand (COD)	200	190
Biological oxygen demand (BOD)	150	38
Hydrocarbons	52	55
Phenol	5	7
Acid (H^+)	47	57
Ammonia (NH_4)	11	10
Na^+	370	250
Mg^{2+}	3	<1
Ca^{2+}	21	1
Al^{3+}	<1	18
Metals, unspecified	48	59
NO_3^-	6	18
Other nitrogen	8	5
Cl^-	340	1,300
F^-	18	1
SO_4^{2-}	49	56
CO_3^{2-}	26	32
Phosphate	1	3
Sulfur/sulfide	5	1
Dissolved organics	27	65
Other suspended solids	2,200	670
Other dissolved solids	350	110
Detergent/oil	68	71

Source: Data from Ref. 2.

produced from reprocessing the HDPE into a usable form are generally far less than those produced in manufacture of virgin resin. Of course, the energy and emissions associated with collecting the HDPE and getting it to the reprocessing facility must be added. The condition of the collected material will have an important effect on the processing requirements, so this must also be considered. At current recovery levels and those foreseeable in the near future, recycling of HDPE results in a net

TABLE 10 Solid Waste Resulting from Production of 1000 kg Extrusion-Blow-Molded HDPE Bottles and Injection-Molded PP Components

Waste type	HDPE Bottles (kg)	PP Components (kg)
Mining waste	57	93
Slag and ash	24	27
Chemical process waste	8.34	11.5
Mixed industrial waste	3	2.4
Construction waste	0.28	0.20
Other and unspecified	1.26	53.7

Source: Data from Ref. 2.

reduction of environmental impact. It is difficult to quantify this impact, however, because it is strongly affected by individual system designs and recovery rates.

Of course, if the recycled HDPE detergent bottles are used for some product other than new detergent bottles, these environmental benefits must be apportioned in a defensible fashion between the detergent bottle system and the manufacturing systems for the products being produced from the recycled bottles. If the recycled HDPE is used in the manufacture of new detergent bottles, these benefits stay within the detergent bottle system.

There is a fairly widespread perception that "bottle-to-bottle" recycling, or closed-loop recycling, is inherently more beneficial than open-loop recycling, where the recycled material may become part of an entirely unrelated product in its subsequent life. However, if the boundaries of the system being analyzed are drawn to include the detergent bottles and these other products, it can be easily seen that, at least to the extent that these new uses displace virgin HDPE resin of equivalent quality, it is immaterial whether the recycled resin is used in detergent bottles or alternative products. The controlling factor is simply the amount of virgin HDPE resin being manufactured. From this perspective, it makes environmental as well as economic sense to use the recycled material in whatever application is most environmentally efficient. Environmental efficiency, in this context, means the application that results in the smallest environmental impact. Here's one simple example: It does not generally make environmental sense to send the recycled HDPE 500 km away to a detergent bottle manufacturing facility if there is a plastic pallet manufacturing facility 10 km away that, if it is not permitted to use the recycled bottles, will use virgin HDPE resin of quality equivalent to that used in the manufacture of detergent bottles. A significant amount of transportation energy can be saved by sending the recycled bottles to the pallet plant and virgin resin to the bottle plant (assuming transport distances for virgin resin to the two manufacturing facilities are equal). Closed-loop recycling can, however, be of great importance in ensuring markets for recycled material. This is particularly significant when recycling systems are still in a developing state and ready markets for the recovered material may not exist. The existence of a guaranteed market for the collected material can provide a very significant incentive for the investment required to get the system operating. It is also likely to be of increasing importance whenever recovery rates increase to the point that existing markets near saturation.

In the United States and other locations where use of multilayer HDPE detergent bottles containing recycled HDPE is routine, the bottles typically contain about 25–30% recycled content by weight. Most often, the origin of this material is recycled milk bottles rather than detergent bottles. When recycled detergent bottles are used, some color sorting is generally required, to ensure that the color of the recycled layer does not show through the thin outside layer of pigmented virgin HDPE, detracting from the bottle's appearance.

Three states in the United States have mandatory recycling requirements for plastic bottles. In California and Oregon, manufacturers of products such as laundry detergent that are packaged in plastic containers must comply with state laws requiring that recycling rate, recycled content, or source reduction requirements be met. In Oregon, all plastic bottles are automatically in compliance because the recycling rate provision calls for a 25% overall recycling rate for plastic containers. The recycling rate for plastic containers in the state has been above 25% ever since the law came into effect. However, the state has issued an official advisory that with the pattern of decreasing plastic recycling rates experienced over the last several years, this may not continue to be the case, beginning perhaps as early as 2003. In California, the overall plastic container recycling rate is below the 25% that is also required for compliance in that state, so most manufacturers must comply by using a minimum of 25% post consumer recycled content in their bottles. For laundry detergent bottles, this has not been an issue, since virtually all manufacturers are using bottles with sufficient recycled content. Wisconsin requires 10% recycled content in plastic bottles for products such as laundry detergent, and again this is not an issue for detergent bottles. The law is reportedly little enforced, in any case.

Accurate recycling rates for HDPE detergent bottles are not readily available. In many places, plastics recycling rates are reported only on an overall basis, not separated by resin type. Even where resin-specific rates are available, separate rates by product type are not usually reported. For the United States, the Environmental Protection Agency issues annual reports on the disposal of municipal solid waste that contain fairly detailed information on packaging, including plastics packaging. The E.P.A. reports information on the production and recycling of unpigmented HDPE milk and water bottles and on "other plastic containers" made of HDPE [3]. Detergent bottles are included in the "other plastic container" segment, as are many other HDPE packages. Recycling rates for HDPE detergent bottles can be assumed to be lower than those for milk and water bottles, since some programs collect only the higher-value, unpigmented bottles. Recycling rates are significantly higher for bottles than for containers such as tubs, which are not accepted by most community recycling programs. Therefore, the recycling rate for HDPE detergent bottles can confidently be assumed to lie between the rate for HDPE milk and water bottles and the rate for other HDPE containers. Figure 3 shows U.S. recycling rates for HDPE milk and water bottles, for other HDPE containers, and for PET soft drink bottles and other PET containers as a comparison.

Recycling of PP from laundry detergent bottles occurs at a considerably lower rate than recycling of HDPE. As discussed, PP occurs mostly in the caps and fitments. Most recycling programs regard caps as undesirable contaminants and instruct participants to remove and discard the caps before placing the bottles in the recycling

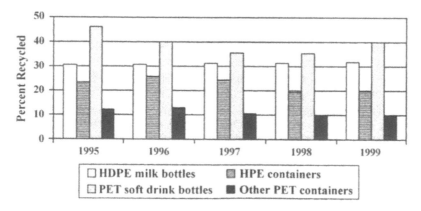

FIG. 3 Recycling rates for HDPE and PET bottles in the United States. (From Ref. 3.)

collection. Therefore, many of the caps are not recovered. The fitments, however, are typically collected with the bottles (which is important since they cannot easily be removed). In addition, a significant fraction of people ignore the instructions to remove the caps, so some caps are also collected with the bottles. If collection involves sorting at curbside, the bottles with caps are likely rejected and not picked up for processing, but if sorting is done at a material recovery facility, it is usual to accept the caps as an unavoidable contaminant. Since systems that can economically separate the PP in the caps or fitments from the HDPE are not generally available, the PP will be ground up with the HDPE and carried along in the finished product. This contamination of pigmented HDPE with PP does result in some limitations on the use of the recycled resin. Manufacturers of drainage pipe, for example, report that they must limit their use of pigmented HDPE because of the PP contamination. While similar problems exist because of PP caps on unpigmented HDPE milk and water bottles, the relative amount of PP contamination is considerably less, and it imposes few restrictions on use of recycled HDPE from that source.

The mixture of colors found in mixed HDPE containers also puts limitations on the use of the recycled resin. When the HDPE collected in typical recycling systems is processed, the result is usually a dull green color. This can readily be tinted black, but it is difficult to achieve other colors that may be desired.

Because just about any color that is desired can be achieved from unpigmented recycled resin, and because of the decreased amount of PP contamination in this resin stream, the unpigmented bottles are of higher value than the pigmented bottles. Therefore, it is usual to separate out the unpigmented HDPE bottles (mostly milk and water bottles) in an initial processing step. Until recently, it was unusual to do any further sorting. However, automatic equipment for color-sorting whole bottles has been developed, and it is being used with increasing frequency in large dedicated recycled plastics processing facilities. The same systems usually provide for separation by resin type as well as by color. This facilitates use of the recycled HDPE detergent bottles in a wider variety of applications.

Effective collection of plastic laundry detergent bottles as well as of other materials for recycling remains challenging. Overall, the collection rate, and thus the recycling rate, for plastic packaging remains well behind the collection rate for a number of other packaging materials (Fig. 4). Several studies have shown that the collection rate for HDPE bottles can be improved if programs target all plastic bottles rather than HDPE and PET bottles alone, which is the typical pattern [4]. The American Plastics Council has endorsed "all plastic bottle" collection and is pushing for expansion of this approach as a significant component in its efforts to improve recycling rates for plastic bottles [5].

With the exception of the deterioration of properties due to the incorporation of PP, HDPE tends to retain its properties well during use and recycling. Recycled HDPE is used in a variety of applications. One of the earliest uses was in the manufacture of agricultural drainage pipe. It can also be used in housewares, toys, pallets, plastic lumber, new plastic bottles, and a variety of other applications. Figure 5 shows the relative proportions of recycled HDPE containers going to various applications. As can be seen, new HDPE containers is now the largest market [6]. Since the PP in the bottles is not generally separated from the HDPE, there is no significant targeted use of PP recovered from laundry detergent bottles. It simply appears as a contaminant in the uses described.

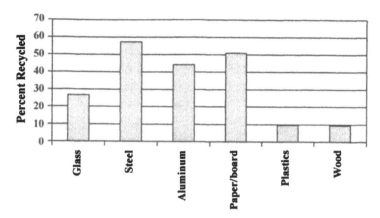

FIG. 4 Recycling rates for packaging materials in the United States, 1999. (From Ref. 3.)

C. Incineration

For materials that are not recycled, incineration offers the benefit of reduction in the amount of space required for disposal of the material. In the case of plastic laundry detergent bottles, nearly all of the volume can be eliminated. If combustion is complete, the only solids remaining will be from incombustible additives, such as certain pigments, plus a small contribution from residues originating in the emission control systems. Most modern incineration facilities, however, do not simply burn the material. They use the heat generated as a source of energy. In this regard, plastic laundry detergent bottles can make a valuable contribution. HDPE and PP are, in essence, petrochemicals, and the feedstock energy embodied in the bottle and other components, which as we have seen is a significant portion of the total energy required to make the bottle system, can be recovered when the bottle and its components are combusted.

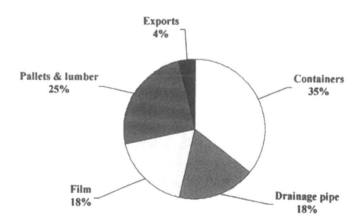

FIG. 5 Uses for recycled HDPE bottles in the United States, 1996. (From Ref. 6.)

Further, HDPE and PP burn relatively cleanly. Neither contain any chlorine, so they do not contribute to the production of chlorinated organics. If combustion is complete, the HDPE and PP themselves will produce only carbon dioxide and water vapor. If pigments and other additives are selected properly, combustion will result in no significant production of pollutants.

Of course, this should not be interpreted to mean that incineration is preferable to recycling. The embodied feedstock energy of the material is not lost when it is re-molded and used again. Incineration should be restricted to those materials that cannot be recycled in an economically and environmentally effective manner, either because they cannot be collected, they have become too contaminated, or they have been recycled so many times that they no longer maintain useful properties.

D. Landfill

Just as plastic detergent bottles can be incinerated safely, they do not usually contain materials that will pose any significant environmental hazards if they are disposed of in landfills. The plastic bottles will be inert in a landfill environment, and will not degrade to any significant degree. This is a disadvantage in one sense, since it means that they will persist indefinitely in the landfill environment and may well still be intact if they are dug up a century later. On the other hand, it is an advantage in that they will not contribute to emissions of methane or other volatile compounds, and they will not release substances that can lead to groundwater pollution.

V. DESIGN GUIDELINES

Proper design of a plastic detergent bottle can significantly reduce its environmental impact. Most obviously, if the bottle can be made with less material, all the impacts associated with production and distribution of the plastic resin will be reduced. Impacts associated with bottle production and distribution and with distribution of the filled container are also likely to be reduced. However, some caution must be exercised. If the bottle's strength is reduced so much that damage increases or that a greater amount of secondary packaging is required in order to successfully distribute the product, then this could easily result in greater overall environmental impact than from a stronger bottle using more resin. There is a point of diminishing returns in reducing bottle weight to reduce environmental impact and a point after which further reduction results in an increase in overall environmental impact. It is for precisely this type of reason that analysis of the environmental impact of a package cannot be looked at in isolation. The whole product/package system must be considered.

Other design features in addition to the bottle weight can either increase or decrease the bottle's environmental impact. Another obvious example is that the bottle should not use heavy metal–based pigments or other additives that can lead to environmental problems. Substitution of more environmentally friendly substances can decrease the environmental impact of the bottle. Such a change will affect impacts associated with the production of those hazardous substances and impacts associated with the disposal of the bottle by landfill or incineration.

There are also design decisions that can significantly affect the recyclability of the bottle, impacting both the ease of recycling and the value of the recovered materials. Several organizations have issued design guidelines for plastic bottles of various types, intended to provide information to help product designers make choices that either facilitate recycling or at least do not cause problems for existing recycling systems. One such organization is the Association of Postconsumer Plastic Recyclers (APR). Design guidelines were also produced by the EPA and state government-funded Plastic Redesign Project.

A. APR Guidelines

The Association of Postconsumer Plastic Recyclers (APR) is, as the name indicates, made up of businesses that recycle plastic bottles collected from consumers. APR has formulated design guidelines for bottles manufactured from various specific plastic resins, intended to facilitate the recycling of those containers.

For pigmented HDPE bottles used for laundry detergent and household chemicals, APR specifies the basic "design for recyclability" guideline as consideration of the general compatibility with the base resin (copolymer HDPE) of any attachment to the bottle or the efficiency of removing the attachment in conventional water-based separation systems that separate components by density. Attachments listed include closures, closure liners, base cups, inserts, labels, pour spouts, handles, sleeves, safety seals, coatings, and layers. Therefore, the recommendation is that any such materials either have a density greater than 1.0 g/cm^3 so that they can be separated or that they be compatible with HDPE.

For closures and closure liners, HDPE, low-density polyethylene (LDPE), and PP are the preferred materials. Preference should also be given to linerless closure systems that leave no residual rings or other attachments on the bottle when the closure is removed. Closures the same color as the bottle are desirable. Metal closures should be avoided.

If tamper evidence or tamper resistance is required, it should be an integral feature of the bottle rather than achieved by the use of sleeves or seals. If sleeves or seals are used, they should be designed to detach completely from the bottle. Shrink sleeves are preferred if sleeves are necessary. The use of PVC sleeves or seals should be avoided.

While the presence of color limits the uses to which recycled resin can be put, the APR was concerned about the use of unpigmented copolymer HDPE bottles. Recyclers use the presence of color to differentiate between homopolymer HDPE bottles, such as for milk and water, which are usually unpigmented, and copolymer HDPE. Therefore, APR generally favors the use of pigmented bottles for laundry detergent.

For labels and adhesives, APR prefers PP, HDPE, LDPE, linear low-density polyethylene (LLDPE), or polystyrene (PS). Paper labels, metallized labels, and PVC labels should be avoided. Label adhesives should be water soluble or water dispersible at temperatures of 140–180°F, for ease of removal. Adhesive use should be minimized.

For printing, inks that do not bleed color when agitated in water should be used so that they do not discolor the recovered HDPE.

APR also addresses layers and coatings, stating that the use of non-HDPE layers should be avoided unless they are compatible with or easily separated from HDPE in conventional recycling systems. It is noted that current systems can tolerate the use of ethylene vinyl alcohol (EVOH) and MXD6 nylon, but APR recommends that the use of non-HDPE resins be minimized to the extent possible.

The use of other attachments, such as handles, inserts, and pour spouts, is discouraged. If they are used, it is recommended that they be made either from materials with a density greater than 1.0 g/cm^3 so that they can be readily separated or from compatible materials such as PP, LDPE, and HDPE. Use of unpigmented homopolymer HDPE is preferred. If PP or LDPE is used, the weight should be limited to less than 5% of the total bottle weight. Additionally, pour spouts should be designed not to leave any product residue and to permit complete emptying of the container. Adhesives used for attachments should also be water soluble or dispersible at temperatures of 140–180°F, and their use should be minimized.

Finally, APR recommends the use of postconsumer HDPE in the bottles whenever possible.

B. Plastic Redesign Project Guidelines

The Plastic Redesign Project was funded by the U.S. Environmental Protection Agency and the states of Wisconsin, New York, and California. It has developed into a broad consortium including a number of state agencies, recycling organizations, and local organizations in addition to EPA. A major goal of the project is to strengthen the economics of local plastics recycling programs by examining new technologies and encouraging industry to voluntarily address the challenges these pose to the plastics recycling infrastructure.

In 1998, the Plastic Redesign Project issued a set of final design recommendations for plastic bottles, written to address all plastic bottles rather than containing specific recommendations for different resins. There is considerable agreement between the two sets of recommendations, but there are some differences.

The major recommendations relevant to HDPE detergent bottles include the statement that all layers in multilayer bottles should be sufficiently compatible so that the postconsumer regrind (PCR) can be sold into high-end-value markets without requiring higher processing costs. Caps, closures, and spouts on HDPE bottles, except for living hinge applications, should be compatible with HDPE so that the PCR can be sold into high-end-value markets, such as film and bottles, without requiring manual removal of caps during processing. Aluminum caps should not be used on plastic bottles, and aluminum inner seals should not be used unless the seal pulls completely off when the consumer opens the bottle. Inks on labels should not bleed onto the plastic flakes. Label adhesives should be water dispersible, or they should be avoided through use of shrink or stretch wraps. PVC and PVDC labels should not be used on non-PVC containers.

C. Alternative Materials

It is certainly possible to consider the use of plastics other than HDPE for laundry detergent bottles. In the United States, however, HDPE and PET are the only plastic bottles that have reasonable recycling rates. In Europe, PVC bottles are also recycled to a certain extent, but there are other environmental considerations that disfavor PVC. Recycling of other resins occurs only to a very small degree. Therefore, to the extent that recyclability is desired, HDPE and PET become the only choices. For a long time, the recycling rate for PET bottles was significantly higher than the rate for HDPE bottles. During the last several years, this gap has closed significantly. In fact, in the United States the recycling rate for PET bottles has consistently fallen over the last 5 years or so, while the recycling rate for HDPE bottles has risen. Therefore, there is no compelling reason to switch from HDPE to PET to improve recyclability.

REFERENCES

1. Tellus Institute. *CSG/Tellus Packaging Study*; Tellus Institute: Boston, 1992.
2. Association of Plastics Manufacturers in Europe. Eco-profiles of Plastics. www.apme.org, 1999.
3. U.S. Environmental Protection Agency. Municipal Solid Waste in the United States: 1999 Facts and Figures. U.S. Environmental Protection Agency: Washington, DC, 2001
4. Perkins, R.; Halpin, B. Resource Recycling. June 2000; 28–33.
5. American Plastics Council. www.plasticsresource.com, 2001.
6. U.S. Environmental Protection Agency. Characterization of Municipal Solid Waste in the United States: 1997 Update. U.S. Environmental Protection Agency: Washington, DC, 1998.

12

European Environmental Safety Legislation on Detergents

JOSÉ LUIS BERNA Petresa, Madrid, Spain

I. INTRODUCTION

The switch in the early 1960s from the poorly degradable surfactant tetra propylene benzene sulfonate (TPBS) to the readily biodegradable surfactant alternative LAS (linear alkyl benzene sulfonate) was the start of a series of regulatory activities in today's European Union aimed at controlling and limiting the use of nondegradable substances used in household products as well as defining the criteria to officially measure and determine the biodegradability of surfactants by approved methods. These activities have in turn generated the development of other regulations, directives, and legislative tools in connection with the production, use, and disposal of detergents and related products.

Surfactants are key ingredients used in detergents, cleaning products, cosmetics, and body care preparations as well as in a broad variety of industrial activities. Most of them are directly or indirectly discharged after use to the environment and hence the need to control and guarantee an adequate environmental behavior of these products. Furthermore, they are large-volume chemicals whose world consumption in 1999 was estimated around 19 million tons [1], including soap, as shown in Figure 1.

The volume of finished detergents is considerably higher, and its per capita consumption in the various European countries has reached a significant level, despite the variability among the various geographical zones, as shown in Figure 2 [2].

II. EU LEGISLATION

Environmental legislation in the European Union has been developing since around 1970 and continues today. Nowadays it includes more than 400 legal acts, such as directives, regulations, decisions, and recommendations. Among them are those that constitute the legislative frame covering detergents and their ingredients.

The information contained in this chapter is believed to be as up to date as possible at the time of its writing. It might not reflect, however, the existing legislation at the time of the publishing of the handbook due to ongoing adaptations of EU legislation.

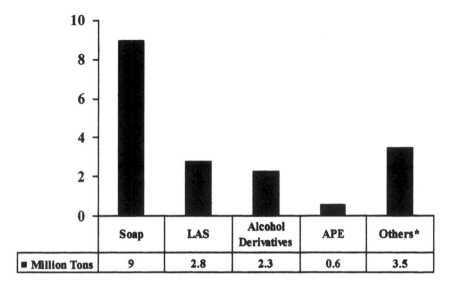

FIG. 1 Surfactant consumption (world estimate, 1999). Includes: APG, SAS, AOS, BABS, Cationics, and Amphoterics. (APG: alkyl polyglucosides; SAS: secondary alkane sulfonates; AOS: alpha olefin sulfonates; BABS: branched alkylbenzene sulfonates; APE: alkyl phenol ethoxylates; LAS: linear alkylbenzene sulfonates.)

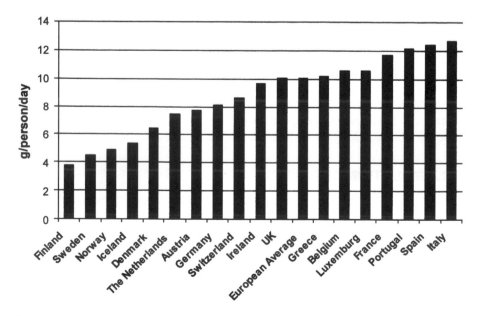

FIG. 2 1998 per capita detergent consumption, by country.

What are the differences between directives, regulations, and decisions?

Directive: Most EU environmental laws are directives, which are a form of law specific to the EU. They are designed to impose obligations on member states and to be sufficiently flexible to take into account different legal and administrative traditions and procedures. The choice and method of aligning the national legal and administrative system are left to the discretion of the member state.

Regulation: A minor portion of the environmental laws in the EU is in the form of regulations. Regulations are directly binding on member states and supersede any conflicting national laws. Member states may not transpose the provisions of regulations into national law, even if the national law is identical to the regulation.

Decision: Decisions are individual legislative acts that are binding in their entirety upon the parties to whom they are addressed. They are usually very specific in nature and are less common in the environmental field.

Among the most important pieces of EU environmental legislation affecting the detergent industry are the following.

- Directives on biodegradation of surfactants, which can be split in two parts: The first one covers the period 1973–2003; the second covers from 2004 onward with the new Detergent Regulation issued in 2004.
- Directives governing the classification, packaging, and labeling of dangerous substances and/or preparations.
- Directives and regulations on the transportation of dangerous goods.
- Directives of biocide products.
- Criteria for Eco-labeling of different products, substances, and materials. These are voluntary legislative tools and therefore are of less importance, since it is up to the concerned industrial groups whether to use the Eco-label system in their products or not.
- Other legislative non-EU tools, such as the OSPARCOM (Oslo Paris Commission) decisions. This is a non-EU regulatory group dealing specifically with matters affecting the pollution and discharges of chemicals to the North Sea Basin. Recommendations adopted by OSPARCOM are binding to the member states and are not officially published in official journals. The minutes of the meetings reflecting the various recommendations suffice to consider them as officially agreed.

The first two groups, biodegradability and dangerous substances, are the most relevant ones for detergents and surfactants, and they are discussed in more detail in the following sections.

III. BIODEGRADATION LEGISLATION IN THE EU

A. Surfactant Biodegradation Legislation (1973–2003)

Until 2003 the legislation affecting surfactants was covered under four directives:

- *Council Directive of 22 November 1973 (73/404/EEC)* on the approximation of the laws of the member states relating to detergents [3]. This can be considered the *Basic Directive*.
- *Council Directive of 31 March 1982 (82/242/EEC)* [4] on the approximation of the laws of the member states relating to methods of testing the biodegradability

of nonionic surfactants and amending Directive 73/404/EEC and Directive 86/94/EEC [5].

* *Council Directive of 22 November 1973 (73/405/EEC)* on the approximation of the laws of the member states relating to methods of testing the biodegradability of anionic surfactants [6].
* *Council Directive of 31 March 1982 (82/242/EEC)* on the approximation of the laws of the member states relating to methods of testing the biodegradability of anionic surfactants [7].

1. Directive 73/404/EEC

The introduction of the Basic Directive (73/404/EEC) refers clearly to a series of effects that could be due to the already observed foaming in surface waters due to the low biodegradability of surfactants being used until then:

> One of the pollution effects of detergents on waters, namely the formation of foam in large quantities, restricts contact between water and air, renders oxygenation difficult, causes inconvenience to navigation, impairs the photo-synthesis necessary to life of aquatic flora, exercises an unfavorable influence on the various stages of processes for the purification of waste waters, causes damage to wastewater purification plants, and constitutes an indirect micro-biological risk due to possible transference of bacteria and viruses.

The directive was also referring to the existing diversity of laws in the various member states relating to biodegradability and consequently the need to harmonize them in order to avoid hindrance to trade.

In Article 2 of the Basic Directive, the limit of biodegradability was set officially for the first time in the EEC:

> Member States shall prohibit the placing on the market and use of detergents where the average level of biodegradability of the surfactants contained therein is less than 90% for each of the following categories: anionic, cationic, non-ionic, and ampholytic.

Although there is no specific mention in the directives, it is understood that the biodegradation under discussion is aerobic.

The requirements set in this directive were ambitious because no testing methods were yet available for cationic and ampholytic surfactants. In fact, it was not until 27 years later that a substantial modification of this directive was agreed to incorporate the testing of all surfactants.

Of special interest is Article 3 of the directive: "No Member State may, on grounds of the biodegradability or toxicity of surfactants, prohibit or restrict or hinder the placing on the market and use of detergents which comply with the provisions of this Directive." Such provisions seems to have been ignored 20 years after the publication of the directive because there are cases of legislation, although voluntary, in certain EU member states, i.e., Eco-labeling, in which the anaerobic biodegradation of surfactants is required as a pass–fail criterion for the acceptability of these products in detergent formulations eligible to carry the label. Surprisingly, however, there are no directives or any other type of legislation dealing with the need to implement the anaerobic biodegradation, the justification for its requirement, or the methods for testing such property. This situation has created considerable debates and criticism between industry experts and regulatory authorities, and it has not been solved yet, despite the existence of well-documented reports [8] indicating that there is no need to use the anaerobic biodegradation criteria.

2. Directive 73/405/EEC

This directive, together with its amendment (82/243/EEC), states that

> Member States shall prohibit the placing on the market and use on their territory of a detergent if the biodegradability of the anionic surfactants contained therein is *less than 80%* determined on a single analysis in accordance with one of the following methods.

Therefore, although in the Basic Directive (73/404/EEC) the minimum level of biodegradability for all type of surfactants was set at 90%, the specific requirement for anionics is fixed now at 80%.

The biodegradability determination must be carried out on the anionic surfactants contained in commercial detergents after alcoholic extraction from such detergents. Directive 73/405/EEC specifies the following official methods for the determination of the biodegradability:

- The OECD method published in the OECD technical report of 19 December 1970 on the "Determination of the Biodegradability of Synthetic Anionic Surfactants" [9].
- The method in use in Germany, established by the "Verordnung über die Abbaubarkeit von Detergentien in Wäsh- und Reiningunsmitteln" of 1 December 1962, published in the *Bundesgesetzblatt* 1962, part 1, p. 698.
- The method in use in France, approved by Decree of 11 December 1970 published in the *Journal Officiel de la République Française*, No. 3 of 5 January 1971 and the experimental standard T 73-260 of February 1971 published by the *Association Française de Normalisation* (AFNOR).

The method developed by the OECD comprises two steps that differ both in principle and in the information they can provide regarding the biodegradability of the surfactants tested:

- A static test of the "open flask" type
- A dynamic test based upon the simulation of the conditions existing in biological sewage treatment works

The Council of the OECD recommended that the system for testing the biodegradability of anionic surfactants be as follows:

- Use of the static test as a "screening test" and acceptance of products whose biodegradability determined by this test exceeds a required percentage
- The use of the dynamic simulation test as a "confirmatory test" for any product that may not have satisfied the "screening test," the results of the "confirmatory test" being the only ones to be taken into consideration in the refusal or acceptance of products not accepted by the "screening test."

This was the system in place (see Fig. 3) until 2003 and the dynamic test was the reference method in the directives of the European Community.

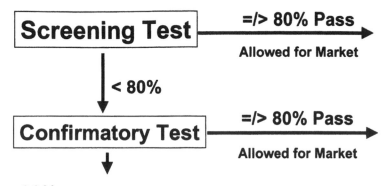

FIG. 3 Biodegradability of surfactants (anionics and nonionics), European scheme until 2003.

In its amendment of 31 March 1982 (82/243/EEC), the directive also includes the method in use in the United Kingdom, the "porous pot test" as described in Technical Report No. 70 (1978) of the Water Research Center and updates the three other previous methods to newer versions:

- OECD method published in the OECD technical report of 11 June 1976 on the "Proposed Method for the Determination of the Biodegradability of Surfactants used in Synthetic Detergents"
- The German method refers now to the one established by the "Verordnung über die Abbaubarkeit von anionischer und nichtionischer grenzflächenaktiver Stoffe in Wash-und Reinigunsmitteln" of 30 January 1977 published in the *Bundesgesetzblatt* 1977, Part I, p. 244, as set out in the regulation amending that regulation of 18 June 1980, published in the *Bundesgesetzblatt* 1980, Part I, p. 706
- The method in use in France, approved by the Decree of 28 December 1977 published in the *Journal officiel de la République Française* of 18 January 1978, p. 514 and 515, and experimental standard T 73-260 of June 1981 published by the Association Française de Normalisation (AFNOR)

A summary of the methods used in these protocols is shown in Table 1.

3. Determination of Biodegradability

The method described in Directives 73/405/EEC, 82/242/EEC, and 82/243/EEC is the same in all cases and is based on the use of a small activated-sludge plant, also

TABLE 1 Methods Used in Directives 82/242 EEC and 82/243 EEC

Screening tests		
1. OECD method	Anionic + nonionic	Pass limit: 80%
2. German method	Anionic + nonionic	Pass limit: 80%
3. French method	Anionic + nonionic	Pass limit: 80%
Confirmatory tests		
1. Porous Pot	Anionic + nonionic	Pass limit: 80%
2. OECD Confirmatory	Anionic + nonionic	Pass limit: 80%

known as Husmann unit, as shown in Figure 4. This type of unit is nowadays widely used in nearly all laboratories working on biodegradability testing. The directive also provides instructions regarding the preparation of synthetic sewage to be used in the experiments as well as the procedure to extract the surfactant portion from the formulated detergents. The biodegradability tests must be conducted on the extracted surfactants after treating the sample with ethanol or, alternatively, on pure surfactant solutions when available. The analytical method to control biodegradability is the determination of the content, in mg/L, of substances active to methylene blue (MBAS) for anionic surfactants and to BIAS (bismuth-active substances) for nonionic surfactants.

The method using the simulated wastewater treatment plant, also known as the "Confirmatory Test," is the reference method in case of dispute for both anionic and nonionic surfactant biodegradation tests.

In the case of anionic surfactants, the method uses a "nondegradable" surfactant as a reference (TPBS: tetra propylene benzene sulfonate) as well as a "biodegradable" standard, which is LAS (linear alkyl benzene sulfonate). The percentage of biodegradation based on MBAS or BIAS is calculated every day on the basis of the content of such active substances, in mg/L, in the raw sewage and in the effluent stream of the units. A typical representation of the results is shown in Figure 5 for LAS and TPBS.

The biodegradation test results are calculated as an arithmetic mean of the data obtained over 21 days after the acclimation period during which biodegradation was regular and the operation of the plant trouble free. In any case the total duration of the tests should not exceed 6 weeks.

B. Comments on Pre-2004 Legislation

The existing legislation described in the previous sections and in force until 2003 contributed significantly to a better environment. However, it contains important shortcomings, which are worthwhile to describe.

- The legislation does not define what a surfactant is. In the legislative framework, surfactants are extracted from commercial detergents following a certain pro-

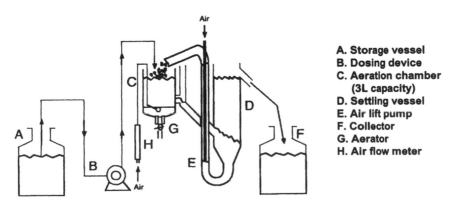

A. Storage vessel
B. Dosing device
C. Aeration chamber
   (3L capacity)
D. Settling vessel
E. Air lift pump
F. Collector
G. Aerator
H. Air flow meter

FIG. 4 Biodegradability determination equipment for the dynamic simulation test.

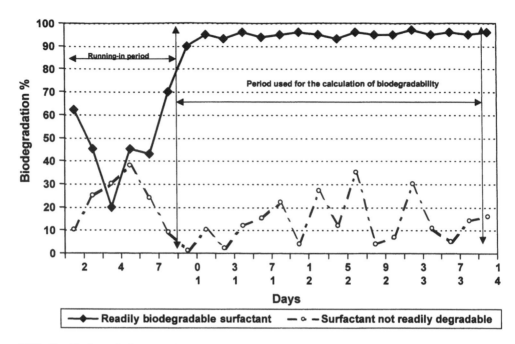

FIG. 5 Biodegradation graph.

cedure. Therefore, anything that can be extracted and shows a reaction with MBAS or BIAS, both nonspecific, is considered a surfactant.

- Cationic and amphoteric surfactants are not covered under the legislation.
- The methods proposed to determine the biodegradation couldn't discriminate between biodegradation and removal. A nonbiodegradable surfactant that is well removed and reaches 80% or higher elimination in a confirmatory test due to sorption will be allowed for commercialization under the legislation.
- The biodegradability determination was limited to "primary" biodegradation, i.e., loss of chemical identity of parent surfactant, such as MBAS (methylene blue–active substances) for anionics, and the potential toxic effects of metabolites formed was not taken into account.

These situations triggered the need for revisions and updating of the biodegradation directives in force until 2003 in the EU.

C. Surfactants—New Biodegradation Legislation (2004): Detergents Regulation

In 1995 the EU DG-III, presently the DG-Enterprise and Information Society, started a series of activities aimed at conducting a considerable revision of the existing legislation on the biodegradation of surfactants. The reasons for concern were that there were no applicable regulations for cationic and amphoteric surfactants and that the existing regulations take no account of metabolites formed during biodegradation. This was particularly important in the case of DTDMAC (di tallow di methyl ammonium chloride) and APEs (alkyl phenol ethoxylates), for which there is an OSPARCOM (Oslo Paris Commission) recommendation to phase them out. In addition to that, the new legislation proposes a testing protocol that includes in the first step the determination of the ultimate biodegradability based on BOD (biological oxygen demand) disappearance or CO_2 evolution. In general the new legislation will:

- Better define a surfactant.
- Complete current directives, including cationics and amphoterics.
- Address the toxicity of undegraded metabolites.
- Facilitate control to make easier current mechanisms that are difficult to implement, namely, the extraction of surfactants from commercial products.
- Address OSPARCOM recommendations to phase out DTDMAC and APEs.
- Integrate biodegradation and labeling requirements of detergents.

Within the new detergent regulation, the EU proposes the following definitions:

Surfactant shall mean any organic substance and/or preparation used in detergents, intentionally added to achieve cleaning, rinsing, and/or fabric softening due to its surface-active properties. It consists of one or more hydrophilic and one or more hydrophobic groups of such a nature and size that it is capable of forming micelles.

Detergent shall mean any substance or preparation containing soaps or other surfactants intended for water-based laundry or dishwashing processes. Detergents may be in any form (liquid, powder, paste, bar, cake, molded piece, and others) and used for household and/or institutional and/or industrial purposes.

Washing shall mean the cleaning of laundry, fabrics, dishes, or kitchen utensils.

Primary biodegradation shall mean the structural change (transformation) of a surfactant by microorganisms resulting in the loss of its surface-active properties due to the degradation of the parent substance and consequent loss of the surface-active property as measured in test methods listed in the annex to the regulation.

Ultimate aerobic biodegradation shall mean the level of biodegradation achieved when the surfactant is totally used by microorganisms in the presence of oxygen resulting in its breakdown to carbon dioxide, water, and mineral salts or any other elements present (mineralization) and new microbial cellular constituents (biomass) as measured in tests listed in the annex to the regulation.

The following products containing soap or other surfactants are also covered within the definition of detergent.

Auxiliary washing preparation, as intended for soaking (prewashing), rinsing, or bleaching clothes, household linen, etc.

Laundry fabric softener, as intended to modify the feel of fabrics in processes that are used to complement the washing of fabrics

Cleaning preparation, as intended for domestic all-purpose cleaners and/or other water-based cleaning surfaces (e.g., materials, products, machinery, mechanical appliances, means of transport and associated equipment, instruments, apparatus)

Other cleaning and washing preparations, as intended for any other water-based processes

The new legislation proposes a new testing scheme based on three steps. The first step uses an ultimate biodegradation test. If the surfactant fails the first step, then a second step must be carried out using the same system as the old legislation: screening + confirmatory test for primary biodegradation. If a detergent contains surfactants for which the level of "ultimate aerobic biodegradation" is less than that stipulated, manufacturers of detergents containing surfactants and/or surfactants for detergents having justified reasons may ask for derogation in accordance with certain provisions. Detergents containing surfactants, for which the level of "primary aerobic biodegradability" is less than that stipulated cannot be granted derogation. The commission will determine the conditions under which the derogation will apply. These conditions will depend on the results of a complimentary risk assessment, which constitutes the third step in the new scheme.

A summary diagram with the proposed three steps is shown in Figure 6. In the first step the ultimate biodegradation test should be conducted on the surfactant itself, not on the extracted material from the finished detergent as in the old legislation. For this test the level of mineralization shall be more than 60% within 28 days as measured according to one of the following tests.

1. The reference method for laboratory opinion on ultimate biodegradability shall be based on EN ISO Standard 14593 (1999) headspace CO_2. Surfactants in detergents shall be considered as biodegradable if passing at least one of the five following tests and if the threshold-pass level of mineralization according to one of them shall be 60% or more in less than 28 days.

   • EN ISO Standard 14593. Water quality, evaluation of ultimate aerobic biodegradability of organic compounds in aqueous medium. Method by

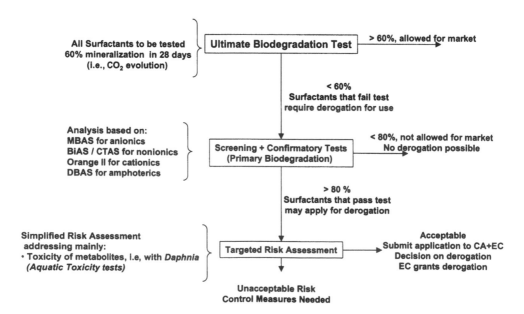

FIG. 6   Diagram for the new EU biodegradability testing of surfactants.

analysis of inorganic carbon in sealed vessels (CO_2 headspace test). Pre-adaptation not used. The 10-day window principle is not applied (Reference method).

- Method of Council Directive 67/548/EEC (CO_2 evolution: modified Sturm test). Preadaptation is not used. The 10-day window principle is not applied.
- Method of Council Directive 67/548/EEC (closed-bottle test). Preadaptation is not used. The 10-day window principle is not applied.
- Method of Council Directive 67/548/EEC (manometric respirometry). Pre-adaptation is not used. The 10-day window principle is not applied.
- Method of Council Directive 67/548/EEC (MITI test, Ministry of International Trade and Industry—Japan). Preadaptation is not used. The 10-day window principle is not applied.

2. Depending on the physical characteristics of the surfactant, other methods can be used for which the threshold limit of 70% is to be considered as equivalent to the 60% on the methods listed in point 1. These methods are:

- Method of Council Directive 67/548/EEC (DOC die-away test). Preadaptation is not to be used. The 10-day window principle is not applied. The threshold limit in this test should be 70% or more in less than 28 days.
- Method of Council Directive 67/548/EEC (modified OECD screening DOC die-away test). Preadaptation is not to be used. The 10-day window principle is not applied. The threshold-pass level in this test shall be 60% or more in less than 28 days.

If a surfactant fails to pass the first step, a second one must be conducted to measure the primary biodegradability using as a reference method the confirmatory test procedure of the OECD method. The level of primary biodegradability shall be more than 80%.

The third step consists of providing information on the surfactants subject to derogation to a local competent authority in the member state. The study should cover the aquatic environmental compartment and shall contain information about potential recalcitrant metabolites formed. Member states will consequently bring the request on derogation to the attention of the commission, where a scientific committee will make a decision based on the information provided.

Information about substances that are ultimately biodegradable must be available from the producers, upon request from the authorities, and the producers are responsible of keeping this information accessible.

In the new legislation, the EU Commission also wants, to include Recommendation 89/542/EEC [10] on the labeling of the detergent and cleaning products indicating on the package the limits of some ingredients. This will apply to household detergents and not to detergents intended to be used in industrial and institutional applications. In this case, equivalent information will be provided by means of technical data sheets or safety data sheets. The following information must appear in legible, visible, and indelible characters on the packaging in which the detergents are put for sale to the consumer:

- The name of the product
- The name or trade name and address or trademark of the party responsible for placing the product in the market

IV. LEGISLATION ON DANGEROUS SUBSTANCES

This legislation is not specific to detergents, although it contains many aspects that affect this group of products. The legislation was implemented first in 1967. Since then it has been continuously updated, revised, and expanded with numerous amendments. It was intended primarily to guarantee a single market by removing any barrier to trade that might stem from national provisions in individual member states legislation.

There are several basic documents:

1. Council Directive 67/548/EEC [11] on the approximation of the laws relating to the classification, packaging, and labeling of dangerous substances
2. Council Regulation 793/93/EEC [12] on the evaluation and control of the risks of existing substances
3. Council Directive 76/69/EEC [13] on the approximation of the laws relating to restrictions on the marketing and use of certain dangerous substances and preparations
4. Council Directive 88/379/EEC [14] on the approximation of the laws relating to the classification, packaging, and labeling of dangerous preparations
5. Directive 1999/45/EC [15] of the European Parliament and of the Council of 31 May 1999 concerning the approximation of the laws, regulations, and administrative provisions of the member states relating to the classification, packaging, and labeling of dangerous preparations

This last document (5) is the most recent one published, for reasons of clarity, because the previous directive (88/379/EEC) was already amended on several occasions. Furthermore, the rules applying to certain dangerous preparations in the various member states exhibit considerable differences in classification, packaging, and labeling matters, creating situations identified in some cases as barriers to trade. This directive, 199/45/EEC, aims at the approximation of the laws, regulations, and administrative provisions of the member states relating to:

- The classification, packaging, and labeling of dangerous preparations
- The approximation of specific provisions for certain preparations that may present hazards, whether or not they are classified as dangerous within the meaning described in such directive

For the purposes of this directive and for the potential application of its provisions to detergents and cleaning products, it is worth highlighting the definitions used.

- *Substances* means chemical elements and their compounds in the natural state or obtained by any production process, including any additive necessary to preserve the stability of the products and any impurity deriving from the process used, but excluding any solvent that may be separated without affecting the stability of the substance or changing its composition.
- *Preparations* means mixtures or solutions comprising two or more substances.

The directive lists the following substances/preparations as "dangerous":

- Explosive substances and preparations
- Oxidizing substances and preparations
- Extremely flammable substances and preparations

- Highly flammable substances and preparations
- Flammable substances and preparations
- Very toxic substances and preparations
- Toxic substances and preparations
- Harmful substances and preparations
- Corrosive substances and preparations
- Irritant substances and preparations
- Sensitizing substances and preparations
- Carcinogenic substances and preparations
- Mutagenic substances and preparations
- Substances and preparations that are toxic for reproduction
- Substances and preparations that are dangerous to the environment

The last one is of particular importance for detergents because the substances dangerous to the environment are described as "substances and preparations which, where they do enter the environment, would or could present an immediate or delayed danger for one or more components of the environment."

The directive does not apply to the following preparations in the finished state intended for the final user:

- Medicinal products for human or veterinary use as defined in Directive 65/65/ EEC
- Cosmetic products as defined in Directive 76/768/EEC
- Mixtures of substances that in the form of waste are covered by Directives 75/ 442/EEC and 78/319/EEC
- Foodstuffs
- Animal feedstuffs
- Preparations containing radioactive substances as defined in Directive 80/836/ Euroatom
- Medical devices that are invasive or used in direct physical contact with the human body

In order to determine the dangerous properties of preparations there is a series of mechanisms to evaluate the hazards by means of:

- Physicochemical properties
- Properties affecting health
- Environmental properties

In general the assessment of properties dangerous to the environment is given as concentration limits expressed as a weight/weight percentage, except for gaseous preparations, which are expressed as a volume/volume percentage and in conjunction with the classification of a substance.

The directive describes procedures to evaluate environmental hazards on the aquatic environment and on the nonaquatic environment. The nonaquatic environment covers two compartments: the ozone layer and the terrestrial environment.

The labeling of the hazard is made by means of *risk phrases* (R phrases) and *safety advice phrases* (S phrases) listed in detail in Tables 2 and 3, respectively. A description of the technical details and limits to define the various phrases is published in Directive 67/548/EEC.

TABLE 2 Risk Phrases

R number	
1. Explosive when dry	34. Causes burns
2. Risk of explosion by shock, friction, fire, or other sources of ignition	35. Causes severe burns
	36. Irritating to eyes
3. Extreme risk of explosion by shock, friction, fire, or other sources of ignition	37. Irritating to respiratory system
	38. Irritating to skin
	39. Danger of very serious irreversible effects
4. Forms very sensitive explosive metallic compounds	40. Possible risk of irreversible effects
5. Heating may cause an explosion	41. Risk of serious damage to eyes
6. Explosive with or without contact with air	42. May cause sensitization by inhalation
7. May cause fire	43. May cause sensitization by skin contact
8. Contact with combustible material may cause fire	44. Risk of explosion if heated under confinement
9. Explosive when mixed with combustible material	45. May cause cancer
10. Flammable	46. May cause heritable genetic damage
11. Highly flammable	47. Not assigned
12. Extremely flammable	48. Danger of serious damage to health by prolonged exposure
13. Not assigned	49. May cause cancer by inhalation
14. Reacts violently with water	50. Very toxic to aquatic organisms
15. Contact with water liberates highly flammable gases	51. Toxic to aquatic organisms
16. Explosive when mixed with oxidizing substances	52. Harmful to aquatic organisms
	53. May cause long-term adverse effects in the aquatic environment
17. Spontaneously flammable in air	54. Toxic to flora
18. In use, may form flammable/explosive vapor–air mixture	55. Toxic to fauna
	56. Toxic to soil organisms
19. May form explosive peroxides	57. Toxic to bees
20. Harmful by inhalation	58. May cause long-term adverse effects in the environment
21. Harmful in contact with skin	59. Dangerous for the ozone layer
22. Harmful if swallowed	60. May impair fertility
23. Toxic by inhalation	61. May cause harm to the unborn child
24. Toxic in contact with skin	62. Possible risk of impaired fertility
25. Toxic if swallowed	63. Possible risk of harm to the unborn child
26. Very toxic by inhalation	64. May cause harm to breastfed babies
27. Very toxic in contact with skin	65. May cause lung damage if swallowed
28. Very toxic if swallowed	66. Repeated exposure may cause skin dryness or cracking
29. Contact with water liberates toxic gas	
30. Can become highly flammable in use	67. Vapors may cause drowsiness and dizziness
31. Contact with acids liberates toxic gas	
32. Contact with acids liberates very toxic gas	
33. Danger of cumulative effects	

TABLE 3 Safety Phrases

S Number	
1. Keep locked up	34. *Not assigned*
2. Keep out of the reach of children	35. This material and its container must be disposed of in a safe way
3. Keep in a cool place	36. Wear suitable protective clothing
4. Keep away from living quarters	37. Wear suitable gloves
5. Keep under (appropriate liquid to be specified by the manufacturer)	38. In case of insufficient ventilation, wear suitable respiratory equipment
6. Keep under (inert gas to be specified by the manufacturer)	39. Wear eye/face protection
7. Keep container tightly closed	40. To clean the floor and all objects contaminated by this material use ... (to be specified by the manufacturer)
8. Keep container dry	
9. Keep container in a well-ventilated area	41. In case of fire and/or explosion, do not breathe fumes
10. *Not assigned*	42. During fumigation/spraying, wear suitable respiratory equipment (appropriate wording to be specified by the manufacturer)
11. *Not assigned*	
12. Do not keep the container sealed	
13. Keep away from food, drink, and animal feedstuffs	43. In case of fire use ... (indicate in thespace the precise type of firefighting equipment. If water increases the risk add: Never use water)
14. Keep away from (incompatible materials to be indicated by the manufacturer)	
15. Keep away from heatc	44. *Not assigned*
16. Keep away from sources of ignition— No smoking	45. In case of accident or if you feel unwell, seek medical advice immediately (show the label where possible)
17. Keep away from combustible material	
18. Handle and open container with care	46. If swallowed, seek medical advice immediately and show this container or label
19. *Not assigned*	
20. When using, do not eat or drink	47. Keep at temperature not exceeding ... °C (to be specified by the manufacturer)
21. When using, do not smoke	
22. Do not breathe dust	
23. Do not breathe gas/fumes/vapor/spray (appropriate wording to be indicated by the manufacturer)	48. Keep wetted with (appropriate material to be specified by the manufacturer)
24. Avoid contact with skin	
25. Avoid contact with eyes	49. Keep only in the original container
26. In case of contact with eyes, rinse immediately with plenty of water and seek medical advice	50. Do not mix with ... (to be specified by the manufacturer)
27. Take off immediately all contaminated clothing	51. Use only in well-ventilated areas
	52. Not recommended for interior use on large surface areas
28. After contact with skin, wash immediately with plenty of ... (to be indicated by the manufacturer)	53. Avoid exposure—Obtain special instructions before use
29. Do not empty into drains	54. *Not assigned*
30. Never add water to this product	55. *Not assigned*
31. *Not assigned*	56. Dispose of this material and its container to hazardous waste or special waste collection point
32. *Not assigned*	
33. Take precautionary measures against static discharges	57. Use appropriate containment to avoid environmental contamination

TABLE 3 Continued

S Number	
58. *Not assigned*	62. If swallowed, do not induce vomiting:
59. Refer to manufacturer/supplier for	Seek medical advice immediately and
information on recovery/recycling	show this container or label
60. This material and its container must be	63. In case of accident by inhalation:
disposed of as hazardous waste	Remove casualty to fresh air and keep
61. Avoid release to the environment.	at rest
Refer to special instructions/Safety	64. If swallowed, rinse mouth with
data sheet	water (only if the person is conscious)

Of particular interest in the directive of dangerous preparations is the fact that indications such as "nontoxic," "nonharmful," "nonpolluting," "ecological," or any other statement indicating that the preparation is not dangerous or likely to lead to underestimation of the dangers of the preparation in question shall not appear on the packaging or labeling of any preparation subject to the directive.

V. LEGISLATION ON BIOCIDES

Directive 98/8/EC of the European Parliament and of the Council of 16 February 1998 concerning the placing of biocide products on the market [16] is an important and extensive document. It concerns:

- The authorization and placing of biocide products on the market
- The mutual recognition of authorizations within the European Community
- The establishment at the European Community level of a positive list of active substances that may be used in biocide products

The directive is not specific to surfactants or detergents, but it concerns, in some cases, substances or ingredients that may also be used in detergents or that have surfactant properties.

For the purpose of this directive, the following definitions apply.

1. *Biocide products*: Active substances and preparations containing one or more active substances, put up in the form in which they are supplied to the user, intended to destroy, deter, render harmless, prevent the action of, or otherwise exert a controlling effect on any harmful organism by chemical or biological means. Annex V of this directive contains a list of three product types, with set of descriptions within each type of product.

2. *Low-risk biocide product*: A biocide product that contains as an active substance(s) one or more of those listed in Annex IA and that does not contain any substance(s) of concern. Under the conditions of use, the biocide product shall pose only a low risk to humans, animals, and the environment.

An active substance cannot be included in Annex IA if it is classified according to Directive 67/548/EEC as any of the following

- Carcinogenic
- Mutagenic

- Toxic for reproduction
- Sensitizing
- Being accumulative and not readily degradable

3. *Basic Substance*: A substance that is listed in Annex IB whose major use is nonpesticide but that has some minor use as a biocide either directly or in a product consisting of the substance and a simple diluent that itself is not a substance of concern and that is not directly marketed for this biocide use. The substances that could potentially enter in Annex IB are the following:

- Nitrogen
- Carbon dioxide
- Ethanol
- 2-Propanol
- Acetic acid
- Kieselgur

4. *Active substance*: A substance or microorganism, including a virus or a fungus, having general or specific action on or against harmful organisms.
5. *Substance of concern*: Any substance, other than the active substance, that has an inherent capacity to cause an adverse effect on humans, animals, or the environment and that is present or is produced in a biocide product in sufficient concentration to create such an effect. Such a substance, unless there are other grounds of concern, would normally be a substance classified as dangerous according to Directive 67/548/EEC.
6. *Harmful organism*: Any organism that has an unwanted presence or a detrimental effect on humans, their activities, or the products they use or produce, or on animals or the environment.
7. *Residues*: One or more of the substances present in a biocide product that remains as a result of its use, including the metabolites of such substances and products resulting from their degradation or reaction.

The inclusion of substances in annexes of the directive may be renewed on one or more occasions for periods not exceeding 10 years. The initial inclusion, as well as any renewed inclusion, may be reviewed at any time if there are indications that any of the requirements is no longer satisfied.

The directive on biocide products is very extensive document and also provides details on many other aspects relevant to the placing in the market of such products as well as to the introduction of substances in the annexes. There are provisions concerning the use of data for second and subsequent applications for authorization of substances, new information requirements for biocide products, derogation from requirements, transitional measures, classification, packaging and labeling of biocides, safety data sheet requirements, advertising of biocide products, information to be provided to poison control centers, and many others.

VI. ECO-LABELING

The European Eco-label award scheme is a voluntary system enabling manufacturers to show on their products and to communicate to their customers that their products comply with the approved criteria. The general requirements for a European Community Eco-label award scheme are established by Regulation EEC 880/92 [17]. The process can be applied to a variety of products and services, among them detergents, and consists of the following steps.

- *Product selection*: This is done with the participation of national competent bodies, industry, environmental and consumer organizations, trade, and retailers as well as any other appropriate group. This forum proposes the criteria to be further discussed in the next step.
- *Criteria setting*: Discussed among experts of the involved groups.
- *Adoption of criteria*. The final proposed criteria in the previous step are discussed by the Consultation Forum and then forwarded to the member states and finally adopted by the European Commission.
- Once approved, the criteria are officially communicated in the Journal.

When a manufacturer or importer is aware that Eco-label criteria are available for his products, the process to apply for the award is as follows.

- Submission of the application by contacting the national competent body, filling the application forms, and providing the necessary tests and details required on the product proving that it complies with the required ecological and performance criteria.
- The national competent body assesses the application, using information provided by the manufacturer and tests results provided by independent laboratories.
- The national competent body informs the European Commission that it intends to award the Eco-label to the manufacturer or importer. All the other member states are informed. If no objections are raised, the national competent body awards the Eco-label.

At this moment there are criteria designed to award the Eco-label to the following products:

Paints and varnishes
Footwear
Textiles
Light bulbs
Bed mattresses
Refrigerators
Detergents (see later)
Washing machines and dishwashers
Copying paper
Soil improver
Personal computers
Tissue paper

The Eco-label for detergents is split into several categories: laundry detergents [18] automatic dish-washing detergents [19], hand dishwashing detergents [20] and all-purpose and sanitary cleaners [21].

In general, the criteria applied in the design of the Eco-label schemes for detergents are aimed at promoting: (1) the reduction of water pollution, both by reducing the quantity of detergents used and by limiting the quantity of harmful ingredients, and (2) the reduction of energy use by promoting low-temperature detergents.

The ecological criteria established for the various detergent family groups include hurdles and scores limiting to a minimum the content of substances and preparations classified as dangerous in detergents that may be awarded the Eco-label. The decision on the criteria to be used for the award is proposed by the selected member state competent body responsible for designing the criteria in cooperation with other interested parties. The criteria are usually revised every four years. In some cases these criteria are in conflict with other existing regulations/legislation that allow the use of products that, under the Eco-label criteria, might be limited or even banned. These situations have generated criticism toward the lack of scientific criteria used in some cases in the preparation of the Eco-label system for detergents and have generated only little acceptability of this scheme among European detergent manufacturers. Of special relevance is the fact that the requirements under Regulation 880/92 clearly establish the need to assess the environmental impact of detergents using the life cycle procedure. However, this is frequently ignored in selecting or proposing criteria for the Eco-label award.

The following definitions are used in the context of the various detergent groups for Eco-label purposes.

- *Laundry detergents* means all laundry detergents in powder, liquid, or any other form, for washing of textiles intended to be used principally in household washing machines.
- *Hand dish-washing detergents* shall mean all detergents intended to be used to wash by hand dishes, crockery, cutlery, pots, pans, or other kitchen utensils.
- *All-purpose cleaners* are detergent products intended for routine cleaning of floors, walls, ceilings, and other fixed surfaces and that are dissolved or diluted in water prior to use.
- *Cleaners for sanitary facilities* are detergent products intended for the routine removal (including by scouring) of dirt and/or deposits in sanitary facilities, such as laundry rooms, bathrooms, showers, toilet, and kitchens. Products that are automatically used when a toilet is flushed, for example, self-dosing products such as toilet blocks, or products for use in a toilet's cisterns, are not included. Products that have no cleaning effect other than calcium carbonate (scale) removal are not included. Disinfectants are not included.

The environmental performance and the fitness for use of each product group shall be assessed according to ecological and performance criteria. Some of these criteria are specific for each product category, while others are similar in the fourth type of cases. In general the criteria cover the following areas.

- *Ecological criteria on ingredients and packaging.* The parameters are calculated and expressed as grams/wash or liters/wash. They are aggregated and assessed as a whole according to an agreed approach. The following are examples of parameters that can be considered:

- Total chemicals content in the formula
- Critical dilution volume—toxicity (CDV—tox.)
- Phosphates content
- Insoluble inorganics content
- Soluble inorganics content
- Organic compounds nondegradable (aerobic)

- Organic compounds nondegradable (anaerobic)
- Biological oxygen demand

The directive provides details concerning the selected criteria, their exclusion hurdles, their weighting factors, the maximum achievable scoring result, as well as the scoring formulae to be used to calculate the score in respect to each criterion.

Of special importance in the discussion of ecological criteria is the document called Detergent Ingredient Data (DID) List, which compiles relevant data about the most widely used ingredients in detergents.

The ecological criteria selection also provides the opportunity to select "other" criteria, in particular the limitation of the presence of some ingredients due to specific properties. The Eco-label scheme for laundry detergents provides the broader example of such limitation.

- The total weight of ingredients that are or may be classified as dangerous to the aquatic environment and assigned the risk phrase R50 (very toxic for aquatic organisms) according to Directive 67/548/EEC shall not be higher than 10 g/wash.
- The total weight of ingredients that are or may be classified as dangerous to the environment and assigned the risk phrase R50 or R53 (may cause long-term adverse effects on the aquatic environment) shall not be higher than 0.25 g/wash.
- Phosphonates shall not exceed 1 g/wash
- Surfactants from the alkylphenol ethoxylates family, perfumes containing aromatic nitro compounds, the complex formation agent EDTA, and ingredients classified as carcinogenic, toxic to reproduction, and mutagenic shall not be used.

The ecological criteria on product packaging state that only primary packaging is considered. The weight of the container shall not exceed 7 g/wash. The cardboard packaging shall be 80% recycled packaging, and the plastic packaging shall be labeled according to ISO 1043.

- *Performance criteria.* The product shall be compared on its washing performance with reference detergents of the same type according to the EU Eco-detergents performance test. The testing shall be performed at the expense of the applicant by laboratories that meet the general requirements laid out in European Norm 45001 or any equivalent system.
- *Consumer information.* The Eco-label system establishes that the following information shall appear on or in the package:

Presort laundry.
Wash with full load.
Follow usage instructions for detergent.
Choose low-temperature washing cycle.
Recommended dosage for normally soiled and for heavily soiled textiles.

VII. GOOD ENVIRONMENTAL PRACTICE FOR HOUSEHOLD LAUNDRY DETERGENTS

The Commission Recommendation of 22 July 1998 [22] reflects the code of conduct elaborated by the European Soap and Detergent Association (AISE), representing over 90% of the detergent and cleaning products in the European Union. The code of conduct was elaborated to improve the information available to consumers to ensure proper usage and in particular the dosing of detergents by more detailed labeling, educational advertising, and other programs aimed at increasing consumer awareness, to save resources and to bring about a direct positive impact on water quality and on the environment in general.

The code focuses on several measures to reduce the impact on several key aspects of the environment, such as (1) decreasing CO_2 emissions through a reduction of the washing temperatures and therefore a decrease in energy consumption, (2) lowering the general environmental impact related to detergents through a reduction in detergent consumption and consequently in detergent packaging, and (3) reducing the percentage of poorly degradable ingredients used in detergents. These measures are translated into the following series of concrete recommendations.

- The total amount of energy used per wash cycle for the product group shall be reduced by 5% by the year 2002 as compared with 1996.
- The consumption per capita of the product group in the European Union should be reduced by 10% by the year 2002 as compared with 1996.
- The consumption per capita of primary and secondary packaging in the product group should be reduced by 10% by the year 2002 as compared with 1996.
- The content of all poorly biodegradable organic ingredients should be decreased by 10% in the product group by the year 2002 as compared with 1996.
- Consumers should be provided with information designed to encourage the correct use of household laundry detergents.

In order to monitor the progress of these recommendations, statistics on detergent consumption will be collected in member states and reported to the European Commission through AISE.

VIII. NATIONAL ECO-LABELING SYSTEMS FOR DETERGENTS

In addition to the Eco-label award system of the European Union for detergents, other systems have been developed separately in other European countries or regions. They have been designed for the same purpose as the EU system, and they also use similar assessment procedures to award the corresponding Eco-label, although they may differ in some specific criteria and the application of other parameters. Among them the most relevant ones are:

German Blue Angel system
Nordic White Swan criteria

In those countries having their own local Eco-label system, the EU program coexists with the local one, and it is up to the producer to apply for one or another system for its products. Being systems of local application, however, they will not be discussed in this chapter.

IX. WATER-QUALITY, AQUATIC-ENVIRONMENT, AND OTHER RELATED LEGISLATION

Although specific directives and regulations, such as those referring to biodegradability, dangerous preparations, and Eco-labelling, are the ones most directly related to surfactants and detergents, there are other legislative tools that indirectly refer to or may affect these products. Some of them deal with water quality, while others refer to the use of wastewater treatment plants, the use of sludge in agriculture, or the protection of groundwater. Among them, the following are worth highlighting.

Drinking water–quality criteria
Groundwater protection
Urban water treatment
Discharges of dangerous substances
Prevention of marine pollution by accidental spills
Sludge directive
Transportation directive

X. OSPAR

The grounding of the ship Torrey Canyon in 1967, and the subsequent release of 117,000 tons of crude oil with disastrous consequences for the environment, proved to be a pivotal point for international cooperation to combat marine pollution in the Northeast Atlantic and stimulated the signing in 1969 of the Agreement for Cooperation in Dealing with Pollution in the North Sea by Oil, also known as the *Bonn Agreement*. The next important development in the growing general awareness of the dangers of pollution of the seas and oceans came with the agreement and signing of the Convention for the Prevention of Marine Pollution by Dumping from Ships and Aircraft: the *Oslo Convention*. The agreement was signed in 1972 and came into force in 1974. Later on it was felt necessary to develop a similar agreement for the prevention of marine pollution by discharges of dangerous substances from land-based sources, watercourses, and pipelines. The negotiations resulted in the so-called *Paris Convention*, which was opened for signing in 1974 and came into force in 1978.

The OSPAR Convention was opened for signing on 1992 and came into force on March 1998. It replaces the Oslo and Paris conventions, but decisions, recommendations, and all other agreements previously adopted under those conventions will continue to be applicable and unaltered in their legal nature unless they are terminated by new measures adopted under the OSPAR Convention. The convention has been ratified by the EU Commission and the following countries: Belgium, Denmark, Finland, France, Germany, Iceland, Ireland, the Netherlands, Norway, Portugal, Spain, Sweden, the United Kingdom, Luxembourg, and Switzerland.

The new OSPAR Convention consists of a series of provisions and, among other things:

- Requires the application of:
    The precautionary principle
    The polluter-pays principle
    Best available techniques (BAT) and best environmental practice (BEP), in
        cluding clean technology
- Provides binding decisions
- Provides the participation of observers
- Establishes rights of access to information about the maritime area of the convention

Within the OSPAR Convention is a series of annexes that deal with specific areas, some of which are of direct interest to the detergent industry:

I. Prevention and elimination of pollution from land-based sources
II. Prevention and elimination of pollution by dumping and incineration
III. Prevention and elimination of pollution from offshore sources
IV. Assessment of the quality of the marine environment
V. Provisions with regard to the protection and conservation of the ecosystems and biological diversity of the maritime area

Future work will be directed in the following main areas:

Hazardous substances
Radioactive substances
Eutrophication

For the purpose of current and future OSPAR activities, it is worth pointing out the definitions used under the convention.

• *Hazardous substances* are those that fall under one of the following categories:

1. *Substances* or groups of substances that are toxic, persistent, and liable to bioaccumulate (PBT).
2. Other substances or groups of substances that are assessed by the commission as requiring a similar approach as substances referred to in category 1, even if they do not meet all the criteria for toxicity, persistence, and bioaccumulation but that give rise to an equivalent level of concern.

• *Substance* means a chemical element or compound in the natural state or obtained by any production process, including any additive necessary to preserve the stability and any impurity deriving from the process used, but excluding any solvent that may be separated without affecting the stability of the substance or changing its composition.
• *Group of substances* means a number of substances where:

1. The substances have been shown to present a similar level of hazard, using internationally accepted criteria.
2. Extrapolation from the assessment of an appropriate sample among a number of substances shows that those substances: (a) require preventive action because of the level of risk they pose to man and the environment, and (b) are sufficiently related in terms of both their physic-chemical properties and their field of application to be jointly managed for the purposes of the strategy.

The aspects of the OSPAR activities of interest for the detergent industry derive from the List of Chemicals for Priority Action, which contains some ingredients used so far in different detergent and cleaning products. The list includes the following products:

Polychlorinated dibenzodioxins (PCDDs)
Polychlorinated dibenzofurans (PCDFs)
Polychlorinated biphenyls (PCBs)
Polyaromatic hydrocarbons (PAHs)
Pentachlorophenol (PCP)

Short-chain chlorinated paraffins (SCCP)
Hexachlorocyclohexane isomers (HCH)
Mercury and organic mercury compounds
Cadmium
Lead and organic lead compounds
Organic tin compounds
Nonylphenol ethoxylates and related substances (NPEs)
Musk xylene
Brominated flame retardants
Certain phthalates—dibutylphtalate and diethylhexylphtalate

From this list, the surfactant group of nonylphenol ethoxylates has been phased out under OSPAR recommendation.

XI. PRECAUTIONARY PRINCIPLE

It is worth mentioning the use of the precautionary principle in some of the regulations and legislation affecting surfactants and detergents, because in some cases this has been the driving factor to justify the actions or regulations proposed. This has given rise to many debates and to mixed, sometimes contradictory, views due to the dilemma of balancing the freedom and rights of individuals, industry, and organizations with the need to reduce the risk of adverse effects on the environment or human, animal, or plant health. The precautionary principle, which is essentially used by decision-makers in the management of risk, should not be confused with the element of caution that scientists apply in their assessment of scientific data.

Although the precautionary principle is not explicitly mentioned in the treaty, except in the environmental field, its scope is far wider and covers those specific circumstances where scientific evidence is insufficient, inconclusive, or uncertain and there are indications through preliminary objective scientific evaluation that there are reasonable grounds for concern that the potentially dangerous effects on the environment or human, animal, or plant health may be inconsistent with the chosen level of protection. It should be noted, however, that the precautionary principle could under no circumstances be used to justify the adoption of arbitrary decisions.

The precautionary principle should be considered within a structured approach to the analysis of risk, which comprises three elements:

Risk assessment
Risk management
Risk communication

The precautionary principle is particularly relevant to the second element, the management of risk.

The recourse to the precautionary principle presupposes that potentially dangerous effects deriving from a phenomenon, product, or process have been identified and that scientific evaluation does not allow the risk to be determined with sufficient certainty. The precautionary principle has been overused or even misused in some particular cases affecting surfactants where there were no objective reasons for concern yet the regulators and decision-makers applied it on the basis of subjective concerns, even ignoring or neglecting available scientific information on such matters.

For this reason the implementation of an approach based on the precautionary principle should start with a scientific evaluation, as complete as possible and, where possible, identifying at each stage the degree of scientific uncertainty. Of particular interest in the communication from the Commission of 2/2/2000 [23] is the fact that where action is deemed necessary, measures based on the precautionary principle should be:

- *Proportional* to the chosen level of protection
- *Nondiscriminatory* in their application
- *Consistent* with similar measures already taken
- *Based on an examination of the potential benefits and costs of action or lack of action* (including, where appropriate and feasible, an economic cost/benefit analysis)
- *Subject to review* in the light of new scientific data
- *Capable of assigning responsibility for producing the scientific evidence necessary* for a more comprehensive risk assessment

Whether or not to invoke the precautionary principle is a decision exercised where scientific information is insufficient, inconclusive, or uncertain and where there are indications that the possible effects on the environment or human, animal, or plant health may be potentially dangerous and inconsistent with the chosen level of protection.

The communication of 2/2/2000 contains guidelines intended to serve as general guidance and not to modify the treaty or secondary European Community legislation. An important objective is to avoid unwarranted recourse to the precautionary principle, which in certain cases could serve as a justification for disguised protectionism. It also stresses the need to clarify a misunderstanding as regards the distinction between reliance on the precautionary principle and the search for zero risk, which in reality is rarely to be found. The search for a high level of health and safety as well as environmental and consumer protection belongs to the framework of the single market, which is a cornerstone of the European Community.

The conclusion of the communication states that it should contribute to reaffirming the European Community's position at an international level in view of the increasing attention given to the precautionary principle. The Commission stresses that the communication is not meant to be the last word; rather, it should be seen as the point of departure for a broader study of the conditions in which risks should be assessed, appraised, managed, and communicated.

XII. WHITE PAPER ON A STRATEGY FOR A FUTURE CHEMICALS POLICY

On February 13, 2001, the European Commission published a document on a strategy for Europe's future chemicals policy under the form of a white paper. The document was to be studied and criticized in the following months by the European Parliament, member states, and stakeholders in order to provide their own input, suggestions, and improvements. The document is still very imprecise, giving rise to more questions than answers. The Commission is now proposing a single regime for new and existing substances (around 30,000) for which the same level of information and testing should be available. This system, called REACH (registration, evaluation, and authorization) requires the registration with national authorities of all chemical substances following preliminary risk assessments by industry. A second step, evaluation, will be needed for high-production-volume substances that will undergo more thorough testing. In addition, a general ban will apply to substances whose known or suspected dangerous properties cause concern. Any of such substances will be subject to specific authorization. The new strategy aims to increase the protection of human health and the environment and is found mainly in the following directives:

• Directive on the Classification, Packaging, and Labeling of Dangerous Substances
• Directive on the Classification, Packaging, and Labeling of Dangerous Preparations
• Regulation on the Evaluation and Control of the Risks of Existing Substances
• Directive on Restrictions on the Marketing and Use of Certain Dangerous Substances and Preparations.

The present system does not provide a way to rapidly restrict the use of a chemical if it is shown to be unsafe. Because of that, the foregoing directives have frequently been criticized for failing to ascertain and address the effect of chemicals on human health or the environment.

The proposal will imply that as of 2012, new and existing substances will be subject to the same registration, testing, and authorization procedures. Testing and permission to use a given chemical depend on the proven or suspected hazardous properties, uses, exposure, and volume produced or imported. The Commission recommends a regime where testing is mandatory for all substances and where the volume of the substance produced triggers the depth of testing. A similar system already applies to substances placed on the EU market after 1981.

The white paper has certainly opened a Pandora's box on the entire issue of chemical hazards, consumer exposure, and damage to the environment. A wide range of interests is affected, since both chemical and product manufacturers must ensure the safety of their products. Because surfactants and detergents are products of large volume and of direct consumer interest, it is reasonable to expect that they will be affected by the future actions derived from the white paper. In this context, however, it is worth mentioning the HERA [24] project.

XIII. THE HERA PROJECT

The HERA (human and environmental risk assessment) Project is a European voluntary initiative launched in 1999, before the white paper, by:

A.I.S.E. (International Association for Soaps, Detergents, and Maintenance Products), representing the formulators and manufacturers of such products
CEFIC (European Chemical Industry Council), representing the suppliers and manufacturers of the raw materials

HERA focuses on ingredients of household products and will provide a common risk assessment framework for such industrial products and show that this process will deliver evaluated safety information on the ingredients used in these products in a speedy, effective, and transparent way. This process is intended to support a risk-based approach to chemicals legislation in the European Union and may serve as pilot for the application of the same process in other sectors and/or geographical areas.

The need for HERA was a voluntary decision by the European Industry because, until recently, safety assessments were carried out on ingredients of detergents by the companies and the results remained, in most cases, confidential to them. Current regulations and public demands require a greater transparency of the information on the potential risks posed by the ingredients of products used in the home.

The surfactant and detergent industry firmly believes that hazard assessment is not sufficient to reach conclusions on the safety of chemicals. Only when the hazard is combined with exposure information is it possible to estimate the risk a chemical may pose in a particular use scenario. Because of that, HERA will approach risk from a definite viewpoint: It will focus on risks to consumers and the environment resulting solely from the use of these chemicals in household cleaning products when used by the consumer. In that respect it is important to understand and realize the progress made in the management of chemicals in the EU as well as the evolution in the various pieces of legislation affecting chemicals, as shown in Figure 7.

The first priority in the implementation of the HERA project was to gain consensus on the procedures within the supplier and formulator companies sponsoring the project and to complete the assessments on three pilot chemicals (alkyl sulfate, zeolites, and fluorescent whitening agent), which were the basis of the first phase of the project. Subsequent phases of the project involve a wider range of substances used in household products.

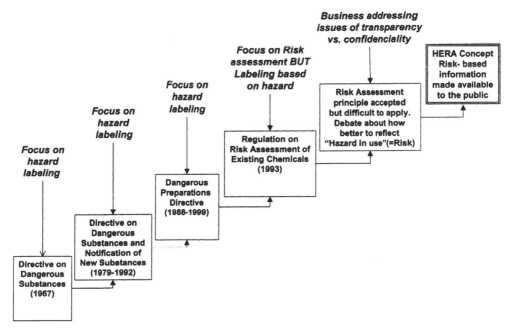

FIG. 7 How management of chemicals and product safety has evolved in the EU.

HERA operates transparently and, after internal industry peer review, the methodology as well as the results of the various substances will be made available to all interested parties through the HERA Web site.

The concept of *risk* versus *hazard*:

- The *hazard* associated with a chemical is its intrinsic ability to cause adverse effects.
- The *risk* is the probability that such effects will occur in the various applications in which the chemical will be used and discharged (exposure scenarios). For a chemical, risk assessment takes into account both the hazards of that chemical and the exposure to it (both human and the environment)

HERA is a large and complex project requiring a substantial working structure to ensure that the necessary knowledge and experience can be brought together from the various parties and expertise involved. Companies and associations involved in the HERA project have committed themselves to an efficient development of the project by agreeing to contribute to the methodology, helping to select the substances to be studied, providing the hazard and exposure information on human health and the environment, talking to shareholders, and, finally, producing and publishing the final HERA assessments. HERA is also forging close links with other, related associations in Japan and the United States to enhance and extend the coverage of the information developed in the assessments.

The basic HERA principles are:

- *Partnership* between supplier and formulator companies
- *Open dialogue with partnerships*
- *Transparency* in all activities
- *No preconceptions* on the outcome of the assessments
- *Sound scientific basis* usage throughout the whole process

REFERENCES

1. CAH Report. Surfactants in Consumer Products. Colin Houston & Associates, New York, 1999.
2. Fox, K. Presentation at XXXI Jornadas of CED, Barcelona (Spain), March 2001.
3. Official Journal of the European Communities. L 347/51 of 17.12.73.
4. Official Journal of the European Communities. L 109 of 24.04.82.
5. Official Journal of the European Communities. L 80 of 25.03.86.
6. Official Journal of the European Communities. L 347 of 17.12.73.
7. Official Journal of the European Communities. L 109 of 24.04.82.
8. Anaerobic Biodegradation of Surfactants. Review of Scientific Information. ERASM, Brussels (Belgium), 1999.
9. OECD Guidelines for Testing of Chemicals—Paris, 1981.
10. Official Journal of the European Communities. L 291 of 10.10.89.
11. Official Journal of the European Communities. L 196 of 16.08.67.
12. Official Journal of the European Communities. D 793 of 15.07.97.
13. Official Journal of the European Communities. L 769 of 27.07.76.
14. Official Journal of the European Communities. L 187 of 16.7.88.
15. Official Journal of the European Communities. L 200 of 30.07.99.
16. Official Journal of the European Communities. L 123 of 24.04.98.

17. Official Journal of the European Communities. L 99 of 11.4.92.
18. Official Journal of the European Communities. L 187 of 20.07.99.
19. Official Journal of the European Communities. L 167 of 02.07.99.
20. Hand Dishwashing Detergents. Ecolabeling Criteria not officially published at the time of printing.
21. All-Purpose Cleaners and Cleaners for sanitary facilities. Eco-labeling Criteria not officially published at the time of printing.
22. Official Journal of the European Communities. L 215 of 01.08.98.
23. Communication from the Commission of 2/2/2000.
24. www.heraproject.com

13

The Application of Environmental Risk Assessment to Detergent Ingredients in Consumer Products

CHRISTINA COWAN-ELLSBERRY, SCOTT BELANGER, and DONALD VERSTEEG The Procter & Gamble Company, Cincinnati, Ohio, U.S.A.

GEERT BOEIJE and TOM FEIJTEL The Procter & Gamble Company, Brussels, Belgium

I. INTRODUCTION

In this chapter, the principles of risk assessment are applied to the assessment and management of risks to the environment. The focus in this chapter is on detergent ingredients in consumer products. These substances need to be carefully assessed because their use and disposal will involve widespread release into the environment.

A number of environmental risk assessment guidelines have been written by governments that describe how to conduct these assessments for industrial chemicals [1–3]. While these guidelines were developed with the aid of risk assessment approaches developed by the consumer products industry [4–11], they are not specific to these types of chemicals. These guidelines cover a wide variety of environmental stressors and are intended to ensure environmental protection by use of conservative approaches. When the chemicals being assessed are released via known pathways and act via known exposure routes, refinements of these guidelines to address these specific pathways and routes are appropriate, especially when they can be supported with case studies. Hence, the information in this chapter represents a refinement of the available guidelines and provides for the protection of the environment from adverse effects of detergent ingredients in consumer products.

A specific issue that arises in environmental, as opposed to human health, risk assessment relates to the question of what is to be protected. It is neither reasonable nor practical to aim to protect every living organism from the possible adverse effects of human activities. Indeed, some activities (e.g., the use of antibiotics, disinfectants, pesticides) are specifically designed to cause harm to certain organisms. Therefore the first step of environmental risk assessment is concerned with identifying the populations, communities, ecological processes, or ecological structure that need to be protected. Other issues that need to be addressed include the extrapolation of adverse effects to the diversity of organisms under the range of conditions encountered in the real world using information largely gleaned from controlled, and inevitably limited, laboratory tests. The approach taken to resolve these issues is discussed and described in this chapter.

II. ENVIRONMENTAL RISK ASSESSMENT GOAL

Environmental risk assessment is generally considered to be the process of defining and quantifying the risk (potential) for effects to occur given specific environmental concentrations in the compartments of interest [12]. Risk management is the process of deciding what actions to take if the assessment indicates a potential risk to the environment [12]. By its nature, detergent ingredient usage is both widespread and diffuse, because these type of consumer products are typically used in the home and then disposed of in wastewater. After treatment and release to the environment, these ingredients can reach the freshwater, marine, and terrestrial compartments. Due to this diffuse distribution in the environment, the goal of risk assessment and risk management is to ensure that adverse effects to the environment do not occur. This is generally accomplished by limiting the volume of ingredients to quantities that do not pose unacceptable risks to the environment.

III. PROBLEM FORMULATION, OR HAZARD IDENTIFICATION

The first component of the environmental risk assessment is sometimes called the problem formulation, or hazard identification, stage. The purpose of this stage is to formulate the objective of the assessment and to identify the critical components needed to carry out the risk assessment. Since this stage establishes the basis for the whole assessment process, including identifying the experimental data that must be collected, it is extremely important that it be done well and thoroughly.

This stage follows a five-step process, which is described in Box 1. The aims are to clearly identify (1) the objective of the assessment, (2) the compartments that are at risk and may need to be protected from the hazard under study, (3) what is already known, and (4) any additional information that will need to be generated. Of particular importance is the identification and documentation of the assessment endpoint. Unlike human health risk assessments, where it is clear that the aim is the protection of human health, environmental risk assessment requires a choice to be made about which species, populations, or communities should be protected. This will not encompass every organism in the environment: Some products are designed specifically to cause harm to some organisms (for example, pesticides, disinfectants), and in general it is not practical to try to protect every individual organism in the environment. However, there may be particular species or particular communities (e.g., those that are rare and endangered) that may be affected and need protection.

STEPS IN POTENTIAL PROBLEM OR HAZARD IDENTIFICATION

Step 1. *Formulate the objective* of the assessment (e.g., to compare the risks from existing technology with those of a promising new technology; to assess the risks arising at the site of production, formulation, or use of a material or chemical; to identify the major risks associated with a product as an input to the identification of potential public relations issues).

Step 2. *Assemble what is known* about the substance for which the assessment is being conducted. Key properties of the substance that should be compiled include its type, structure, anticipated uses, and physical, chemical, toxicological, and degradation properties. An initial estimate should also be made of the potential emissions of the substance to the environment. Where the necessary physical/chemical and biodegradation data are not available for the substance of interest, data on a close structural analogue may be used, since at this stage a qualitative understanding of the substance's fate and partitioning in the environment is all that is required. Additional critical pieces of information relate to the environmental compartments into which the substance is released, the relative proportions released to each compartment, the total annual emissions of the substance from all sources into each compartment, and the matrix in which this release will occur. Any concerns about the accuracy of the physical/chemical, degradation and emissions data can be addressed when more definitive data are available in the subsequent fate assessment.

Step 3. *Conduct a pathway analysis* to focus efforts on the most critical environmental compartment(s). The objective of the pathway analysis is to determine the fate and flow of the substance after it is released to the environment during production, use, and disposal and to identify which environmental compartments will be exposed to the substance.

Step 4. Based on the results of the pathways analysis, identify those characteristics of the environmental compartments or ecosystems that could be affected by exposure to the substance, and *choose the assessment endpoint* or set of assessment endpoints. The assessment endpoint, an explicit statement of the environmental characteristics to be protected, is chosen on the basis of its ecological relevance and societal value. The assessment endpoint may range from specific important species, e.g., an endangered species, to the general characteristics and functions of an ecosystem, e.g., energy flow. Its choice is a critical element of the assessment process. It is therefore important to communicate clearly the chosen assessment endpoint and the reasons for the choice to any internal and external decision-makers.

Step 5. Based on the results of the first four steps, identify the critical components that are either missing or that need to be further refined in the effects and exposure assessments, and describe how this information will be used to accomplish the objective of the assessment identified in Step 1.

Box 1

The identification of such target species, populations, and/or communities needs to reflect their ecological importance and societal value.

For consumer product chemicals, the problem formulation, or hazard identification, process is simplified by the similarity among ingredients, the fact that the release-and-exposure scenario does not change for each ingredient, and a general consensus on what the endpoint is that needs to be protected. Consumer products and the detergent ingredients that they contain are typically used in the home, disposed of into municipal sewers, treated in municipal wastewater treatment plants, and released to the environment, where they are diluted. In the environment, sorptive and degradative fate processes continue to act to reduce the detergent ingredient's concentration. Based on their partitioning and degradation characteristic, consumer product compounds are most commonly found in the pelagic or sediment compartments of freshwater and marine systems or in soil (via sludge disposal on land). Differences in physical and chemical properties will lead to an emphasis on one or another of these compartments for additional information on exposure concentration and effects. The environmental endpoint that is chosen for protection from risk is the structure and function of the ecological systems in which the detergent ingredient is found. This protection of the structure and function does not imply the protection of every species or organism but rather the ability of the system to support normal energy flow and production of economically important species and population, e.g., sport or commercial fisheries. Because of the similarities among detergent ingredients, the problem formulation and risk identification phase of the risk assessment process for new ingredients is not necessarily repeated unless the new compounds appear to present unique issues.

IV. ENVIRONMENTAL EXPOSURE ASSESSMENT

The aim of environmental exposure assessment is twofold: (1) to identify in which environmental compartments exposures are expected to occur, and (2) to estimate the concentration of the chemical being assessed in the exposed environmental compartments (resulting in predicted environmental concentrations, PECs). For consumer product compounds, the environmental compartments that are usually considered are surface water, aquatic sediment, and agricultural soil (terrestrial compartment).

A. Exposure Pathway and Compartments of Concern

In the environmental exposure assessment, it is essential to define the target compartments for the exposure assessment. In particular, it is necessary to define what is being exposed and also how long the exposure will last. For detergent chemicals, the "down-the-drain" pathway is the single most important exposure scenario. After use in the household, detergent ingredients are released into the domestic wastewater stream. Before discharge into surface water, a wastewater treatment step usually occurs. During treatment, a fraction of the substance is degraded, another fraction is adsorbed onto the wastewater treatment biosolids (sludge), another is volatilized, and the remaining fraction is discharged with the treated effluent. The fraction of the chemical discharged into the surface water may further (bio)degrade, or it may partition into the aquatic sediment. Finally, the fraction adsorbed onto biosolids may be applied to agricultural soil when the sludge is used as a fertilizer. Because the household use of detergent ingredients is fairly constant over time, the exposure to the environment can be considered a continuous process (i.e., the emission can be assumed constant over time).

For the surface water compartment, the relationships between chemical-related information (laboratory fate testing, physicochemical data, mathematical modeling) and the actual fate and distribution in the field are well understood. There is, however, a more limited understanding of the partitioning, bioavailability, fate, and transport in the sediment compartment and of the bioavailability and leaching in soil, which in turn limits the accuracy of exposure assessments in these compartments. This typically leads to the use of more conservative assumptions in the prediction of environmental concentrations in the sediment and soil compartments than for water.

B. Spatial Scale—Regional and Local

Environmental exposure can be considered at different spatial scales. Close to emission sites, the local chemical concentrations will be the highest. Further away, concentrations decrease, due to dilution, partitioning into other environmental compartments, and chemical degradation processes.

Using advanced geographical information system (GIS) approaches, in combination with mathematical fate and dilution models, these spatial exposure patterns can be simulated and analyzed (e.g., Refs. 13 and 14).

Because these tools are available for only a few locations, so-called "generic" exposure assessment calculation techniques that differentiate between "local" and "regional" exposures have been used in most assessments. The local exposure calculation estimates the exposure next to emission sources (e.g., wastewater effluent) and takes into account both the chemical influx due to the local emission and any background concentrations already present in the environment. The regional exposure calculation, on the other hand, focuses on an "average" situation that occurs remotely from individual emission points. The regional exposure concentrations are often considered to represent background levels. The regional calculation usually requires considering the fate, transport, and distribution of the chemical among the different environmental media (air, water, sediment, soil, and biota).

Due to the widespread use (and thus release to the environment) of detergent chemicals, the regional exposure concentration cannot be ignored. For specific point-source emissions (e.g., related to a specific chemical production facility, occurring only in a limited number of plants), the local emission will typically by far exceed any regional exposure estimates, making a local assessment more relevant. For "down-the-drain" substances, on the other hand, the emissions are distributed more or less uniformly across a region, which leads to a more important contribution of the regional exposure component in the overall local/regional equation. However, when the detergent ingredient is readily biodegradable, the regional exposure will be negligible compared to the local exposure. On the other hand, if the ingredient is poorly biodegradable, the environmental background levels may be in the same order of magnitude as the local emissions. Hence, for poorly degradable substances a regional exposure assessment (as well as the local exposure) is used in the risk assessment.

C. Exposure Assessment Calculations

1. Tiered Approach

The fate of a substance in the environment depends on several factors: the properties of the environmental compartment into which it is released, the matrix within which it is released, its physical/chemical properties, and its potential for and rates of degradation (biodegradation, photolysis, hydrolysis, etc.).

The way these concepts are applied in exposure assessment can vary from large simplifications to very high levels of detail (Table 1, Fig. 1). The tools used for

TABLE 1 Tiered Exposure Assessment for Local PEC Estimation

Phase	Data needed to calculate PEC	Experimental approach
Tier I	Removal = 0	No removal/degradation information is required
	Release or emission	Calculated
	Dilution factor	Between 1 and 10
Tier II	Removal	Literature search for data from structurally similar substances; QSAR estimates from chemical/physical data; or assumed to be zero. Estimated from chemical/physical data combined with screening-level biodegradation and environmental partitioning test data
	- Biodegradation	O_2 uptake or CO_2 generation to assess mineralization in the matrix of interest. Die-away testing to estimate removal
	- Environmental partitioning	Utilize material characterization parameters to predict partitioning behavior
	Emission or release	Mostly calculated
	Dilution factor	Between 1 and 10, or indicative for geography/site.
Tier III	Removal	Estimated from chemical/physical data or preferably from experimental laboratory data
	- Biodegradation	Continuous activated-sludge system used to assess mineralization, primary biodegradation, identify biodegradation intermediates, and model actual treatment. Soil and sediment systems are used for biodegradation assessment in the terrestrial and benthic environments. Measured rates from respirometric tests or from primary degradation measured in die-away test
	- Chemical/physical degradation	Either standardized or specialized tests to assess if a material undergoes oxidation, hydrolysis, or photolysis
	- Environmental partitioning	Estimated from chemical/physical data or preferably from realistic experimental laboratory data
	Emission or release	Calculated or measured
	Dilution factor	Between 1 and 10, or indicative for geography/site
Tier IV	Monitoring data	Field studies or laboratory ecosystem tests to verify previous predictions

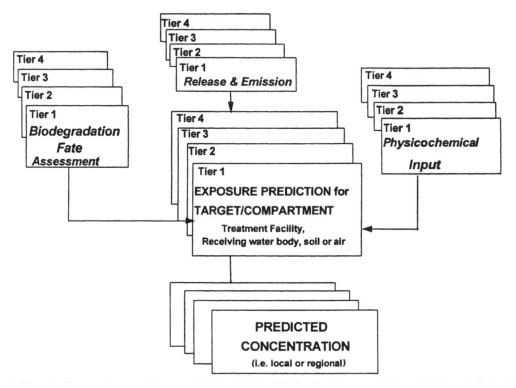

FIG. 1 Tiered environmental exposure assessment.

exposure analysis range from simple hand calculations at lower tiers, to "rules of thumb" and multimedia or generic environmental models [9,15,16] to sophisticated GIS-linked site-specific models [13,14] and field monitoring at the highest tiers. All these methods have the following steps in common: (1) estimation of the "down-the-drain" emission, (2) removal and partitioning calculations in wastewater treatment, and (3) calculation of fate in the receiving compartment (water, sediment, soil). This process is illustrated in Figure 1.

Early tiers rely on fate and distribution models for the environmental compartment of interest, use of predicted or measured physicochemical descriptors, and screening biodegradation and sorption data to estimate the PEC. These assessments can be limited to a local exposure calculation, assuming conservative emission, removal, and dilution factors. The required chemical-specific information is very limited (i.e., tonnage, hydrophobicity, biodegradability).

In later tiers more realistic test systems and/or monitoring data are used to derive more reliable PECs. The degree of realism, or representativeness of the test system and test concentration, increases with each tier, and the uncertainty associated with extrapolating from experimental laboratory results to generic or specific environmental compartments and conditions correspondingly decreases. At these higher tiers, gradually more and more accurate, rather than estimated, chemical information is needed, and more complex models and/or more realistic laboratory tests are applied. For example, specific kinetic studies can be conducted to establish exact biodegradation rates in the environmental compartment of interest. In combination with sorption data, these rates can be used to run fate models, such as wastewater treatment plant simulation models, to obtain a more accurate estimate of the exposure concentrations. If more information is needed, for example, to estimate the removal of the chemical in a wastewater treatment plant, laboratory simulation tests with benchtop treatment plants (e.g., OECD 303 Continuous Activated Sludge test) can be used. This is illustrated in Figure 2.

Finally, at the highest tier, field monitoring can be conducted. For example, chemical removal in treatment plants can be measured, and these numbers can be used to override removal values obtained using models or laboratory tests. Alternatively, the entire exposure calculation can be replaced by monitoring data, when measured effluent concentrations or river water concentrations are used instead of PECs.

It should be noted that the lower-tier assessments are more generic, whereas the higher-tier assessments are typically directed toward understanding the exposure at more specific locations or in more specific conditions. To develop a monitoring-based exposure assessment that is valid for a large area, an extensive monitoring

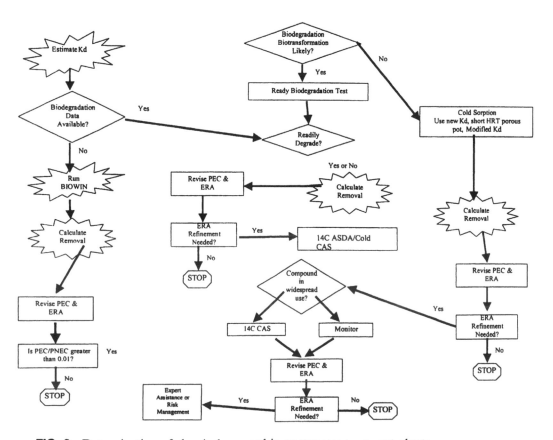

FIG. 2 Determination of chemical removal in wastewater treatment plants.

program is required, covering a wide range of locations, to ensure statistically valid results.

2. Estimation of "Down-the-Drain" Emissions

Based on the total down-the-drain tonnage of a substance used in a region, the average per capita emission rate can be calculated, assuming a uniform use of the substance across the population. This substance "mass flux" can then be converted to a concentration in domestic sewage using the daily per capita water consumption. This is illustrated in Box 2.

For risk assessments covering a large area (e.g., the entire European Union), it is appropriate to take into account the regional variations in chemical use. Typically, a worst-case assumption (i.e., highest product consumption) is used. Specifically for European use of detergents, it was found that to cover the regions with the most intensive consumption, the average per capita product use in the EU should be multiplied by a factor of 1.85 [17]. In addition, local variability (within a single region) may also be considered. In Ref. 17 it is argued that the worst-case local emissions will not exceed twice the average local emission.

3. Estimation of Removal and Partitioning in Wastewater Treatment

Environmental exposure assessments typically focus on activated-sludge wastewater treatment plants, because normally this degree of treatment is installed to meet

---

EXAMPLE ENVIRONMENTAL EXPOSURE ASSESSMENT APPROACHES FOR CHEMICAL SUBSTANCE DISPOSED OF "DOWN THE DRAIN"

The general equation for calculating the exposure concentration in the aquatic environment for substances disposed down the drain and released to the environment via wastewater treatment plants is as follows:

$$PEC\ local = I(1 - R)/D$$

where I is the concentration of the chemical in the influent to the wastewater treatment plant (WWTP), R is the fraction removed during wastewater treatment, and D is the dilution factor of the WWTP euent in surface water.

The concentration of a consumer product chemical in wastewater (I) is calculated by assuming that the amount of the chemical used daily per capita is dissolved in the daily per capita wastewater flow. The equation is

$$I = Q/(365PW)$$

where Q is the annual mass of the chemical used in the region of interest, P is the population of that region, W is the daily per capita wastewater flow, and 365 is the number of days in a year, to convert to daily concentration. Alternatively, Q can be calculated from the amount of a consumer product used annually (A) and the fraction of this chemical (F) in the product of interest (i.e., $Q = AF$).

Box 2

surface water quality regulations. The removal of a substance in activated-sludge wastewater treatment plants can be obtained in several ways, at different tiers of refinement.

1. *Model calculation.* At the first tier, the SimpleTreat model [18] or the AS-TREAT model [19,20] can be applied. These models predict the fractions of a substance going to air, water, and sewage sludge as well as the fraction degraded. Furthermore, a concentration of the substance on sewage sludge is also calculated. These calculations are based mainly on the volatility, sorptivity, and biodegradability of the substance. Either defaults or measured values can be used for the parameters describing these properties.

2. *Laboratory simulation studies.* Removal data observed in CAS (OECD 303) or similar (e.g., porous pot) simulation units can be used to obtain a more accurate estimate of the removal. Typically, these studies allow only the fraction removed to be determined. Thus, the differentiation between sorbed, degraded, and volatilized chemical cannot be determined using these laboratory tests. If it is important to know the relative distribution of these various fate processes, the relative fractions obtained with the mathematical model can be used to fit the measured removal to the relative contributions of the various fate processes (i.e., % removed = % degraded + % to sludge + % volatilized).

3. *Field monitoring.* Field-monitoring studies can be used to provide even more accurate removal estimates, provided the monitoring data are extensive enough to be considered sufficiently representative of the general situation in the geography covered by the assessment.

Finally, it should be noted that normally the removal efficiencies are based on parent material disappearance, unless the degradation results in metabolites that are known to be persistent and/or more toxic than the parent compound.

4. Calculation of Fate and Distribution in Receiving Compartments

To calculate regional exposure, multimedia or "unit-world" models are most often used. These models contain a simple representation of the world and estimate the effect of partitioning, transformation, and intermedia transfer processes on the fate of the chemical within this environment (Fig. 3). Within the environment, fate is controlled by two factors, the inherent physical, chemical, and transformation properties of the substance and the nature of the environment into which the substance is released [21]. The environmental properties can be measured or set to values that represent average air, water, soil, and sediment environments, as described by Mackay [22] and Mackay et al. [23] or as developed in the EU Technical Guidance Documents on Risk Assessment [3]. The quantitative predictions of multimedia models result in regional estimates of background concentrations in all environmental compartments.

Local aquatic exposure can be calculated using a simple dilution calculation, taking into account the effluent concentration, the dilution factor (ratio of river flow to effluent flow), and the upstream background concentration in the river. For the terrestrial compartment (agricultural soil), a similar "dilution" model can be used to calculate PECs immediately after sludge application. However, for a more relevant exposure assessment in the terrestrial environment, often the chemical concentration

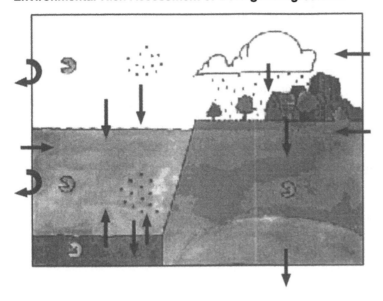

FIG. 3 Diagrammatic representation of fate and distribution analysis. (Modified from www.scienceinthebox.com.)

is calculated after a certain period (3–6 months), thus allowing chemical biodegradation and biotransformation processes to occur.

Local PECs can be representative of generic situations (e.g., average or realistic worst-case dilution), but they can also be site specific through the addition of site-specific information. When multiple site-specific assessments are performed, exposure assessments can be viewed in a probabilistic context. For example, in the United States, population, per capita water usage, the type of treatment process, and dilution factors are known for a large number (>15,000) of wastewater treatment plants. Use of treatment-type-specific removal values, together with this information, allows the estimation of river concentration immediately below the mixing zone of each wastewater treatment plant [24]. These numbers can be further statistically processed according to the specific exposure assessment's needs.

V. ENVIRONMENTAL EFFECTS ASSESSMENT

The primary objective of the environmental effects assessment is to determine the concentration of the substance of interest in receiving environmental compartments that will result in minimal or no effects on the chosen assessment endpoint. This concentration, which as stated previously is aimed to protect the structure and function of the ecosystem, is denoted the predicted no-effect concentration (PNEC). Determining the PNEC involves predicting what is likely to happen to the populations and communities in the ecosystem(s) of interest in the real world. Data typically come from models (i.e., QSARs) and the results of tests carried out in the laboratory. The rationale and brief descriptions of the methodologies frequently used are presented later.

The three environmental compartments traditionally evaluated in effects assessments of consumer product compounds are the aquatic, sediment, and terrestrial compartments. The relationships between laboratory testing and effects in the field are best understood for the aquatic compartment. Currently, equilibrium partitioning is used to extrapolate toxicity values from the pelagic to the sediment compartment. This allows the extensive database from the pelagic compartment to be used to protect sediment biota. In the terrestrial compartment, effects assessments are conducted using best available test methods involving terrestrial plants and animals.

As with other stages of risk assessment, effects assessment can be divided into multiple tiers. Early tiers rely on modeled and acute toxicity data to estimate the PNEC. Later tiers (3 and 4) use chronic and ecosystem-level data to derive PNECs. The degree of realism, or representativeness of the test conditions to the real world, increases with each tier, and the uncertainty associated with extrapolating from experimental results to the PNEC correspondingly decreases. This extrapolation is represented by the application or assessment factors.

Application factors (AFs) account for uncertainties such as differences between the species to be protected and the species on which the tests have been carried out, differences between observed immediate effects of short-term exposure (acute toxicity) and the longer-term effects of prolonged exposure (chronic toxicity) likely to be of more concern in the field, and extrapolations from the levels of exposure used in laboratory toxicity tests to the levels likely to be found in the field [25]. AFs are greatest when PNECs are derived from models and acute toxicity data and least when they are derived from chronic toxicity and model ecosystem data (see Fig. 4). They are typically expressed as factors of 10, and, while there are differences in the precise factors used, there is remarkable consistency in the factors selected across geographical boundaries [2,3,26,25]. The AFs recommended here are derived from years of risk assessment experience with consumer product compounds and reflect the fact that consumer product chemicals are typically not highly chlorinated or bioactive and do not contain heavy metals. They represent a compromise of AFs used in the United States, Europe, and Canada. In a tiered testing approach, risk is evaluated after each tier and a decision made whether the risk is acceptable or unacceptable based on the current data. If the risk is unacceptable, there are two choices: (1) conduct additional testing to better understand the potential risk and (2) consider managing the risk by using a lower volume of the chemical in question.

Effects assessments should be conducted for each of the environmental compartments in which the substance will reside, as identified in the pathway analysis. In the following sections, a more detailed description is provided of the effects assessment for the aquatic compartment, to illustrate the approach. Similar principles apply to conducting effects assessments for the sediment and soil compartments.

A. Tier I: QSAR or Analogue Estimates of Effects

The QSAR tier is relevant for situations in which there is little or no effects information regarding the substance of interest or closely related compounds. This is the typical case for a new compound or existing compound with new-found use that lacks specific data. Effects assessments at this tier make use primarily of known effects data for analogous compounds or estimations based on quantitative structure–activity relationship (QSAR) models (see Box 3). The majority of QSARs predict

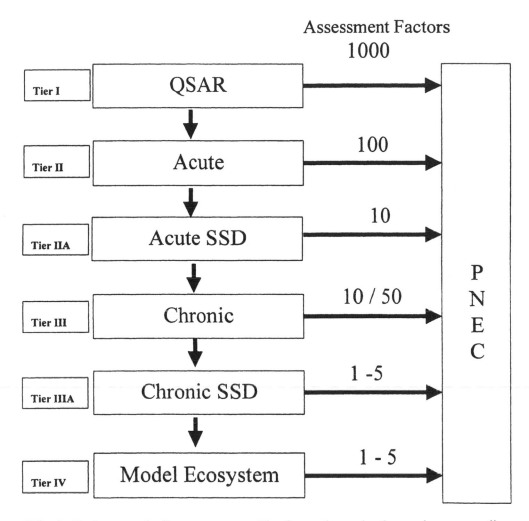

FIG. 4 Environmental effects assessment. The figure shows the tiers and corresponding assessment factors in environmental effects assessments. SSD refers to the single-species distribution approach to calculating a PNEC and can be applied at either the acute and/or chronic tiers when sufficient data are available.

acute toxicity. Most QSARs are based on the octanol–water partition coefficient (log K_{ow}), and their domain and application remain limited. Other QSARs may use physical/chemical parameters, such as molecular weight, chain length, and critical micelle concentration, but these approaches are still under development.

The first step in the responsible use of QSARs is to understand the chemical domain that the set of compounds used to establish the QSAR represents and to ensure that the chemical of interest falls within this domain. Where significant overlap exists, the QSAR being used can be expected to provide a reliable estimate of the toxicity of the test compound. However, if the chemical of interest does not fall into the domain of the compounds used to derive the QSAR, other methods should be considered to predict toxicity [27].

QUANTITATIVE STRUCTURE–ACTIVITY RELATIONSHIPS (QSARs)

The objective of QSAR (quantitative structure–activity relationship) modeling is to obtain an estimate of key physical/chemical/toxicological parameters for use in risk assessments prior to the initiation of experimental work.

Several sources of QSAR models are used routinely, such as Numerica and SRC (Syracuse Research Corp., Syracuse, NY) for physical/chemical information. Both QSAR models require CAS (Chemical Abstract Service) number or a SMILES notation to make a prediction. Information that can be obtained from the QSAR models includes the following: octanol–water partition coefficient, boiling point, vapor pressure, molecular weight, water solubility, melting point, heat of vaporization, acid dissociation constant, parachor, molar volume, molar refraction, Henry's law constant, aquatic toxicity estimates, biodegradation half-life, bioconcentration factor, sorption coefficient (K_d(s)) to sediment, sludge, and soil, hydrolysis half-life, phytoxicity estimate, and mutagenic assessement.

Box 3

Due to the level of uncertainty in estimating a PNEC from a QSAR, assessment factors typically used in this tier are high. Predicted no-effect concentrations (PNECs) derived using QSAR or analogue data are generally based on an assessment factor of 1000 for the protection of structure and function of aquatic ecosystems. However, depending on the relevance and adequacy of the QSARs, lower assessment factors may be used. This tier has the advantage of eliminating animal usage but may be skipped if there are no data on material analogues or if a relevant structure–activity relationship is not available. In this case, the ecotoxicologist will have to measure effects in order to estimate a PNEC.

B. Tier II: Acute Toxicity

The purpose of tiers II and III is to determine the toxicity of the substance empirically in acute and chronic toxicity tests. Acute toxicity tests typically involve the survival of organisms over a short duration (hours to days), while chronic toxicity tests measure growth, survival, and reproduction during a sensitive part of the organism's life cycle. A specific duration cannot be put on a chronic toxicity test due to the differences in the life history of different species; however, a chronic test must involve a significant portion of the organism's life span. Algae pose an issue in testing nomenclature because the standard algal population growth test has a duration of 3–4 days. The duration of the test (i.e., longer than the doubling time of a cell) and the fact that cell replication causes the algal population size to increase by orders of magnitude make the algal test a chronic test. However, since there is no acute algal test that is well accepted, the EC_{50} from the test is used as an acute value in tier II, and the NOEC (no-observed-effect concentration) or EC_{20} value from the same test is used as the chronic value in tier III.

Tier II PNECs are based on measurements of the concentrations at which 50% of the population suffer some defined acute effect, like immobility (EC_{50}) or mortality (LC_{50}), in short-term toxicity tests. The species used in acute toxicity tests have typically been selected because of their sensitivity and ease of testing/culturing. Since the overall goal of effects testing is to extrapolate laboratory effects to field effects, a variety of species representing different classes of organism and levels of complexity is used (see Box 4). Typically, an algae, an invertebrate, and a fish are tested and the lowest LC_{50} or EC_{50} data used to derive the PNEC. In this tier, a factor of 100 is applied to the lowest of these toxicity values. Table 2 presents species currently used in routine tests. Table 3 lists guidelines for the most common aquatic acute tests.

Because multiple acute toxicity values are generated, the use of a factor of 100 to derive a PNEC is no longer appropriate. While the use of lower and lower assessment factors as more and more data are collected would seem appropriate, this would be difficult to support with data confirming the validity of each extrapolation factor. Further, questions on how to value data on less sensitive taxa would be difficult to address. Hence, we use a probabilistic approach to incorporating data from five or more acute toxicity tests. This single-species sensitivity distribution approach has been most widely used to understand chronic toxicity [28,29] and is further discussed in Section V.E. Since acute and chronic data are separated by an assessment factor of 10 and the chronic $SSD_{0.05}$ (concentration predicted to exceed the EC_{10-20} or NOEC for 5% of species with 50% confidence) has an assessment factor of 1, an assessment factor of 10 is applied to the acute $SSD_{0.05}$ (Fig. 4).

TYPES AND LEVELS OF SPECIES USED FOR ACUTE (TIER II) TOXICITY TESTING

Algal tests
Algal tests provide surrogate information regarding effects of chemicals on phytoplankton and periphyton in the field. Phytoplankton are algae and small plants that are suspended or floating in water, and they are found in lakes, large rivers, and oceans. Periphyton are attached algal communities that are found in shallow areas of lakes, rivers, and estuaries. Phytoplankton are typically tested rather than periphyton.

Invertebrate tests
Invertebrate tests provide surrogate information for free-swimming and sediment-dwelling macroinvertebrates. Daphnids are the most common invertebrates tested to date. They are free-swimming, filter-feeding organisms that are distributed worldwide and have been shown to adequately represent the sensitivities of the broad array of invertebrates that exist in freshwater. The mysid shrimp is the analogue to the daphnid in marine systems.

Fish tests
As with the invertebrates, fish tests provide information regarding fish effects in the field. Fathead minnows and zebra fish are the most commonly tested species.

Box 4

TABLE 2 Examples of Species Routinely Used for Acute Aquatic Toxicity Tests

Major Taxa	Common name	Species name
Fish	Fathead minnow	*Pimephales promelas*
	Rainbow trout	*Oncorhynchus mykiss*
	Bluegill	*Lepomis macrochirus*
	Zebra danio	*Danio rerio*
Invertebrate	Water flea	*Daphnia magna*
	Water flea	*Daphnia pulex*
	Water flea	*Ceriodaphnia dubia*
	Amphipod	*Gammarus* sp.
	Midge	*Chironomus* sp.
	Rotifer	*Brachionus* sp.
Algae	Green algae	*Selenastrum capricornutum*
	Green algae	*Scenedesmus subspicatus*
	Green algae	*Scenedesmus quadricauda*
	Blue-green algae	*Microcystis aeruginosa*
	Diatom	*Navicula pelliculosa*

C. Tier III: Chronic Tests

While tiers I and II typically assess substantial and acute effects (e.g., survival) over a limited part of the organisms life cycle, tier III evaluates more subtle effects over greater (longer) parts of the organisms life cycle. While survival is also considered, so are a variety of sublethal endpoints (e.g., growth and reproduction) because

TABLE 3 Selected Guidelines for Acute Toxicity Testing

Major Taxa	Method	Ref.
Algae	OECD 201 Growth Inhibition Test	46
	Algal Acute Toxicity Test	47
	E1218-97 Standard Guide for Conducting 96-h Toxicity Tests with Microalgae	48
Invertebrates	OECD 202 *Daphnia* sp. Acute Immobilization Test and Reproduction Test	46
	Daphnid Acute Toxicity Test	49
	E-729-96 Standard Guide for Conducting Acute Toxicity Tests on Test Materials with Fishes, Macroinvertebrates, and Amphibians	48
Fish	OECD 203 Fish Acute Toxicity Test	46
	Fish Acute Toxicity Test	50
	E-729-96 Standard Guide for Conducting Acute Toxicity Tests on Test Materials with Fishes, Macroinvertebrates, and Amphibians	48

these endpoints are often more sensitive than mortality. When data from this tier are used to derive a PNEC, low levels of effects such as the EC_{20} for growth or reproduction are evaluated. The same species identified in tier II are typically used in chronic toxicity tests (see Box 5). Typical standard test methods are listed in Table 4.

Historically, the NOEC, or no-observable-effect concentration, has been used to summarize chronic toxicity test results. However, the NOEC suffers from a number of defects, as described in Refs. 30–33, including:

1. The NOEC must be one of the test concentrations. If these concentrations are widely spaced, this may cause the NOEC to be substantially underestimated (i.e., lower than expected) or the LOEC to be substantially overestimated.
2. The NOEC will depend upon the sensitivity of the test system, which, in turn, depends in part on the number of replicates used and on replicate-to-replicate

---

TYPES AND LEVELS OF SPECIES USED FOR CHRONIC (TIER III) TOXICITY TESTING

Algal tests
For algae, the NOEC may be determined from the same toxicity test from which the EC_{50} was determined because this is a multigeneration test. EC_{20} values are based on the extent of population growth inhibition as well as the rate of growth during the test period.

Invertebrate tests
For invertebrates, survival, growth, and reproduction are typically monitored during chronic toxicity tests. Daphnids (e.g., *Daphnia magna*, *Ceriodaphnia dubia*) are the invertebrate species most commonly used in chronic tests. Daphnids reproduce parthenogenetically–females produce neonates without need of fertilization. Growth of daphnids can be measured as well as the number of neonates produced during a test. Typical tests are conducted such that three batches, or generations, of neonates are produced. For example, *Daphnia magna* and *Ceriodaphnia dubia* require 21 and 7 days, respectively, to achieve three generations. When reproduction is the most sensitive endpoint, EC20 values are based on the growth and extent of neonate production during the test period.

Fish tests
As with the invertebrates, survival, growth, and reproduction are the key attributes considered in evaluating chronic toxicity to fish. Full life cycle tests are not commonly conducted, due to extensive labor needs and cost. Studies conducted from the late 1970s to the mid-1980s showed that full life cycle results were based primarily on effects noted during the early life stages, the embryo and larval stages. Hence, most "chronic" tests with fish conducted today are early-life-stage tests, where the embryo and/or larval stages are exposed to the test material. For fathead minnows these tests run from 28 to 32 days, whereas rainbow trout studies are conducted for at least 90 days, due to lower temperatures and the slower development of this species. Effects on hatching and larval development are also measured.

Box 5

TABLE 4   Selected Guidelines for Chronic Toxicity Testing

Taxa	Method	Ref.
Invertebrates	E-1193-97 Standard Guide for Conducting *Daphnia magna* Life-Cycle Toxicity Tests	48
	E 1295-89 Standard Guide for Conducting Three-Brood Renewal Toxicity Tests with *Ceriodaphnia dubia*	48
	Daphnid Chronic Toxicity Test	51
Fish	E 1241-98 Standard Guide for Conducting Early Life-Stage Toxicity Tests with Fishes	48
	Fish Early Life Stage Toxicity Test	52
	OECD 210 Fish Early Life Stage Toxicity Test	46

variability. This has the undesired consequence of rewarding poor experimentation with higher NOECs.

3. Information on the concentration–response curve and the variability in the data is lost. A compound with a steep concentration–response curve poses a different threat to the test species than one with a shallow concentration–response curve.
4. The NOEC is a misnomer, because effects can be present but not statistically significant.

To eliminate these problems in summarizing the results of toxicity studies, suitable dose–response models have been developed to interpolate a *benchmark dose*, a dose producing a given level of change, from controls. Setting this level of change low enough would ensure that the population as a whole and their function in the community are not adversely affected. The key advantages of the concentration–response approach include:

- The estimated benchmark concentration has known biological effects.
- Decisions on the benchmark concentration and the acceptable environmental concentration can be based on biological criteria.
- Estimation of the benchmark concentration is not biased by the selection of alpha risk level, test concentrations, or number of replicates.
- Variability in the data does not bias the benchmark concentration.
- Quantitative information on the variability or confidence in the endpoint and information on the steepness of the concentration response curve is available.

Based on this rationale, the NOEC has been replaced by the EC_x in this risk assessment approach. While not every risk assessor uses the EC_x, likely due to the difficulties in identifying an appropriate value of x, we greatly prefer this parameter.

A key issue regarding use of the concentration–response approach in aquatic toxicology is selecting an appropriate EC_x level (i.e., what x should be). Concentration–response models allow estimation of an infinite number of effective concentrations. But for hazard assessment, biological judgment is needed to select the value corresponding to an acceptable level of biological effect. No consensus currently exists within the scientific community, and the question has not received much attention. Ideally, one would consider information about the species and its life history, the endpoint measured (reproduction, survival, growth, r intrinsic rate of population increase), test design (e.g., duration), and the slope of the concentration–response curve in deriving x. Alternatively, ecosystem-level modeling of toxicant behavior and effects similar to that performed by Bartell et al. [34] could be used to establish an appropriate level for x for a given ecosystem. Until such a time, we have decided to use an EC_{10} value for organisms with long generation times (e.g., salmon) and an EC_{20} value for organisms with rapid generation times (e.g., daphnids, algae). Where a NOEC currently exists and the data are not readily available to estimate an EC_x, it is recommend that the NOEC continue to be used and that it be considered equivalent to the EC_x. As with the selection of x, this is a practical solution to this issue.

As a general rule, an application factor of 10 is applied on the lowest EC_x value derived from chronic tests to protect the structure and function of aquatic ecosystems and to derive a PNEC (Fig. 4). In situations where up to four chronic values (from different species) are known, the application factor is applied to the lowest chronic value. In cases where there are repeated tests with a particular species, the geometric mean of the EC_x values can be used to represent that species value. With chronic databases that have EC_x values on at least five different species, a statistical extrapolation method may be used to estimate the PNEC (Fig. 4). Statistical extrapolation methods have been used successfully to relate diverse chronic toxicity data to tier IV (field, mesocosm) effects, particularly for surfactants [28,29,35]. Novel tests are often conducted in this tier to provide increased understanding of the properties that influence bioavailability and toxicity. This information is valuable for more realistic PEC-to-PNEC comparisons and to help bridge tier IV tests with lower-tier data. This topic is further discussed in Section V.E.

D. Tier IV: Aquatic Ecosystem Studies

While single-species tests serve an efficient purpose in estimating a PNEC, they have inherent limitations for predicting ecosystem responses, because they do not assess multispecies interactions (e.g., predation, competition), fundamental ecosystem processes (e.g., energy flow, nutrient cycling), or the importance of assimilative capacity. Therefore, field data or simulated ecosystem (i.e., mesocosm) data may be needed to fully ensure low risk in the aquatic compartment. Tests to obtain these data should reflect the essential elements of ecosystem structure and function and be sufficiently long to cover relevant portions of key organisms' life cycles.

Since tier IV tests are usually expensive and time consuming, they are likely to be conducted primarily on substances with PEC/PNEC ratios near or above 1 (after tier III assessment), when there are perceived risk issues, or when related compounds have demonstrated community-level effects that are more sensitive than single-species-level effects.

There are two categories of tier IV tests: experimental ecosystems and field biological monitoring. Experimental systems simulate ecosystem realism, whereas field tests monitor for exposure and effects under existing conditions in the field. Experimental ecosystems are confined systems that can be manipulated to accommodate a variety of hypotheses and statistical designs. As with previous tier tests, one goal of these tests is to estimate the PNEC. The nearness to reality is dependent upon several factors, such as test duration, complexity, location, chemical/biological evaluations, and statistical rigor [36].

Field tests can be conducted to answer site-specific questions. These types of tests may be conducted above and below targeted discharge points. Field tests are difficult to conduct because of the difficulties of differentiating the effects of the substances of interest from those of other potential stressors in the system and because of other uncontrollable factors, such as the unpredictability of the weather and its effects on sampling, habitat, and biological structural changes. To overcome the latter problem, field tests are often long-term endeavors. Nevertheless, the prime advantage of field tests is that they are carried out on real systems, increasing the certainty of the results in application to other communities. The statistical rigor of field biomonitoring is often challenged but can be overcome through careful design.

This tier uses application factors between 1 and 5, depending on the purpose for which the data were collected and the uncertainty of the predicted no-effect concentration (PNEC). For site-specific applications, there is no need for an application factor if field test data are used (i.e., application factor = 1). However, if these data are to be used to extrapolate to other ecosystems, factors such as test system complexity and sensitivity need to be evaluated before deciding the appropriate application factor. Laboratory-scale ecosystems often do not span the broad array of species used in large-scale model ecosystem tests and are not discussed here for their applicability to risk assessment.

E. Use of Statistical Approaches to Estimate SSD$_{0.05}$

In the statistically based approach, multiple single-species toxicity data are viewed as a sensitivity distribution (Fig. 5). This distribution has a mean and a standard deviation allowing appropriate statistical procedures to be used to estimate a concen-

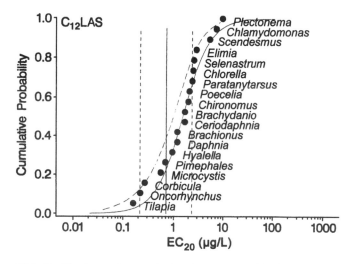

FIG. 5 Single-species sensitivity distribution. This is an example of a single-species sensitivity distribution using data for C$_{12}$LAS. The SSD$_{0.05}$, which is the point at which the effects assessment is conducted, is lower than 95% of single-species EC$_{20}$ values and, hence, is protective of 95% of species.

tration that is less than 90% or 95% of the data. Selection of an appropriate concentration would then confer protection on a large percentage of the species (i.e., 95% of the species). Currently, the convention is to protect 95% of species from acute or chronic effects. The point at which 95% of species are protected against acute or chronic effects is referred to as the $SSD_{0.05}$acute or $SSD_{0.05}$chronic. The advantages of this approach are:

- The approach values additional data (because more data are collected, confidence in the true $SSD_{0.05}$ increases, but the $SSD_{0.05}$ does not necessarily get lower).
- Differences between compounds are reflected in the shape and location of the distribution; these difference are reflected in the distribution's statistics.
- Changes in environmental concentrations can be directly associated with changes in the risk posed to the ecosystem.
- Confidence limits can be generated for the $SSD_{0.05}$, providing additional value in the risk assessment.

The limitations of the statistical approach include:

- The assumption that a specific level of protection of single species will confer a similar level of protection of the ecosystem
- The need to select a finite percentage of species to be protected in order to set the $SSD_{0.05}$ and, hence, a finite number of species that may be impacted
- Statistical questions around the selection of the most appropriate distribution
- Questions around the inclusion of species in this procedure

The importance of these limitations in using the statistical approach to protect the ecosystem was investigated by Versteeg et al. [28]. The authors compared the PNEC derived from multiple chronic single-species toxicity tests with model ecosystem and field-level-derived PNECs for 11 diverse substances. The derivation of the single-species PNEC followed the procedures specified in this guideline. Overall, there was excellent agreement between mean model ecosystem NOECs and single-species effect data. Mean model ecosystem data corresponded with effects on 10–30% of the species, as predicted by the single-species sensitivity distributions. This is likely due to differences in bioavailability between the two types of studies and should not be interpreted to mean that 30% of the species can be affected without impacting the ecosystem. The conclusion from the paper was that the single-species-based statistical approach to setting the PNEC can be used to confer protection on ecosystems in general.

VI. ENVIRONMENTAL RISK CHARACTERIZATION

Environmental risk assessment finally involves the development of a predicted no-effect concentration (PNEC) and its comparison with the predicted exposure concentration (PEC). This is usually done by calculating the risk quotient, the ratio of the PEC to the PNEC. If the risk quotient is less than 1, then the substance is likely to have little potential to have an adverse effect on the assessment endpoint; if the risk quotient is greater than 1, then action may be required, either to obtain better data or to manage the use and/or release of the substance. This risk characterization is an iterative process by which the risk assessment proceeds through progressively higher tiers of testing. It is best accomplished by involving the risk manager in the risk characterization of each iteration, providing the risk manager with the opportunity to have an input into the design and direction of subsequent analyses. In cases where environmental goals and criteria are not clearly defined, however, there is a danger that an iterative risk assessment process may continue indefinitely and exhaust valuable time and resources. For this reason, decision points and criteria should be clarified to the fullest extent possible, and the resources made available for the assessment should be commensurate with the benefits of the technology (and substance) and the magnitude (on a spatial and a temporal scale) of the potential risks involved.

VII. CASE STUDIES

As described in the introduction, environmental risk assessments of detergent ingredients in consumer products have been conducted for over 20 years. Through this time the approach has been further and further refined and more accurate test methods developed to estimate the characteristics of the type of chemicals that influence their fate and represent their potential effects. Although the general approach described herein has been followed for all of these assessments, because of the unique characteristics of the chemicals assessed and the particular environmental concern that is being addressed, the approaches are typically tailored to meet the specific needs of the assessment. To better understand how the general principles are applied to specific detergent ingredients, the reader is referred to specific publications that describe these assessments, which have ranged from QSAR-based screening assessments on 2100 fragrance materials [37] to detailed assessments on individual ingredients. For example, one of the most extensive assessments conducted was one that focused on the assessment of four of the major detergent surfactants [i.e., linear alkyl benzene sulfonates (LAS), alcohol ethoxylates (AE), alcohol ethoxylated sulfonates, (AES), and soap] in the Netherlands. The publications include exposure and fate predictions [38], effects assessment [39], and overall risk assessment [11]. Other representative assessments include a cationic surfactant, DTDMAC [40], chelator [41], phosphonates [42], and LAS [43]. In addition, there are case studies that illustrate how the approach can be applied to the terrestrial environment [44] and the marine environment [45]. Furthermore, the soap and detergent industry in Europe (AISE) is involved in an extensive assessment of the major detergent ingredients used in Europe. These environmental exposure assessments can be found on their Web site (www.heraproject.com). Finally, the Procter & Gamble Company has developed a Web site where the approach used to conduct environmental risk assessments is described and the results of specific assessments are presented (www.scienceinthebox.com).

REFERENCES

1. US EPA. *Guidelines for Ecological Risk Assessment. EPA/630/R-z95/002F*; U.S. Environmental Protection Agency: Washington, DC, 1998.
2. Environment Canada. Environmental Assessments of Priority Substances Under the Canadian Environmental Protection Act, Guidance Manual Version 1.0, EPS 2/CC/3E,

Chemicals Evaluation Division, Commercial Chemical Evaluation Branch: Ottawa, Canada, 1997.

3. EC. Technical Guidance Documents in support of directive 96/67EEC on risk assessment of new notified substances and Regulation (EC) No. 1488/94 on risk assessment of existing substances (Parts I, II, III, and IV). EC catalogue numbers CR-48-96-001, 002, 003, 004-EN-C, Office of Official Publications of the European Community, 2 rue Mercier, L-2965, Luxembourg, 1996.

4. Duthie, J.R. In Aquatic Toxicology and Hazard Assessment: Seventh Symposium, ASTM STP 634; Mayer, F.L. Hamelink, J.L., Eds.; American Society for Testing Materials: Philadelphia, 1977, 17–35.

5. Kimerle, R.A.; Levinskas, G.J.; Metcalf, J.S.; Scharpf, L.G. In *Aquatic Toxicology and Hazard Assessment, Seventh Symposium, ASTM STP 634*; Mayer, F.L. Hamelink, J.L., Eds.; American Society for Testing Materials: Philadelphia, 1977; 36–43.

6. Cairns, J.; Dickson, K.L.; Maki, A.W. Hydrobiologia 1979, *64*, 157–166.

7. Beck, L.W.; Maki, A.W.; Artman, N.R.; Wilson, E.R. Regul. Toxicol. Pharmacol. 1981, *1*, 19–58.

8. Holman, W.F. In *Aquatic Toxicology and Hazard Assessment: Fourth Conference, ASTM STP 737*; Branson, D.R. Dickson, K.L., Eds.; American Society for Testing and Materials: Philadelphia, 1981; 159–182.

9. Cowan, C.E.; Versteeg, D.J.; Larson, R.J.; Kloepper-Sams, P.J. Regul. Toxicol. Pharmacol. 1995, *21*, 3–31.

10. Fendinger, N.J.; Versteeg, D.J.; Weeg, E.; Dyer, S.; Rapaport, R.A. Environmental Chemistry of Lakes and Reservoirs. Advances in Chemistry Series No. 237; Baker, L.A., Ed.; American Chemical Society: Washington, DC, 1994; 527–557.

11. Feijtel, T.C.J.; van de Plassche, E. Environmental risk characterization of four major surfactants used in the Netherlands. Report Number 679101 025, September 1995. Bilthoven, National Institute of Public Health and Environmental Protection; The Netherlands, 1995.

12. Suter, G.W. *Ecological Risk Assessment*; Lewis: Chelsea, MI, 1993, 538 pp.

13. Feijtel, T.C.J.; Boeije, G.; Matthies, M.; Young, A.; Morris, G.; Gandolfi, C.; Hansen, B.; Fox, K.; Holt, M.; Koch, V.; Schröder, R.; Cassani, G.; Schowanek, D.; Rosenblom, J.; Niessen, H. Chemosphere 1997, *34* (11), 2351–2374.

14. Cowan, C.E.; White, C.E.; Merves, M.L.; Capara, R.J.; Gullotti, M.J. *66th Annual Conference and Exposition*; Water Environment Federation: Alexandria, VA, 1993, 351–358.

15. Feijtel, T.C.J.; Veerkamp, W.; Koch, V.; Niessen, H. Toxicol Modeling 1995, *1* (1), 5–19.

16. Mackay, D.; Patterson, S.; Kicsi, G.; DiGuardo, A.; Cowan, C.E. Environ. Toxicol. Chem. 1996, *15* (9), 1618–1626.

17. AISE and CEFIC. HERA—Human and Environmental Risk Assessment on Ingredients of European Household Cleaning Products. Guidance Document Methodology. 2002. www.heraproject.com.

18. Struijs, J.; Stoltenkamp, J.; van de Meent, D. Water Res. 1991, *25* (7), 891–900.

19. McAvoy, D.C.; Shi, J.; Schecher, W.D.; Rittmann, B.E. ASTREAT: a model for calculating chemical loss within an activated-sludge treatment system. Version 1.0. Users Manual; Procter & Gamble: Cincinnati, OH, July 1997.

20. Lee, K.C.; Rittmann, B.E.; Shi, J.C.; McAvoy, D.C. Water Environ. Res. 1998, *70* (6), 1118–1131.

21. Cowan, C.E.; Mackay, D.; Feijtel, T.C.J.; Van de Meent, D.; di Guardo, A.; Davies, J.; Mackay, N. In *Multimedia Fate Models—A Vital Tool for Predicting the Fate of Chemicals*; SETAC Press: Pensacola, FL, 1995.

22. Mackay, D. *Multimedia environmental models. The fugacity approach*; Lewis: Chelsea, MI, 1991.

23. Mackay, D.; Paterson, S.; Shiu, W.Y. Chemosphere 1992, *24*, 695–717.

24. Versteeg, D.J.; White, C.E.; Cowan, C.E. In *Environmental Toxicology and Risk Assessment: Modeling and Risk Assessment, Sixth Volume, STP 1317*; Dwyer, F.J. Doane, T.R., Hinman, M.L., Eds.; American Society for Testing Materials: Philadelphia, 1997, 82–97.

25. Zeeman, M.; Gilford, J. In *Environmental Toxicology and Risk Assessment. ASTM 1179*; Landis, W.G. Hughes, J.S., Lewis, M.A., Eds.; American Society for Testing Materials: Philadelphia, 1993, 7–21.

26. OECD. Report of the OECD Workshop on the Extrapolation of Laboratory Aquatic Toxicity Data to the Real Environment. OECD Environment Monographs No. 59. OCDE/GD (92)169. Paris: Organization for Economic Cooperation and Development, 1992.

27. ICCA. (Q)SARS for Human Health and the Environment. Workshop on Regulatory Acceptance. Brussels: CEFIC Research and Science, 2002.

28. Versteeg, D.J.; Belanger, S.E.; Carr, G.J. Environ. Toxicol. Chem. 1999, *18* (6), 1329–1346.

29. Posthuma, L.; Suter, G.W. II; Traas, T.P. *Species Sensitivity Distributions in Ecotoxicology*; Lewis: Boca Raton, FL, 2002.

30. Crump, K.S. Fund. Appl. Toxicol. 1984, *4*, 854–871.

31. Stephan, C.E.; Rogers, J.W. In *Aquatic Toxicology and Hazard Assessment: Eighth Symposium. ST 891*; Bahner, R.C. Hansen, D.J., Eds.; American Society for Testing Materials: Philadelphia, 1985; 328–338.

32. Barnthouse, L.W.; Suter, G.W.; Rose, A.E.; Beauchamp, J.J. Environ. Toxicol. Chem. 1987, *6*, 811–824.

33. Bruce, R.D.; Versteeg, D.J. Environ. Toxicol. Chem. 1992, *11*, 1485–1494.

34. Bartell, S.M.; Gardner, R.H.; O'Neill, R.V. In *Aquatic Toxicology and Hazard Assessment: Tenth Symposium. ST 891*; Adams, W.J. Chapman, G.A., Landis, W.G., Eds.; American Society for Testing and Materials: Philadelphia, 1988, 261–274.

35. Aldenberg, T.; Slob, W. Ecotoxicol. Environ. Saf. 1993, *25*, 48–63.

36. Belanger, S.E. Ecotoxicol. Environ. Saf. 1997, *36*, 1–16.

37. Salvito, D.T.; Senna, R.J.; Federle, T.W. Environ. Toxicol. Chem. 2002, *21* (6), 1301–1308.

38. Feijtel, T.C.J.; Struijs, J.; Matthijs, E. Environ. Toxicol. Chem. 1999, *18*, 2645–2652.

39. van de Plassche, E.J.; de Bruijn, J.; Stephenson, R.R.; Marshall, S.J.; Feijtel, T.C.J.; Belanger, S.E. Environ. Toxicol. Chem. 1999, *18*, 2653–2663.

40. Versteeg, D.J.; Feijtel, T.C.J.; Cowan, C.E.; Ward, T.E.; Rapaport, R.A. Chemosphere 1992, *24* (5), 641–662.

41. Jaworska, J.S.; Schowanek, D.; Feijtel, T.C.J. Chemosphere 1999, *38*, 3597–3625.

42. Jaworska, J.S.; van Genderen-Takken, H.; Hanstveit, A.; van de Plassche, E.; Feijtel, T. Chemosphere 2002, *47*, 655–665.

43. Feijtel, T.C.J.; Webb, S.F; Matthijs, E. Food Chem. Toxicol. 2000, *38*, 43–50.

44. de Wolf, W.; Feijtel, T.C.J. Chemosphere 1998, *36*, 1319–1343.

45. Temara, A.; Carr, G.; Webb, S.; Versteeg, D.; Feijtel, T.C.J. Marine Pollution Bull. 2001, *42* (8), 635–642.

46. OECD. OECD Guidelines for the Testing of Chemicals. Volume 1. Paris: Organization for Economic Cooperation and Development, 1993.

47. US EPA. Short-Term Methods for Estimating the Chronic Toxicity of Effluents and Surface Waters to Freshwater Organisms. EPA-821-R-02-013; U.S. Environmental Protection Agency: Washington, DC, 2002.

48. ASTM. Biological Effects and Environmental Fate: Biotechnology: Pesticides 2000, Vol. 11.05. American Society for Testing and Materials: Philadelphia, 2000.

49. US EPA. Ecological Effects Test Guidelines: OPPTS 850.1010. Aquatic Invertebrate Acute Toxicity Test, Freshwater Daphnids. EPA712-C-96-114; U.S. Environmental Protection Agency: Washington, DC, 1996.

50. US EPA. Ecological Effects Test Guidelines: OPPTS 850.1075. Fish Acute Toxicity Test,

Freshwater and Marine. EPA712-C 96-118; U.S. Environmental Protection Agency: Washington, DC, 1996.

51. US EPA. Ecological Effects Test Guidelines: OPPTS 850.1300. Daphnid Chronic Toxicity Test. EPA712-C-96-120; U.S. Environmental Protection Agency: Washington, DC, 1996.

52. US EPA. Ecological Effects Test Guidelines: OPPTS 850.1400. Fish Early Life-Stage Toxicity Test. EPA712-C-96-121; U.S. Environmental Protection Agency: Washington, DC, 1996.

14

Surfactants in the Environment

Fate and Effects of Linear Alkylbenzene Sulfonates (LAS) and Alcohol-Based Surfactants

LUCIANO CAVALLI* Sasol Italy, Milan, Italy

I. INTRODUCTION

Surfactants are the basic active ingredients of any detergent used in both domestic and industrial applications. They are manufactured in large quantities [1], used ubiquitously by many people and industries, and disposed of after use into the environment. Worldwide the annual consumption of surfactants cannot be determined exactly because of the lack of knowledge of their content in commercial products and because market estimates are limited only to some specific application sectors. At any rate, the worldwide annual consumption of surfactants in 2000 can reasonably be estimated to have been about 10–11 million tons, excluding soap, whose estimation is around 9 million tons [2]. Only a few types of surfactants are currently used in large quantities on the market. Excluding soap, which is definitely the most widely used anionic surfactant, the market is dominated by linear alkylbenzene sulfonates (LAS) and alcohol derivatives, which comprise alcohol sulfates (AS), alcohol ether sulfates (AES), and alcohol ethoxylates (AE) (Table 1). A multitude of other different surfactants are present on the market, some relevant in volume, such as alkylphenolethoxylates (APE), lignin, and petroleum sulfonates, and others important as specialities, such as cationics and amphoterics. A breakdown by geography and applications shows that nearly 60% of surfactants, excluding soap, are consumed in the developed countries of Europe, North America, and Japan, and more than 50% are used in houscholds (Figs. 1 and 2).

All these substances can enter the environment and have the potential to impact microbial, plant, and animal life. This environmental impact depends on inherent properties of the substance relevant to its fate in the environment and to its effects as well (e.g., biodegradation, ecotoxicity) and on the conditions of the

---

* *Current affiliation*: UNICHIM, Milan, Italy.

TABLE 1 World Surfactant Consumption, Million Tons (Mt), Year 2000, Excluding Soap

Surfactant	World		W. Europe	
	Mt	%	Mt	%
LAS	3.2	30	0.39	18
Alcohol derivatives	2.5	23	0.79	36
APE	0.7	7	0.07	3
Other	4.3	40	0.95	43
Total	10.7	—	2.20	—

Source: Refs. 1 and 2.

receiving environmental compartments (e.g., STP conditions, dilution factors in rivers, soil type). Surfactants enter above all the aquatic environment by several routes, which include direct discharge into rivers and discharge into the water from sewage treatment plants (STPs). In the most environmentally developed areas of Europe and the United States, the waste streams are treated by STPs, which reduce significantly the load of surfactants to the receiving surface waters. Surfactants enter the terrestrial compartment by irrigation with wastewaters or by sewage sludge application to agricultural land. Surfactants, however, may also reach the environment, both surface waters and soil, when used as emulsifying, dispersing agents in the manufacture and distribution of fertilizers and pesticides in agriculture.

As shown, the highest-volume surfactants currently on the market, excluding soap, are LAS and alcohol-based derivatives, namely, AE, AS, and AES (Fig. 3). These surfactants, by the way, are among the few for which laboratory and environmental monitoring studies have been conducted using specific analytical methodology. Only the use of specific analytical techniques, on the other hand, allows insight into the fate and removal mechanisms of substances that enable valid exposure data in the environment to be obtained. These specific studies have been quite extensive for LAS. In contrast, analogue studies for AE, AS, and AES have not been so rigorous as for LAS so far and are still limited by the lack of application of sensitive and selective analytical techniques.

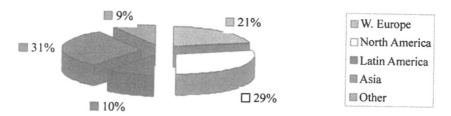

FIG. 1 World surfactant consumption, by region. Total consumption: 10.7 million tons (Mt); excluding soap: about 9 Mt.

FIG. 2 World surfactant consumption, by end use. Total consumption: 10.7 million tons (Mt); excluding soap: about 9 Mt.

Commercial LAS is a mixture of closely related isomers and homologues generated in the manufacture of the raw material, linear alkylbenzene (LAB), each containing an aromatic ring sulfonated at the *para* position and attached to a linear alkyl chain, typically from 10 to 13 carbon units, at any position except the terminal carbons [3,4]. The European LAS cut has an average alkyl carbon number of about 11.6. A new LAS, alternative to the conventional one, with superior claimed performance, induced by the presence of a modified alkyl structure, was recently proposed at the research level [5,6], testifying to how much this surfactant is still alive.

AE is a class of nonionic surfactants composed of a long-chain fatty alcohol with an ether linkage to a chain of ethylene oxide (EO) units. Commercial-grade materials usually have an alkyl chain length normally from 12 to 18 carbon units, with the EO chain typically averaging between 2 and 10 EO units. The most commonly used AE category is the C_{12-15}, EO_{6-9}. The fatty alcohols may be derived from vegetables or petroleum sources and are primarily linear or methyl mono-branched [7,8].

AES is an anionic class of surfactants derived by sulfation of AE materials. The commercial-grade products normally consist of an alkyl chain length of 12–15

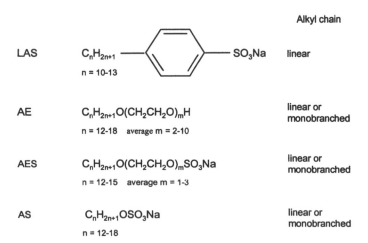

FIG. 3 Commercial linear alkylbenzene sulfonate (LAS) and alcohol-based surfactants, namely, alcohol ethoxylate (AE), alcohol sulfate (AS), and alcohol ethoxy sulfate (AES), presently on the market.

carbon units, with the EO group typically averaging between 1 and 3 EO units. About 20–50% of the mixture has no EO unit and represents an alkyl sulfate (AS) component [8].

AS is a class of anionic surfactants derived by sulfation of fatty alcohols. The commercial products have an alkyl chain length typically in the range of 12–18 carbons [8].

This chapter will deal with the preceding classes of surfactants only.

II. FATE AND EFFECTS IN THE ENVIRONMENT

The environmental fate and effects of LAS and of the alcohol-based surfactants (AE, AS, and AES) have been studied since the appearance of the products on the market. A schematic general picture of the fate of surfactants in the environment is shown in Figure 4. The environmental information on LAS is founded on an extensive database, larger than for any other surfactant, which dates back to more than 35 years ago and includes both aquatic and terrestrial compartments. Analogous information on AE, AS, and AES is not so extensive and normally includes only the aquatic compartment.

Here the available environmental information on these surfactants both in aquatic and terrestrial compartments will be summarized, referring mainly to the latest studies, those carried out after the introduction of sensitive analytical techniques necessary to detect and monitor selectively their concentrations in the environment and in both laboratory and field experiments.

A. Aquatic Compartment

1. Biodegradation

Biodegradation is the dominant removal mechanism for surfactants discharged into waters [9,10]. It reduces the chemical's dispersion through the environment and its

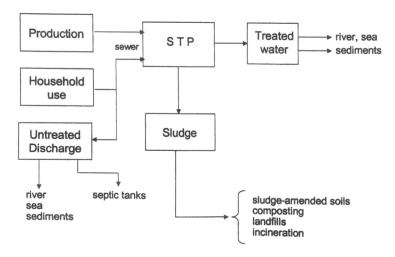

FIG. 4 Schematic of the fate of surfactants in the environment.

concentration in specific habitats, and it is a variable key in determining its exposure. Surfactants can be degraded aerobically and anaerobically by microorganisms in the presence of water. As to the degradation behavior of surfactants under anaerobic conditions, see Section II.A.1(c). Aerobic biodegradation, however, is of primary importance when one considers surfactants that are discharged into sewage and eventually into rivers and the sea. A distinction has to be made between primary and ultimate biodegradation. *Primary* biodegradation is the attack on the surfactant molecule, with loss of interfacial activity and of toxicity vs. aquatic organisms due to a change of the molecule structure. It can be determined by nonspecific analyses detecting anionic surfactants (LAS, AS, AES), as methylene blue–active substances (MBAS), and nonionics (AE), as bismuth-active substances (BiAS). Alternative, and better, analytical methods specific to the single surfactant can be used. *Ultimate* biodegradation is the complete conversion of the surfactant into inorganic substances (e.g., H_2O, CO_2, Na_2SO_4) and the incorporation of its constituents into the biomass of microorganisms. This biodegradation can generally be measured by indirect methods via removal of dissolved organic carbon (DOC), consumption of O_2, and evolution of CO_2. Several test methods to measure the extent and rate of biodegradation exist and have been used, most of them adopted and numbered in the OECD (Organization for Economic Cooperation and Development) guidelines [11]. They can be divided into two major categories: screening and confirmatory or simulation tests. The description and the use of both screening and simulation tests as well as the proposals for improving scientific understanding of biodegradation were deeply discussed at a Setac—Europe workshop [12]. A critical review study on new biodegradability test methods was also commissioned by the EU Commission [13]. Screening tests are usually static and include the methods for ready biodegradability (OECD 301 A–F), also employed for determining the primary biodegradation, and the inherent biodegradability methods (OECD 302 A–C). A degree of primary biodegradation (at least 80%) is the pass level for the acceptance of surfactants on the market. A degree of ultimate degradation (at least 70%, as measured by DOC removal, or 60%, as measured by O_2 uptake or by CO_2 evolution) has to be achieved to consider that the biodegradability of a substance in the environment is complete. Simulation tests attempt to simulate an environmental compartment and include, for example, the confirmatory continuous activated-sludge (CAS) method, in which the process occurring in an STP is approximated (OECD 303 A). A list of updated biodegradation values for LAS, AS, AES, and AE, as recently reported by the EU Commission [14] is given in Table 2, where data for primary biodegradation at the screening and confirmatory levels and for ultimate biodegradation at the ready, inherent, and simulation levels, are summarized.

The screening tests, despite their importance in European legislation for the registration and environmental classification of substances, utilize conditions far from the real world, e.g., too high ratios of test substance to biomass and non-acclimated inocula [15], and thus are inadequate for exposure assessment purposes. To this end the rate of parent disappearance, namely, that of the primary biodegradation, rather than that of the ultimate biodegradation, is the most relevant parameter in case of surfactants. In fact, the disappearance of the parent molecule (primary biodegradation) results in the loss of interfacial activity and toxicity of surfactants in the environment [16]. Degradation rates derived from most existing screening studies, conducted over days and weeks, are usually much slower than those obtained under realistic conditions, as observed in monitoring and continuous

TABLE 2 Biodegradation Data for LAS and Alcohol-Based Surfactants in the OECD Test Methods

Surfactant	Primary biodegradation[a] %		Ultimate biodegradation[b] %		
	Screening	Confirmatory	Ready	Inherent	Simulation
LAS	>99	>99	50–88	95–98	80->95
AS	99	98–99	62–100	86–98	96–97
AES	99	96–98	41–100	96–97	89–99
AE	98–99	—	72–88	—	94–100

[a] As measured by MBAS and BiAS or by application of specific analytical methodologies.
[b] As measured by DOC removal or by O_2 uptake and CO_2 evolution.
Source: Ref. 14.

activated-sludge studies, which show extensive removals. The application of specific and sensitive analytical methodologies to the biodegradation studies is important because it allows us to work in the low concentration range and helps to approach the conditions of the real world, also becoming very informative. The biodegradation of products is more accurate and reliable, the degradation process is characterized in better detail, the presence of intermediates or biotransformed products can be considered, and kinetic data directly relevant for exposure assessment are eventually provided. Specific analytical methodologies based on high-performance liquid chromatography (HPLC) and gas chromatography or liquid chromatography coupled with mass spectrometry (GC/MS, LC/MS) have been developed for LAS and alcohol-based surfactants. Their applications are becoming more common, and several studies have already been published. A further improvement, but complex and expensive, is the use of ^{14}C materials in a biodegradation experiment. In this case the biodegradation conditions can be brought quite near to those of the real world, operating at realistic ratios between test material and biomass and at the low concentrations present in the environment. A comprehensive approach has been devised coupling a realistic test system with specific radioanalytical techniques that measure the evolution of $^{14}CO_2$ by liquid scintillation counting (LSC) and the biodegradation residues by radio thin-layer chromatography (Rad-TLC) [17]. Intermediates or biotransformed products, besides the intact material, could also be evaluated, providing additional information on the mechanism of biodegradation.

Next, recent biodegradation studies carried out applying specific or innovative analytical methodologies are summarized.

(a) LAS. To determine LAS and its biodegradation intermediates in laboratory tests and in the environment, HPLC, coupled to UV or to fluorometric detection, remains the most commonly used technique in many laboratories. Different methodologies based on this technique have been developed since its first introduction [18] up to one of the latest proposals [19]. The relative biodegradation rates with respect to the alkyl chain length and the initial concentration of the surfactant have also been presented [20,21].

A better understanding of the LAS biodegradation process [22] has been achieved using pure cultures of bacteria [23]. Commercial LAS is believed to be completely aerobically degraded not by single organisms but by a community of organisms with a three-tier structure.

The two most important commercial LAS products, depending on the two manufacturing catalyzed (HF and AlCl3) processes of linear alkylbenzene (LAB), the LAS precursor, were tested in a continuous activated-sludge (CAS) system, according to the OECD 303 A method [24]. Specific analytical methodologies, using HPLC as the final detection technique to determine the residual amount of LAS as well as that of its biodegradation intermediates, the sulfophenyl carboxylates (SPCs), were applied to effluents (LAS: 0.08–0.28 mg/L; SPC: 0.2–0.65 mg/L), and sludge (LAS: 0.1–0.45 mg/g), getting a full mass balance in the CAS tests. This allowed a specific assessment of the primary and ultimate biodegradation of LAS comparable to those attained in STPs. The primary biodegradation is always greater than 99% and the ultimate one is in the 95–98% range, higher than that measured via DOC (85%) [24].

Commercial LAS, at least in Europe, is a C_{10}–C_{13} alkyl chain product with linearity between 88% and 98%, depending on the manufacturing process of LAB [3,4]. The nonlinear components of LAS, on average 2–12%, are mainly dialkyltetralin sulfonate (DATS) and methyl-substituted alkylbenzene sulfonate (iso-LAS). DATS can be present in considerable amounts only in the LAS derived from the AlCl3-catalyzed LAB process. The use of this LAS, however, is diminishing in the market and is practically zero in Europe [25]. iso-LAS consists almost exclusively of single methyl substitutions in the alkyl chain, which are different from the highly branched structure of dodecylbenzene sulfonate (DBS), in which the alkyl chain is derived from propylene tetramer [26]. The biodegradation characteristics of both DATS and iso-LAS have been investigated using specific analytical methodology. As to DATS, it was shown that it is totally biodegradable if the biodegradation process is allowed to proceed long enough at realistic environmental conditions [27]. Using a screening biodegradation test (OECD 301 B conditions), the fate of LAS and of its nonlinear components was followed [28]. The biotransformation intermediates of DATS, mono- and dicarboxylates, were identified and monitored by LC/MS. The intact material, DATS, disappeared completely after about 15 days, whereas the intermediates, as expected, were not mineralized at the mild biodegradation conditions of the screening test and leveled off at about 40% of the initial DATS concentration. As to iso-LAS, it appeared that it was more biodegradable than DATS. The biodegradation intermediates of iso-LAS could not be distinguished from those of LAS [28]. Even if more testing is necessary, an overview study on iso-LAS indicates that the methyl substitution does not limit its biodegradation [29]. Two model compounds with the methyl substitutions, one on an alkyl carbon and the other on the benzylic carbon, showed a complete primary biodegradation in a screening test (OECD 301 E) and 75–90% ultimate biodegradation in a simulation CAS test (OECD 303 A) [30]. Specific analyses using HPLC and NMR were carried out to measure and monitor both intact materials and corresponding SPCs and to elucidate the structure of SPCs [30]. Other iso-LAS models, labeled with ^{14}C, were studied with radio techniques in CAS tests, showing comparable good ultimate biodegradation (80–90%) but also giving some poor results (about 12%), in the latter case due to a possible specific quaternary carbon effect [27]. Two different methyl substitutions, in fact, are likely to occur; one on "odd" carbons and the other on "even" carbons with respect to the end of the alkyl chain. The latter is expected not to interfere with a β-oxidation pathway, whereas the former is expected to slow it down [29]. iso-LAS with various midchain methyl substitutions have a rate ($k = 0.1$– 0.7 h^{-1} namely, $t_{0.5} = 6.9$–1 h) and extent of primary biodegradation (about 96%) comparable to those of commercial LAS in SCAS, CAS, and activated-sludge die-away studies [29]. Problematic structures for the biodegradation seem to be confined only to some iso-LAS with quaternary carbons in the alkyl chain. The nonlinear components of LAS were also involved in a study carried out on commercial LAS products using a laboratory trickling-filter test [31,32]. Five to ten percent of refractory organic carbon (ROC) was found at the end of the test and assigned to nondegraded metabolites, originated from the nonlinear components of LAS, mainly from DATS [31]. This lack of complete degradation could be attributed to a microbial mixture present in the trickling-filter system, different and selectively unbalanced from that existing in a CAS system [33]. A commercial LAS with a content of 6.5% iso-LAS and of ≤0.5% DATS was tested in a prolonged and "living" biodegradation test, where the product was continuously added for 80 days [33]. The final residue of this test was characterized by HPLC and NMR, showing that no accumulation of iso-structures, intact iso-LAS, or corresponding metabolites had occurred and concluding that the nonlinear components of LAS, at least the iso-LAS, had to mineralize at rates comparable to those of the linear ones [33].

The kinetics of primary and ultimate biodegradation of LAS in activated sludge were carefully assessed, applying an innovative laboratory die-away test method to radiolabeled materials [34,35]. $^{14}CO_2$ evolution was determined by LSC, and the parent molecule and metabolites were analyzed by RAD-TLC. First-order rate constants were in the range of 0.96–1.10 h^{-1} range ($t_{0.5} = 0.63$–0.72 h) for the primary biodegradation and 0.50–0.53 h^{-1} ($t_{0.5} = 1.31$–1.39 h) for the ultimate biodegradation. Biodegradation kinetics of LAS were also derived under realistic discharge conditions in river water, applying the same testing procedure [36]. LAS was dosed at 75 μg/L into a river water amended with 1% activated sludge to simulate the mixing zone of a river. Both primary and ultimate biodegradation rates in these conditions were about 10–15 times lower than those found in activated sludge [34,36].

C_{12} LAS, in comparison with C_{12} EO$_8$, was tested in a die-away test using instream river waters as the test medium and HPLC analysis to follow the biodegradation. The C_{12} LAS standard showed $t_{0.5} = 11$–14 h for the primary biodegradation and $t_{0.5} = 2$–4 d for the ultimate biodegradation [37].

A general kinetic model for degradation of surfactants was proposed [38]. To validate the biodegradation model, the kinetics of LAS in river water [39] and in seawater [38] were obtained using a die-away test method. In river water, different concentrations (5 and 20 mg/L) of a commercial LAS at different temperatures (7–25°C) were tested. The experiments at low temperature needed a long period of acclimation (up about 20 days). Those at higher temperature scarcely involved a period of induction. The extent of the primary biodegradation (>99%) was never affected by the various conditions, whereas the rate was affected. Even in the most severe conditions (7°C, 20 mg/L of LAS), primary biodegradation was always complete (>99%). At LAS concentrations over 10 mg/L, the biodegradation process is slowed down, likely as a result of a bacteriostatic effect of the surfactant. The dependence of the biodegradation kinetics on temperature has to be related to the metabolism of the microorganism. At 10 mg/L and at 25°C, LAS disappeared completely in about 6 days [39]. Using seawater, two surfactants, LAS and AS, were tested in the laboratory at 20 mg/L concentration and at two temperatures (5 and 20°C for LAS and only at 20°C for AS) [38]. It was observed that LAS biodegradation strongly depends on temperature. At 5°C the metabolic process of microorganisms is considerably inhibited; however, one should notice that the microorganisms used originated from the seawater of Cadiz bay, where temperatures as low as 5°C are not found even in winter months and, thus, they might not be acclimated to the low temperature of the biodegradation experiment. At 20°C, LAS disappeared completely in 12–15 days. At the same temperature, AS disappeared in about 3 days [38]. No significant seasonal variation in the biodegradation of LAS was found in laboratory experiments using waters collected from three different types of ponds at different periods of the year [40]. This was interpreted as the result of acclimatization of the pond microbial community. Another work on LAS biodegradation in salty water describes an experiment where LAS at 15 mg/L concentration is added to 3 L of water collected from a briny pond (salinity 9.5 g/L) in a reactor kept at 21°C, under continuous stirring and oxygenation air: 0.5 L/min) [41]. Primary biodegradation corresponded to a first-order rate constant $k = 0.46$ d^{-1} ($t_{0.5} = 1.5$ d).

The function and structure of microbial communities capable of degrading gray waters (waters produced by human residences, excluding toilet wastes), with the addition of variable amount of LAS as surfactant, were studied in a continuous-flow reactor operating in the temperature range of 30–62°C [42,43]. LAS at the concentration of about 25 mg/L started to inhibit the microbial activity. Community complexity was inversely related to temperature. The microbial community in the reactor adapted to the operating temperature. At high temperatures (53 and 62°C) high activity was restricted to a narrow temperature range (± 7°C), whereas at low temperature (30 and 44°C) high activity extended over temperatures ± 12–15°C from the optimum [43].

Few studies describe surfactant behavior and fate in groundwater. Biodegradation studies generally are not conducted under the oxygen-limited conditions that exist in some contaminated groundwaters. Detailed investigations in this area were recently carried out on LAS for a range of dissolved oxygen concentrations using both column laboratory and field tracer tests [44,45]. Biodegradation increased with increasing dissolved oxygen concentration. At comparable dissolved oxygen concentration, LAS removal rates were generally two to three times greater in laboratory tests than in field experiments. These differences could be attributed to the difference both in LAS concentration in the experiments (13 mg/L in laboratory tests and 20 mg/L in field experiments) and in the temperature (12°C in field experiments compared to 25°C in laboratory tests). Partial LAS removal and the appearance of LAS metabolites in the moderately aerobic transition zone (sewage-contaminated) demonstrate that biodegradation occurred. However, observation of no LAS biodegradation in zones where oxygen was present indicated that aerobic conditions are not the only prerequisite for LAS biodegradation [45]. Under anaerobic conditions (<0.1 mg/L of dissolved oxygen), in sewage highly contaminated zones, or suboxic zones, no biodegradation of the LAS mixture was observed during a 45-day test. First-order rate constants of 0.002–0.08 d^{-1} ($t_{0.5} = 9$–346 d) were measured in the low-oxygen region (dissolved oxygen around 0.5–1.0 mg/L) of the described continuous field tracer test [45]. These rates were comparable to those estimated using a pulsed field tracer test ($k = 0.002$–0.09 d^{-1}) [44].

(b) Alcohol-Based Derivatives.   HPLC as the final detection technique was used to measure the biodegradation of a commercial AE obtained from an oxo-alcohol with a C_{12}–C_{15} alkyl chain, 44% linear and 56% mono-branched, and with an average ethoxylation degree of 7 [24]. The AE concentrations of effluents and sludge of a biodegradation CAS test, OECD 303 A, operating at steady-state conditions, were measured and found in the range of 5–35 μg/L and at about 200 mg/kg, respectively. From the mass balance of the test, the calculated primary biodegradation was greater than 99%, with a DOC removal of 92.7% [24].

Specific advanced analytical methodologies based on HPLC were also applied to compare the biodegradation process of the linear and monobranched components present in commercial AE products [46–48]. Residual AE and biodegradation intermediates such as the neutral polyethylene glycols and various carboxylated derivatives were identified and quantified. The alkyl chain side, linear or monobranched, has a role in determining the biodegradation rates and mechanism of the AE products. Monobranched components exhibit biodegradation rates slower than the linear ones. In addition, linear and short alkyl (methyl, ethyl) monobranched AE components follow predominantly a mechanism that leads to the formation of polyethylene glycols and that can be interpreted in terms of a primary biodegradation path where the scission of the alkyl-ether bond is the first step [49]. AE with longer alkyl branches, on the contrary, biodegrades, mainly with the formation of carboxylated intermediates with the carboxylic group on the hydrophobic chain. Highly branched AE were also examined [48]. The biodegradation in this case is slower than that of linear and monobranched AE materials and seems to proceed mainly by hydrolytic oxidation of the polyethoxy chain, with formation of carboxylated derivatives with the carboxylic group on the hydrophilic moiety. Advanced analytical methodologies based on LC/MS with an electrospray were applied to the biodegradation of a monobranched AE model molecule under screening test conditions [50]. The study allowed the elucidation of various aspects of the biodegradation pathway of this AE molecule, identifying that the major biodegradation route was the initial formation of carboxylated intermediates generated via ω- and β-oxidation of the main alkyl chain. Results somewhat different from the previous ones on the effect of branching over the biodegradation of alcohol-based surfactants were obtained in other studies [51]. AE and AS commercial products as well as some corresponding model structures, both linear and branched, were taken into consideration, measuring their biodegradation in ready OECD tests and in CAS sewage treatment tests. The data demonstrated that the biodegradation of those derivatives with a monobranched 2-alkyl side group in the alkyl chain was comparable to that of the corresponding totally linear surfactants. The length of the 2-alkyl branch also appeared to have a negligible effect on the extent of biodegradation. Tertiary carbon atoms, whether present as proximal or distal to the functional group, had little effect. Quaternary carbon atoms, in contrast, seriously impeded biodegradation [51].

Various AE homologues were assessed in a batch activated-sludge die-away test using ^{14}C radiolabeled materials and applying an innovative testing procedure in order to determine the effect of the EO unit number and alkyl chain length on the biodegradation as well as on their sorption onto sludge [52]. All AE showed very rapid biodegradation, with first-order rates corresponding to $t_{0.5}$ around 1 min for the primary biodegradation and around 0.5 h for the ultimate biodegradation. Applying the same testing procedure, representative ^{14}C radiolabeled materials of AE, AS, and AES were tested under realistic discharge conditions in river water, simulating the mixing zone downstream of an STP outfall [36]. The biodegradation kinetics of these surfactants were determined at 75-μg/L concentration under conditions simulating the mixing zone below an STP outfall. All alcohol-based surfactants tested showed first-order rates with $t_{0.5}$ in the range of 0.2–0.7 h for the primary biodegradation and 1–18 h for the ultimate biodegradation [36].

The biodegradation rate was also studied for some standard AE samples (C_{12}, C_{13}, and C_{14} alkyl chain derivatives), both linear and monobranched, using OECD 301 E and river water die-away tests [37,53]. HPLC analysis and instream river waters as test medium, sampled far away from STP outfalls, were used. Primary biodegradation rates, as expected, were slower than those observed in previous studies [36]. Linear AE components showed first-order rates, with $t_{0.5}$ in the range of 4.5–14 h, whereas the monobranched AE components were slower to biodegrade, with $t_{0.5}$ in the range of 19–58 h [37,53]. Two AE mixtures were also tested under continuous-flow activated-sludge conditions, applying recently developed analytical methodologies based on the indirect tensammetric (ITM) technique [54]. At the biodegradation steady-state conditions, the authors found a high AE primary biodegradation, 96.8% and 96.6%, and measured the consistent presence of free fatty alcohols and polyethylene glycols, whose concentrations were 30–100% of those theoretically expected by a central fission mechanism of AE [49].

The seasonal temperature effect as well as algal sensitivity were investigated during stream mesocosm studies carried out over a 5-year period for three alcohol-based derivatives, AE, AS, and AES, using ^{14}C materials [55]. Seasonal temperatures ranged from 0 to 28°C over all studies. The average decrease in temperature for most of the studies was about 12°C. The overall results indicate that the mineralization of these surfactants in the realistic environmental conditions of the studies, where the various algal species are acclimated following the natural temperature fluctuations, was at least maintained and often increased during significant seasonal decreases in temperature. These findings differ from laboratory studies, which show a positive correlation between temperature and biodegradation. In the laboratory, however, it is likely that the drop of the temperature in the batch testing system prevents the natural acclimation of the microorganism community [55].

Biodegradation kinetic data for LAS and alcohol-based surfactants including data from monitoring studies (see Section II.A.4) are summarized in Table 3.

(c) Anaerobic Biodegradation. Most biodegradation processes take place in the presence of oxygen (aerobic conditions). Biodegradation, however, can also proceed, even if at a lower rate, in the absence of oxygen (anaerobic conditions). Surfactants entering the environment are mostly exposed to aerobic conditions, and only a part of them can potentially reach some environmental compartments that are permanently or temporarily anaerobic, such as STP digester sludge and sediments. That results from the specific physical-chemical properties of surfactants that are strongly adsorptive onto solids. Anaerobic biodegradation, even if it does not have the same

TABLE 3 Biodegradation Kinetic Data for LAS and Alcohol-Based Surfactants at Room Temperature

Surfactant	Type of measurement	Biodegradation rate half-life time, $t_{0.5}$ (h)		Refs.
		Primary	Ultimate	
LAS	Laboratory test with:			
	activated sludge	0.6–0.7	1.3–1.4	34, 35
	river water[a]	ca. 10	ca. 20	36
	river water[b]	11–14	2–4 (days)	37, 53
	seawater	36	—	41
	groundwater	9–346 (days)	—	44, 45
	Monitoring in river	0.9–23	—	36, 91, 127, 128, 130
AE	Laboratory test with:			
	activated sludge	ca. 0.02	0.5	52
	river water[a]	0.2–0.7	1–18	36
	river water[b]	4.5–58	—	37, 53
AS/AES	Laboratory test with:			
	activated sludge	—	—	—
	river water[a]	0.2–0.7	1–18	36
	Monitoring in river	ca. 1	—	110

[a] River water at STP outfalls.
[b] Instream river water.

environmental relevance as aerobic biodegration, is, however, an important element of the exposure assessment of surfactants. An overview of the anaerobic biodegradation issue in general and of its relevance for surfactants in particular has recently been prepared by industry [56,57]. The report consists of a comprehensive compilation of the available literature data on the fate and biodegradability of surfactants under anaerobic conditions. It deals first with discussions on anaerobic compartments and existing methods to measure anaerobic biodegradation and then with the interpretation of the data and criteria to evaluate the importance of anaerobic biodegradation [56]. The results for LAS and alcohol-based surfactants obtained in laboratory tests conducted under anaerobic conditions are summarized in Table 4. As with aerobic biodegradation tests, anaerobic biodegradation methods include both laboratory screening and simulation tests, the former characterized by a high test-substance-to-biomass ratio, usually of the batch type, the latter by test conditions approaching the real world. Most of the tests measure ultimate biodegradation, determining the final gas products, CO_2 and CH_4 [58]. ^{14}C-radiolabeled materials or specific analytical methods are also used, but generally they are applied for simulation tests and in the case of research studies [59,60].

LAS, as shown in Table 4, doesn't degrade anaerobically [61–63]. In oxygen-limited conditions, however, which are likely to occur in the real environment, it was demonstrated that LAS biodegradation can initiate, with the formation of sulfophenyl carboxylates (SPCs), and then continue in anaerobic conditions [64–66]. Academically it was demonstrated that desulfonation reactions, catalyzed by anaerobic bacteria, exist in nature [67]. These reactions, which involve the assimilation of

TABLE 4 Typical Anaerobic Biodegradation Results for LAS and Alcohol-Based Surfactants

Surfactant	Test type	Biodegradation results[a] %	Refs.
LAS, C_{10-13}	Screening/simulation	0	61–63
LAS, C_{10-13}	Research studies	14–100[b]	64, 65, 69, 70, 72
AS, C_{12-18}	Simulation	80–90	59, 74
AS, C_{12-18}	Screening	65–88	75
AES, C_{12-18}, EO_{2-3}	Screening/simulation	>75	59, 62
AE, C_{9-15}, EO_{5-23}	Screening	>70	63, 76–78
AE, C_{12-18}, EO_{7-23}	Simulation	>80	60, 76

[a] Ultimate biodegradation over about 2 months.
[b] Studies conducted under particular anaerobic conditions (i.e., sulfate-limited conditions). In these cases primary biodegradation was measured.

sulfur, seem to be widespread: alkyl- and aryl-sulfonates were desulfonated [68]. It was shown that LAS was also subjected to a quantitative desulfonation [69]. These reactions, however, occur in sulfur-limited conditions, with the sulfonates as the sole source of sulfur for the growth of the anaerobic bacteria. There is no evidence, thus, of how relevant the phenomenon can be under real-world conditions.

In 2000, using the ECETOC screening test [58] under various modified conditions, specific HPLC analysis were applied to measure LAS concentration and thus to assess its behavior under anaerobic conditions [70]. While no significant biogas formation from LAS was measured, in agreement with previous studies, a relevant disappearance of LAS, even up to ca. 90%, was claimed to exist in the mass balance of the tests after 8 months of incubation. It was shown that this primary biodegradation process could have already started in the STP digester from which the testing sludge was sampled [70]. Transformation of LAS, by the way, was also documented under certain anaerobic conditions with selected appropriate inocula [71] and in continuously stirred tank reactors (CSTR) [72]. The degree of transformation varied between 14% and 25%, depending on the bioavailability of LAS.

The anaerobic inhibition potentials, expressed as the 50% inhibition concentration, IC_{50}, were measured while conducting experiments in an upflow anaerobic sludge blanket (UASB) type of reactor [73]. Inhibition started only at LAS concentrations higher than 25 mg/L in all cases, corresponding to about 17 g/kg of LAS in sludge on a dry basis. IC50 of LAS was found to be in the range of 40–150 mg/L, corresponding to a concentration of LAS in sludge of about 27–100 g/kg, a value much larger than those usually found in STP sludge (1–10 g/kg).

As to alcohol-based surfactants, anaerobic biodegradation mechanisms of linear AE were studied in experiments carried out on a pure AE, C_{12} EO_8, single ethoxymer, and a commercial AE, C_{12} EO_9, with an average EO unit number of 9, using anoxic sewage sludge [79]. Specific HPLC analytical methods were applied. During the first stage of the degradation, AE with shortened EO units were released and identified. No PEG products were observed, supporting the conclusion that the central ether-bond scission, which is known to occur predominantly in the aerobic pathway [49], is very improbable in the anaerobic microbial attack. From the results of these studies, the anaerobic biodegradation of AE appears to proceed at first with the cleavage of the terminal EO unit, releasing acetaldehyde stepwise, and then shortening the EO chain until the lipophilic moiety is reached [79].

2. Exposure Assessment

The exposure assessment of surfactants includes the aquatic compartment as the primary environmental target. To conduct an accurate exposure assessment of surfactants as well as of any chemicals discharged down the drain, it is necessary to estimate:

- Concentrations in raw sewage based on usage volume and dilution
- Effluent concentrations based on removal in sewers and during sewage treatment
- Concentrations in surface waters after the mixing zone, based on effluent dilution

The final concentrations are usually calculated by modeling, using distribution coefficients and biodegradation rate constants [80]. They can be predicted with increasing accuracy by various mathematical models nowadays. Removal during sewage treatment can be estimated using, e.g., the SIMPLETREAT [81] or the WWTREAT [82] models. Mathematical models of the "McKay level III" type [83] can be used instead to estimate the exposure in rivers. One can describe the exposure of the aquatic compartment close to the source of emission, called the local predicted environmental concentration (PEC), namely, worst-case estimates. Alternatively, the exposure assessment may be developed taking into consideration the fate, transport, and distribution of the chemical into different media (water, sediments, soil) that are far from the source of emission (regional PEC, namely, background estimates). To this purpose, models such as HAZCHEM [84] and EUSES [85] have been used. An improved and accurate aquatic exposure prediction has recently been developed and validated, the GREAT-ER [86,87], which, different from other models, takes account of spatial and temporal variability in regional infrastructure and of river flows and chemical emissions.

The experimental measurement of surfactants concentrations and of their removal in the various environmental compartments by using specific analytical methodologies, however, is the approach necessary to validate the various modelings and is correctly viewed as the definitive means of obtaining exposure data to perform valid risk assessments.

Next we will consider only the experimental aspects of the exposure assessment, summarizing the known measured data obtained by specific determinations of LAS and alcohol-based surfactants in monitoring studies. Several specific monitoring data, in fact, have recently become available in the literature that, as expected, are heavily biased toward LAS, with alcohol-based derivatives having received less attention.

3. Removal in Sewage Treatment Plants

(a) LAS. The first removal occurs in sewer systems, where the surfactants are ordinarily discharged before entering a sewage treatment plant (STP). There are indications that the sewer is more than just a transportation system for the sewage; it is also a reactor where biodegradation, precipitation and sorption processes are

initiated. LAS was demonstrated to be removed in sewers by some field studies in Spain [88], Italy [89], and the Netherlands [90] up to 68%. More recently in-sewer LAS removals were also documented in monitoring studies to develop and validate the aquatic chemical exposure model, GREAT-ER. The removals averaged between 48% and 56% in the UK studies [91] and were about 62% in the Italian study [92]. In-sewer removal can vary strongly and depends on the length of the sewer, the travel time, and the degree of microbiological activity present in the sewer. Laboratory studies have demonstrated that all surfactants can be significantly reduced in sewers [93].

Concentrations of LAS in raw sewage and STP effluents with the corresponding LAS removal have been measured by several authors in different countries. Table 5 is a summary of the most recent results. Early studies have already shown that LAS could be extensively removed in STP [89,94–97]. In Europe in recent years, several monitorings were programmed and carried out. Pilot LAS studies, as a part of a concerted European-wide project to perform environmental risk assessment of surfactants, were carried out in the United Kingdom [98], Germany [99], the Netherlands [100], Spain [101], and Italy [102]. The major insights gained from these studies were summarized [103] and were the base to set up an accurate and extensive monitoring program over seven STPs in the Netherlands, not only for LAS but also for other major surfactants on the market [90]. The LAS effluent concentrations in all STPs averaged 39 μg/L, with an average removal of 99.2%. The measured alkyl chain length was 11.6. All these results were used to estimate the exposure concentrations for the river water compartment in the Netherlands [104].

TABLE 5 Average Measured Concentrations of LAS in STP Raw Sewage and Effluent

STP type[a]/country	Concentration		Removal (%)	Refs.
	Raw sewage (mg/L)	Effluent (μg/L)		
(as) DE	7.0	67	99.2	99, 103
(as) UK	15.1	10	99.9	98, 103
(as) NL	4.0	9	99.8	95, 103
(as) E	9.6	140	98.5	101, 103
(as) I	4.6	68	98.5	102, 103
(as) NL	3.4–8.9	19–71	98.0–99.6	90
(as) DE	2.1–5.1	16–42	98.0–99.9	109
(as) DE	0.6–2.2	5	99.2–99.3	110
(as) I	3.1–8.4	13–115	97.8–99.4	105
(as) I	—	3–9	>99	129
(as) UK/NL	5.6–11.4	4–80	99.1–99.9	108
(tf) UK/NL	6.8–9.7	530–580	91.5 94.5	108
(tf) UK	2.5–4.4	80–360	90.4–97.1	91, 263
(as) USA	3.8–5.4	<5–7	98.0–>99.9	107
(tf) USA	1.8–6.1	73–1500	72.2–98.6	107
(as) USA	3.0–7.7	3–86	99.5	106
(tf) USA	2.7–6.4	140–2300	82.9	106

[a] (as): activated sludge; (tf): trickling filter.

On the basis of mass balance studies [89,95,96,102], the split for LAS in the activated-sludge (as) STPs was found to be as follows: 80–90% degraded, 10–20% into sludge, and about 1% released to surface waters.

In Italy, modified analytical methodologies based on LC/MS detection were employed over five STPs to monitor, over different periods of the year, not only LAS but also the nonlinear components, DATS and iso-LAS, and their corresponding biodegradation intermediates [105]. The average removal of DATS and iso-LAS (95.1–96.9%) is equivalent to that of LAS (98.6%), whereas the disappearance of the biodegradation intermediates generated by the nonlinear components appears to be less prominent (46–65%). The LAS effluent concentration was found to be in the range of 13–115 µg/L. The corresponding DATS and iso-LAS effluent concentrations were in the range of 7–54 µg/L and 2–29 µg/L, respectively. Different biodegradation intermediates were identified and estimated to total about 300–600 µg/L [105].

A new analytical procedure based on a derivatization electron capture GC/MS method was developed to measure at high sensitivity and selectivity the low concentrations of LAS and DATS and of their biodegradation intermediates in environmental samples [106]. This work was an extension of a previous study that reported environmental concentrations of LAS at 50 domestic STP sites in the United States [97]. This new analytical procedure was applied to monitor 10 STPs in the United States, half of the activated-sludge (as) type and the other half of the trickling-filter (tf) type. The average removal of LAS in the (as) plants was 99.5%, whereas that of the (tf) plants averaged 82.9%, in agreement with what was already reported in a previous US study [97]. DATS, present at 0.28–0.54 mg/L in influents and at 1–152 µg/L in effluents, was shown to be removed 95% in (as) plants and 63.2% in (tf) plants. The biodegradation intermediates originating from LAS were well removed, 97%, whereas those from DATS were only removed about 59% in all STPs [106]. Their residual presence in effluents was in the range of 19–196 µg/L. In the United States, with the purpose of getting realistic STP removal rates for LAS as well as for other major surfactants on the market, a monitoring study was conducted over 10 other STPs, six (tf) and four (as) plants [107]. The overall average influent value for LAS was 4.4 ± 1.5 mg/L, with an average alkyl length of 11.9. The final effluent concentrations of the same surfactant were quite variable for the (tf) plants (750 ± 570 µg/L) and below or near the detection limit (<5 µg/L) for all (as) plants. The removal averaged 82.3% for the (tf) plants and was 99.9% for the (as) plants [107].

An equivalent study (in the United Kingdom/Netherlands), involving both (tf) and (as) plants was reported in 2000 for some cationics, also taking into account LAS as the reference surfactant [108]. The more efficient (as) plants removed LAS to an extent greater than 99%, whereas (tf) plants showed lower and more variable removal rates, with average values in the range of 72–98%. Still, in the United Kingdom, for four STPs, all (tf) type, the average LAS concentration in flow-proportional composite crude sewage was found to be 3.65 mg/L and in the final effluents 240 µg/L [91,263]. The average removal rate throughout all STPs was 92.9% higher than rates reported for analogous plants in the United States [107].

In Germany several studies were carried out monitoring LAS in STP influents and effluents and in the corresponding receiving rivers [109]. The average LAS elimination in the STPs involving biological treatment was 99.2%, with average LAS concentrations of 3.3 mg/L in influents and 25 µg/L in effluents. One STP with sewage treated only mechanically was also investigated. In this case the average LAS was 2260 µg/L in the effluent and 570 µg/L downstream (2.5–11.5 km) in the receiving river [109]. Another monitoring program, covering different periods of the year, dealing not only with LAS but also with other surfactants (i.e., AS, AES), was conducted over two STPs and river reaches receiving the treated waters [110]. In both cases the LAS elimination rates were on average 99.2–99.3%, with an average concentration in the effluents of around 5 µg/L. The LAS concentrations in the rivers just upstream and downstream of the STP outfall were of the same order of magnitude [110].

(b) Alcohol-Based Surfactants. AE and AES were tested in a sewage die-away study to estimate their kinetic removal and fate in sewers [93]. Radiolabeled material of both products was added to raw sewage collected from a municipal STP, with the biodegradation process monitored via thin-layer chromatography with radio scanner (TLC/RAD). The disappearance of the parent AE molecule as a function of time corresponded to a half-life time of 3 hours, whereas that of AES was approximately 4 hours. During degradation both AE and AES were shown to turn into transient polar intermediates, polyethylene glycol carboxylates and sulfates, respectively, which rapidly mineralized to carbon dioxide. The laboratory findings for AES were also confirmed by a field study where the AES concentration present in a STP raw sewage was measured by LC/MS and compared with the concentration expected according to the known surfactant consumption, the served population, and average flow rates [93]. AES removal in the sewer during travel to the STP was calculated to be approximately 47%.

Only a few environmental monitoring studies have used specific analytical methodologies to determine the alcohol-based surfactants. Apart from a few early and preliminary determinations on AE [111] and AES [112], the most recent data are summarized in Table 6. The extensive monitoring program conducted in the Netherlands to evaluate the risk assessment of priority detergent surfactants included the alcohol-based surfactants, namely, AE, AS, and AES, in addition to LAS and soap [90]. Seven STPs were considered and monitored. AE was determined in influent by a specific HPLC methodology [113] and in effluent by HPLC coupled to thermospray mass spectrometry (LC/TS/MS) [114]. AS and AES were determined by HPLC coupled to ion spray mass spectrometry (LC/IS/MS) in both influent and effluent [115]. The effluent concentrations over all STPs averaged 6.2 µg/L for AE, 6.5 µg/L for AES, and 5.7 µg/L for AS, with an average removal of 99.8%, 99.6%, and 99.2%, respectively. The average alkyl chain length of AE in the effluent was 13.3, with an average EO number of 8.2, similar to that of the original commercial material. The analyses indicated that 50% of AE was associated with the suspended solids. As to AES, the measured alkyl chain length in the effluent averaged 12.5 (12.3 for AS), with an average EO value of 3.4 [90].

In Germany, two STPs and the corresponding river reaches receiving the treated waters were monitored for AS and AES over different periods of the year [110]. The effluent concentrations of both AS and AES were in the range of 1–4 µg/L, with an experimental average removal of 98.75%, which underestimates the real elimination in the STPs of these surfactants, likely to exceed the 99% limit by far. The AS and AES concentrations in the receiving river reaches were, most of the time,

TABLE 6 Average Measured Concentrations of Alcohol-Based Surfactants in STP Raw Sewage and Effluent

Surfactant	STP type[a]/ Country	Concentration		Removal (%)	Refs.
		Raw sewage (mg/L)	Effluent (µg/L)		
AE	(as) NL	1.6–4.7	2.2–13	99.6–99.9	90, 113
AE	(as) USA	3.1–3.7	8–235[b]	90.4–99.7	107
AE	(tf) USA	0.7–3.4	49–509[b]	79.4–96.5	107
AE	(tf) USA	0.7–2.7	50–114	92.6–95.7	116
AE	(as) USA	3.2–3.7	11–71	98.1–99.6	116
AE	(as) I	—	5–15	—	129
AS	(as) NL	0.1–1.3	1.2–12	99.0–99.6	90
AS	(as) DE	0.14–0.38	2–4	98.6–98.9	110
AES	(as) NL	1.2–6.0	3–12	99.0–99.6	90
AES	(as) DE	0.09–0.36	1–4	98.9	110
AES	(as) USA	0.15–1.0	4–18	97.7–98.2	107
AES	(tf) USA	0.1–2.18	32–164	69.7–96.7	107
AES	(rbc) USA	0.57	88	—	115

[a] (as): activated sludge; (tf): trickling filter; (rbc): rotating biological contractor.
[b] The high concentration figures are anomalous because of hydraulic overloading and bulking sludge problems.

very low (ca. 1 µg/L), close to the detection limit. For AES, in one river reach, a half-life time of about 1 hour was estimated [110].

In the United States an analytical method based on GC after derivatization with hydrogen bromide was developed to determine AE concentrations in environmental samples [116]. The method was applied to four STPs in the United States, two (tf) and two (as) plants. The AE alkyl homologues C_{12} and C_{14} in the influent were more abundant than the C_{13} and C_{15} ones. This is not unexpected, because much of the C_{12} and C_{14} are derived from oils of renewable sources, while C_{13} and C_{15} are derived from petrochemical oils. Effluent concentrations are independent of influent concentrations and remain relatively constant, at approximately 11–71 µg/L for the (as) plants and 50–114 µg/L for the (tf) plants. Total AE removal ranged from 92% for a trickling-filter plant to as high as over 99% for an activated-sludge plant [116]. Ten other STPs, six (tf) and four (as) plants, were monitored in the United States for AE and AES [107]. The overall average influent AE concentrations were 2.6 ± 0.9 mg/L and the average effluent AE concentrations, excluding some anomalous results, were 106 µg/L in the (tf) treatment and 30 µg/L in the (as) treatment. The overall average removal for AE was 89.9% by (tf) treatment and 97.0% by (as) treatment. Concentrations of AES in STPs influents ranged quite a lot, going from 0.1 to 2.18 mg/L. The final effluent concentrations ranged from 4 µg/L in (as) plants to 164 µg/L in (tf) plants, with an average removal of 98.0% and 83.5% in (as) and (tf) plants, respectively. The average alkyl chain length and EO unit numbers were found to be almost equal in both influent and effluent, namely, about 14 and 1.7, respectively [112]. Background concentrations of AES in grab samples of an STP influent and effluent, used to develop and validate a specific method for this surfactant, were also reported, namely, 0.57 mg/L for the influent and 88 μg/L for the effluent [115].

4. Monitoring in Surface Waters

(a) LAS. Estimates of the LAS exposure in the Dutch rivers were obtained by modeling on the basis of the extensive monitoring program carried out on the STPs of this country [90] and in Europe [103]. Considering a conservative in-stream loss rate of 0.029 h^{-1} ($t_{0.5}$ = 24 h) for surfactants, similar to in-stream BOD removal rates, the average PEC of LAS in Dutch rivers was calculated to be 3.7 μg/L [104,117]. Similarly, in the United States the information obtained by monitoring a series of STPs was used to predict that 50% of the receiving waters immediately below the mixing zones of the 11,500 U.S. STPs have on a LAS concentration below 2 μg/L, even under low-flow conditions [107].

Average measured LAS concentrations in surface waters, mainly in rivers, upstream, and downstream STPs, and instream, in different countries are given in Table 7.

Early studies have already monitored LAS in small reaches of rivers [118] and in groundwater [119]. In recent years in Europe, extensive monitoring programs have been carried out for LAS in the United Kingdom [98], Germany [99], the Netherlands [100], Spain [101], and Italy [102], showing values ranging from about 2 μg/L to

TABLE 7 Average Measured Concentrations of LAS in Surface Waters[a]

Country	Concentration (μg/L)		Refs.
	Upstream of the STP	Downstream of the STP or instream[b]	
DE	9	11	99, 103
UK	30–130	9–47	98, 103
NL	<2.1–2.9	<2.1–7.1	100, 103
E	27	30	101, 103
I	8.5	9, 6	102, 103
DE	3	5	109
DE	63	570 (dd)	109
DE	—	4–6	110
I	1.5	36 (0.1 km), 0.9 (27 km)	105
USA	—	0.1–10.3; 0.1–2.8	121
USA	3–110	2–94	106
E	—	bay w.: 10–30	122, 124
E	—	estuarine w.: 138 (top), 14 (1 m deep)	125
I	177	187[c]; lagoon w.: 2.5–8.4; open sea: 1.9	126
UK	5	416 (tf), 33 (4.8 km)	127
I	5–50	100–200 (dd), 10–50 (20 km)	92, 128
I	—	16–19	129
DE	7–8	12–30	130

[a] In rivers unless stated otherwise.
[b] (dd): direct discharge; (tf): tricking filter.
[c] Downstream mechanical-biological STP.

47 µg/L. In Germany the monitoring activity was particularly intensive, covering different sites and environmental situations, including one with poorly treated discharge [109]. The LAS concentrations in rivers downstream of the outfalls of well-operated STPs were in the range of 4–6 µg/L [109,110] and at about 570 µg/L in the case of poorly treated discharges [109]. The monitoring of the river reaches provided clear evidence that the elimination rates for LAS, as well as that for the other surfactants investigated, was considerable: the half-life for LAS was in the range of 1–3 hours [110]. The overall data, compared with those obtained by modeling, were considered representative of the river situation in Germany and were used to assess the LAS impact on the aqueous environment related to the year 1993 in Germany [120]. Risk is low when STPs are present and well operated; risk is present if sewage treatment is lacking or insufficient [120]. The use of LAS has decreased since 1993 in Germany. A simple estimate came to the conclusion that the average concentration of this surfactant in German rivers would be 25% less in 1995 [120].

LAS, DATS, as well as their biodegradation intermediates were analyzed in U.S. river sites receiving the treated discharges of 10 STPs by applying a new, sensitive analytical procedure based on the electron capture GC/MS method [106]. The ranges of LAS, DATS, and biodegradation intermediates in the receiving waters downstream of these STPs were: LAS 2–94 µg/L, DATS 1–23 µg/L, and biodegradation intermediates 1–159 µg/L. Using analytical procedures based on LC/MS, similar measurements were performed in the receiving waters of one Italian river [105]. LAS, DATS, iso-LAS, and biodegradation intermediates were measured at 0.1 km and 27 km downstream of one STP. The concentration of each of the sulfonates was around 18–36 µg/L near the STP outfall (0.1 km) and 0.9–1.8 µg/L 27 km downstream. The total concentration of the corresponding biodegradation intermediates in the same sites was about 1 order of magnitude higher than that of the sulfonates [105].

In the United States an extensive sampling program was carried out in the main stream of the Mississippi River and its tributaries, with three cruises in summer, fall, and spring collecting totally 515 grab samples [121]. LAS was detected in 21% of the main stream samples at concentrations in the range of 0.1–10.3 µg/L and in 15% in the tributary samples in the range 0.1–2.8 µg/L. LAS was analyzed by GC/MS after derivatization. Dissolved LAS was present mainly downstream of the sewage outfalls of the major cities. Sorption and biodegradation are the principal processes affecting dissolved LAS. Dilution was the major factor affecting LAS concentration in the river [121].

Studies were conducted on the fate and behavior of LAS in the Spanish littoral environment of the Cadiz Bay in a shallow narrow channel (3–6 m deep, ca. 20 km long) subjected to strong tidal currents, into which untreated urban wastewaters are discharged [122]. The concentration of suspended solids (SS) was quite high, 90–157 mg/L in the channel water measured at one end of the channel. Most of LAS was adsorbed on these SS. The highest LAS concentration was recorded in the part of the channel nearest to the untreated urban effluent discharge point (1.7 mg/L and 5.9 mg/L as dissolved and adsorbed LAS, respectively). In the Cadiz Bay waters at both ends of the channel there was an appreciable reduction in concentration due to biodegradation, adsorption into sediments, and dilution produced by tidal inflows. The total LAS measured here was in the range of 10–30 µg/L [122], similar to the values observed in the outer zone of Tokyo Bay [123]. Worth mentioning is the evidence of the strong sorption character of surfactants at the water–atmosphere interface. In zones close to the urban effluent discharge point a steep vertical gradient in LAS concentration was observed, with values in the top water (3–5 mm deep) some orders of magnitude higher than those in deep waters (>0.5 m deep). LAS biodegradation intermediates, SPC, with long alkyl chain up to 11 carbon atoms were also detected in the marine environment of Cadiz Bay [124]. This finding is evidence that LAS biodegradation can be a slow process in seawaters that are deficient in oxygen and highly contaminated with various organic substrates. SPC reached the maximum concentration values (about 120 µg/L) in the channel about 3 km from the urban effluent discharge point [124]. Estuarine water samples as well as some sediments taken from a river near Santander were also analyzed [125]. Different LAS concentrations were found at different water depths having different salinity. In the top layer (0–5 mm deep) total LAS was at about 138 µg/L, with 1:5:5:2 homologue distribution ratio from C_{10} to C_{13}, whereas 1–2 m deep the concentrations were one order of magnitude less. SPC were detected but definitely at lower concentration values, namely, around 10 µg/L in the top layer [125].

A monitoring study on LAS and their biotransformation products was carried out in the central lagoon of Venice, Italy [126]. NPE were also investigated. The study included 13 sites, one in the open sea (2 km offshore), two along the terminal reach of a river entering the lagoon upstream and downstream of a mechanical STP, and the other 10 inside the lagoon. Monitoring was done twice monthly for 6 months from January to September 1994. The average concentrations of LAS found in the monitoring campaigns were: 1.9 µg/L (open sea), 177 and 187 µg/L (river, upstream and downstream of one STP), and 2.5–8.4 µg/L, with a yearly average of 4.3 µg/L (lagoon). SPC intermediates were also measured and were found to be approximately twice times the LAS concentration. Taking into account these monitoring results of both LAS and SPC and considering the annual LAS loading (630 kg/d) to the lagoon and the mean water renewal (10 d) in the lagoon (132 km^3), the annual disappearance of LAS by biodegradation and sedimentation was calculated. It turned out to be approximately 90%, whereas its mineralization was estimated at about 72%. Remarkable seasonal differences, namely, efficient loss of LAS by biodegradation, were noticed in the hot period of the year, late spring and summer [126].

An extensive monitoring program on LAS was performed at the European level to provide the specific data required to calibrate and validate the GREAT-ER model [86,87]. LAS removal was measured in several STPs and in specific rivers in the United Kingdom, Italy, and Germany. Nearly 3000 samples from STP effluents and river-monitoring sites were collected over a period of about 2 years and analyzed for LAS but also for boron and other water-quality parameters. In the UK study, four catchment areas of the Yorkshire region were chosen, proposing various objectives [91]. A small stream receiving a high effluent level of LAS from a trickling-filter STP was considered in detail [127]. The LAS concentration was measured upstream and then at seven sites downstream of the effluent STP outfall. The decrease in LAS concentration downstream is caused by dilution, degradation, and settling out of SS. The values were: 5 µg/L (upstream), 416 µg/L (mixing zone), and 33 µg/L downstream, at the end of the reach, 4.8 km away from the effluent outfall. The study showed a LAS removal that can be described by a first-order rate constant of 0.31 h^{-1} ($t_{0.5}$ = 2.2 h). This removal was somewhat faster than ammonia removal and twice as fast as BOD5 removal [127]. Measurements were also made in other UK rivers, observing different and slower removal kinetics, k = ca. 0.07 h^{-1}, with $t_{0.5}$ = 9.6–10.4 h [91,263]. In these latter cases the approximate average LAS concentration in the river reaches, however, was almost an order of magnitude less, namely, around 20 µg/L [91,263]. In Italy, two extended campaigns were carried out over a small river, about 30 km long, north of Milan [92,128]. The river catchment is characterized by the presence of a mixture of intensive industrial and residential activity. Two (as) STPs operate in the study area. The presence of intermittent direct discharges from one of the two STPs, which was undersized and located near the upstream end of the monitoring zone, proved to have a very strong influence on the river water quality, generating a pronounced diurnal cycle in the concentration of pollutants along the river. All that required the adoption of specific sampling methodologies in order to understand the dynamics of the pollution behavior in the receiving water body. The results from the composite samples showed a near-exponential decrease in the mean LAS concentration with distance, 120 µg/L, just downstream of the charged STP, and 27 µg/L at the end of the river reach, 26 km downstream. The collected data suggested an average LAS removal rate constant of 0.052 h^{-1} ($t_{0.5}$ = 13.3 h). The regional LAS PEC, produced by aggregating all local LAS data and calculated according to the GREAT-ER model, was 49 µg/L [92,128]. The situation of this Italian river reach was recently upgraded, removing the intermittent input of untreated sewage. The improvement effect could be simulated using GREAT-ER and then documented by monitoring [129]. LAS dropped from an average of 120 µg/L to 16 µg/L downstream of that charged STP; the effluents of the upgraded STP were measured in the range of 3–9 µg/L [129]. In Germany a small tributary stream, about 20 km long, that meets the River Rhine near Düsseldorf was considered [130]. The hydrological conditions of the stream are significantly affected by the effluents of three (as) STPs. The monitoring study was closely linked to the GREAT-ER project [86,87]. The measured LAS concentration near the springs of this tributary was at about 7–8 µg/L, even if the origin of the presence of the surfactant there could not be understood. The maximum measured concentration was always less than 30 µg/L. The mean LAS value at the output of the tributary, at the confluence with the Rhine, was 27 µg/L. Comparison between measured LAS data and those calculated by GREAT-ER modeling along the entire river clearly showed that the residual presence of LAS in the river is determined mainly by two parameters, the degree of connectivity to STPs and their efficiency, namely, the elimination rate of the surfactant in STPs. A further observation is that degradation is far more important than sedimentation. In addition, high variability of the LAS elimination was shown, k = 0.03–0.8 h^{-1} ($t_{0.5}$ = 0.9–23 h), which was highly correlated with water-quality parameters such as TOC and ammonium [130].

Also taking into account other, less recent field studies [109,110,131], we find that quite a wide range of LAS removal rate constants in rivers exists, with k between 0.006 and 1.71 h^{-1}, namely, half-lives in the range of 0.4–115 h. Presence of biomass, especially biofilms, was shown to be determinant of the removal rate, particularly in small rivers. A new biodegradation modeling concept considering microbial activity both in biofilm and in bulk water was proposed [132]. The model was calibrated using experimental data obtained in artificial river experiments conducted in the context of the GREAT-ER project [127]. The model was able to predict the LAS removal measured in the field study, a calculated removal rate of 0.25 h^{-1} ($t_{0.5}$ = 2.8 h) vs. an experimentally field removal rate of 0.31 h^{-1} ($t_{0.5}$ = 2.2 h).

(b) Alcohol-Based Surfactants. Estimates of the exposure of the alcohol-based surfactants, namely, AE and AES, in Dutch rivers were given by mathematical modeling as part of the comprehensive aquatic risk assessment carried out for the major surfactants on the market [117]. Assuming a conservative instream removal rate of surfactants similar to that of BOD_5, $k = 0.029 \ h^{-1}$ ($t_{0.5} = 24$ h), the average PEC of AE and AES in the rivers was calculated to be 0.5 µg/L and 1.2 µg/L, respectively [104,117]. In the United States, a similar project predicted that 50% of the receiving waters immediately below the mixing zone of the 11,500 existing STPs had concentrations of AE and AES below 4 µg/L under low-flow conditions [107]. Few experimentally measured data of alcohol-based surfactants in surface waters exist in the literature. These are summarized in Table 8.

AE concentrations, using a GC-based methodology, were measured upstream and downstream of four STPs in the United States [116]. Total AE was in the range of 17–26 µg/L upstream and always less than 37 µg/L in the receiving stream below the STP outfalls, despite the low surface water dilution. In Italy, over the same small river north of Milano monitored for LAS on behalf of the GREAT-ER project [92,128], a monitoring study on AE covering different periods of the year was also carried out [48,129]. The river water quality was strongly affected by an STP that discharged partly untreated sewage. Composite samples had AE concentration in the range of 78–258 µg/L just downstream (300 m) of this STP and 29–54 µg/L about 10 km downstream. After the recent upgrading of the STP, with raw sewage fully treated, the AE concentration downstream of the STP dropped to 8–9 µg/L. The corresponding STP effluents showed an average AE concentration of 5–15 µg/L [129]. These AE figures have to be considered conservative because of the HPLC-based methodology employed.

Some AES concentrations in surface waters have been reported as well. Accurate monitoring requires a sensitive and sophisticated analytical methodology for this surfactant that quantifies individual AES components without interference from non-AES species. A method using ion spray LC/MS technique to determine individual AES components has been developed [115]. Applying the method to a river water the background concentration of AES was found to be 5–10 µg/L, showing that the components with EO units equal to zero, namely AS, were the majority.

TABLE 8 Average Measured Concentrations of Alcohol-Based Surfactants in Surface Waters

| Surfactant | Country | Concentration (µg/L) | | Refs. |
		Upstream of the STP	Downstream of the STP[a]	
AE	USA	17–24	18–37	116
AE	I	—	78–258 (dd), 29–54 (10 km)	48, 129
AE	I	—	8–9	129
AS	DE	—	2–10	110
AES	DE	—	1–2	110
AS/AES	USA	5–10	—	115

[a] (dd): direct discharge.

AS and AES were monitored in two German river reaches receiving STP treated sewage [110]. Concentrations of both surfactants were in the range of 1–4 µg/L 100 m downstream of the STP outfalls. Monitoring along the river reaches showed that the biodegradation rate of these surfactants was high, with half-lives on the order of 1 h [110].

5. Toxicity in Water

(a) Single-Species Testing. Aquatic toxicity data of some commercial LAS and alcohol-based surfactants are summarized in Table 9. The values of LAS, AE, and AES are taken from the BKH reports, a comprehensive review of existing environmental information on these surfactants prepared in 1993–94 on behalf of their risk assessment in the Netherlands [133]. The total records of toxicity data collected in BKH reports amount to 749 for LAS [134], 388 for AE [135], and 91 for AES [136]. As to AS, because data were not collected by BKH, reference was made to the IPCS report [137] and to the BUA report [138]. The toxicity database for LAS appears very rich, whereas those for AE, AES, and AS, even if not as extensive, are still remarkable. The overall toxicity data are quite spread and in general cover several taxonomic groups. Intra- and interspecies variability is large for all surfactants. That is true particularly in the case of algae. This is due to the chemical heterogeneity of surfactants, which are mixtures of different isomers and homologues, and also to differences in test design as well as to the diversity of species sensitivity. Because of that, to present a simplified picture of the aquatic toxicity of these surfactants and for comparison purposes, Table 9 shows only average data relative to *Daphnia magna*, as representative of crustacea, to *Pimephales promelas*, as representative of fish, to algae, and to marine species. Data of EC_{50} (median effect concentration) after 48-h exposure, LC_{50} (median lethal concentration) after 96-h exposure, and NOEC (no-observable-effect concentration) at 21-d exposure were extracted from the aforementioned reports. They were selected for only the commercially most representative products: LAS, with C_{10-13} alkyl chain, the so-called European cut, with

TABLE 9 Average Measured Aquatic Toxicity of LAS and Alcohol-Based Surfactants[a]

Species	LAS^b (mg/L)	AE^b (mg/L)	AES^b (mg/L)	AS^c (mg/L)
Invertebrates: *Daphnia magna*				
EC_{50}	5.0	1.1	6.4	10–27
NOEC	0.7	0.2	0.4	16.5, $(0.2–0.6)^d$
Fish: *Pimephales promelas*				
LC_{50}	2.9	1.5	4.2	5–38
NOEC	1.1	0.2	0.1, 0.9	$(1.7)^e$
Algae, EC_{50}	0.05–32	0.09–30	3.5–10	9–65
Marine species, NOEC	0.025–0.12	2.7–48	—	0.29–0.73

[a] Data refer to the standard commercial products: LAS, C_{10-13}; AE, C_{12-15} EO_{6-9}; AES, C_{12-16} EO_{1-3}; AS, C_{16-18}.
[b] Data from BKH reports, 1993–94 (134–136).
[c] Data from IPCS report, 1996 (137) and from BUA report (138).
[d] Data for *Ceriodaphnia dubia* (139).
[e] Datum for *Brachydanio rerio* (138).

average $C_{11.6}$ carbon number; AE, with C_{12-15} alkyl chain and average EO_{6-9} units; AES, with C_{12-16} alkyl chain and average EO_{1-3} units; AS, with C_{16-18} alkyl chain. Several data for *Daphnia m.* and *Pimephalas p.* of the preceding surfactants, LAS, AE, and AES, exist in the BKH reports [134–136]. Their toxicity figures in Table 9 are averages of various independent records, with the exception of the NOEC of AES for fish, for which only two data were found. As to AS, one NOEC (16.5 mg/L) for *Daphnia m.* is given in IPCS and BUA reports [137,138]. A NOEC of 0.2–0.6 mg/L for *Ceriodaphnia dubia* [139] is, however, a better representative toxicity figure for invertebrates. One NOEC of 1.7 mg/L for fish, *Brachydanio rerio*, and a biocenotic NOEC of 0.55 mg/L were reported for this surfactant AS, C_{16-18} [137,138]. Toxicity data of all surfactants for algae are variable, which justifies the report of their range. Data on marine species are presented in a similar way, since only a few values are available in the literature.

To illustrate how toxicity depends on chemical structure, Table 10 shows the various measured acute and chronic aquatic toxicity of LAS homologues for a species of crustaceous and fish found in the literature [134]. The strong positive dependence of the alkyl chain length on toxicity is evident. The structure–toxicity relationship of AE to aquatic organisms is more complex. It was shown in fact that toxicity increases with increasing alkyl chain length and with decreasing EO units [140]. As a general rule, when EO units remain the same, an increase in the alkyl chain length increases toxicity. Conversely, though less marked, when the average alkyl chain length remains constant, an increase in the EO chain units lowers toxicity. More simply, an increase in toxicity typically occurs with an increase in hydrophobicity of the surfactant mixture. This rule is valid in general for all surfactants.

An alternative valuable list of aquatic toxicity of surfactants was recently developed in the context of the European Eco-labeling. That list is a part of the Detergent Ingredient Database (DID) of the ecological criteria set up to award the European Eco-labeling of detergents. It was developed not for risk assessment purposes, but to have a comparative and relative ranking of the toxicity of detergent ingredients. The DID was set up by a team of experts coordinated by the German Ministry of the Environment. It was published first in 1995 [141] and then updated in 1999 [142]. The toxicity values, called in this case LTE (long-term effect), were based on the lowest existing measured NOEC data, when available, and then relatively refined altogether by expert judgment. LTE data for the most commonly used LAS

TABLE 10 Average Measured Aquatic Toxicity (mg/L) of LAS Homologues

Alkyl chain	Crustacea (*Daphnia magna*)		Fish (*Pimephales promelas*)	
	Acute	Chronic	Acute	Chronic
C_{10}	29.5	9.8	57.5	14
C_{11}	21.1	—	21.9	7.2
C_{12}	5.9	4.9	6.6	1.08
C_{13}	2.6	0.57	1.8	0.12
C_{14}	0.7	0.1	0.5	0.05
Commercial, $C_{11.6}$	5.0	0.7	2.9	1.1

Source: Ref. 134.

and alcohol-based surfactants are summarized in Table 11. Two LAS, the European, cut with $C_{14} < 1\%$, and that with an average higher molecular weight ($C_{14} > 1\%$), not used in Europe but more common in the United States, six AE with different alkyl chain length and EO units, two AS at high and low molecular weight, and one AES are given. As shown, with the exception of AS, for which there is a general lack of toxicity experimental data, the hydrophobicity rule on toxicity is followed in each surfactant group. In particular, it is worth noting that for the most important and most used AE category, the C_{12-15} EO_{6-9}, two different toxicity values were identified, a low one, when the average alkyl C number is under 14, and a high one, when the average alkyl C number is over 14.

Other relevant and more recent toxicity studies on LAS and alcohol-based surfactants are considered next.

The toxicity effects of LAS, cationic surfactants (CS), and their complexes (LAS-CS) were examined in *Dunaniella* sp., a living green alga, by NMR (nuclear magnetic resonance) analysis of glycerol and in *Chlorella p.*, another alga, by growth inhibition [143–145]. The glycerol content decreased quantitatively with increased concentration of the surfactants and of their complexes. The 24-h EC_{50} of LAS and CS in free form were in the range of 0.38–18 mg/L, whereas that of LAS-CS complexes, at molar ratios of 1:1 and 2:1, were in the range of 1.3–13 and 5.6–9.6 mg/L, respectively, expressed as LAS-equivalent concentration. It is evident that not only LAS but also its complexes with cationics might exert toxic effects on organisms in the environment.

A series of laboratory studies and a 19-d mesocosm study were carried out on blue mussel, *Mytilus edulis*, larvae and plantigrades exposed to LAS [146]. This organism lives in near-shore environments and estuarine waters, where surfactants can be present because of untreated urban and industrial discharges. In the

TABLE 11 Long-Term Effect (LTE) Values of LAS and Alcohol-Based Surfactants

Surfactant group	LTE (mg/L)[a]
LAS, C_{10-13}, "European cut"	0.3
LAS, C_{10-14}	0.12
AS, C_{12-15}	0.1
AS, C_{16-18}	0.55
AES, C_{12-15} EO_{1-3}	0.15
AE, C_{12-15} EO_{2-6}	0.18
AE, C_{12-15} ($C_{av} < 14$) EO_{6-9}	0.24
AE, C_{12-15} ($C_{av} > 14$) EO_{6-9}	0.17
AE, C_{12-18} EO_9	0.2
AE, C_{16-18} EO_{2-6}	0.03

[a] Long-term effect (LTE) values based on the lowest existing measured NOEC data as reported in the Detergent Ingredients Database (DID) of the ecological criteria set up for the Eco-labeling of detergents.
Source: Refs. 141 and 142.

laboratory, larvae showed a 50% mortality at 3.8 mg/L of LAS after 96-h exposure. The swimming behavior was affected at 0.8 mg/L of LAS. During the mesocosm experiment, the larvae population showed a dramatic decrease in abundance within 2 d at 0.08 mg/L of LAS [146].

The toxicity of LAS on the common goby, *Pomatoschistus microps*, a fish that inhabits estuarines, fjords, and shallow near-shore waters, was investigated [147]. The acute toxicity, LC_{50}, after a 96-h static exposure, was established at 2.6 mg/L. Physiological responses were observed after prolonged exposure (28 d) to an LAS concentration of 0.05–1.0 mg/L. Growth and respiration were significantly affected at 0.1 mg/L of LAS.

Acute toxicity effects, after a 48-h static exposure, of various LAS with average molecular weight (MW) in the 232–262 range, two AS, C_{12-14} and C_{16-18}, and one AES, C_{12-14} $EO_{2.35}$, were measured comparatively for *Daphnia m.* [148]. The EC_{50} of the LAS blends varied between 9.5 and 2.8 mg/L going from the low-MW to the high-MW product. The EC_{50} of the two AS were 5.7 and 4.2 mg/L and that of AES was 7.1 mg/L. A chronic test (21 d) on *Daphnia m.* was also performed for the most common commercial LAS (MW = 242), yielding a NOEC of 1.25 mg/L. The toxicity effect is proportionally related to the hydrophobicity of the surfactant alkyl chain. The 2-phenyl LAS homologues with even alkyl chains in the C_8–C_{20} range were synthesized and measured for their acute toxicity, 48-h EC_{50}, to *Daphnia m.* [149]. The EC_{50} values were between 108 mg/L (2-phenyl C_8) and 0.7 mg/L (2-phenyl C_{20}). Correlations between the theoretical log K_{OW} (octanol/water partition coefficient), experimental CMC (critical micelle concentration), and EC_{50} data were established [149].

LAS homologues, C_{10}–C_{14}, and one biodegradation intermediate, SPC C_{11}, were studied to test their toxicity to embryos and larvae of seabream [150]. Lethality of 100% was observed in the range of 0.1–0.25 mg/L for the C_{13} and C_{14} LAS homologues, whereas lethal effects started at 5 mg/L for the C_{10}–C_{12} homologues. No mortality was observed for the SPC C_{11} up to a concentration of 10 mg/L. Other LAS homologues, C_{11} and C_{13}, were studied by means of flow cytometry analysis for their toxicity to four marine microalgae species [151]. The GTI_{50} (growth inhibition test) was found to be in the range of 1.38–13.37 mg/L for LAS C_{11} and 0.18–1.23 mg/ L for LAS C_{13}. The toxicity increased by a factor of 10, increasing the alkyl chain with two carbons.

As for the alcohol-based surfactants, a few important works have recently appeared. Three AE, C_{9-11} EO_6, C_{12-13} $EO_{6.5}$, and C_{14-15} EO_7, were tested for their chronic toxicity to *Daphnia m.* after a 21-d flow-through laboratory exposure, measuring the two sensitive parameters: survival and reproduction [152]. The survival NOECs of the three AE were 2.77, 1.75, and 0.79 mg/L, respectively, whereas the reproduction NOECs were 2.77, 0.77, and 0.79 mg/L. These survival NOECs were in good agreement with those for cladoceran densities (4.35, 1.99, and 0.33–0.55 mg/L) obtained in stream mesocosm studies (30-d exposure) of the same AE [153–155].

Nine commercial-grade AE with different average alkyl chain length and EO content were studied for their acute toxicity to *Pimephales p.* and *Daphnia m.* [156]. The blends investigated were the following: C_{9-11} EO_6, C_{9-11} EO_8, C_{11} EO_7, C_{11} EO_9, C_{12-13} $EO_{4.5-6}$, $C_{12.13}$ $EO_{4.5-6}$, C_{12-13} $EO_{6.5}$, C_{12-15} EO_{12}, and C_{14-15} EO_{13}. The response of each species to the AE was generally similar. The LC_{50} (96-h) values of the various AE for the fish were in the range of 0.96–8.5 mg/L and those of EC_{50} (48 h) for the crustaceous were between 0.46 and 6.7 mg/L. Quantitative structure–activity relationship (QSAR) models were developed from the data. Calculation leads to the following linear models:

$$\log LC_{50} = 4.35 - 0.34(ACL) + 0.05(EO) \qquad \text{for } \textit{Pimephales p.}$$

$$\log EC_{50} = 4.23 - 0.38(ACL) + 0.10(EO) \qquad \text{for } \textit{Daphnia m.}$$

where (ACL) is the average alkyl length, (EO) is the ethoxy number, and $L(E)C_{50}$ is expressed in μmol/L [156].

The toxicity and teratogenic effects of commercial surfactants of the AE and AES categories were investigated on an aquatic amphibian normally used as a model animal for both embryological and larval development [157]. The acute effects and malformations were measured, the latter via light and electron microscopy. The effects as documented by the 72-h LC_{50} results (4.59 mg/L for AE and 6.75 mg/L for AES) were in general higher for AE than for AES.

Eighteen different compounds synthesized in the laboratory, namely, six AS with single alkyl chains from C_{12} to C_{18}, 12 AES components with single alkyl chains from C_{12} to C_{15} and single EO units from 1 to 8, and one AES mixture of four AES pure structures, were studied for their acute and chronic toxicity, using *Ceriodaphnia dubia* via a novel flow-through method [158]. As for the acute LC_{50} after a 48-h exposure, the values were in the range of 0.16–5.55 mg/L for AS, of 0.78–167.3 mg/L for AES, and at 4.96 mg/L for the AES mixture. As for the chronic NOEC after a 7-d exposure, the values were in the range of < 0.06–0.88 mg/L for AS and 0.06–6.25 mg/L for AES and at 0.05 mg/L for the AES mixture. Acute LC_{50}, expressed in mole/L, increased with increased alkyl chain length (ACL) and decreased with increased EO units (EO), as shown in the following significantly determined linear regression for AES:

$$\text{Log } LC_{50} = (ACL)0.3216 + (EO)0.3049 - 0.8379$$

The 7-d chronic toxicity was found to be related to the structure by parabolic relationships. NOEC values expressed as mole/L can be represented by the following general quadratic equation:

$$NOEC = (ACL)^{2}0.128 - (ACL)3.767 + (EO)0.152 + 21.182$$

This equation gives the best mathematical fit. But even so, only about 70% of the variation in toxicity results could be explained. Other models were tried but yielded worse coefficients of determination [158].

A flow-through laboratory microcosm was developed to assess the effects of toxicants on natural algae [159]. AS and AES samples were tested and the observed NOECs for a 28-d exposure were 553 and 608 μg/L, respectively.

Responses of some invertebrates and fish to AS, C_{12}, and to an AES mixture, C_{14-15} $EO_{2.17}$, were investigated applying flow-through toxicity tests and using nonstandard species, which were, however, of regional ecological importance [160]. The invertebrate *Corbicula f.* was the most sensitive organism, for which NOEC values of 418 and 75 μg/L were found for AS and AES, respectively.

(b) Field Model Studies.   Also called mesocosm studies, they have been used effectively to assess the environmental effects of surfactants. A variety of field model

ecosystem studies, namely, studies that are not single-species laboratory tests but involve simultaneously several communities of organisms in simulated realistic conditions, have been performed for LAS and alcohol-based surfactants.

Early studies concluded that the lower limit of NOECs from the various field studies with both high and low taxonomic groups was realistically restricted between 0.25 mg/L and 0.50 mg/L for LAS [161,162]. In-depth investigation on the LAS commercial product was recently carried out in two different facilities located in the United Kingdom and the United States [163,164]. Most of the NOECs measured in these studies were found at >0.12 mg/L. Only two species, *Baetis* sp. and *Gammarus p.*, appeared to be rather sensitive, and apparently their NOECs reduced from 0.12 mg/L to 0.06 mg/L and to 0.03 mg/L, respectively, in the rifle zone, not in the pool of the stream, when the exposure time was prolonged from 28 to 56 days [164]. However, uncertainties were associated with these high toxicity data, observed only at the end of the extended 56-day study, which were not confirmed by a long-term (107 days) laboratory toxicity study performed specifically on *Gammarus p.* [165]. An integrated model stream ecosystem fate and effects study over a C_{12} LAS homologue, with a high content (35.7%) of its most hydrophobic and toxic 2-phenyl isomer, was completed in 2002 [166]. A NOEC of 0.27 mg/L was found for large and sensitive communities of organisms using a long-term exposure of 56 days. A critical review of all mesocosm studies available for LAS was also given, concluding that the NOEC of 0.27 mg/L was the most reasonable and substantive measure of no effects of this surfactant in the aquatic ecosystems [166].

Various commercial-type AEs were field tested. Three AE, namely, C_{9-11} EO_6 [167,168], C_{12-13} $EO_{6.5}$ [155], and C_{14-15} EO_7 [169], were tested in stream facilities for an exposure period of 30 days. Another AE, C_{12-15} EO_9 [170], was tested for a period of 56 days. The analysis of the surfactants was done by HPLC after an extraction and a cleanup procedure. HBr fission followed by GC/MS was used for AE, C_{12-15} EO_9. NOECs for these AE were found in the range of $0.73->11.24$ mg/L, of $<0.32-5.15$ mg/L, of $0.08->0.55$ mg/L, and of $0.07->0.74$ mg/L, respectively. The overall model ecosystem NOECs of these AE, derived from ecologically significant endpoints, were concluded, thus, to be 0.73, 0.28, 0.08, and 0.07 mg/L, respectively. Stream mesocosm and laboratory results for these AE were in general quite similar to NOECs in the same range. The toxicity effects, particularly of AE, C_{14-15} EO_7, on fish were reported in another study, conducted both in the laboratory and an outdoor stream mesocosm [171]. In the laboratory the LC_{50} for fish after a 96-h exposure was found to be 0.65–0.77 mg/L. In the stream mesocosm, where the fish were exposed for 30 d, the NOEC for the survival was 0.16 mg/L. A provisional study on C_{12-15} EO_7 [172] put forward NOECs in the range of 70–100 µg/L for a critical endpoint, namely, some macro-invertebrates. The mesocosm studies clearly demonstrated a relationship between AE alkyl chain length and biological response [173]. Results from laboratory chronic studies are similar to those derived from these mesocosm tests. Mesocosm and laboratory studies when conducted in similar water-quality conditions produce similar results [173].

Another AE, a noncommercial C_{12-15} EO_6, was field tested in 2000, assessing several communities over an exposure period of 2 months [174]. Invertebrate populations and communities were quite sensitive to AE, showing adverse effects in the range of 36–760 µg/L. These data indicated a model ecosystem NOEC of 13 µg/L for this AE. This AE appears more toxic than others. At the ecosystem level, a good relationship between calculated K_{OW} (octanol/water partitioning coefficient), as indicator of hydrophobicity, with the various AE NOEC conclusions present in the literature was shown to be still valid. The high toxicity of the last AE, $C_{12-15} EO_6$, is underpredicted according to the foregoing relationship and a clear reason for that was not understood, even if some experimental reasons may exist to justify this observation. In this context it is worth reporting a study where the potential effects of phase behavior of AEs were considered important to develop tests to assess their toxicity effects and to interpret the corresponding results [175]. That is particularly true for studies of some commercial AEs that, being a mixture of several surface-active species (over 100, depending on alkyl chain and EO unit distribution), may have a high content of poorly soluble components. For optimal dosing during the toxicity tests, the stock solutions of the AE should be single-phase micelle solutions, which may be difficult to achieve, particularly for the most hydrophobic AE systems. Turbidity and video microscopy measurements were used to suggest the right temperature regime for a variety of AE to get homogeneous micelle solutions [175].

In any case, despite the existence of some high-toxicity data in one AE surfactant group, the collective information on fate and effects measured both in the laboratory and in field systems was considered enough to state that AE in general pose a low risk to the environment [174].

AS and AES were studied in two distinct mesocosm experiments for their ecotoxicological effects on stream communities (macro-invertebrates) [176]. Assessment was conducted over an 8-week exposure in the summer–fall seasons. The AS was a pure C_{12} compound and the AES was a mixture of components with a 50:50 distribution of C_{14} and C_{15} alkyl chains and an average EO unit number of 2.17. The two chemicals were measured in the stream by specific techniques, GC-FID for AS [177] and LC-MS for AES [115]. Analytical losses due to rapid sulfate biodegradation were minimized using individual sample recovery factors [178]. A conservative ecotoxicological NOEC of 224 µg/L was found for AS. AES observations were highly consistent with those of AS: a mesocosm adverse NOEC of 251 µg/L was, in fact, found for AES. This finding is somewhat surprising because AS and AES in these parallel studies differ by an average alkyl chain length of 2.5 carbons. Structure–activity relationships would have predicted AES to be more toxic than AS in this case. It was postulated that confounding indirect effects are present and responsible for the lower AS NOEC, whose value, thus, can be considered conservative [176].

The overall picture of the toxicity results from field studies is summarized in Table 12.

6. Risk Assessment

Assessment of the aquatic effects of a surfactant requires the derivation of a PNEC (predicted no-effect concentration) of that product in the environment. Because toxicity depends on the chemical structure of the surfactant, the literature toxicity data for each group of surfactants have to be normalized to the typical structures present in the environment. That is what was done for the aquatic risk assessment of the major surfactants present on the market, namely, LAS, AE, AES, and soap, carried out in the Netherlands [117,133]. On the basis of the results from the monitoring studies in Dutch surface waters, the following average structures were assumed present in the aquatic compartment: LAS: $C_{11.6}$; AE: $C_{13.3} EO_{8.2}$; AES:

TABLE 12 Toxicity Results from Field Studies with LAS and Alcohol-Based Surfactants

Test substance	Lowest observed NOEC (µg/L)	Refs.
LAS, C_{10-13}	250–500	161, 162
LAS, C_{10-13}	120	163
LAS, C_{10-13}	$(30)^a$	164
LAS, C_{12}	270	166
AE, C_{9-11} EO_6	730	167, 168
AE, C_{12-13} $EO_{6.5}$	280	155
AE, C_{14-15} EO_7	80	169
AE, C_{14-15} EO_7	160	171
AE, C_{12-15} EO_7	70–100	172
AE, C_{12-15} EO_9	70	170
AE, C_{12-15} EO_6	$(13)^a$	174
AES, C_{14-15} $EO_{2.17}$	251	176
AS, C_{12}	224	176

^{a} Experimental uncertainties; see text.

$C_{12.5}$ $EO_{3.4}$. Short-term and long-term toxicities, refined by statistical extrapolation [179], were normalized to the average structures using accepted QSARs [140] and then compared with NOECs from field studies available at the time. PNECs were derived that are the 95th percentile of the distribution of all NOECs at 50% confidence level. A final refinement of the PNECs was then made on the basis of expert judgment, when necessary. These PNECs were compared with PECs in surface waters, derived from surfactant concentration results obtained by monitoring a set of representative STPs [90]. PECs were 50–100 times lower than PNECs. Only soap had a PEC about equal to PNEC. That was due to a lack of an effective toxicity database for soap. Since the substance is considered no more toxic than other surfactants, there is no reason for environmental concern about it. So the assessment carried out in the Netherlands concluded that the risk for the aquatic compartment from the use of LAS, AE, and AES as well as of soap is low, of course, if a regime of well- and proper STP functioning exists. PECs, PNECs, and their corresponding ratios, considering also that instream removals might occur in rivers, derived from the Dutch risk assessment of the mentioned surfactants, are summarized in Table 13. More recently LAS appeared on a list of high-production-volume chemicals that the U.S. EPA was interested in sponsoring in the OECD SIDS program. An Industry coalition, lead by CLER (Council for LAB/LAS Environmental Research), has prepared a SIDS dossier that was submitted to the EPA and is at present under review [180].

A separate environmental risk assessment for AS has not been carried out so far. Its toxicity is comparable to that of other anionic surfactants, such as LAS [137,138]. So there is no immediate concern about environmental risk, considering also its relatively lower consumption volume and its positive environmental fate characteristics. As for AS, an improved new structure presenting a small number of methyl branches in the alkyl chain, with high solubility and water hardness

TABLE 13   Aquatic Risk Assessment Conclusions for LAS, AE, and AES in The Netherlands

Surfactant	PNEC (µg/L)	PEC^a (µg/L)	PEC^a/PNEC	PEC^b (µg/L)	PEC^b/PNEC
LAS, $C_{11.6}$	250	9.2	0.04	3.7	0.02
AE, $C_{13.3} EO_{8.2}$	110	1.3	<0.01	0.5	<0.01
AES, $C_{12.5} EO_{3.4}$	400	2.9	<0.01	1.2	<0.01

[a] Calculated with no instream removal.
[b] Calculated with a first-order instream removal of $t_{0.5} = 24$ h.
Source: Ref. 133.

tolerance, was recently developed at the research level and environmentally characterized [181]. Several model mixtures with average carbon numbers between $C_{14.5}$ and C_{18} were tested for their biodegradability and in acute toxicity studies with fish, invertebrates, and algae. One model mixture was also tested in chronic studies. Fish species (LC50 = 0.25–0.51 mg/L) were in general more sensitive than invertebrates (EC50 = 1.2–1.3 mg/L) and algae (EC50 = 1.8–23 mg/L). Chronic toxicity was evaluated for the C_{16-17} model mixture. As expected, fish, represented by the fathead minnow (28-d NOEC = 0.04 mg/L), was more sensitive to this model than *Daphnia m.* (21-d NOEC = 0.12 mg/L) or *Selenastrum c.* (96-h NOEC = 1.8 mg/L). All models tested were readily biodegradable, with 80% biodegradation at the end of the 10-day window [181].

In September 1999 the European industry set up a project called HERA (Human and Environmental Risk Assessment), with the aims to provide a common risk assessment framework for household detergent and cleaning products and to deliver evaluated safety information on the detergent ingredients in a speedy, effective, and transparent way (www.heraproject.com/RiskAssessment.cfm). A first task for HERA was to propose a refinement of the TGD methodology [182] to estimate the amount of a chemical input to an STP at the regional and local levels [183]. On the basis of actual detergent consumption and of monitoring data through Europe it was demonstrated that, in the worst scenario, the chemical releases in the standard EU region have to be considered 7% of the total EU tonnage, and not 10% as assumed by the TGD. In addition, at the local level, the HERA conclusion is that, even for the worst possible case, any STP always receives less than 1.5 times the regional average load, instead of 4.0 times as assumed presently by the TGD. These experimentally based release factors will be used by HERA as new default values in the TGD to evaluate the targeted environmental risk assessment of several detergent ingredients (LAS and AS are included) in any important regional and local areas in Europe [183].

7. Bioconcentration

Information on bioconcentration is essential to assess the environmental risk of surfactants. Hydrophobicity is the driving force for bioconcentration. The longer the alkyl chain of the surfactant, the higher are the hydrophobicity and the bioconcentration factor, BCF, expressed in L/kg and defined as the aquatic organism/water concentration ratio. Log K_{OW} (octanol/water partitioning coefficient) is used in general as a measure of the hydrophobicity of a chemical. However, log K_{OW} cannot be experimentally measured for surfactants because of their surface-active properties, but only approximately calculated [184]. When calculations are applied to LAS, for example, log K_{OW} predicts a BCF of 100–1000 in water, with alkyl chain length being an important factor [185].

The bioconcentration of surfactants has not yet received much attention. Most of the early experimental studies, as shown by a critical review [186], are not appropriate because the analytical methods usually employed to measure the surfactant concentration were based on radioanalysis. That doesn't allow us to distinguish between the parent compound and its biotransformed products. Hence, the measured concentrations overestimate the parent surfactant concentration present in the aquatic organism body; consequently, the aquatic organism/water concentration ratio, CR, consistently overestimates the true bioconcentration [186]. For example, in the case of the commercial LAS $C_{11.6}$, a CR of 269 and 480 L/kg were found for fish and *Daphnia m.* [187]. In the case of AE in the range of C_{12-14} EO_{4-8}, CRs from 222 to 799 were found for a few aquatic organisms [188,189]. These CRs consistently overestimated the true BCF. They, however, demonstrated that the bioconcentration of surfactants increased with increasing length of the alkyl chain, depending on the composition of the aqueous phase, and that surfactant biotransformation occurred in the aquatic organism [186].

This lack of a reliable database on the BCF behavior of surfactants stimulated an in-depth research program, commissioned by industry and carried out at the University of Utrecht (NL), which focused the attention on the two most important commercial surfactant categories, LAS and AE [190]. These two surfactants were studied by employing a flow-through test system, in line with the OECD guidelines, using *Pimephales p.* as test fish. Single homologues and isomers representative of the commercial LAS and a set of AE components with distinct alkyl chains and EO units were synthesized and then tested, determining the uptake and elimination rates. Specific analysis of the individual surfactant components in the water phase and in the fish body showed that these chemicals very quickly reach a steady-state concentration in the fish body, namely, an equilibrium between uptake rate and elimination rate, ca. 3 d for LAS and ca. 1 d for AE. BCF data for the tested LAS standards ranged between 2 and 990 L/kg [191] and allowed the determination of the BCF of the various homologues and of the internal and external isomers present in an LAS mixture. Positive relationships between BCF and uptake rate constants were also observed with estimated log K_{OW}. The BCF potentials of two LAS mixtures, that of the commercial product and that typically present in the water phase of a river [121], were calculated using the average of all BCF determinations carried out on the LAS standards and showed to be 66 and 16 L/kg, respectively (190. p. 175). This difference is due to the LAS composition in surface waters, which has on average ca. 1 C atom less than that of the commercial product, due to alkyl chain switch to shorter homologues induced by the higher biodegradation rate of the longer homologues. In order to deepen the understanding of the LAS bioconcentration behavior, the BCF of selected external LAS components (2-phenyl isomers) was measured using the rainbow trout as fish and also taking into consideration the internal organs of the animal [192]. The BCF with trout increased with increasing length of the alkyl chain and was significantly lower than that measured with *Pimephales p.* [191]. The internal LAS redistribution in the fish body appeared to be controlled by perfusion, and uptake was demonstrated to occur via the gills rather than the skin. In general BCF was apparently not related to the fish lipid content, and its hydrophobicity dependence was affected by water hardness [192].

An isomer mixture of an LAS C_{12} homologue was used as test material in an experimental stream mesocosm to study isomer partitioning in biological matrices [193]. Applying new analytical methodologies based on HPLC/MS, it was observed that external 2,3-phenyl LAS isomers were preferentially bioconcentrated by biological organisms, fish, clams, and snails. Incidentally, no enrichment of specific LAS isomers was observed in sediments and suspended soils.

The BCF behavior of a single ^{14}C-labeled AE product, ^{14}C-C_{13} EO_8 [194], and that of a series of standard AE compounds [190] were measured. The BCF of the parent AE from the radiolabeled standard was found to be 32 L/kg, whereas the fish-to-water steady-state concentration ratio, based on radioactivity measures, was 224 L/kg. That testifies to the importance of biotransformation as a process contributing to the reduction of the bioconcentration potential of surfactants and demonstrates that the radioactivity of biotransformed products is partly incorporated in the fish body. BCF data for the tested AE standards ranged between < 5 and 390 L/kg. BCF as well as uptake rate constants increased with increasing length of the alkyl chain and decreasing length of the EO units. In contrast a decrease of the elimination rate constants with increasing alkyl chain length and decreasing EO units was found [190]. Similar to what was performed for LAS, the BCF potential of a typical commercial AE mixture with average composition of $C_{13.3}$ $EO_{9.1}$, fully characterized by specific analysis [114], was calculated, using the average of all BCF determinations carried out on the AE standards, and a conservative estimate of 142 L/kg was obtained (190, p. 175). The bioconcentration potential of this surfactant family can be predicted to remain largely unaltered in surface waters after STP treatment, because its STP removal rate is uniform over all components, as shown by the concentration distribution in STP effluents, which is rather similar to that of the commercial product (see Section II.A.3(b)).

The BCF for both LAS and AE was found, as already said, to be dependent on hydrophobicity. In addition, biotransformation contributes significantly to the elimination of surfactants in fish. However, both surfactants differ in their respective relationships. Extrapolation, thus, from LAS and AE data to other classes of surfactants, such as AS and AES, is regarded as difficult.

On the basis of these BCF studies and derived oral PEC data for surfactants, a comprehensive sensitivity analysis was developed to address also the secondary poisoning effects to predatory species and man [195]. Even with a number of conservative assumptions relating to bioconcentration, oral PEC derivation, and exposure potential, for the major commercial surfactants, LAS and AE as well as AS and AES, the BCF potential of concern for a secondary poisoning would be at minimum 10,000 L/kg, one to two orders of magnitude higher than that found experimentally for LAS and AE [195,196]. Hence, the bioconcentration behavior of the discussed commercial surfactants is believed not to represent an issue of concern either to human health or to aquatic safety.

B. Terrestrial Compartment

Waste streams, at least in the most environmentally developed countries, are treated by STPs, which reduce significantly the load of surfactants to the receiving surface waters, as described in the previous paragraphs. Surfactants can also end up, however, in the sewage sludge produced by STPs, as shown in Figure 4. They were found, in fact, to occur in sludge at relatively high concentrations, often at levels exceeding 1 g/kg dw (dry weight). If sludge is then used in agriculture, surfactants are introduced onto soil. Surfactants can also enter the terrestrial environment, even if to a minor extent, by other routes, such as leakage of sewer lines, presence of septic tanks, use of sewage effluents as irrigation water, processing of fertilizers, and distribution of pesticides in agriculture, where surfactants serve as emulsifying and dispersing agents for remediation of contaminated soil. Compared to the aquatic environment (see Section II.A), the terrestrial environment has received considerably less attention [197–199]. The only surfactant for which a relevant database exists is LAS. The terrestrial environmental data on alcohol-based surfactants, on the contrary, are scarce, and what little information there is related mainly to AE.

1. Surfactants in Sludge

Sorption is the most important mechanism that affects the fate and effects of surfactants in the terrestrial environment. Through sorption to sludge, surfactants enter soil. Partition coefficients, K_d, expressed in L/kg as the ratio of adsorbed fraction to the dissolved fraction, vary quite a bit for surfactants in soil and sediments, usually between a few units to a few thousand [185]. Overall, K_d seems to be higher for sediments than for soil. This could be due to a more organic content in sediments. Sorption is a complex process and depends on soil composition and the hydrophobicity of the molecule. In the case of sludge, for example, K_d in the 600–5000 range was reported for LAS, with homologues with longer alkyl chains having higher adsorptive character [185]. Sorption of LAS was shown to correlate significantly with the organic content of soils and also to be strongly affected by pH, decreasing with increasing pH [200]. Sorption in general can influence the residence time of surfactants in soil and, consequently, their biodegradation as well as the expression of their toxicity.

Table 14 summarizes the concentrations of LAS and alcohol-based surfactants in sludge as reported in the literature [198,199,201,202]. The concentration of LAS is quite high due to its sorption to primary STP sludge, precipitation of its Ca salts, and the lack of biodegradation in an anaerobic digester. Apart from LAS, only AE was measured in sludge. Its concentration depends on its high adsorptive character; it is lowered, however, by its potential to biodegrade anaerobically in the STP digesters.

As for LAS, several data are available, obtained in different countries. They are related to two different categories of sludge, aerobic and anaerobic sludge, depending on the stabilization used by the STP. Aerobic sludge has an LAS concentration of no more than 500 mg/kg, whereas anaerobic sludge was found in the range of <1000–30,000 mg/kg, typically between 1000 and 10,000 mg/kg, depending on the surfactant content in the sewage, the STP operating conditions, and the water hardness. An overview of LAS in Danish sludge was recently presented at a workshop in Copenhagen [203]. Six hundred samples of sludge were analyzed by accredited laboratories, with every STP in the country being examined at least once a year [204]. Only 64% of the samples contained LAS. LAS concentration ranged between <50 mg/kg to more than 7000 mg/kg. More than 85% of the samples were below the present cutoff value of 1300 mg/kg, set up for the sewage sludge to be used in agriculture by Danish Statutory Order [205].

TABLE 14 Average Measured Concentrations of LAS and AE in STP Sewage Sludge

Surfactant	Country/sludge	Concentration (mg/kg dw)	Refs.
LAS	CH anaerobic	2900–11900	206
LAS	CH anaerobic	5500	207
LAS	D aerobic	182–432	18
LAS	D anaerobic	1330–9930	18
LAS	E aerobic	100–500	95
LAS	E anaerobic	7000–30,200	95
LAS	E anaerobic	12,100–17,800	208
LAS	I anaerobic	11,500–14,000	89
LAS	I anaerobic	6000 ± 1200	102
LAS	NL primary	3400–5930	100
LAS	NL aerobic	205	100
LAS	EU aerobic/anaerobic	210–9400	103
LAS	DK aerobic	11–<500	209
LAS	DK anaerobic	1000–16,100	209
LAS	DK anaerobic	3700–5100	227
LAS	DK aerobic/anaerobic	<50–7000	204
LAS	USA anaerobic	4660 ± 1540	210
LAS	USA aerobic	152 ± 120	97
LAS	USA anaerobic	10,460 ± 5170	97
AE	EU anaerobic	<700	211, 212
AE	DK anaerobic	150–200	213

As for AE, monitoring exercises for different anaerobic sludges were carried out in different countries [211,212]. AE never exceeded 700 mg/kg in a German ring test [211]. In a European study involving several STPs in five countries, AE content at the inlet of the digesters was found to be in the range of 550–2950 mg/kg, and at the outlet it never exceeded 470 mg/kg, with an average removal in the anaerobic digesters of 82% [212]. Other data for AE were also obtained in some Danish anaerobic sludge and were found to be in the range of 150–200 mg/kg [213]. As for other major surfactants, measurements in sludge exist only for SAS (secondary alkane sulfonates), APE (alkylphenol ethoxylates), and soap [198,199,201,202]. By way of comparison, the highest concentration of one surfactant ever found in sludge is that of soap, 18,800–51,900 mg/kg [214].

Sludge, prior to its use in agriculture, is normally stored at the STP facility for several months, from 6 to 12 months. The biodegradation of surfactants as well as of any chemical substance in sludge during storage in big compact piles can be difficult, if aerobic conditions in the mass are not ensured and if there is a lack of microbial activity. This is the conclusion from a sludge stability study where two sludges kept in containers with LAS content of approximately 2000 and 4000 mg/kg were kept under observation for 6 months [215]. The levels of LAS as well as of any other chemicals, for example, AE and soap, which were measured at levels of 150–200 mg/ kg and 660–670 mg/kg [213], remained practically stable for the entire observation period. If the sludge, on the contrary, is piled in relatively small field stacks or laid down on the soil, biodegradation of LAS in sludge could be quick, with half-lives times ranging from <10 to 50 days, as actually observed [201,216]. Recently, in fact, it was found that LAS concentration in the bulk of aged dry sludge could drop by 74% as compared to a freshly produced sludge [218].

LAS concentration in sludge can drastically be reduced by an aerobic post-treatment of the sludge and by composting. Aerobic stabilization experiments, performed using wastewater sludge from an anaerobic digester in sequencing batch reactors, indicated that LAS was reduced by 50% and 90% in 3 and 9 hours, respectively [215]. Recent composting studies in both the laboratory [215] and the field [219] also demonstrated that not only can LAS be totally removed, but other contaminants of sludge can also be effectively reduced [220].

The availability of LAS in sludge seems to be affected by water hardness, as documented by sorption/desorption experiments carried out on a standard treated sludge [221].

2. Surfactants in Sludge-Amended Soil

With regard to soil, only LAS measurements are known. There has been a growing database about LAS from laboratory and field studies in recent years. These investigations were usually not performed in relevant agricultural conditions. Consequently, after application of sludge the initial LAS concentration in soil was often quite high and corresponded to an unrealistic sludge application rate. The observed disappearance of LAS is always relatively quick, particularly in the initial period after sludge application and has a kinetic rate, with $t_{0.5}$, half-life, usually in the range of 10–30 days, with some exceptions above and below the range. The measured half-lives in studies with ^{14}C material, dealing, thus, with ultimate biodegradation, were in the range of 13–26 days [222]. Experimental LAS concentration data measured in soil with corresponding biodegradation rates are summarized in Table 15 [198,199, 201,202]. To complement the existing literature data, other recent laboratory and field data were produced in Denmark. Greenhouse experiments with sludge application rates in the range of 0.4–90 t/ha were carried out under the Danish Environmental Research Program, to study the uptake and biodegradation of LAS in a soil–sludge–plant system [227]. In all experiments, no uptake of LAS from plants was

TABLE 15 Average Measured Concentrations of LAS in Sludge-Amended Soil

Country	Concentration (mg/kg dw)		Biodegradation rate ($t_{0.5}$, days)	Refs.
	Initial	Typical[a]		
CH	45	5	9, 87	223
D	16, 27	—	26, 13	222
E	16, 53	0.3	26, 33	95
E	22.4	0.7, 3.1	—	208
UK	max. 66	0–20	7–22	224
UK	max. 145	0–8	3–30	225
USA	max. 250	1–7	18–26	226
DK	max. 234	<0.2–86	12–44	218, 227
DK[b]	1	0.1	25–40	215

[a] Typical values after a test period.
[b] Model simulation for the typical agricultural situation in Denmark.

observed (detection limit: 0.5 mg/kg). Degradation of LAS was stimulated by the presence of plant growth. The initial measured LAS concentration in the soil, at the highest sludge application rate of 90 t/ha, with LAS content in sludge of 3700–5100 mg/kg, were 227–334 mg/kg. With a sludge application rate at or above 6 t/ha, LAS was still present in the soil at harvest time (19–85 days, depending on the plants), with a concentration that was 4–36% of the initial one. With a sludge application rate below 6 t/ha, close to those used in agriculture, LAS concentration in soil at harvest time was below the detection limit, 0.2 mg/kg [227]. In agreement with these findings, field experiments with a sludge application rate of 4.4 t/ha showed that, after a 4-month growth of oat, LAS concentration was below the detection limit of 0.2 mg/kg [227]. If sludge application is carried out according to the prevailing agricultural rules, LAS concentration in all soils was found to be under 1 mg/kg [218]. In all cases the estimated half-life of LAS in sludge-amended soil was in the range of 12–44 days [227]. In the context of a ring test among five different laboratories to evaluate the methodology to measure LAS in environmental samples of sludge and soil, standard samples of soils and sludge were carefully analyzed. Average LAS concentrations of 3.7 mg/kg and of 4108 mg/kg were found for a standard sludge-amended soil and for a sludge, respectively [217]. Despite the use of different analytical procedures, repeatability and reproducibility among laboratories on the LAS measurements were within 20%. A model simulation study indicated that 98–99% of LAS present in sludge-amended soil degraded within one year after the sludge application [215]. LAS is not expected to accumulate in soil. Model simulations indicated that typical LAS concentration in soil was in the range of 0.1–1 mg/kg, with a higher concentration immediately after sludge application (sludge application rate: 2 t/ha/year; LAS in sludge: 1000 mg/kg). In the worst case of sludge with LAS content of 16,100 mg/kg, simulations indicated that LAS was in the range of 1–10 mg/kg, with higher concentrations immediately after sludge application [215].

In Germany, the behavior of anionic surfactants as LAS in sandy soils with low amounts of organic matter and amended with sludge was examined under real field conditions, performing field and lysimeter studies [228]. LAS was mobile in these soils. It could be detected down to a depth of 30–40 cm. No leaching of LAS, however, was pointed out. LAS was investigated in all depths and was shown to biodegrade rapidly. So only biodegradation is responsible for the LAS disappearance. The half-lives were in the range of 3–7 days [228]. The homologue profile of LAS in percolating water was also described, showing that homologues with longer alkyl chains were retained more strongly than those with shorter alkyl chains [223,224].

Surfactants at levels present in sludge do not increase the risk to the environment of other xenobiotic compounds possibly present in sludge or soil [229]. Their concentration in soil, at any time after sludge application, is always too low to contribute significantly to the mobilization of possible hydrophobic organic compounds [230].

3. Prediction of Surfactant Concentration in Soil

The EU sludge Directive 86/278/EEC, "Sludge to land," is going to be revised in the near future. A single application of sludge per year of 3 t/ha/year, equivalent to an application rate of 0.3 kg/m^2/y, is likely to be adopted. According to the EU TGD

[182], other exposure data for the soil are: soil depth of 0.2 m and soil density of 1500. An updated EU TGD was to have appeared in 2003 (http://ecb.jrc.it/existing-chemicals/): the principles for estimating exposure concentrations in soil as described later are still valid. Thus initial concentration of a chemical just after sludge application, C_{soil} (0), in the beginning of the growing season can be calculated by the following expression:

$$C_{soil} \text{ (0) (mg/kg)} = (C_{sludge} \times APP_{sludge})/(DEPTH_{soil} \times RHO_{soil})$$

where C_{sludge} is the concentration in dry sludge, mg/kg, $APPL_{sludge}$ is the dry sludge application rate, kg/m^2/y, $DEPTH_{soil}$ is the mixing depth of soil, m, and RHO_{soil} is the bulk density of soil, kg/m^3.

Following the TGD, for exposure of the endpoints, the concentration of a chemical, namely, the local PEC in soil, needs to be averaged over a certain time period, t, which is 30 days for the terrestrial ecosystem, PEC_{soil}, and 180 days for crops for human consumption, $PEC_{agric.soil}$. For surfactants it is reasonable to assume that in topsoil, because of aerobic conditions, they degrade at a rate with a half-life equivalent to that observed for LAS, namely, $t_{0.5} = 10$–30 days. The average local concentration of surfactants in soil over t days is, then, derived by applying the following expression:

$$\text{PEC in soil (mg/kg)} = 1/kt \times C_{soil} \text{ (0)} \times (1 - e^{-kt})$$

where k is the first-order rate constant for biodegradation on the topsoil, d^{-1}, ($k = 0.693/t_{0.5}$), t is the average time, (30 days for soil, 180 days for agricultural soil), and C_{soil} (0) is the initial concentration of the surfactant after sludge application, mg/kg.

On the basis of these parameters, local PECs of LAS and of AE in soil are calculated, assuming a biodegradation rate in soil equivalent to $t_{0.5} = 30$ days. As an example, data are reported next for some typical surfactant concentrations in sludge, as experimentally found and reported in Table 14.

Surfactant	Concentration in sludge (mg/kg)	C_{soil} (0) (mg/kg)	PEC_{soil} (mg/kg)	$PEC_{agric.soil}$ (mg/kg)
LAS	500	0.5	0.36	0.12
LAS	1,000	1.0	0.72	0.24
LAS	10,000	10.0	7.2	2.4
AE	700	0.7	0.5	0.16

These calculated PEC values are conservative. They were calculated considering the worst-case situations of sludge application rate and biodegradation in soil. In addition, the possibility of chemical degradation in sludge during storage was disregarded and delay periods after the application of sludge before PNEC operates were not taken into consideration.

The approach in Denmark of calculating local PEC_{soil} values according to the national sludge legislation [205] sticks to the C_{soil} (0), namely, to the initial concentration of the surfactant after sludge application, without taking into account biodegradation in sludge and soil and delay periods before the PNEC is to be applied.

4. Toxicity in Soil

Toxicity test methods for the species living in the terrestrial compartments are not as well developed as those for the aquatic compartment. In addition toxicity can be influenced by the different routes of exposure, by the chemical characteristics of the surfactant, and by the type of organism involved [230]. Few reliable data were produced and reported in the literature in the recent past [197,231]. However, chronic data in adequate numbers became, available only recently and only for LAS. A large number of new long-term toxicity data, in fact, were produced in both the laboratory and the field on the effects of LAS in soil, namely, toxicities for soil fauna, soil plants, soil microorganisms, and microbial soil processes [199,232–234]. All data were presented and discussed in an international workshop in Denmark [203] and in a world surfactant Congress [235,236]. These new data, together with those previously published [231], for a total of 23 chronic data points, were assembled and used by applying a statistical extrapolation method [237] to derive the various PNEC values for the different taxonomic groups:

Species	PNEC (mg/kg)	EC_{10} (mg/kg)
Soil fauna	4.6	6
Plants	5.3	9
Animals and plants	4.6	6
Microorganisms	—	<8

For this assessment, EC_{10} values were generally preferred for calculations to NOEC values, and the calculations were performed on data from experiments with Na LAS in soil without sludge amendment. The final PNEC of LAS was set at 4.6 mg/kg. The salt speciation of LAS, the soil type, the use of sludge, and the recovery rate of the ecosystem were also included in the evaluation [232,238]. In the final assessment, all tested microbial processes or functions [234,239,240] would be protected by this PNEC. This is the lowest definite figure combining all effect data on LAS available in the literature and produced in the latest studies, and it represents protection of 95% of species with 50% confidence [232,235]. Field observations are also available [232,235].and were discussed at the Danish workshop [203]. On their basis, LAS-containing sludge didn't produce any adverse effect on soil ecosystem after sewage sludge treatment, even at an LAS concentration that had shown significant effects in the laboratory. It could be concluded that an LAS dosage of 10–50 kg/ha or an average soil concentration of 5–15 mg/kg is unlikely to cause long-term adverse effects on the agricultural soil ecosystem [232,235]. Other conclusions were: The bioconcentration in plants, if ever, is at low level, as substantiated by field studies [226], and LAS toxicity depends only on one nonspecific mechanism, namely, perturbation of membranes, which implies that only small differences in toxicity between species are expected [232,235].

Other field studies were conducted to evaluate the possible adverse effects of LAS on the soil ecosystem [216]. Strings of aerobic sludge with a low LAS content (69 mg/kg), spiked with different and large amounts of LAS (0, 10, and 50 g/kg), were applied to soil. The field was sowed with oat and then analyzed on a weekly basis for up to 5 months for the microbial parameters and the LAS content. The population size of *Enchytreids*, important and sensitive soil invertebrates, did not recover to normal size in the LAS-spiked sludges during the 5-month study period. An EC_{50} of ca. 4 g/kg LAS for the *Enchytreids* in sludge strings was deduced. No toxic effects were reported for the adjoining soil compartment. A potential ammonium oxidation (PAO) inhibition was found to be present in the LAS-spiked sludge fraction during the first 3–4 weeks. Activity, however, returned to normal values after that period. In the adjoining soil fractions this activity was only partially inhibited. However, due to the high bioavailability of LAS spiked in sludge, higher than that of LAS usually present in sludge, and the high LAS doses applied, these results overestimate the risk associated with amendment of LAS-containing sludge to agricultural soil. The conclusion seems that LAS applied via sludge doesn't constitute a serious threat to organisms residing in soil. The findings of this work [216], however, deserve further attention at the much lower LAS concentrations regularly encountered in sludge.

5. Risk Assessment

The risk characterization of a chemical in soil implies a simple comparison of the local predicted environmental concentration, PEC, with the predicted no-effect concentration, PNEC, normally expressed as the risk ratio, PEC/PNEC, at a given time, for an averaging time of at least 30 days. The typical local PEC in soil for some concentrations of LAS and AE in sludge were calculated, according to TGD [182], and reported in Section II.B.3. As for LAS, the conclusive PNEC that protects the terrestrial organisms is 4.6 mg/kg dw [232,235], as described in Section II.B.4. Calculations to achieve the safe LAS PEC in soil of 4.6 mg/kg are shown shortly. These calculations were done by considering the mixing of LAS in 3 tons of sludge in 1 hectare of soil, 0.2 m deep and at a density of 1500, applying an averaging time 30 and 180 days with a conservative half-life of LAS in soil of 30 days, as described in Section II.B.3:

C_{sludge} (mg/kg)	C_{soil} (0) (mg/kg)	PEC_{soil} (mg/kg)	$PEC_{agric.soil}$ (mg/kg)
6,400	6.4	4.6	—
19,400	19.4	—	4.6
20,800	20.8	15	—

As shown, to reach a PEC_{soil} and a $PEC_{agric.soil}$ of 4.6 mg/kg it is necessary to have LAS content in sludge of 6,400 and 19,400 mg/kg in sludge, respectively. Calculations are also reported for an LAS PEC_{soil} of 15 mg/kg, a concentration that soil can reach through sludge amendment without showing any chronic effects in its ecosystem, as pointed out by field studies [232,235]. In this case an LAS content in sludge of 20,800 mg/kg would still appear to represent a safe concentration. A comparison of the measured concentrations in sludge shown in Table 14 would indicate that only in the extreme case of some sludge in Spain would such a limit be exceeded. These conclusions are in line with what is presented in the recent literature [202,236].

The risk assessment conclusions on LAS at the workshop in Denmark [203] for the specific Danish situation are summarized next. The initial PEC in soil, just after sludge application, was typically equal to 0.7 mg/kg and in the worst case equal to 6.9 mg/kg. As for PNEC, 5 mg/kg LAS in soil was defined as the scientific value, substantiated by a great deal of data, whereas a value of 1 mg/kg would be regarded as conservative and a value of 10 mg/kg would represent the real world and imply an acceptance of some risk.

PEC (mg/kg)		PNEC (mg/kg)		
Typical	Worst case	Conservative	Scientific	Real world
0.7	6.9	1	5	10

A risk characterization of alcohol-based surfactants has not been carried out so far, because in the literature terrestrial information on these products is still poor. Only for AE were some concentration data in sludge derived, as shown in Table 14. The low conservative PEC$_{soil}$ that was calculated for this surfactant (Section II.B.3) is unlikely to cause any effect in the soil ecosystem.

6. Surfactants in Sediments

Surfactant residues present in STP effluents (1–2% of the total surfactant load) enter river waters both dissolved in water and strongly adsorbed on suspended soils [241]. These suspended solids are the major constituents that form sediments near the STP outfall, downstream along a river and then at the river mouth when entering the sea. In rivers, the structure of the sediment will be affected by the velocity, turbulence, and volume of the river flow, which in turn causes the transport of sediment particles either by suspension or by sliding and rolling. Marine sediments in coastal areas are essentially similar to freshwater sediments. Surfactants adsorbed on sediment surface layers are easily removed because they biodegrade rapidly at more or less the same rate as the product dissolved in river water. That was demonstrated for LAS in laboratory experiments and confirmed by field-monitoring studies [242]. The half-lives in aerobic sediments were in the range of <1–5.9 days. River sediments are aerobic only in the upper layers, but they can become anaerobic a few centimeters below the surface. In estuarine and coastal areas, sediments usually become anaerobic as shallow as a few millimeters below the surface. To determine the biodegradation of surfactants on natural sediments in a suitable way, an in situ experimental approach, coupled with mathematical simulations, has recently been presented [243]. Sediments, poorly contaminated by LAS (<1 mg/kg), were collected from a pond having relatively high salinity (ca. 10 g/L). They were then artificially spiked by 1-(p-sulfophenyl) dodecane up to a concentration of 60 mg/kg (wet basis) and distributed in a series of specific modules that were finally implanted on the bottom of the pond (ca. 2 m deep), close to the location of the original sediment sampling. The biodegradation of the standard surfactant was followed for a period of 52 days, analyzing its concentration decrease in two different layers, one near the surface (0–4 cm) and the other in the 4 to 14-cm layer. The half-lives of the surfactants were about 49 d in the top layer and about 173 d in the bottom sediment layer. These biodegradation rates, obtained in the in situ experimental approach, were slower than those noted for laboratory tests using the same pond water ($t_{0.5}$ = 1.5 d) [41] and for the same sediment kept under aerobic conditions ($t_{0.5}$ = 2.9 d) [244]. The in situ experiment showed that biodegradation could occur only in layers or superficial sediments (up to ca. 10 cm deep), which could easily be reworked and oxygenated by hydrodynamism and bioturbation processes [243].

(a) Monitoring. Few papers are present in the literature dealing with the specific monitoring of surfactants in sediments. Toxicity data on sediment organisms are also rare. Most of the information available today refers to LAS and APE [245]. Some data on AE in river sediments have also become available [246]. Typical average measured concentrations in sediments are summarized in Table 16.

As for freshwater sediments, European monitoring studies were carried out in five countries in rivers upstream and downstream of activated-sludge STPs. LAS concentrations were in the range of 0.2–5.3 mg/kg [103]. In Germany, dated sediment layers (1939–1991) of the Lippe River were studied for their LAS concentration [247]. These concentrations ranged from 0 to 3.3 mg/kg. In Switzerland LAS concentrations were determined by GC-MS [248] to be between 0.2 and 0.7 mg/kg in lake sediment layers deposited between 1980 and 1994 [249]. After a sharp LAS concentration peak in a layer deposited around 1970, a continual decrease in LAS has been noted, even though LAS usage has increased almost twofold in Switzerland since 1970. That corresponded to a period when STPs were introduced, which drastically reduced the

TABLE 16 Average Measured Concentrations of LAS and AE in Sediments

Surfactant	Sediment: country	Concentration (mg/kg)	Refs.
	Freshwater:		
LAS	EU	0.2–5.3	103
LAS	D	0–3.4	247
LAS	CH	0.2–0.7	248, 249
LAS	I	0.3–4.7	246
LAS	DK	0	250
LAS	USA	5–11, 174[a]	251
LAS	USA	0.3–3.8	97, 251
LAS	USA	<1, 20[a]	121
AE	I	0.1–1.0	246
	Marine:		
LAS	J, I	5–17	252–254
LAS	E	1–2. >20[a]	255
LAS	E	0.1	208
LAS	F	0.5–6.0	257
LAS	DK	0.8, 2.9, 22	256

[a] Downstream of a polluted discharge.

loading of LAS onto sediments [249]. In Denmark a monitoring survey from 1990 to 1998 for 57 substances on various sediments did not detect LAS in any sediment, despite its consistent presence both in STP influents and sludge (130–6500 mg/kg) [250]. In Italy, composite sediment samples of the Po, the major Italian river, were collected from 10 reaches (mostly 20–40 km long) along the watercourse (645 km), close to the springs, downstream of the principal tributaries and at the mouth [246]. LAS and AE were specifically determined. As for LAS, the concentration ranged between 0.3 and 4.7 mg/kg. AE concentration was three to four times lower than that of LAS, namely, 0.1–1.0 mg/kg, but, because of the analytical methodology, it has to be considered conservative. As expected, the highest surfactant concentration was found in sediments downstream of the polluted tributaries [246]. United States monitoring studies reported results similar to the European ones. In one real-world monitoring study a high LAS level of 174 mg/kg was found [251]. This value is an exceptional worst-case result, found just below the outfall of a poorly functioning STP in which the effluent contributed a large fraction of the receiving river volume. Downstream of the same STP outfall, the LAS concentration in sediments dropped to 5–11 mg/kg [251]. United States monitoring studies reported average sediment LAS levels of 0.3–3.8 mg/kg below the outfalls of activated-sludge STPs [97]. An in-depth monitoring investigation on the Mississippi River showed that 31 of the 32 sediments sampled and analyzed had an LAS concentration of less than 1 mg/kg [121]. One exception: A sediment with 20 mg/kg LAS was found in a drainage canal carrying undiluted effluent from the STP of a large city.

As far as marine sediments are concerned, LAS levels of 5–17 mg/kg were reported for highly polluted estuarine sediments in Tokyo Bay [252], in Teganuma Lake [253], and in the Venice Lagoon [254], close to outfalls of probably untreated sewage. LAS was undetectable farther out to sea. In a shallow canal 20 km long adjoining the Bay of Cadiz in Spain, LAS in the top layer of the sediment in the central part of the canal close to an untreated discharge was definitely higher than that reported for the contaminated parts of Tokyo Bay and the Venice Lagoon [255]. At the two exits of the canal the LAS concentration decreased sharply (1–2 mg/kg). Other monitoring in Spain reported an LAS content of 0.1 mg/kg in sediments off a seacoast near an underwater sewage discharge pipe [208]. In Denmark sediment samples near 65 STPs were collected from December 1996 to March 1997 and examined for 110 substances [256]. LAS was found only on samples collected near 24 of the 65 STPs. The highest LAS concentration was found at the bottom of three fjords at levels of 22.0, 2.9, and 0.8 mg/kg [256]. In France LAS was found in the sediments of a pond, the salinity of which was 10 g/L, at levels in the range of 0.5–6.0 mg/kg [257]. The highest concentrations were near the mouth of a river entering the pond.

(b) Toxicity in Sediments. Toxicity data on surfactants for sediment organisms are lacking. Long-term toxicity tests exist only for LAS. As for microbial processes, LAS content usually encountered in freshwater sediments should not be inhibitory. That is suggested the by the low-toxicity data known for aerobic bacteria [245] and by the fact that no inhibition has ever been found in STP digesters, even at the unusual LAS level in sludge of 30 g/kg [95]. As for toxicity to higher organisms, an 80-day LAS chronic test found that the NOEC for a freshwater mussel was over 750 mg/kg [16,258]. The lowest NOEC in sediment, 319 mg/kg, was observed for the larvae of the benthic organism *Chironomous riparus* (midge) [16]. A study with saltwater

mussels (*Mytilus*) showed that the only effect observed was an increase in the rate of feeding, at 281 mg/kg [258]. Other, similar toxicity tests were also performed with some mollusc species, such as *Unio elongatus*, *Anodonta cignea*, and *Mytilus galloprovincialis*, exposed to sediments with much higher LAS concentration (200 mg/kg), showing that these animals could survive well without any apparent effects [258]. A long-term study was performed with a tubificid species, *Branchiura sowerbyi*, a benthic filter-feeding organism, which was exposed for 220 days to a sediment having an irreversibly adsorbed LAS concentration of 26 mg/kg [259]. No effects were observed.

Chronic studies were conducted with *Lumbriculus variegatus* and *Caenorhabditis elegans* [260]. Using radiolabeled material, a 28-d NOEC = 81 mg/kg and a 3-d NOEC = 100 mg/kg were derived, respectively.

Multivariate statistical techniques were applied to evaluate marine sediment quality in the Gulf of Cadiz [261,262] and in San Francisco Bay [262]. This multivariate tool revealed groupings of relationship between organic and inorganic chemical concentrations in sediments and biological effects that allowed the proposing of sediment-quality guidelines for the two areas taken into consideration. Among chemicals considered important for their adverse biological effects, LAS was mentioned, but they were considered of relevance only for the Gulf of Cadiz. No or trivial adverse biological effects were set up at ≤12.8 mg/kg LAS, whereas major adverse effects are expected only at ≥62 mg/kg LAS. So uncertainty would exist in the range of 12.8–62 mg/kg concentration [262].

Only for LAS could a preliminary screening approach of risk characterization in sediments, as expressed by the PEC/PNEC ratio, be put forward tentatively. As for the PNEC, as shown, long-term toxicity results for more than two species are available, with the lowest NOEC on the most sensitive tested species at values of about 100 mg/kg. Following the EU TGD [182], an application factor of 10 could be applied in this case to derive the PNEC of LAS in sediment, which, thus, could be placed at the value of about 10 mg/kg. As for the PEC, available measured data in freshwater sediments were generally found in the range of <1–5.3 mg/kg in both Europe and the United States. In marine sediments, LAS was found on average in the range of 1–2 mg/kg only in some polluted parts of estuaries, lagoons, bays, and fjords. Some higher concentrations were registered downstream of polluted STP outflows and in proximity of the mouth of rivers and canals. The PEC/PNEC ratio, thus is lower than 1 in most cases, indicating that this surfactant should pose little risk to sediment organisms. There are, however, documented situations where LAS concentrations in sediments are near this PNEC. There is a need, therefore, to refine the effect assessment and to characterize the contaminant/O_2 content ratio and the biology of the corresponding sites. AE is the only alcohol-based surfactant for which some concentration data in sediments are available. These values are lower than those for LAS by more than three times in the same monitored sites [246]. Although no long-term toxicity data are available for this surfactant, it is unlikely that these low AE concentrations are harmful to the sediment organisms.

III. SUMMARY

Worldwide consumption of surfactants in 2000 was estimated to about 10–11 million tons, excluding soap. LAS as well as AE, AS, and AES constitute more than 50% of total consumption, and the alcohol-based surfactants in particular are consumed mostly in developed countries. These surfactants were found in relatively high concentrations in raw sewage water in the range of 1–15 mg/L. When the sewage was properly treated in activated-sludge (as) STPs, they were on average removed 98–99% or more, decreasing their content in the STP-treated waters to residual total concentrations from 3 µg/L to a maximum of 140 µg/L. Their concentrations were further decreased by dilution in the receiving waters and by biodegradation and adsorption processes. In river waters each of them could be found from zero up to 50 µg/L downstream of some (as) STPs and up to 20–30 µg/L in some polluted coastal marine waters. All these surfactants are aerobically degraded, with average half-lives in river waters typically of 10–15 h for LAS and 0.2–1 h for alcohol derivatives. Under anaerobic conditions, alcohol-based surfactants degrade effectively, whereas LAS degrades slowly and partially only at the level of primary biodegradation. LAS, however, can totally be desulfonated in sulfur-limited environments. LAS was found at a relatively high concentration in STP sludge, typically from 1 to 10 g/kg. In comparison, AE had a concentration in sludge of 0.1–0.7 g/kg.

Because of sludge use in agriculture, LAS was found in sludge-amended soil, particularly just after sludge application, at a relatively high level (>100 mg/kg) when high sludge application rates, unrealistic for the usual agricultural practice, were used. In typical and realistic agricultural applications (sludge application rate <0.5 kg/m^2/y), however, its initial concentration was 1–10 mg/kg, which then decreased quickly, with half-lives of 10–30 days at levels over 1 mg/kg at harvest time. In these actual situations, neither accumulation in soil nor uptake from plants could experimentally be documented. In freshwater and marine coastal sediments LAS was found mostly from 0 to 6 mg/kg, with some exceptions of higher concentrations downstream of polluted discharges. When experimentally measured in freshwater sediments, AE was found at 0.1–1 mg/kg, three to four times lower than LAS. Toxicity data for aquatic organisms are abundant and well documented. Long-term toxicity results in both laboratory and fields studies when carried out in similar water-quality conditions were identical and were found for the commercial products in the range of 0.1–1.0 mg/L. Two exceptions with a higher toxicity for some invertebrate organisms were observed in mesocosm studies, run with a prolonged exposure time, for LAS, C_{10-13}, and AE, C_{12-15} EO$_6$. Experimental uncertainties came to light in these mesocosm studies, and, at least for LAS, the high toxicity value was not confirmed in laboratory. Final PNECs in the aquatic environment were derived in the Dutch environmental risk assessment and they were fixed at 0.25 mg/L for LAS, 0.11 mg/L for AE, and 0.4 mg/L for AES. Bioconcentration data for fish of LAS and AE model molecules were experimentally obtained, allowing for calculation of the bioconcentration factor (BCF) potentials in river waters of the commercial products, LAS (16 L/kg) and AE (conservative estimate: 142 L/kg).

Long-term toxicity data for soil are available only for LAS. A large number of new data were produced in both laboratory and field studies for soil fauna, plants, and microorganisms. A final PNEC of LAS in soil was derived in the Danish environmental risk assessment and set up at 4.6 mg/kg.

The environmental risk assessment (ERA) of surfactants for the aquatic compartment carried out in the Netherlands concluded that the risk from the use of LAS, AE, and AES was low, as shown by the PEC/PNEC ratios found, always equal to or below 0.02.

Terrestrial risk assessment (TRA) of LAS in sludge-amended soil carried out in Denmark concluded that this surfactant in the long run and under realistic sludge application in agriculture didn't pose a threat to ecosystems, as shown by the PEC/PNEC ratios, typically always below 1 and approaching 1 only exceptionally and only immediately after sludge application. It is likely that alcohol-based surfactants, because of their low content in sludge, even if not documented, do not represent any risk to terrestrial life.

A risk assessment in sediments could be approached at a screening level only for LAS. PEC/PNEC ratios in both freshwater and marine coastal sediments were lower than 1 in most cases, approaching 1 only in a few sites downstream of polluted discharges.

ACKNOWLEDGMENTS

My thanks to G. Cassani, D. Calcinai, L. Valtorta, and A. Zatta for providing me with comments and updated literature. I am also indebted to J.L. Berna, J. Heinze, and T.W. Federle, who supplied me with information on unpublished studies.

REFERENCES

1. Dolkemeyer, W. Surfactants on the eve of the third millenium. Challenges and opportunities. 5th World Cesio Congress, V. 1, Florence, Italy, May–June 2000; 38 pp.
2. Houston, C. Detergent alkylates. World markets, 1995–2000. CAHA 2000.
3. Cavalli, L.; Clerici, R.; Radici, P.; Valtorta, L. Tenside Surf. Det 1999, *36*, 254–258.
4. Valtorta, L.; Radici, P.; Calcinai, D.; Cavalli, L. Riv. Ital. Sost. Grasse LXXVII: 2000; 73–76.
5. Bhore, N.A.; Cimini, R.J.; Propp, J.L.; McWilliams, J.P.; Schild, R.L.; Elangovan, T.; Rajashankar, R.; Thiyagarajan, K. 5th World Cesio Congress, V. 1, Florence, Italy, May–June 2000; 147–151.
6. Burckett St. Laurent, J.; Connor, D.; Cripe, T.; Kott, K.; Scheibel, J.; Stidham, R.; Reilman, R. 5th World Cesio Congress, V. 1, Florence, Italy, May–June 2000; 716–722.
7. Körnig, W. Production of alkoxylated surfactants. 5th World Cesio Congress, V. 1, Florence, Italy, May–June 2000; 11–23.
8. McKendrick, C.B. Alcohol based surfactants. 5th World Cesio Congress, V. 1, Florence, Italy, May–June 2000; 133–137.
9. Karsa, D.R.; Porter, M.R. *Biodegradability of Surfactants*; Blackie Academic & Professional: Glasgow, 1995.
10. Scott, M.J.; Jones, M.N. Biochimica et Biofisica Acta 2000, *1508*, 235–251.
11. OECD. Revised guidelines for testing chemicals. Paris, 1993.
12. SETAC-Europe. Biodegradation kinetics. Generation and use of data for regulatory decision making; Brussels, April 1997.
13. European Commission DGIII. Study on new biodegradability test methods for surfactants in detergency. Europlus, ETD/96/500210, September 1997.
14. European Commission DGIII. Study on the possible problems for the aquatic environment related to surfactants in detergents. WRc, EC 4294, February 1997.
15. Ruffo, C.; Fedrigucci, M.G.; Valtorta, L.; Cavalli, L. Riv. Ital. Sost. Grasse LXXVI: 1999; 277–283.
16. Kimerle, R.A. Tenside Surf. Det. 1989, *26*, 169–176.

17. Federle, T.W. In *Biotechnology in the Sustainable Environment*; Sayler, R., Ed.; Plenum Press: New York, 1997; 223–232.
18. Matthijs, E.; DeHenau, H. Tenside Surf. Det. 1987, *24*, 193–199.
19. Sarzin, L.; Arnoux, A.; Rebouillon, P. J. Chromatogr. 1997, *760*, 285–291.
20. Moreno, A.; Ferrer, J.; Bravo, J.; Berna, J.L.; Cavalli, L. Tenside Surf. Det. 1998, *35*, 375–378.
21. Perales, J.A.; Manzano, M.A.; Sales, D.; Quiroga, J.M. Bull. Environ. Contam. Toxicol. 1999, *63*, 94–100.
22. Schöberl, P. Tenside Surf. Det. 1989, *26*, 86–94.
23. Cook, A.M.; Hrsak, D. 5th World Cesio Congress, V. 2, Florence, Italy, May–June 2000; 1387–1396.
24. Cavalli, L.; Cassani, G.; Lazzarini, M. Tenside Surf. Det. 1996, *33*, 158–165.
25. ECOSOL. Sector Group of CEFIC, Statistics, Brussels, 2001.
26. Sweeney, W.A.; Anderson, R.G. J. Am. Oil Chem. Soc. 1989, *66*, 1844–1849.
27. Nielsen, A.M.; Britton, L.N.; Beall, C.E.; McCormick, T.P.; Russell, G.L. Environ. Sci. Technol. 1997, *31*, 3397–3404.
28. DiCorcia, A.; Casassa, F.; Crescenzi, C.; Marcomini, A.; Samperi, R. Environ. Sci. Technol. 1999, *33*, 412–4118.
29. Dunphy, J.; Federle, T.W.; Itrich, N.; Simonich, S.; Kloepper-Sams, P.; Scheibel, J.; Cripe, T.; Matthijs, E. 5th World Cesio Congress, V. 2, Florence, Italy, May–June 2000; 1489–1497.
30. Cavalli, L.; Cassani, G.; Lazzarini, M.; Maraschin, C.; Nucci, G.; Berna, J.L.; Bravo, J.; Ferrer, J.; Moreno, A. Toxic. Environ. Chem. 1996, *54*, 167–186.
31. Kölbener, P.; Baumann, U.; Leisinger, T.; Cook, A.M. Environ. Toxicol. Chem. 1995, *14*, 561–579.
32. Kölbener, P.; Ritter, A.; Corradini, F.; Baumann, U.; Cook, A. Tenside Surf. Det. 1996, *33*, 149–156.
33. Cavalli, L.; Cassani, G.; Lazzarini, M.; Maraschin, C.; Nucci, G.; Valtorta, L. Tenside Surf. Det. 1996, *33*, 393–398.
34. Federle, T.W.; Itrich, N.R. Environ. Sci. Technol. 1997, *31*, 1178–1184.
35. Kaiser, S.K.; Peng, C.; Namkung, E.; Gledhill, D.W.; Nuck, B.A.; Federle, T.W. WEFTEC; Singapore, 1998; 257–264.
36. Itrich, N.R.; Federle, T.W. Setac Meeting; Vancouver, 1995.
37. Cassani, G.; Lazzarin, M.; Nucci, G.; Cavalli, L. Setac Meeting; Washington, DC, November 1996.
38. Quiroga, J.M.; Perales, J.A.; Romero, L.I.; Sales, D. Chemosphere 1999, *39*, 1957–1969.
39. Perales, J.A.; Manzano, M.A.; Sales, D.; Quiroga, J.M. Internat. Biodeterior. Biodegrad. 1999, *43*, 155–160.
40. Nishihara, T.; Hasebe, S.; Nishikawa, J.; Kondo, M. J. Appl. Microbiol. 1997, *82*, 441–447.
41. Sarrazin, L.; Arnoux, A.; Rebouillon, P.; Monod, J.L. Toxicol. Environ. Chem. 1997, *58*, 209–216.
42. Konopka, A.; Zakharova, T.; Oliver, L.; Turco, R.F. J. Ind. Microb. Biotechnol. 1997, *18*, 235–240.
43. Konopka, A.; Zakharova, T.; LaPara, T.M. J. Ind. Microb. Biotechnol. 1999, *23*, 127–132.
44. Krueger, C.J.; Barber, L.B.; Metge, D.W.; Field, J.A. Environ. Sci. Technol. 1998, *32*, 1134–1142.
45. Krueger, C.J.; Radakovich, K.M.; Sawyer, T.E.; Barber, L.B.; Smith, L.R.; Field, J.A. Environ. Sci. Technol. 1998, *32*, 3854–3961.
46. Marcomini, A.; Zanette, M.; Pojana, G.; Suter, M.J.F. Environ. Toxicol. Chem. 2000, *19*, 549–554.

47. Marcomini, A.; Zanette, M.; Pojana, G.; Carrer, C.; Cavalli, L.; Cassani, G.; Lazzarin, M. Environ. Toxicol. Chem. 2000, *19*, 555–560.

48. Marcomini, A. 5th World Cesio Congress, V. 2, Florence, Italy, May–June 2000; 1380–1386.

49. Tidswell, E.C.; Russel, N.J.; White, G.F. Microbiology 1996, *142*, 1123–1131.

50. DiCorcia, A.; Crescenzi, C.; Marcomini, A.; Samperi, R. Environ. Sci. Technol. 1998, *32*, 711–718.

51. Battersby, N.S.; Kravetz, L.; Salanitro, J.P. 5th World Cesio Congress, V. 2, Florence, Italy, May–June 2000; 1397–1407.

52. Itrich, N.R.; Kerr, K.M.; McAvoy, D.C.; Federle, T.W. Setac Meeting; Charlotte, NC, 1998.

53. Cassani, G.; Cavalli, L.; Lazzarin, M.; Nucci, G.; Marcomini, A. XXVIII CED Congress; Barcelona; May 1998.

54. Szymanski, A.; Wyrwas, B.; Swit, Z.; Jaroszynski, T.; Lukaszewski, Z. Water Res. 2000, *34*, 4101–4109.

55. Lee, D.L.; Guckert, J.B.; Belanger, S.E.; Feijtel, T.C.J. Chemosphere 1997, *35*, 1143–1160.

56. Anaerobic biodegradation of surfactants. Review of scientific information. AISE/CESIO Report, 1999.

57. Berna, J.L.; Battersby, N.; Cavalli, L.; Fletcher, R.; Guldner, A.; Schovanek, D.; Steber, J. Tenside Surf. Det. 2001, *38*, 86–93.

58. Evaluation of anaerobic biodegradation. ECETOC. Technical report No. 28, 1988.

59. Nuck, B.A.; Federle, T.W. Environ. Sci. Technol. 1996, *30*, 3597–3603.

60. Steber, J.; Wierich, P. Water Res. 1987, *21*, 661–667.

61. Steber, J.; Wierich, P. Tenside Surf. Det. 1989, *26*, 406–411.

62. Steber, J. Textilveredulung 1991, *26*, 348–354.

63. Federle, T.W.; Schwab, B.S. Water Res. 1992, *26*, 123–127.

64. Larson, R.J.; Rothgeb, T.M.; Shimp, R.J.; Word, T.E.; Ventullo, R.M. J. Am. Oil Chem. Soc. 1993, *70*, 645–657.

65. Heinze, J.; Britton, L. Anaerobic biodegradation: environmental relevance. In *3rd World Conference on Detergents: Global Perspectives*; Cahn, A., Ed.; AOCS Press: Champaign, IL, 1994; 235–239.

66. Leon, V.M.; Gonzalez-Mazo, E.; Forja Pajares, J.M. Environ. Tox. Chem. 2001, *20*, 2171–2178.

67. Cook, A.M. Tenside Surf. Det. 1998, *35*, 52–56.

68. Denger, K.; Cook, A.M. Arch. Microbiol. 1997, *167*, 177–181.

69. Denger, K.; Cook, A.M. J. Appl. Microbiol. 1999, *86*, 165–168.

70. Prats, D.; Rodriguez, M.; Llamas, J.M.; DeLaMuela, M.A.; Ferrer, J.; Moreno, A.; Berna, J.L. 5th World Cesio Congress, V.2, Florence, Italy, May–June 2000; 1655–1658.

71. Angelidaki, I.; Mogensen, A.S.; Ahring, B.K. Biodegradation 2000, *11*, 377–383.

72. Angelidaki, I.; Haagensen, F.; Ahring, B.K. 5th World Cesio Congress, V.2, Florence, Italy, May–June 2000; 1551–1557.

73. Sanz, J.L.; Rodriguez, M.; Amils, R.; Berna, J.L.; Ferrer, J.; Moreno, A. Riv. Ital. Sost. Grasse LXXVI: 1999; 307–311.

74. Salanitro, J.P.; Diaz, L.A. Chemosphere 1995, *30*, 813–830.

75. Steber, J.; Gode, P.; Guhl, W. Soap-Cosmetics Chemical Spec 1998, *64*, 44–50.

76. Wagener, S.; Schink, B. Water Res. 1987, *21*, 615–622.

77. Madsen, T.; Rasmussen, H.B.; Nilson, L. Chemosphere 1995, *31*, 4243–4258.

78. Siegfried, M.; Mueller, M.T.; Baumann, U. Anaerobic degradation and toxicity of AE in anaerobic screening test systems. 4th World Cesio Congress; V. 3, Barcelona, Spain, May 1996; 261–275.

79. Huber, M.; Meyer, U.; Rys, P. Environ. Sci. Technol. 2000, *34*, 1737–1741.

80. Feijtel, T.C.J.; Webb, S.F.; Matthijs, E. Food Chem. Toxicol. 2000, *38*, S43–S50.
81. Struijs, J.; Stoltenkamp, J.; Van de Meent, D. Water Res. 1991, *25*, 891–900.
82. Cowan, C.E.; Larson, R.J.; Feijtel, T.C.J.; Rapaport, R.A. Water Res. 1993, *27*, 561–573.
83. McKay, D.; Paterson, S.; Cheung, B.; Brock Neely, W. Chemosphere 1985, *14*, 335–374.
84. ECETOC. Special report No. 28, 1994.
85. EUSES, European Union System for the Evaluation of Substances. European Chemicals Bureau, 1997.
86. Feijtel, T.; Boeije, G.; Matthijs, M.; Young, A.; Morris, G.; Gandolfi, C.; Hansen, B.; Fox, K.; Holt, M.; Koch, V.; Schroder, R.; Cassani, G.; Schowanek, D.; Rosenblom, J.; Niessen, H. Chemosphere 1997, *34*, 2351–2373.
87. Schowanek, D.; Fox, K.; Holt, M.; Schroeder, F.R.; Koch, V.; Cassani, G.; Matthies, M.; Boeije, G.; Vanrolleghem, P.; young, A.; Morris, G.; Gandolfi, C.; Feijtel, T.C.J. Water Sci. Technol. 2001, *43*, 179–185.
88. Moreno, A.; Ferrer, J.; Berna, J.L. Tenside Surf. Det. 1990, *27*, 312–315.
89. Cavalli, L.; Gellera, A.; Landone, A. Environ. Toxicol. Chemistry 1993, *12*, 177–1788.
90. Matthijs, E.; Holt, M.S.; Kiewiet, A.; Rijs, G.B.J. Environ. Toxicol. Chem. 1999, *18*, 2634–2644.
91. Holt, M.S.; Daniel, M.; Buckand, H.; Fox, K.K. 5th World Cesio Congress, V. 2, Florence, Italy, May–June 2000; 1358–1369.
92. Gandolfi, C.; Facchi, A.; Whelan, M.J.; Cassani, G.; Tartari, G.; Marcomini, A. 5th World Cesio Congress, V. 2, Florence, Italy, May–June 2000; 1370–1379.
93. Matthijs, E.; Debaere, G.; Itrich, N.; Masscheleyn, P.; Rottiers, A.; Stalmans, M.; Federle, T.W. Water Sci. Technol. 1995, *31*, 321–328.
94. Takada, H.; Ishiwatari, R. Environ. Sci. Technol. 1987, *21*, 875–883.
95. Berna, J.L.; Ferrer, J.; Moreno, A.; Prats, D.; Ruiz Beria, F. Tenside Surf. Det. 1989, *26*, 101–107.
96. Painter, H.A.; Zabel, T. Tenside Surf. Det. 1989, *26*, 108–115.
97. McAvoy, D.C.; Eckhoff, W.S.; Rapaport, R.A. Environ. Toxicol. Chem. 1993, *12*, 977–987.
98. Holt, M.S.; Waters, J.; Comber, M.H.I.; Armitage, R.; Morris, G.; Newbery, C. Water Res. 1995, *29*, 2063–2070.
99. Schöberl, P.; Klotz, H.; Spilker, R.; Nitschke, L. Tenside Surf. Det. 1994, *31*, 243–252.
100. Feijtel, T.C.J.; Matthijs, E.; Rottiers, A.; Rijs, G.B.J.; Kiewier, A.; de Nijs, A. Chemosphere 1995, *30*, 1053–1066.
101. Sanchez Leal, J.; Garcia, M.T.; Tomas, R.; Ferrer, J.; Bengoechea, C. Surf. Det. 1994, *31*, 253–256.
102. DiCorcia, A.; Samperi, R.; Bellioni, A.; Marcomini, A.; Zanette, M.; Lemr, K.; Cavalli, L. Riv. Ital. Sost. Grasse LXXI: 1994; 467–475.
103. Waters, J.; Feijtel, T.C.J. Chemosphere 1995, *30*, 1939–1956.
104. Feijtel, T.C.J.; Struijs, J.; Matthijs, E. Environ. Toxicol. Chem. 1999, *18*, 2645–2652.
105. DiCorcia, A.; Capuani, A.; Casassa, F.; Marcomini, A.; Samperi, R. Environ. Sci. Technol. 1999, *33*, 4119–4125.
106. Trehy, M.L.; Gledhill, W.E.; Mieure, J.P.; Adamove, J.E.; Nielsen, A.M.; Perkins, H.O.; Eckhoff, W.S. Environ. Toxicol. Chem. 1996, *15*, 233–240.
107. McAvoy, D.C.; Dyer, S.D.; Fendinger, N.J.; Eckhoff, W.S.; Lawrence, D.L.; Begley, W.M. Environ. Toxicol. Chem. 1998, *17*, 1705–1711.
108. Waters, J.; Lee, K.S.; Perchard, V.; Flanagan, M.; Clarke, P. Tenside Surf. Det. 2000, *37*, 161–171.
109. Schöberl, P. Tenside Surf. Det. 1995, *32*, 25–35.
110. Schröder, F.R. Tenside Surf. Det. 1995, *32*, 492–497.
111. Gledhill, W.H.; Huddleston, R.L.; Kravetz, L.; Nielsen, A.M.; Sedlak, R.I.; Vashon, R.D. Tenside Surf. Det. 1989, *26*, 276–282.

112. Neubecker, T.A. Environ. Sci. Technol. 1985, *19*, 1232–1236.
113. Kiewiet, A.T.; Parsons, J.R.; Govers, H.A.J. Chemosphere 1997, *34*, 1795–1801.
114. Evans, K.A.; Dubey, S.T.; Kravetz, L.; Dzidic, I.; Gumulka, J.; Mueller, R.; Stork, J.R. Anal. Chem. 1994, *66*, 699–705.
115. Popenoe, D.D.; Morris, S.J.; Horn, P.S.; Norwood, K.T. Anal. Chem. 1994, *66*, 1620–1629.
116. Fendinger, N.J.; Begley, W.; McAvoy, D.C.; Eckhoff, W.S. Environ. Sci. Technol. 1995, *29*, 856–863.
117. Vandepitte, V.; Feijtel, T.C.J. Tenside Surf. Det. 2000, *37*, 35–40.
118. Takada, H.; Ogura, N.; Ishiwatari, R. Env. Sci. Technol. 1992, *26*, 2517–2523.
119. Field, J.A.; Barber, L.B.; Thurman, E.M.; Moore, B.L.; Lawrence, D.L.; Peake, D.A. Environ. Sci. Technol. 1992, *26*, 1140–1148.
120. Greiner, P.; Six, E. Tenside Surf. Det. 1997, *34*, 250–255.
121. Tabor, C.F.; Barber, L.B. Environ Sci. Technol. 1996, *30*, 161–171.
122. Gonzalez-Mazo, E.; Forja, J.M.; Gomez-Parra, A. Environ. Sci. Technol. 1998, *32*, 1636–1641.
123. Kikuchi, M.; Tokai, A.; Yoshida, T. Water Res. 1986, *20*, 643–650.
124. Gonzalez-Mazo, E.; Honing, M.; Barceló, D.; Gomez-Parra, A. Environ Sci. Technol. 1997, *31*, 504–510.
125. Leon, V.M.; Gonzalez-Mazo, E.; Gomez-Parra, A. J. Chromatogr. A 2000, *889*, 211–219.
126. Marcomini, A.; Pojana, G.; Sfriso, A.; Quiroga Alonso, J.M. Environ. Toxicol. Chem. 2000, *19*, 2000–2007.
127. Fox, K.; Holt, M.; Daniel, M.; Buckland, H.; Guymer, I. Sci. Total Environ. 2000, *251/252*, 265–275.
128. Whelan, M.J.; Gandolfi, C.; Bischetti, G.B. Water Res. 1999, *33*, 3171–3181.
129. Marcomini, A., Pojana, G. Private communication, 2001.
130. Schulze, C.; Matthies, M.; Trapp, S.; Schröder, F.R. Chemosphere 1999, *39*, 1833–1852.
131. Takada, H.; Muton, K.; Tomita, N.; Miyadzu, T.; Ogura, N. Water Res. 1994, *28*, 1953–1960.
132. Boeije, G.M.; Schowanek, D.R.; Vanrolleghem, P.A. Water Res. 2000, *34*, 1479–1486.
133. Van de Plassche, E.J.; de Bruiju, J.H.M.; Stephenson, R.R.; Marshall, S.J.; Feijtel, T.C.J.; Belanger, S.E. Environ. Toxicol. Chem. 1999, *18*, 2653–2663.
134. Consulting Engineers. The use of existing toxicity data for estimation of the maximum tolerable environmental concentration of LAS. Part I: main report. Part II: data lists. Delft (NL), May 1993.
135. Consulting Engineers. Environmental data review of AE. Final report. Delft (NL), October 1994.
136. Consulting Engineers. Environmental data review of AES. Data lists. Delft (NL), January 1994.
137. Environmental Health Criteria 169, 15–244. WHO; Geneva, 1996.
138. BUA Report 189 (August 1996), Fatty alkyl sulfates. S. Hirzel, 1998.
139. Dyer, S.D.; Lauth, J.R.; Morral, S.W.; Herzog, R.R.; Cherry, D.S. Environ. Tox. Water Quality 1997, *12*, 295–303.
140. Roberts, D.W. Sci. Total Environ. 1991, *109/110*, 557–568.
141. Official Journal of Eur. Communities. L 13/9/1995, *217*, 14–30.
142. Official Journal of Eur. Communities. L 20/7/1999, *187*, 52–69.
143. Utsunomiya, A.; Watanuki, T.; Matsushita, K.; Tomita, I. Environ. Toxicol. Chem. 1997, *16*, 1247–1254.
144. Utsunomiya, A.; Watanuki, T.; Matsushita, K.; Tomita, I. Chemosphere 1997, *35*, 1215–1226.
145. Utsunomiya, A.; Watanuki, T.; Matsushita, K.; Nishina, M.; Tomita, I. Chemosphere 1997, *35*, 2479–2490.

146. Hansen, B.; Fotel, F.L.; Jensen, N.J.; Wittrup, L. Marine Biology 1997, *128*, 627–637.

147. Christiansen, P.D.; Brozek, M.; Hansen, B.W. Environ. Toxicol. Chem. 1998, *17*, 2051–2057.

148. Verge, C.; Moreno, A. Tenside Surf. Det. 2000, *37*, 172–175.

149. Lopez-Serrano, I.; Bravo, J.E.; Moreno, A. XXXI CED Congress; Barcelona, March 2001; 39–48.

150. Hampel, M.; Ortiz, J.B.; Sarasquete, C.; Moreno, A.; Blasco, J. 5th World Cesio Congress, V. 2, Florence, Italy, May–June 2000; 1617–1626.

151. Hampel, M.; Sobrino, C.; Lubian, L.; Blasco, J. 5th World Cesio Congress, V. 2, Florence, Italy, May–June 2000; 1627–1636.

152. Gillespie, W.B.; Steinriede, R.W.; Rodgers, J.H.; Dorn, P.B.; Wong, D.C.L.; Wiley: New York, 1999, *14*, 293–300.

153. Gillispie, W.B.; Rodgers, J.H.; Crossland, N.O. Environ. Toxicol. Chem. 1996, *15*, 1418–1422.

154. Gillispie, W.B.; Rodgers, J.H.; Dorn, P.B. Aquatic Toxicol. 1997, *37*, 221–236.

155. Dorn, P.B.; Rodgers, J.H.; Gillespie, W.B.; Lizotte, R.E.; Dunn, A.W. Environ. Toxicol. Chem. 1997, *16*, 1634–1645.

156. Wong, D.C.L.; Dorn, P.B.; Chai, E.Y. Environ. Toxicol. Chem. 1997, *16*, 1970–1976.

157. Cardellini, P.; Ometto, L. Ecotox. Environ. Safety, B 2001, *48*, 170–177.

158. Dyer, S.D.; Stanton, D.T.; Lauth, J.R.; Cherry, D.S. Environ. Toxicol. Chem. 2000, *19*, 608–616.

159. Belanger, S.E.; Rupe, K.L.; Lowe, L.L.; Johnson, D.; Pan, Y. Environ. Toxicol. Water Qual. 1996, *11*, 65–76.

160. Belanger, S.E.; Rupe, K.L.; Bausch, R.G. Bull. Environ. Contam. Toxicol. 1995, *55*, 751–758.

161. Lewis, M.A.; Pittinger, C.A.; Davidson; Ritchie, C.J. Environ. Toxicol. Chem. 1991, *10*, 1803–1812.

162. Fairchild, J.F.; Dwyer, F.J.; La Point, T.W.; Burch, S.A.; Ingersoll, C.G. Environ. Toxicol. Chem. 1993, *12*, 1763–1775.

163. Tattersfield, L.J.; Holt, M.; Girling, A.G.; Mitchell, G.C.; Pearson, N.; Ham, L. The fate and effects of LAS in outdoor artificial streams and pools. Shell Research 1995. *unpublished results*.

164. Tattersfield, L.J.; Mitchell, G.C.; Holt, M.; Girling, A.G.; Pearson, N.; Ham, L. LAS: fate and effects in outdoor artificial streams and pools. An extended study. Shell Research 1996. *unpublished results*.

165. ERASM Report. Long-term toxicity of LAS on *Gammarus* sp. Brussels, April 2000.

166. Belanger, S.E.; Bowling, J.W.; Lee, D.M.; LeBlanc, E.M.; Kerr, K.M.; McAvoy, D.C.; Christman, S.C.; Davidson, D.H. Ecotox. Environ. Safety 2002, *52*, 150–171.

167. Dorn, P.B.; Rodgers, J.H.; Dubey, S.T.; Gillespie, W.B.; Lizotte, R.E. Ecotoxicology 1997, *6*, 275–292.

168. Harrelson, R.A.; Rodgers, J.H.; Lizotte, R.E.; Dorn, P.B. Ecotoxicology 1997, *6*, 321–333.

169. Dorn, P.B.; Rodgers, J.H.; Dubey, S.T.; Gillespie, W.B.; Figueroa, R.A. Ecotoxicol. Environ. Safety 1996, *34*, 196–204.

170. Tattersfield, L.J.; Young, L.J.; Pearson, N.; Davies, E.H. NEODOL 25-9: fate and effects in outdoor artificial streams and pools. Shell Research Report, SBER. 95. 013, July 1996.

171. Kline, E.R.; Figueroa, R.A.; Rodgers, J.H.; Dorn, P.B. Environ. Toxicol. Chem. 1996, *15*, 997–1002.

172. Tattersfield, L.J.; Holt, M.; Mitchell, G.C.; Pearson, N. Effects of NEODOL 25-7 in outdoor artificial streams. Shell Research Report, SBER. 95. 001,1995.

173. Dorn, P.B.; Tattersfield, L.J.; Holt, M.; Dubey, S.T. 4th World Cesio Congress, V. 3, Barcelona, June 1996; 97–212.

174. Belanger, S.E.; Guckert, J.B.; Bowling, J.W.; Begley, W.M.; Davidson, D.H.; LeBlanc, E.M.; Lee, D.M. Aquatic Toxicol. 2000, *48*, 135–150.

175. Raney, K.H. Colloids Surfaces A: Physicochem. Eng. Aspects 2000, *167*, 151–164.

176. Belanger, S.E.; Meiers, E.M.; Bausch, R.G. Aquatic Toxicol. 1995, *33*, 65–87.

177. Fendinger, N.J.; Begley, W.M.; McAvoy, D.C. Eckhoff. Environ. Sci. Technol. 1992, *26*, 2498–2693.

178. Guckert, J.B.; Walker, D.D.; Belanger, S.E. Environ. Toxicol. Chem. 1996, *15*, 262–269.

179. Aldenberg, T.; Slob, W. Ecotoxicol. Environ. Safety 1993, *25*, 48–63.

180. Heinze, J. CLER, USA. Private communication.

181. Shi, J.; Kloepper-Sams, P.J.; Giolado, S.T.; Federle, T.W.; Versteeg, D.J.; Belanger, S.E. 5th World Cesio Congress, V. 2, Florence, Italy, May–June 2000; 1525–1531.

182. Technical Guidance Document (TGD) in support of Commission Directive 93/67/EEC and Commission Regulation No. 1488/94, EU Chemical Bureau, 1996.

183. Fox, K. XXXI CED Congress, Barcelona, March 2001; 213–226.

184. Roberts, D.W. 5th World Cesio Congress, V. 2, Florence, Italy, May–June 2000; 1517–1524.

185. Painter, H.A. Anionic Surfactants. Handbook Environ. Chem. 1992, *3*, 2–88.

186. Tolls, J.; Kloepper-Sams, P.; Sijm, D.T.H.M. Chemosphere 1994, *29*, 693–717.

187. Comotto, R.M.; Kimerle, K.A.; Swisher, R.D. In *Bioconcentration and Metabolism of LAS by Daphnids and Fatheads minnows. Aquatic Toxicology*; Marking, L.L., Kimerle, R.A., Eds.; ASTM: V. STP, 1979; 667 pp.

188. Bishop, W.E.; Maki, A.W. A critical comparison of bioconcentration test methods. In *Aquatic Toxicol*; Eaton, J.G., Parish, P.R., Hendricks, A.C., Eds.; ASTM: V. STP, 1980; 707 pp.

189. Wakabayashi, M.; Kikuchi, M.; Sato, A.; Yoshida, T. Ecotoxicol. Environ. Safety 1987, *13*, 148–163.

190. Tolls, J. Bioconcentration of Surfactants. Thesis, Utrecht University, June 1998.

191. Tolls, J.; Haller, M.; De Graaf, I; Thijssen, M.A.T.C.; Sijm, D.T.H.M. Environ. Sci. Technol. 1997, *31*, 3426–3431.

192. Tolls, J.; Haller, M.; Seinen, W.; Sijm, D.T.H.M. Environ. Sci. Technol. 2000, *34*, 304–310.

193. Morral, S.W.; Begley, W.M.; Rawlings, J.M.; Eckhoff, W.S.; Versteeg, D.J. 5th World Cesio Congress, V. 2, Florence, Italy, May–June 2000; 1468–1474.

194. Tolls, J.; Sijm, D.T.H.M. Environ. Toxicol. Chem. 1999, *18*, 2689–2695.

195. Komber, M.H.I.; de Wolf, W.; Cavalli, L.; Van Egmond, R.; Steber, J.; Tattersfield, L.; Priston, R.A. Chemosphere. 2003, *52*, 23–32.

196. Steber, J. Chimica oggi, October 2000; 83–88.

197. Kloepper-Sams, P.; Tors, F.; Feijtel, T.C.J.; Gooch, J. Sci. Tot. Environ. 1996, *185*, 171–185.

198. De Wolf, W.; Feijtel, T.C.J. Chemosphere 1998, *36*, 1319–1343.

199. Jensen, J. Sci. Tot. Environ. 1999, *226*, 93–111.

200. Fytianos, K.; Voudrias, E.; Papamichali, A. Chemosphere 1998, *36*, 2741–2746.

201. Cavalli, L.; Valtorta, L. Tenside Surf. Det. 1999, *36*, 22–28.

202. Cavalli, L.; Valtorta, L.; DeWolf, W.; Feijtel, T.C.J. TRA for LAS in Sludge-Amended Soil. IAWQ Conference: Athens, Oct. 3–15, 1999.

203. SPT Workshop in coordination with the Danish EPA, April 19–20, 1999. LAS Risk Assessment for Sludge-Amended Soils. Copenhagen, September 8, 1999.

204. Brink, M. Monitoring Data in Sludge. The DK Plant Directorate. Presented at Ref. 203.

205. Statutory Order No. 823, Ministry of the Environment and Energy, Denmark, 16/9/1996.

206. McEvoy, J.; Giger, W. Naturwissenschaften 1985, *72*, 429–431.

207. Marcomini, A.; Giger, W. Tenside Surf. Det. 1988, *25*, 226–229.

208. Prats, D.; Ruiz, F.; Vaquez, B.; Zarzo, D.; Berna, J.L.; Moreno, A. Environ. Toxicol. Chem. 1993, *12*, 1599–1608.

209. Torslov, J.; Samsoe-Petersen, L.; Rasmussen, J.O.; Kristensen, P. VKI Report No. 366, 1997.

210. Rapaport, R.A.; Hopping, W.D.; Eckhoff, W.S. 8th SETAC Meeting; Pensacola, FL, 1987.

211. Klotz, A. AE in sewage treatment sludges. Results of a German ring test. "Analytica" Conference, April 22, 1998.

212. Matthijs, E.; Burford, M.D.; Cassani, G.; Comber, M.H.I.; Eadsforth, C.V.; Haas, P.; Klotz, H.; Spilker, R.; Waldhoff, H.; Wingen, H.-P. Paper to be submitted.

213. Cassani, G. Condea Augusta, Centro Ricerche Paderno D. (Mi), Italy. Private Communication, 2001.

214. Moreno, A.; Bravo, J.; Ferrer, J.; Bengoechea, J. J. Am. Oil Chem. Soc. 1993, 70, 667–671.

215. Madsen, T.; Winther-Nielsen, M.; Rasmussen, D. 5th Cesio Congress, V. 2, Florence, Italy, May–June 2000; 1428–1432.

216. Brandt, K.K.; Krogh, P.H.; Cassani, G.; Soerensen, J. 5th Cesio Congress V. 2, Florence, Italy, May–June 2000; 1590–1597.

217. Mortensen, G.C.; Cassani, G.; Verge, C.; Volfing, M.; Bennetzen, S. 5th World Cesio Congress, V. 2, Florence, Italy, May–June 2000; 558–1565.

218. Carlsen, L.; Metzon, M.B.; Kjelsmark, J. Sci. Total Environ. 2002, 290, 225–230.

219. Prats, D.; Rodriguez, M.; Muela, M.A.; Llamas, J.M.; Moreno, A.; Ferrer, J.; Berna, J.L.; Naylor, G.C.; Nielsen, A.M. 5th Cesio Congress, V. 2, Florence, Italy, May–June 2000; 1475–1488.

220. Petersen, P.H. Rambol. Degradation of xenobiotics by composting. Presented at Ref. 203.

221. Garcia, M.T.; Ribosa, I.; Sanchez Leal, J. 5th Cesio Congress, V. 2, Florence, Italy, May–June 2000; 1447–1456.

222. Figge, K.; Schöberl, P. Tenside Surf. Det. 1989, 26, 122–128.

223. Marcomini, A.; Capel, P.D.; Lichtensteiger, Th.; Brunner, P.H.; Giger, W. J. Environ. Qual. 1989, 18, 523–528.

224. Waters, J.; Matthijs, E. Tenside Surf. Det. 1989, 26, 129–135.

225. Holt, M.S.; Matthijs, E.; Waters, J. Water Res. 1989, 23, 749–759.

226. Ward, T.E.; Larson, R.J. J. Ecotoxicol. Environ. Safety 1989, 17, 119–130.

227. Mortensen, G.K.; Egsgaard, H.; Ambus, P.; Jensen, E.S.; Groen, C. J. Environ. Quality 2001, 30, 1266–1270.

228. Küchler, T.; Schnaak, W. Chemosphere 1997, 35, 153–167.

229. Fox, K.K.; Chapman, L.; Solbé, J.; Brennand, V. Tenside Surf. Det. 1997, 34, 436–441.

230. Haigh, S. Sci. Tot. Environ. 1996, 185, 161–170.

231. Jensen, J.; Folker-Hausen, P. DMU Report No 47, Denmark, 1995.

232. Jensen, J.; Lokke, H.; Holmstrup, M.; Krogh, P.H.; Elsgaard, L. Environ. Tox. Chem. 2001, 20, 1690–1697.

233. Holmstrup, M.; Krogh, P.H. Environ. Tox. Chem. 2001, 20, 1673–1679.

234. Elsgaard, L.; Petersen, S.O.; Debosz, K. Environ. Tox. Chem. 2001, 20, 1656–1663.

235. Loekke, H.; Holmstrup, M.; Jensen, J. 5th World Cesio Congress, V. 2, Florence, Italy, May–June 2000; 1439–1446.

236. Solbé, J.; Berna, J.L.; Cavalli, L.; Feijtel, T.J.; Fox, K.K.; Heinze, J.; Marshall, S.J.; DeWolf, W. 5th World Cesio Congress, V. 2, Florence, Italy, May–June 2000; 1433–1438.

237. Wagner, C.; Lokke, H. Water Res. 1991, 25, 1237–1242.

238. Holmstrup, M.; Krogh, P.H.; Lokke, H.; De Wolf, W.; Marshall, S.; Fox, K. Environ. Tox. Chem. 2001, 20, 1680–1689.

239. Elsgaard, L.; Petersen, S.O.; Debosz, K.; Kristiansen, I.B. Tenside Surf. Det. 2001, 38, 94–97.

240. Elsgaard, L.; Petersen, S.O.; Debosz, K. Environ. Tox. Chem. 2001, *20*, 1664–1672.
241. Westall, J.C.; Chem, H.; Zhang, W.; Brownawell, B.J. Environ. Sci. Technol. 1999, *33*, 3110–3118.
242. Larson, R.J.; Rothgeb, T.M.; Shimp, R.J.; Word, T.E.; Ventullo, R.M. J. Am. Oil Chem. Soc. 1993, *70*, 645–657.
243. Sarrazin, L.; Schembri, T.; Rebouillon, P. Toxicol. Environ. Chem. 1999, *72*, 113–125.
244. Sarrazin, L.; Limouzin, Y.; Rebouillon, P. Toxicol. Environ. Chem. 1999, *69*, 487–498.
245. Painter, H.A.; Zabel, T.F. Review of environmental safety of LAS. WRc report London, UK, 1988.
246. Cavalli, L.; Cassani, G.; Viganò, L.; Pravettoni, S.; Nucci, G.; Lazzarin, M.; Zatta, A. Tenside Surf. Det. 2000, *37*, 282–288.
247. Schöberl, P.; Spilker, R. Tenside Surf. Det. 1996, *33*, 400–403.
248. Reiser, R.; Toljander, H.O.; Giger, W. Anal. Chem. 1997, *69*, 4923–4930.
249. Reiser, R.; Toljander, H.O.; Albrecht, A.; Giger, W. In *Molecular Markers in Environmental Geochemistry*; Eganhouse, R.P., Ed.; ACS Symposium Series 671; Washington, DC, 1997; 196–212.
250. Aarhus Report. Investigation on xenobiotics from wastewater into environmental circulation in Aarhus county. Denmark, September 1998.
251. Rapaport, R.A.; Eckhoff, W.S. Environ. Toxicol. Chem. 1999, *9*, 1245–1257.
252. Takada, H.; Ogura, N. Mar. Chem. 1992, *37*, 257–273.
253. Amano, K.; Fukushima, T.; Nakasugi, O. Hydrobiologia 1992, *235–236*, 491–499.
254. Marcomini, A.; Pavoni, P.; Sfrisio, A.; Orio, A.A. Mar. Chem. 1990, *29*, 307–323.
255. Gonzalez-Mazo, E.; Quiroga, J.M.; Sales, D.; Gomez-Parra, A. Toxicol. Environ. Chem. 1997, *59*, 77–87.
256. Lillebælt Report. The Lillebælt Cooperation. Denmark, December 1998.
257. Sarrazin, L.; Arnoux, A. Toxicol. Environ. Chem. 1998, *65*, 163–171.
258. Bressan, M.; Brunetti, R.; Castellato, S.; Fava, G.C.; Giro, P.; Marin, M.; Negrisolo, P.; Talandini, L.; Thoman, S.; Tosani, L.; Turchetto, M.; Campesan, G.C. Tenside Surf. Det. 1989, *26*, 148–158.
259. Casellato, S.; Aiello, R.; Negrisolo, P.A.; Seno, M. Hydrobiologia 1992, *232*, 169–173.
260. Webb, S.; Comber, S.; Stuart, S.; Höss, S. . SETAC Congress, Vienna, 2001.
261. DelValls, T.A.; Forja, J.M.; Gomez-Parra, A. Ciencias Marinas 1998, *24*, 127–154.
262. DelValls, T.A.; Chapman, P.M. Ciencias Marinas 1998, *24*, 313–336.
263. Holt, M.S.; Fox, K.K.; Daniel, M.; Buckland, H. The Science of the Total Environment 2003, *314–316*, 271–288.

15

The Environmental Safety of Alkylphenol Ethoxylates Demonstrated by Risk Assessment and Guidelines for Their Safe Use

CARTER G. NAYLOR*   Consultant, Austin, Texas, U.S.A.

I. INTRODUCTION

Formulators of cleaning products have the responsibility not only to provide the consumer with products of high performance, but also to ensure that the products cause no harm to humans and the environment. Assessment of the safety of household and commercial cleaning products is an integral part of modern product development and maintenance. The health and environmental properties of each ingredient must be known prior to its inclusion in a cleaner; information should be publicly available or provided by manufacturers.

This chapter is a case study of a family of popular, widely used surfactants about which much public information is contradictory; they have come under intense scrutiny following claims of environmental and human health hazard. The technical case for their safety and the scientific methods for determining their safety are discussed in this chapter. It presents the environmental risk assessment of alkylphenol ethoxylate surfactants (APE) as a case study of the process used to establish systematically and rigorously the low degree of risk posed by widely used surfactants when they are handled and disposed of properly.

When criticisms of APE began appearing in the technical literature in the mid-1980s, the manufacturers undertook a comprehensive program of environmental research to address the unfavorable reports. At that time there was little known about the fate of APE in the environment and their potential for causing adverse effects. Since then there has been a great deal of research conducted on both fate and effects, and a reassuring picture of the risk they pose to the environment has emerged.

---

*Retired from Huntsman Corporation, Austin, Texas, U.S.A.

Formal risk assessment of a substance requires extensive knowledge of its exposure to organisms and environments at risk, its fate in the environment, and the nature of the effects the substance is capable of causing. The more information that is available, the more robust is the risk assessment and the less need for added "safety factors" that compensate for uncertainty. The risk assessment for APE in the aquatic environment has been completed and forms the basis of guidelines for their safe use and disposal.

II. APE SURFACTANTS

A. Chemistry of APE Surfactants

Most of the APE in commerce are based on nonylphenol (NPE, about 80% of the APE market) and octylphenol (OPE, most of the remaining 20%) [1]. Minor APE include dodecylphenol and dinonylphenol ethoxylates. NPE and OPE are manufactured by the same chemical process steps of acid-catalyzed alkylation of phenol with olefin followed by base-catalyzed reaction of the alkylphenol with ethylene oxide. Most octylphenol is made using di-isobutylene, a branched eight-carbon olefin; nonylphenol production uses the branched nine-carbon olefin mixture termed *nonene* or propylene trimer. The major positional AP isomer is para ($\geq 90\%$), while the ortho isomer is 10% or less. Figure 1 shows one of the several major isomers. NP is a complex mixture of at least 20 nonyl group isomers [2]. Most of those have a quaternary carbon attachment to the phenol ring.

The AP ethoxylates, being oligomeric with repeating oxyethylene units, consist of ethoxylate groups with a distribution close to that of Poisson (bell-shaped curve, Fig. 2). The greater the number of EO units, the more water soluble the APE becomes. Complete water solubility of NPE is reached with an EO number of about 7, that of OPE with about 6 EO. The most commonly used ethoxylate has on average 9–10 EO; it is shown in Figure 2.

OP consists almost entirely of a single octyl group isomer, (1,1,3,3-tetramethylbutyl)phenol.

APE surfactants are highly surface active; that is, they lower the interfacial tension between water and other liquids, air, and solids very effectively. They are especially suited for cleaning, emulsifying, dispersing, and wetting applications [3].

B. Product CAS Identification

APE products are known by a variety of names and Chemical Abstract Service Registry numbers, as shown in Table 1. All five CASR numbers represent the same

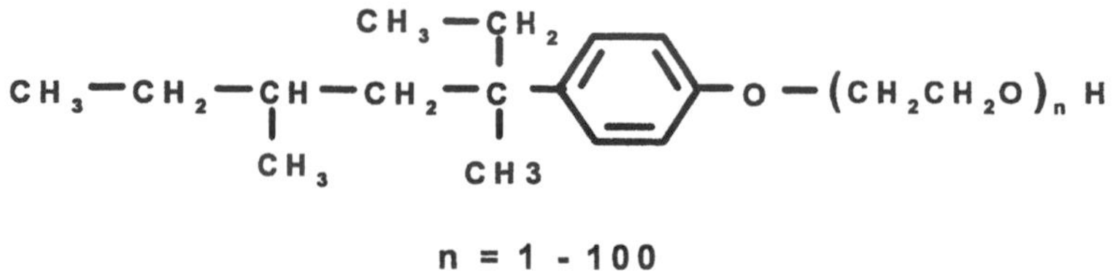

$$CH_3-CH_2-CH-CH_2-C \overset{\overset{\displaystyle CH_3-CH_2}{|}}{\underset{\underset{\displaystyle CH3}{|}}{}} \! \left\langle\!\!\!\!\bigcirc\!\!\!\!\right\rangle \!\!-O-(CH_2CH_2O)_n H$$

$$\underset{\displaystyle CH_3}{|}$$

n = 1 - 100

FIG. 1 Structure of nonylphenol ethoxylate. (From Ref. 2.)

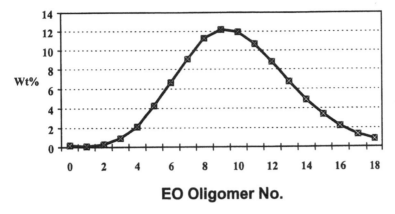

FIG. 2 Nonylphenol ethoxylate (NPE9) oligomer distribution.

commercially relevant NPE of EO number greater than 1 and so may be considered equally valid, and all ten NPE synonyms are equivalent. Similarly for the three OPE CAS numbers and seven synonyms.

C. Uses of APE Surfactants

APE surfactants have been in commercial use for more than 50 years. Production worldwide is about 700 Kt/year [4a], of which about 250 Kt/year is in North America [4b] and about 120Kt/year is in western Europe [4c]. The largest uses for NPEs are in cleaning products and processes: household and institutional detergents, hard surface

TABLE 1 NPE and OPE Nomenclature and CASR Numbers

CASR No.	NPE Synonym
9016-45-9	Nonylphenol ethoxylate
26027-38-3	Nonylphenol polyethylene glycol ether
37205-87-1	Nonylphenoxypolyethoxyethanol
68412-54-4	Nonylphenol polyethylene oxide
127087-87-0	Nonylphenol polyglycol ether
	Nonylphenoxypoly(oxyethylene)ethanol
	p-Nonylphenol polyethylene glycol ether
	α-(p-Nonylphenyl)-ω-hydroxypoly(oxyethylene)
	Poly(oxy-1,2-ethanediyl), α-(isononylphenyl)-ω-hydroxy
	Ethoxylated nonylphenol

CASR No.	OPE Synonym
9036-19-5	Octylphenoxy poly(ethoxy)ethanol
9002-93-1	*tert*-Octylphenoxy poly(ethoxy)ethanol
68987-90-6	*tert*-Octylphenoxy poly(oxyethylene)ethanol
	Octylphenoxy poly(oxyethylene)ethanol
	p-*tert*-Octylphenoxy poly(ethoxy)ethanol
	α-[(1,1,3,3,-Tetramethylbutyl)phenyl]-ω-hydroxypoly(oxy-1,2-ethanediyl)
	α-(p-Octylphenyl)-ω-hydroxypoly(oxyethylene), branched

cleaners, pulp and paper manufacture, paper recycling, textile scouring, and metal cleaning. Other important uses include pesticide emulsification and dispersion, emulsion polymerization, additives for plastics, lubricants, and fuels. The major application of OPE is emulsion polymerization.

Modern laundry detergents are sophisticated formulations designed to cleanse clothing and other fabric of many different soils under a wide range of conditions. The surfactants must be economical, easy to formulate, compatible with the other components, and effective Typically, combinations of nonionic and anionic surfactants are used to get optimum cleaning performance. NPE are the most cost-effective nonionic surfactants available to the formulator, so detergents based on NPE can be the most economical for the consumer.

D. NPE and Their Biodegradation Intermediates

Figure 3 illustrates the numerous initial steps NPE pass through as they are biodegraded. The intermediates shown are all surface active. The polyether chain of

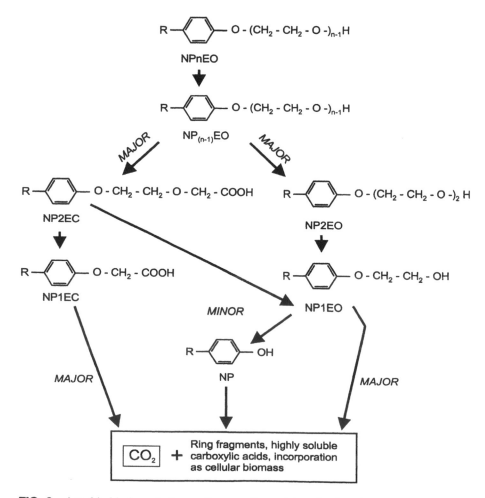

FIG. 3   Aerobic biodegradation pathways of nonylphenol ethoxylates.

NPE is shortened stepwise, probably via the carboxylate at each step. NP and the shortest NPE oligomers, NPE1 and NPE2, are hydrophobic and adsorb onto the organic solids of sludges and sediments. The carboxylates are hydrophilic, so they tend to remain in solution. Under oxic condition (oxygen present), very little NP is formed. The aromatic ring is oxidized and cleaved while the ethoxy or ether carboxylate group is still attached to the ring.

III. ENVIRONMENTAL CONSIDERATIONS FOR APE DISPOSAL AFTER USE

A. Treatment

After a cleaning or other processing operation is finished, the water containing the detergent and soil is discharged as wastewater. Unless it is treated to remove the organic load (the surfactants and soil), it is a potential pollution threat to the aquatic environment. The simplest and least satisfactory wastewater treatment is mechanical settling and flotation of insoluble solids, commonly called *primary* treatment. Typically, little organic matter is removed and little biodegradation takes place, so discharge of primary effluent to receiving waters can be a serious source of pollution. Surfactants are highly toxic to fish and other aquatic fauna. Unless they are biodegraded before being discharged, they may harm aquatic life.

Impounding primary effluent for a time enables microorganisms to consume dissolved and suspended organic matter. The most common design for biological (*secondary*) treatment is termed *activated sludge*, in which the microbial population has high density and is well aerated and the hydraulic residence time is sufficient for extensive removal of the organic load by biodegradation. Less efficient are trickling-filter plants and passive lagoon systems. The surfactants most commonly used in cleaning products are readily broken down during biological treatment. This eliminates their toxicity.

B. Biodegradation Testing

Laboratory test methods are commonly used to measure biodegradation of surfactants. APE have been examined extensively for many years. Published data are highly variable; the extent of degradation has been reported [1] from 0% to 100%. Test conditions and assay methods differ widely, even for standard test protocols, so cautious interpretation of test results is necessary. (In this chapter, *primary biodegradation* is defined as the disappearance of surface-active parent materials and degradation intermediates, and *ultimate biodegradation* is mineralization and uptake into biomass.) Most reliable are tests using microbial inocula well acclimated to APE and validated analytical methods specific for APE and their biodegradation intermediates [5]. They demonstrate the high biodegradability of APE and their metabolites and allow the following generalizations:

- AP/APE will be found in sewage.
- Much less AP/APE will be found in wastewater after biological treatment.
- AP/APE will be found in waters receiving untreated sewage.
- Little or no AP/APE will be found in waters receiving treated wastewater.

C. Field Assessment of APE Biodegradation/Treatment

Results from laboratory tests indicate a chemical's treatability in secondary treatment and persistence in receiving waters. But neither rates of degradation nor the extent of mineralization (conversion to CO_2) in the laboratory are accurate quantitative measures of "real-world" behavior. Field-monitoring studies at treatment plant sites are necessary to confirm the chemical's environmental fate and concentration during and after treatment. The more concentration data, the greater the confidence in the relevance of the data. The field sites should be characterized by determining the type of treatment received by the wastewater and the impact from wastewater discharges on receiving waters. "Hot spots" of high pollution are thereby distinguished from more representative sites.

D. Toxicity Testing

Toxicity tests are routinely run in the laboratory under controlled and well-defined conditions. Acute tests measure the concentrations lethal to organisms, and chronic tests measure effects such as growth rate, reproduction rate, and development. These tests are artificial in significant ways, such as the mode of chemical dosing and how the system is maintained. To develop a comprehensive picture of the chemical's toxicity it is necessary to test many species of organisms representing several taxonomic orders, using different acute and chronic exposure scenarios. In the case of APE the various surface-active chemical species formed as biodegradation intermediates and present in wastewater should be included with the parent surfactant in the testing program. Their profiles of toxicity define the potential hazard posed by APE.

IV. RISK ASSESSMENT OF APE SURFACTANTS IN THE ENVIRONMENT

The following discussion focuses on the aquatic environment because enough information is available for a formal risk assessment. A parallel assessment of the terrestrial environment has not yet been completed, but the available data suggest that risks are comparable to those in the aquatic environment.

A. Elements of a Risk Assessment

Risk assessment is the methodical process of determining the net risk to the environment from all the identified hazards and exposure to those hazards. *Hazard* is an adverse effect that can occur, while *risk* is the likelihood that it will and is dependent on the concentrations to which an environmental receptor is *exposed*. The *degree of risk* is the ratio of exposure to the hazard threshold.

The main elements of the aquatic environmental risk assessment of APE are summarized in the following outline.

1. Environmental concentrations of APE.

    a. Measured in enough separate locations where exposure is likely for statistical validity.
    b. Accepted protocols for sample collection and handling and validated methods for analysis of APE and intermediates are used.

   c. Default or assumed values for exposure that tend to overestimate actual measured concentrations should not be used.

2. APE removal in wastewater treatment.

   a. Laboratory tests are used to determine the potential for APE to biodegrade.
   b. Field studies should include as wide a variety of treatment plant designs, wastewater sources, organic loading, climates, and seasonal conditions as possible.
   c. Default or assumed values for treatability that tend to underestimate actual removal rates should not be used.

3. Toxicity tests conducted on as wide a range of aquatic species as possible.

   a. Literature data from studies meeting quality standards should be included but not those from studies of dubious validity.
   b. Criteria for acceptance are:

      i. Adherence to well-defined protocols.
      ii. Properly identified and well-characterized test chemicals.
      iii. Well-studied test organisms receiving proper care and husbandry.
      iv. Appropriate and well-documented toxicity endpoints.

   c. When sufficient data for enough species and families are available, toxicity thresholds can be developed that are protective of the environment. No added "safety factor" needs to be applied; it would not significantly improve environmental quality while unnecessarily biasing the assessment. Safety factors are appropriate if the toxicity database is not large.

4. Statistical analysis of the toxicity hazards posed by APE and the range of their actual concentrations.

   a. A commonly accepted risk probability standard is protection of 95% of resident species in order to maintain ecological vitality.
   b. Risk assessment is performed by dividing the relevant exposure concentrations to toxicity thresholds. The result is referred to as a *hazard quotient* (HQ). HQ values of less than 1 are desirable.

B. Environmental Fate of APE Surfactants

Concentrations of APE in the aquatic environment are minimized by degradation before entering surface waters. An understanding of the capacity fo microbes to break down APE, through studies of biodegradation and wastewater treatment, must accompany the exposure assessment.

1. APE Biodegradability

As mentioned earlier, APE biodegradability measurements vary widely in the literature. Only recently has the picture become clear as biodegradation and analytical test methods have improved and actual treatment plant studies have been reported. Standard laboratory methods for assessing biodegradability are widely practiced and are an essential first step [6]. The first tier of methods, for "ready biodegradability," measure the production of CO_2 or the uptake of oxygen in dilute solution, and can be made more useful by also measuring the consumption of organic carbon and the concentration change of the test chemical. If a test chemical does not meet the stringent and rather arbitrary "pass" criteria, it does not imply environmental persistence. It may be subjected to a more vigorous second-tier test for "inherent biodegradability" that more closely simulates treatment plant conditions.

APE and their major intermediates have been examined by the Organization for Economic Cooperation and Development (OECD) 301B test for ready biodegradability [5,7]. Table 2 summarizes the results. The two octylphenoxy carboxylates (OPEC) passed as readily biodegradable, the other materials as inherently biodegradable. The remaining dissolved organic carbon was presumed to consist of ring-opened and other degradation intermediates no longer surface active. The carbon missing from the balance of CO_2, DOC, and residual test materials was presumed to be incorporated into the biomass. Both OP and NP were nondetectable and extensively mineralized by the end of the test, showing clearly that they are highly biodegradable and refuting claims of their recalcitrance.

A similar test method, the ISO/DIS-14593, was used on NPE9. It reached 70% conversion to CO_2 and left 12% DOC and 3% NPE, passing the test as readily biodegradable and further establishing the facile biological breakdown of NPE [5].

2. APE Treatability

The extensive removal of APE by wastewater treatment has been well documented [8–10a,b]. Several activated-sludge treatment plants were selected for examination of treatment effectiveness because they were reported or suspected to have operating problems due to NPE (Table 3). Their rates of NPE removal (decrease of concentration from influent to effluent) were uniformly high, even in winter, except for one of the pulp mills. Its inability to adequately handle the load of organic material from the pulping process was reflected in its much lower rate of NPE removal. Research in Switzerland showed a direct correlation between treatment plant operating performance, as measured by removal of organic carbon, and NPE treatability [11].

NPE are effectively treated in on-site septic systems [12]; in a 2-year study, the extent of NPE removal by septage leach fields was shown to exceed 99.9% within the leach zones without any contamination of groundwater beneath the field.

TABLE 2   APE Biodegradation in the OECD 301B Test

Test material	Conversion to CO_2, % after 35 days	DOC remaining, %	Test material remaining, %	Biodegradability rating
NP	48.0	3.4	0.0	Inherent
OP	69.6	2.9	0.0	Inherent
NPE9	78.7	3.0	1.5	Inherent
OPE9	81.9	2.4	2.7	Inherent
NPE1.5	57.5	7.9	0.2	Inherent
OPE1.5	64.0	3.3	0.5	Inherent
NPE1C	59.0	Not analyzed	Not analyzed	Inherent
NPE2C	64.6	Not analyzed	Not analyzed	Inherent
OPE1C	71.7	Not analyzed	Not analyzed	Ready
OPE2C	80.0	Not analyzed	Not analyzed	Ready

TABLE 3 Removal of NPE by Wastewater Treatment Plants

Plant location, month	Operating problems	NPE removal rate, %
North Carolina, two plants; Jan., April, May, July	Excessive foaming, wastewater from furniture industry	96–99%
North Carolina, two plants; May	Excessive foaming, wastewater from textile industry	95% to 98%
Wisconsin; March, August	None, wastewater from cleaning product manufacturer	92% in winter, 97% in summer
Wisconsin; April	None, wastewater from paper recycling mill	99%
Alaska, two plants; June, September	None at one pulp mill, second mill seriously overloaded	98% at first plant, 84% at second plant

C. Exposure: APE Environmental Concentrations

A nationwide monitoring survey of NPE concentrations in U.S. rivers was conducted so that the data could be used to construct a statistical model of United States rivers [13]. Thirty river sites known to receive wastewater discharge, and thus expected to have NPE present, were chosen randomly from the U.S. EPA database of river reaches. About 100 samples each of water and sediment, three each per site plus some duplicates, were analyzed for NP and NPE; high quality assurance standards were met.

Table 4 summarizes the ranges of concentrations found of NP, NPE1, and NPE2 (the first ethoxylate oligomers) and the total of NPE3-17 in water, and NP and NPE1 in sediment. It is clear from these results that a small number of the river water samples accounted for the highest concentrations, and most of the samples had nondetectable or very low concentrations. Ten rivers had almost all the high concentrations, while the other 20 were devoid of NPE or nearly so.

An alternate way of presenting the data [14] is as log-normal curves superimposed on frequency distribution plots, not shown here. All views of the data show the same pattern: most locations with minimal or no measurable concentrations of NPE in the water and the sediments, and a small number of locations seriously polluted. The highest levels of NP, the most toxic biodegradation intermediate, were 0.64 µg/L or ppb in water and 2.96 mg/kg or ppm (basis dry weight) in sediment. NP was the dominant intermediate in sediments but a minor one in water (average 3.4% of NPE in rivers with total NPE >1.0 µg/L) [13].

An extensive monitoring study of a river heavily impacted by industrial and municipal wastewater [15,16] provided a test of the Thirty-River U.S. national statistical model. The lower Fox River of Wisconsin is home to about 20 paper mills, a population of 250,000, and 21 wastewater treatment plants. Concentrations of NPE and degradation intermediates were within the scope of the national model.

D. Toxicity of APE Surfactants

1. Aquatic Toxicity

The amount of information on the toxicity of APE, in particular NPE, is very extensive. The largest part of the data is on nonylphenol aquatic toxicity [17–19], but

TABLE 4 Thirty-River Study: Water and Sediment Concentrations

Water	NP	NP-1EO	NP-2EO	NP-3-17EO	
Method detection limit (concentration, μg/L)	0.11 ppb	0.067 ppb No. of water samples	0.063 ppb	1.6 ppb	Samples
<MDL	69	66	59	<MDL	77
MDL–0.3	19	26	28	MDL–5	13
0.3–0.6	9	6	10	5–10	8
0.6–1.2	1	0	4	10–15	3
Average value, μg/L	0.12	0.09	0.1	Average	2.0
Max. value found	0.64	0.6	1.2	Max. value	14.9

Sediment	NP	NP-1EO
Method detection limit concentration, μg/kg	2.9 ppb	2.3 ppb No. of sediment samples
<MDL	16	36
MDL–10	26	30
10–100	24	14
100–1000	12	4
1000–3000	3	0
Average value, μg/kg	162	18
Max. value found	2,960	175

enough studies with NPE are available for a rigorous evaluation of both acute and chronic toxicity as functions of the degree of ethoxylation. NP is more toxic than the ethoxylates, so NP toxicity provides a conservative estimate of NPE toxicity. Less is known about the toxicity of the ether carboxylates; they are receiving more attention, so they will be better understood in the future as new studies are reported.

Table 5 summarizes the range of acute toxicity values (either LC50—concentrations lethal to 50% of the subjects, or EC50—concentrations causing an adverse effect on 50% of the subjects) for NP, water-insoluble NPE, water-soluble NPE and NPEC from the published literature. While the ranges overlap and are very broad, covering 1–3 orders of magnitude, it can be seen that NP is more toxic than NPE.

Some aquatic organisms are widely used to assess the toxicity of chemicals because they are easily grown in the laboratory, they have provided a large amount of data on many chemicals, and the data are commonly accepted by regulatory agencies. While such data are convenient to obtain and compare, it should be recognized that they might not accurately represent toxicity to other organisms, so data from uncommon organisms give them added confidence. Table 6 summarizes NPE toxicity toward five common test species; the data merely show the relative effects of the NPEs and biodegradation intermediates. The test conditions, chronic endpoints, and duration are not given here but are available, along with data from many other reported studies. (Chronic NOEC is the maximum concentration having no observable effect.)

TABLE 5 Acute Aquatic Toxicity Ranges of NPE and Metabolites

	Range of acute values, µg/L		
	Invertebrates	Fish	Algae
NP	21–3,000	17–3,000	27–1,500
NPE1-6	110–5,000	1,300–5,400	6,000–500,000
NPE7-12	1,200->1,000,000	1,000–13,800	210–1,000,000
NPEC	990–14,000	2,000–9,600	

While there are many gaps in the table, a number of trends can be noted:

- Toxicity varies among aqutic species and test chemicals.

  - NP is more toxic than NPE and NPE1C.
  - NP is about twice as toxic as NPE1-2.
  - NP is about 10 times as toxic as NPE8-9 or NPEC.
  - NPE8-9 and NPE1C have similar toxicity.

- The thresholds for chronic effects are much lower than for acute effects.
- Toxicity values for these aquatic species are fairly representative of literature data on many other species. In fact these species are more sensitive than most.

TABLE 6 Selected Aquatic Toxicity Values for NPE and Metabolites

	Acute LC50 or EC50, µg/L	Chronic NOEC, µg/L
	Daphnia magna (water flea, freshwater)	
NP	190	24
NPE2	148	
NPE9	14,000	10,000
NPE1C	14,000	11,000
	Mysidopsis bahia (mysid shrimp, marine)	
NP	43	3.9
NPE1-2	110	7.7
NPE9	1,200	
NPE1C	9,400	
	Pimephales promelas (fathead minnow)	
NP	135	7.4
NPE9	4,600	1,000
NPE1C	2,000	
	Oryzias latipse (killifish)	
NP	1,400	
NPE1	3,000	
NPE8	11,600	
NPE1C	9,600	
	Selenastrum capricornutum (green alga)	
NP	410	

2. Endocrine Disruption

A form of toxicity receiving enormous attention is the disruption of hormone-mediated biological processes within organisms. A number of man-made chemicals display hormone-mimicking, in particular estrogen-like, properties. Nonylphenol and octylphenol and their ethoxylates, along with many other suspected endocrine-disrupting chemicals, are being studied extensively in fish, mammals, and aquatic invertebrates and have been reviewed [19,20].

Both NP and OP are capable of expressing weak estrogenic effects in vitro (cell culture) and in vivo (intact organism). The concentrations required are orders of magnitude higher than those of natural synthetic estrogen hormones, approaching acutely toxic levels in vitro and above chronic toxicity levels in vivo. There is currently much debate over the significance of the effects measured and their relevance to environmental conditions.

Many of the published in vivo studies of NP and OP have been on the sexual development of fish. Two studies [21,22] illustrate the effects seen and the difficulty of interpretation. The first study exposed male trout to NP and OP for three weeks and measured their testicular growth rate and blood concentration of the egg yolk protein vitellogenin. The maximum no-effect levels for inhibition of testicular growth were 20 µg/L and 44 µg/L, respectively, and for vitellogenin 5.0 and 1.6 µg/L, respectively. Thus there was no correlation between the biochemical marker (vitellogenin) and developmental effects. The second study was an extended (six-month) exposure of Japanese medaka to OP and their reproductive success. No clear correlation was found between OP concentration and effects; no effects were seen at 10 µg/L and no consistent effects up to 100 µg/L.

A related study identified the compounds in treated wastewater that caused the observed elevation of vitellogenin in English river fish [23]. The only active substances were natural estrogens of human origin. NP was present in the wastewater but was not in the active fraction. The same pattern was found in the United States [24a,b].

Since actual levels of NP and OP are well below 1 µg/L in rivers except in highly polluted areas of poor or no wastewater treatment, natural estrogen is the likely cause of estrogenic effects observed near treated wastewater discharges.

Studies on mammals for endocrine effects have shown OP and NP to be at most very weakly estrogenic and the common ethoxylates to be inactive [25–27]. It may be concluded that the health risk to mammals, including humans, from exposure to NPE is insignificant.

3. Toxicity Thresholds

When all the reliable data are pooled, they may be treated statistically as distributions. The U.S. EPA has compiled the database of NP toxicity and calculated statistical NP concentration values protective of aquatic ecosystems [18,28] and issued draft water-quality criteria. The proposed freshwater "chronic value" is 5.9 µg/L NP, the threshold below which the aquatic environment is considered safe. NPE and NPEC were not included in the assessment.

Environment Canada recently prepared a similar database but included NPE and NPEC concentration data by applying "toxic equivalency quotient" (TEQ) factors, relating them to NP [29] with the assumption that the weighted toxicities are additive. The TEQ weight factors used were 0.5 for NPE1-2, 0.005 for NPE9, and 0.005 for NPEC.

Figure 4 (based on Figures 9 and 16 of Ref. 29) presents the effect data from a multitude of aquatic toxicity studies on NP, NPE, and NPEC [19] in order of increasing value versus the cumulative percentage of the total database. The plot includes chronic toxicity in addition to the more numerous data from acute tests, the values (LC50s) converted to estimated chronic values assuming an acute-to-chronic ratio of 4. The 5% dashed arrow represents the statistical 95% protection level, in this case about 5 μg/L.

The plot on the left side of the graph displays the measured NP/NPE TEQ concentrations [30] in rivers, lakes, and estuaries, with effluents from municipal wastewater treatment plants divided by 10 to estimate concentrations in the receiving waters of effluent-dominated streams. This is an informative way of displaying surface water concentration data from disparate sources published since 1982; the individual data points each represent a set of data. About 20% of the TEQ plot falls above the statistical 95% protection level. Those 20% are data from seven papers—four studies of one polluted river, the Glatt in Switzerland, two of the most documented "hot spots" of pollution in the United Kingdom, the River Aire and the Tees Estuary, and one of a Spanish river heavily impacted by industry and human population, the Llobregat. All of the next 20% (seven data points) of the TEQ plot are data from the Glatt River and other polluted Swiss streams. This illustration demonstrates that levels of NPE and metabolites in surface water are below the threshold of concern, except at otherwise-polluted sites (Section IV.B).

The three vertical lines within the chart represent the concern thresholds of the European Union (PNEC — predicted no-effect concentration = 0.33 μg/L NP), Environment Canada (ENEV = estimated no-effect value = 1.0 μg/L TEQ), and U.S. EPA (draft Water Quality Criteria chronic value for fresh water = 5.9 μg/L NP). The Environment Canada 95% species protection value of 10 μg/L, derived from a

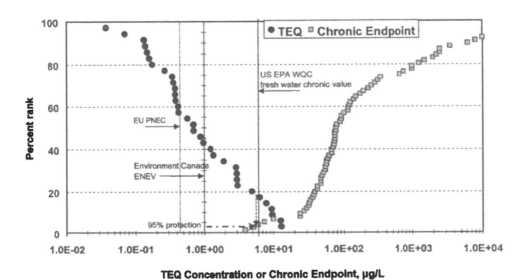

FIG. 4 Toxic equivalency quotient (TEQ) concentrations and toxicity values of NPE and metabolites. (From Ref. 29.)

statistical calculation similar to the U.S. EPA's, was divided by a 10-fold "assessment factor" to cover uncertainties and give the ENEV, which has a protection value approaching 100%. The EPA did not apply any extra factor because of the abundance of assessment data. The EU assessment is discussed later in the chapter.

V. USING WATER-QUALITY CRITERIA
A. Canada

Environment Canada, in the course of developing its NPE risk assessment, identified a small number of problematic sites within Canada where wastewater treatment is missing or inadequate so that receiving waters could be adversely affected. TEQ concentrations in receiving water estimated from effluent analyses exceeded the ENEV of 1 μg/L in these cases. For this reason NPE and NP have been declared "CEPA toxic" under the definition of the Canadian Environmental Protection Act. The link between treatment and compliance with the ENEV will likely be a tool for improving water quality in Canada. Better wastewater treatment will solve many water-quality problems besides APE contamination. Replacing NPE with other surfactants in institutional cleaning formulations or industrial processes to lower effluent TEQs would not achieve that goal. Hence the toxic declaration does not mean product bans or use restrictions, but environmental management that will ensure compliance with the guidance.

B. United States

The EPA's Water Quality Criteria (WQC) document was finalized in early 2004. The EPA has not considered NPE as a priority. In its own risk assessment [31] of NP, the agency stated, "Nationwide, nonylphenol does not pose a significant risk to aquatic organisms but there are rivers that can be impacted"; that is, there may be pollution hot spots. When the criteria are issued, they may be adopted by the states as standards.

The slightly different computational methods of the two federal environment agencies reflect similar approaches to environmental protection. Both are scientific, quantitative, and risk based. The difference between the WQC and ENEV is of less significance than the main purpose of establishing them: defining safe levels and providing industries and municipalities with means to monitor their water-quality performance.

Biological wastewater treatment is the key to compliance with the emerging standards.

C. European Union

The European Union has produced a risk assessment [32] of nonylphenol using procedures lacking the objectivity and rigor of those just discussed. In the absence of monitoring data, default worst-case estimates were used to calculate "predicted environmental concentrations" (PECs). The extensive monitoring data from North America were not used, and the scant European data that existed were mostly from pollution hot spots. Reliance on the older biodegradation literature led to the assumptions of poor treatability of NPEs and NP being the major degradation product of NPEs, now known to be incorrect. The toxicity data used were much the same as those used by Environment Canada and the U.S. EPA; the PNEC calculation initially gave a similar value, about 3.3 μg/L. Then a 10-fold assessment factor was applied, giving the official PNEC of 0.33. The PEC/PNEC ratio (the hazard quotient), after PEC was exaggerated and PNEC was minimized, exceeded 1 in most scenarios.

The three approaches to risk assessment are summarized in Figure 5. The bars represent the ranges of NPE TEQ concentrations divided by the respective toxicity thresholds (the hazard quotients) used by the U.S., Canada, and the EU. It is readily seen that the U.S. EPA gives NPE a low priority, Environment Canada gives it medium priority, and the EU gives it high priority. The EU, after concluding that the NPE environmental risk is high and pervasive, now is planning to institute APE product bans in most applications rather than allow treatment or other pollution management measures [33].

D. Toronto, Canada

More questionable from a scientific point of view is the local action by the city of Toronto, Canada, to adopt a de facto ban on industrial and institutional uses of NPE-containing products. (Household cleaning products are not included.) The Sewer Use By-Law [34] of 2000 bans the discharge of wastewater containing more than traces of NPEs to the sewer system. Risk assessment was not used to demonstrate an environmental problem before this measure was introduced; the action appears to have been solely political. Environment Canada monitoring of the city's treatment plants showed them to be functioning well enough that their effluent ENEVs were not exceeded [35].

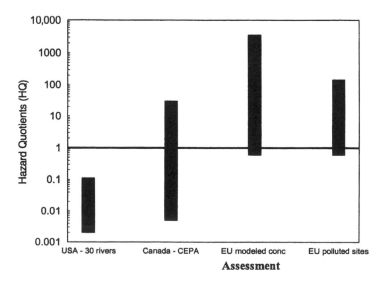

FIG. 5 The use of hazard quotients by the U.S., Canada, and the EU for NPE risk. (From Refs. 13,18,29.)

VI. GUIDELINES FOR USE AND DISPOSAL OF APE SURFACTANTS

Formulators and consumers of cleaning products can help ensure the continued availability of APE surfactants by following these guidelines for responsible use, handling, and disposal.

- Train workers to follow safe handling practices.
- Avoid overuse and misuse.
- Minimize product waste and loss.
- Ensure proper down-the-drain disposal of all wastewater to a biological treatment facility.
- Encourage and support improvements to local wastewater treatment plants or to their construction if they do not yet exist.
- Provide these handling practices and other relevant safety information to customers and consumers.
- Be informed on APE regulatory issues and scientific facts.
- Contact the APE Research Council www.aperc.org) with questions.

ACKNOWLEDGMENTS

The author is grateful to APE Research Council colleagues for valuable comments and suggests for the manuscript, S.G. Gilbert of Dow (formerly Union Carbide), Dr. C.A. Staples of Assessment Technologies, Inc. and Dr. J.E. Heinze of John Adams Associates.

REFERENCES

1. Talmage, S.S. *Environmental and Human Safety of Major Surfactants—Alcohol Ethoxylates and Alkylphenol Ethoxylates, Part II.* Soap and Detergent Association, Lewis Publishers: Boca Raton, FL, 1994.
2. Wheeler, T.F.; Heim, J.R.; LaTorre, M.R.; Janes, A.B. J. Chromatographic Sci. 1997, *35* (1), 19–30.
3. Rosen, M.J. *Surfactants and Interfacial Phenomena,* 2nd Ed.; Wiley: New York, 1989.
4a. Summary of the report Higher Alcohols: Market Forecast to 2010, from the Web site of Colin A. Houston Associates: http://www.colin-houston.com/Press_Releases/Press_Releases.htm.
4b. US International Trade Commission Publication No. 2933, Synthetic Organic Chemicals 1994, Table 3.1.
4c. Data from CEPAD (European Council for Alkylphenols and Derivatives), a CEFIC sector group, Brussels.
5. Staples, C.A.; Naylor, C.G.; Williams, J.B.; Gledhill, W.E. Environ. Toxicol. Chem. 2001, *20* (11), 2450–2455.
6. OECD Guidelines for the Testing of Chemicals, vol. 2, OECD Bookshop, Paris, France, http://oecdpublications.gfi-nb.com/cgi-bin/oecdbookshop.storefront, 1993.
7. Staples, C.A.; Williams, J.B.; Blessing, R.L.; Varineau, P.T. Chemosphere 1998, *38* (9), 2029–2039.
8. Naylor, C.G. Soap Cosmetics Chem. Specialties 1992, *68* (8), 27–31.
9. Naylor, C.G. Industrial Wastewater Sept/Oct 1996; 61–64.
10a. Naylor, C.G. Textile Chemist Colorist 1995, *27* (4), 29–33.

10b. Naylor, C.G. AATCC Rev. 2001, *1* (3), 34–36.

11. Ahel, M.; Giger, W.; Koch, M. Water Res. 1994, *28* (5), 1131–1142.

12. Naylor, C.G.; Huntsman, B.E.; Solch, J.G.; Staples, C.A.; Williams, J.B. Proceedings: CESIO 5th World Surfactants Congress; 2000; Vol. 2, 1645–1653.

13. Naylor, C.G.; Mieure, J.P.; Adams, W.J.; Weeks, J.A.; Castaldi, F.J.; Ogle, L.D.; Romano, R.R. J. Am. Oil Chem. Soc. 1992, *69* (8), 695–703.

14. Weeks, J.A.; Adams, W.J.; Guiney, P.D.; Hall, J.F.; Naylor, C.G. Proceedings: CESIO 4th World Surfactants Congress; 1996; Vol. 3, 276–291.

15. Naylor, C.G.; Williams, J.B.; Varineau, P.; Webb, D. Proceedings: CESIO 4th World Surfactants Congress; 1996; Vol. 4, 378–391.

16. Field, J.; Reed, R. Environ. Sci. Technol. 1996, *30* (12), 3544–3550.

17. Staples, C.A.; Weeks, J.; Hall, J.F.; Naylor, C.G. Environ. Toxicol. Chem. 1998, *17* (12), 2470–2480.

18. US EPA. Ambient Aquatic Life Water Quality Criteria: Nonylphenol, EPA Contract No. 68-C6-0036, Work Assignment No. B-05, draft of September 1998.

19. Servos, M.R. Water Quality J. Canada 1999, *34* (1), 123–177.

20. Nimrod, A.; Benson, W.H. Crit. Rev. Toxicol. 1996, *26*, 335–364.

21. Jobling, S.; Sheahan, D.; Osborne, J.A.; Matthiessen, P.; Sumpter, J.P. Environ. Toxicol. Chem. 1996, *15* (2), 194–202.

22. Gray, M.A.; Metcalfe, C.D. Aquat. Toxicol. 1999, *46*, 149–154.

23. Desbrow, C.; Routledge, E.J.; Brighty, G.C.; Sumpter, J.P.; Waldock, M. Environ. Sci. Technol. 1998, *32* (8), 1549–1558.

24a. Environmental News Section. Environ. Sci. Technol. 1998, *32* (1), 8A.

24b. Snyder, S.A.; Keith, T.L.; Verbrugge, D.A.; Snyder, E.M.; Gross, T.S.; Kannan, K.; Giesy, J.P. Environ. Sci. Technol. 1999, *33* (16), 2814–2820.

25. Cunny, H.C.; Mayes, B.A.; Rosica, K.A.; Trutter, J.A.; Van Miller, J.P. Regul. Toxicol. Pharmacol. 1997, *29*, 172–178.

26. Williams, J.; Brady, A.M.; Lewis, R.W.; Hughes, L. Proceedings: Fourth CESIO World Surfactants Congress; 1996; Vol 3, 34–41.

27. Tyl, R.W.; Myers, C.B.; Marr, M.C.; Brine, D.R.; Fail, P.A.; Seely, J.C.; Van Miller, J.P. Regul. Toxicol. Pharmacol. 1999, *30* (2), 81–95.

28. Stephen, C.E.; Mount, D.I.; Hansen, D.J.; Gentile, J.H.; Chapman, G.A.; Brungs, W.A. Guidelines for Deriving Numerical National Water Quality Criteria for the Protection of Aquatic Organisms and Their Uses; National Technical Information Service: Springfield, VA, 1985 (Report PB85-227049).

29. Canadian Environmental Protection Act Priority Substance List Assessment Report: Nonylphenol and Its Ethoxylates, Environment Canada and Health Canada, draft for public comments, March 2000.

30. Bennie, D.T. Water Quality J. Canada 1999, *34* (1), 79–122.

31. Rodier, D. US EPA Office of Pollution Prevention and Toxics, CSRAD, RM-1 Document for para-Nonylphenol, October 2, 1996.

32. EU Risk Assessment of Nonylphenol. Draft of September 1999, latest version available.

33. Final Report, Nonylphenol Risk Reduction, commissioned by the UK Department of the Environment, Transport and the Regions, for the EU, Risk and Policy Analysts, Lodden, Norfolk, UK, Sept. 1999.

34. New sewer use bylaw no. 457-2000 passed by Toronto City Council on July 6, 2000. Available at: http://www.city.toronto.on.ca/involved/wpc/nbylaw.htm.

35. Bennie, D.T.; Sullivan, C.A.; Lee, H.-B.; Maguire, R.J. Water Quality J. Canada 1998, *33* (2), 231–252.

16
Estrogenic Effects of the Alkylphenol Ethoxylates and Their Biodegradation Products

KAREN L. THORPE * **and CHARLES R. TYLER**   Exeter University, Exeter, Devon, United Kingdom

I. INTRODUCTION

Observations of reproductive abnormalities in humans and wildlife over the last few decades have raised concerns about the potential for environmental contaminants to disrupt the endocrine system [1–7]. These *endocrine-disrupting*, or *endocrine-active*, chemicals are believed to interfere with the function of endogenous hormones, in particular the female sex steroid estradiol-17β (E2), and in turn to cause alterations in normal growth, development, and/or reproduction.

It has been known for more than 60 years that simple alkylphenols (AP) can bind to the estrogen receptor (ER) and act as weak oestrogens [8,9]. Concerns about their potential environmental estrogenic activity were not raised, however, until 1991, when it was reported that *para*-nonylphenol (NP; a contaminant in the antioxidant additive in polystyrene, which is released by leaching) was capable of promoting the proliferation of estrogen-dependent MCF-7 breast cancer cells [10]. Nonylphenol is a biodegradation product of a group of nonionic surfactants known as the alkylphenol polyethoxylates (APEO), which are used as ingredients in domestic and industrial cleaning products and in many industrial processes. The presence of the APEO in the environment and the weak estrogen-like activity of their biodegradation products have prompted a number of investigations into the potential reproductive effects of these and other groups of surfactants. It has been demonstrated, however, that the estrogenic activity of the surfactants is restricted to the biodegradation products of the APEO [11]. Parent anionic, nonionic, cationic, and amphoteric surfactants do not themselves possess the ability to bind the ER [11], neither do the sulfophenyl carboxylates, which are the biotransformation metabolites of the linear akylbenzene sulfonates [11,12].

Current affiliation: AstraZeneca UK Limited, Brixham, Devon, United Kingdom.

This chapter considers the potential estrogenic effects of the biodegradation products of the APEO in the environment. Information on the occurrence and estrogenic effects of the major biodegradation intermediates of the APEO in the aquatic environment and their estrogenic effects in fish, mammals, and invertebrates exposed under laboratory conditions is reviewed. Because the majority of the APEO are discharged to the aquatic environment via sewage treatment work (STW) effluents, they are likely to be present as complex mixtures with other estrogens. In the final section of this chapter, a case study is included to consider the potential for interactive effects of AP with other estrogenic chemicals.

For more extensive reviews on some of the aspects discussed in this chapter, readers are referred to the other chapters of this handbook as well as the comprehensive reviews provided by Nimrod and Benson [13], Bennie [14], and Servos [15].

II. CONCENTRATIONS OF THE ALKYLPHENOLIC CHEMICALS IN THE AQUATIC ENVIRONMENT

It is estimated that around 60% of the APEO used every year in the United States [16] enter the aquatic environment, following their disposal in wastewater. Nearly all of this receives biological treatment before discharge into rivers and other water bodies. Alkylphenol ethoxylates discharged to modern STW plants generally undergo aerobic activated-sludge treatment that routinely removes more than 90% of the APEO [17,18]. The parent APEO are transformed into 1- or 2-mol APEO, alkylphenol ether carboxylates (APEC), and AP [16,17,19,20]. Normally the AP occur in minute amounts compared with the APEO and the APEC [20]. Although these metabolites are often described as persistent [17], evidence has recently been provided that shows the extensive biodegradability of APEO and their biodegradation intermediates. Staples et al. [21] derived ultimate biodegradation half-lives of 7–28 days for APEO and their biodegradation intermediates, in aerobic conditions. Despite this apparent lack of persistence in the environment, the biodegradation intermediates of the parent APEO have been measured in STW effluents discharging into the environment. In a study of effluents from STWs in the United Kingdom, AP1EO, AP2EO, and AP were identified as major estrogenic contaminants in effluents from Keighley STW, which discharges into the River Aire [22,23]. Along the Aire, wool-scouring plants use APEO in large amounts to wash grease from the fleeces, and the liquor is then discharged via Keighley STW into the river. In 1994, the mean concentration of APEO discharging into the River Aire was 950 μg/L, and the concentration of NP exceeded 300 μg/L [23]. Following commissioning of a new treatment plant by a wool-scouring operation, the mean concentration of APEO discharging into the River Aire was 450 μg/L and the concentration of NP was consistently less than 100 μg/L [23]. The APEO have also been identified (measured concentrations of NP and NP1EO of 3 μg/L and 45 μg/L, respectively) as contributors to the estrogenic activity in effluents from Howdon STW, which is a major STW discharging into the industrialized and urbanized Tyne Estuary [24]. In effluents from STWs receiving primarily domestic waste in the United Kingdom, the APEO have generally been discounted as major contributors to the estrogenic activity [25,26]. As an example, in effluent from the Chelmsford STWs (an STW receiving primarily domestic waste in South East England), NP was measured at concentrations ranging from 1.2 to 2.7 μg/L and from <0.2 to 8.9 μg/L for NP(1 + 2)EO

[26]. In effluents in Europe, NP has been measured at concentrations ranging from 36 to 202 µgL in Switzerland [27] and at concentrations ranging from 0.7 to 9.7 µg/L in Italy [28]. In effluents from 18 different STWs in Baden-Württemberg, southwestern Germany, NP was measured at an average concentration of 0.19 µg/L [29]. Little information is available on the concentrations of the NPEC in the environment, despite the fact that they are more likely to be present in the water column due to their higher water solubility than the AP. In Europe, NP1EC has been measured in effluents at concentrations ranging from 1.5 to 3.9 µg/L in Italy [28] and at concentrations up to 5.8 µg/L in Germany [29]. In the United States, in 15 paper mill effluents discharging into the Fox River (near Green Bay, WI), the total concentration of NPEC (NP1EC–NP4EC) ranged from below detection (<0.4 µg/L) up to 1300 µg/L but was typically less than 100 µg/L [19]. The concentrations of NPEC measured in six STW effluents discharging into the same river ranged from 140 to 270 µg/L. In all cases, NP2EC was the dominant oligomer, constituting 72% and 54% of the total NPEC in paper mill and STW effluents, respectively [19].

Analysis of surface waters has revealed that concentrations of the APEO residues are typically low. In the United Kingdom, Blackburn et al. [30] found that in 75% of rivers and estuaries, concentrations of total aqueous APEO (NP, OP, NPEO, and NP2EO) were below the limit of detection (<0.1–0.6 µg/L), but measurable concentrations were recorded in the River Aire (15–76 µg/L), the River Mersey (6–11 µg/L), and the Tees Estuary (up to 76 µg/L). In rivers in Switzerland, the average total concentrations of the lipophilic metabolites, NP, NP1EO, and NP2EO, ranged from 0.05 to 0.3 µg/L, with NP being the most abundant [31]. The total concentration of NPnEC in the Glatt River was found in the range from 0.5 to 3 µg/L [31]. In the state of Baden-Württemberg, Germany, 4-NP was measured in most (70%) of the samples at concentrations up to 0.458 µg/L, and 4-t-OP was present in 87% of samples at concentrations up to 0.189 µg/L [32]. In harbor and coastal areas in Spain, NPEO have been measured at concentrations up to 11 µg/L, with average concentrations ranging between 0.3 and 5 µg/L [33]. Nonylphenol was found in 47% of these water samples at concentrations between <0.15 and 1.0 µg/L, OP was detected (0.3 µg/L) at only one site, near the outflow of a chemical plant, and NP1EC concentrations were below the limit of detection [33]. In Israel, APEO were measured at concentrations ranging from 12.5 to 75.1 µg/L in rivers and at concentrations ranging from 4.2 to 25.0 µg/L in the eastern Mediterranean seacoastal water, at locations close to where sewage-containing rivers/streams flow into the sea [34]. In a survey of 30 U.S. rivers receiving wastewater discharges, 70% of the sites had concentrations of NP of 0.1 µg/L or less [18]. The total concentration of NPEC in the Fox River (USA), downstream of the paper mill effluents, was 13.5 µg/L [19]. In an additional five out of eight rivers sampled in the United States, NPECs were measured at concentrations ranging from 1.4 to 6.3 µ/L [19]. In all those rivers containing measurable concentrations of the NPEC, NP2EC was the dominant oligomer and was present at between two and seven times the concentration of NP1EC [19].

Concentrations of AP tend to be higher in sediments than in the water column, due to their lipophilic nature and their poor biodegradation under anaerobic conditions [35,36]. In heavily contaminated sediments in the United Kingdom, concentrations of APEO have been measured at between 1600 and 9050 µg/kg for NP, between 125 and 3970 µg/kg for NP1EO, and between 30 and 340 µg/kg for OP (dry weight), in the Tees Estuary [24]. At a less contaminated UK site (Tyne Estuary), concentrations of APEO measured in sediments ranged from 30 to 80 µg/kg for NP, from 160 to 1400 µg/kg for NP1EO, and from 2 to 20 µg/kg for OP (dry weight) [24]. In sediment samples collected in Spain, concentrations of NPEO and NP ranged between 35 and 620 µg/kg and from <5 to 1000 µg/kg, respectively [33]. Octylphenol was found in only three of these Spanish samples at concentrations ranging between 17 and 145 µg/kg [33]. In Germany, sediment concentrations of NP and OP were measured at between 10 and 259 µg/kg and from <0.5 to 9 µg/kg (dry weight), respectively [32].

Alkylphenolic chemicals have also been measured in the body tissue of wild fish taken from rivers and estuaries polluted with the biodegradation products of the APEO. In flounder (*Platichthys flesus*) from the Tyne Estuary in the United Kingdom, NP was detected in muscle and liver tissues at concentrations between 5 and 60 ng/g wet weight, and NP1EO was detected in liver tissues at concentrations between 190 and 940 ng/g [24]. In the Tees Estuary, NP was detected in muscle and liver tissues from the flounder at concentrations between 30 and 180 ng/g [24]. In the River Aire in the United Kingdom, wild gudgeon (*Gobio gobio*) contained up to 0.8 µg/g NP in their muscle tissue, and wild chub (*Leuciscus cephalus*) contained up to 9.5 µg/g NP1EO + NP2EO in their liver tissue [30]. In a single common carp, living in Las Vegas Bay of Lake Mead (NV, USA), NP and NP1EO were detected at concentrations of 184 and 242 ng/g wet weight, respectively [37]. The concentrations of the APEOs measured in the wild fish are generally higher than those measured in the surrounding waters, as a consequence of their tendency to bioconcentrate in animals. For wild fish from the River Aire, Blackburn et al. [30] calculated bioconcentration factors (BCFs) of approximately 50 and 475 for NP and NP1EO + NP2EO, respectively. These values are consistent with BCFs of 13–408 for NP, 3–300 for NP1EO, and 3–326 for NP2EO observed in fish from the surface waters in the Glatt Valley, Switzerland [38]. In rainbow trout (*Oncorhynchus mykiss*) exposed to 65 µg/L NP for 3 weeks in the laboratory, Blackburn et al. [30] observed a BCF of between 90 and 125. Octylphenol has also been shown to bioconcentrate in fish; a BCF of 471 was estimated for an exposure via the water for 10 days [39]. The ability of NP and its ethoxylates and OP to bioconcentrate in fish needs to be realized when considering the biological activity/ potency from lab-based studies in the following sections (especially for studies based on short-term exposures).

III. ESTROGENIC EFFECTS OCCURRING IN THE AQUATIC ENVIRONMENT: ASSOCIATION WITH ALKYLPHENOLIC CHEMICALS

Although the presence of the APEO and their degradation products in the aquatic environment has been clearly demonstrated, the extent to which they contribute to endocrine activity in the environment is at present unknown. A number of studies have established a clear link between exposure to STW effluent and effects on the endocrine system and/or reproductive physiology of fish. These effects include induction of the estrogenic biomarker vitellogenin (VTG) in the plasma of fish (in both Europe and the United States) [22,23,26,40–46], reduced testicular growth in male rainbow trout, *Oncorhynchus mykiss* [23], an inhibition in both testicular growth and secondary sexual characteristics in male fathead minnow, *Pimephales promelas* [46], intersexuality in wild populations of roach, *Rutilus rutilus* [47], and gudgeon, *Gobio gobio* [48], in UK rivers and feminization of the reproductive duct in the early life stages of the roach [49]. Given that the vitellogenic response is an estrogen-dependent process and that reductions in testicular growth [50] and secondary sexual characteristics [51] and feminization of the reproductive duct [52] and formation of oocytes in the testis of males [53] can be induced by exposure to estrogens, attempts to identify the causative agents in effluents have focused on estrogens and estrogen-like chemicals. Toxicity identification and evaluation (TIE) studies to isolate and identify the major estrogenic chemicals present in STW effluents have shown that in domestic effluents the most active (estrogenic) fraction (>80% total activity in domestic effluent) contains the natural and synthetic steroidal estrogens [25,26,29,46]. In most effluents studied, alkylphenolic compounds make little contribution to the observed estrogenic activity. Alkylphenolic compounds are present, however, at much higher concentrations in effluents from STWs that receive trade influents and may, therefore, be significant contributors to the estrogenic activity in rivers receiving large volumes of effluents from industrial sources. In an effluent from Keighley STW (UK), which treats between 6% and 10% trade effluent from a variety of sources, including wool-scouring and other textile industries, Sheahan et al. [22] determined that the major estrogenic fractions contained NP (63 μg/L), NP(1 + 2)EO (230 μg/L), estrone (18 ng/L), estradiol (4 ng/L), and ethynylestradiol (<0.1 ng/L). Using potency estimates and estradiol-equivalent values they determined that the combined estrogenic equivalent of the AP together with the steroids would induce a vitellogenic response (10 mg/mL) similar to that observed in male rainbow trout placed at the point of effluent discharge for a 3-week period. However, the observed vitellogenic response at 2 and 5 km downstream was higher than expected, using estimated estradiol equivalents, based on the concentrations of the chemicals present at these sampling points [22]. This higher-than-expected estrogenic response was attributed to the presence of more soluble estrogenic compounds, such as the NPEC. The results of this investigation have provided the first evidence of a close link between the presence of alkylphenolics in the environment and an estrogenic response (VTG induction). Furthermore, the magnitude of the vitellogenic response was observed to decrease between 1994 and 1995, concomitant with decreasing concentrations of the APEO in the effluent [23], supporting the evidence for a link between alkylphenolics and estrogenic responses in fish.

IV. ESTROGENIC EFFECTS IN FISH

The most conclusive evidence for the estrogenic effects of the APEO and their degradation products is provided by laboratory studies employing fish. The most widely studied biological response in fish exposed to environmental estrogens is the induction of the hepatic phospholipoglycoprotein vitellogenin (VTG). In fish, VTG induction is specifically an estrogen-dependent process, normally restricted to females; however, immature females and male fish possess the machinery for VTG production [54], and exposure to estrogens (and their mimics) can trigger VTG synthesis via the ER [55–57]. A number of investigations have clearly demonstrated the ability of the alkylphenolic degradation products to bind the ER and induce VTG production in a concentration-dependant manner. Reported lowest-observed-effect concentrations (LOECs) for induction of VTG by 4-NP range from 4 to 51.7 μg/L, depending on the fish species, age, and sex employed and the duration of exposure. In adult male swordtails [58] and sheepshead minnows [59], exposed to NP for 3 and 5 days, respectively, the observed LOECs for induction of VTG were 4 and 5.4 μg/L. In rainbow trout, reported LOECs range from 6.4 µg/L in immature females exposed for 2 weeks [60] and 8.3 µg/L in adult females exposed for 6 weeks [61] to 10 µg/L in immature females and males exposed for 3 days [62] and up to 20 µg/L in adult males exposed for 3 weeks [63]. In adult fathead minnows, LOECs for induction of VTG after exposure to 4-NP for 3 weeks, range from 8.1 µg/L in males to 51.7 µg/L in females [51]. For OP, LOECs vary between 4.8 µg/L [63] and 10 µg/L [64] in adult male rainbow trout exposed for 3 weeks and up to 100 µg/L in adult roach exposed for 3 weeks [64]. Induction of VTG has also been observed in adult male rainbow trout exposed for 3 weeks to 30 µg/L of either NP1EC or NP2EO [63]. The ability of the AP to stimulate VTG production suggests that they are acting as estrogens through binding the ER.

The AP have also been shown to affect tissue growth and tissue morphology. Ashfield et al. [65] exposed juvenile female rainbow trout to NP, OP, NP2EO, or NP1EC (each at 1, 10, 50 µg/L) from hatch for 35 days, followed by 431 days in clean water. Modifications in growth (increases or decreases in weight) were observed in fish sampled at the end of the experiment. Treatment with NP (30 µg/L) also increased the weight of the ovary relative to the body weight of the fish (the gonadosomatic index; GSI). NP1EC treatment caused a significant reduction of the GSI at 1 and 10 µg/L (although no effect was observed with the highest dose). Neither OP nor NP2EO caused significant changes in the GSI (as recorded at the end of the study) [65]. In contrast to the results of Ashfield et al. [65] for NP, Harris et al. [61] observed a decrease in GSI in adult female rainbow trout exposed for 18 weeks to 85.6 µg NP/L. Inhibitions in gonadal growth have also been reported in adult male rainbow trout exposed to NP, OP, NP1EC, or NP2EO (30 µg/L for 3 weeks) [63]. A concentration of 54.3 µg NP/L completely inhibited testicular growth in trout [63]. Reductions in the GSI were also observed in male eelpout (*Zoarces viviparous*) following intraperitoneal injections with NP (10 and 100 µg/g body weight/week) [66,67]. In these experiments, histological examination of the testis from the eelpout revealed that NP treatment also affected the Sertoli cells, causing degeneration of the germ cells during active spermatogenesis [66,67]. Treatment with NP also resulted in a reduction in the amount of milt in the eelpout testis during late spermatogenesis [66,67]. Similar effects on the testis were observed in male eelpout exposed to E2, indicating that these effects of NP were a consequence of the estrogenic activity of NP [66,67]. Effects of NP on gonad development have been reported in Japanese medaka, *Oryzias latipes*, [68]. Induction of testis-ova was observed in 50% and 86% of male medaka exposed to 50 and 100 µg/L of NP, respectively (where the exposure was from hatch for 3 months). The sex ratio at the highest concentration was also biased toward females relative to the controls [68]. In a life cycle study in the medaka, exposure of the parent (F_0) generation to 51.5 µg NP/L from 24 days postfertilization to 104 days posthatch resulted in a sex ratio skewed toward female, with 40% of the fish having testis-ova [69]. In this study, induction of testis-ova was also observed in 20% of the fish exposed to 17.7 µg NP/L. Fecundity was unaffected in females exposed to 4-NP at concentrations between 4.2 and 17.7 µg/L. In the F_1 generation, there were no effects on hatching success, posthatch mortality, or growth, but the sexual differentiation at 60 days posthatch was affected. Induction of testis-ova in the gonads of the F_1 fish was observed in 10% and 25% of the fish in the 8.2 and 17.7 µg/L treatments, respectively [69]. Exposure to 100 µg NP/L has also been shown to affect the histopathology of the testis and suppress spermatogenesis in adult male swordtails, *Xiphophorus helleri* [58].

Effects of OP, a mixture of NP(1 + 2)EO, and 4-*tert*-pentylphenol (TPP) on the testis have also been reported. Induction of testis-ova was observed in 29% of male medaka exposed from 3 to 100 days posthatch to 100 μg OP/L [70] and in one of 30 males exposed from 1 to 90 days posthatch to 100 μg NP(1 + 2)EO/L [71]. Exposure of adult male guppies (*Poecilia reticulata*) to OP (100 and 300 μg/L) for 30 days resulted in an increased number of sperm cells in the ejaculates and inhibited testis growth [72]. Exposure of sexually immature juvenile male summer flounder, via injection with OP at 100 mg/kg, also resulted in reduced testicular size [73]. Exposure of genetically male carp for 60 days during the period of sexual differentiation to 4-*tert*-pentylphenol (TPP) nominally at 100, 320, and 1000 μg/L resulted in a very poorly developed testis with an oviduct in most fish where spermatogenesis was severely inhibited. In some individuals (two out of six), exposure to the highest TPP concentration induced the formation of ovo-testis and severely inhibited spermatogenesis [52]. In a further investigation, Gimeno et al. [74] demonstrated that the formation of the oviduct in male carp following exposure to TPP for the period encompassing sexual differentiation was a permanent effect.

Exposure to NP has also been shown to reduce the prominence of androgen-dependent secondary sexual characteristics, indicating a feminizing or demasculinizing effect. Kwak et al. [58] observed a concentration-related inhibition of the sword growth in 30-day-old juvenile male swordtails exposed to NP at concentrations between 0.2 and 20 μg/L for 60 days. In adult male fathead minnows exposed for 21 days to NP (at concentrations between 0.65 and 57.7 μg/L), there was a concentration-related reduction in the thickness of the dorsal fatpad and a reduction in the number of nuptial tubercles at the highest (57.7 μg/L) exposure concentration [51].

For pair-breeding fathead minnows, exposure to 4-NP reduced fecundity (number of eggs, frequency of spawning) in a dose-related manner, with an LOEC between 8.1 and 57.7 μg/L [51]. Concentrations of NP at or above 48 μg/L inhibited reproduction completely [51]. Reduced male courtship activity and a reduced fertilization success was observed when unexposed female medaka were mated with male medaka that had been exposed to 25 and 50 μg OP/L from 1 day to 6 months posthatch [75]. Various developmental problems were also observed in the F1 generation upon hatch from the matings with males exposed to 10 and 25 μg OP/L [75].

The results of these studies demonstrate that the AP can induce a range of estrogenic responses in fish, resulting in alterations in normal growth, development, and reproduction. With the exception of effects on VTG induction and secondary sexual characteristics, which occur at environmentally relevant concentrations of the AP, most of these effects were observed only at relatively high exposure concentrations.

V. ESTROGENIC EFFECTS IN MAMMALS

Investigations into the potential endocrine-disrupting effects of the biodegradation products of the APEO in mammalian systems have focused on NP and OP. For both NP and OP, different test species, life stages, and duration and route of exposures have been employed, making it difficult to compare and contrast the reported results. The effects of both NP and OP are therefore considered separately, according to the exposure route used, injection versus oral exposure.

The most widely accepted and "classical" assay for the measurement of estrogenicity in mammals is the measurement of growth in the reproductive tract of sexually immature female rodents [76–78]. Using this assay, both NP, administered by intraperitoneal injection (1 mg/animal; [79]), and OP, administered by subcutaneous injection (10 mg; [80]) have been shown to induce estrogen-dependent uterine growth in immature female rats in a manner similar to E2. Further investigations into the effects of NP and OP have demonstrated other altered female reproductive parameters. In neonatal female rats given daily subcutaneous injections of 500 mg NP/kg/day from postnatal day 1 to day 5, development of the ovaries was disrupted, with an increased atresia of the follicles, a decrease in the corpora lutea, and abnormal luminal epithelial cells in the uterus [81]. An altered estrous cycle and abnormal reproductive function were also observed [81]. In female pups exposed via injection to 1 mg OP the day after birth and in adults exposed via injection to 20 or 40 mg OP, persistent vaginal estrous was observed [82].

Effects on the development of the reproductive tract have also been reported in immature male rats exposed to NP via injection. Decreases in the size of the testes, epididymis, seminal vesicle, and ventral prostate were observed in neonatal male rats exposed to NP (8 mg/kg/day) via intraperitoneal injection for 15 days [83]. Macroscopic and microscopic alterations of the gonads, characterized by marked edema or atrophy of the testes and accessory sex organs, and suppression of spermatogenesis were observed in neonatal male rats given daily subcutaneous injections of 500 mg NP/kg/day from postnatal day 1 to day 5 [81]. No abnormalities in locomotor activity, sperm motion, or plasma testosterone concentrations were observed [81]. Chronic administration of OP (80 mg) to adult male rats via subcutaneous injection suppressed luteinizing hormone (LH), follicle-stimulating hormone (FSH), and testosterone secretion and enhanced prolactin secretion [84]. Reductions in the number of sperm and effects on the sizes, weights, and histological structures of the testes, epididymides, ventral prostate glands, seminal vesicles, and coagulating glands were also observed [85]. Additional evaluation of sperm morphology revealed marked increases in the proportions of head and tail abnormalities from exposed males [85]. These results for OP were similar to those obtained for estradiol valerate, suggesting that the effects of OP were a consequence of an estrogenic activity [84,85].

Exposure of adult Sprague-Dawley rats for 90 days to *para*-NP via the diet (at concentrations ranging from 200 to 2000 ppm; 15–150 mg/kg/day) did not induce any treatment-related effects on endocrine organs, estrous cycling, or sperm measurements [86]. However, using the same exposure regime and test concentrations, Chapin et al. [87] observed indications of estrogenic activity in female rats of this strain when exposed as pups over three generations, with accelerated vaginal opening and disruption of the estrous cycle, at the highest exposure concentrations. No consistent reproductive changes were observed in the males. Administration of higher concentrations of 4-NP (400 mg/kg) via the diet, however, resulted in decreased testicular mass and epididymal mass and total cauda epididymal sperm count in adult male Sprague-Dawley rats [88]. Similar effects were also observed in males exposed to *p*-NP (250 mg/kg) via the diet during fetal life and for the duration of lactation [89]. In a two-generation reproduction study with *t*-OP in rats exposed orally to 0.2–2000 ppm, there were no treatment-related effects in reproductive measurements, reproductive organs, or extensive evaluations of sperm measurements in three generations of males (F_0, F_1, and F_2), leading the authors to conclude that exposure to OP via the diet produces no estrogenic activity in the rat [90].

Exposure of Wistar rats in utero to 15 or 75 mg/kg/day of 4-*n*-NP on gestation days 11 through 18 resulted in a dose-dependent reduction in the absolute weight of the right epididymis, but this effect was not significant when related to the body weights [91].

The effect of intrauterine exposure to t-OP (10 or 1000 μg t-OP/kg body weight) on reproductive parameters in the pig over three generations demonstrated that OP extended pregnancy length and induced cervical proliferation in the F_0 adults and accelerated onset and reduced litter size in F_1 females [92]. This suggests that exposure of sows to OP during pregnancy might interfere with fetal maturation or induction of parturition as well as the prepubertal stimulation of follicular growth and fertility in the F_1 generation. Farrowing rate of the F_0 and F_1 female pigs and the number of stillborn piglets, sex ratio, and birth weight of F_1 and F_2 offspring were not affected by OP. Octylphenol did not increase ERα activity in the maternal part of the placenta of F_0 sows. The morphology of genital organs of F_1 and F_2 offspring as well as semen production in F_1 boars were unaffected.

VI. ESTROGENIC EFFECTS IN INVERTEBRATES

Although invertebrates represent a large proportion of the known species in the animal kingdom, there has been very little investigation into the effects of endocrine-active chemicals on their growth, development, or reproduction. This is partially due to the fact that their hormonal systems are poorly understood in comparison with vertebrates. The best-documented example of the effects of an AP on invertebrates is provided by Oehlmann et al. [93], who investigated the responses of the freshwater ramshorn snail, *Marisa cornuarietis*, and the marine dogwhelk, *Nucella lapillus*, to OP exposure. Exposure of the ramshorn snail over two generations to OP (nominally 1 μg/L) induced a complex syndrome of alterations in the F_0 and F_1 females referred to as *superfemales*. Affected specimens were characterized by the formation of additional female organs, an enlargement of the accessory pallial sex glands, gross malformations of the pallial oviduct section resulting in an increased female mortality, and a massive stimulation of oocyte and spawning mass production. Similarly, superfemales were observed in the marine dogwhelk following exposure to OP for 3 months (nominally 1 μg/L). In contrast with the ramshorn snail, no oviduct malformations were found in the dogwhelk; this was attributed to species differences in the gross anatomical structure of the pallial oviduct. Additionally, a lower percentage of exposed male dogwhelks had ripe sperm stored in their vesicular seminalis, and the males exhibited a reduced penis length and prostate gland when compared with controls. The results of this study demonstrate that the prosobranch snails are sensitive to OP, which causes endocrine disruption at environmentally relevant concentrations.

Nonylphenol has also been shown to have effects on invertebrates, but at much higher concentrations than for OP. Exposure to 100 μg NP/L altered the uptake and metabolism of testosterone and decreased fecundity in *Daphnia magna* [94] and altered the number of female offspring produced by *Daphnia galeata mendotae* [95]. Prenatal exposure of *Daphnia galeata mendotae* to 10 μg NP/L affected the morphological development of the offspring [95]. The significance of these effects of APs for wild populations of invertebrates has not been determined.

VII. MECHANISMS OF ACTION OF ALKYLPHENOLIC CHEMICALS

The results of the laboratory investigations demonstrate that the biodegradation products of the APEOs are able to elicit effects on the endocrine system of fish, mammals, and invertebrates; however, the mechanisms by which they mediate these effects are largely unknown. Most estrogen mimics are thought to exert their effects through binding the ER(s); however, some chemicals have been found to exert estrogenic effects via indirect mechanisms, such as interfering with synthesis or degradation of endogenous steroids. Given that both induction of VTG synthesis in fish and increases in uterine weight in mammals are estrogen-dependent processes, the ability of the alkylphenolic compounds to stimulate these processes, in a manner similar to E2, supports the hypothesis that they mediate at least some of their effects through competitive binding to the ER(s). Furthermore, it has been demonstrated that coadministration with an antiestrogen blocks the stimulatory effect of the alkylphenolics on both VTG synthesis in fish [96] and uterine weight in mammals [79]. In neonatal rats, the inhibitory effects of NP on testis and male accessory organs were also blocked by concomitant treatment with ICI 182,780 (an ER antagonist), suggesting that NP acts on the male reproductive tissues through the ER(s) [83]. Despite this strong evidence for a direct effect of the alkylphenolics on the ER, it has been suggested that the estrogenic effects of NP in fish may also be induced indirectly, by increasing endogenous E2 concentrations through stimulation of aromatase [97]. However, this is not supported by the findings of Harris et al. [61], who observed reduced plasma E2 concentrations in adult female rainbow trout exposed to 85.6 µg NP/L. Furthermore, it was reported in 2000 that female rats dosed with OP had lowered serum E2 concentrations, despite showing signs of estrogenic stimulation [98]. Decreased plasma E2 concentrations have also been observed in juvenile atlantic salmon [99] following exposure to low (1 and 5 mg/kg), but not high (25 and 125 mg/kg) concentrations of NP. These decreases in plasma E2 concentrations may be indicative of an inhibition of E2 synthesis. A direct inhibitory effect of NP on E2 synthesis has not been reported, but Nakajin et al. [100] observed that 4-t-PP, 4-t-OP, and 4-NP behave as inhibitors of cortisol synthesis in human adrenocortical cells, through inhibition of various steroidogenic enzymes. This may in turn result in decreases in E2 synthesis. Similarly, in an in vitro assay using immature Leydig cells from rat testis, the stimulatory effects of OP on testosterone synthesis were not affected by coadministration with ICI 182,780, suggesting that although OP is a hormonally active agent, some of its actions are distinct from those of E2 and are not mediated through the ERα or -β pathways [101,102]. Murono et al. [102] proposed that the observed effects on testosterone formation in cultured rat precursor and immature Leydig cells were due to inhibition of the steroidogenic enzymes responsible for the conversion of cholesterol to testosterone. Further investigation suggests that the 17α-hydroxylase/c17-20-lyase step, which converts progesterone to androstenedione, is inhibited by OP [101]. Decreases in plasma E2 concentration may also be due to an increase in the activity of the steroid-metabolizing enzymes. Arukwe et al. [99] demonstrated that NP causes variations in the activity of the steroid hydroxylases and reduces the activity of the cytochrome P450 isozymes and conjugating enzyme levels in juvenile atlantic salmon exposed via injection to NP. They suggested that NP has a dual activity in regulating reproductive hormones by increasing the action of steroid-metabolizing enzymes at low concentrations and decreasing the action of these enzymes at higher concentrations.

Estrogens and endocrine-active chemicals may also exert effects on reproduction via the hypothalamo–pituitary–gonad axis. Estrogen receptors occur in the brain (hypothalamus) and pituitary gland, where they play important roles in controlling luteinizing hormone (LH) and follicle-stimulating hormone (FSH) synthesis and secretion. These hormones in turn control gonad development in both males and females. Blake and Boockfor [84] observed that chronic administration of OP (80 mg) to adult male rats via subcutaneous injection suppressed LH, FSH, and testosterone secretion and enhanced prolactin secretion. Similarly, Harris et al. [61] observed that NP suppressed FSH and LH secretion in adult female rainbow trout exposed via the water.

VIII. INTERACTION OF ALKYLPHENOL ETHOXYLATES WITH THE ESTROGEN RECEPTOR: STRUCTURE–ACTIVITY RELATIONSHIPS

Although it has been demonstrated that the alkylphenolic chemicals may have multiple sites of action, acting directly via the ER or indirectly through affecting the steroidogenic enzymes or the hypothalamo–pituitary–gonad axis, investigations into structure–activity relationships have concentrated on interactions with the ER. Collectively these investigations have demonstrated that for effects mediated via the ER, estrogenicity is dependent upon both the size and the structure of the alkylphenolic degradation product. It has been found that AP require a mass equivalent to the B ring of E2 in order to be active [9]. Among the active degradation products, which include the AP, APEO with a low degree of ethoxylation, and their carboxylic acid derivatives, estrogenic potency decreases with increasing EO chain length [9,103–105]. The estrogenic activity of the APEO also appears to be associated with the *para* position of the alkyl side chain: when this side chain is moved *ortho* or *meta* to the phenolic group, activity is lost [9,103]. Hydrophobicity of the alkyl group is also important for estrogenic activity, for the introduction of a hydroxyl, carboxyl, or nitro group renders the parent compound inactive [9]. Each of these investigations has employed the use of in vitro assays, considering only the ability of the APEO to bind the ER. Further investigatons are required to determine the structure–activity relationships of the APEO for other mechanisms of action.

IX. CHEMICAL MIXTURES

The laboratory evidence presented to date for the estogenic/reproductive effects of the alkylphenolic chemicals has been based on exposures to single chemicals. In the environment though, wildlife, especially in the aquatic environment, are unlikely to be exposed to the alkylphenolic chemicals in isolation, but rather to a cocktail of chemicals. Indeed TIE approaches on effluent discharges into surface waters have determined that the estrogenic fraction may contain a mixture of natural and synthetic steroids and a combination of APEO [22,26]. In most cases, especially when toxicity is used as an endpoint, mixtures of chemicals have been shown to be additive in their effects to fish and other aquatic organisms [106–109]. In a small number of cases, however, deviations from additivity have been identified. Greater than additive effects on toxicity have been reported in the microcrustacean *T. brevicomis*, exposed to mixtures of metals and insecticides [110], and in the midge *Chironomus tentans*, exposed to combinations of atrazine with organophosphate pesticides [111], while combinations of atrazine with mevinophos or methoxychlor were less than additive in their toxicity to the midge [111]. These interactive effects may occur through one chemical reaching with another, physically or chemically, to affect the availability of either chemical to the organism; through one chemical affecting the absorption, transportation, metabolism, accumulation, or excretion of another and thus influencing the quantity of either chemical available at the site or sites of action; or through one chemical altering the tissue sensitivity to another chemical. This potential for chemicals to interact to modify the effects of each other highlights the need to consider the issue of mixtures when assessing the estrogenicity of the APEO in the environment. As yet, relatively little is known about the combined effects of mixtures of estrogenic chemicals; however, a limited number of in vitro and in vivo screening assays have been used to assess the effects of mixtures of AP with other environmental pollutants. In vitro investigations have considered the effects of combinations of AP and phthalates in competitively binding the ER [112] and the activity of an equimolar mixture of bisphenol-A, 4-*t*-OP, and NP in inducing cell proliferation in the MCF-7 assay [113]. In vivo screening assays have assessed the activity of combinations of OP and butylbenzylphthalate (BBP) or OP and E2 on regulation of the ER and induction of zona radiata protein in mixed-sex juvenile rainbow trout receiving intraperitoneal injections of OP, BBP (both at 5 and 50 mg/kg b.w.), or E2 (1 and 10 mg/kg) [114]. The effect of a mixture of NP and BPA on VTG mRNA expression and inhibition of sword growth in the swordtail has also been investigated [58]. In each of these in vitro and in vivo investigations, the mixtures were observed to be more potent than the individual chemicals at the concentrations tested, demonstrating that mixtures of estrogenic chemicals are more active than the individual chemicals. None of these investigations, however, used the appropriate mathematical models to determine whether the observed effect was additive or more than or less than additive. To determine the magnitude of the observed interaction requires the use of mathematical models to compare the observed mixture concentration–response curve with a theoretical curve calculated from the concentration–response curves for the individual chemicals. There are two main analytical models for defining the expected effects of mixtures of agents: the model of concentration addition (CA) [115] and the model of response addition [116]. The first model, CA, assumes that the compounds act via a similar mechanism in producing an effect, while the second model, response addition, assumes that they act via independent pathways. Using these models, if expectations are met for a mixture of xenestrogens, the combined effect is described as additive; but if a mixture of xenestrogens is more potent than would be predicted from the estrogenic activity of the individual chemicals, the combination effect is called *synergistic*. If it is less effective, the combination effect is labeled *antagonistic*. Describing the combination effect is important if assessments of the estrogenic potency are to be made for a mixture.

The following case study considers the estrogenic activity of an environmentally relevant binary mixture of E2 and NP, using an in vivo VTG screening assay in the rainbow trout [117]. Given that all the available evidence shows that both E2 and NP act via the ER to induce VTG synthesis, the model of CA was used to assess the estrogenic activity of the binary mixtures of the test chemicals. A simulation technique termed *bootstrap* [118] was used with the model of CA to construct a 95% confidence belt around the line of prediction. This provides a statistical basis to determine whether deviations from expectation were significant (those that fall outside the 95% confidence of the predicted curve), or simply due to natural variation in the biological response (those that fall within the 95% confidence of the predicted curve).

X. ESTROGENIC ACTIVITY OF A BINARY MIXTURE OF 4-NONYLPHENOL AND ESTRADIOL-17β—A CASE STUDY

A full description of this work is provided in Thorpe et al. [60]. Briefly, juvenile female rainbow trout (10.47 ± 0.71 g; $n = 24$) were exposed for 14 days to nominal concentrations of E2 at 2.4, 4.2, 7.5, 13.5, and 24.0 ng/L and NP at 2.4, 4.2, 7.5, 13.5, and 24.0 μg/L and to fixed ratio binary mixtures of E2 + NP at concentrations of 4.2 ng/L + 4.2 μg/L, 7.5 ng/L + 7.5 μg/L, and 13.5 ng/L + 13.5 μg/L, respectively ($n = 12$/treatment). The ratio used for the mixture was selected on the basis of earlier work, in which E2 was found to be approximately 1000-fold more potent than NP [117]. Each experiment included a dilution water control and a methanol solvent control. Plasma VTG concentrations were determined in a subset of fish ($n = 24$) sampled on day 0 and in all fish on day 14. Concentrations of E2 and NP were determined in the individual aquaria on days 0, 7, and 14 of exposure using GCMS and HPLC, respectively.

For the description of the concentration–effect relationships for the individual test compounds and for the binary mixtures, a four-parameter logit regression model was used, defined as:

$$f(x) = \theta_{min} + \frac{\theta_{max} - \theta_{min}}{1 + \exp(-\theta_1 - \theta_2 * \log_{10}(x))} \qquad \text{if } x > 0$$

where x = concentration and $f(x)$ = mean effect. The model parameter θ_{min} describes the minimal mean effect (control response), θ_{max} the asymptotical maximal effect, θ_1 is termed the *location* parameter, and θ_2 characterizes the *steepness* of the concentration–response relationship. The experiments were not designed to determine maximal effects, so estimation of θ_{max} contains a high degree of statistical uncertainty. Due to heterogeneous nonrandom variabilities in the replicated data, each model was fitted using the estimation method of generalized least squares [118]. To fulfill the statistical prerequisite of symmetrically distributed effect data for this estimation method, the plasma VTG concentrations were log 10-transformed. The expected concentration–response relationships of the binary mixtures were determined using the model of CA [115]. The "bootstrap" methodology [118] was employed with the model of CA to determine the statistical accuracy of the predicted combined effects.

All results are described using the actual measured concentrations of the test chemicals (shown in Table 1). The concentrations of VTG in the plasma of juvenile female fish at the onset of the experiments were 500 ± 80 ng/mL. There were no detectable increases in plasma VTG concentrations in either the dilution water control or in the solvent control fish after the 14-day exposure period in any of the experiments ($p > 0.05$).

TABLE 1 Mean Measured Tank Concentrations of NP and Estradiol in Individual and Binary Mixtures over 14 Days

Mean measured concentration estradiol (ng/L)			Mean measured concentration nonylphenol (µg/L)		
Nominal	Individual	Mixture	Nominal	Individual	Mixture
2.4	2.3 ± 0.0		2.4	1.8 ± 0.4	
4.2	4.7 ± 0.2	4.9 ± 0.3	4.2	3.6 ± 0.3	3.3 ± 0.4
7.5	7.2 ± 0.2	7.4 ± 0.2	7.5	6.1 ± 0.7	6.1 ± 1.1
13.5	13.3 ± 0.4	12.6 ± 0.2	13.5	10.2 ± 0.6	10.2 ± 1.1
24.0	21.3 ± 0.3		24.0	12.2 ± 3.8	

Data as means ± SEM (n = 3).
Source: Ref. 57.

Estradiol-17β (concentrations ranging from 2.3 to 21.3 ng/L) and NP (concentrations ranging from 1.8 to 12.2 µg/L) produced concentration-dependent increases in plasma VTG (Fig. 1), with LOECs of 4.7 ng/L (plasma VTG concentration of 4100 ± 940 ng/mL, $p < 0.05$) and 6.1 µg/L (plasma VTG concentration of 1450 ± 230 ng/mL, $p < 0.05$) for E2 and NP, respectively. The mixture of E2 and NP, at a fixed 1:1000 ratio, also produced a concentration-dependent increase in plasma VTG, with the lowest mixture concentration tested (4.9 ng/L E2 and 3.3 µg/L NP) inducing a 17-fold increase in VTG concentration (8720 ± 4370 ng/mL, $p < 0.05$). When comparing the observed VTG induction data for the mixtures with the expected mixture effects according to the model of concentration addition, the mixtures of E2 and NP were shown to act in a concentration-additive manner (Fig. 1). The 95% confidence belt of the fitted concentration–response relationship for the observed mixture data overlapped with the 95% confidence bootstrap belt for the calculated mean of CA, for the whole effect range tested. The concentration-additive behavior of the mixture was also observed at concentrations in the mixture below that for the LOEC of the individual chemicals. This demonstrates the need to consider the presence of low concentrations of the AP in the environment when assessing estrogenicity. In many cases, concentrations of NP in the environment are below the threshold of response for the individual chemical, which may have resulted in the AP's being eliminated as potential contributors to any observed estrogenic activity. The data from this experiment, however, suggest that concentrations of alkylphenolic chemicals may be contributing to vitellogenic effects observed in wild or caged fish in many environments receiving estrogenic contaminants.

Laboratory investigations have demonstrated that NP may also mediate effects on a number of other estrogenic parameters in fish, including inhibition of secondary sexual characteristics and induction of testis-ova in males and an impairment of reproduction. Although the concentrations of the individual AP required to produce these estrogenic effects are generally much higher than those measured in many surface waters, the ability of the estrogenic chemicals to produce additive effects at environmentally relevant concentrations must be considered. Further work is required, however, to determine whether the alkylphenolics, which have been shown to have multiple mechanisms of action on the endocrine system, are also additive in their effects on these biological systems.

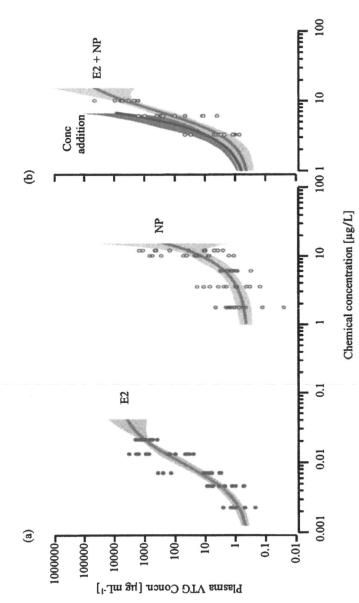

FIG. 1 Plasma vitellogenin (VTG) concentrations in female juvenile rainbow trout exposed to (a) estradiol-17β (E2) (closed circles) and 4-*tert*-nonylphenol (NP) (open circles), and (b) fixed-ratio binary mixtures (1:1000) of E2 and NP (open circles). In some cases the plasma VTG concentrations were very similar between fish within a treatment so not all data points are visible. For each of these exposures the 95% confidence belts (light grey-shaded regions) of the fitted concentration–response relationships (grey lines) are shown. The expected vitellogenic response for the binary mixture, calculated using the model of concentration addition, is shown as a black line (b) with a 95% confidence bootstrap belt (dark grey-shaded region). (From Ref. 57.)

XI. CONCLUSIONS ON ALKYLPHENOLIC CHEMICALS AND ESTROGENIC CONTAMINATION IN WASTE- AND SURFACE WATERS

Laboratory studies have demonstrated that the alkylphenolic chemicals are able to induce a range of estrogenic responses in fish, mammals, and invertebrates, but these effects are generally observed only at high concentrations, compred with those measured in most environments. It should be emphasized, however, that most of the laboratory studies were relatively short in duration and did not consider the ability of the alkylphenolics to bioconcentrate/bioaccumulate. In addition, most laboratory studies do not consider the potential for the APEO to interact in an additive manner with other estrogenic chemicals present in the environment. The presence of APEO biodegradation products in the aquatic environment has been clearly demonstrated. These chemicals enter the environment via effluents discharging from STWs receiving industrial waste, particularly from textile industries, and can be measured in surface waters and fish tissues downstream from STWs. The highest proportion of the APEO concentrate in the sediments. The APEO have been identified as the key estrogenic contaminants in highly polluted rivers (e.g., the rivers Aire and Calder in the United Kingdom) receiving industrial effluents from textile industries. In most environments, however, they are unlikely to represent the major estrogenic contaminant, although their potential to contribute to the total estrogenic load should not be discounted, particularly in environments receiving municipal/industrial effects. Effective wastewater treatment could reduce the concentrations of the parent APEO in the environment, but as highlighted by Bennie [14] this may be a double-edged sword, with the degradation products, i.e., AP, AP1EO, and AP2EO, having greater environmental impacts than their precursors.

REFERENCES

1. Howell, W.D.; Black, D.A.; Bortone, S.A. Copeia 1980, 676–681.
2. Fry, D.M.; Toone, C.K. Science 1981, *213*, 922–924.
3. Reijnders, P.J.H. Nature 1986, *324*, 456–457.
4. Gibbs, P.E.; Pascoe, P.L.; Burt, G.R. J. Mar. Biol. Assoc. U.K. 1988, *68*, 715–731.
5. Carlsen, E.; Giwercman, A.; Keiding, N.; Skakkebaek, N.E. Br. Med. J. 1992, *305*, 609–613.
6. Guillette, L.J.; Gross, T.S.; Masson, G.R.; Matter, J.M.; Percival, H.F.; Woodward, A.R. Environ. Health Perspect. 1994, *102*, 680–688.
7. Jobling, S.; Nolan, M.; Tyler, C.R.; Brighty, G.; Sumpter, J.P. Environ. Sci. Technol. 1998, *32*, 2498–2506.
8. Dodds, E.C.; Lawson, W. Proc. R. Soc. Lond. Ser. B—Biol. Sci. 1938, 222–232.
9. Mueller, G.C.; Kim, U-H. Endocrinology 1978, *102*, 1429–1435.
10. Soto, A.M.; Justicia, H.; Wray, J.W.; Sonnenschein, C. Environ. Health Perspect. 1991, *92*, 167–173.
11. Routledge, E.J.; Sumpter, J.P. Environ. Toxicol. Chem. 1996, *15*, 241–248.
12. Navas, J.M.; Gonzalez-Mazo, E.; Wenzel, A.; Gomez-Parra, A.; Segner, H. Mar. Pollut. Bull. 1999, *38*, 880–884.
13. Nimrod, A.C.; Benson, W.H. Crit. Rev. Toxicol. 1996, *26*, 335–364.
14. Bennie, D.T. Water Qual. Res J. Canada 1999, *34*, 79–122.
15. Servos, M.R. Water Qual. Res J. Canada 1999, *34*, 123–177.

16. Naylor, C.G. Soap Cosmet. Chem. Spec. 1992, *68*, 27–31, 72.
17. Ahel, M.; Giger, W.; Koch, M. Water Res. 1994, *28*, 1131–1142.
18. Naylor, C.G. Text Chem. Color 1995, *27*, 29–33.
19. Field, J.A.; Reed, R.L. Environ. Sci. Technol. 1996, *30*, 3544–3550.
20. Fujita, M.; Ike, M.; Mori, K.; Kaku, H.; Sakaguchi, Y.; Asano, M.; Maki, H.; Nishihara, T. Water Sci. Technol. 2000, *42*, 23–30.
21. Staples, C.A.; Naylor, C.G.; Williams, J.B.; Gledhill, W.E. Environ. Toxicol. Chem. 2001, *20*, 2450–2455.
22. Sheahan, D.A.; Brighty, G.C.; Daniel, M.; Kirby, S.J.; Hurst, M.R.; Kennedy, J.; Morris, S.; Routledge, E.J.; Sumpter, J.P.; Waldock, M.J. Environ. Toxicol. Chem. 2002, *21*, 507–514.
23. Sheahan, D.A.; Brighty, G.C.; Daniels, M.; Jobling, S.; Harries, J.E.; Hurst, M.R.; Kennedy, J.; Kirby, S.J.; Morris, S.; Routledge, E.J.; Sumpter, J.P.; Waldock, M.J. Environ. Toxicol. Chem. 2002, *21*, 515–519.
24. Lye, C.M.; Frid, C.L.J.; Gill, M.E.; Cooper, D.W.; Jones, D.M. Environ. Sci. Technol. 1999, *33*, 1009–1014.
25. Desbrow, C.; Routledge, E.J.; Brighty, G.C.; Sumpter, J.P.; Waldock, M. Environ. Sci. Technol. 1998, *32*, 1549–1558.
26. Rodgers-Gray, T.P.; Jobling, S.; Morris, S.; Kelly, C.; Kirby, S.; Janbakhsh, A.; Harries, J.E.; Waldock, M.J.; Sumpter, J.P.; Tyler, C.R. Environ. Sci. Technol. 2000, *34*, 1521–1528.
27. Stephanou, E.; Giger, W. Environ. Sci. Technol. 1982, *16*, 800–805.
28. Dicorcia, A.; Samperi, R.; Marcomini, A. Environ. Sci. Technol. 1994, *28*, 850–858.
29. Spengler, P.; Korner, W.; Metzger, J.W. Environ. Toxicol. Chem. 2001, *20*, 2133–2141.
30. Blackburn, M.A.; Kirby, S.J.; Waldock, M.J. Mar. Pollut. Bull. 1999, *38*, 109–118.
31. Ahel, M.; Molnar, E.; Ibric, S.; Giger, W. Water Sci. Technol. 2000, *42*, 15–22.
32. Bolz, U.; Hagenmaier, H.; Korner, W. Environ. Pollut. 2001, *115*, 291–301.
33. Petrovic, M.; Fernandez-Alba, A.R.; Borrull, F.; Marce, R.M.; Mazo, E.G.; Barcelo, D. Environ. Toxicol. Chem. 2002, *21*, 37–46.
34. Zoller, U.; Hushan, M. Water Sci. Technol. 2001, *43*, 245–250.
35. Ball, H.A.; Reinhard, M.; McCarty, P.L. Environ. Sci. Technol. 1989, *23*, 951–961.
36. Johnson, A.C.; White, C.; Bhardwaj, L.; Jurgens, M.D. Environ. Toxicol. Chem. 2000, *19*, 2486–2492.
37. Snyder, S.A.; Keith, T.L.; Naylor, C.G.; Staples, C.A.; Giesy, J.P. Environ. Toxicol. Chem. 2001, *20*, 1870–1873.
38. Ahel, M.; McEvoy, J.; Giger, W. Environ. Pollut. 1993, *79*, 243–248.
39. Ferreira-Leach, A.M.R.; Hill, E.M. Analusis 2000, *28*, 789–792.
40. Purdom, C.E.; Hardiman, P.A.; Bye, V.J.; Eno, N.C.; Tyler, C.R.; Sumpter, J.P. Chem. Ecol. 1994, *8*, 275–285.
41. Harries, J.E.; Janbakhsh, A.; Jobling, S.; Matthiessen, P.; Sumpter, J.P.; Tyler, C.R. Environ. Toxicol. Chem. 1999, *18*, 932–937.
42. Harries, J.E.; Sheahan, D.A.; Jobling, S.; Matthiessen, P.; Neall, M.; Sumpter, J.P.; Taylor, T.; Zaman, N. Environ. Toxicol. Chem. 1997, *16*, 534–542.
43. Harries, J.E.; Sheahan, D.A.; Jobling, S.; Matthiessen, P.; Neall, P.; Routledge, E.J.; Rycroft, R.; Sumpter, J.P.; Tylor, T. Environ. Toxicol. Chem. 1996, *15*, 1993–2002.
44. Folmar, L.C.; Denslow, N.D.; Rao, V.; Chow, M.; Crain, D.A.; Enblom, J.; Marcino, J.; Guillette, L.J. Environ. Health Perspect. 1996, *104*, 1096–1101.
45. Larsson, D.G.J.; Hallman, H.; Forlin, L. Environ. Toxicol. Chem. 2000, *19*, 2911–2917.
46. Hemming, J.M.; Waller, W.T.; Chow, M.C.; Denslow, N.D.; Venables, B. Environ. Toxicol. Chem. 2001, *20*, 2268–2275.
47. Jobling, S. Pure Appl. Chem. 1998, *70*, 1805–1827.

48. van Aerle, R.; Nolan, M.; Jobling, S.; Christiansen, L.B.; Sumpter, J.P.; Tyler, C.R. Environ. Toxicol. Chem. 2001, 20, 2841–2847.

49. Rodgers-Gray, T.P.; Jobling, S.; Kelly, C.; Morris, S.; Brighty, G.; Waldock, M.J.; Sumpter, J.P.; Tyler, C.R. Environ. Sci. Technol. 2001, 35, 462–470.

50. Panter, G.H.; Thompson, R.S.; Sumpter, J.P. Aquat. Toxicol. 1998, 42, 243–253.

51. Harries, J.E.; Runnalls, T.; Hill, E.; Harris, C.A.; Maddix, S.; Sumpter, J.P.; Tyler, C.R. Environ. Sci. Technol. 2000, 34, 3003–3011.

52. Gimeno, S.; Gerritsen, A.; Bowmer, T.; Komen, H. Nature 1996, 384, 221–222.

53. Piferrer, F.; Donaldson, E.M. Aquaculture 1989, 77, 251–262.

54. Leguellec, K.; Lawless, K.; Valotaire, Y.; Kress, M.; Tenniswood, M. Gen. Comp. Endocrinol. 1988, 71, 359–371.

55. Vanbohemen, C.G.; Lambert, J.G.D.; Vanoordt, P. Gen. Comp. Endocrinol. 1982, 46, 136–139.

56. Sumpter, J.P.; Jobling, S. Environ. Health Perspect. 1995, 103, 173–178.

57. Mackay, M.E.; Lazier, C.B. Gen. Comp. Endocrinol. 1993, 89, 255–266.

58. Kwak, H.I.; Bae, M.O.; Lee, M.H.; Lee, Y.S.; Lee, B.J.; Kang, K.S.; Chae, C.H.; Sung, H.J.; Shin, J.S.; Kim, J.H.; Mar, W.C.; Sheen, Y.Y.; Cho, M.H. Environ. Toxicol. Chem. 2001, 20, 787–795.

59. Hemmer, M.J.; Hemmer, B.L.; Bowman, C.J.; Kroll, K.J.; Folmar, L.C.; Marcovich, D.; Hoglund, M.D.; Denslow, N.D. Environ. Toxicol. Chem. 2001, 20, 336–343.

60. Thorpe, K.L.; Hutchinson, T.H.; Hetheridge, M.J.; Scholze, M.; Sumpter, J.P.; Tyler, C.R. Environ. Sci. Technol. 2001, 35, 2476–2481.

61. Harris, C.A.; Santos, E.M.; Janbakhsh, A.; Pottinger, T.G.; Tyler, C.R.; Sumpter, J.P. Environ. Sci. Technol. 2001, 35, 2909–2916.

62. Lech, J.J.; Lewis, S.K.; Ren, L.F. Fundam. Appl. Toxicol. 1996, 30, 229–232.

63. Jobling, S.; Sheahan, D.; Osborne, J.A.; Matthiessen, P.; Sumpter, J.P. Environ. Toxicol. Chem. 1996, 15, 194–202.

64. Routledge, E.J.; Sheahan, D.; Desbrow, C.; Brighty, G.C.; Waldock, M.; Sumpter, J.P. Environ. Sci. Technol. 1998, 32, 1559–1565.

65. Ashfield, L.A.; Pottinger, T.G.; Sumpter, J.P. Environ. Toxicol. Chem. 1998, 17, 679–686.

66. Christiansen, T.; Korsgaard, B.; Jespersen, A. Mar. Environ. Res. 1998, 46, 141–144.

67. Christiansen, T.; Korsgaard, B.; Jespersen, A. J. Exp. Biol. 1998, 201, 179–192.

68. Gray, M.A.; Metcalfe, C.D. Environ. Toxicol. Chem. 1997, 16, 1082–1086.

69. Yokota, H.; Seki, M.; Maeda, M.; Oshima, Y.; Tadokoro, H.; Honjo, T.; Kobayashi, K. Environ. Toxicol. Chem. 2001, 20, 2552–2560.

70. Gray, M.A.; Niimi, A.J.; Metcalfe, C.D. Environ. Toxicol. Chem. 1999, 18, 1835–1842.

71. Metcalfe, C.D.; Metcalfe, T.L.; Kiparissis, Y.; Koenig, B.G.; Khan, C.; Hughes, R.J.; Croley, T.R.; March, R.E.; Potter, T. Environ. Toxicol. Chem. 2001, 20, 297–308.

72. Toft, G.; Baatrup, E. Ecotoxical Environ. Saf. 2001, 48, 76–84.

73. Zaroogian, G.; Gardner, G.; Horowitz, D.B.; Gutjahr-Gobell, R. Aquat. Toxicol. 2001, 54, 101–112.

74. Gimeno, S.; Komen, H.; Venderbosch, P.W.M.; Bowmer, T. Environ. Sci. Technol. 1997, 31, 2884–2890.

75. Gray, M.A.; Teather, K.L.; Metcalfe, C.D. Environ. Toxicol. Chem. 1999, 18, 2587–2594.

76. Bulger, W.H.; Kupfer, D. Am. J. Ind. Med. 1983, 4, 163–173.

77. Eroschenko, V.P. Anat. Rec. 1981, 199, A79.

78. Galey, F.D.; Mendez, L.E.; Whitehead, W.E.; Holstege, D.M.; Plumlee, K.H.; Johnson, B. J. Vet. Diagn. Invest. 1993, 5, 603–608.

79. Lee, P.C.; Lee, W. Bull. Environ. Contam. Toxicol. 1996, 57, 341–348.

80. Bicknell, R.J.; Herbison, A.E.; Sumpter, J.P. J. Steroid. Biochem. Mol. Biol. 1995, 54, 7–9.

81. Nagao, T.; Saito, Y.; Usumi, K.; Nakagomi, M.; Yoshimura, S.; Ono, H. Hum. Exp. Toxicol. 2000, *19*, 284–296.
82. Blake, C.A.; Ashiru, O.A. Proc. Soc. Exp. Biol. Med. 1997, *216*, 446–451.
83. Lee, P.C. Endocrine 1998, *9*, 105–111.
84. Blake, C.A.; Boockfor, F.R. Biol. Reprod. 1997, *57*, 255–266.
85. Boockfor, F.R.; Blake, C.A. Biol. Reprod. 1997, *57*, 267–277.
86. Cunny, H.C.; Mayes, B.A.; Rosica, K.A.; Trutter, J.A.; VanMiller, J.P. Regul. Toxicol. Pharmacol. 1997, *26*, 172–178.
87. Chapin, R.E.; Delaney, J.; Wang, Y.F.; Lanning, L.; Davis, B.; Collins, B.; Mintz, N.; Wolfe, G. Toxicol. Sci. 1999, *52*, 80–91.
88. de Jager, C.; Bornman, M.S.; van der Horst, G. Andrologia 1999, *31*, 99–106.
89. de Jager, C.; Bornman, M.S.; Oosthuizen, J.M.C. Andrologia 1999, *31*, 107–113.
90. Tyl, R.W.; Myers, C.B.; Marr, M.C.; Brine, D.R.; Fail, P.A.; Seely, J.C.; Van Miller, J.P. Regul. Toxicol. Pharmacol. 1999, *30*, 81–95.
91. Hossaini, A.; Dalgaard, M.; Vinggaard, A.M.; Frandsen, H.; Larsen, J.J. Reprod. Toxicol. 2001, *15*, 537–543.
92. Bogh, I.B.; Christensen, P.; Dantzer, V.; Groot, M.; Thofner, I.C.N.; Rasmussen, R.K.; Schmidt, M.; Greve, T. Theriogenology 2001, *55*, 131–150.
93. Oehlmann, J.; Schulte-Oehlmann, U.; Tillmann, M.; Markert, B. Ecotoxicology 2000, *9*, 383–397.
94. Baldwin, W.S.; Graham, S.E.; Shea, D.; LeBlanc, G.A. Environ. Toxicol. Chem. 1997, *16*, 1905–1911.
95. Shurin, J.B.; Dodson, S.I. Environ. Toxicol. Chem. 1997, *16*, 1269–1276.
96. Toomey, B.H.; Monteverdi, G.H.; Di Giulio, R.T. Environ. Toxicol. Chem. 1999, *18*, 734–739.
97. Giesy, J.P.; Pierens, S.L.; Snyder, E.M.; Miles-Richardson, S.; Kramer, V.J.; Snyder, S.A.; Nichols, K.M.; Villeneuve, D.A. Environ. Toxicol. Chem. 2000, *19*, 1368–1377.
98. Yoshida, M.; Katsuda, S.; Ando, J.; Kuroda, H.; Takahashi, M.; Maekawa, A. Toxicol. Lett. 2000, *116*, 89–101.
99. Arukwe, A.; Forlin, L.; Goksoyr, A. Environ. Toxicol. Chem. 1997, *16*, 2576–2583.
100. Nakajin, S.; Shinoda, S.; Ohno, S.; Nakazawa, H.; Makino, T. Environ. Toxicol. Pharmacol. 2001, *10*, 103–110.
101. Murono, E.P.; Derk, R.C.; de Leon, J.H. Reprod. Toxicol. 2000, *14*, 275–288.
102. Murono, E.P.; Derk, R.C.; de Leon, J.H. Reprod. Toxicol. 1999, *13*, 451–462.
103. Jobling, S.; Sumpter, J.P. Aquat. Toxicol. 1993, *27*, 361–372.
104. White, R.; Jobling, S.; Hoare, S.A.; Sumpter, J.P.; Parker, M.G. Endocrinology 1994, *135*, 175–182.
105. Routledge, E.J.; Sumpter, J.P. J. Biol. Chem. 1997, *272*, 3280–3288.
106. Matthiessen, P.; Whale, G.F.; Rycroft, R.J.; Sheahan, D.A. Aquat. Toxicol. 1988, *13*, 61–75.
107. Alabaster, J.S.; Calamari, D.; Dethlefsen, V.; Konemann, H.; Lloyd, R.; Solbe, J.F. Environ. Top. 1994, *6*, 145–205.
108. Walker, M.K.; Cook, P.M.; Butterworth, B.C.; Zabel, E.W.; Peterson, R.E. Fundam. Appl. Toxicol. 1996, *30*, 178–186.
109. Bailey, H.C.; Miller, J.L.; Miller, M.J.; Wiborg, L.C.; Deanovic, L.; Shed, T. Environ. Toxicol. Chem. 1997, *16*, 2304–2308.
110. Forget, J.; Pavillon, J.F.; Beliaeff, B.; Bocquene, G. Environ. Toxicol. Chem. 1999, *18*, 912–918.
111. PapeLindstrom, P.A.; Lydy, M.J. Environ. Toxicol. Chem. 1997, *16*, 2415–2420.
112. Knudsen, F.R.; Pottinger, T.G. Aquat. Toxicol. 1999, *44*, 159–170.
113. Korner, W.; Hanf, V.; Schuller, W.; Kempter, C.; Metzger, J.; Hagenmaier, H. Sci. Total. Environ. 1999, *225*, 33–48.

114. Knudsen, F.R.; Arukwe, A.; Pottinger, T.G. Environ. Pollut. 1998, *103*, 75–80.
115. Berenbaum, M.C. Pharmacol. Rev. 1989, *41*, 93–141.
116. Bliss, C.I. Ann. Appl. Biol. 1939, *26*, 585–615.
117. Thorpe, K.L.; Hutchinson, T.H.; Hetheridge, M.J.; Sumpter, J.P.; Tyler, C.R. Environ. Toxicol. Chem. 2000, *19*, 2812–2820.
118. Scholze, M.; Boedeker, W.; Faust, M.; Backhaus, T.; Altenburger, R.; Grimme, L.H. Environ. Toxicol. Chem. 2001, *20*, 448–457.

17

The Survival and Distribution of Alkylphenol Ethoxylate Surfactants in Surface Water and Groundwater

The Case of Israel—Is There an Environmental Problem?

URI ZOLLER Haifa University–Oranim, Kiryat Tivon, Israel

I. INTRODUCTION, SCIENTIFIC BACKGROUND, AND ENVIRONMENTAL RELEVANCE

The occurrence and persistence of anthropogenic pollutants in the environment showing estrogenic-endocrine modulating effects in aquatic organisms is a "hot" issue of major health- and environment-related concern worldwide [1–6]. A well-known group of the potential endocrine-disrupting chemicals (EDCs), the so-called "endocrine disrupters," are the nonionic surfactants alkylphenol ethoxylates (APEOs, Fig. 1). This significant group of surfactants, 7% of the world surfactant consumption ($\sim 7 \times 10^6$ tons annually [8]), is used in detergent formulations, mostly as components of institutional, industrial, but also household cleaning products. Also they are used as emulsifying wetting agents in agricultural applications. Some of them, especially the nonionic branched-chain nonylphenol and octylphenol ethoxylates (NPEOs and OPEOs, respectively) [9], constitute environmentally persistent organic pollutants (POPs), since their biodegradation under natural environmental conditions is very slow and quite often incomplete for a long period of time [9–11].

Moreover, many of their degradation metabolites are more toxic to aquatic organisms than are the parent molecules, and these metabolites remain in the affected environmental compartment for a long period of time [12]. Both the parents and their metabolites are known to elicit estrogenic response (i.e., capable of mimicking or antagonizing the action of steroid hormones) in both mammals and fish [13–15]. Indeed, they were found in tissues of mature and juvenile fish (flounder), indicating an environmental/estrogenic exposure [16], apparently to wastewater discharges.

R-C$_6$H$_4$-O-(CH$_2$CH$_2$O)$_n$H

1

n = 4-16

R = a mixture of branched-
   chain C$_9$H$_{19}$(mainly)
   and C$_8$H$_{17}$; C$_6$H$_4$ = Ph

FIG. 1 Chemical structure of representative alkylphenol-based ethoxylates (APEOs).

Nonylphenol (NP), octylphenol (OP), and nonyl- and octylphenol polyethoxylates (NPEOs and OPEOs) have previously been found to be estrogenic [17] and to occur in wastewater effluents and discharges from sewage treatment works [10,12,16,18–19]. Therefore, their reaching surface water and, ultimately, groundwater is to be expected; indeed, their environmental persistence has been demonstrated [18,20–25].

Thus, the nature and extent of possible effects of EDCs on human health and wildlife is receiving growing attention from the scientific community, regulatory agencies, and the public at large, particularly with respect to their capability of affecting the endocrine system. Driven by this increasing awareness of possible EDC impact, a recent worldwide effort is stimulating innovative, multidisciplinary research that addresses scientific uncertainties concerning the potential adverse endocrine-disrupting effects of these chemicals. In this respect, the environmentally related most critical question to be answered is: Which of the anthropogenic APEO pollutants and their metabolites and/or degradation products can definitely be characterized as environmental endocrine disrupters [26] and in what concentration level in terms of acute and, more important, chronic toxicities.

II. ENVIRONMENTAL HEALTH RISK POTENTIAL OF APEOs

APEOs are found in all aqueous compartments around the world [6–13,18–27] and are considered to constitute a significant environmental and health hazard. Their survival, homologic distribution, and accumulation in aqueous and solid environmental compartments constitute a problem of short- and long-range, economical, technological, and sociopolitical implications. Both APEOs and their metabolites have been identified by the World Health Organization (WHO) and the EPA as primary pollutants, primarily in the aquatic environment [28–29].

Acute toxicity of nonionic surfactants to aquatic organisms occurs at concentrations usually greater than 100 ppb. However, long-term environmental effects of the APEOs and their metabolites on organisms are still poorly known and are limited to few species. Studies show that surfactants present in aquatic environments may significantly decrease population size, reduce reproduction success, affect larvae survival rates and swimming performance, and inhibit testicular growth [10–12]. Studies of the toxicity of nonylphenol, a product of APEOs biodegradation, to aquatic organisms, including invertebrates, fish, and algae, have demonstrated "no effect" in concentrations ranging from 6.7 ppb for *Mysidopsis bahia* reproduction to 24 ppb for *Daphnia magna* [30]. Our own recent study [6] demonstrated a decrease in egg production of zebrafish (*Danio rerio*) in freshwater containing APEOs within the concentration range of 10–75 ppb.

Servos [31] pointed out that data on sublethal (chronic) effects of APEOs are scarce and that more data about the chronic effects of APEOs are needed. Lower concentrations and/or chronic exposure may cause sublethal damage that cannot be detected by LD_{50} or short-term tests–based estrogenic effects. Aquatic organisms, especially if one considers bioaccumulation in the animals' tissues, may interfere with cell-membrane functions and mucous layers [32]. Such interferences may affect membrane permeability and interfere mainly in osmoregulation and neural function. Indeed, some effects, such as a decrease in swimming activity, changes in gill permeability, gill structural damage, and decrease in growth and reproduction of the exposed organisms were reported (for review, see Refs. 33 and 34). Nonetheless, data as well as knowledge on physiological effects are still lacking.

Endocrine-disruptive activity of APEOs, their metabolites, and their homologues have been shown to induce vitelogenin synthesis in male trout (and other fish), inhibit testicular growth and sex changes (from males to females), and reduce fecundity and offspring survival and development [35–39]. The results of our preliminary study in this respect is complementary and in agreement with these findings [6]. Although there is still not enough evidence to support the hypothesis that exposure to endocrine-disrupting/estrogenic chemicals is a global environmental human health problem, the following relevant facts have been established.

1. Estrogenic activity of domestic sewage treatment work effluents was shown to occur at levels capable of producing biological effects in fish exposed to this water [40].
2. Many of the xenoestrogens, such as the APEOs, and their degradation products, such as the carboxylated metabolites [27,41], enter the aquatic environment by means of discharge from municipal sewage treatment works, industrial effluents, and untreated sewage and can occur in effluents and rivers from nanogram to mg L^{-1} levels [10,22,40,42], as well as in groundwater within the range of 0–24 ppb [43].
3. Although no clearcut evidence of endocrine disruption was demonstrated, a recent study with 4-nonylphenol, a persistent metabolite of APEOs, has indicated that this compound and some of its short-chain polyethoxylates may have toxic effects at environmentally realistic concentrations of 0.01–10 ppb [15,17]. As previously mentioned, our own results [6] corroborate these findings and their implications (see Section IV).

It is now clear that APEOs as well as their homologues and metabolites constitute a potential environmental health hazard. In chronic toxicity tests, no-observable-effect concentrations (NOECs) are as low as 6 ppb in fish and 3.7 ppb in invertebrates. There is an increase in the toxicity of both NPEOs and OPEOs with decreasing EOs chain length. NPECs (nonylphenol ethoxy-carboxylates) and OPECs are less toxic than the corresponding APEOs and have acute toxicities similar to those of APEOs if they have 6–9 EO units [31]. Alkylphenols (APs) and APEOs bind to the estrogen receptor, resulting in the expression of several responses, both in vitro and in vivo, as mentioned earlier. The available literature suggests that the ability of APs and APEOs to bioaccumulate in aquatic biota in the environment is low to moderate. Thus, with respect to the latter, bioconcentration factors (BCFs) and bioaccumulation factors (BAFs) in biota, including algae, plants, invertebrates, and fish, range from 0.9 to 3400 ppb. The potential of OPs and OPEOs to bioaccumulate is expected to be similar to that of the corresponding NPs and NPEOs [31].

Studies from the 1990s and later with APEOs have indicated reproductive abnormalities in populations of marine organisms [44,45], such as inhibition of barnacle settlement [15], effect on the reproduction of blue crab [46] and other aquatic invertebrates [47], and reduction in egg production of zebra fish [6]. These results support the contention that exposure of wildlife to environmentally persistent endocrinic/estrogenic chemicals can result in deleterious reproductive consequences.

Uppgard et al. [48] have developed a multiderivate-quantitative structure–activity relationship (M-QSAR) model, according to which the most important physicochemical variables are the molecular hydrophobicity, the hydrophilic–lipophilic balance, the critical packing parameter with respect to whether the hydrophobe is branched, and the critical micelle concentration. The bottom-line conclusion evolved from this model is that surfactant toxicity tends to increase with increasing alkyl chain length, which, in turn, corresponds to an increase in hydrophobicity but a decrease in water solubility. Indeed, the acute toxicity (LD_{50}) of LAS (linear alkylbenzene sulfonates), AOS (alpha-olefin sulfonates), AS (alkyl sulfonates), and MES (methyl ester sulfates) was shown to increase with increasing alkyl chain length [7] and those of APEOs with decreasing hydrophilic ethoxylate chain length [31]. Be this as it may, the concentration levels and the homologic distribution of the APEOs and their metabolites, determined by their persistence in the aquatic environments, constitute a major "limiting factor" of their environmental impact.

Thus, a systematic conceptual approach is needed in order to analyze and assess quantitatively, the effects and the biochemical-ecological impact of the actual concentrations and distribution of the APEOs in the aquatic environment. The development of such a comprehensive systematic approach should consider all chemical, biochemical, and physical parameters associated with the hydrological, chemical, and biological processes involved in the transport, changes, chemical, and biochemical degradation, and transformation processes of these surfactants, in order to estimate their health risk potential as well as their effects on environmental biological webs.

III. ENVIRONMENT–DETERGENT FORMULATION RELATIONSHIPS: THE PROBLEM

Sustainable development is a key demand in our world of finite resources and endangered ecosystems. Given the environmental imperatives, the potential ecotoxicological risk of anthropogenic chemicals, and the limited economic feasibility of advanced large-scale treatment and remediation technologies, the currently emerging corrective-to-preventive paradigm shift in production, development, consumption, and disposal is unavoidable. Surfactants, the dominant components in detergent formulations, constitute a particular issue of concern since they enter the aquatic and terrestrial compartments of the environment as such or as (bio-)degradation products. A research-based systematic understanding of the distribution, pathways, biodegradation, survival, and effects/consequences of these surfactants is the key for appropriate, integrated relevant environmental policy, regulation, and education to be implemented worldwide.

While the anionic and nonionic surfactants, having a linear alkyl chain, are biodegradable and classified as "soft," those of the branched-chain type, having tertiary and quarternary carbon atoms causing them to resist biodegradation by microorganisms in natural environments, are classified as "hard" and nonbiodegradable (Fig. 1). The less biodegradable a surfactant is, the more of an environmental problem it constitutes. The branched-chain nonylphenol-based nonionic ethoxylates (NPEOs)—the most used alkylphenol-based ethoxylates (APEOs) in detergent formulations [9]—are environmentally persistent pollutants, because their biodegradation is very slow and quite often incomplete for a long period of time. Furthermore, their degradation in wastewater treatment plants or in the environment generates more persistent shorter-chain APEOs, some of which can mimic natural hormones and in their concentration levels in aquatic environments may disrupt endocrine functions in wildlife and potentially in humans [10,31,42,49]. Typical compositions of household liquid and powder laundry formulations are provided in Table 1 [22].

One way or another, all kinds of surfactants and/or their degradation products as well as all the other components of consumed/used detergents find their way into either man-made sewage systems or receiving natural surface water and groundwater. Therefore, in the 1960s the detergent industry switched from the nonbiodegradable branched-chain dodecylbenzene sulfonate (DDBS or ABS-alkylbenzene sulfonate) mostly used (at that time) to the biodegradable linear alkylbenzene sulfonate (LABS or LAS). Nevertheless, excluding human secretions, anthropogenic surfactants are the major man-made organic contaminant in raw sewage waters and municipal influents.

TABLE 1 Typical Compositions (%) of "Classical" Household Liquid and Powder Laundry Detergent Formulations

Ingredients	Liquid	Powder
Surfactants (anionics and nonionics)	25–45	20–35
Builder-sequestrants (polyphosphates, zeolites, chelating agents)	5–10	15–45
Alkaline components (silicates, sodium carbonate, etc.)	—	10–30
Alcohols/glycols, and/or coupling agents	5–15	—
Bleaching agents (perborates) and bleach activators	—	0–30
C.M.C., enzymes, foam regulators, optical brighteners, perfume, etc.	1–5	2–7
Balance:		
Water	40–55	5–10
Sodium sulfate	—	10–30

Source: Ref. 22.

In the 1980s, typical concentrations of anionic and nonionic surfactants in Europe and Israel sewage influents were within the range of 9–11 and 0.8–2.5 mg/L, respectively [21,22,50,51], whereas those of the APEO in industrial influents (paper mills) in the United States were within the range of 1.3–12.2 mg/L [52].

The emerging threat of irreversible pollution of all kinds of receiving waters [19,50] has turned the issue of water resources and their availability, quality control, and management, pollution, health concerns, treatment, and economics into a pressing, complex global environmental problem par excellence. Studies show, repeatedly, that the penetration of persistent organic substances into soil and ground-water is much more substantial than was previously believed [53]. The immense quantity of detergents used in household, institutional, and industrial applications and, consequently, the surfactants' presence and survival in both surface water and groundwater, alkylphenol-based nonionics in particular, give rise to much concern with respect to the latter's direct and indirect effects on the environment [19,27,50] and the possible potential risk to people's health due to their estrogenic properties [13,54].

In spite of many years of efforts worldwide to prevent pollution of surface water and groundwater bodies, studies have indicated that both are contaminated, partic-ularly by the nonbiodegradable ("hard") nonionics and their metabolites [22,24,55,56]. Significantly, the APEOs are still extensively used without legislative restrictions in several Mediterranean as well as European countries. Moreover, until the early 1990s, the "hard" APEOs were the most commonly used nonionic surfac-tants in Israeli domestic laundry detergent formulations, a situation with no parallel in other Western industrial countries.

IV. THE CASE OF THE APEOs IN ISRAELI SURFACE WATER AND GROUNDWATER

A. Concentration Profiles and Homologic Distribution of APEOs

1. Background/Conceptualization, Water Resources, Supply, Treatment, and APEOs

Israel as a country with a high standard of living and, hence, typified by a high consumption of detergents will be used here as an illustrative case study. It is located in a semiarid region, thus experiencing an extreme shortage of water supplies. Of about 5 $\times 10^8$ m^3 of the annually produced sewage—containing in the 1980s about 9–12 ppm of anionic (mainly LABS) and 1–3 ppm of nonionic (mainly APEOs) surfactants (approx. 85:15 ratio [22])—about 27% and 45% of the total quantity are used, following secondary treatment, or directly, for aquifer recharge and agricultural irrigation, respectively. Since, practically speaking, (1) only secondary treatment is available for sewage effluents in the country, (2) APEOs are quite biodegradation resistant due to their branched-carbon alkyl chain, and (3) about two-thirds of the nonionic surfactants used in Israel until the early 1990s were of the "hard" APEOs type, these nonionic surfactants and/or their metabolites reach surface water and groundwater. Thus, for example, until the start of operation of the combined mechanical-activated sludge-SAT (soil-aquifer treatment) SHAFDAN national sew-age treatment project, about 1700 kg of anionic (mainly LAS) and about 350 kg of nonionic (mainly APEOs) surfactants were discharged daily into the coastal Mediterranean Sea water of Tel Aviv.

The Israeli APEO surfactant case may be considered a well-investigated example, typical of other countries confronted with (superficially conceived as) conflicting interests in the socio-techno-environmental context and problems that require sustainable solutions. As will be demonstrated, hard nonionic surfactants and/or their metabolites, the origin of which are either domestic, institutional, industrial, or agricultural wastewater and/or surface streams polluted by municipal sewage, reach both surface water and groundwater. Neither the existing sewage treatment facilities nor naturally occurring biodegradation processes in the receiving surface water systems or physical soil adsorption appear to be capable of avoiding this environmental pollution.

The relevant systematically conceptualized surfactant/wastewater/environment (water) supply cycle is schematically illustrated in Figure 2. Table 2 provides a snapshot of the effectiveness of the sewage treatment plants operating in the country in removal of the APEOs present in the treated influents [19,22]. Although the data in

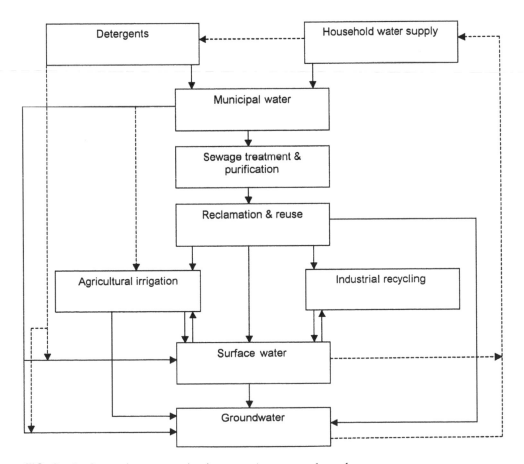

FIG. 2 Surfactant/wastewater/environment/water supply cycle.

TABLE 2 Effectiveness of Nonionic Surfactants' (Mainly APEOs) Removal[a] in Representative Primary and Secondary Treatment Plants in Israel in the 1980s

Location of treatment site	Type of treatment	Nonionic detergent concentration[a]		Removal efficiency[b] (%)	Effluent destination
		Influents	Effluents		
Tiberias	Precipitation/ oxidation ponds	1.8	0.5	72	Agricultural irrigation (partly)
Haifa	Precipitation/ aerated oxidation	1.2	0.25	79	Agricultural irrigation and industrial recycling
Hadera	Secondary oxidation	1.0	0.3	70	Mediterranean Sea
Dan Region[c]	Mechanical/ biological/ chemical	2.35	0.45	81	Aquifer recharge (mainly)

[a] Figures are mg/L.
[b] In terms of primary degradation (determined by the modified SDA-CTAS method) (Ref. 25).
[c] Data of the Dan Region "Shafdan" Project prior to 1987 (Ref. 19).
Source: Refs. 19 and 22.

Table 2 are about 20 years old, the same practical level of removal effectiveness of APEOs still holds in most of these still-operating sewage treatment plants (STPs), and therefore they may be reliably cited whenever necessary.

Since in the 1980s the hard nonionic APEOs made up about 10% of the surfactants found in Israeli wastewaters, at least trace amounts of them were expected to be found in water wells adjacent to wastewater-containing rivers and streams. Thus, two groups of three wells, each located not equidistantly in the vicinity of polluted streams (each pumps water from a different depth), were sampled. The first group (designated D.I., I.B., and M.C.) is located near the Hadera stream, and its water is pumped from the Pleistocene sandy aquifer. The second group (designated A.N., S.T., and Z.A.) is located near the Hadas stream, and its water is pumped from a cenomanianturonian aquifer. Water samples were collected simultaneously from the wells and the corresponding sewage-containing streams at the shortest distance from the former, and their nonionic surfactant content was determined.

Data concerning the characteristics of each well (distance from stream, depth, and hydraulic head between the stream and the well) and the concentrations of nonionic surfactants found in both the wells and the corresponding streams are given in Table 3 [55]. The concentrations found in the wells surveyed constitute a matter of concern not only because the surviving APEOs found their way into the groundwater, but also because some of their metabolites/degradation products, particularly the more aquatically persistent short–EO chain APEOs, are known for their higher toxicity [31] and longer-term persistence in wastewaters, particularly under anaerobic conditions [18,49,56].

TABLE 3 Nonionic Surfactant Concentrations in Water Wells Adjacent to Rivers (Streams) Polluted by Municipal Sewage

| Stream/well | Well characteristics | | | Nonionic concentrations (mg/L)[a] | |
	Distance[b]	Depth (m)	Hydraulic head	Well	Stream[c]
Hadera					
D.I.	40	61	12.9	0.78	1.7
I.B.	40	50	11.7	0.77	2.6
M.C.	160	50	12.5	0.12	1.6
Hadas					
A.N.	150	91	20.3	0.44	2.4
S.T.	150	156	21.0	0.48	2.1
Z.A.	300	80	20.6	0.27	2.0

[a] Expressed in terms of a weighted average mixture of nonylphenol-based polyethoxylated nonionic surfactants. Determined during 1985–86.
[b] From the adjacent polluted stream.
[c] Refers to concentrations found in the spot nearest the corresponding well.
Source: Refs. 19 and 22.

Our results suggest that approximately 23% of the concentration found in the wells' water, compared with the initial concentration in the adjacent streams (average: 0.48:2.07 mg/L), probably represents the lower limit of nonbiodegradable, persistent nonionic surfactants reaching groundwaters under similar conditions.

All of these results, historically typical for the case of the APEOs in Israeli surface water and groundwater, are in accord with the findings concerning the behavior, occurrence/survival, and elimination of APEOs and their metabolites in rivers and groundwater in Switzerland [18,27,56,57]. Thus, the concentrations of 18–25 ppb of NPs (1 or 2) EOs or NPs (1 or 2) EC found in rivers and 0.1–1 ppb in groundwater in Switzerland in the mid-1990s [57], about 10 years after the use of NPEOs in laundry detergents was banned, are "comparable" with the range of concentrations of the parent APEOs found in Israeli sewage-containing rivers and adjacent water wells in the mid-1980s (Table 3).

Based on the data in Table 3, the mean velocity of movement of the surfactant front(s) was calculated to be 0.42 and 0.40 m/d for the most distant wells, M.C. (Hadera stream) and Z.A. (Hadas stream), respectively. This means that it takes the nonionic surfactants about 75 and 38 days, respectively, to reach these wells from the corresponding streams. Clearly this long period of time is insufficient for the complete chemical and/or biological degradation of the "hard" APEOs, and therefore their penetration into aquifers and groundwater is unavoidable [55,57].

In view of these results, which indicated that nonionic surfactants do penetrate into groundwater, additional selected water wells, representing both "ordinary" and "red regions" (areas with a high risk of wastewater penetration into aquifers), were sampled and the concentrations of the nonionic surfactants (mainly APEOs) determined [21]. Also, a few observation and recovery wells at the infiltration site of the SAT stage of the Dan project were sampled and their nonionic surfactant (mainly APEOs) content determined [21]. Representative results are given in Table 4. The concentrations of nonionic surfactants found in the water wells in the "red regions" are very high, an order of magnitude or more higher than those found in surface effluents in Europe [58] or the United States [59] and higher than the European and EPA limits for just anionic or all kinds of surfactants. The potential health risk, i.e., human intake via drinking water, is thus obvious. Although the high concentrations found may be due to a local "accident-type" pollution problem (e.g., leakage from adjacent untreated raw sewage water inappropriately handled), other possibilities, such as "hydrological shortcuts" [60], cannot be excluded. Also, the APEOs concentration dependence on the season (the "dilution effect"), particularly in semiarid regions of low precipitation levels, should always be taken into consideration.

Interestingly, in spite of the overall 97% efficiency of the combined activated-sludge/soil aquifer treatment in removing the nonionic surfactants initially present in the raw sewage influents and the 92.5% removal efficiency of the soil/aquifer treatment (SAT), alone at the SHAFDAN (Dan region) project, concentrations of 22–25 ppb of "hard" nonionic surfactants were found in the reclaimed water used for irrigation after recovery. This level is about one order of magnitude below the old European standard of 200 ppb for anionic surfactants and the corresponding EPA maximum level of 500 ppb for foaming agents.

Although the SAT was shown to be 70–90% efficient with respect to biological oxygen demand, chemical oxygen demand, total organic carbon, and anionic surfactant removal, the following were observed [61]:

1. There was an increase in the organic parameters of the reclaimed water in 1989 compared with those of 1988.
2. There was some increase in the anionic surfactant concentrations in the production wells in one of the recharge/recovery sites.
3. Phenolic compounds were not satisfactorily removed by the SAT.

Therefore, although the concentrations of nonionic surfactants in several water wells in non-"red regions" throughout the country were recently found to be within 0–2 ppb

TABLE 4 Nonionic Surfactant Concentrations in Water Wells in Northern and Central Israel

Location	Depth (m)	Concentration (mg/L)[a]	
		March–April 1990	May–June 1990
Tut[b]	110	0.10	0.37
Guaton[b]	180	0.066	1.60
Kefar Yasif[b]	120	0.073	1.80
Dan Region[c]	—[d]	—	0.024
Dan Region[c]	—[d]	—	0.022

[a] An average of three determinations.
[b] Water well located in a "red region."
[c] Observation well 061: after an activated-sludge process followed by soil aquifer treatment (SAT).
[d] A few dozen meters.
Source: Ref. 22.

[22], the satisfactory removal of nonionic surfactants from sewage influents even by the most sophisticated treatment facilities available in the country remains an open question. Hence, the previously accumulated data in the case of Israel point to the persistence of the organic "hard" nonionic surfactants (APEOs) in the subsurface. The effectiveness of surfactant removal in the existing primary and secondary treatment processes in the country was determined to be within the range of 85–97% and 69–81% for the anionic and nonionic surfactants, respectively [19,22]. Therefore, treated and untreated influents containing ca. 0.3–0.5 and 1.1–2.4 ppm of nonionic surfactants, respectively, are expected to reach receiving soils or waters in the environment: Neither the existing treatment facilities nor naturally occurring biodegradation in the subsurface and/or physical soil adsorption processes are capable of avoiding this contamination. These data of the 1980s are insufficient to provide a full picture of the contamination of the Israeli aquifers by nonionic surfactants. They indicate, however, that nonionic surfactants do penetrate into surface water resources and aquifers (in spite of the various kinds of treatment whenever and wherever applied) and, consequently, find their way to reclaimed water. The presence of "hard" nonionic surfactants in fresh/potable and reused water is an issue of major environmental and health risk concern. Since the assurance of adequate water supply to the growing and developing population and economy in the country constitutes a continuous major problem, the reuse of sewage water as part of the Israeli national water supply system is unavoidable. The problem of conceptualizing the country's water resources contamination by APEOs is, therefore, conditioned by the predictions of Israel's water consumption and supply given in Table 5. The use of reclaimed effluents for irrigation and aquifer recharge should be carried out with caution. The residuals of nonionic surfactants in surface water and groundwater constitute a problem that requires either their removal from or, at least, a substantial reduction of their concentrations in wastewaters.

Groundwater is quite different from other environmental aqueous systems, because it is characterized by a slow environmental response. Consequently, groundwater is susceptible to long-term contamination by organic ingredients, such as APEOs which have a high degree of hydrophobe branching and therefore constitute environmentally persistent organic pollutants (POPs). Although (1) the soft-type surfactants are the ones most used, (2) effective secondary-type wastewater treatment processes constitute common practice in highly industrialized Western countries, (3) adsorption and other physicochemical processes in soil do remove surfactants from downward-infiltrating wastewater, and (4) the slow advance of wastewater in the unsaturated zone enables biodegradation processes of organic material to continue, low-level concentrations of "hard" surfactants are expected to be found in groundwater. Their level(s), homologic distribution, and persistence should be closely monitored to facilitate a rational action to be taken accordingly.

2. Concentration and Homologic Distribution in Surface Water and Groundwater

Grab sampling of selected "representative" sewage-containing rivers, groundwater wells, and the Mediterranean Sea coastal water was followed by reverse-phase HPLC determination of the total APEO concentration (mainly nonyl/octyl phenolethoxylates) in the samples [51,58,62] as well as their local (at the sampling site) homologic distribution by normal HPLC [63].

TABLE 5 Water Consumption and Supply in Israel in the Years 2000–2005 (in and to All Sectors)—Mm³/year

		2000	2001	2002	2003	2004	2005
Rigid consumption excluding agriculture	Domestic consumption	710	750	760	770	770–90	770–810
	Industrial consumption[a]	95	98	101	104	107	110
	Palestinian consumption[b]	30	34	38	42	46	50
	Supply to Jordan	55	55	55	55	55	55
	Total consumption (without agriculture)	890	847	854	901	928	955
Water to agriculture	Total water to agriculture[c]	1,065	990	990	990	1,150	1,150
	Effluents	265	265 + 15	265 + 40	265 + 65	265 + 90	265 + 120
Total	Natural waters to agriculture	800	710	685	660	795	765
Total consumption of natural water		1,690	1,557	1,559	1,561	1,723	1,720
Potential of natural water (average)		1,160	1,555	1,555	1,555	1,555	1,555
Additional potable water	Desalination of brackish water	0	6	12	18	24	30
	Desalination of seawater[d]	0	0	0	15	165	165
	Import from Turkey	0	0	20	40	40	40
Total inventory of natural water + additions		1160	1561	1,587	1,628	1,784	1,790
Balance for aquifer replenishment		−530	4	28	67	61	70

[a] Without brackish water in the Dead Sea.
[b] In the Palestinian Authority excluding Gaza Strip (exploitation from the western aquifers).
[c] Not including occasional flood waters (~31 Mm³/year in the last 20 years).
[d] Additional desalination of 100 mm³/y, beginning 2004.
Source: Based on a publication of the Office of National Infrastructures–Water Commission, Israel, 2001.

The APEOs loads of the sewage-containing rivers/streams, which enter into the Mediterranean Sea at the sites of the coastal seawater sampling, are given in both Figure 3 and Table 6 [6,24,43]. The APEOs concentrations in Israel's groundwater wells are given in Table 7 [6,43]. The APEOs concentrations in seawater at the estuaries of the coastal rivers are given in Table 8 [6,24,43]. The homologic distributions of the APEO in Israeli rivers, coastal/river estuaries and offshore Mediterranean Sea water of the country are given in Figure 4 and Table 9 [6]. The bottom line: The APEO nonionic surfactant concentrations of the country's rivers/streams (clearly containing wastewater), groundwater, and Mediterranean coastal seawater are 12–75, trace–20,

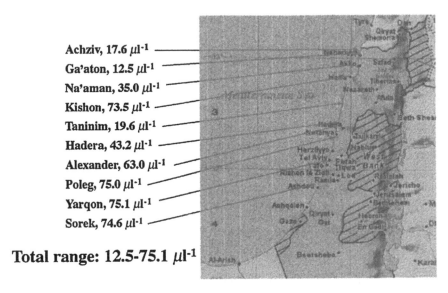

Achziv, 17.6 μl^{-1}
Ga'aton, 12.5 μl^{-1}
Na'aman, 35.0 μl^{-1}
Kishon, 73.5 μl^{-1}
Taninim, 19.6 μl^{-1}
Hadera, 43.2 μl^{-1}
Alexander, 63.0 μl^{-1}
Poleg, 75.0 μl^{-1}
Yarqon, 75.1 μl^{-1}
Sorek, 74.6 μl^{-1}

Total range: 12.5-75.1 μl^{-1}

FIG. 3 APEOs in Israeli freshwater (rivers/streams).

and 4–25 µg/L, respectively. The homologic profiles of the APEOs in the sampled sites revealed a homologic distribution, percentage-wise, within the range of 1–10, somewhat *skewing* toward the more toxic shorter-chain ethoxylates (Fig. 4) [6]. Thus, for example, 2.8–4.6% (of the total concentration) of the shortest homologs-$APEO_{1-3}$ means survival concentrations of 2.1–3.4 µg/L of these homologs, which are known to be more toxic than the longer-chain ones [16,31,49]. Such levels of short-chain APEOs in the aquatic environments of semiarid Mediterranean regions constitute an issue of health-related concern [64–66].

All of these findings are in full agreement with the previously collected and interpreted data locally (in Israel) [19,22,43] and elsewhere (e.g., Ref. 57). Also, they

TABLE 6 Concentrations of Nonionic (APEOs) Surfactants in Israeli Sewage-Containing Rivers

River	APEOs concentration (µg/L)
Achziv	17.6
Gaaton	12.5
Naaman	35.0
Kishon	73.5
Taninim	19.6
Hadera	43.2
Alexander	63.0
Poleg	75.0
Yarkon	75.1
Sorek	74.6

TABLE 7 Concentrations of Nonionic (APEOs) Surfactants in Israeli Groundwater (Selected Representative Water Wells)

Well location	Adjacent to river (upstream)	APEOs concentration (μg/L)
Metzer K.	Hadera	11.9
Ma'anit K.B.	Hadera	Trace
Nazlet Issa	Hadera	20.2
MEK Yad Hana	Alexander	0
Tul Karem A.Q.K.	Alexander	7.6
P. Taibe D.N.	Poleg	8.5
P. Kafr Qasem I.	Yarkon	2.7
Nakhshonim	Yarkon	8.7
Mek Rosh Ha'Ayin 2	Yarkon	Trace
P. Kfar Sirkin	Yarkon	6.4

simultaneously confirm a similar contemporary state of affairs in, e.g., Switzerland [27,67] in spite of a partial ban on the use of APEOs there, and reinforce a similar problem conceptualization/perspective in other semiarid countries/continents, e.g., Australia [68].

B. Environmental Concentration Effect of APEOs on Fish Reproduction

The potential endocrinological effect of the actual concentrations of the APEO found in Israel's surface water and groundwater (Tables 6–8) was studied via exposing zebrafish (*Danio rerio*) to these "real-world" environmental concentrations of an industrially produced and extensively used commercial APEOs mixture (marlophen 810).

Indeed, egg production by exposed zebrafish was significantly reduced after 8, 16, and 18 days in 75, 25, and 10 μg/L of the APEOs, respectively, and continued to decrease statistically significantly until day 20 to 89.6, 84.7, and 76.9% of the baseline

TABLE 8 Concentrations of Nonionic (APEOs) Surfactants in the Coastal Water of the Eastern Mediterranean Sea (Israel)

Mediterranean–river/stream junction	APEOs concentration (μg/L)
Achziv	5.9
Gaaton	4.2
Naaman	11.7
Kishon	24.5
Taninim	6.5
Hadera	14.4
Alexander	15.2
Poleg	25.0
Yarkon	25.0
Sorek	24.9

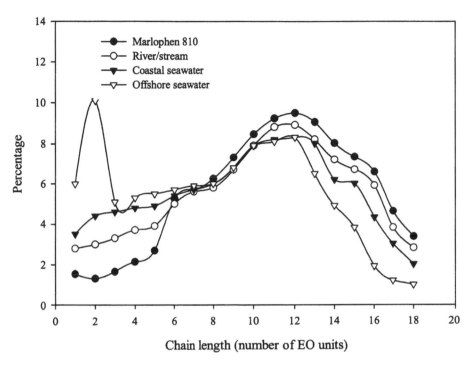

FIG. 4 Typical (compared with a reference commercial product) homologic distribution profile of APEOs (% of total concentration) in surface water.

levels, respectively, when the APEOs were removed from the water (Table 10). No significant reduction in egg production was detected after 20 days' exposure to 0 (control) and 5 µg/L of APEOs [6].

Our overall results suggest that the APEOs concentrations found in Israel's coastal rivers, groundwater, and seawater may be detrimental to aquatic fauna. Also, they should be evaluated as potential environmental/health (endocrine disrupting) risks. Further, the effect of the APEOs is both time and concentration positively related, so long-term exposure to APEOs may affect fish reproduction in lower concentrations than those examined here (<5 µg/L). Thus, the draft EPA water-quality guidelines for NP [freshwater, 6 g/L (4-day average) and 26 µg/L (1-hour average)] appear to be rather liberal. This is especially true while keeping in mind that endocrine-disrupting chemicals were found to affect not only egg production rates, but also other stages in fish and invertebrate reproduction processes [37].

C. Is There an Environmental Problem?

The hard nonionic APEO surfactants and their biodegradation/degradation metabolites/products, the origin of which are either domestic, industrial, or agricultural wastewater, reach rivers/streams as well as the Mediterranean Sea and infiltrate into groundwater in Israel and persist there. Neither the existing sewage treatment facilities nor naturally occurring biodegradation processes in the receiving surface water systems appear to be capable of avoiding this penetration.

TABLE 9 APEOs Homologic Distribution (%) in Israeli Surface Waters (2000/2001)

No. of EO units	Rivers/streams	Coastal seawater	Offshore seawater
1	2.8	3.5	6.0
2	3.0	4.4	10.1
3	3.3	4.6	5.1
4	3.7	4.8	5.3
5	3.9	4.9	5.5
6	5.0	5.4	5.7
7	5.6	5.7	5.9
8	5.8	6.0	6.0
9	6.7	6.8	6.8
10	7.9	7.9	7.9
11	8.8	8.2	8.1
12	8.9	8.3	8.3
13	8.2	8.0	6.5
14	7.2	6.2	4.9
15	6.7	6.0	3.8
16	5.9	4.3	1.9
17	3.8	3.0	1.2
18	2.8	2.0	1.0

The levels of the APEOs, particularly those of their shorter-chain homologues found in these aquatic environments, constitute an issue of major environmental long-term health risk, particularly in semiarid countries with limited water resources in which the "hard" environmentally persistent APEOs are still in extensive use domestically and/or industrially-institutionally.

In view of (1) the estrogenic/endocrinic potential of the anthropogenic APEO nonionic surfactants, (2) the recently reported findings of estrogenic activity of domestic sewage treatment wastewater effluents in European countries (e.g., the

TABLE 10 Effect of 20 Days' Exposure to APEOs on Egg Production of Zebrafish (*danio rerio*)

APEOs concentration (μg/L)	Time of exposure until significant rate reduction (days)	Egg production after 20 days of exposure (% of baseline)
0	—	98.5 ± 0.5
5	—	101.3 ± 2.4
10	18	89.6 ± 3.9
25	16	84.7 ± 3.9
75	8	76.9 ± 2.2

United Kingdom, (3) the recently reported results that indicate that NPs are ubiquitous in food [64], (4) the levels of NPEOs and their metabolites found in Switzerland surface waters and groundwaters [27,57,67] and their risk assessment [66], (5) the case of the APEOs in Israel here presented, and (6) the direct effect of APEOs on zebrafish egg production found by us, it appears that there is a potential problem. There is still not enough evidence to conclusively support the hypothesis that exposure to endocrine-disrupting/estrogenic chemicals is a global environmental health problem. Yet, neither existing sewage treatment facilities nor naturally occurring biodegradation appears to be capable of keeping the APEOs and/or their metabolites from reaching and persisting in natural water resources. It appears, therefore, that there is a related potential environmental and health risk problem.

REFERENCES

1. Kaiser, J. Science 2000, *260*, 695–697.
2. Legler, J.; Broekhof, J.L.M.; Brouwer, A.; Lanser, P.H.; Murk, A.J.; van der Saag, P.T.; Vethaak, A.D.; Wester, P.; Zivkovic, D.; van der Burg, B. Environ. Sci. Technol. 2000, *34*, 4439–4444.
3. Renner, R. Environ. Sci. Technol. 2000, *68A*.
4. Sole, M.; de Alda, M.J.; Castillo, M.; Porte, C.; Ladegaard-Pedersen, K.; Barcelo, D. Environ. Sci. Technol. 2000, *34*, 5076–5083.
5. Keith, T.L.; Snyder, S.A.; Naylor, C.G.; Staples, C.A.; Summer, C.; Kannan, K.; Giesy, J.P. Environ. Sci. Technol. 2001, *35*, 10–13.
6. Zoller, U.; Plaut, I.; Hushan, M. Water Sci. Technol. 2003 (accepted for publication).
7. Cavalli, L. Chapter 14, this book.
8. Hager, C.D. In *Annual Surfactant Review*; Karsa, D.R., Ed.; Academic Press: Sheffield, UK, 1998; 1–19.
9. Naylor, C.G. Chapter 15 in this book.
10. Britton, L.N. J. Surfact. Deter. 1998, *1*, 109–117.
11. Maguire, R.J. Water Qual. Res. J. Canada 1999, *34* (1), 37–78.
12. Thiele, B.; Gunther, K.; Schwuger, M.J. Chem. Rev. 1997, *97*, 3247–3272.
13. Jobling, S.; Sumpter, J.P. Aquat. Toxicol. 1993, *27*, 361–372.
14. Jobling, S.; Nolan, M.; Tylor, C.R.; Brighty, G.; Sumpter, J.P. Environ. Sci. Technol. 1996, *32*, 2498–2506.
15. Billinghurst, Z.; Clare, A.S.; Fileman, T.; McEvoy, J.; Readman, J.; Depledge, M.H. Mar. Pollut. Bull. 1998, *36*, 833–839m.
16. Lye, C.M.; Frid, C.L.J.; Gill, M.E.; Cooper, D.W.; Jones, D.M. Environ. Sci. Technol. 1999, *33*, 1009–1014.
17. Jobling, S.; Sheahan, D.; Osborne, A.; Mathiessen, P.; Sumpter, J.P. Environ. Toxicol Chem. 1996, *15*, 194–202.
18. Ahel, M.; Giger, W.; Koch, M. Environ. Res. 1994, *2* (1–2), 29–38.
19. Zoller, U. In *Groundwater Contamination and Control*; Zoller, U., Ed.; Marcel Dekker: New York, 1994; 273–292.
20. Naylor, C.G. Chem. Spec. 1992, *68*, 27.
21. Zoller, U. J. Environ. Sci. Health 1992, *A27*, 1521–1533.
22. Zoller, U. Toxicol. Environ. Chem. 1998, *66*, 145–157.
23. Zoller, U. Pollution of the coastal aquifer by nonionic surfactants and the implications to the mountain aquifer: Is there a problem? Proceedings of the Annual Meeting of the Isr. Assoc. Wat. Resources (Hebrew), Acre; 2000; 89–93.

24. Zoller, U.; Hushan, M. Water Sci. Technol. 2000, *42* (1/2), 325–330.
25. Zoller, U.; Romano, R. J. Am. Oil Chem. Soc. 1984, *61*, 971–976.
26. Sadik, A.O.; Witt, D.M. Environ. Sci. Technol. 1999, *33*, 368A–374A.
27. Ahel, M.; Molnar, E.; Ibric, S.; Giger, W. Water Sci. Technol. 2000, *42* (7–8), 15–22.
28. US Environmental Protection Agency. Special Report on Environmental Endocrine Dissruption. An Effects Assessment and Analysis, 1997.
29. Fed. Reg. (US), 1992, *57*, 246.
30. Ekelund, R.; Bergman, A.; Granmo, A.; Berggren, M. Environ. Pollut. 1990, *64*, 107.
31. Servos, M.R. Water Qual. Res. J. Canada 1999, *34* (1), 123–177.
32. Heath, A.G. *Water Pollution and Fish Physiology*; CRC Lewis: London, 1995.
33. Lewis, M.A. Water Res. 1991, *25*, 101–113.
34. Lewis, M.A. Water Res. 1992, *26*, 1013–1023.
35. Dreze, V.; Monod, G.; Cravedi, J.-P.; Biagianti-Risbourg, S.; Le Gac, F. Ecotoxicology 2000, *9* (1–2), 93–103.
36. Nice, H.E.; Thorndyke, M.C.; Morritt, D.; Steele, S.; Crane, M. Mar. Pollut. Bull. 2000, *40* (6), 491–496.
37. Harris, J.E.; Runnals, T.; Hill, E.; Harris, C.A.; Maddix, S.; Sumpter, J.P.; Tyler, C.R. Environ. Sci. Technol. 2000, *34*, 3003–3011.
38. Gisey, J.P.; Pierens, S.L.; Snyder, E.M.; Miles-Richardson, S.; Kramer, V.J.; Snyder, S.A.; Nichols, K.M.; Villeneuve, D.A. Environ. Toxicol. Chem. 2000, *19* (5), 1368–1377.
39. Kelly, S.A.; Di Giulio, R.T. Environ. Toxicol. Chem. 2000, *19* (10), 2564–2570.
40. Routledge, E.J.; Sheahan, D.; Desbrow, C.; Brighty, G.C.; Waldock, M.; Sumpter, J.P. Environ. Sci. Technol. 1998, *32*, 1559–1565.
41. Field, J.A.; Read, R.L. Environ. Sci. Technol. 1996, *30*, 3544–3550.
42. Naylor, C.G.; Mieure, J.P.; Adams, W.J.; Weeks, J.A.; Castaldi, F.J.; Olge, L.D.; Romano, R.R. Alkylphenols Alkylphenol Ethoxylates Rev. 1998, *1* (1), 32–43.
43. Zoller, U. J. Environ. Eng., 2003 (accepted for publication).
44. Colborn, T., Clement, C., Eds.; *Advance Modern Environ Toxicol*; Princetown Scientific: Princetown, US, 1992; Vol. 12.
45. Colborn, T.; von Saal, F.S.; Soto, A.M. Health Perspect. 1993, *103*, 81–85.
46. Lee, R.F.; F & T Noone. Mar. Environ. Res. 1995, *39*, 151–154.
47. LeBlanc, G.A.; Bain, L.J. Health Perspect. 1997, *105S*, 65–80.
48. Uppgard, L.-L.; Lingren, A.; Sjostrom, M.; Wold, S. J. Surfact. Deterg. 2000, *3* (1), 33–41.
49. Ying, G.-G.; Williams, B.; Kookana, R. Environ. Inter. 2002, *28* (3), 215–226.
50. Zoller, U. J. Am. Oil Chem. Soc. 1985, *62*, 1006–1008.
51. Ahel, M.; Giger, W. Anal. Chem. 1985, *57*, 2584–2590.
52. Naylor, C.G.; Mieure, J.P.; Adams, W.J.; Weeks, J.A.; Castaldi, F.J.; Ogle, L.D.; Romano, R.R. J. Oil Chem. Soc. 1992, *69* (8), 695–703.
53. Zoller, U., Ed.; *Groundwater Contamination and Control*; Marcel Dekker: New York, 1994.
54. Purdon, C.E.; Hardiman, P.A.; Bue, V.J.; Eno, N.C.; Tyler, C.R.; Sumpter, J.P. Chem. Ecol. 1994, *8*, 275–285.
55. Zoller, U.; Ashash, G.; Ayali, G.; Shafir, S.; Azmon, B. Environ. Inter. 1990, *26*, 301–306.
56. Ahel, M.; Giger, W.; Schaffner, C. Water Res. 1994, *28* (5), 1143–1152.
57. Ahel, M.; Schaffner, C.; Giger, W. Water Res. 1996, *30* (1), 37–46.
58. Ahel, M.; Giger, W. Anal. Chem. 1989, *57*, 1577–1583.
59. Gledhill, W.E.; Haddleston, R.L.; Kravet, L.; Nielsen, A.M.; Sellack, R.I.; Vashon, R.D. Tensides Surfact. Deterg. 1989, *26*, 275–281.
60. Goldberg, L.C.; Melloul, A.J.; Zoller, U. J. Environ. Manage 1996, *46*, 311–326.
61. Kanarek, A.; Aharoni, A.; Sherer, D.; Michail, M.; Goldman, B. Groundwater recharge with municipal effluent. Dan Region Project—Stage 2. Tahal Water Planning for Israel and Mekorot Water Company; Tel Aviv, 1989.
62. Marcomini, A.; Giger, W. Anal. Chem. 1987, *59*, 1709–1715.

63. Kubeck, E.; Naylor, C.G. J. Am. Oil Chem. Soc. 1990, *67*, 400–403.
64. Guenther, K.; Heinke, V.; Thiele, B.; Kleist, E.; Prast, H.; Reacker, T. Environ. Sci. Technol. 2002, *36* (8), 1676–1680.
65. Soto, A.M.; Sonnenschein, G.; Chung, K.L.; Farmandez, M.F.; Olea, N.; Olea, S.F. Environ. Health Perspec. 1995, *103*, 113.
66. Fenner, K.; Kooijman, C.; Schringer, M; Hungerbühler, K. Environ. Sci. Technol. 2002, *36* (6), 1147–1154.
67. Ahel, M.; Giger, W.; Molnar, E.; Ibric, S. Croatia Chem. Acta 2000, *73* (1), 209–227.
68. Ying, G.-G.; Kookana, R.S. Awa. Wat. 2002, *29* (6), 42–48.

18

Ecology and Toxicology of Alkyl Polyglycosides

ANDREAS WILLING, HORST MESSINGER,
and WALTER AULMANN Cognis Deutschland GmbH & Co. KG,
Düsseldorf, Germany

I. ECOLOGICAL PROPERTIES AND ENVIRONMENTAL IMPACT OF ALKYL POLYGLYCOSIDES

Alkyl polyglycosides (APG) are nonionic surfactants produced in quantities of about 80,000 t/a [1]. Their main use is in laundry detergents and hard-surface cleaners [2] and in personal care products [3]. All these applications have in common that the products are discharged into the domestic wastewater after use. Therefore, the majority of the APG will directly (in case of direct discharge) or indirectly (after passage through a sewage treatment plant) enter the aquatic environment. Another use of APG is in agricultural applications [4], and this fraction will be exposed primarily to the terrestrial compartment. Thus, APG are products of high environmental relevance, which leads to the question of their environmental compatibility. According to the broadly accepted risk assessment scheme for chemicals [5], environmental compatibility requires proof that the use of the chemical will not result in environmental exposure levels higher than the predicted ecotoxicological no-effect concentration (the PEC/PNEC concept). To assess the environmental impact of APG, knowledge of their aquatic and terrestrial toxicity as well as of their environmental fate is necessary. The ecotoxicity of surfactants is linked to their surface activity (detergency) and to the resulting effects on biological membranes [6], i.e., surfactants generally have a significant ecotoxicity. However, the ecotoxicity of surfactants depends on the exact structure and can vary by more than two orders of magnitude. The environmental fate of nonionic surfactants is determined primarily by their biodegradability, because biodegradation is the most prominent mechanism for the ultimate removal of water-soluble organic chemicals from the aquatic and terrestrial environments. Thus, fast and complete biodegradability is the most important presupposition for an environmentally compatible surfactant.

APG is a generic term for a group of structurally closely related substances that differ in the alkyl chain distribution and in the degree of polymerization (DP). Usually the reaction conditions are chosen so that APG with an average DP of 1.2–

1.7 are formed. If the alkyl chains (usually linear chains derived from renewable resources) are in the range of 8–14 carbon atoms, these products are water soluble. Water-soluble APG are by far the most important products because they are good surfactants [7]. Less important are the APG with 16 or more carbon atoms in the alkyl chain. Long-chain alkyl polyglycosides are water insoluble, if the DP is below 2 and water soluble if the DP is above 5. For the ease understanding, this chapter is focused on three representative APG, i.e., the short-chain (technical) $C_{8/10}$-APG, the medium-chain (technical) $C_{12/14}$-APG, and the long-chain water-soluble C_{16-18}-APG. The properties of the "mixed" APG, e.g., $C_{10/16}$-APG, can be approximated by interpolation. The following section will summarize the ecological properties of these APG and will evaluate the environmental impact of this interesting class of surfactants.

A. Environmental Fate and Biodegradability of Alkyl Polyglycosides

The environmental fate of nonionic surfactants is determined primarily by their biodegradability, because biodegradation is the most prominent mechanism for the ultimate removal of water-soluble organic chemicals from the aquatic and terrestrial environments. Thus, fast and complete biodegradability is the most important presupposition for an environmentally compatible surfactant. The biodegradability of organic substances can be determined in a broad variety of different test systems. An overview of the test methods can be found in Ref. 8. To make the test results from different laboratories comparable, however, it is advisable to use internationally standardized test systems, like the OECD 300 series methods [9], and to apply the "principles of good laboratory practice" [10]. Therefore, the degradability data cited herein were chosen according to these guidelines whenever possible. The basic test designs for the evaluation of the biodegradability of organic substances can be grouped into two categories: screening tests and simulation tests. The screening tests are simple, discontinuous tests that investigate the biodegradability of the test substance under highly artificial but very stringent conditions, especially with regard to the inoculum and the nutrient status. The result of these tests, namely, the results of the OECD 301 A–F tests, are used for classification and labeling purposes, e.g., according to the EU Dangerous Substance Directive, 92/69/EEC [11], and the German Water Hazard Classification scheme [12]. If, under these stringent conditions, a substance turns out as readily biodegradable, this allows the general conclusion to be drawn that the substance will be easily biodegradable under environmental conditions too [9]. The simulation tests are continuous tests designed to mimic the frame conditions of a municipal sewage treatment plant. They are more realistic with regard to the elimination of chemicals from the wastewater, and the results of the simulation tests, therefore, build the basis for the deviation of the predicted environmental concentration (PEC) in the framework of an environmental risk assessment.

1. Primary Biodegradability of Alkyl Polyglycosides

Primary biodegradability for surfactants is defined as the initial cleavage of a molecule leading to the loss of the physicochemical properties of the parent compound, i.e., the loss of the surface activity. Because the ecotoxicity of surfactants is linked to their surface activity (detergency) and the resulting effects on biological membranes

[6], primary biodegradation generally leads to a detoxification. Based on this perception, surfactants intended for use in detergents and cleaning products are regulated in the Euopean Union. According to the Detergent Law [13] for such applications, the only surfactants permitted are those that are at least 90% (80% in the screening test) primary biodegradable, determined as removal of, respectively, methylene blue- or Bismuth–reactive substances in the OECD conformatory test (OECD 303 A). Methylene blue is a substance group-specific reagent forming colored complexes with anionic surfactants, and the Bismuth reagent is substance-group-specific reagent for alkoxylated nonionic surfactants. Although alkyl polyglycosides are nonionic surfactants, they are not alkoxylated and, therefore, cannot be determined by the routine analytical procedure used to monitor the primary degradation of alkoxylated nonionic surfactants. Thus, the European legal requirements on the primary biodegradability of surfactants intended for use in detergents and cleaning products do not apply to this group of surfactants. Nevertheless, the primary biodegradability of $C_{12/14}$-APG has been studied in the OECD conformatory test using a substance-specific analytical method (HPLC) [14]. Under these conditions the medium-chain APG exhibited 99.5% or greater primary biodegradation, determined as removal of the parent compound (Table 1). It is also reasonable to assume that the short-chain ($C_{8/10}$-APG) and the long-chain ($C_{16/18}$-APG) APG underlie a rapid primary biodegradation.

2. Ultimate Biodegradability of Alkyl Polyglycosides

The assessment of the environmental fate of chemicals normally starts with screening tests for ultimate biodegradability. These tests determine the ultimate biodegradation of the test compound, i.e., the microbial transformation of the organic test substance into the final products of the degradation process, such as carbon dioxide, water, and assimilated bacterial biomass, a process often referred to as *mineralization*. These discontinuous tests are characterized by their stringency, which is attributable to the low bacterial concentration, the fact that only nonadapted inoculum is to be used, and to the fact that the test substance serves as the sole carbon source in the test.

The short-chain alkyl polyglycoside ($C_{8/10}$-APG) was tested in two different ultimate biodegradability screening tests, the closed-bottle test (OECD 301 D) and the modified OECD screening test (OECD 301 E) [14]. The results obtained (Table 2) show a very high degree of ultimate biodegradation within the 28-day period. In the closed-bottle test, which is the most exacting of all the OECD 301 tests [15], the substance exhibited mineralization levels of 81% (test substance concentration 2 mg/L) and 82% (test substance concentration 5 mg/L), far exceeding the

TABLE 1 Primary Degradation Test Data of APG

Method	Test substance			Remark/conclusion
	$C_{8/10}$-APG, DP = 1.4	$C_{12/14}$-APG, DP = 1.4	$C_{16/18}$-APG, DP = 1.1	
OECD confirmatory test	n.d.	>99.5%	n.d.	Removal of parent compound (HPLC)
Surface water simulation test	n.d.	>98%	n.d.	Removal of parent compound (HPLC)

TABLE 2 Ultimate Degradation Test Data of APG

Method	Test substance			Remark/conclusions
	$C_{8/10}$-APG, DP = 1.4	$C_{12/14}$-APG, DP = 1.4	$C_{16/18}$-APG, DP = 1.1	
OECD 301 A	n.d.	95–96%[a] (DOC)	n.d.	"Ready biodegradable"
OECD 301B	n.d.	78% (CO2)	n.d.	
OECD 301 D	81–82%[a] (BOD/COD)	73–88%[a] (BOD/COD)	86–92%[a] (BOD/COD)	According to OECD classification
OECD 301 E	94%[a] (DOC)	90–93%[a] (DOC)	n.d.	
ISO 10708	n.d.	89% (BOD/COD)	n.d.	

[a] 10-d window fulfilled.

OECD limit value for ready biodegradability (60% BOD/COD). In the Modified OECD screening test (OECD 301 E), $C_{8/10}$-APG exhibited 94% DOC removal. In order to ensure that the high DOC removal is not overly influenced by physical elimination processes, TOC removal was also analyzed. This parameter enables the contribution of mineralization to the biodegradation level to be evaluated. The data (88% TOC removal) show that the high DOC removal is due mainly to the mineralization of the alkyl polyglycoside underlying the ready ultimate biodegradability of this surfactant. In addition, the kinetic of the degradation was so fast that the limit value of 60% was exceeded within 10 days after the onset of the degradation reaction, i.e., $C_{8/10}$-APG fulfills the 10-day-window criterion and can therefore be regarded as readily biodegradable. The medium-chain alkyl polyglycoside ($C_{12/14}$-APG), which with regard to production volumes is the most important product, was tested in several ultimate biodegradability screening tests. In the DOC die-away test (OECD 301 A), $C_{12/14}$-APG achieved 95–96% DOC removal (66–81% with regard to TOC, indicating the presence of insoluble material). In the modified Sturm test (OECD 301 B), $C_{12/14}$-APG was 78% mineralized to carbon dioxide within the 28-day test period (no value for the 10-d window given). In the closed-bottle test (OECD 301 D), the substance exhibited mineralization levels of 73% (test substance concentration 2 mg/L) and 88% (test substance concentration 5 mg/L). In the modified OECD screening test (OECD 301 E), the substance achieved 90–93% DOC removal (56–82% with regard to TOC). The results obtained (Table 1) show a very high degree of ultimate biodegradation within the 28-day period, which far exceeds the OECD limit value for ready biodegradability (60% BOD/COD). In addition, the kinetics of the degradation was so fast that the limit value of 60% was exceeded within 10 days after the onset of the degradation reaction; i.e., $C_{12/14}$-APG fulfills the 10-day-window criteria and can therefore be regarded as readily biodegradable. In addition to the OECD 301 tests, the biodegradability of the medium-chain-length APG has been studied in other standardized screening tests for ultimate biodegradability [14]. In the two-phase closed-bottle test (BODIS test, ISO 10708), $C_{12/14}$-APG exhibited 89% BOD/COD after 28 days, thus confirming the results from the OECD 301 tests. The long-chain alkyl polyglycoside was tested in the closed-bottle test (OECD 301 D). With mineralization levels of 86% BOD/COD (test substance concentration 2 mg/L) and 92% BOD/COD (test substance concentration 5 mg/L), the substance (experimental product, $C_{16/18}$-APG, DP = 1.1, water insoluble) exhibited high degradation levels comparable to those of the water-soluble APG [16]. Although the article focuses primarily on the commercially most important linear alkyl polyglycosides, the mention of the ecological properties of the branched isomers may be of interest as well. 2-Ethylhexyl alkyl polyglycoside (*iso*-C_8-APG) was tested in the closed-bottle test (OECD 301 D). With mineralization levels of 61% BOD/COD (test substance concentration 2 mg/L) and 64% BOD/COD (test substance concentration 5 mg/L) and the fulfillment of the 10-d-window criteria the substance (experimental product, 2-ethylhexyl-APG, DP = 1.23) exhibited degradation levels comparable to those of the linear APG [17]; i.e., it can be regarded as readily biodegradable.

To sum up, the results of the various screening tests for ultimate biodegradability show unequivocally that alkyl polyglycosides are readily biodegradable and, therefore, will undergo rapid and ultimate biodegradation in the environment, according to the conclusions of the OECD [8].

3. Degradation Pathway for Alkyl Polyglycoside: Exclusion of the Formation of Recalcitrant Degradation Metabolites

The degradation of surfactants can lead to the formation of recalcitrant (stable) degradation metabolites. A well-known example is nonylphenol-10-ethoxylate (NP-10). NP-10 degradation occurs via initial shortening of the EO chain from the free end. When the number of remaining EO moieties has fallen below five, the degradation rate slows down significantly (probably caused by a steric hindrance from the neighboring aromatic ring), leading to the accumulation of the more hydrophobic, and thus more ecotoxic, NP-1 to NP-3 derivatives. As a consequence of the reduced degradation rates of the recalcitrant metabolites, it is likely that such surfactants will fail to surpass the limit values for ready biodegradability in the corresponding OECD screening tests for ultimate biodegradability. However, even in the cases where a substance has surpassed the limit values for ready biodegradability (60% BOD/COD or CO_2 formation, respectively, DOC removal), the test design of the OECD screening tests cannot unequivocally exclude the possibility of the formation of small but stable degradation metabolites. This uncertainty is related to the fact that in the OECD screening tests the test substance is the only carbon source. Therefore, the microorganisms can grow and proliferate only if they use the test substance as carbon source to increase their biomass. Thus, it is feasible to assume that the difference between the amount of test substance initially added to the test (e.g., 100 mg TOC) and the amount mineralized after 28 days (e.g., 80 mg TOC) ended up in the anabolism of the microorganisms (e.g., 20 mg TOC), i.e., to increase their biomass. Although it is likely that this assumption generally holds true, the possibility of the formation of small recalcitrant degradation metabolites cannot be ruled out in cases where OECD screening test results of under 100% are obtained. To address the question of the completeness of degradation, which is of interest especially for high-production-volume chemicals like detergent surfactants, the so-called "metabolite test" has been developed [18]. The "metabolite test" is in principle a DOC-based sewage treatment plant simulation test (coupled-units test,

OECD 303 A) for ultimate biodegradation that has been modified specifically to detect the buildup of recalcitrant degradation metabolites. This is achieved by recycling the plant effluent daily to the continuous activated-sludge unit and replenishing it with nutrients ("synthetic sewage") and the test substance. In this test the formation of a stable metabolite will lead to a continuous increase of the DOC in the system, which can easily be detected analytically. The medium-chain-length alkyl polyglycoside has been investigated in the "metabolite test." $C_{12/14}$-APG revealed a constant DOC removal rate of 102 ± 2% under these conditions [19]. This result rules out the possibility of any recalcitrant metabolite's being formed during the biodegradation of the test substance. Based on this result and taking into consideration the excellent comparability of the test results for $C_{8/10}$-APG, $C_{12/14}$-APG, and $C_{16/18}$-APG from the ultimate biodegradability studies, it can be concluded that the alkyl polyglycoside structure as such is completely biodegradable without the formation of recalcitrant degradation metabolites. This conclusion is confirmed by a recent study of the biodegradation pathway of alkyl polyglycosides [20]. Rhine River water was spiked with 1 mg/L each of octyl-, decyl-, and dodecyl-β-monoglycoside and continuously pumped through a fixed-bed reactor so that a stable biofilm could develop. The temporal evolution of the test substance concentration was monitored by liquid chromatography–electrospray mass spectrometry (LC-ES/MS). The alkyl polyglycosides were degraded in this system with a half-life of about 18 h. In addition, the test-filter samples were investigated with respect to the occurrence of possible metabolites. In principle, two degradation pathways are feasible: (1) ω-degradation of the alkyl chain, leading to "polyglycoside alkanoic acids," and (2) cleavage of the glycosidic bond, leading to the formation of glucose and fatty alcohol. Because the search for the masses of the corresponding polyglycoside octanoic acid ($m/z = 322$), polyglycoside decanoic acid ($m/z = 350$), and polyglycoside dedecanoic acid ($m/z = 378$) as well as for related "polyglycoside alkanoic acids" using LC-ES/MS was unsuccessful, it can be concluded that the ω-degradation pathway is unlikely. In case of alternative pathway [2], the liberated glucose is rapidly further metabolized via pyruvate, whereas the fatty alcohol is oxidized to the corresponding fatty acid. Because fatty acids are biochemically rather inert, they are subsequently reacted with coenzyme A to the corresponding (activated) thiolesters. The fatty acid thiolesters are then channeled into the classical fatty acid degradation pathway, i.e., into the β-oxidation cycle [21]. Again, LC-ES/MS has been used to detect the free fatty acids, however, without success. The lack of detection of the free fatty acids in the degradation solution can be traced back to a fast intracellular metabolism. In conclusion it can be stated that alkyl polyglycosides are most likely degraded via pathway (2), i.e., via an initial cleavage of the glycosidic bond and a subsequent rapid intracellular metabolization of the glucose and the fatty alcohol.

4. Biodegradation of Alkyl Polyglycosides in Surface Waters

It is feasible to assume that readily biodegradable substances such as alkyl polyglycosides are biodegraded in natural surface waters. However, no standardized test method exists for this endpoint at the moment. A simulation method for determining biodegradation in surface waters based on a river flow model has recently been developed and applied to several surfactants, including $C_{12/14}$-APG [22]. It has been shown that the described method can be regarded as a suitable simulation model that can be used to determine the biodegradation of substances in surface waters. Using this model it is possible to estimate not only the degree of degradation but also the flow time, and thus the distance, within which this degree of degradation is achieved in a natural surface water. For $C_{12/14}$-APG a degradation rate of 98% (primary degradation) has been determined in the surface water simulation model. This result is in good agreement with the degradation rate of 99.5–99.8% determined for $C_{12/14}$-APG in the OECD confirmatory test [23], i.e., in a sewage treatment plant simulation test. The corresponding half-lives for the primary degradation of $C_{12/14}$-APG were 0.3–0.7 hours under the surface water simulation test conditions. In conclusion, it can be stated that alkyl polyglycosides are easily biodegraded in surface waters.

5. Biodegradability of Alkyl Polyglycosides Under Marine Conditions

Alkyl polyglycosides are produced mainly in North America, as well as in Europe, and are routinely shipped in bulk to customers all over the world. Therefore, it cannot be excluded that marine environments might eventually get exposed to alkyl polyglycosides. In contrast to the vast numbers of freshwater degradation studies, very few studies are available that address the biodegradability of surfactants and especially of alkyl polyglycosides under marine conditions. This is probably related to the fact that with regard to the use pattern of surfactants, the most obvious exposed compartment is the continental (fresh) watershed. In addition, it is not unreasonable to assume that the substances that turned out in an OECD screening test for ultimate biodegradability as readily biodegradable are easily biodegradable under all environmental conditions, including the marine environment. This assumption has recently been tested for two groups of surfactants using inocula from different fresh- and saltwater environments [24]. As a typical example for a European surface water, the Rhine River was chosen. As representative of the marine environment, samples from North Sea coastal waters as well as from the open sea were taken. The degradation rates of the surfactants (LAS and a fatty alcohol ethoxylate) using these environmental samples as inocula were then compared to the degradation rates under standard OECD screening test conditions, i.e., with the effluent from a municipal sewage treatment plant (containing 10^4–10^6 colony-forming units per liter) serving as inoculum. To summarize the results: The fatty alcohol ethoxylate achieved degradation values after 28 days almost identical with those of the control, sodium benzoate. However, under standard conditions, the results (79–84% BOD/COD) were about 30–40% higher than with the inocula from the marine environment (40–55%). LAS was degraded somewhat slower, but exhibiting a similar pattern (60–70% BOD/COD under standard conditions, compared to 25–35% with marine environmental inocula). Between the different environmental inocula, however, there was no significant difference in the degradation rates. This clearly indicates that surfactants that are found to be readily biodegradable under OECD screening test conditions are easily, although eventually more slowly, degraded under marine conditions, too. Although this study did not include alkyl polyglycosides, the general validity of its conclusion is confirmed by an independent study of a Japanese research team. This team has studied the ultimate biodegradability of a number of surfactants (including APG and LAS) in seawater using a comparable method, i.e., a modified OECD 301 D test [25]. Under these conditions APG was degraded much faster than LAS, and indeed the fastest of all surfactants tested. This clearly indicates that alkyl polyglycosides, which turned out in all kinds of biodegradation tests as one of the most easily degradable substances, will easily be biodegradable in the marine environment, too.

6. Biodegradability of Alkyl Polyglycosides Under Anaerobic Conditions

As explained in the Section I.A.2 alkyl polyglycosides are easily ultimately biodegradable under aerobic conditions. In regard to the application areas and the corresponding discharge pathway for APGs, i.e., via municipal sewage treatment plants, oxidative cleavage (aerobic biodegradation) is certainly the primary pathway for the mineralization of these surfactants. Nevertheless, it cannot be excluded that small amounts of APG could reach anaerobic compartments (e.g., sediment, flooded soils, deep groundwater) unaltered. Therefore, knowledge of the biodegradability in the absence of oxygen is required to assess the fate of alkyl polyglycosides in anaerobic environmental compartments. The anaerobic biodegradability of the short-chain ($C_{8/10}$-APG) and of the medium-chain. ($C_{12/14}$-APG) APG have been studied in the ECETOC screening test [26]. The ECETOC test, which has been developed by an expert group under the umbrella of the European Center for Ecotoxicology and Toxicology of Chemicals (ECETOC), evaluates the ultimate biodegradability of a test substance under anaerobic conditions using the parameter of CH_4/CO_2 formation. Its stringency is comparable to the OECD screening tests mentioned earlier. Under these conditions, $C_{8/10}$-APG exhibited $95 \pm 22\%$ mineralization within 56 days [27]. The medium-chain alkyl polyglycoside ($C_{12/14}$-APG) under the same conditions achieved $84 \pm 15\%$ mineralization within 56 days [28]. Based on these results both APG types can be regarded as easily biodegradable under anaerocic conditions, and it is quite feasable to assume that the stucturally very closely related long-chain alkyl polyglycoside ($C_{16/18}$-APG) is easily biodegradable under anaerobic conditions, too. It is, however, noteworthy that the kinetics of anaerobic degradation are somewhat slower compared to degradation under aerobic conditions (see the time scale of the tests). The faster degradation under aerobic conditions is related to the high chemical reactivity of molecular oxygen. On the basis of the predictive value of the ECETOC screening test result for the real environmental situation, it can be concluded that alkyl polyglycosides will undergo ultimate biodegradation in anaerobic municipal and household digester tanks, so significant contaminations of river sediments and soils (widespread use of sewage digester sludges as agricultural fertilizers) by alkyl polyglycosides from this source is unlikely. Accordingly, alkyl polyglycosides can join the ranks of surfactants with optimal biodegradation properties, in that they are ultimately biodegradable under all environmental conditions.

B. Elimination of Alkyl Polyglycosides in Sewage Treatment Plants

The degradation (screening) tests discussed in Sections I.A.2 to I.A.5 are designed to determine the biodegradability of chemicals under very stringent but artificial conditions, especially with regard to the inoculum and the nutrient status. As mentioned, the result of these tests, namely, the results of the OECD 301 A–F tests, are used for classification and labeling purposes, e.g., according to the EU Dangerous Substance Directive, 92/69/EEC [29], and the German Water Hazard Classification scheme [12]. However, in the context of an environmental risk assessment, data from tests that mimic the real exposure scenarios more closely are preferred. As in most developed countries, wastewater is not discharged directly (untreated), but is purified in sewage treatment plants; the (predicted) environmental concentrations (PECs) in surface waters are dependent on the efficiency of the treatment plants to remove (eliminate) the chemicals from the wastewater. The removal of the chemicals from the wastewater is attributed mainly to two mechanisms, adsorption onto the sludge and biodegradation. Because the simulation tests have a much higher inoculum and are run in the presence of plenty of nutrients, they have a significantly higher degradation potential as compared to the screening tests.

1. Primary Elimination of Alkyl Polyglycosides (Parent Compound) in Simulation Tests

The removal of the medium-chain alkyl polyglycoside ($C_{12/14}$-APG) in terms of primary biodegradation was determined in the OECD confirmatory test using substance-specific HPLC analysis with electrochemical detection [30]. The principal test method is laid down in the EU Directive 82/242/EEC [13]. The test substance was dosed continuously with 10 mg C/L and 20 mg C/L, respectively. The hydraulic retention time of the model plant was adjusted at 3 h. In both cases the removal rate of the $C_{12/14}$-APG was very high from the start of the test, i.e., already exceeding 98% during the 1-week working-in period of the model plant. During the 3-week evaluation phase, virtually no $C_{12/14}$-APG parent compound was detectable in the effluent. Based on the limit of detection of 0.2 ppm, the surfactant was removed to 99.5–99.8% from the wastewater during passage through the model plant. At the same time, the degradation of the so-called synthetic wastewater representing the main organic substrate continuously added to the model plant was 94–96%, based on DOC removal. This shows impressively that $C_{12/14}$-APG present in the sewage will have no detrimental effect on the degrading organisms or on the purification efficiency of sewage treatment plants.

2. Ultimate Bioelimination of Alkyl Polyglycosides in Simulation Tests (DOC Removal)

The elimination of the medium-chain alkyl polyglycoside ($C_{12/14}$-APG) from synthetic wastewater was studied in terms of DOC removal in the coupled-units test (OECD 303 A) at a test substance concentration of 10 mg C/L [31]. With a hydraulic retention time of 3 hours and in the presence of a surplus of easily biodegradable by-substrates, the surfactant exhibited a constantly high $89 \pm 2\%$ DOC removal within the 39-day test period. This result ensures that the high elimination of the $C_{12/14}$-APG parent compound observed in the OECD confirmatory test is connected with an extensive removal of dissolved organic carbon under realistic sewage treatment plant conditions (Table 3).

Thus, the investigations into the fate of $C_{12/14}$-APG in model wastewater treatment plants prove that this surfactant is removed to a very high degree from the water phase, resulting in extremely low concentrations of the parent compound in the plant effluent and, accordingly, to a very low predicted environmental concentration (PEC). In addition these studies have shown that the organic matter (DOC) of alkyl polyglycosides resp. of their degradation products are also removed to great extent from wastewater by the biological treatment in modern sewage treatment

TABLE 3 Elimination of APG in Sewage Treatment Plant Simulation Tests

Method	Test substance			Remark/conclusions
	$C_{8/10}$-APG, DP = 1.4	$C_{12/14}$-APG, DP = 1.4	$C_{16/18}$-APG, DP = 1.1	
OECD confirmatory test	n.d.	99.5–99.8% (parent compound)	n.d.	Primary degradation determined by substance-specific HPLC analysis
OECD 303 A (coupled-units test)	n.d.	89 ± 2% (DOC)	n.d.	Ultimate bioelimination

plants (for information on degradation products resp. the completeness of degradation, see Section I.A.3).

C. Ecotoxicological Properties of Alkyl Polyglycosides

According to the broadly accepted risk assessment scheme for chemicals [5], environmental compatibility requires proof that the use of the chemical will not result in environmental exposure levels higher than the predicted ecotoxicological no-effect concentration (the PEC/PNEC concept). To assess the environmental impact of APG, knowledge of their aquatic and terrestrial toxicity as well as of their environmental fate is necessary. The ecotoxicity of surfactants is linked to their surface activity (detergency) and to the resulting effects on biological membranes [6]; i.e., surfactants generally have a significant ecotoxicity. However, the ecotoxicity of surfactants depends on the exact structure and can vary by more than two orders of magnitude. To make the test results from different laboratories comparable, it is advisable to use internationally standardized test systems, like the OECD 200 series methods [32], and and the principles of good laboratory practice [10]. Therefore, the ecotoxicological data cited herein were chosen according to these guidelines whenever possible. The ecotoxicological evaluation covers the acute, subchronic, and chronic effects of alkyl polyglycosides toward organisms living in the aquatic and terrestrial environments.

1. Acute Aquatic Toxicity of Alkyl Polyglycosides

Tests for acute aquatic toxicity are designed to determine the half-maximum-effect concentrations of chemicals to aquatic organisms upon short-term exposure. Acute (short-term) exposure is defined as an exposure that comprises 10% or less of the life span of an organism. Because the endpoints observed are rather severe effects (e.g., death of the organism), relatively high test substance concentrations are employed. Data from acute toxicity tests are widely used as the basis for environmental classification and labeling, e.g., in the context of the EU Dangerous Preparation Directive and the German Water Hazard Classification Scheme. Chronic toxicity data are required in the context of an environmental risk assessment.

(a) Acute Toxicity of Alkyl Polyglycosides Toward Freshwater Organisms. The acute (short-term exposure) effects of technical, linear alkyl polyglycosides have been

comprehensively studied for the short-chain ($C_{8/10}$-APG) and the medium-chain ($C_{12/14}$-APG) alkyl polyglycosides [14] and less comprehensively for the long-chain $C_{16/18}$-APG. For $C_{8/10}$-APG in several independent studies with laboratory samples as well as with representative samples taken out of production (DP = 1.4–1.6), LC_{50} values for acute fish toxicity in the range of 100–200 mg/L were observed [33–35]. As the benchmark for acute fish toxicity of $C_{8/10}$-APG (DP = 1.4), an LC_{50} = 101 mg/L has been defined [14], determined according to GLP in the 96-h OECD 203 test with zebra fish (*Brachydanio rerio*). Similarly, for 24-h acute invertebrate toxicity, EC_{50} values between 100 and over 500 mg/L were observed [36,37]. This reflects the intra- and interlaboratory variations of the biological test systems. Under the prolonged 48-h-exposure conditions of method OECD 202/1, $C_{8/10}$-APG exhibited EC_{50} = 20 mg/L against the water flea (*Daphnia magna*). Therefore, as the benchmark for acute invertebrate toxicity of $C_{8/10}$-APG, EC_{50} = 20 mg/L has been defined [14]. The acute (subchronic) toxicity of $C_{8/10}$-APG for aquatic plants has been determined in the 72-h OECD 201 test with the unicellular algae *Scenedesmus subspicatus* (= *Desmodesmus subspicatus*) [38]. Under these test conditions, $C_{8/10}$-APG exhibited ErC_{50} = 47 mg/L (EbC_{50} = 21 mg/L). In the context of the global harmonization scheme for classification and labeling of chemicals, the environmental classification with regard to the endpoint algae toxicity should be based on the ErC_{50} value. The rationale for using the ErC_{50} value and not the usually more sensitive EbC_{50} value is that the 72-h algae test is, scientifically speaking, not an acute toxicity test, but rather a chronic test, because the exposure period is long compared to the life span of the test organisms (within the 72-h test period the algae multiply several times). However, as the benchmark for algae toxicity of $C_{8/10}$-APG, the more sensitive parameter has been defined, i.e., EC_{50} = 20 mg/L [14]. Another type of short-chain alkyl polyglycoside is the branched 2-ethylhexyl alkyl polyglycoside (*iso-C_8-APG*). The acute fish toxicity of *iso-C_8-APG* has been determined under standard semistatic conditions in a 96-h test with zebra fish (*Brachydanio rerio*) [39]. Under these conditions the branched APG exhibited LC_{50} = 558 mg/L. In the 48-h acute invertebrate toxicity test with the water flea (*Daphnia magna*), the branched APG exhibited EC_{50} = 557 mg/L [40]. The algae toxicity of *iso-C_8-APG* has been determined in two different algae species according to test method OECD 201. With the species *Kirchneria subcapitata* (synonymous with *Raphidocelis subcapitata* = *Selenastrum capricornutum*), a 72-h E_rC_{50} = 1,543 mg/L was determined [40], whereas with the species *Scenedesmus subspicatus* a comparable 72-h E_rC_{50} = 1,800 mg/L (E_bC_{50} = 720 mg/L) was observed [40]. In conclusion it can be stated that the linear short-chain ($C_{8/10}$-APG) alkyl polyglycoside is only moderately resp. weakly toxic for freshwater organisms and that the branched (*iso-C_8-APG*) short-chain alkyl polyglycoside is even an order of magnitude less toxic (Table 4).

For different $C_{12/14}$-APG samples from the laboratory as well as out of production (DP = 1.3–1.6), LC_{50} values for acute fish toxicity between 2.5 and 5.0 mg/L were observed under GLP in the 96-h OECD 203 test with zebra fish (*Brachydanio rerio*) [40,41]. As the benchmark for the acute fish toxicity of $C_{12/14}$-APG (DP = 1.4), LC_{50} = 3.0 mg/L was defined [14]. For the 48-h acute invertebrate toxicity against the water flea (*Daphnia magna*), EC_{50} values between 7 and 14 mg/L were observed [40,42]. Because the first study was conducted under GLP and with analytical confirmation of the test substance concentration throughout the study, EC_{50} = 7.0 mg/L can be used as the representative value (benchmark) for acute

TABLE 4 Ecotoxicological Data of APG Toward Freshwater Organisms

Endpoint	Method	$C_{8/10}$-APG, DP = 1.4	$C_{12/14}$-APG, DP = 1.4	$C_{16/18}$-APG, DP = 5.4
			Test substance	
Acute toxicity				
Fish	OECD 203 (96-h, LC50)	101 mg/L	3.0 mg/L	2.6 mg/L (48 h)
Daphnia	OECD 202/1 (48-h, EC50)	20 mg/L	7.0 mg/L	n.d.
Algae	OECD 201 (72-h, E_rC50)	47 mg/L	10.0 mg/L	n.d.
Bacteria/Sludge	OECD 209 (30-min, EC0)	n.d.	200 mg/L	n.d.
Subchronic/chronic toxicity				
Fish	OECD 204 (4-week, NOEC)	n.d.	1.8 mg/L	n.d.
Daphnia	OECD 202/2 (21-d, NOEC)	n.d.	1.0 mg/L	n.d.
Algae	OECD 201 (72-h, NOEC)	5.7 mg/L	2.0 mg/L	n.d.
Bacteria	DIN 38412, part 8 (16-h, NOEC)	1,700 mg/L	5,000 mg/L	n.d.

invertebrate toxicity of $C_{12/14}$-APG [14]. The acute (subchronic) toxicity of different samples of $C_{12/14}$-APG for aquatic plants was determined in the 72-h OECD 201 test with two unicellular algae species, *Kirchneria subcapitata* and *Scenedesmus subspicatus* [40,43,44]. Under these conditions, $C_{12/14}$-APG exhibited E_rC_{50} values between 10 and 25 mg/L (E_bC_{50} = 6 and 10 mg/L). In conclusion it can be stated that the acute aquatic toxicity data of $C_{12/14}$-APG for (eukaryotic) freshwater organisms are in the range of typical surfactants used in detergents (i.e., LC/EC_{50} = 1–10 mg/L). Nevertheless, a comparison of the current LC/EC_{50} values with the corresponding data from other nonionic surfactants of this alkyl chain length (e.g., fatty alcohol + 7EO) revealed that the alkyl polyglycoside has a relatively favorable ecotoxicity profile, especially with regard to the effects on invertebrates and aquatic plants. In addition, the acute toxicity of $C_{12/14}$-APG against sewage sludge [40] resp. against representative microorganisms of the sludge from sewage treatment plants (*Pseudomonas putida*) [45] has been investigated in a oxygen uptake inhibition test (OECD 209, 30 min). Under these conditions up to 200 mg/L (sludge, highest concentration tested) resp. 500 mg/L (*P. putida*) did not inhibit the respiration rate of the corresponding microorganisms, i.e., EC_0 = 200 mg/L. This result has relevance for the proper functioning of sewage treatment plants. As long as the alkyl polyglycoside concentration in the influent does not exceed 200 mg/L there is no (APG-related) adverse effect on the efficieny of the plant to be expected. For comparison, the concentration of alkyl polyglycoside in the influent of sewage treatment plants is expected to be three to four orders of magnitude lower, i.e., < 20 μ/L [46].

The ecotoxicity of long-chain alkyl polyglycosides has been studied less comprehensively, because these products do not play a significant economic role. A $C_{16/18}$-APG sample prepared in the laboratory (glucose:fatty alcohol, tallow, 1:2 reaction product, DP = 5.4) has been studied with regard to acute fish and invertebrate toxicity. In a fish toxicity study with golden orfe (*Leuciscus idus*), a 48-h LC_{50} value of 2.6 mg/L was observed (German standard method DIN 38412, part 15) [47]. Taking into consideration the short exposure time of this study compared to the studies with short- and medium-chain alkyl polyglycosides and, in addition, the high degree of polymerization of the sample, this result fits into the general structure/toxicity relationship developed for APG (see later). In the 24-h toxicity study with the water flea (*Daphnia magna*) according to the German standard method DIN 38412, part 11, the test substance exhibited an EC_{50} value of over 1000 mg/L [48]. This result contradicts the fish toxicity result and the invertebrate toxicity data for the other APG. Because it is stated in the test report that the test substance was water soluble but yielded "black impurities resembling charcoal," the result is likely to be an artifact.

A clear structure/toxicity relationship is observed when comparing the data of the short-chain-length $C_{8/10}$-APG with those of the medium-chain-length $C_{12/14}$-APG of the same degree of polymerization (DP = 1.4). The fish toxicity increased by more than an order of magnitude, and the daphnia and algae toxicity increased by about a factor of 3 between the short-chain and the medium-chain alkyl polyglycoside. Therefore, it can be stated in general that the acute toxicity increased significantly with increased alkyl chain length, i.e., with the molecular hydrophobicity of the surfactant. This rather coarse structure/toxicity relationship has recently been refined [49]. The authors investigated the dependencies of the acute toxicity of freshwater shrimps and rotifers (i.e., of invertebrates) with regard to different structural elements (chain length, branching, DP) and physicochemical variables (HLB, CMC, surface tension, contact angle, etc.) of 34 alkyl polyglycosides. The properties with the strongest influences on the toxicity of the surfactants were the critical micelle concentration (CMC), wetting, contact angle, and the number of carbon atoms in the alkyl chain. This multivariate quantitative structure–activity relationship (M-QSAR) might be useful for the development of new, low-toxicity alkyl polyglycoside surfactants with optimized technical properties. With regard to the representative short-chain ($C_{8/10}$-APG) and medium-chain ($C_{12/14}$-APG) alkyl polyglycosides, EC_{50} values of 129 mg/L and 9.8 mg/L have been determined for the freshwater shrimp *Thamnocephalus platyurus*, clearly demonstrating, again, the increase of toxicity with increased chain length [49].

(b)   Acute Toxicity of Alkyl Polyglycosides Toward Marine Organisms.   Although the main application areas for alkyl polyglycosides are in consumer products such as detergents and personal care products, alkyl polyglycosides might be used for industrial application as well, due to their technical properties. One specific industrial application area for alkyl polyglycosides is in the offshore oil production. Here APG are used in drilling, e.g., to clean the borehole in between the use of different drilling fluids. For the area of the North Sea, the use of offshore chemicals has to be approved for the envisaged national sector (e.g., Netherlands, UK) by the competent authorities [50]. The basis for the approval is exotoxicological (and environmental fate) data on the chemicals. Therefore, there is a need to determine the ecotoxicity toward marine species, and some data regarding saltwater species are now available for alkyl polyglycosides (Table 5).

TABLE 5 Ecotoxicological Data of APG Toward Marine Organisms

Endpoint	Method	Test substance		
		$C_{8/10}$-APG, DP = 1.4	$C_{12/14}$-APG, DP = 1.4	$C_{16/18}$-APG, DP = 5.4
Acute toxicity				
Fish (*Scophthalmus*)	OECD 203 adapted (96-h, LC50)	97 mg/L	n.d.	n.d.
Invertebrates (*Acartia*)	ISO/PARCOM (48-h, EC50)	32 mg/L	6.25 mg/L	n.d.
Algae (*Skeletonema*)	ISO/PARCOM (72-h, EC50)	20 mg/L	2.3 mg/L	n.d.

The ecotoxicity of the short-chain alkyl polyglycoside has been studied with regard to fish, invertebrate, and algae toxicity. Acute fish toxicity has been studied according to the OSPARCOM protocol [51] against the marine fish *Scophthalmus maximus* (OECD 203, adapted to seawater conditions). $C_{8/10}$-APG exhibited a 96-h LC_{50} of 97 mg/L, i.e., a value almost identical to the value determined for freshwater species [52]. The acute invertebrate toxicity of $C_{8/10}$-APG has been tested against the marine copepod *Acartia tonsa* according to the ISO/PARCOM guidelines [53]. Under these conditions $C_{8/10}$-APG exhibited a 48-h EC_{50} of 32 mg/L [54]. The toxicity of $C_{8/10}$-APG against the marine algae *Skeletonema costatum* has been determined according to the corresponding ISO/PARCOM guidelines [55]. Under these conditions $C_{8/10}$-APG exhibited a 72-h EC_{50} of 20 mg/L [56]. The ecotoxicity of the medium-chain alkyl polyglycoside has been studied with regard to invertebrate and algae toxicity. The acute invertebrate toxicity of $C_{12/14}$-APG has been tested against the marine copepod *Acartia tonsa* according to the ISO/PARCOM guidelines [53]. Under these conditions $C_{12/14}$-APG exhibited a 48-h EC_{50} of 6.25 mg/L [57]. The toxicity of $C_{12/14}$-APG against the marine algae *Skeletonema costatum* has been determined according to the corresponding ISO/PARCOM guidelines [55]. Under these conditions $C_{12/14}$-APG exhibited a 72-h EC_{50} of 2.3 mg/L [58]. In conclusion it can be stated that marine species show the same degree of sensitivity against the short-chain alkyl polyglycoside as freshwater species and a comparable or slightly higher sensitivity against the medium-chain alkyl polyglycoside.

2. Long-Term Aquatic Toxicity of Alkyl Polyglycosides

Whereas data from acute toxicity tests are required to assess the effects of a single environmental exposure, e.g., in case of an accident and, accordingly, for classification and labeling purposes, with regard to an environmental risk assessment, chronic (long-term) toxicity data are much preferred, because these data substantially improve the validity of the predicted no-effect concentration (PNEC). The chronic toxicity of the short-chain ($C_{8/10}$-APG) alkyl polyglycoside has been investigated in the 16-h bacterial cell multiplication inhibition (DIN 36412, part 8) test with *Pseudomonas putida*, as well as in the 72-h algae cell multiplication inhibition test with *Scenedesmus subspicatus* according to OECD 201. For the chronic bacterial toxicity a NOEC of 1,700 mg/L has been observed. In the study with algae the increase of biomass (E_bC_0) was the most sensitive parameter. From this study a

NOEC of 5.7 mg/L has been established for $C_{8/10}$-APG [59]. In the same test system for the medium-chain ($C_{12/14}$-APG) alkyl polyglycoside, a NOEC of 2.0 mg/L has been established [60]. Thus, as already observed in the acute toxicity tests, the medium-chain alkyl polyglycoside is about three times more toxic for aquatic organisms than is the short-chain alkyl polyglycoside. This, together with the higher production volume of the medium-chain alkyl polyglycoside, is the rationale behind the focus of most of the expensive long-term studies on the $C_{12/14}$-APG. The (sub)chronic fish toxicity of $C_{12/14}$-APG has been studied in an intermittent flow-through, 4-week prolonged test with zebra fish (*Brachydanio rerio*) according to an adapted OECD 204 procedure [61]. Under these conditions the growth of the fish was not affected up to a test substance concentration of 3.2 mg/L, although food intake already seemed to be somewhat affected (slowed down) at 1.8 mg/L. This was, however, not considered an adverse effect because the growth of the fish was not retarded. The most sensitive parameter under the test conditions employed was the mortality of the fish, resulting in an NOEC of 1.8 mg/L. The chronic invertebrate toxicity of $C_{12/14}$-APG has been studied in two independent studies according to OECD 202/2 using *Daphnia magna* [37,62]. The exposure time in both tests was 21 days, and the endpoints evaluated were reproduction rates and mortality. Mortality turned out to be the more sensitive endpoint, and the corresponding NOEC values were 1.5 mg/L (reproduction) and 1.0 mg/L (mortality), respectively. The chronic toxicity of the medium-chain ($C_{12/14}$-APG) alkyl polyglycoside has been investigated in the 16-h bacterial cell multiplication inhibition test with *Pseudomonas putida* (DIN 36412, part 8). For the chronic bacterial toxicity, NOEC $\geq$ 5,000 mg/L (highest concentration tested) has been observed. As a conclusion it can be stated that the ecotoxicological effect concentrations observed upon chronic exposure of aquatic organisms to alkyl polyglycosides are in the same range as the acute effect concentrations, based on the comparison of the NOEC and the EC_0 values). This is indicative of the fact that alkyl polyglycosides exert their adverse ecotoxicological effects due primarily to their reactions with biological membranes, but that they do not have a significant cumulative systemic toxicity (Table 4).

3. Biocenotic Toxicity of Alkyl Polyglycosides

Biocenotic toxicity tests are designed to determine the threshold of toxicity using a community of interdependent species as a very sensitive bioindicator. $C_{12/14}$-APG was accordingly tested in a river flow model system [14]. In this system during the 4-week working-in period (without test substance dosage), a stable and well-comparable biocenosis comprising 19 different species of algae, protozoans, and small metazoans had established in the two parallel tracks (each consisting of eight tanks). Subsequently, for a 4-week period the surfactant was continuously added to the effluent of the first tank of one of the river flow tracks, resulting in an influent concentration of 5 mg/L in the second tank. A biological analysis of the species and their abundance was made once a week in each tank in comparison to the control tank. In addition, the surfactant concentration was analytically determined three times a week. During the whole test period a significant biocenotic difference between the dosed and the control tanks was observed only in the initially dosed tank (tank 2). As in tank 3, the APG concentration (determined by substance-specific HPLC analysis) was between 2.1 and 3.0 mg/L; 2.1 mg/L is established as the biocenotic NOEC. A comparison shows that the biocenotic and single-species NOEC values are comparable, indicating, again, the favorable ecotoxicological profile of alkyl polyglycosides. Additionally, the reversibility of the toxic effects was tested after the close of the APG dosage by biological analysis of the biocenosis recovery in tank 2. Within 5 days the biocenotic difference between the dosed and the control track had diminished. Thus, APG permitted a faster recovery of the aquatic biology than most other surfactants [63]. As a conclusion of the biocenotic toxicity study it can be stated that even a significantly high local environmental exposure of the more toxic medium-chain ($C_{12/14}$-APG) alkyl polyglycoside would not have long-term consequences to the biocenosis.

4. Terrestrial Toxicity of Alkyl Polyglycosides

Because surfactants could become exposed to terrestrial environments as well, e.g., when they are used as inerts in pesticide formulations or when sludge from sewage treatment plants is used as a fertilizer for agricultural soils, the terrestrial toxicity of alkyl polyglycosides is of some environmental relevance too. Following the rationale outlined in Section II.C, the terrestrial toxicity of alkyl polyglycosides has been evaluated primarily for the more toxic $C_{12/14}$-APG. The terrestrial toxicity of $C_{12/14}$-APG has been determined for earthworms as well as for higher plants. In a 14-day test with the soil-dwelling worm species *Eisenia foetida* using artificial soil homogeneously spiked with the test substance, no mortality has been observed up to the highest test substance concentration employed (654 mg/kg). Thus, a NOEC $\geq$ 654 mg/kg has been established for earthworms. The toxicity of $C_{12/14}$-APG against higher plants has been determined according to test method OECD 208 using three different plant genera, i.e., monocotyledons (oat = *Avena sativa*), dicotyledons (turnip = *Brassica rapa*), and tomato (= *Lycopersicum esculentum*). The test substance was homogeneously mixed with the soil and the germination success and the growth of the developing plants was evaluated over a time period of 21 days compared to an untreated control. No adverse effects have been observed up to the highest test substance concentration employed (654 mg/kg). Thus, a NOEC $\geq$ 654 mg/kg has been established for terrestrial plants. In addition, the ecotoxicity of alkyl polyglycosides toward marine sediment–inhabitating organisms has been investigated. Tests were conducted according to the OSPARCOM method with the sediment reworker *Corophium volutator* [64]. The sediment was homogeneously spiked with the test substance and the 10-day survival rate was monitored. Under these conditions $C_{8/10}$-APG exhibited a LC_{50} of over 433 mg/kg soil (highest concentration tested) [65], whereas the $C_{12/14}$-APG exhibited a LC_{50} of 1,300 mg/kg soil [66]. In conclusion it can be stated that alkyl polyglycosides are only weakly to scarcely toxic to terrestrial resp. sediment inhabitating organisms. The lower toxicity of alkyl polyglycosides to terrestrial organisms compared to aquatic organisms is probably related to the adsorption of the APG onto the soil ("buffering capacity" of the soil), as indicated by the increasing K_{oc} values of the more toxic (hydrophobic) APG homologues. The following K_{oc} values have been calculated from the log P_{ow} data for the individual homologues of $C_{8/10}$-APG: C8 + 2Glc: K_{oc} = 1.62; C10 + 2Glc: K_{oc} = 5.23; C8 + 1Glc: K_{oc} = 31.9; C10 + 1Glc: K_{oc} = 104.3 [67] (Table 6).

D. Environmental Impact of Alkyl Polyglycosides

The environmental compatibility assessment of chemicals is generally based on the comparison of the predicted environmental concentration (PEC) and the predicted

TABLE 6 Ecotoxicological Data of APG Toward Terrestrial Organisms

		Test substance		
Endpoint	Method	$C_{8/10}$-APG, DP = 1.4	$C_{12/14}$-APG, DP = 1.4	$C_{16/18}$-APG, DP = 5.4
Soil dwellers				
Earthworm (*Eisenia foetida*)	UBA proposal 1984 (14-d, NOEC)	n.d.	654 mg/L[a]	n.d.
Higher plants				
Oat (*Avena sativa*)	OECD 208 (21-d, NOEC)	n.d.	654 mg/L[a]	n.d.
Turnip (*Brassica rapa*)	OECD 208 (21-d, NOEC)	n.d.	654 mg/L[a]	n.d.
Tomato (*L. esculentum*)	OECD 208 (21-d, NOEC)	n.d.	654 mg/L[a]	n.d.

[a] Highest concentration tested.

no-effect concentration (PNEC), the latter being derived from the earlier-discussed ecotoxicological endpoints [5]. An additional environmental safety aspect to be taken into account may be the endocrine-modulating properties. However, such an activity can be excluded for alkyl polyglycosides [68]. The PEC estimation (exposure assessment) as well as the derivation of the PNEC should be based on realistic worst-case assumptions; i.e., the scenarios employed for the evaluation of the environmental fate and the extrapolation of ecotoxicological data from the laboratory have to rely on stringent but realistic conditions (Table 7).

TABLE 7 PEC/PNEC values for the Aquatic and Terrestrial Compartments

Compartment	Scenario	PEC	PNEC	PEC/PNEC ratio	Conclusion
Aquatic compartment	German surface waters, consumer product scenario A[a]	0.05 mg/L	0.1 mg/L	0.05	No environmental risk
	German surface waters, consumer product scenario B[b]	0.016 mg/L	0.1 mg/L	0.02	No environmental risk
Terrestrial compartment	Agricultural soil fertilized with German digester sludge[a]	0.15 mg/L	6.5 mg/L	0.023	No environmental risk

[a] See Section I.D.1.
[b] *Source*: Ref. 8.

1. Environmental Risk Assessment for the Aquatic Compartment

As the scenario for the aquatic compartment, the area of Germany was arbitrarily chosen, and it was assumed that the relevant environmental exposure originates from the use of alkyl polyglycoside-containing household products, e.g., detergents and cleaning products. Because not all possible relevant exposure scenarios are taken into account, the assessment is not considered a comprehensive risk assessment, but rather an initial resp. a targeted risk assessment. The biggest problem for a realistic risk assessment is to get reliable data on the exposure situation. In this regard, realistic assumptions about the amount of alkyl polyglycosides used in these down-the-drain applications are needed, i.e., identification of the corresponding product categories, their use volumes, and the concentration of APG in these products. An overview of the different hard-surface cleaner and laundry detergent product categories containing alkyl polyglycosides and an indication of the typical APG concentration ranges in these products can be found in a recent monography [2]. Accordingly, alkyl polyglycosides are—due to their excellent skin compatibility—used primarily in liquid or gelly cleaning products that might come in direct contact with consumers' skin, e.g., manual dishwashing detergents, all-purpose cleaners, and—less important—liquid detergents. The concentrations of alkyl polyglycosides in these products span a wide range from ca. 2 to 10%. On average the concentration is under 5%. The use of alkyl polyglycosides in detergent powders is also possible, although it is less common. To be on the safe side it is assumed that one-quarter of the heavy-duty powder detergents contain on average 5% alkyl polyglycosides. The area of the German Federal Republic is used as the boundary for the derivation of the $PEC_{local,aquatic}$ because Germany is representative of western Europe and a good database exists on the consumption of detergents and cleaning products in Germany. According to a recent survey conducted by the German Cosmetics and Detergents Association (IKW), 27,900 tons of powder detergents (category 1a) and 25,700 tons of liquid detergents and cleaners (categories 1b–6) were used in 2000 in Germany [69]. Assuming that 25% of the powder detergents contain on average 5% alkyl polyglycosides and that the liquid products contain on average 5% alkyl polyglycosides too, about 2000 t of alkyl polyglycosides are disposed of via wastewater annually. Based on a population of 81 million in Germany and an average daily per capita water consumption of 200 L [70], an average waste water concentration of 0.34 mg/L is calculated. Recent monitoring data from Eichhorn et al., although from Spain (see later), indicate that this calculation, though formally not including the personal care products, like shampoos and cosmetics, is still conservative. Eichhorn et al. have determined the concentration of alkyl polyglycosides in the influent of three different sewage treatment plants in Spain and found concentrations of only 6.5–16 µg/L [46]. Since APG removal in sewage treatment plants is about 90% regarding DOC removal (more than 99% regarding primary degradation), it is expected that the APG-derived effluent DOC concentration will be less than 50 µg/L. Based on an average effluent/river water dilution factor of 10, a $PEC_{local,aquatic}$ of 5 µg alkyl polyglycoside/L (refering to DOC) in river waters is calculated (assuming no adsorption onto suspended matter).

The counterpart of the exposure analysis, the estimation of the PNEC for alkyl polyglycosides, is based on the worst-case assumption that all APG is represented by ($C_{12/14}$-APG) alkyl polyglycoside, which is the more ecotoxic type. The NOEC values of $C_{12/14}$-APG in the acute as well as in the chronic bacterial toxicity tests are two to three orders of magnitude above the concentrations expected for the sewage treatment plant influent (200 mg/L vs. 0.34 mg/L). This is in line with the observation of high elimination rates (for APG as well as for the carbon removal) of the model sewage treatment plant operated with APG concentrations of 36 mg/L (equivalent to 10 mg C/L), proving the conclusion that the use of APG-containing products will not have adverse effects on the proper functioning of municipal sewage treatment plants.

According to the standard methodology of the EU Guidance Document on environmental risk assessment [70], the river water PNEC of alkyl polyglycosides was deduced from the NOEC of the most sensitive species among the aquatic organisms tested in subchronic/chronic tests by application of an appropriate safety factor. In the case of $C_{12/14}$-APG, the invertebrates are the most sensitive of the three species tested (fish, daphnia, algae), with a NOEC of 1.0 mg/L, obtained from the 21-day life cycle study with *Daphnia magna*. This NOEC has to be divided by a safety factor of 10, yielding a $PNEC_{aquatic}$ = 100 µg/L. The resulting PEC/PNEC ratio of 0.05 indicates that no adverse environmental impact is to be expected from the use of alkyl polyglycosides in consumer products. Similar PEC/PNEC ratios for APG have been derived by others; e.g., Schöberl and Scholz estimated a PEC/PNEC ratio of 0.02 for German surface waters [8]. Therefore, it can be stated that the use of alkyl polyglycosides in down-the-drain consumer products is—due to the excellent bio-degradability and relatively low aquatic toxicity of these surfactants—without doubt environmentally compatible. This proof of the environmental safety of alkyl poly-glycosides is also valid in circumstances where discharge of untreated sewage may occur, because the ratio of the sewage influent monitoring data (< 20 µg/L) with the $PNEC_{aquatic}$ (100 µg/L) = 0.2 is still below 1, i.e., below the threshold for concern. In addition, investigation of the biocenotic toxicity in the model river system has shown that even an APG concentration of 5 mg/L would lead to only a slight biocenotic effect at the immediate inflow site but would not cause any adverse effect dowstream, due to the ready biodegradation of the surfactant.

2. Environmental Risk Assessment for the Terrestrial Compartment

The exposure scenario for terrestrial risk assessment assumes that digester sludge is used as an agricultural fertilizer. Based on the discussed down-the-drain use scenario for APG-containing consumer products yielding a sewage treatment plant influent concentration of 0.34 mg/L, the sewage plant sludge concentration of alkyl poly-glycoside was calculated. Calculation was done using the HAZCEM mathematical model [71], indicating a sludge concentration of 266 mg APG/kg. Assuming that on average 70% of the adsorbed alkyl polyglycoside is degraded anaerobically during the sludge digestion process (see Section I.A.6), a digester sludge concentration of about 100 mg/kg can be expected. Use of such digester sludge for agricultural purposes at an application rate of 0.5 kg sludge/m² and applying the appropriate default values given in the EU Technical Guidance Document would result in an APG concentration of 0.15 mg APG/kg in the top 20-cm layer of the soil. Thus, a $PEC_{local,soil}$ of 0.15 mg/kg is obtained.

The derivation of the $PNEC_{terrestial}$ of alkyl polyglycosides is based on the worst-case assumption that all APG in the sludge is represented by the more toxic, medium-chain ($C_{2/14}$-APG) alkyl polyglycoside. The terrestial NOEC values of $C_{12/14}$-APG have been determined in two species from different trophic levels (NOEC = 654 mg/kg for higher plants as well as for earthworms). According to the EU TGD, a safety factor of 100 has to be applied to derive the PNEC from this NOEC. Therefore, a $PNEC_{soil}$ = 6.5 mg/kg is established for alkyl polyglycosides. The corresponding $PEC_{local,soil}/PNEC_{terrestrial}$ ratio of 0.023 indicates that the use of APG-containing digester sludge does not pose an environmental risk for the terrestrial compartment.

II. TOXICOLOGY OF ALKYL POLYGLYCOSIDES

Since product safety is increasingly regarded as part of product quality, basic toxicological and ecotoxicological data have become increasingly important with regard to the handling, storage, transport, and even marketing of new products. This, in turn, has stimulated the collation of comprehensive data on the acute and systemic toxicology of alkyl polyglycosides [72]. Toxicological data addressing all relevant endpoints allow meaningful safety evaluations and risk assessments for all kinds of consumer-use scenarios and therefore provide valuable tools for both formulators and regulatory authorities. As an important component of modern surfactant formulations, a legitimate demand for supplementary data has resulted in considerable efforts to demonstrate the innocuousness of this important class of surface-active substances.

Therefore, a tiered toxicological test program for alkyl polyglycosides was undertaken in compliance with acknowledged guidelines, especially OECD Guidelines for the Testing of Chemicals and the principles of good laboratory practice [73].

A. Acute Oral Toxicity

By definition, acute toxicity describes the short-term adverse effects of single or multiple doses of the substance in question. This includes dermal and oral applications, along with inhalation and more specific routes, such as intraperitoneal and intravenous injection, although the latter two are of negligible relevance for surfactants used in consumer products.

The acute oral and dermal toxicity of various types of APG, representing various linear alkyl chain length distributions and numbers of glucosidic repeating units, were studied in two strains of rat, Wistar and Sprague–Dawley.

Assuming low toxicity, primary limit tests were performed with alkyl polyglycosides, using only a single high dose (2000 mg/kg b.w.) in one group of test animals. These dosages were chosen to comply with the regulatory requirements in Europe [74] and the United States [75] used to define individual classification limits for chemical substances. In Europe, a chemical substance with an LD_{50} over 2000 mg/kg b.w. is not subject to classification, whereas in the United States the level is set at 5000 mg/kg b.w. A compilation of test results for different APG types is given in Table 8.

No deaths were reported for test animals, even after the application of relatively high amounts of alkyl polyglycosides. However, symptoms of toxicity were recorded, although some of these were attributed to the administration of a chemical displaying a certain irritancy (see Section II.C) rather than effects specific to APG. These included slight depression, urine stains, saliva stains, and reddening around

TABLE 8 Acute Oral Toxicity of APG

	C8/10	C12/14	C10/16
Degree of polymerization	1.6	1.5	1.6
Active substance	50	60	50
Limit dose (mg/kg b.w.)	> 5000	> 2000	> 5000
Test animals	Sprague–Dawley rat, 5 male/5 female	Wistar rat, 2 male/2 female	Sprague–Dawley rat, 5 male/5 female
Mortalities	0/10	0/4	0/10
Animals with gross necropsy findings	0/10	0/4	0/10
Ref.	102	103	104

the eye. In contrast, after necropsy no macroscopic effects were seen in any rat at any dose level.

Chemical structure appeared to have little influence on the acute oral toxicity of APG. The products tested covered a broad range of linear C8–C16, with degrees of polymerization between 1.5 and 1.6. For the dose range used, none of the test animals showed any adverse effects in response to APG challenge, irrespective of fine chemical structure. This may be interpreted as demonstrating their acute innocuity.

Summarizing these findings, accidental ingestion of alkyl polyglycosides does not represent an acute health risk, and APG need not be subject to the appropriate labeling required by European or U.S. law.

B. Acute Dermal Toxicity

As a general rule, chemical substances are more readily absorbed via the oral than the dermal route, because the latter's stratum corneum presents an effective barrier to absorption of many substances. Nevertheless, exceptions to this rule are known, by which percutaneous absorption of some compounds—mainly hydrophobic—is almost complete [76]. Since alkyl polyglycosides are used as surfactants in home care products and thus involve unavoidable skin contact, there is a legitimate need for an investigation of APG toxicity after dermal application.

Five male and five female New Zealand white rabbits received a single limit dose of 2000 mg/kg b.w. of C8/C10 alkyl polyglycoside applied dermally [77]. Clinical signs recorded in individual animals during the observation period of 14 days included mild to moderate irritant effects, fecal stains, yellowing around the application site, emaciation, nasal discharge, and lacrimation. One animal died of Tyzzer's desease, confirmed by microscopic examination. At the end of the 14-day observation period, all surviving animals were killed. Gross necropsy revealed spotty areas of hemorrhage on the lungs of five rabbits [78].

The same test protocol was applied using C10–16 alkyl polyglycoside as test substances. The clinical signs observed included slight depression, hunched posture, mild to marked erythema, and marked desquamation in all animals. No deaths occurred and no pathological findings were reported after autopsy [79].

Although no treatment-related deaths occurred, it was apparent that longer-C-chain APG, while systemically less toxic than shorter chain APG, were locally more of an irritant. This observation is supported by several in vivo studies discussed later (see Section II.C). Data on the acute dermal toxicity of APG are summarized in Table 9.

According to EU guidelines for acute dermal application, alkyl polyglycosides need not be classified as toxic.

C. Dermal Irritation

Dermal irritation/corrosion is defined as the production of reversible/irreversible changes to the skin following application of a test substance. OECD Guideline 404 is the international standard method for investigating dermal irritation, using rabbits as test species. The test substance is applied under occlusive or semiocclusive conditions to one flank of the animals and remains in situ for 4 hours. Treated areas are scored 24, 48, and 72 hours after removal of the patch. If necessary, the reversibility of skin effects is checked during the postobservation period.

Table 10 summarizes the results of skin irritation tests undertaken according to OECD Guideline 404, using linear alkyl polyglycosides as the test substance. Analysis of the data reveals a clear structure–response relationship. Short-chain (C8/10) alkyl polyglycosides are significantly less of an irritant to the skin than their corresponding long-chain (C10/16 or C12/16) homologues. According to European legislation, $C_{8/10}$ APG need not be labeled below the usual maximum concentration in technical products (70% active substance), whereas C10/16 and C12/16 APG require labeling as a skin irritant if their concentration exceeds 30%. Furthermore, the primary dermal irritation index (PDII) and mean values for the 24-, 48-, 72-hour erythema data reveal an almost linear increase with increasing concentration, a relationship independent of the degree of polymerization. Even mixtures of short-chain and long-chain APG fit this scenario well, allowing meaningful extrapolations for labeling issues (Fig. 1). A product tested at both pH 7 and pH 11 revealed that irritating properties were almost unaffected by alkalinity [80]. Systemic toxicity after a single dermal application in rabbits was not recorded in any of the tests with linear alkyl polyglycosides. Furthermore, it is worth noting that dermatological studies have shown alkyl polyglycosides used as secondary surfactants significantly enhance the skin compatibility of primary surfactants [81].

TABLE 9 Acute Dermal Toxicity of APG

	C8/10	C10/16
Degree of polymerization	1.6	1.6
Active substance	50%	50%
Limit dose (mg/kg b.w.)	>2000	>2000
Test animals	New Zealand white rabbit, 5 male/5 female	New Zealand white rabbit, 5 male/5 female
Mortalities	1/10 (died due to Tyzzer's disease)	0/10
Animals with gross necropsy findings	5/10	0/10
Ref.	78	79

TABLE 10 Data from In Vivo Skin Irritation Studies with Linear APG

Chain length	Degree of polymerization	Active substance	PDII	Positive responder: irritation score 2			Mean values 24/48/72		Remarks	Cognis reference
				Erythema (absolute)	Erythema (%)	Edema (absolute)	Erythema	Edema		
C8/10	1.6	70%	0.8	1/3	33	0/3	0.9	0.0	No labeling, no signs of systemic toxicity	R 9300407
C8/10	1.6	35%	1.3	1/3	33	2/3	1.1	0.3	No labeling, no signs of systemic toxicity	R 93 00408
C8/10	Not specified	15%	0.0	0/5	0	0/5	0.0	0.0	No labeling	TBD 870150
C10/16	1.4	60%	4.6	4/4	100	4/4	2.9	2.1	R38, occlusive application	TBD 880089
C12/16	1.4	50%	3.7	3/3	100	3/3	2.2	1.6	R 38, no signs of systemic toxicity	R 9400459
C12/16	1.4	50%	3.0	3/3	100	1/3	2.1	0.9	R 38, no signs of systemic toxicity	R 9300116
C12/16	1.6	50%	3.0	3/3	100	1/3	1.9	1.1	R 38, no signs of systemic toxicity	R 9300178
C12/16	1.4	30%	2.8	5/6	83	3/6	1.7	0.9	No labeling, no signs of systemic toxicity	R 9400460
C10/16	1.4	20%	0.4	1/4	25	2/4	1.2	1.3	No labeling, no signs of systemic toxicity	TBD 880405
C8/10 + C12/16	Not specified	35% C12/16, 21% C8/10	2.7	3/3	100	2/3	1.8	0.8	No labeling, no signs of systemic toxicity, occlusive application	R 9300115

Semiocclusive patches, unless otherwise specified.

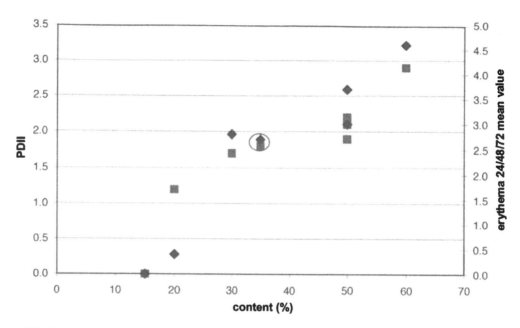

FIG. 1 Concentration–response relationship for C10/16 and C12/16 alkyl polyglycosides. ◆ = PDII, ■ = mean value 24/48/72 h for erythema, ○ = mixture containing 35% C10/16 APG.

D. Mucous Membrane Irritation (Eye Irritation)

Since accidental eye contact with irritating substances may lead to irreversible changes in the cornea and even loss of sight, specific legal requirements for labeling have been established. Because alkyl polyglycosides display significant skin-irritating properties, an in vitro pretest was done to avoid unnecessary animal testing. The HET-CAM test is a screening method suitable for detection of mucosa irritation. HET-CAM is the hen's egg test—chorionallantois membrane (a fine, highly vascularized membrane facilitating oxygen diffusion). Determination of such parameters as hemorrhagia, lysis, and coagulation within membrane vessels over time results in the determination of the quotient of irritation index Q, a measure of the irritant properties of the test substance, compared to a benchmark chemical (generally 0.5% SDS, sodium dodecylsulfate).

C10/16 alkyl polyglycosides showed only slight reactions at pH 7 and pH 11.5. In alkaline solutions, Q reached 55% of the benchmark chemical, whereas the neutralized product caused irritation scores as low as 19% of the benchmark [82]. At the same time, C8/10 alkyl polyglycosides were significantly more irritating than 0.5% SDS and were evaluated as being highly irritating to the eye.

An in vivo study was carried out according to OECD Guideline 405. An aliquot of 0.1 mL aqueous solution of C12/16 alkyl polyglycoside (50% active substance) was introduced into the eyes of four albino rabbits and left for 24 hours. Eyes were scored 24, 48, and 72 hours after application. The 24-/48-/72-hour mean scores were 0.5 for the cornea, 2.08 for the conjunctival erythema, and 0.25 for the iris. The conjunctivae displayed moderate to strong reactions that did not abate completely within 21 days in two animals. In addition, one of these animals showed slight but persistent corneal reactions that did not disappear within the observation period of 21 days [83]. Due to these nonreversible corneal effects, C10/16 alkyl polyglycosides require labeling as presenting the "risk of serious damage to eyes" (R 41).

Nevertheless, data from validated animal tests show that diluted and neutralized formulations of C12/16 alkyl polyglycoside are only moderate eye irritants and so are not subjected to classification or labeling with regard to the content in APG. A dose of 0.1 mL of lauryl glucoside (neutralized, 12.5% active substance) was introduced into the eyes of six New Zealand white rabbits. Observations of ocular irritation were recorded 24, 48, and 72 hours following introduction of the test article. Additional readings were made 4 and 7 days afterwards. No irritant reactions were reported in any of the six animals. Very slight corneal reactions, which persisted for less than 48 hours, were observed in one animal. Medium to mild irritation of the conjunctiva occurred in all rabbits. These irritations were reversible within the 7-day observation period in all but one animal. The remaining redness of the conjunctiva in the single rabbit was very slight and was expected to disappear completely the following day. Applying the evaluation scheme according to European Directive 67/548/EEC, C12/16 APG (neutralized) need not to be labeled as an eye irritant at concentrations of 12.5% or less, which underlines that specific adverse effects depend on applied dose or concentration.

E. Skin Sensitization

After repeated dermal contact with a chemical substance, type IV sensitization (contact dermatitis) may result when an excessive immunological response is induced. After a characteristic delay, dermatological symptoms such as pruritis, erythema, edema, and papules may occur in humans sensitized against a specific antigen.

The acknowledged standard for investigation of the allergenic potential of chemical compounds is OECD Guideline 406, which comprises a selection of closely related protocols. In a first step (induction phase), the test substance is applied several times intracutaneously (Magnusson Kligman protocol) and/or epidermally (Buehler protocol) to guinea pigs. Ideally, a mildly irritating concentration is used in order to push allergenic properties. The Magnusson and Kligman protocol additionally uses a bacterial mixture ("Freund's adjuvant") during intradermal induction to improve the sensitivity of the method. Following a rest interval of two weeks, the challenge phase is initialized by applying a nonirritating concentration of the test substance epidermally. Signs of hypersensitivity are scored 24 and 48 hours after the challenge treatment.

As shown in Table 11, none of the tests with C12/16 alkyl polyglycosides displayed significant potential for sensitization, either with the Buehler method or the Magnusson–Kligman protocol. Only a very weak skin reaction was observed in one animal at the limit concentration for irritation of 20%, in the Buehler test [84]. Because a single irritation also occurred during the induction phase at this concentration, this effect is not considered indicative of sensitization. In addition, the Buehler protocol requires at least 15% positive responder animals before classifying a substance as sensitizer. Furthermore, the absence of sensitizing properties was confirmed by the results of the more stringent Magnusson–Kligman test (intradermal induction with adjuvant) [85].

TABLE 11 OECD 406 Sensitization Tests with C12/16 APG

Protocol	Concentration for induction	Concentration for challenge	Number of positive responders	Classified as sensitizer to skin?	Ref.
Buehler	Epidermal: 20%	Epidermal: 20%	1[a]/20	No	84
Magnusson–Kligman	Intracutaneous: 1% Epidermal: 60%	Epidermal: 10%	0/20	No	85

[a] Very weak skin reaction.

Because animal tests revealed no sensitizing properties, a wide range of alkyl polyglycosides were tested on humans in a human repeated insult patch test (HRIPT) using the Shelanski protocol. Two-tenths mL of the test substance (active ingredient: 5%) was applied to the upper back of 48 volunteers and kept there for 24 hours. This procedure was repeated 10 times for 3 weeks. Following a 2-week rest interval, the panel was challenged again with the same test solution and evaluated for skin reactions 24 and 48 hours after application. Although the test products covered a broad range from C8 to C16 alkyl polyglycosides in different ratios and derived from different raw materials (fatty alcohol from natural and synthetic sources), none of the products induced any skin reaction indicative of sensitization in any volunteer [86], supporting the animal study results.

F. Mutagenicity

Mutagenicity is the ability of chemicals to induce structural or chemical changes in the DNA of living cells. It is obvious that such changes may affect basic life processes, and not only of the generation subject to exposure: If germ cells are targeted, even the offspring may suffer from hereditary diseases, whereas the worst effect to the host is the risk of developing cancer. It has been shown that many carcinogenic chemicals are also mutagenic, underlining the importance of mutagenicity testing. These methods are subdivided into two groups: gene mutations in the proper sense and chromosomal aberrations.

1. Gene Mutation

Gene mutations can be investigated by the *Salmonella typhimurium* reversion ("Ames") test. The international standard test method is OECD Guideline No. 471 (consistent with EU guideline No. B14). Mutation is detected by reversion of histidine-auxotrophic bacteria to prototrophy. Auxotrophic *Salmonella typhimurium* strains are unable to grow on a histidine-deficient medium, whereas the prototrophic mutant synthesizes its own histidine and proliferates. The conversion rate of mutant colonies can easily be detected and is a measure of mutagenicity. In order to simulate the metabolic activity present in mammals, test plates are partly incubated with an enzyme fraction obtained from rat liver cells pretreated with the enzyme-inducing agent Aroclor 1254.

Alkyl polyglycosides were tested on *Salmonella typhimurium* strains TA 98, TA 100, TA 1535, TA 1537, and TA 1538 in two independent experiments, both with and without S9 mix metabolic activation. Test concentrations were 8, 40, 200, 100, and 5000 µg/L (1st test) and 11.1, 33.3, 100, 300, and 900 µg/plate (2nd test), respectively. Solutions of the alkyl polyglycoside were prepared in bidistilled water immediately prior to use. The direct plate incorporation assay was used.

Toxic effects were noted starting from 900 µg/plate. Compared with concurrent negative controls, no precipitations or enhanced revertant rates were observed in all strains tested in the presence or absence of metabolic activation. 4-Nitro-o-phenyl-enediamine, 9-aminoacridine, and sodium azide controls were positive without S9 activation, and the aminoanthracene control was positive with S9 activation. Since alkyl polyglycosides failed to induce reverse mutations, they must be regarded as nonmutagenic in this test system [87].

2. Chromosome Mutations

Chromosome mutations are characterized by structural chromatid changes (aberrations) that, for example, can be detected by the in vitro cytogenetics test in mammalian cell lines. The international standard test method is OECD Guideline No. 473 (consistent with EU guideline No. B10), in which cells are exposed to the test substance in the presence and absence of the metabolic activating S9 system (see Section II.F.1). Chromosomes are prepared in posttreatment mitotic cells arrested in metaphase. Preparations are microscopically examined for chromosomal structural aberrations, e.g., breaks, translocations, inversions, or deletions. A significant increase in aberrant cells compared with concurrent and historical controls is indicative of clastogenic activity of the test substance in vitro.

Cultured chinese hamster V79 lung fibroblasts were exposed repeatedly to C10/16 alkyl polyglucosides every 4 hours. For analysis of chromosomal aberrations and based on the results of a pretest, maximum concentrations of 160 µg/mL with and 16 µg/mL without metabolic activation were applied to the test system. Seven, 20, and 28 hours after starting treatment, cells were fixed and prepared for chromosome analysis. Toxic effects, indicated by reduced mitotic indices below 50% compared to control, were observed starting from 5 µg/mL in the *absence* of S9 mix, as shown in an experiment for determination of the plating efficiency. In the *presence* of S9 mix, alkyl polyglycosides exhibit clear signs of toxicity at 100 µg/mL and above.

No biological effects, with respect to aberration induction, were observed at any time, either with or without S9 activation. Aberration counts were comparable to controls. The reference mutagens acting as positive controls produced a significant increase in cells with chromosomal aberrations (ethylmethane sulfonate without metabolic activation, and cyclophosphamide with metabolic activation). Therefore, it can be stated that C10/16 alkyl polyglycosides display no clastogenic activity under the set of conditions of this test design [88].

G. Subchronic Toxicity

Whereas acute exposure to chemical substances generally is restricted to accidental events or misuse, in daily life chronic exposure presents the more subtle effects. Repeated contact with small amounts over a long period of time may well result in the accumulation of the substance in certain organs and subsequent damage to tissue and its physiological functions. These effects are of major concern because incremental changes often are not recognized as being associated with individual exposure and therefore may evade detection. Consequently, detergents, with which consumers unavoidably have repeated contact, need to be tested for subchronic, i.e., cumulative, toxicity. Test methods for subchronic effects investigate toxicity after repeated dosing of a substance for a period not exceeding 10% of the test animal's life span.

C12/16 alkyl polyglycosides were investigated in compliance with OECD Guideline 408 for subchronic toxicity. Ten male and 10 female Sprague–Dawley rats per dosage group were fed daily with 250, 500, and 1000 mg/kg b.w./day by gavage. An untreated group served as control [89]. All dosage groups were screened daily for clinical signs and mortality, whereas body weight gain and food and water intake were recorded weekly. Hematological and biochemical parameters were examined at weeks 6 and 13. Before animal necropsy at the end of the study, ophthalmological investigations were done in groups 1 and 4. All animals were subject to histopathological examinations. An additional group of five male and five female animals in dose groups 1 and 4 served as control for the reversibility of possible systemic or local effects.

The deaths of one male animal in the low-dose group and one female animal in the medium-dose group during the study were not substance related. No significant or dose-responsive changes in body weight gain or relative organ weights were seen in any dose group. Absolute gonad weights were reduced in all male test groups, but were not considered to be treatment related due to the lack of dose response, which was further supported by unchanged relative gonad weights. Furthermore, the one-generation screening test (see later) confirmed that C12/16 alkyl polyglycosides do not influence the gonad weight in Sprague–Dawley rats.

All biochemical parameters remained within the typical range of the control animals. Hematological changes (increased number of thrombocytes) in group 4 at week 6 were considered neither significant nor compound related. A dose-dependant and slowly reversible irritation and ulceration of the forestomach mucosa was noted in groups 3 and 4, a local effect induced by a well-known irritant substance. In contrast, systemic toxicity was not seen in any dose group.

Therefore, 1000 mg/kg b.w./day is the no-observed-adverse-effect level (NOAEL) dose for systemic toxicity after oral admission, whereas the no-observed-effect concentration (NOEC) for local compatibility is deduced as 2.5 % active substance.

Another study for subacute toxicity is available for C8–10 APG. The test substance (60% active substance) was applied openly to the intact skin of rabbits (New Zealand white) at doses between 0.06 and 3 g/kg b.w. day, respectively. Each dose group consisted of six male and six female animals, which received aliquots of 2 mL/kg once daily over a treatment period of 14 days. After each daily 6-h exposure time, the application sites were wiped. A control group was treated with only water. The rabbits were observed daily for clinical signs and skin irritation. Body weights were determined on days 5 and 9 of treatment, whereas hematological and clinical chemical parameters were determined prior to dosing and on termination of the study. Absolute and relative organ weights, necropsy observations, and histopathology of the testes and accessory sex glands were evaluated at the end of treatment. Doses at and above 1.5 g/kg b.w. day induced severe skin irritation after repeated application as well as several changes in hematological and clinical parameters and testicular degeneration. These effects were at least partly attributed to stress and inflammation due to the severe irritation caused by the test substance. Minimal to mild skin irritation was seen in dose groups starting from 0.54 g/kg b.w. day, whereas no clinical, haematological, or organ changes were reported at this dose. At and below 0.18 g/kg b.w. day, none of the described adverse effects was seen, which consequently was defined as the $NOAEL_{dem}$ after repeated dermal application [90]. The NOAEL for *systemic* effects can be set at 0.54 g/kg b.w.

H. Toxicity to Reproduction and Developmental Toxicity

Reproductive toxicity, in the broader sense defined by the EU guidelines, encompasses adverse effects either on the development of the progeny or on the sexual functions. The latter may be manifested by an influence on spermatogenesis and oogenesis, hormonal balance, or impairment of reproductive performance, such as sexual behavior, including libido and weaning.

1. Impairment of Hormonal Functions

The potential risk of chemical substances interacting with the sexual hormone system can be investigated in in vitro screening assays. Of these, the E-screen and the reporter gene assay are commonly used. Both assays are considered very sensitive [91,92]. Although these in vitro assays have some limitations, e.g., pharmacokinetics is not covered, exposure conditions are rather artificial, positive findings in these assays are considered reliable indications of a potential hormonal (estrogenic) activity and usually trigger further in vivo assays. In contrast, substances inactive in these assays are considered unlikely to be deleterious to the endocrine systems.

The E-screen determines induction of cell proliferation (mitogenic effect) in an estrogen-dependent human breast tumor cell line. A wide range of concentrations of the test substance is compared to a negative control (medium) and to reference hormones. A C12/16 alkyl polyglycoside concentration range of 0.1–10,000 nmol were tested in this assay. Whereas estradiol (physiological estrogen) starting at a concentration of 0.1 nmol clearly induced proliferation, compared to the concurrent control, no effects were noted with APG concentrations up to 10^5 times higher. The sensitivity of the assay is further supported by findings using the very weak estrogen bisphenol A, which showed a clear positive and concentration-dependent effect in this assay, starting at approximately 10 nmol. In the reporter gene assay, the induction of luciferase in stable transfected MCF-7 cells is used as an indicator for estrogenicity.

Again, whereas a clear positive and concentration-dependent induction of luciferase was noted for the physiological estrogen estradiol starting at about 0.1 nmol, no effects were seen with C12/16 alkyl polyglycosides up to a 100,000 times higher test concentration.

Based on these two sensitive in vitro assays, there is no indication that C12/16 alkyl polyglycosides might act as an endocrine modulator [68].

2. Impairment of Sexual Functions

Apart from definitive guidelines, a specifically designed screening test was developed within the OECD between 1990 and 1992 for the assessment of reproductive toxicity, which has since been listed as an accepted OECD Guideline, No. 421. It can be used to provide screening information on possible reproductive effects and development. In this one-generation screening assay, the test substance is applied to male and female rats prior to mating throughout the gestation and lactation period until postpartum day 3. The total duration of the study is 53 days. During the study, parameters of general toxicity, such as clinical signs, food consumption, and body weight gain, are recorded in the parental generation and pups. Moreover, effects related to reproduction, such as estrous cycle, mating performance, pregnancy rates, and the number of resorptions, are registered. Pup losses are recorded and the filial generation is examined for toxic effects.

C10/16 alkyl polyglycosides were investigated in male and female rats following daily administration by oral gavage of 100-, 300-, and 1000-mg/kg/day doses [68]. Each group contained 10 male and 10 female F0 rats. Dams were allowed to litter and rear their offspring to post partum day 4. No effects indicative of general toxicity were observed in parental animals. Relative and absolute weights of testes, epididymides, and seminal vesicles did not differ between test and control animals. A marginal reduction in absolute and relative prostate weights was seen in all treated males compared to the control group, being significant for the low-dose group only and not dose dependent. Therefore, this finding was not considered to be substance related. With regard to reproductive parameters, no test substance–related symptoms were observed. Mean litter weights, mean pup weights, sex ratios, and gestation periods were similar among groups. Some variations in prebirth loss were seen in the high-dose group without being dose dependent or significantly different to the controls. Pre-weaning clinical signs showed no treatment-related effects in pups, nor did necropsy reveal any effects in decedent or Fl pups. Macroscopic examination revealed no difference between treated and control animals. On the basis of the results obtained in this one-generation screening assay, a NOAEL of 1000 mg/kg/day for reproductive effect can be deduced.

3. Developmental Toxicity

Developmental toxicity is the property of a chemical that causes permanent structural or functional abnormalities. It can be investigated in a "segment II study." The international standard test method is OECD Guideline No. 414: The test substance is administered in graduated doses for at least that part of the pregnancy covering the period of organogenesis. Clinical condition and reaction to treatment are recorded at least once daily. Shortly before the expected partum date, dams are sacrificed, uteri removed, and the contents examined for embryonic or fetal deaths and live fetuses. Dams are examined macroscopically and the fetuses subjected to skeletal and visceral examinations. Maternal and reproduction data are recorded allowing further insight in maternal toxicity (MT).

C12/16 alkyl polyglycosides were tested in a segment II study at dose levels of 0, 100, 300, and 1000 mg/kg-day b.w./day in pregnant CD rats. The substance was administered daily by gavage from day 6 to day 15 of gestation. A standard dose volume of 10 ml/kg body weight was used. Clinical condition and reaction of treatment were recorded at least once daily. The maternal body weights were reported for days 0, 6, 16, and 20 of gestation [93].

All dams tolerated the applied dose levels of up to 1000 mg/kg without lethality. Maternal body weight gain, a very sensitive indicator for unspecified toxicity, was not affected by the treatment. For maternal toxicity the following no-adverse-effect-level (NOAEL) can be deduced:

$$NOAEL_{MT} = 1000 \text{ mg/kg b.w.}$$

All females had viable fetuses. Pre- and postimplantation loss as well as mean numbers of resorption were not affected by treatment. All parameters were comparable with control group animals. Skeletal and visceral investigations did not detect any treatment-related malformations. For embryo/fetotoxicity and teratogenicity (EFT/TER) the following NOAEL can be deduced from this study:

$$\text{NOAEL}_{EFT/TER} = 1000 \text{ mg/kg b.w.}$$

Both developmental toxicity and teratogenicity NOAEL [68] exceed, by far, any reasonably foreseeable exposure to alkyl polyglycosides in daily life. Assuming acknowledged safety factors for intra- and interspecies differences, no risk of teratogenic effects in humans induced by APG can be expected. Furthermore, these surfactants give no reason for concern regarding toxicity to maternal reproductivity.

I. Toxicokinetics and Metabolism

Toxicokinetic studies examine absorption, desorption, metabolism, and excretion of chemicals. The distribution between selected organs enables the assessment of metabolic fate as well as providing valuable data with regard to the bioavailablity of individual substances.

Two types of radiolabeled alkyl-β-glucosides, representing both the short-chain and long-chain homologues of commercial APG, were applied by gavage to female NMRI mice. Two hours after treatment the mice were killed and relevant organs were radioanalyzed to determine specific organ distribution.

Stomach, intestines, liver, and kidney showed the highest levels of radioactivity for both compounds. Using simple extraction methods, it was shown that alkyl polyglycosides are readily cleaved into glucose and fatty alcohol, which is further oxidized to the corresponding fatty acid and partly incorporated in normal fat metabolism. Octyl-β-D-[U-^{14}C] glucosides were rapidly and almost completely transformed into hydrophilic metabolites during intestine and liver passage, whereas [1,–^{14}C]-hexadecyl-β-D-glucosides showed a much greater tendency toward lipophilic metabolism, resulting, e.g., in preferential identification in the liver of radiolabeled palmitoyl glycerides. These findings are underlined by the fact that β-oxidation occurs more easily in medium-chain fatty acids than in long-chain fatty acids [94,95].

It could be shown that alkyl polyglycosides are transformed into physiologically occurring metabolites that chemically behave in the same way as their native counterparts, further supporting the results of the subchronic study (see Section II.G) and proving again the innocuousness of this important class of surfactants.

III. CONCLUSION

The ecological, substance-inherent properties of alkyl polyglycosides have been studied comprehensively. (Linear) alkyl polyglycosides are fast and completely biodegradable under all environmental conditions (aerobically and anaerobically, in freshwater as well as in seawater) and easily removed from the wastewater in sewage treatment plants. The ecotoxicological properties of the alkyl polyglycosides are strongly dependent on the structure. The short-chain $C_{8/10}$-APG is less toxic than the medium-chain $C_{12/14}$-APG, and APGs with branched alkyl chains are less toxic than the corresponding linear APGs. The acute aquatic toxicity of the more toxic $C_{12/14}$-APG is in the same range as for most surfactants; i.e., $C_{12/14}$-APG is toxic for aquatic organisms (LC/EC$_{50}$ values between 1 and 10 mg/L). Chronic exposure to C$_{12/14}$-APG is relatively well tolerated by aquatic organisms, with the NOEC values being almost identical with the EC$_0$ values observed in the acute tests. This is indicative of the fact that alkyl polyglycosides—due to their surface activity—act primarily on the cell membranes of the outer surface of the organisms but do not exert systemic toxic effects.

Preliminary risk assessments for the aquatic (PEC/PNEC$_{aquatic}$ = 0.02) and terrestrial compartments (PEC/PNEC$_{terrestrial}$ = 0.023) based on the corresponding German use volumes of the year 2000 have shown that the use of alkyl polyglycosides in down-the-drain consumer products does not have a detrimental environmental impact. In conclusion, alkyl polyglycosides prove to be a promising class of modern nonionic surfactants that has been comprehensively investigated in terms of its ecological properties and that exhibits an excellent environmental compatibility.

Alkyl polyglycosides of different chain length have been investigated in depth for their potential health hazard. All studies reported comply with current legal requirements for the classification and labeling of chemical substances or were performed in accordance with acknowledged protocols. Consequently, the quality of data allows scientifically sound risk assessments for a variety of application scenarios, for both occupational and personal care use.

Undiluted APG displayed very low acute toxicity (LD50 in the range of several grams per kilogram of body weight) but proved to be skin irritants and severe eye irritants. Since the latter data are only applicable to undiluted APG, they give no indication of the effects of surfactant formulations used in daily life. Quite the reverse, it is known that APG as cotensides enhance skin compatibility of formulations significantly.

Furthermore, APG showed neither sensitizing properties nor genotoxic effects, either in vitro or in vivo, in a variety of different test systems. No endocrine-modulating effects have been observed in an E-screen and a reporter gene assay.

Several authors, estimating exposure to surfactants via relevant sources such as drinking water, food, and dental hygiene products, have produced figures for daily intake ranging from 0.3 to 3 mg/person day [97–102]. Considering a standardized body weight of 60 kg, these figures result in a "worst-case" daily dose of 0.05 mg/kg b.w., far below the NOAEL for repeated oral uptake, teratogenicity, foetotoxicity, and maternal toxicity (1000 mg/kg b.w. each) indicated by different studies with APG. In view of a margin of safety of at least 1000, it can be concluded that APG are not cumulatively toxic to humans.

Lacking any irreversible effects due to exposure at marketable concentrations, APG represent no risk for the consumer or general public, whereas the intrinsic hazards of the raw material, occurring exclusively during manufacture, can be countered effectively by personal protective measures.

REFERENCES

1. Hill, K.H.; Rhode, O. Fett/Lipid 1999, *101* (1), 25–33.
2. Andree, H., Hessel, J.F., Krings, P., Meine, G., Middelhauve, B., Schmid, K., Hill, K.H., von Rybinski, W., Stoll, G., Eds.; Alkyl Polyglycosides—Technology, Properties, and Applications; Weinheim: Verlag Chemie, 1997, 99–130.
3. Tesmann, H., Kahre, J., Hensen, H., Salka, B.A., Hill, K.H., von Rybinski, W., Stoll, G.,

Eds.; Alkyl Polyglycosides—Technology, Properties, and Applications; Weinheim: Verlag Chemie, 1997; 71–98.

4. Aleksejczyk, R.A. ASTM Spec Tech Publ 1993, *12*, 22–32.
5. Commission Regulation (EC) No. 1488/94 of June 28, 1994. Off J Europ Comm: L161, June 29, 1994.
6. Partearroyo, M.A.; Pilling, S.J.; Jones, M.N. Com. Biochem. Physiol. 1991, *100C* (3), 381–388.
7. Eskuchen, R., Nitsche, M., Hill, K.H., von Rybinski, W., Stoll, G., Eds.; Alkyl Polyglycosides—Technology, Properties, and Applications; Weinheim: Verlag Chemie, 1997; 9–22.
8. Schöberl, P.; Scholz, N. Surfactant Sci. Ser. 2000, *91*, 331–363.
9. OECD Guidelines for the Testing of Chemicals. Vol 1, Sec 3: Degradation and Accumulation. OECD, Paris, 1993.
10. Organization For Economic Cooperation and Development—OECD (1981).
11. Commission Directive (92/69/EEC). European Community (1992), OJ L 383-29.12.1992.
12. German Ministry for Environmental Affairs (1999): Allgemeine Verwaltungsvorschrift zum Wasserhaushaltsgesetz über die Einstufung wassergefährdender Stoffe in Wassergefährdungs klassen (VwVwS) vom 17, Mai 1999.
13. EEC (1982): Council Directive of 31 March 1982 (82/242/EEC). Off. J. Europ. Comm. L 109, April 22, 1982.
14. Steber, J.; Guhl, W.; Stelter, N.; Schröder, F.R. Tenside Surf. Det. 1995, *32*, 515–521.
15. EEC (1986): Guidance Document of the Competent Authorities for the Implementation of Directive 79/831/EEC, Doc. XI/861/86, 1986.
16. Cognis Deutschland. Unpublished results, Protocol No. 33. Page/Assay 789. Duesseldorf, 1986.
17. Cognis Deutschland. Unpublished results, Report No. R 9800392. Duesseldorf, 1998.
18. Gerike, P.; Holtmann, W.; Jasiak, W. Chemosphere 1984, *13*, 121–141.
19. Cognis Deutschland. Unpublished results, Report No. R 9601746. Duesseldorf, 1996.
20. Eichhorn, P.; Knepper, T.P. J. Chrom. A 1999, *845*, 221–232.
21. Karlson, P. Biochemie für Mediziner und Naturwissenschaftler; Stuttgart: Georg Thieme Verlag, 1977.
22. Schöberl, P.; Guhl, W.; Scholz, N.; Taeger, K. Tenside Surf. Det. 1998, *35*, 279–285.
23. Cognis Deutschland. Unpublished results, Report No. R 9400449. Duesseldorf, 1994.
24. Cognis Deutschland. Unpublished results, Report No. R 9701182. Duesseldorf, 1997.
25. Toshima, Y.; Tsugukuni, T.; Takashima, F. Nippon Kaisui Gakkaishi 1996, *50* (1), 18–22.
26. Birch, R.R.; Biver, C.; Campagna, R.; Gledhill, W.E.; Pagga, U.; Steber, J.; Reust, H.; Bontinck, W.J. Chemosphere 1989, *19*, 1527–1550.
27. Cognis Deutschland. Unpublished results, Report No. R 9602229. Duesseldorf, 1996.
28. Cognis Deutschland. Unpublished results, Report No. R 9400894. Duesseldorf, 1994.
29. Commission Directive (92/69/EEC), European Community (1992), OJ L 383-29.12.1992.
30. Cognis Deutschland. Unpublished results, Report No. R 9400449. Duesseldorf, 1994.
31. Cognis Deutschland. Unpublished results, Protocol No. 300/5. Page/Assay 2221/2222. Duesseldorf, 1986.
32. OECD Guidelines for the Testing of Chemicals, Vol 1, Sec 2: Effects on Biotic Systems OECD, Paris, 1993.
33. Cognis Deutschland. Unpublished results, Report No. RE 920189. Duesseldorf, 1992.
34. Cognis Deutschland. Unpublished results, Report No. TBD 892144. Duesseldorf, 1989.
35. Cognis Deutschland. Unpublished results, Report No. TBD 852387. Duesseldorf, 1985.
36. Cognis Deutschland. Unpublished results, File 407/4. Duesseldorf, 1984.
37. Garcia, M.T.; Riboss, L.; Campos, E.; Leal, J.S. Chemosphere 1997, *35* (3), 545–556.
38. Cognis Deutschland. Unpublished results, Report No. R 9400074. Duesseldorf, 1994.

39. Madsen, T.; Petersen, G.; Seireo, C.; Torlov, J. J. Am. Oil Chem. Soc. 1996, *73* (7), 929–933.
40. Cognis Deutschland. Unpublished results, Report No. R 9800015. Duesseldorf, 1998.
41. Cognis Deutschland. Unpublished results, Report No. R 9400701. Duesseldorf, 1994.
42. Cognis Deutschland. Unpublished results, Report No. R 9500085. Duesseldorf, 1995.
43. Cognis Deutschland. Unpublished results, Report No. R 9400443. Duesseldorf, 1994.
44. Cognis Deutschland. Unpublished results, Report No. R 9400738. Duesseldorf, 1994.
45. Cognis Deutschland. Unpublished results, Report No. R 9500541. Duesseldorf, 1995.
46. Eichhorn, P.; Petrovic, M.; Barcelo, D.; Knepper, T.P. Vom Wasser 2000, *95*, 245–268.
47. Cognis Deutschland. Unpublished results, File 401/12. Duesseldorf, 1983.
48. Cognis Deutschland. Unpublished results, File 407/4. Duesseldorf, 1984.
49. Uppgard, L.; Sjöström, M.; Wold, S. Tenside Surf. Det. 2000; 131–142.
50. OSPAR Decision 2000/2 on a Harmonized Mandatory Control System for the Use and Reduction of the Discharge of Offshore Chemicals, 2000.
51. OECD 203 Guideline as adopted by OSPARCOM (1995) for marine testing of off-shore chemicals, 1995.
52. Cognis Deutschland. Unpublished results, Report No. C 0100640-22. Duesseldorf, 2001.
53. ISO Proposal to TC147/SC5/WG2: Determination of acute lethal toxicity to marine adult copepods, 1990.
54. Cognis Deutschland. Unpublished results, Report No. C 0100640-23. Duesseldorf, 2001.
55. ISO/PARCOM: Toxicity test with marine unicellular algae: technical support document for the ISO DP 10253 Standard Method, 1990/1991.
56. Cognis Deutschland. Unpublished results, Report No. C 0100640-24. Duesseldorf, 2001.
57. Cognis Deutschland. Unpublished results, Report No. C 0100640-11. Duesseldorf, 2001.
58. Cognis Deutschland. Unpublished results, Report No. C 0100640-12. Duesseldorf, 2001.
59. Cognis Deutschland. Unpublished results, Report No. R 9400074. Duesseldorf, 1994.
60. Cognis Deutschland. Unpublished results, Report No. R 9400738. Duesseldorf, 1994.
61. Cognis Deutschland. Unpublished results, Report No. R 9500527. Duesseldorf, 1995.
62. Cognis Deutschland. Unpublished results, Report No. R 9500594. Duesseldorf, 1995.
63. W. Guhl. German Environmental Protection Agency: Report No. 10603057, 1989.
64. OSPARCOM (1995). Sediment bioassay using an amphipod *Corophium* sp., 1995.
65. Cognis Deutschland. Unpublished results, Report No. C 0100640-25. Duesseldorf, 2001.
66. Cognis Deutschland. Unpublished results, Report No. C 0100640-13. Duesseldorf, 2001.
67. Cognis Deutschland. Unpublished results, Report No. C 0100011-2. Duesseldorf, 2001.
68. Aulmann, W.; Koehl, W.; Pitterman W. Publication in preparation, 2004.
69. IKW (2000). Industrieverband Körperpflege- und Waschmittel e. V. Frankfurt/M., 2000.
70. EEC (1994). Risk Assessment of existing substances. Technical Guidance Document. European Commission DG XI. Brussels, 1994.
71. ECETOC (European Center for Ecotoxicology and Toxicology of Chemicals). HAZ-CHEM, Special Report No. 8, ECETOC, Brussels, 1994.
72. Aulmann, W.; Sterzel, W.; Hill, K.H. von Rybinski, W., Stoll, G., Eds.; Alkyl Polyglycosides—Technology, Properties, and Applications; Weinheim: Verlag Chemie, 1997; 151–167.
73. Organization For Economic Cooperation and Development—OECD (1981).
74. European Directive 67/548/EEC.
75. OSHA's Hazard Communication (HAZCOM) Standard (29 CFR 1910.1200).
76. S. Pfeiffer, P.; Pflegel, H.H.; Borchert. Grundlagen der Biopharmazie: Pharmacokinetik, Bioverfügbarkeit, Biotransformation. Weinheim: Verlag Chemie, 1984.
77. Test protocol according to 40 CFR 152 (Federal Insecticide, Fungicide, and Rodenticide Act).
78. Cognis Deutschland. Unpublished results, Report no. TBD EX 0323. Duesseldorf, 1987.
79. Cognis Deutschland. Unpublished results, Report no. TBD EX 0449. Duesseldorf, 1987.

80. Cognis Deutschland. Unpublished results, Reports RT 930138 and RT 930139. Duesseldorf, 1993.
81. Balaguer, F.; Castan, P.; Coll, J.; Reig, J.; Recasens, M.M.; Prat, A.; Pelejero, C. Chim. Oggi 2000, *18*, 23–25.
82. Cognis Deutschland. Unpublished results, Report no. R 9500934. Duesseldorf, 1995.
83. Cognis Deutschland. Unpublished results, Report no. 880160. Duesseldorf, 1988.
84. Cognis Deutschland. Unpublished results, Report No. R 94 00208. Duesseldorf, 1994.
85. Cognis Deutschland, Unpublished results, report No. TBD 900290. Duesseldorf, 1990.
86. Cognis Deutschland. Unpublished results, Report no. R 97 00189. Duesseldorf, 1993.
87. Cognis Deutschland. Unpublished results, Report no. R 9300209. Duesseldorf, 1993.
88. Cognis Deutschland. Unpublished results, Report no. R 94 00243. Duesseldorf, 1995.
89. Cognis Deutschland, unpublished results, report no. TBD 89 0161. Duesseldorf, 1989.
90. Rohm and Haas. Internal Reports 83P-533, 82P-250, and 85P-144. Spring House, PA, 1985–1986.
91. Soto, A.M.; Sonnenschein, C.; Chung, K.L.; Fernandez, M.F.; Olea, N.; Olea-Serrano, M.F. The E-SCREEN assay as a tool to identify estrogens: an update on estrogenic environmental pollutants. Environ. Health Perspect 1995, *103*, 113–122.
92. Gutendorf, B.; Westendorf, J. Comparison of an array of in vitro assays for the assessment of the estrogenic potential of natural and synthetic estrogens, phytoestrogens and xenoestrogens. Toxicology 2001, *166* (1–2), 79–89.
93. Cognis Deutschland. Unpublished results, Report no. R 96 00504. Duesseldorf, 1996.
94. Scheig, R. Hepatic metabolism of medium chain fatty acids. In *Medium-chain tri-glycerides*; Senior, J.R., Ed.; University of Pennsylvania Press: Philadelphia, 1968; 39–49.
95. Petit, D.; Raisonnier, A.; Amit, N.; Infante, R. Lack of induction of VLDL apoprotein synthesis by medium chain fatty acids in the isolated rat liver. Ann. Nutr. Metab. 1982, *26*, 279.
96. Swisher, R.D. Arch. Environ. Health 1968, *17*, 232–246.
97. Borneff, J. Arch. Hyg. Bakt. 1957, *141*, 578.
98. Sterzel, W. Anionic Surfactants: Biochemistry, Toxicology, Dermatology; Marcel Dekker: New York, 1992; 411–417.
99. Wedell, H. Fette, Seifen, Anstrichmittel 1966, *68*, 551–556.
100. Schmitz, J. Tenside Detergents 1973, *10*, 11–13.
101. Krüger, R. Seifen, Öle, Fette, Wachse 1960, *86*, 289–292.
102. Cognis Deutschland. Unpublished results, Report no. TBD EX 0321. Duesseldorf, 1987.
103. Cognis Deutschland. Unpublished results, Report no. TBD 860296. Duesseldorf, 1986.
104. Cognis Deutschland. Unpublished results, report no. TBD EX 0448. Duesseldorf, 1990.

19

Biodegradation of Cationic Surfactants

An Environmental Perspective

C. G. VAN GINKEL   Akzo Nobel Chemicals Research Arnhem, Arnhem, The Netherlands

I. INTRODUCTION

Surface-active agents—or surfactants—are partly water soluble, or hydrophilic, and partly lipohilic (soluble in lipids, or oils). Surfactants are classified according to the nature of the hydrophilic part of the molecule. Cationic surfactants carry a positive charge in the hydrophilic moiety. Chemically they have four alkyl groups linked directly to the nitrogen atom through covalent bonds. The alkyl groups may be alike or different, saturated or unsaturated, and may contain ester linkages. Finally, cationic surfactants may be aromatic compounds. The nitrogen atom plus the attached alkyl groups form the positively charged moiety of the molecule. Primary, secondary, and tertiary long-chain amines are considered quasicationic, because in acid solutions the nitrogen can be protonated. Finally, ethoxylated quaternary ammonium salts and polyoxyethylene alkylamines are sometimes regarded as cationic surfactants [1]. Agriculture provides tallow and vegetable oils like coconut oil and palm-kernel oil, which have been traditionally used as raw material for the soap industry and are now also used for the production of fatty amine derivatives. In addition, cationic surfactants derive their hydrophobic chains from crude oil via intermediates. Important cationic surfactants are shown in Figure 1.

Cationic surfactants are used in liquid and powdered laundry detergents to enhance cleaning, softening, and static control in fabric conditioners, and thickening in hard-surface cleaners. Cationic surfactants therefore constitute an important and widely used class of surfactants. Because they are used in detergents, cationic surfactants are released into the environment. Environmental acceptability of chemicals is greatly justified, if the environmental concentration or the period of exposure is significantly reduced by biodegradation. Biodegradation of cationic surfactants is therefore important in any consideration of their environmental safety.

Consequently, the biodegradability of cationic surfactants has become almost as important as the actual performance of these substances. Different methods are

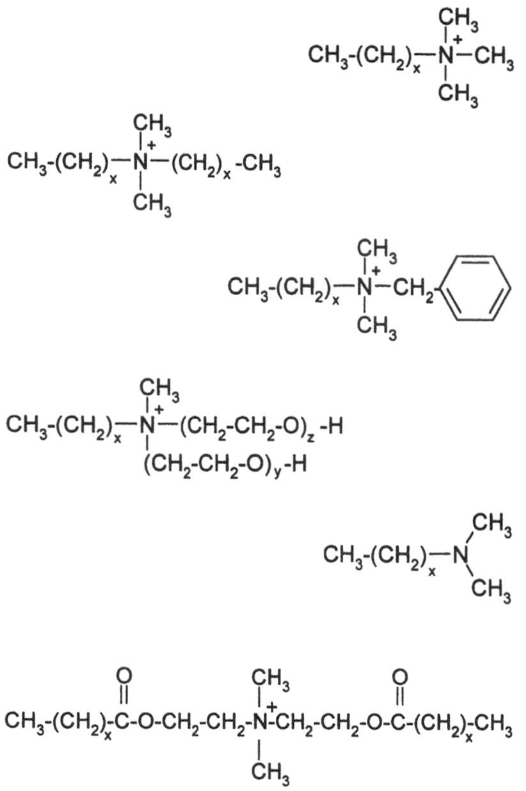

$$CH_3\text{-}(CH_2)_x\text{—}\overset{\overset{\displaystyle CH_3}{|}}{\underset{\underset{\displaystyle CH_3}{|}}{N^+}}\text{—}CH_3$$

$$CH_3\text{-}(CH_2)_x\text{—}\overset{\overset{\displaystyle CH_3}{|}}{\underset{\underset{\displaystyle CH_3}{|}}{N^+}}\text{—}(CH_2)_x\text{-}CH_3$$

$$CH_3\text{-}(CH_2)_x\text{—}\overset{\overset{\displaystyle CH_3}{|}}{\underset{\underset{\displaystyle CH_3}{|}}{N^+}}\text{—}CH_2\text{—}\langle\bigcirc\rangle$$

$$CH_3\text{-}(CH_2)_x\text{—}\overset{\overset{\displaystyle CH_3}{|}}{\underset{\underset{\displaystyle (CH_2\text{-}CH_2\text{-}O)_y\text{-}H}{}}{N^+}}\text{—}(CH_2\text{-}CH_2\text{-}O)_z\text{-}H$$

$$CH_3\text{-}(CH_2)_x\text{—}N\overset{\diagup CH_3}{\diagdown CH_3}$$

$$CH_3\text{-}(CH_2)_x\text{-}\overset{\overset{\displaystyle O}{\|}}{C}\text{-}O\text{-}CH_2\text{-}CH_2\text{-}\overset{\overset{\displaystyle CH_3}{|}}{\underset{\underset{\displaystyle CH_3}{|}}{N^+}}\text{-}CH_2\text{-}CH_2\text{-}O\text{-}\overset{\overset{\displaystyle O}{\|}}{C}\text{-}(CH_2)_x\text{-}CH_3$$

FIG. 1 Structures of some important cationic surfactants used in detergents.

available for assessing the biodegradation of chemicals under aerobic conditions. First, screening tests discriminate readily biodegradable compounds from inherent and persistent chemicals. These screenings or ready biodegradability tests include the closed-bottle test, MITI I test, and the Sturm test. All these tests are usually performed within a 28-day period. Positive ready biodegradability test results indicate that the test substance is biodegraded without problems. However, ready biodegradability tests tend to underestimate the potential for biodegradation in the environment. When the result of a ready biodegradability test is negative or dubious, inherent biodegradability tests or simulation tests are required. They include the semicontinuous activated sludge (SCAS) test and the continuous activated sludge (CAS), or confirmatory test. The amount of carbon dioxide produced, the oxygen consumed, or the removal of the parent compound or its carbon during microbial metabolism determines the extent of biodegradation in biodegradability tests. The biodegradation of surfactants in these tests has been reviewed extensively [2,3].

In addition to qualitative and quantitative testing for microbial activity by oxygen uptake, carbon dioxide evolution, and DOC (dissolved organic carbon) removal, there are ways to identify biodegradation pathways through metabolic studies. Metabolic studies are an important tool in environmental microbiology to show complete mineralization, i.e., conversion of the organic compounds into carbon dioxide, water, and mineral salts of any other element present in the organic compounds. Nowadays, the complete mineralization of surfactants, avoiding the formation of recalcitrant and/or toxic products, is a major issue [4]. This follows findings demonstrating that biodegradation of branched alkylphenol ethoxylate chain occurs via ethoxylate shortening, leading mainly to nonylphenol monooxyethylene, nonylphenol dioxyethylene, nonylphenol acetic acid, and nonylphenol ethoxyacetic acid [5,6]. These metabolites are more toxic than the parent compound.

Studies in environmental microbiology often emphasize the search for microorganisms isolated after selective enrichment. Use of these pure cultures, in dense cell suspensions, facilitates analysis of substrates, intermediates, and products. Over the last decade, considerable advances have been made in our understanding of the strategies underlying biochemical degradation of cationic surfactants using these pure culture studies. Up until now, pure cultures of microorganisms capable of metabolizing cationic surfactants were primarily investigated to establish biodegradation pathways [7].

The objective of this chapter is to review biodegradability test results and biodegradation pathways and to relate test results and metabolic studies to assess the environmental fate of cationic surfactants more accurately.

II. BIODEGRADATION PATHWAYS

A. Benefit of Metabolic Studies

Early research aimed at assessing the potential of cationic surfactants to biodegrade in ecosystems and biological treatment plants was carried out primarily using biodegradability tests [2]. Such tests are inherently limited since only the parent compound, carbon content, oxygen consumption, or carbon dioxide formation is used as determinant. Moreover, in these tests only 50–90% is mineralized by microorganisms utilizing the test compound as a source of carbon and energy, while the remaining part is converted into new biomass. This new biomass is mineralized, in the long-term ecological sense. This phenomenon makes it impossible to demonstrate unambiguously complete mineralization in standard biodegradability tests with or without radiolabeled material. Mineralization without formation of toxic intermediates during the biodegradation of cationic surfactants can be demonstrated by determining catabolic pathways using pure cultures of microorganisms. Based on the structure of quaternary ammonium salts, three biodegradation pathways are plausible: (1) hydroxylation of the far end of the alkyl chains, followed by β-oxidation, progressing toward the hydrophilic moiety, (2) a central fission of the molecule, giving rise to a hydrophobic part and a hydrophilic part, and (3) an attack on the hydrophilic part.

B. Alkyltrimethylammonium Salts

In order to examine the biodegradation pathway of alkyltrimethylammonium salts, a bacterium capable of utilizing decyltrimethylammonium salts as sole source of carbon and energy was isolated [8]. Using GC-MS, 9-carboxynonyl- and 7-carboxyheptyltrimethylammonium chloride were detected in the supernatant liquor during growth of the isolate on this quaternary ammonium salt. Identification of these carboxyalkyltrimethylammonium salts suggests that the metabolism of decyltrimethylammonium chloride includes ω- and β-oxidation of the alkyl chain. However, respiration experiments reported indicate a C_{alkyl}–N cleavage yielding trimethylamine and decanal [8]. In 1992, van Ginkel et al. [9] isolated a pure culture that grows on hexadecyltrimethylammonium chloride. This bacterium, tentatively identified as a *Pseudomonas* sp., was obtained through enrichment in a continuous culture inoculated with activated sludge. Growth of the *Pseudomonas* sp. on alkyltrimethylammonium salts with alkyl chain lengths ranging from C_{12} to C_{18} was achieved in batch cultures in the presence of silica gel. The isolate therefore displays a broad substrate specificity with respect to the alkyl chain length. Growth on hexadecyltrimethylammonium chloride is accompanied by the excretion of trimethylamine into the medium. Cell-free extracts catalyze an $NADH/O_2$-linked formation of trimethylamine in the presence of hexadecyltrimethylammonium chloride. The first step in the degradation pathway of alkyltrimethylammonium chloride is therefore a fission at the C_{alkyl}–N bond by a monooxygenase. For this, the C_{alkyl}–N bond is probably activated by the monooxygenase-mediated addition of a hydroxyl group to the α-carbon. Subsequently, C_{alkyl}–N bond cleavage occurs by chemical hydrolysis into trimethylamine and hexadecanal. Monooxygenases are common enzymes initiating the biodegradation of many organic compounds. The aldehyde formed is converted into an alkanoic acid by an alkanal dehydrogenase. The alkanoic acid is then activated with coenzyme A by an alkanoyl CoA synthetase. The thiol ester formed is degraded by β-oxidation, yielding a succession of acetyl-CoA units as the fatty acid is progressively shortened by C_2 units. The acetyl-CoA is assimilated through the tricarboxylic acid cycle and glyoxylate shunt [10]. It was shown in the *Pseudomonas* sp. that the enzymes degrading quaternary ammonium salts are constitutive. The proposed biodegradation pathway associated with the *Pseudomonas* sp. is detailed in Figure 2.

Another demonstration of the C_{alkyl}–N fission as the first degradation step of alkyltrimethylammonium salts with a mixed culture of microorganisms was presented by Nishiyama et al. [11]. During degradation of tetradecyltrimethylammonium salt by activated sludge, trimethylamine accumulated transiently in the culture fluid to a concentration of about 12 μmol/L (approximately 25% of the added quaternary ammonium salt). The oxidation of the alkyl chain generates trimethylamine.

C. Dialkyldimethylammonium Salts

Long-chain dialkyldimethylammonium salts (C_{14}–C_{18}) are not readily biodegradable [2]. The number of alkyl chains attached to the hydrophilic group could be a hindrance to the scission of the C_{alkyl}–N bond [12,13]. Separation of the first alkyl chain from the nitrogen of fatty amine derivatives with more than one alkyl chain was therefore thought to be difficult. To assess the capacity of microorganisms to attain alkyl chains of dialkyldimethylammonium salts, the metabolic basis of the growth on dialkyldimethylammonium salts by microorganisms was studied. To this end, a bacterial strain was isolated from activated sludge by growth on didecyldimethylammonium chloride as a sole source of carbon and energy [14]. This strain, identified as an *Achromobacter* sp., was examined for growth on a variety of fatty amine derivatives. Of those tested, alkyltrimethylammonium chloride (C_{10}–C_{18}), dialkyldimethylammonium chloride (C_{10}–C_{18}), and decyldimethylamine served as growth substrates. Utilization of dioctadecyldimethylammonium salt by the *Achromobacter* sp. could be demonstrated only in a study with columns packed with silica gel. Dioctadecyldimethylammonium

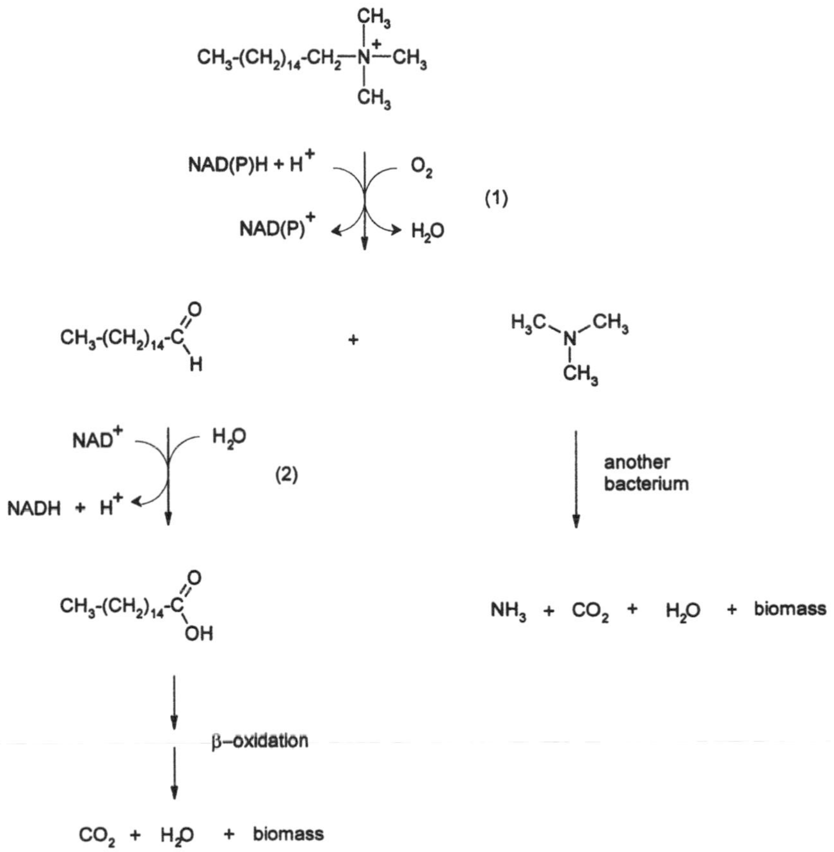

FIG. 2 Degradation of hexadecyltrimethylammonium in a *Pseudomonas* sp. (1) monooxygenase, and (2) alkanal dehydrogenase.

salt desorbed from the silica gel particles was metabolized within a few days by *Achromobacter* sp. present in the column. A decline in the effluent concentration of a sterile column serving as a control was not observed [14]. Decanal, decanoic acid, and acetic acid chosen from the postulated metabolites also serve as growth substrates. No growth was detected on methylamines. When considering methylamines, the oxygen consumption by a washed cell suspension of *Achromobacter* sp. grown on didecyldimethylammonium chloride was not significantly different from the endogenous respiration. Oxygen consumption increased significantly when decyldimethylamine, decanal, decanoic acid, and acetic acid were supplied to a washed cell suspension. Finally, acetate-grown cells showed significant oxygen consumption in the presence of didecyldimethylammmonium and decyltrimethylammonium salt. This proves that the enzymes responsible for the degradation of dialkyldimethylammonium salts are constitutively expressed. *Achromobacter* sp. does not exhibit a high substrate specificity with respect to the length of the alkyl chains. It is therefore very unlikely that the slow degradation of poorly water-soluble dialkyldimethylammonium salts is to be attributed to the absence of competent organisms. Their failure to biodegrade as readily as didecyldimethylammonium salt can probably be attributed to its limited bioavailability (see later). Cultures of *Achromobacter* sp. growing on didecyldimethyl-ammonium chloride accumulate dimethylamine as breakdown product (unpublished Akzo Nobel results).

A *Pseudomonas fluorescens* was also isolated from an activated-sludge sample using didecyldimethylammonium chloride as sole carbon and energy source [15]. This bacterium was screened for growth on a variety of cationic surfactants, and growth was observed on alkyltrimethylammonium, alkylbenzyldimethylammonium, and dialkyldimethylammonium salts. Biodegradation of didecyldimethylammonium chloride by *Pseudomonas fluorescens* is probably initiated by the activity of a mono-oxygenase, effectively cleaving a $C_{alkyl}-N$ bond, as shown by the degradation of didecyldimethylammonium chloride in cell-free extract in the presence of NADH and NADPH. Products from the monooxygenase-mediated reaction were not detected in cell-free extracts. Transient accumulation of decyldimethylamine during didecyldi-methylammonium chloride degradation suggests that this tertiary amine is the first intermediate. Accumulation of dimethylamine was not stoichiometrically, as found with the *Achromobacter* sp. Loss of dimethylamine was attributed either to volatiliza-tion from the medium or to oxidation by the *Pseudomonas fluorescens* strain. Figure 3 shows the ability of microorganisms to catalyze $C_{alkyl}-N$ fissions, thereby forming alkanals that can enter the common pathways of metabolism.

D. Alkylbenzyldimethyl Ammonium Salts

The degradation of alkylbenzyldimethylammonium salts has not been studied in pure cultures. In an attempt to substantiate that alkylbenzyldimethylammonium chloride–degrading microorganisms first utilize the alkyl chain, closed-bottle tests inoculated with adapted and unadapted sludge were conducted (unpublished Akzo Nobel results). Adaptation to an organic compound reduces the lag period, i.e., the time required for the development of an effective population of microorganisms [16]. Adapted mixed cultures degrading decylbenzyldimethylammonium chloride were obtained by enrichment in an SCAS unit. The acclimatized sludge from the SCAS unit was able to utilize alkylbenzyldimethylammonium chloride, *N,N*-dimethylbenzyl-amine, and dimethylamine without lag period. Appreciable lag phases were detected with these compounds using unacclimatized sludge (Fig. 4). For decylamine, lag phases in the closed-bottle test inoculated with adapted and unadapted sludge were both on the order of hours. Similarly, no consistent differences were found with decyldimethylamine in closed-bottle tests. This suggests that decyldimethylamine and decylamine are not intermediates in the degradation route of decylbenzyldimethyl-ammonium chloride. Consequently, exposure of sludge to decylbenzyldimethylam-monium chloride probably selects for three microorganisms that utilize the alkyl chain, the aromatic moiety, and dimethylamine. The degradation of the alkyl chain of alkylbenzyldimethylammonium chloride probably precedes the breakdown of the aromatic moiety (Fig. 3).

E. Alkylamines

The long alkyl chains of primary, secondary, and tertiary alkylamines are excellent sources for microbial growth, as demonstrated in ready biodegradability tests [17]. A

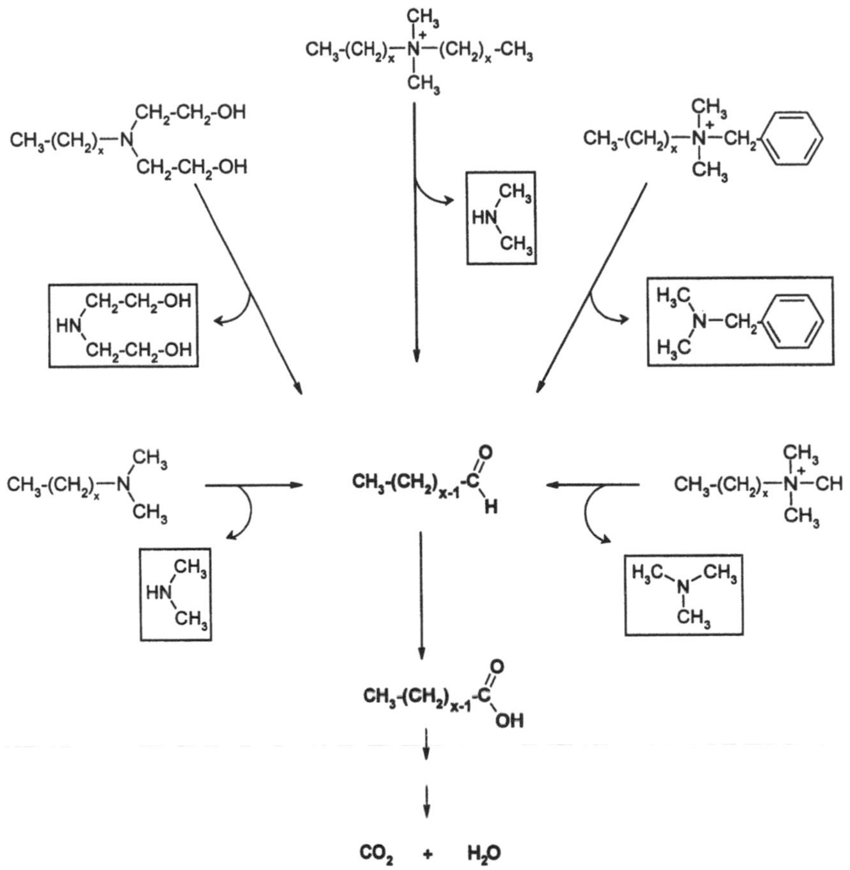

FIG. 3 General degradation pathway of cationic surfactants. The alkanals formed through the oxidation of the α-carbon of the alkyl chain enter one channel of metabolism, namely, β-oxidation.

bacterial strain was isolated from activated sludge by using dodecyldimethylamine as sole source of carbon and energy to study the biodegradation of fatty amines in more detail [18]. This organism was identified as a *Pseudomonas* sp., and its potential for the degradation of several fatty amine derivatives and potential intermediates was elucidated. The lengths of the alkyl chain sustaining growth of the *Pseudomonas* sp. is not limited to C_{12} because a range of C_8–C_{18} alkyl chains supports growth. In addition to alkyldimethylamines, the strain utilizes dodecanal, dodecanoic acid, and acetic acid as carbon sources. Dimethylamine does not support growth. These results again point toward the utilization of the alkyl chain. Biodegradation of dodecyldimethylamine begins with the cleavage of the alkyl chain with a dehydrogenase reaction. The oxidation of the aldehyde to a fatty acid is carried out by another dehydrogenase. The alkanoic acid is converted into acetyl-CoA units, which can be used for energy-yielding reactions and biosynthesis. Alkanals and alkanoic acids are also key intermediates in the metabolism of other alkylamines. Although the oxidation

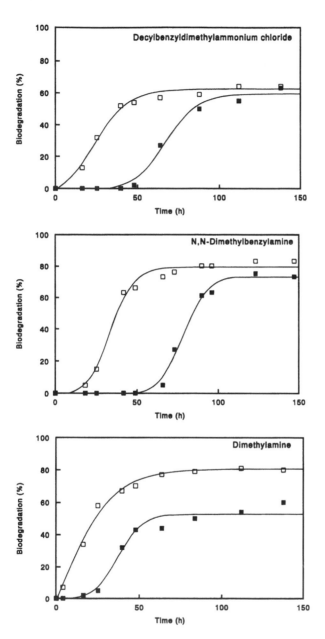

FIG. 4 Biodegradation of decylbenzyldimethylammonium chloride, N,N-dimethylbenzyl-amine, and dimethylamine in closed-bottle tests inoculated with unadapted sludge (■) and sludge adapted to alkylbenzyldimethylammonium chloride (□).

of the alkyl chain may lead directly to full mineralized hydrophiles (i.e., NH_3), in many instances methylamines remain [17].

F. Ethoxylated Fatty Amines

To understand the metabolic basis of growth on ethoxylated fatty amines by microorganisms, the pathway of alkylbis(2-hydroxyethyl)amines was studied [19]. To this end, a bacterium was isolated for its ability to use octadecylbis(2-hydroxyethyl)amine as sole carbon and energy source. Octadecylbis(2-hydroxyethyl)amine is also metabolized after an initial cleavage of the C_{alkyl}–N bond. A dehydrogenase converts octadecylbis(2-hydroxyethyl)amine into octadecanal and diethanolamine. The dehydrogenase requires DCPIP as artificial electron donor and appeared to be constitutive. Another dehydrogenase, effectively oxidizing the alkanal, produces dodecanoic acid. The route of the fatty acid degradation is by β-oxidation. Diethanolamine is not attacked by the isolate. The elucidated reaction sequence is shown in Figure 3. Closed-bottle test results obtained with ethoxylated fatty amines with various numbers of oxyethylene groups indicate that biodegradation of all ethoxylated fatty amines is initiated by the degradation of the alkyl chain [20].

G. Quaternary Ammonium Salts Containing Ester Bonds

Fatty acid esters of quaternary ammonium compounds with polyalcoholic groups (ester quats) are used extensively as softener. Although metabolic studies are not reported, degradation pathways of ester quats can be formulated, using research with mixed cultures and radiolabeled chemicals. Activated sludge extensively degraded ^{14}C-methyl-labeled di-(tallow fatty acid) ester of di-2-hydroxyethyldimethylammonium chloride (DEEDMAC) during 28 days of incubation [21]. Solid-phase extraction of quaternary ammonium salts, followed by thin-layer chromatography with radiochemical detection, clearly showed that the disappearance of DEEDMAC was sequentially followed by the appearance of tallow fatty acid ester of di-2-hydroxyethyldimethylammonium chloride and then di-2-hydroxyethyldimethylammonium chloride. Degradation of DEEDMAC was not detected in abiotic controls. The absence of abiotic degradation suggests that the hydrolysis of the ester bonds is biologically mediated. These results are consistent with the removal and degradation of fatty acids groups preceding mineralization of di-2-hydroxyethyldimethylammonium chloride [21]. The proposed pathway for DEEDMAC is presented in Figure 5.

The use of three ^{14}C-labeled di-(hardened tallow fatty acid)ester of 2,3-dihydroxypropyltrimethylammonium chloride (DEQ) provides information on the degradation pathway of DEQ. Incubation with river water demonstrated >80% of all ^{14}C-labeled DEQs was mineralized to carbon dioxide. In these die-away experiments, the microorganisms consumed the labeled DEQs within 10 days. The rate of disappearance varied with the location of the radiolabel. DEQ labeled in the fatty acid moiety was rapidly degraded without a detectable lag period. Mineralization of ^{14}C-methyl– and ^{14}C-dihydroxypropyl–labeled DEQ was preceded by a lag period, followed by extensive degradation. The mineralization of ^{14}C-methyl– and ^{14}C-dihydroxypropyl–labeled DEQ following that of the ^{14}C fatty acid–labeled DEQ supports the removal of fatty acid moieties prior to the degradation of the quaternary ammonium compound [22].

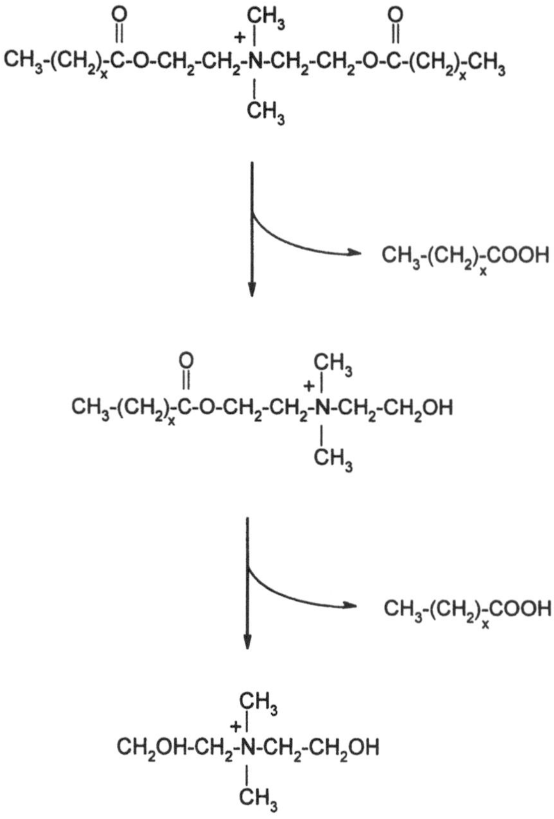

$$CH_3\text{-}(CH_2)\overset{}{\underset{x}{}}\text{-}\overset{\overset{\displaystyle O}{\|}}{C}\text{-}O\text{-}CH_2\text{-}CH_2\text{-}\overset{\overset{\displaystyle CH_3}{\underset{}{|}}}{\overset{+}{N}}\text{-}CH_2\text{-}CH_2\text{-}O\text{-}\overset{\overset{\displaystyle O}{\|}}{C}\text{-}(CH_2)\overset{}{\underset{x}{}}\text{-}CH_3$$

$$CH_3\text{-}(CH_2)_x\text{-}COOH$$

$$CH_3\text{-}(CH_2)_x\text{-}\overset{\overset{\displaystyle O}{\|}}{C}\text{-}O\text{-}CH_2\text{-}CH_2\text{-}\overset{\overset{\displaystyle CH_3}{\underset{\underset{\displaystyle CH_3}{|}}{|}}}{\overset{+}{N}}\text{-}CH_2\text{-}CH_2OH$$

$$CH_3\text{-}(CH_2)_x\text{-}COOH$$

$$CH_2OH\text{-}CH_2\text{-}\overset{\overset{\displaystyle CH_3}{\underset{\underset{\displaystyle CH_3}{|}}{|}}}{\overset{+}{N}}\text{-}CH_2\text{-}CH_2OH$$

FIG. 5 Microbial degradation pathway of DEEDMAC. The hydrolysis of the ester bonds giving rise to fatty acids and diol quaternaries is probably a general mechanism for ester quats.

H. Degradation of Hydrophilic Moieties

The key step in the pathways of cationic surfactants is the C_{alkyl}–N bond cleavage, or the hydrolysis, of the ester bond. The C_{alkyl}–N bond is cleaved by either a monooxygenase or a dehydrogenase. Hydrolysis of ester quats results in the release of fatty acids. Microbial metabolism of cationic surfactants therefore concurs with one biochemical strategy for the degradation. That is, the channeling of these structurally diverse fatty amine derivatives to fatty acids. Once the initial enzymatic degradative step has occurred, the cationic surfactant loses its toxicity. Hydrophilic products of initial transformation subsequently have to be broken down by other microorganisms because alkyl chain-degrading microorganisms usually do not degrade hydrophilic moieties.

Oxidation of the alkyl chain of ethoxylated fatty amines results in the production of polyethoxylated amines. A microorganism isolated from activated sludge was able to utilize diethanolamine as sole source of carbon and energy. Diethanolamine is degraded through ethanolamine, ethanolamine-*O*-phosphate to end products such as ammonium and acetate [23]. Other polyoxyethyleneamines have not been studied in detail. The biodegradation of alkylbenzyldimethylammonium salts is also initiated by a central cleavage, resulting in the formation of dimethylbenzylamine. Only a closed-bottle test result demonstrates that microorganisms metabolize this polar moiety. The alkyl chains of alkylmethylamines, monoalkyltrimethylammonium salts, and dialkyl-dimethylammonium salts are metabolized by pure cultures to give methylamines, which accumulate in the medium. Degradation of the methylamines by methylo-trophic microorganisms have been studied extensively [24].

Microorganisms metabolize trimethylamine by cleavages of the C_{methyl}–N bonds. Microorganisms employ a direct route by synthesizing trimethylamine dehy-drogenase [25,26]. This enzyme responsible for the initial degradation of trimethyl-amine yields dimethylamine and methanal. Other microorganisms possess a monooxygenase and a trimethyl-N-oxide aldolase catalyzing the conversion of trimethylamine into dimethylamine via trimethylamine-N-oxide [25,27]. Removal of a methyl group from dimethylamine was shown to be accomplished by secondary amine monooxygenase, yielding methanal and methylamine [25,28,29]. Finally, the cleavage of the C–N bond of methylamine can be achieved by a primary amine dehydrogenase [30,31]. The methanal formed is channeled through the central path-ways of methylotrophic microorganisms. Hydrolysis of di-(hardened tallow fatty acid) ester of 2,3-dihydroxypropyltrimethylammonium chloride results in the formation of 2,3-dihydroxypropyltrimethylammonium chloride. 2,3-Dihydroxypropyltrimethyl-ammonium chloride was biodegraded by a *Pseudomonas putida* strain to completion. This compound served as growth substrate, and surplus nitrogen was excreted as ammonium by the isolate [32].

I. Consortia of Microorganisms

Complete biodegradation of cationic surfactants requires the concerted action of at least two microorganisms because a single organism usually lacks the full complement of enzymatic capabilities. Slater and Lovatt [33] reviewed several studies in which single isolates and mixed microbial communities degraded organic compounds. In many of these studies the substrates were degraded more extensively by mixed cultures than by single isolates. The mechanisms for the enhanced biodegradation of mixed cultures are the provision of specific nutrients, removal of growth-inhibiting products, and the combined metabolic attack on the substrate. Degradation of decyltrimethyl-ammonium chloride by a pure culture of a *Xanthomonas* sp. was capable only in the presence of yeast extract. The unidentified cofactor present in yeast extract required for growth of the *Xanthomonas* sp. was produced by the other bacterium in the consortium [8]. This is an example of a consortium of microorganisms based on the supply of a nutrient. However, the main mechanism for biodegradation of surfactants by mixed cultures is the combined metabolic attack.

The limited metabolic capacities of pure cultures of microorganisms utilizing the moieties of the surfactants point to the requirement of consortia to degrade cationic surfactants completely. Consortia are required for the complete degradation of not only cationic surfactants. Examples of other surfactants degraded by consortia comprising at least two microorganisms are linear alcohol ethoxylates and linear alkylbenzene sulfonates [7].

To fully explore the hypothesis of concerted metabolic activity by at least two microorganisms to degrade fatty amine derivatives, comparative studies were per-formed with a two-membered reconstructed community [34]. A strain of *Burkholderia cepacia* utilizing dodecyldimethylamine as carbon and energy source released dimethylamine quantitatively. Correspondingly, four isolates that could utilize dimethylamine as sole source of carbon were unable to degrade the tertiary fatty amine. One dimethylamine-utilizing isolate was identified as a *Stenotrophomonas maltophilia*. A two-membered culture consisting of *Burkholderia cepacia* and *Stenotrophomonas maltophilia* is capable of complete mineralization of dodecyldimethylamine, as demonstrated by the carbon and nitrogen balances [34]. Batch culture experiments revealed that the two-membered culture consisting of *Burkholderia cepacia* and *Stenotrophomonas maltophilia* is based on a commensalistic relationship under carbon-limited conditions. Under nitrogen-limited conditions, this relationship is transformed from a commensalistic to a mutualistic one (Fig. 6). A two-membered

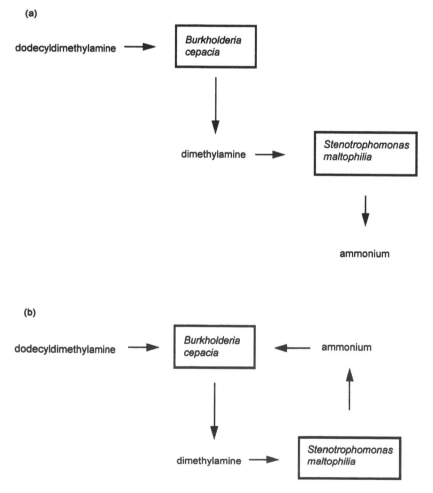

FIG. 6 A two-membered culture consisting of *Burkholderia cepacia* and *Stenotrophomonas maltophilia* growing on dodecyltrimethylamine as sole source of carbon and energy in (a) a commensalistic relationship, which is altered into (b) a mutualistic relationship under nitrogen-limited conditions.

culture is therefore imperative for growth on dodecyldimethylamine under nitrogen-limited conditions, whereas a pure culture of *Burkholderia cepacia* is capable of growth on dodecyldimethylamine under carbon-limited conditions [34].

III. BIODEGRADABILITY TESTS

A. Biodegradability Testing

Numerous standardized tests for different situations are used for assessing the biodegradation of chemicals. There are three testing method levels: ready biodegradability, inherent biodegradability, and simulation tests. In addition to analytical methods, biodegradation can be assessed using a number of chemically nonspecific methods. These include dissolved organic carbon (DOC) depletion, biological oxygen consumption, and carbon dioxide formation.

Ready biodegradability tests are primarily used for regulatory purposes. Test results are employed, for instance, to assess the environmental risks of chemicals. The closed-bottle test, Sturm test, and MITI I test have been specified to determine the ready biodegradability of organic compounds. Biodegradation is monitored, using oxygen consumption or carbon dioxide production as parameter, for 28 days. Biodegradation is expressed as the percentage of the carbon dioxide formation to the theoretical carbon dioxide evolution or the biological oxygen consumption to the theoretical oxygen demand.

Ready biodegradability tests use batch cultures, and initial concentrations of the test substance are in the range of 2–100 mg/L. The test compound is added to a mineral salts medium and is exposed to a very small inoculum of microorganisms. In these tests, biodegradation can be brought about only by microorganisms that are capable of growing on the test substance. A chemical that passes a ready biodegradability test is expected to biodegrade rapidly under most aerobic conditions.

The relationship between substrate carbon assimilation and oxidation to carbon dioxide is referred to as *yield* or *metabolic efficiency*. It is determined by the substrate's energy content, by the degrading organisms' biochemical pathways, and by growth conditions. During aerobic metabolism in ready biodegradability tests, the biomass yield ranges from 0.1 to 0.5 g biomass per gram of substrate. The biomass-incorporated carbon does not consume oxygen or produce carbon dioxide during the test period. Biodegradation percentages of > 60 are therefore interpreted as complete or near-complete biodegradation.

Failure to achieve biodegradation in ready biodegradability tests may be due either to recalcitrance of the test substance or to the severe conditions imposed during the testing. Inherent biodegradability tests, therefore, are to satisfy conditions required to enable biodegradation, thereby making it possible to assume that any failure to observe biodegradation is due to recalcitrance. Any technique may be employed to demonstrate inherent biodegradability, but only three test methods have become standard. The three standardized methods corresponding to the demands of inherent biodegradability assessment are MITI II, Zahn–Wellens, and SCAS tests. In SCAS units, which are operated on a 24-hour cycle, the test substance is fed once a day to activated sludge, mimicking biological treatment. Zahn–Wellens and SCAS tests measure overall DOC removal via both biodegradation and adsorption to sludge, and the MITI II test assesses biodegradation through oxygen consumption. Finally, two reactor types can be used to simulate biological treatment, i.e., the porous-pot reactor and an activated-sludge reactor with external clarifier. These reactors are used to measure biodegradability in the CAS (confirmatory) test. In the CAS test, a specific concentration of surfactant spiked to sewage is continuously introduced into the reactor and the overflow effluent is analyzed for DOC or surfactant. The use of specific analysis and DOC implies that observed disappearance may be due to biodegradation and/or sorption onto solids. These tests are designed to simulate activated-sludge treatment.

B. Apparent Recalcitrance in Ready Biodegradability Tests Due to Inhibition

The use of cationic surfactants as antibacterial agents represents their first important application. The primary mechanism by which cationic surfactants are toxic to microorganisms is through disruption of their cell membrane [35]. Cationic surfactants are therefore inhibitory to microorganisms capable of degrading these compounds. Consequently, many experiments with cationic surfactants yield negative results due to toxicity [2,8,36]. Substrate inhibition resulting from high initial concentrations in ready biodegradability tests is therefore a very important barrier to overcome if the ready biodegradability of cationic surfactants is to be demonstrated.

Due to an initial concentration of 20 mg/L, octadecyltrimethylammonium salt is not biodegraded in a Sturm test [37]. However, examination of mineralization with radiolabeled octadecyltrimethylammonium chloride at a concentration of 10 μg/L resulted in extensive biodegradation. Biodegradation was determined by measuring the formation of $^{14}CO_2$ from the ^{14}C-labeled test substance. After 10 days, already >70% of the radioactivity was detected as $^{14}CO_2$. Larson [37] has also demonstrated the potential of anionic surfactants to circumvent quaternary ammonium salt toxicity during ready biodegradability testing. The formation of a cationic/anionic complex reduces the concentration of octadecyltrimethylammonium salt in the water phase. Evidently, the inhibitory effect of the substrate at 20 mg/L was prevented, as shown by a biodegradation percentage of >70% obtained after an incubation period of 28 days.

Oxygen consumption by activated sludge in the absence and presence of 2 mg/L hexadecyltrimethylammonium chloride as obtained in a closed-bottle test illustrates the toxicity of this cationic surfactant. Inhibition of the respiration of the activated sludge was detected during a test period of 28 days [38]. The toxicity of hexadecyltrimethylammonium chloride may be reduced through the addition of 2 g silica gel in the closed bottles. In this test, a lag period of only one week was observed, and hexadecyltrimethylammonium chloride was biodegraded >60% at day 28 (unpublished Akzo Nobel results). Adsorption on to silica gel removes cationic surfactants from the aqueous phase and thus may enhance biodegradation. However, adsorption may also hinder biodegradation by making the compound unavailable [39].

In natural waters, 50–80% of the dissolved organic matter is made up of fulvic and humic acids [40]. These humic substances have an anionic surfactant-like structure, containing both hydrophilic and hydrophobic domains. Because of this character, humic substances react with cationic surfactants, resulting in a reduction of toxicity to aquatic organisms [41]. Tallow(hydrogenated)benzyldimethylammonium salt showed no biodegradation at day 28 in the closed-bottle test. In contrast, by adding 2.5 mg/L of humic acid, the pass level of 60% required for classification as readily biodegradable, was achieved within 28 days.

It is clear that the concentration of cationic surfactants in the aqueous phase should not exhibit inhibitory effects in ready biodegradability tests in order to assess their biodegradation. Concentrations of 1 mg/L of the test chemical are allowed in the closed-bottle test, which is very low compared to the initial concentrations used in other ready biodegradability tests. The closed-bottle test is therefore a superior test for cationic surfactants. The addition of an anionic surfactant, silica gel, or humic acid further minimizes the initial concentration of quaternary ammonium salts in the aqueous phase, allowing proper biodegradation testing of cationic surfactants.

The toxicity accounts for the widespread failure to observe biodegradation of water-soluble quaternary ammonium compounds in ready biodegradability tests. Under environmental conditions, cationic surfactants are unlikely to pose a toxicity risk to microorganisms because these compounds will be present in the environment in the microgram-per-liter range. Consequently, negative results obtained in ready biodegradability tests described in the open literature are ignored in the next subsection, given that more appropriate tests with either anionic surfactant, silica gel, or humic acids are available. Provided they are successful, these test results are used irrespective of the fact that tests with negative results exist.

C. Compilation of Ready Biodegradability Test Results

The rationale behind the ready biodegradability tests is that any chemical passing these tests would be rapidly broken down during sewage treatment and in most aerobic ecosystems. For such an approach to be valid, tests have to be extremely stringent. The stringency of the ready biodegradability tests is ensured primarily by precluding the use of acclimated microorganisms, even though this is a natural phenomenon, and by minimizing the initial biomass concentration, which delays the onset of biodegradation and limits the microbial diversity. The 10-day time window concept has been introduced in ready biodegradability tests as a simple parameter to assess the rate of biodegradation. In order to pass the test, over 60% biodegradation has to be achieved within a period of 10 days immediately following the attainment of 10% biodegradation. Biodegradation of a hydrophilic product of cationic surfactants cannot begin until significant amounts of this product are formed by alkyl chain–degrading microorganisms. Due to the sequential degradation of the moieties of cationic surfactants, the requirement of the 10-day window for classification as readily biodegradable is stricter than intended. Sequential degradation does not occur during sewage treatment because products are continuously being released and support a permanent population of competent microorganisms. Because the fulfillment of the 10-day time window has limited relevance, a cationic surfactant is classified as readily biodegradable if at least 60% of initially added substrate, measured as oxygen consumption or carbon dioxide evolution, is removed after 28 days by unacclimatized microorganisms. This view is adopted in the new EU legislation on surfactants used in detergents [42].

A biological oxygen consumption of >70% of the theoretical oxygen demand required by octyltrimethylammonium chloride was attained in a 10-day period in the MITI I test [43]. For octadecyltrimethylammonium chloride complexed with an anionic surfactant, the carbon dioxide formation accounted for approximately 80% of the theoretical carbon dioxide production of the substrate within 28 days [37]. With humic acid present, biodegradation of octadecyltrimethylammonium chloride started almost immediately, reaching 60% biodegradation within 14 days (unpublished Akzo Nobel results). These satisfactory ready biodegradability test results and the scientific evidence that a single microorganism can degrade all alkyltrimethylammonium salts through a joint biodegradation pathway lead to the conclusion that all alkyltrimethylammonium salts are readily biodegradable.

In closed-bottle tests, on day 28 levels of degradation of >70% (didecyldimethylammonium chloride) and >60% (didodecyldimethylammonium bromide complexed with humic acid) were observed (unpublished Akzo Nobel results). These results show that both dialkyldimethylammonium salts can be classified as readily biodegradable. After 28 days, negligible biodegradation percentages of long-chain dialkyldimethylammonium salts (C_{14}–C_{18}) were achieved in closed-bottle tests. However, in prolonged closed-bottle tests, extensive biodegradation was noted for dicocodimethylammonium chloride and di(hydrogenated)tallowdimethylammonium chloride (DHTDMAC), with biodegradation percentages of >60 [13].

Degradation of polyoxyethylene(15)cocomethylammonium chloride was sustainable but did not exceed 30% after 28 days of incubation. In the prolonged closed-bottle test, >70 % of the theoretical amount of oxygen was measured after 120 days [3]. Polyoxyethylene(15)cocomethylamine, a structurally related compound, is also inherently biodegradable, as shown in the prolonged closed-bottle test [20]. Octadecylbis(2-hydroxyethyl)amine was >70% degraded within 28 days in the closed-bottle test [19,20].

Krzeminzki et al. [44] and Masuda et al. [43] have reported high biodegradation percentages for alkylbenzyldimethylammonium salts under aerobic conditions. After 10 days, >60% of the theoretical oxygen uptake of alkylbenzyldimethylammonium chlorides with alkyl chains ranging from C_8 to C_{14} were found in MITI I tests [43]. Longer alkyl chains of the quaternary ammonium salts increasingly impair biodegradation in MITI I tests [43]. The capacity of microorganisms to degrade alkylbenzyldimethylammonium salts was also assessed using closed-bottle tests. A biodegradation of >60% with decylbenzyldimethylammonium chloride was conclusively demonstrated within 6 days (Fig. 4). Using the closed-bottle test, tallow(hydrogenated)benzyldimethylammonium chloride complexed with humic acid degraded >60% of the theoretical oxygen demand (unpublished Akzo Nobel results). These ready biodegradability test results clearly indicate that alkylbenzyldimethylammonium salts (C_{10}–C_{18}) are readily biodegradable.

For primary fatty amines (C_8–C_{18}), biodegradation of >60% were obtained after an incubation period of 12 days [12]. The secondary fatty amines also exhibited considerable oxygen consumption [12]. Finally, alkyldimethylamines (C_{12}–C_{18}) studied were readily degraded, achieving >60% biodegradation in the MITI I and closed-bottle tests [12,17]. Closed-bottle tests inoculated with several sources from natural habitats, including soils, ditch water, marine water, and river waters, gave biodegradation of >60%, enabling ready biodegradability classifications of didodecylmethylamine and dodecyldimethylamine [17]. Proof that the alkyl chains of fatty amines can be degraded through β-oxidation by a single microorganism and the high biodegradation percentages permit classification of alkylamines and alkyldimethylamines as readily biodegradable.

Biodegradation of the ester quats and their hydrolysis products has been extensively studied using Sturm tests. Di-(tallow fatty acid) ester of di-2-hydroxyethyldimethylammonium chloride (DEEDMAC) is readily biodegradable, because approximately 80% of the theoretical carbon dioxide formation was produced in 28 days [21]. Puchta et al. [45] reported that N-methyl-N,N-bis(2-($C_{16/18}$-acyloxy)ethyl)-N-2-hydroxyethylammonium methosulfate is readily biodegradable. The biodegradation percentages for tris(hydroxyethyl)ammonium salt spanned a range of 76–94%. In another Sturm test, 85% of di-(hardened tallow fatty acid) ester of 2,3-dihydroxypropyltrimethylammonium chloride (DEQ) degradation could be accounted for as carbon dioxide. 2,3-Dihydroxypropyltrimethylammonium chloride produced was biodegraded to carbon dioxide, with a level of 80% [22]. These Sturm tests measured sufficient degradation to conclude that this ester quat and its hydrolysis product are readily biodegradable.

Appropriate biodegradation testing of cationic surfactants demonstrates that most cationic surfactants should be classified as readily biodegradable. Table 1 lists the cationic surfactants classified as readily biodegradable.

D. Two-Phase Biodegradation in Ready Biodegradability Tests

Knowledge of the sequential degradation of both moieties of fatty amine derivatives allows a better interpretation of biodegradation curves obtained in ready biodegradability tests. A two-phase degradation of octadecylbis(2-hydroxyethyl)amine was identified in a closed-bottle test using a special funnel to enable multiple measurements in a single bottle [19]. The degradation of the alkyl chain started at day 8 and was probably completed at day 15. After day 15 there was a second lag period of approximately one week, followed by further degradation to >60% at day 28. The second lag was required while microorganisms acclimatize to diethanolamine formed between days 8 and 15.

The biodegradation curve obtained with polyoxyethylene(15)tallowamine is also characterized by two growth phases. Closed-bottle test results obtained with ethoxylated fatty amines with various numbers of oxyethylene groups suggest rapid mineralization via an initial oxidation of the alkyl chain [20]. The intermediates formed, viz. ethoxylated secondary amines, were biodegraded slowly. Finally, two-phase growth on polyoxyethylene(15)cocomethylammonium chloride was demonstrated in a prolonged closed-bottle test [3]. Initially the alkyl chain of this cationic surfactant is probably oxidized. This oxidation is followed by a phase of slow oxidation of the hydrophilic moiety. According to the evidence presently available, microorganisms readily oxidize the alkyl chains of fatty amine derivatives. Although chemically linked, the hydrophobic alkyl chain(s) and hydrophilic moiety of a cationic surfactant should be classified separately. An overview representing this approach is given Table 1.

E. Removal in Biological Treatment Systems

A key factor involved in controlling the environmental levels of cationic surfactants is the efficiency of wastewater treatment systems. Biological treatment systems constitute

TABLE 1 Biodegradation Classification of Various Cationic Compounds and Their Hydrophobic and Hydrophilic Moieties

Cationic compound	Both moieties	Alkyl chain	Hydrophilic moiety	
			Identity	Classification
Alkyltrimethylammonium salts (C_{10}–C_{18})	Ready	Ready	Trimethylamine	Ready
Dialkyldimethylammonium salts (C_{10}–$_{12}$)[a]	Ready	Ready	Dimethylamine	Ready
Alkylbenzyldimethylammonium salts (C_{10}–C_{18})	Ready	Ready	N,N-Dimethylbenzylamine	Ready
Di-(tallow fatty acid) ester of di-2-hydroxyethyldimethyl ammonium chloride	Ready	Ready	Dihydroxyethyldi-methylamimonium salt	Ready
N-Methyl-N,N-bis(2-$C_{16/18}$-acyloxy)ethyl)-N-(2-hydroxyethyl ammonium methosulfate	Ready	Ready	Trihydroxyethyl-methylammonium salt	Ready
Di-(hardened tallow fatty acid)ester of 2,3-dihydroxypropyl trimethylammonium chloride	Ready	Ready	2,3-Dihydroxypropyl-trimethylammonium salt salt chloride	Ready
Polyoxyethylene(15)alkylamines	Inherently	Ready	Ethoxylated amines	Inherently
Alkylbis(2-hydroxyethyl)amines	Ready	Ready	Diethanolamine	Ready
Alkyldimethylamines	Ready	Ready	Dimethylamine	Ready

[a] Long-chain cationic surfactants are not classified as readily biodegradable due to their limited bioavailability.

common practice in industrialized countries. These treatment systems use naturally occurring microorganisms to convert organic compounds. The dominant feature of biological treatment is the feedback of most of the sludge containing the microorganisms to the aeration tank. This is usually achieved by removing the sludge from the treated wastewater via settling. The aeration tank serves to encourage rapid adsorption, uptake, and especially oxidation of the compounds present in the wastewater. Excess sludge is wasted from the system to maintain the appropriate concentration of activated sludge. This process is known as activated-sludge treatment. Activated-sludge treatment is usually a secondary process, treating an influent, which has already been subjected to primary settling, to remove grosser solid materials.

Adsorption of cationic surfactants to solids plays an important role in biological treatment systems because these compounds readily adsorb onto particles, which are mostly negatively charged. Non-ready biodegradability therefore does not equate with inefficient operation of biological treatment plants, but unfortunately this view is still being advanced. According to a monitoring study in full-scale activated-sludge plants, approximately 10% was already removed in the primary settling tank [46]. Huber [47] reported removal of cationic surfactants in primary settling tanks ranging from 20% to 40%. Consequently, only part of the cationic surfactants present in raw sewage is passed onto the secondary activated-sludge treatment (biological).

More than 95% of quaternary ammonium salts passed onto aeration tanks are immediately removed by means of adsorption on activated-sludge particles. Larson and Vashon [48] concluded that adsorption of quaternary ammonium salts does not render these compounds unavailable to biodegradation, pointing to the importance of both biodegradation and adsorption as removal mechanisms. The overlap of biodegradation and adsorption in activated-sludge systems has been studied with radiolabeled DHTDMAC in order to estimate the extent of biodegradation in activated-sludge plants [49]. Both adsorption and biodegradation contribute to the efficient removal of DHTDMAC from the wastewater, as demonstrated in a semibatch reactor fed daily with a synthetic sewage without removing effluent from the reactor. A sharp decrease of DHTDMAC in the water phase is the result of adsorption. The subsequent biodegradation of this dialkyldimethylammonium salt is shown by a $^{14}CO_2$ recovery of approximately 40%. The almost complete removal of this dialkyldimethylammonium salt in activated-sludge plants is due to an initial rapid adsorption onto waste activated-sludge solids. The adsorbed DHTDMAC, in turn, is at least partly biodegraded in sewage treatment plants. Because of their strong adsorption and lower biodegradation rate, DHTDMAC was found at relatively high concentrations on the activated-sludge particles.

Adsorption of octadecyltrimethylammonium salt is also rapid, because removal of more than 98% of the cationic surfactant added to the SCAS unit during the aqueous phase was achieved within 90 minutes. To show unequivocally whether or not biodegradation of alkyltrimethylammonium salts was also responsible for the removal, an experiment with ^{14}C-radiolabeled octadecyltrimethylammonium chloride was performed. The extent of biodegradation of octadecyltrimethylammonium chloride ranged from 60% to 90%. In the SCAS test, $^{14}CO_2$ was produced within a few hours. [50]. This rate of biodegradation is sufficiently fast to support biodegradation in activated-sludge plants. This finding is in agreement with the ready biodegradability of alkyltrimethylammonium salts. Alkylbenzyldimethylammonium salts are also readily biodegradable (Table 1). The expected biodegradability of alkylbenzyldi-

TABLE 2   Removal/Biodegradation of Cationic Surfactants in Continuous-Flow
Activated-Sludge Systems (CAS) and Semicontinuous Activated-Sludge (SCAS) units

Test compound	Test	Removal (%)	Ref.
Hexadecyltrimethylammonium bromide	CAS	91–98	52
Hexadecyltrimethylammonium bromide	CAS	98–99	53
Hexadecyltrimethylammonium bromide	CAS	100	54
Octadecyltrimethylammonium chloride	SCAS	98	50
Didecyldimethylammonium chloride	CAS	95	52
Dioctadecyldimethylammonium chloride	CAS	95	52
Dioctadecyldimethylammonium chloride	CAS	91–93	55
Dodecylbenzyldimethylammonium chloride	CAS	96	52
Tetradecylbenzyldimethylammonium chloride	CAS	> 70	56
Cocobenzyldimethylammonium chloride	CAS	94	57
Di-(tallow fatty acid) ester of di-2-hydroxyethyldimethylammonium chloride	CAS	> 99	21
N-Methyl-N,N-bis(2-($C_{16/18}$-acyloxy)ethyl)-N-hydroxyethylammonium methosulfate	CAS	> 90	45
Ditallow ester of 2,3-dihydroxypropanetrimethylammonium chloride	CAS	> 99	22

methylammonium salts was evident in an SCAS test performed with a ^{14}C-radio-labeled compound. After an adaptation period of a few days, 80% of the ^{14}C added to the SCAS unit was converted to [14]CO_2 [44]. Therefore, under normal operating conditions, the major part of alkyltrimethylammonium salts and alkylbenzyldimethylammonium salts adsorbed on activated-sludge particles reaching the bioreactor is likely to be removed by biodegradation.

A substance judged as readily biodegradable will be removed from wastewater in biological treatment systems [51]. However, there is no evidence to support the opposite approach because there are many examples of chemicals not classified as readily biodegradable that are completely removed during biological wastewater treatment. In these cases, biological treatment simulations such as CAS or SCAS tests are appropriate to establish the removal in biological treatment systems. The CAS test in particular is thought to give realistic results similar to full-scale treatment. Results obtained in the CAS and SCAS tests with cationic surfactants are summarized in Table 2. These studies of cationic surfactant removal in activated-sludge systems demonstrate extensive removals.

IV.   KINETICS

A.   Half-Lives and Growth Rate

Simkins and Alexander [58] integrated six kinetic biodegradation models, three for growth and three for nongrowth situations. In addition, Schmidt et al. [59] developed kinetic models for the metabolism of organic compounds that do not support growth. The assignment of the appropriate models requires the measurement of accurate depletion curves, which are usually not available. Moreover, determination of kinetic constants is also difficult, due to the nature of surfactants, i.e., two substrates chemically linked together. Estimating kinetic parameters from biodegradation curves is possible and valid only when a single chemical substance is studied. When two compounds are involved, mixed kinetics can be expected. Moreover, these models assume that the substrate is bioavailable and ignore toxicity, both of which are important for cationic surfactants. Nevertheless, two kinetic approaches used to quantify biodegradation rates, i.e., maximum growth rate (μ_{max}) and first-order rate constant (k), are described because of their importance.

The first-order rate constant is an important parameter used to estimate the concentration of a chemical in natural ecosystems. First-order kinetics is observed when the rate of biodegradation of a chemical is directly proportional to the concentration. Mathematically, first-order degradation can be expressed as

$$C_t = C_0 e^{-kt}$$

where C_0 is the initial concentration, C is the concentration at time t, and k is the first-order rate constant. Using the first-order rate constant, a half-life of a chemical can be calculated via this equation:

$$t_{1/2} = \ln 2 / k$$

Only half-lives of alkyltrimethylammonium and dialkyldimethylammonium salts are reviewed, because the hydrophilic parts, i.e., methylamines, constitute a minor part of these surfactants. Kinetics obtained with ^{14}C-methyl–labeled quaternary ammonium salts is valid only when the degradation of methylamines is not rate limiting, which is probably true. Degradation of both hexadecyltrimethylammonium and octadecyltrimethylammonium salts in river water was rapid when determined using radiolabeled material. Rate constants were first order, with half-lives of 3 days. Distearyldimethylammonium choride underwent biodegradation in river water, with a half-life of 14 days. Biodegradation of the monoalkyltrimethylammonium salts in freshwater sediments was demonstrated, with predicted half-lives of 3 days. The half-life for distearyldimethylammonium chloride in the presence of sediment was 5 days [37,60]. More recently, alkyltrimethylammonium salts degraded with an estimated half-life of less than 10 hours using river water. The degradation of alkyltrimethylammonium salts was also examined in sediments and sludge-amended soils. Degradation in sediment and soil took place, with half-lives of 1–5 and approximatley 28 days, respectively [61]. Rapid removal of alkyltrimethylammonium salts by biodegradation was confirmed in a stream dosing study [61].

The toxic effects of cationic surfactants are probably negligible in the closed-bottle test due to the low initial concentration. When degraded by activated sludge in closed-bottle tests, water-soluble quaternary ammonium salts can have a half-life varying from less than one day to a few weeks. For instance, the estimated half-life for biodegradation of didecyldimethylammonium chloride in water inoculated with activated sludge was 0.5 days (Fig. 7).

The growth rate of the competent microorganisms is needed to generate an estimate of the efficiency of a biological treatment plant, which is determined primarily by the sludge retention time (SRT). The relationship between microbial growth rate

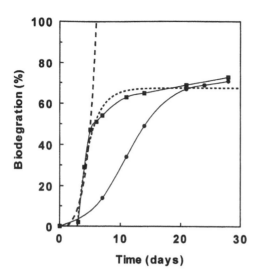

FIG. 7 Biodegradation of didecyldimethylammonium chloride in the absence (■) and presence of silica gel (●) in closed-bottle tests. The exponential dotted line represents the fit used to calculate the maximum growth rate using only the biodegradation percentages at days 3, 4, and 5. The half-life is calculated with the dotted logarithmic curve obtained with biodegradation percentages of days 3 to 28.

and SRT is given as $\mu = 1/\text{SRT}$ [62,63]. Processes linked to growth are usually described mathematically by the Monod equation:

$$\mu = \mu_{max} S/K_s + S$$

where μ is the specific growth rate, μ_{max} is the maximum specific growth rate, S is the substrate concentration, and K_s is a constant that represents the affinity for the substrate. Monod assumes that the compound sustains growth and is the only source of carbon. Because ready biodegradability tests are in essence batch-enrichment cultures, a growth rate may be estimated from the curves. Assuming that the competent microorganisms in the inoculum of ready biodegradability tests manifest a K_s much smaller than the initial substrate concentration, a logarithmic oxygen depletion or carbon evolution curve can be expected. Using closed-bottle test results, maximum growth rates of microorganisms capable of growth on the alkyl chains of alkyltrimethylammonium and water-soluble dialkyldimethylammonium salts were estimated to range from 0.1 to 0.5 day^{-1}. The growth curve obtained with didecyldimethylammonium chloride allows an estimation of 0.3 day^{-1} as maximum growth rate (Fig. 7). Accurate maximum growth rates are determined for a limited number of bacteria capable of utilizing the alkyl chains of fatty amine derivatives. A *Pseudomonas* capable of growth on hexadecyltrimethylammonium chloride degraded this compound at a maximum rate of 3 day^{-1} [9]. A maximum growth rate of 5 day^{-1} was established for an isolate utilizing didecyldimethylammonium chloride as sole source of carbon and energy [14]. Competent microorganisms degrading an organic compound will be retained in an activated-sludge plant only if its maximum growth rate is greater than the sludge wastage rate [63]. The sludge wastage rate in biological treatment systems ranges from 0.02 to 0.15 day^{-1}. Under these conditions, alkyl chain–degrading microorganisms maintain themselves easily. Consequently, removal by biodegradation in biological treatment systems has to be extensive.

B. Bioavailability

The chemical structure may have an indirect impact on how well the substrate will be biodegraded. Because some cationic surfactants are water insoluble, bioavailability is perhaps the key to determining their biodegradation potential. Bioavailability is defined as that fraction that is readily accessible to microbial degradation and exists in a dynamic equilibrium between the aqueous and solid phases. Organic compounds are probably bioavailable only when present in the aqueous phase, although unifying principles on how bioavailability affects biodegradation have not been established [64].

Due to the water insolubility and the capacity to adsorb, long-chain dialkyldimethylammonium salts such as DHTDMAC are not expected to be readily available to organisms for biodegradation. In the prolonged closed-bottle test, biodegradation of DTHDMAC resulted in a linear curve, indicating that desorption of the substrate is rate limiting to biodegradation [13]. Oxygen consumption by didecyldimethylammonium-grown cells of the *Achromobacter* sp. after supplying dialkyldimethylammonium salts with varying alkyl chain lengths also indicates that only dissolved dialkyldimethylammonium salts are mineralized. The substrate respiration in the presence of dioctadecyldimethylammonium salt was minor, whereas the water-soluble didecyldimethylammonium chloride was oxidized at high rates [14].

Bioavailability and toxicity are proposed as the main characteristics responsible for the results observed in closed-bottle tests with water-soluble cationic surfactants. Many water-soluble cationic surfactants are toxic in biodegradation tests [2,8,36]. Inhibitory effects of cationic surfactants are prevented in ready biodegradability tests with anionic surfactants, humic acid, and silica gel. However, it should be realized that adsorption of didecyldimethylammonium chloride onto silica gel decreases the bioavailability to such an extent that the degradation kinetics become more or less linear instead of logarithmic. In the absence of silica gel, a "logarithmic" growth curve was observed after a lag period of 3 days (Fig. 7).

DHTDMAC, a water-insoluble cationic surfactant, has no toxic effects on the microorganisms in the closed-bottle test at 2.0 mg/L [13]. This is most likely due to the limited bioavailability. In addition, DHTDMAC did not inhibit biogas formation even at 100 mg/L [65]. Inhibition of biogas formation by cationic surfactants decreases with longer alkyl chain lengths, supporting the view that nonbioavailable cationic surfactants do not exert toxic effects [66]. Water-insoluble cationic surfactants probably have no impact on microbial communities in the environment. Reduced bioavailability is also responsible for reduced toxicity of cationic surfactants to other organisms [41,61].

A new method was developed to test the hypothesis that water-insoluble dialkyldimethylammonium salts are available to degradative microorganisms when desorbed. To eliminate the influence of the limited bioavailability in biodegradability testing, a flow-through system consisting of a storage vessel with aerated river water, a pump, a column with a quaternary ammonium salt adsorbed on silica gel, and a collecting vessel was used [14]. Aerated river water was pumped through the column-packed silica gel particles with adsorbed quaternary ammonium salts. River water fed

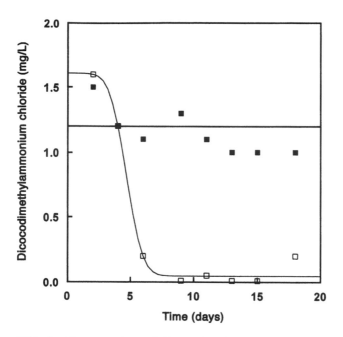

FIG. 8 Concentration of dicocodimethylammonium chloride in effluents of flow-through units fed with river water (□) and sterilized river water (■). The columns contained silica gel with adsorbed dicocodimethylammonium chloride (25 mg/g silica gel).

to the column used as control was sterilized to exclude microbial action. Under steady-state conditions, dicocodimethylammonium salts was eluted from sterile columns. In columns fed with river water, dicocodimethylammonium chloride can be degraded when desorbed from the silica gel. The biodegradation of dicocodimethylammonium chloride was evident from the marked decline in the quaternary ammonium salt concentration in the effluent after a few days (Fig. 8). Adsorbed long-chain dialkyldimethylammonium salts (C_{14}–C_{18}) resist biodegradation for extended periods of time in spite of the presence of many microorganisms capable of degrading these compounds. Negative biodegradability results in standard screening tests are therefore often interpreted as evidence of recalcitrance of long-chain dialkyldimethylammonium salts. In reality, adsorption of these salts to surfaces and the limited water solubility greatly reduces degradation rates. Realistic assessment of the biodegradation potential of nonbioavailable compounds should be based not only on well-known ready and inherently biodegradability tests. An understanding of adsorption/desorption onto solids is essential for interpreting the fate of these chemicals.

V. CONCLUSIONS

Human activity has provided cationic surfactants with the potential to sustain microbial growth. Microorganisms have probably evolved enzymes to degrade cationic surfactants readily because these compounds contain chemical structures in common with those occurring naturally, i.e., alkyl groups and methylamines. Ready biodegradability tests and metabolic studies have demonstrated that alkyltrimethylammonium chloride, dialkydimethylammonium chloride, alkylbenzyldimethylammonium chloride, ester quats, ethoxylated quaternary ammonium salts, and tertiary, secondary, and primary fatty amines are degraded completely by consortia of microorganisms. All research to date identifies hydrophilic moieties as degradation intermediates. The formation of hydrophilic moieties is achieved through one main pathway, namely, cleavage of the C_{alkyl}–N bonds or hydrolyzing the ester bonds. The central fission of the cationic surfactants immediately results in a detoxification due to loss of surfactancy. The possibility that initial conversions of cationic surfactants yields products with increased toxicity can be excluded. The occurrence of incomplete degradation and the accumulation of toxic intermediates are important factors in judging the environmental risks caused by surfactants.

Only a few long-chain cationic surfactants are slowly degraded in ready biodegradability tests, because of poor bioavailability (Table 1). Sorption of cationic surfactants to solid phases renders them less bioavailable to microorganisms. Microorganisms use these cationic surfactants as they dissolve in water. The rate of desorption restricts the biodegradation. Toxicity assessment of poor water-soluble compounds depends largely on the conditions under which the measurements are made. The mere presence of a chemical in an organism's environment is no guarantee that the chemical will be toxic. Indeed, water-insoluble cationic surfactants do not display toxic effects on microorganisms, and toxicity of water-soluble cationic surfactants is reduced by the presence of humic acids and silica gel. Recently, Alexander [67] argued that the risks of nonbioavailable compounds have been exaggerated, because the effect of bioavailability on toxicity is not reflected in current toxicity test methods.

Public concern about the presence of man-made chemicals in the environment is based on their toxicity. However, it should be realized that many compounds produced by living organisms are highly toxic. During the course of evolution, such toxic compounds came to serve as sources of carbon and energy for microorganisms. Biodegradation of toxic, naturally occurring organic compounds is therefore part of the carbon cycle. In this cycle, an enormous variety of organic compounds are produced through reduction of carbon dioxide, driven primarily by solar energy. Microbial degradation is an essential part of the carbon cycle in the continual synthesis, transformation, and biodegradation of many organic compounds. To arrive at a steady-state situation in ecosystems, a balance must exist between the rates of synthesis and biodegradation. The breakdown of cationic surfactants into carbon dioxide and water—thus becoming a part of the carbon cycle and enabling low environmental steady-state concentrations—is ideal. Most cationic surfactants are readily biodegraded, allowing low steady-state concentrations in the environment. This strongly indicates that the risks to the environment of cationic surfactants are very limited.

REFERENCES

1. Sanders, H.L.; Braunwarth, J.B.; McConnell, R.B.; Swenson, R.A. J. Am. Oil Chem. Soc. 1969, *46*, 167–170.
2. Swisher, R.D. *Surfactant Degradation*; Marcel Dekker: New York, 1987.

3. van Ginkel, C.G. *Biodegradability of Surfactants*; Porter, M.R., Karsa, R.D., Eds.; Blackie Academic & Professional: London, 1995; 183–203.

4. de Henau, H.; Masscheleyn, P. Larivista Italiana Sostanze Grasse 1997, *74*, 459–460.

5. Brunner, P.H.; Capri, A.; Marcomini, A.; Giger, W. Water Res. 1988, *22*, 1465–1472.

6. White, G.F. Pest. Sci. 1993, *37*, 159–166.

7. van Ginkel, C.G. Biodegradation 1996, *7*, 151–164.

8. Dean-Raymond, D.; Alexander, M. Appl. Environ. Microbiol. 1977, *33*, 1037–1041.

9. van Ginkel, C.G.; van Dijk, J.B.; Kroon, A.G.M. Appl. Environ. Microbiol. 1992, *58*, 3083–3087.

10. Ratledge, C. *Biochemistry of Microbial Degradation*; Ratledge, C., Ed.; Kluwer Academic Press: Dordrecht, 1994; 89–141.

11. Nishiyama, N.; Toshima, Y.; Ikeda, Y. Chemosphere 1995, *30*, 593–603.

12. Yoshimura, K.; Machida, S.; Masuda, F. J. Am. Oil Chem. Soc. 1980, *57*, 241–338.

13. van Ginkel, C.G.; Kolvenbach, M. Chemosphere 1991, *23*, 281–289.

14. vanGinkel, C.G.; Venema, M.A.; Geurts, M.G.J. Proceedings of the 5th World Surfactants Congress, Florence, Italy, 2000; 1408–1413.

15. Nishihara, T.; Okamoto, T.; Nishiyama, N. J. Appl. Microbiol. 2000, *88*, 641–647.

16. van Ginkel, C.G.; Haan, A.; Luijten, M.L.G.C.; Stroo, C.A. Ecotox. Environ. Saf. 1995, *31*, 218–223.

17. van Ginkel, C.G.; Pomper, M.A.; Stroo, C.A.; Kroon, A.G.M. Tenside Surf. Det. 1995, *32*, 355–359.

18. Kroon, A.G.M.; Pomper, M.A.; van Ginkel, C.G. Appl. Microbiol. Biotechnol. 1994, *42*, 134–139.

19. van Ginkel, C.G.; Kroon, A.G.M. Biodegradation 1993, *3*, 435–443.

20. van Ginkel, C.G.; Stroo, C.A.; Kroon, A.G.M. Tenside Surf. Det. 1993, *30*, 213–216.

21. Gionlando, S.T.; Rapaport, R.A.; Larson, R.J.; Federle, T.W.; Stalmans, M.; Masscheleyn, P. Chemosphere 1995, *30*, 1067–1083.

22. Waters, J.; Kleiser, H.H.; How, M.J.; Barratt, M.D.; Birch, R.R.; Fletcher, R.J.; Haigh, S.D.; Hales, S.G.; Marchall, S.J.; Pestall, T.C. Tenside Surf. Det. 1991, *28*, 460–467.

23. Williams, G.R.; Callely, A.G. J. Gen. Microbiol. 1982, *128*, 1203–1209.

24. Large, P.J. Xenobiotica 1971, *1*, 457–467.

25. Colby, J.; Zatman, L.J. Biochem. J. 1973, *132*, 101–112.

26. Boulton, C.A.; Crabbe, M.J.C.; Large, P.J. Biochem. J. 1974, *140*, 153–163.

27. Large, P.J.; Boulton, C.A.; Crabbe, M.J.C. Biochem. J. 1972, *128*, 137p–138p.

28. Myers, P.A.; Zatman, L.J. Biochem. J. 1971, *121*, 10p.

29. Eady, R.R.; Large, P.J. Biochem. J. 1969, *111*, 37P–38P.

30. Eady, R.R.; Large, P.J. Biochem. J. 1968, *106*, 245–255.

31. Hampton, D.; Zatman, L.J. Biochem. Soc. Trans. 1973, *1*, 667–668.

32. Kaech, A.; Egli, T. System Appl. Microbiol. 2001, *24*, 161–252.

33. Slater, J.H.; Lovatt, D. *Microbial Degradation of Organic Compounds*; Gibson, D.T., Ed.; Marcel Dekker: New York, 1984; 439–485.

34. Kroon, A.G.M.; van Ginkel, C.G. Environ Microbiol 2001, *3*, 131–136.

35. Hugo, W.B. SCI Monograph No. 19, Surface-Active Agents in Microbiology London Soc Chem Ind, 1965; 67–82.

36. Mackrell, J.A.; Walker, J.R.L. Int. Biodeterior Bull 1978, *14*, 77–83.

37. Larson, R.J. Residue Rev. 1983, *85*, 159–171.

38. van Ginkel, C.G.; Stroo, C.A. Ecotox. Environ. Saf. 1992, *24*, 319–327.

39. Alexander, M. *Biodegradation and Bioremediation*; Academic Press: New York, 1994, 149–158.

40. Buffle, J. *Complexation Reactions in Aquatic Systems: An Analytical Approach*; Ellis Horwood: Chichester, UK, 1988.

41. Suffet, I.H.; Javfert, C.T.; Kukkonen, J.; Servos, M.R.; Spacie, A.; Williams, L.L.; Noblet, J.A. *Bioavailability*; CRC Press: Boca Raton, FL, 1994; 73–108.
42. EU revision of detergent legislation, http://europa.eu.int/comm/enterprise/chemicals/detergents/index.htm.
43. Masuda, F.; Machida, S.; Kanno, M. Studies on the biodegradability of some cationic surfactants. Proceedings of VII International Congress on Surface-Active Substances 4, Moscow, 1976; 129–138.
44. Krzeminski, S.F.; Martin, J.J.; Brackett, C.K. Houshold Pers. Prod. Ind. 1973, *10*, 22–24.
45. Puchta, R.; Krings, P.; Sandkühler, P. Tenside Surf. Det. 1993, *30*, 186–191.
46. Topping, B.W.; Waters, J. Tenside Surf. Det. 1982, *19*, 164–169.
47. Huber, L.H. J. Am. Oil Chem. Soc. 1987, *61*, 377–382.
48. Larson, R.J.; Vashon, R.D. Dev. Ind. Microbiol. 1983, *24*, 425–434.
49. Sullivan, D.E. Water Res. 1983, *17*, 1145–1151.
50. Games, L.M.; King, J.E.; Larson, R.J. Environ. Sci. Technol. 1982, *16*, 483–488.
51. Brown, D. *Biodegradability of Surfactants*; Porter, M.R., Karsa, R.D., Eds.; Blackie Academic & Professional: London, 1995; 1–27.
52. Gerike, P. Tenside Surf. Det. 1982, *19*, 162–164.
53. Brown, D. The assessment of biodegradability. In A consideration of possible criteria for surface active substances. Proceedings of VII International Congress on Surface-Active Substances 4, Moscow, 1976; 44–57.
54. Pitter, P.; Svitalkova, J. Sb VSChT 1961, *52*, 25–42.
55. May, A.; Neufahrt, A. Tenside Surf. Det. 1976, *13*, 65–69.
56. Fenger, B.H.; Mandrup, M.; Rohde, G.; KjaerSorensen, J.C. Water Res. 1973, *7*, 1195–1208.
57. Janicke, W.; Hilge, G. Tenside Surf. Det. 1979, *16*, 117–122.
58. Simkins, S.; Alexander, M. Appl. Environ. Microbiol. 1984, *47*, 1299–1306.
59. Schmidt, S.K.; Simkins, S.; Alexander, M. Appl. Environ. Microbiol. 1985, *50*, 323–331.
60. Larson, R.J. *Curr. Perspect. Microb. Ecol*; Klug, M.J., Adinarayana, R.J., Eds.; Proceedings Int Symp 3rd, 1984; 677–686.
61. Woltering, D.M.; Bishop, W.E. *The Risk Assessment of Environmental and Human Health Hazards: A Textbook of Case Studies*; Pautenbach, D.J., Ed.; Wiley: New York, 1989; 345–389.
62. Birch, R.R. J. Chem. Tech. Biotechnol. 1991, *50*, 411–422.
63. Pirt, S.J. *Principles of Microbe and Cell Cultivation*; Blackwell Scientific: Oxford, 1985.
64. Mihelcic, J.R.; Lueking, D.R.; Mitzell, R.J.; Stapleton, J.M. Biodegradation 1993, *4*, 141–153.
65. Garcia, M.T.; Campos, E.; Sanchez-Leal, J.; Ribosa, I. Chemosphere 2000, *41*, 705–710.
66. Garcia, M.T.; Campos, E.; Sanchez Leal, J.; Ribosa, I. Chemospere 1999, *38*, 3473–3483.
67. Alexander, M. Environ. Sci. Technol. 2000, *34*, 4259–4265.

20

Biodegradability of Amphoteric Surfactants

ANDREAS DOMSCH and KLAUS JENNI Degussa Goldschmidt
Personal Care, Essen, Germany

I. INTRODUCTION

Compounds with both acidic and alkaline properties, i.e., the ability to release or bind protons, are described as amphoteric electrolytes or ampholytes. In acidic solutions they form cations, in alkaline solutions anions, and in the mid-pH range *zwitterions*, i.e., molecules with two ionic groups with equivalent charges. Molecules to which this principle is applied on the hydrophilic group and which, at the same time, contain a hydrophobic fatty chain are known as *amphoteric surfactants*.

In this chapter, amphoteric surfactants, or in short amphoterics, are described. There are two main groups manufactured and used commercially: real amphoterics and betaines. The key functional groups are the more or less quaternized nitrogen, derived from an amine, and the carboxylic group. The carboxylic group can be replaced by the sulfonate or the phosphate group, resulting in sulfobetaines and phosphobetaines, respectively.

Amphoterics are surfactants with an ionic charge. Depending on the pH value, they can change between anionic character, the isoelectric neutral stage, and the cationic character. But in the isoelectric neutral stage, they are ionic substances and not nonionics.

The first main group of amphoterics is the group of real amphoterics; i.e., they have one ionic group that determines the character of the structure. Depending on the pH value, either anionics or cationics are formed. At a high pH value the carboxyl group is ionized and a more anionic surfactant results; at a low pH value the amino group is ionized and a cationic surfactant resembles a quaternary ammonium compound. Between these extremes there is a certain pH range in which the molecule has a neutral charge. This pH is the isoelectric range, depending on the alkalinity of the nitrogen atom and the acidity of the carboxylic function in the given structure.

The betaines are the second group among the amphoterics. Because of the fully quaternized nitrogen, they are present in the form of *zwitterions* only. They are inner salts; i.e., they consist of two functional ionic groups with opposite electric charge in one molecule. The difference from the real amphoterics is that an increase in pH does

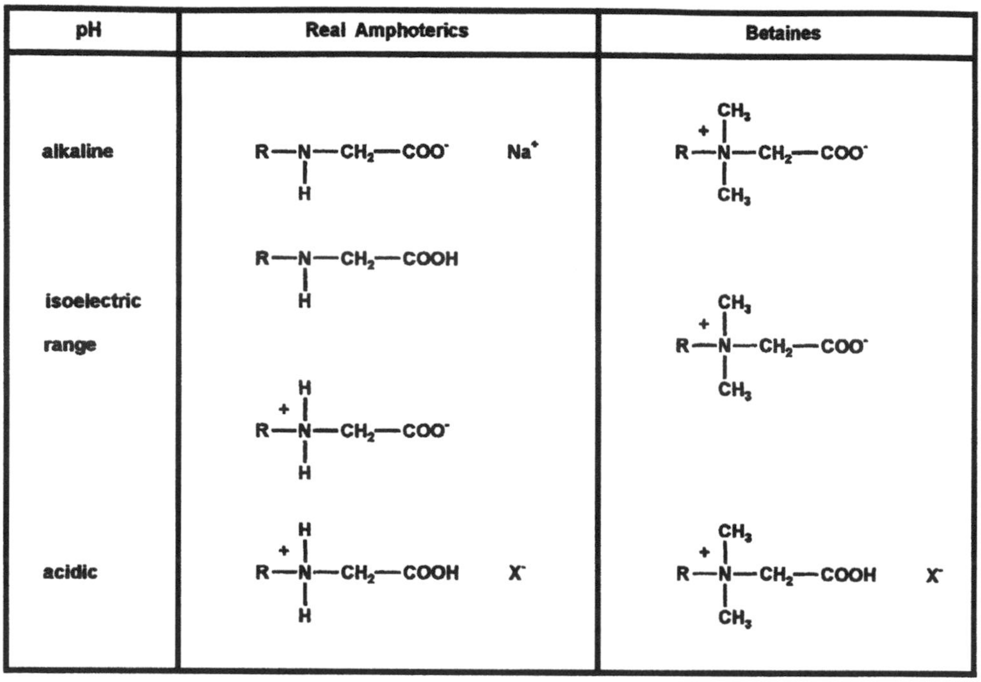

pH	Real Amphoterics	Betaines
alkaline	R—N—CH$_2$—COO$^-$ Na$^+$ | H	CH$_3$ $^+$| R—N—CH$_2$—COO$^-$ | CH$_3$
isoelectric range	R—N—CH$_2$—COOH | H H $^+$| R—N—CH$_2$—COO$^-$ | H	CH$_3$ $^+$| R—N—CH$_2$—COO$^-$ | CH$_3$
acidic	H $^+$| R—N—CH$_2$—COOH X$^-$ | H	CH$_3$ $^+$| R—N—CH$_2$—COOH X$^-$ | CH$_3$

FIG. 1 Influence of pH value on the structure of amphoterics.

not give anionic properties and the quaternization of the nitrogen is independent on the pH value. The structure of real amphoterics and betaines, as well as the influence of the pH value on it, is shown in Figure 1.

All betaines can be regarded as derivatives of "betaine"—*N,N,N*-trimethylglycine—a natural substance occurring in the sugar beet, *Beta vulgaris*. In the simplest case, one of the methyl groups is replaced by a long alkyl chain, forming the alkyl betaines.

Because of the less polar structure, compared to anionics, amphoterics are dermatologically mild surfactants. They can form complexes with anionic surfactants and are able to reduce their irritative properties. Therefore the main use is as mild surfactants in cosmetics and toiletries or hand dishwashing liquids.

But beside mildness, amphoterics are important surfactants in some detergents, especially in light-duty detergents and special wool care products. Special surface-active properties are the reason for these applications: Amphoterics foam strongly and have an excellent capacity to disperse or emulsify oils and fats. They are very effective cleaning agents, even in extreme pH ranges.

II. STRUCTURAL ELEMENTS AND BIODEGRADATION IN GENERAL

The fatty chain is degraded by the mechanism of ω- or β-oxidation, as described, e.g., for linear alkylbenzene sulfonate [1]. This mechanism is generally valid for all surfactants based on natural oils and fats. After ω-oxidation, a low-molecular-weight hydrophilic molecule remains. In case of ω-oxidation of alkyl betaine (structure 1 in Fig. 3), betaine (*N,N,N*-trimethylglycine) results. For this naturally occurring compound the degradation to methane, carbon dioxide, and ammonia is described [2].

Different authors are examining the ability of microorganisms to decompose biological compounds containing quaternary nitrogen, like betaine, choline, carnitine, and butyrobetaine. With this the decomposition of alkyl betaines and related compounds can be understood, assuming that these metabolites are a part of the pathway.

In contrast to anionic and nonionic surfactants, there is no simple analytical method available to determine the surface-active properties, like methylene blue reaction or bismuth reaction for anionics and nonionics, respectively. This could be the reason why very limited information about the primary degradation is published. There is only a polarographic method with which to analyse the active content of amphoterics after a biodegradation process [3]. Specifically for alkylamidobetaines, the orange II method is recommended [5,22]. In the case of polycarboxyglycinates, HPLC has been used to measure the loss of active substance during degradation, which is regarded as primary degradation.

Amphoterics in general have a certain antimicrobiological activity because of the quaternized nitrogen. This has to be taken into consideration for all biodegradation tests. Acclimatization to the bacteria is therefore necessary in order to get correct figures for the ultimate degradability.

Comparing the different types of amphoterics described later, it can be assumed that every type is degraded more or less according to the same mechanism of decomposition, if the conditions of the test are constant as well as the lipophilic part of the molecule. Results on the basis of the OECD test 302 B (Zahn–Wellens test) are shown in Figure 2 for alkylamidobetaine (structure II), alkylamphoacetate (structures IX and X), and alkylamphopropionate (structures XVI and XVII), all with the same cocoalkyl chain as the lipophilic part of the molecule, illustrating nearly the same degradation kinetics.

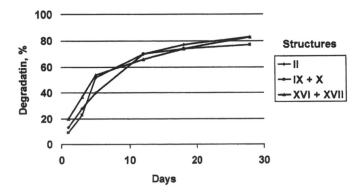

FIG. 2 Amphoterics—Biodegradation.

III. ALKYL BETAINES

A. Chemical Structure

Alkyl betaines can be regarded as derivatives of betaine (*N,N,N*-trimethylglycine) in which one methyl group is replaced by a long alkyl chain. In the preparation process, equivalent quantities of long-chain tertiary amines and sodium chloroacetate are heated in aqueous solution at 70–80°C for several hours. The general formula of the reaction is given in Figure 3. The final product is an aqueous solution of the alkyl betaine and contains an equimolar amount of sodium chloride. For most applications this is accepted. It is also possible to carry out the reaction in an alcoholic solution. In this case sodium chloride is precipitated and can be removed by filtration. Another possibility is removal via electrodialysis. But both processes are avoided because they are more complicated as the alternative route for obtaining salt-free amphoterics. This is described in Section VIII.

The R group in the molecular structure can be the alkyl chain derived from coconut oil or palm kernel oil (C_8–C_{18}), tallow (C_{16}–C_{18} and $C_{18'}$), palm oil (C_{16}–C_{18} and $C_{18'}$), or pure C_{12}-chain. This gives the different types of commercial products.

B. Properties, Application

Alkyl betaines have very good stability against water hardness and are excellent lime soap–dispersing agents. They are also stable at very low and high pH values and compatible with anionic, cationic, and nonionic compounds, especially surfactants. There is complex formation between alkyl betaines and anionic surfactants, which depends upon the isoelectric point. The amphoterics are also stable in systems containing high amounts of electrolytes.

With these properties and good detergency, the alkyl betaines are widely used in hard-surface cleaners and special textile detergents.

C. Primary Degradation

The primary degradation of alkyl betaines (structure I in Fig. 3) can be observed by measuring the loss of the surface-active substance during biodegradation using a polarographic method [3]. The result is given in Table 1. In general, alkylbetaines show good primary degradability.

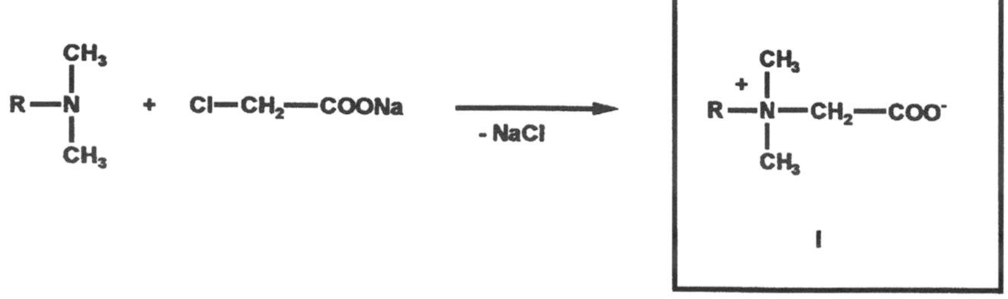

FIG. 3 Reaction scheme of alkyl betaines.

TABLE 1 Alkyl Betaines—Biodegradation

Substance	Method of degradation	Analysis	Degradation (%)	Evaluation of degradation	Ref.
I, R = C_{12}	CBT	Polarography	100	Primary	4
I, R = C_{12}	CBT	O_2	55	Nonreadily	4
I, R = C_{12}	STURM	CO_2	91	Readily	4
I, R = C_{14}	CBT	O_2	58	Nonreadily	4
I, R = C_{14}	STURM	CO_2	84	Readily	4
I, R = C_{16}	CBT	O_2	45	Nonreadily	4
I, R = C_{16}	STURM	CO_2	84	Readily	4
I, R = C_{14-15}	CBT	O_2	52	Nonreadily	4
I, R = C_{14-15}	STURM	CO_2	81	Readily	4
I, R = cocoalkyl	CBT	BOD_{30}/COD	57	Nonreadily	18
I, R = cocoalkyl	MOST		>70	Readily	Henkel
I, R = cocoalkyl	CBT	BOD_{30}/COD	>60	Readily	Henkel
I, R = cocoalkyl	MOST	DOC	95	Readily	Clariant

I = structure I in Fig. 3; CBT = closed-bottle test; MOST = Modified OECD screening test 301 E; STURM = STURM test; OECD 303 A = OECD method 303 A = coupled-units test; O_2 = oxygen uptake; CO_2 = CO_2 evolution; BOD_{30} = biological oxygen demand in 30 days; COD = chemical oxygen demand; Henkel = Safety Data Sheet of Dehyton AB 30, Henkel KGaA, Düsseldorf; Clariant = Clariant GmbH, Frankfurt, private communication.

D. Ultimate Degradation

The degradation of alkyl betaines (structure I) is described with the closed-bottle test as well as with the STURM test [4]. The data are shown in Table 1. A small difference, but probably not a significant one, seems to be between the natural derived C-chain (C_{14}) and the synthetic one (C_{14-15}): The natural chain degrades in a given time to a higher extent. But this difference is minute, i.e., within the normal variation range of results from a biological test system. As a conclusion, alkyl betaines can be regarded as nonreadily biodegradable, but after a certain time of adaptation they are readily biodegradable too.

IV. ALKYLAMIDO BETAINES

A. Chemical Structure

The synthesis of alkylamido betaines is carried out in two steps. The first step is the condensation of a fatty acid or their esters (especially the methyl ester or the corresponding triglyceride) with dimethylaminopropyl amine. The second step is the reaction of this intermediate with sodium chloroacetate. The reaction product is an aqueous solution of alkylamidopropyl betaine containing the equimolar amount of sodium chloride. The sodium chloride can be removed, but very seldom for commercial products. The reaction scheme is given in Figure 4.

The link between the lipophilic and the hydrophilic part is an amide functional group, corresponding to the amide link in proteins. This results in different properties compared with alkyl betaines. The standard products in the range of alkylamido betaines are produced on the basis of R = cocoalkyl (see structural formula II in Fig. 4), mixed coco/oleoalkyl derivative, or pure C_{12}-derivative.

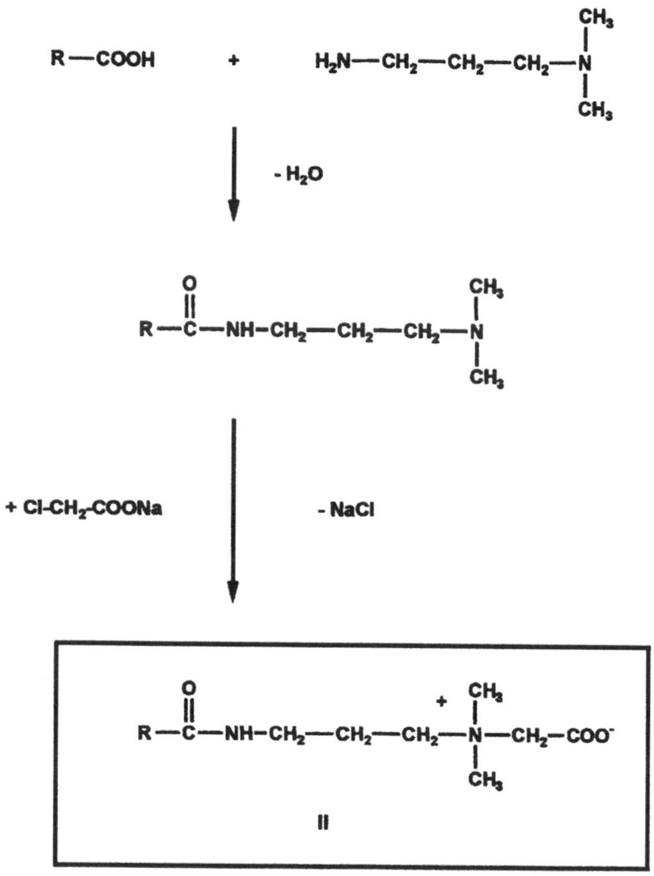

FIG. 4 Reaction scheme of alkylamido betaines.

B. Properties, Application

The basic properties are similar to those of the alkyl betaines; i.e., the alkylamido betaines show stability against electrolytes, acids, alkali, and water hardness. But the dermatological behavior is much better. Therefore the main use is in cosmetics and toiletries, but also in hand dishwashing liquids and special textile detergents.

The addition of alkylamido betaines to anionic surfactants, mainly sodium lauryl ether sulfates, increases the viscosity of the blend. With all of these properties, the alkylamido betaines are now the second most important surfactant in shampoos, shower gels, and foam baths or liquid soaps of higher quality.

C. Primary Degradation

The degradation of alkylamido betaines (structure II in Fig. 4) can be observed by measuring the loss of surface-active substance during biodegradation using the orange II method [5]. The result is given in Table 2. In general, alkylamido betaines show good primary degradability. The orange II method was recently updated [22], but no new results for primary biodegradation are reported.

D. Ultimate Degradation

The inherent biodegradation of cocoamidopropyl betaine (structure II in Fig. 4) was determined with the Zahn–Wellens test OECD 302 B, res. according to the German Standard DIN 38 412 Part 25—static test. The result for the cocoalkyl derivative is given in Table 2. In conclusion, the alkylamido betaines can be regarded as inherently biodegradable.

E. Anaerobic Degradation

If a surfactant is discharged to a recipient, it may end up in the sediment of the bottom layers, where an anaerobic degradation takes place. Because of this, in recent years both aerobic degradation and anaerobic degradation of chemical compounds are examined. This test was carried out according to the ECETOC method. The result for the cocoalkyl derivative is given in Table 2. In conclusion, the alkylamido betaines can be regarded as anaerobically biodegradable.

V. SULFOBETAINES AND HYDROXYSULFOBETAINES

A. Chemical Structure

In the past, the true sulfobetaines have been obtained by the reaction of tertiary amines with propane sultone (structure III in Fig. 5). However the sultones are regarded as carcinogenic; therefore this way is no longer used.

Today the usual procedure for preparation of sulfobetaines is the reaction of tertiary amines with chlorohydroxypropane sulfonic acid. The latter is obtained by the reaction of epichlorohydrin with sodium hydrogensulfite. The reaction scheme is shown in Figure 5.

This product group (structures IV and V in Fig. 5) should be described correctly as hydroxysulfobetaines to separate them very clearly from the former version, described in the first paragraph. Commercial products are available as hydroxysulfobetaines with R = cocoalkyl and synthetic C_{14-15}-alkyl (structures IV and V in Fig. 5) as an aqueous solution with molar equivalents of sodium chloride.

For two special sulfobetaines (structures VI and VII in Fig. 6), the biodegradation is described in the literature [6], but no commercial products of this type are available. For the structures see Figure 6.

Another sulfobetaine is based on ethoxylated primary amine, quaternized and sulfosuccinated (see structure VIII in Fig. 6).

B. Properties, Application

With good wetting properties and good stability against electrolytes, the hydroxysulfobetaines are especially of interest for household products. In the past their use was limited, but now there is increased interest in the use of these products.

C. Primary Degradation

The primary degradation of sulfobetaines (structure III) and hydroxysulfobetaines (structure IV) was determined by the loss of polarographic activity [3] in the closed-

TABLE 2 Alkylamidopropyl Betaine—Biodegradation

Substance	Method of degradation	Analysis	Degradation (%)	Evaluation of degradation	Ref.
II, R = cocoalkyl	OECD Conf.	Orange II	10 d: 98	Primary	
II, R = cocoalkyl		Orange II	7 h: 80–100	Primary	5
II, R = cocoalkyl	CBT (OECD 301 D)	BOD_{30}/COD	30 d: 84	Readily	18, IUCLID
II, R = cocoalkyl	CBT	BOD/COD	7 d: 30	Readily	19
			14 d: 51		
			21 d: 80		
			28 d: 82		
II, R = cocoalkyl	STURM (OECD 301B)		20 d: 100	Readily	IUCLID
II, R = cocoalkyl (basis fatty acid)	OECD 301 D	BOD	1 d: 8	Readily	Z & S, 1998
			7 d: 27		
			14 d: 64		
			21 d: 77		
			28 d: 86		
II, R = cocoalkyl (basis coconut oil)	OECD 301 D	BOD	1 d: 2	Readily	Z & S, 1999
			3 d: 30		
			5 d: 43		
			7 d: 54		
			10 d: 72		
			14 d: 84		
			21 d: 88		
			28 d: 94		
II, R = cocoalkyl	MOST (OECD 301E)	DOC	14 d: 90	Readily	IUCLID
			28 d: 100		
II, R = cocoalkyl	MOST		>70	Readily	Henkel

Compound	Test method	Measurement	Result	Classification	Source
II, R = cocoalkyl	MOST	DOC	90–94	Readily	Hoechst
II, R = soy alkyl	MOST	DOC	71	Readily	Hoechst
II, R = cocoalkyl	OECD 302 B	COD	7 d: 84 28 d: 99	Inherently	IUCLID
II, R = cocoalkyl	OECD 302 B	COD	3 h: 4 24 h: 13 3 d: 28 5 d: 40 12 d: 70 18 d: 77 28 d: 83	Inherently	
II, R = cocoalkyl	OECD 302 B	COD	1 d: 24–27 2 d: 67–75 3 d: 78–85 6 d: 95 7 d: 91–100	Inherently	Z & S
II, R = cocoalkyl	OECD 303 A	DOC removal	97	Ultimately	IUCLID
II, R = cocoalkyl	OECD 303 A	DOC removal	30 d: 71	Ultimately	IUCLID
II, R = cocoalkyl	Anaerobic degradation ECETOC	TIC	42 d: 25–61		
II, R = cocoalkyl	Anaerobic degradation ECETOC	TIC	42 d: 43		IUCLID
II, R = cocoalkyl	Anaerobic degradation ECETOC	TIC	22d: 80	Biodegradable	

II = structure II in Fig. 4; OECD Conf. = OECD confirmatory test; OECD 302 B = OECD method 302 B = Zahn–Wellens test; = Static test DIN 38412, part 25; OECD 303 A = OECD method 303 A = coupled-units test; CBT = closed-bottle test; MOST = modified OECD screening test 301 E; STURM = shake flask CO_2 evolution system (STURM test); COD = chemical oxygen demand; BOD_{30} = biological oxygen demand in 30 days; DOC = dissolved organic carbon; TIC = total inorganic carbon; Orange II = orange II method; Henkel = Safety Data Sheet of Dehyton K, Henkel KGaA, Düsseldorf; Hoechst = Hoechst AG, Frankfurt, private communication; Z & S = Zschimmer & Schwarz GmbH, Lahnstein, private communication; IUCLID = International Uniform Chemical Information Database, ECB, Ispra, Italy.

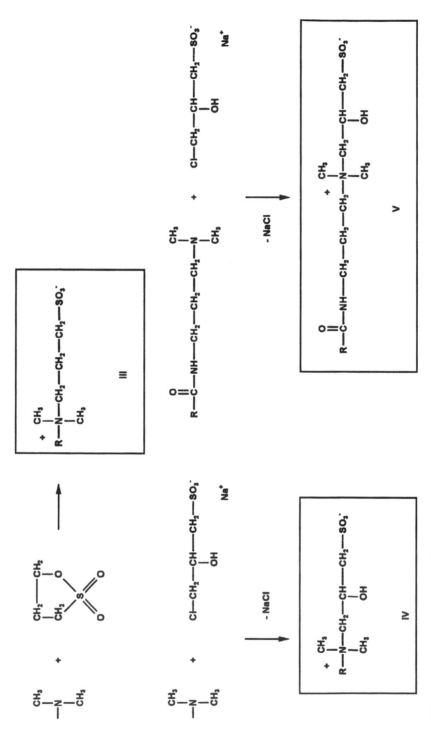

FIG. 5 Reaction scheme of sulfobetaines and hydroxysulfobetaines.

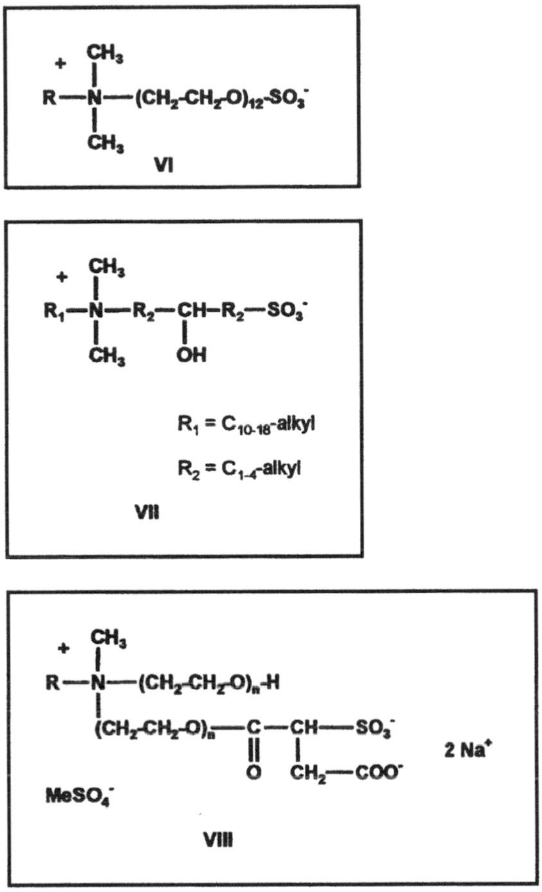

FIG. 6 Structure of special sulfobetaines.

bottle test. Alkyl sulfobetaine as well as alkyl hydroxysulfobetaine are described [4]. The data are given in Table 3. The primary degradation appears to be complete.

D. Ultimate Degradation

The ultimate biodegradation for both groups (structures III, IV, and V) is measured with the closed-bottle test and the STURM test. Data obtained are given in Table 3. Obvious differences can be recognized between sulfobetaines and hydroxysulfobetaines. The former products have only a very limited degradation, but the hydroxysulfobetaines show a better degradation behavior nearer to that of the standard betaines. In conclusion, it can be stated that sulfobetaines as well as hydroxysulfobetaines are not readily biodegradable but may considered "inherently biodegradable." Therefore additional research and development work is needed to overcome this situation.

TABLE 3   Sulfobetaines and Hydroxysulfobetaines—Biodegradation

Substance	Method of degradation	Analysis	Degradation (%)	Evaluation of degradation	Ref.
III, R = C_{12}	CBT	Polarography	90	Primary	4
III, R = C_{16}	CBT	Polarography	97	Primary	4
IV, R = C_{14-15}	OECD 301 E	Polarography	96	Primary	4
IV, R = C_{14-15}	CBT	Polarography	100	Primary	4
III, R = C_{12}	CBT	O_2	25	Nonreadily	4
III, R = C_{12}	STURM	CO_2	49	Nonreadily	4
III, R = C_{16}	CBT	O_2	26	Nonreadily	4
III, R = C_{16}	STURM	CO_2	56	Nonreadily	4
IV, R = C_{14-15}	CBT	O_2	40	Nonreadily	4
IV, R = C_{14-15}	STURM	CO_2	40	Nonreadily	4
V, R = cocoalkyl	STURM	CO_2	33	Nonreadily	
V, R = cocoalkyl	STURM	DOC	47	Nonreadily	
V, R = cocoalkyl	ECETOC	DIC	57–68	Anaerobic degradable	
VI, R = C_{16}	Batch (no detailed information)	COD	33 d: 20	Nonreadily	6
		DOC	33 d: 25	Nonreadily	
		DOC	1 d: 32	Nonreadily	
VII, R = cocoalkyl	Batch (no detailed information)	CO_2	42 d: 41	Nonreadily	6
		DOC	42 d: 49	Nonreadily	
		DOC	1 d: 57	Nonreadily	
VIII, R = cocoalkyl	OECD 302 B	DOC	28 d: 41	Noninherently	6

III = sulfobetaine, structure III in Fig. 5; IV, V = hydroxysulfobetaine, structures IV and V in Fig. 5; VI, VII, VIII = structures VI, VII, VIII in Fig. 6; CBT = closed-bottle test; OECD 301 E = OECD screening test; STURM = STURM test; OECD 302 E = OECD method 302 B; Batch = batch activated sludge; O_2 = oxygen uptake; CO_2 = CO_2 evolution; COD = chemical oxygen demand; DOC = dissolved organic carbon; DIC = dissolved inorganic carbon.

Two additional substances (structures VI and VII) are described in the literature [6]. Both are evaluated as nonreadily biodegradable.

In a patent application [7], 18 different noncommercial ethoxylated types of sulfobetaines are described. For all structures the degradation is measured by the STURM test. Only if the ester function is a part of the molecule could ready biodegradability be confirmed. Ether linkages and higher degrees of ethoxylation did not show a good degradability. The sulfobetaine based on the quaternized ethoxylated amine (structure VIII) has a very limited biodegradation, like that of the traditional sulfobetaine.

E.   Anaerobic Degradation

This test was carried out according to the ECETOC method. The result for the hydroxysulfobetaines is given in Table 3. In conclusion, the hydroxysulfobetaines can be regarded as anaerobically biodegradable.

VI. ALKYLAMPHOACETATES

A. Chemical Structure

This group of surfactants is based on fatty alkyl imidazolines obtained by the condensation of fatty acids or their esters (methyl esters or triglycerides of fatty acids) with aminoethylethanol amine. The reaction scheme is given in Figure 7.

As intermediate, an imidazoline ring structure is formed that was formerly assumed to be present in the final product. But intensive analysis leads to the interpretation that the ring structure is opened by the influence of hydrolyzing conditions [8–11]. With 1 mole sodium chloroacetate, the monoacetate (and with an excess of sodium chloroacetate the diacetate), will be formed. Numerous possible reaction products are described as a result of the first and the second steps of the reaction [8–10,12–15]. Therefore the usual commercial products are complex mixtures.

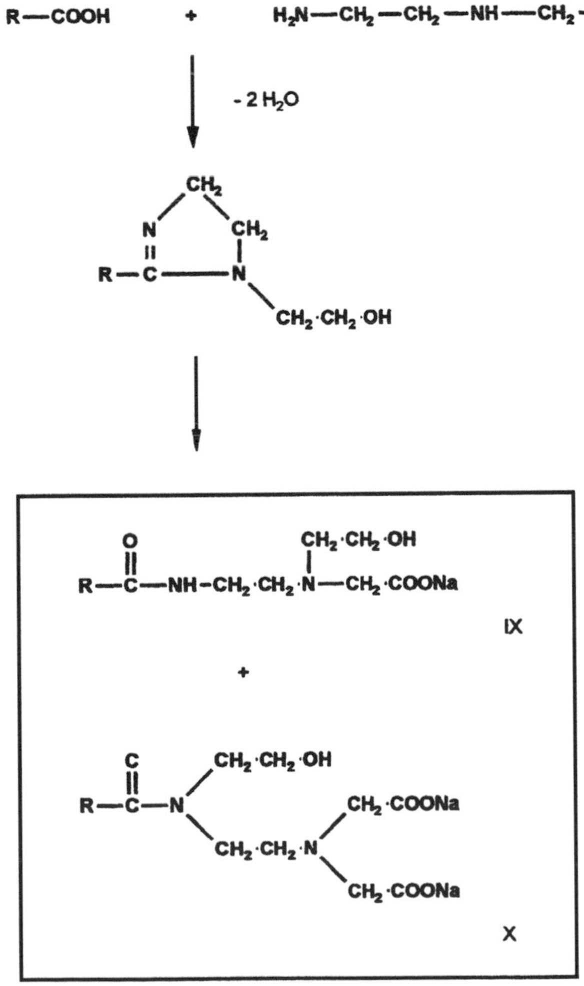

FIG. 7 Reaction scheme of alkylamphoacetates.

Because of the reaction with sodium chloroacetate, alkylamphoacetates contain the corresponding equimolar amount of sodium chloride. The commercial products are aqueous solutions.

Commercial products are available with a broad range of lipophilic groups. In most of the products, R in the formula corresponds to cocoalkyl, but tallowalkyl and palm oil alkyl are also possible. Depending on the molar ratio of the sodium chloroacetate, different degrees of "quaternization" at the nitrogen can be observed. The chemical description of the products differs appreciably from author to author in the literature. Alkylamphoacetates have been described as alkylamino carboxylic acids, hydroxyalkyl alkylamidoethyl glycinates, carboxyglycinates, amphoglycinates, imidazoline derivatives (because of the intermediate), cocoamphodiacetate, or cocoampho(mono)acetate.

B. Properties, Application

Alkylamphoacetates are stable against acidic and alkaline pH values; also, the foaming behavior is not influenced by change in the pH value. There is good stability in systems with a high content of electrolytes, with hard water, and with lime soaps. They are also compatible with anionic, cationic, and nonionic surfactants. But the most important property is the extreme mildness on skin and mucous membranes as compared to other types of surfactants.

Alkylamphoacetates are used in formulating high-quality toiletries, especially baby baths, shower gels, and liquid soaps. In combination with anionic surfactants they improve the mildness of a given basic surfactant. Alkylamphoacetates are regarded as the modern generation of milder surfactants used in personal care products.

C. Ultimate Degradation

The biodegradation of cocoamphodiacetate (mixture of structures IX and X) was first determined with the closed-bottle test (OECD 301 D). The results for the cocoalkyl derivative are given in Table 4. These data show mainly a ready biodegradability at the borderline in terms of "ready." The consequence is to evaluate these substance again with a higher advanced test of the test hierarchy.

The inherent biodegradation has been determined with the Zahn–Wellens test (OECD 302 B, res. according to the German Standard DIN 38 412 Part 25—static test). The results for the cocoalkyl derivative are given in Table 4. These results show that the alkylamphoacetates can be regarded as inherently biodegradable. The same result can be found for the 1:1 mixture of cocoamphodiacetate with sodium lauryl sulfate (see Table 4). The complex between the amphoteric and the anionic surfactant does not significantly influence the biodegradation.

D. Anaerobic Degradation

The anaerobic degradation test of sodium cocoampho(mono)acetate is carried out according to the ECETOC method. The result for the cocoalkyl derivative is given in Table 4. In conclusion, the alkylamphomonoacetates can be regarded as anaerobically biodegradable.

TABLE 4 Cocoamphoacetate—Biodegradation

Substance	Method of degradation	Analysis	Degradation (%)	Evaluation of degradation	Ref.
IX + X, R = cocoalkyl	CBT	BOD_{30}/COD	66	Readily	18
	CBT	BOD_{30}/COD	>60	Readily	Henkel
	MOST	DOC	>70	Readily	Henkel
IX + X, R = lauryl	OECD 301 B	CO_2	68	Not readily	Clariant, 2001
IX + X, R = cocoalkyl	OECD 302 B	COD	3 h: 3 24 h: 9 3 d: 23 5 d: 52 12 d: 70 18 d: 74 28 d: 77	Inherently	
	OECD 302 B	DOC	6 h: 17 1 d: 38 2 d: 69 3 d: 76 4 d: 80 7 d: 81 9 d: 79	Inherently	Z & S
IX + X, R = cocoalkyl 1:1 blended with sodium laurylether sulfate	OECD 302 B	COD	3 h: 0 24 h: 3 2 d: 15 5 d: 22 10 d: 47 15 d: 59 20 d: 68 28 d: 80	Inherently	

IX, X = structures IX and X in Fig. 7; OECD 301 D = closed-bottle test; OECD 301 B = modified Sturm test; CBT closed-bottle test; MOST = modified OECD screening test 301 E; OECD 302 B = Zahn–Wellens test; = static test DIN 38412, Part 25; BOD_{30} = biological oxygen demand in 30 days; COD = chemical oxygen demand; DOC = dissolved organic carbon; Henkel = Safety Data Sheet of Dehyton G, Henkel KGaA, Düsseldorf; Z & S = Zschimmer and Schwarz, Lahnstein, private communication; Clariant = Clariant GmbH, Frankfurt, private communication.

VII. POLYCARBOXYGLYCINATES

A. Chemical Structure

This special group of glycinates consists of the reaction products of a fatty polyamine with sodium chloroacetate [17,18]. The general structure is given in Figure 8.

The polyamine used will have a distribution of amino groups with an average value of four amino groups, which gives the structure shown in Fig. 8. After reaction with sodium chloroacetate, an equimolar quantity of sodium chloride will be produced and will be present in aqueous solutions of commercial polycarboxyglycinates. The R group in the structure can be the alkyl chain derived from coconut oil, oleic acid, or

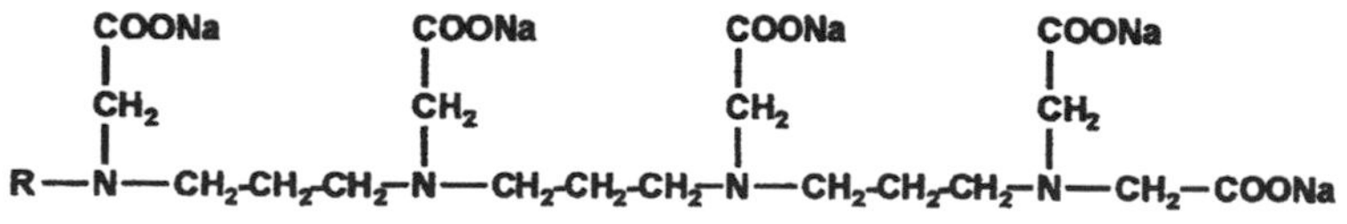

FIG. 8 Structure of polycarboxyglycinates.

tallow. The main product in this range is produced on the basis of an alkyl chain derived from tallow.

B. Properties, Application

From the application point of view, this type of amphoteric is used especially in detergents because of high detergency, high sequestration ability, and dispersibility. This is combined with low irritation rates on skin and mucous membranes and the ability to stabilize enzymes in detergents. In personal care products they act as very mild surfactants.

C. Primary Degradation

Results for the primary biodegradability of tallow polycarboxyglycinate (structure XI) are given in Table 5. The method used was the coupled-units test (according to OECD Method 303 A) combined with HPLC to analyze the content of tallow polycarboxyglycinate. The primary degradability can be confirmed.

D. Ultimate Degradation

The ultimate degradation of tallow polycarboxyglycinate (structure XI) was determined with the closed-bottle test (OECD Method 301 D) and the modified SCAS test (OECD Method 302 A). The results of the measurements are given in Table 5. Based on the data given it could be concluded that these materials are at least inherently biodegradable but may even be readily biodegradable.

TABLE 5 Polycarboxyglycinates—Biodegradation

Substance	Method of degradation	Analysis	Degradation (%)	Evaluation of degradation	Ref.
XI, R = tallowalkyl	CUT	HPLC	>90	Primary	16, 17
XI, R = tallowalkyl	CBT	BOD/COD	5 d: 72,5	Readily	16, 17
XI, R = tallowalkyl	OECD 302 A	DOC	80	Inherently	16, 17

XI = structure XI in Fig. 8; CUT = coupled-units test; CBT = closed-bottle test = OECD 301 D, 5-day version; OECD 302 A = modified SCAS test; BOD = biological oxygen demand; COD = chemical oxygen demand; DOC = dissolved organic carbon.

VIII. ALKYLAMPHOPROPIONATES

A. Chemical Structure

Normally amphoterics contain equimolar quantities of sodium chloride because of the reaction with sodium chloroacetate. The removal of this is technologically complicated. For certain applications, where the presence of chlorides has to be avoided, an alternative product group of salt-free amphoterics is formed by the reaction of acrylic acids or its derivatives.

A salt-free carboxyethyl betaine is obtained by the electrophilic addition of acrylic acid, methyl acrylate, or ethyl acrylate on primary or secondary amines. This is shown in Figure 9.

Depending on the amount of acrylic acid or its ester, the mono- and di-adducts are obtained. The first reaction products are the mono-adducts, especially if a carbonate-free amine is heated with an equimolar amount of methyl acrylate at 100°C, followed by vacuum distillation to remove the excess of acrylate. The ester adduct, a by-product, is hydrolyzed with either alkali or acid. The substances formed can also be described as alanin derivatives (for mono-adducts) or propionates (especially for di-adducts).

Another possible preparation of alkylamphopropionates is the addition of methyl acrylate on N-2-hydroxyethyl-N-2-hydroxyalkyl-β-alanin. The reaction is described in Figure 10.

A second group of salt-free betaines is produced by the addition of acrylic acid, methyl acrylate, or ethyl acrylate on the reaction product of aminoethylethanolamine and fatty acids [13]. The intermediate imidazoline ring is synthesized as described in Section V for alkylamphoacetates. The synthesis in principle is shown in Figure 11.

The first group of alkylamphopropionates is available on the basis of R = cocoalkyl, lauryl, or stearyl. The second group (amide type) is represented by R = cocoalkyl. A special substance in the alkylamphopropionates group—but not as salt-

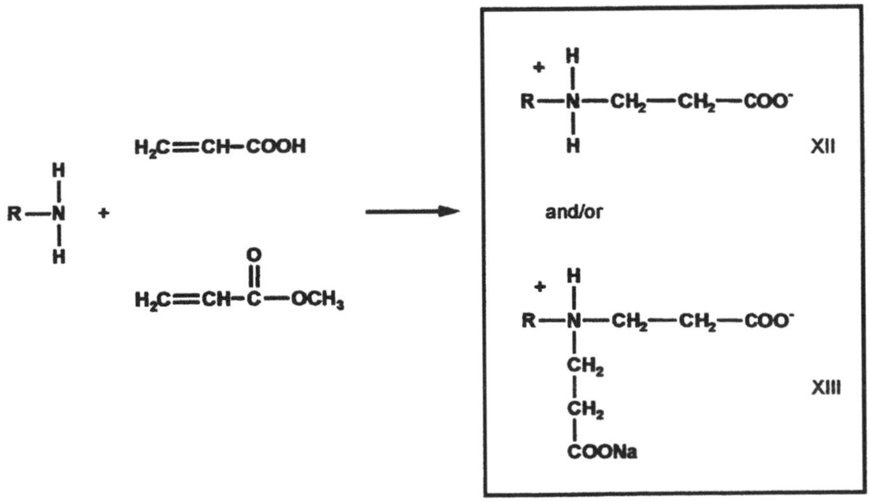

FIG. 9 Reaction scheme of alkylamphopropionates on the basis of amines (I).

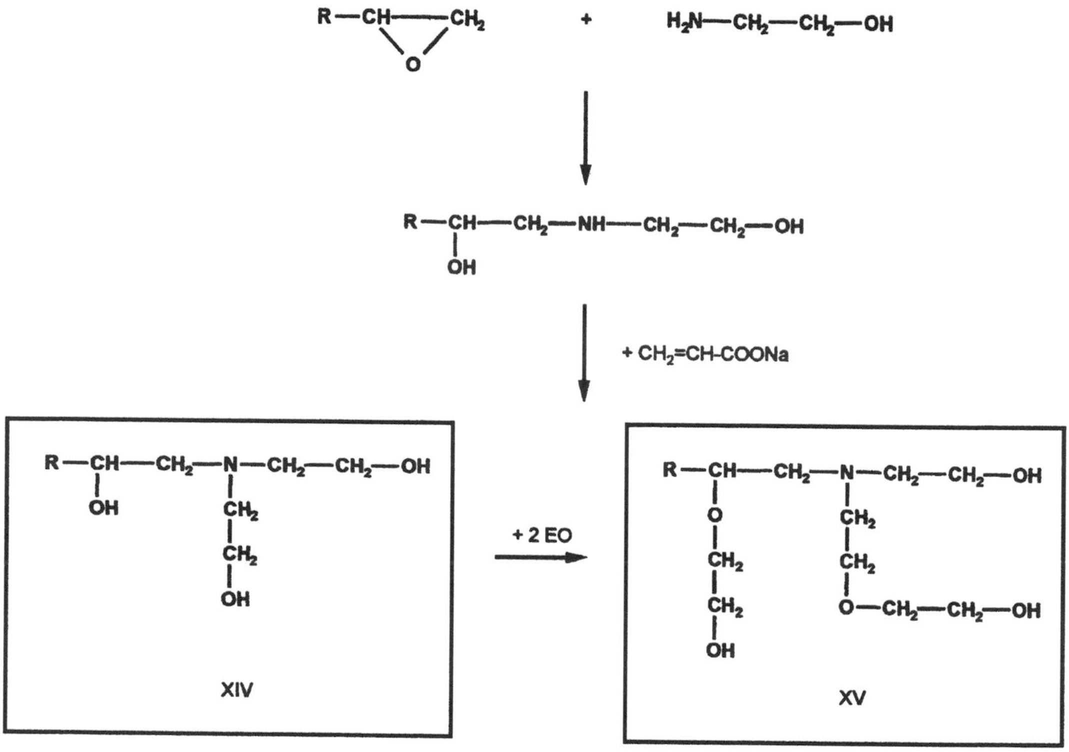

FIG. 10 Reaction scheme of alkylamphopropionates on the basis of amines (II).

free amphoteric—is the reaction product of a tertiary amine with ethylchloroacetate in the first step, later reacted with aminoethylethanol amine and finally with acrylic acid. The structure is shown in Figure 12.

B. Properties, Application

All alkylamphopropionates are free of electrolytes, but all have a very good tolerance against high concentrations of salts or electrolytes. In some cases this could be higher than for other types of amphoterics. But most important is the high stability at very low and very high pH values. The dermatological properties of alkylamphopropionates are quite good.

The main use of alkylamphopropionates is in household products and other types of cleaners containing high amounts of acids or alkalis. They are also used in special cosmetic formulations where the presence of sodium chloride has to be avoided.

C. Ultimate Degradation

Data are given for degradation of alkylamphopropionate on the basis of a primary amine (structures XII and XIII) and N-alkyl-β-alanin derivatives (structures XIV and XV). The methods used were an activated-sludge test and a shake culture test. The results are presented in Table 6.

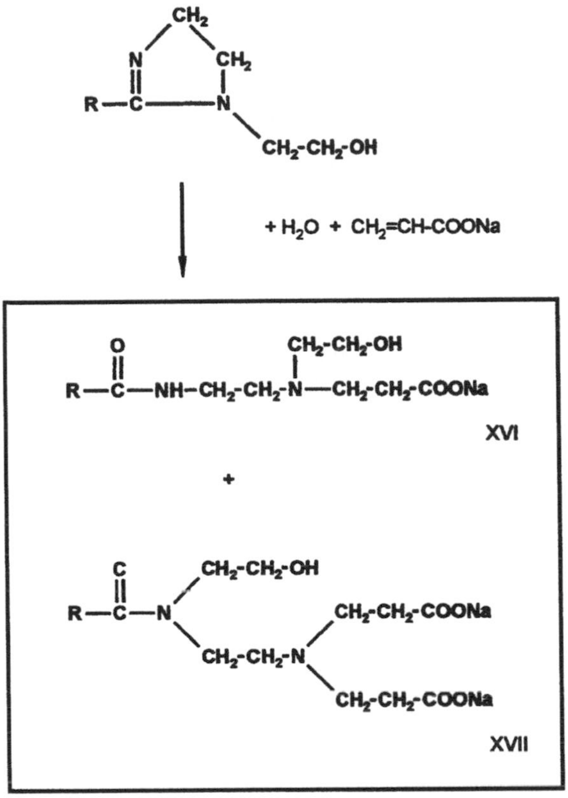

FIG. 11 Reaction scheme of alkylamphopropionates on the basis of amides.

The degradation of C_{12}-β-alanine (structure XII) and N-stearyl-β-amino dipropionic acid (structure XIII) was described on the basis of an activated-sludge test [20] and shows a good degradability. The degradation of N-(2-hydroxyethyl)-N-(2-hydroxyalkyl)-β-alanine (structure XIV) and its ethoxylated derivative (structure XV) occurs to a high degree. The results are part of Table 6.

The inherent biodegradation has been determined for the alkylamphopropionate on the basis of an amidoamine (mixture of structures XVI and XVII) with the

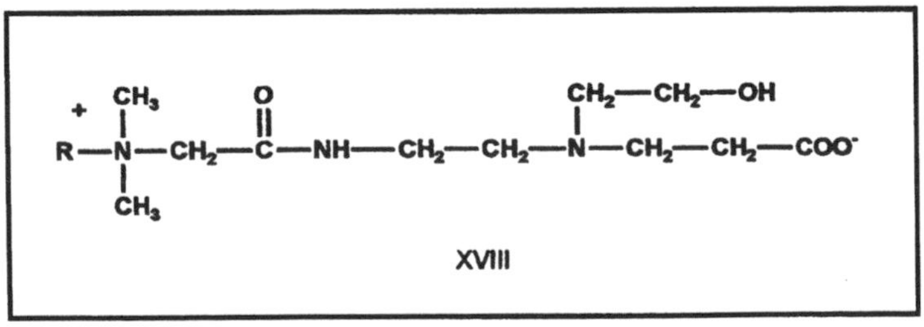

FIG. 12 Structure of cocodimethylacetamido betaine.

TABLE 6 Alkylamphopropionates—Biodegradation

Substance	Method of degradation	Analysis	Degradation (%)	Evaluation of degradation	Ref.
XII, R = C_{12-18}	MOST	DOC	79	Readily	Hoechst
XII, R = C_{12}	SCAS	Surface tension	30 d: >95	Primary	20
XIII, R = C_{18}	SCAS	Surface tension	30 d: >95	Primary	20
XIV, R = C_{12-14}	SCT	No information	8 d: 98		21
XV, R = C_{12-14}	SCT	No information	2 d: 72.2 8 d: 95.5		21
XVI + XVII, R = capryl	STURM	CO_2		Readily	
XVI + XVII, R = cocoalkyl	OECD 302 B	COD	2 h: 0 1 d: 20 2 d: 37 5 d: 54 12 d: 66 18 d: 74 28 d: 86	Inherently	
XVII, R = cocoalkyl	OECD 302 B	COD	2 h: 1 24 h: 26 2 d: 33 5 d: 45 12 d: 50 28 d: 70	Inherently	

XII = C_{12}-β-alanine, structure XII in Fig. 9; XIII = C_{18}-β-amino dipropionic acid, structure XIII in Fig. 9; XIV = hydroxyalkyl-β-alanine, structure XIV in Fig. 10; XV = ethox. hydroxyalkyl-β-alanine, structure XV in Fig. 10; XVI, XVII = amides based alkylamphopropionates, structures XVI and XVII in Fig. 11; XVIII = cocodimethylacetamido betaine, structure XVIII in Fig. 12; MOST = modified OECD screening test 301 E; SCAS = semicontinuous activated-sludge test; SCT = shake culture test (Japanese method JIS K-3363-1976); STURM = shake flask CO_2 evolution system (STURM Test); OECD 302 B = Zahn–Wellens test; DOC = dissolved organic carbon; COD = chemical oxygen demand; CO_2 = CO_2 evolution; Hoechst = Hoechst AG, Frankfurt, private communication.

Zahn–Wellens test (OECD 302 B, resp. according to the German Standard DIN 38 412 Part 25—static test). The results for the imidazoline-derived cocoalkyl derivatives (structures XVI and XVII) are given in Table 6. With the degradability of 83% after 28 days, this group of alkylamphopropionates can be classified as inherently biodegradable.

Cocodimethylacetamido betaine (structure XVIII) was also tested for inherent biodegradability with the Zahn–Wellens test. Despite a certain antimicrobial activity, the degradation of this substance can be compared with that of standard betaines and can be stated as inherently biodegradable.

ACKNOWLEDGMENTS

The use of unpublished data kindly given by Dr. Volker Martin of Zschimmer and Schwarz GmbH, Lahnstein, Germany, Alwin K. Reng of the former Hoechst AG, Frankfurt, Germany, and Dr. Peter Klug of Clariant GmbH, Frankfurt, Germany, is greatly acknowledged. Dr. Hans-Juergen Koehle, Degussa Performance Chemicals, Steinau, Germany, contributed significantly to the chemistry of amphoterics.

REFERENCES

1. Steber, J.; Berger, H. Biodegradability of anionic surfactants. In *Biodegradability of Surfactants*; Karsa, D.R., Potter, M.R., Eds.; London: Blackie Academic and Professional, 1995.
2. Greenberg, L.L.M. *Metabolic Pathways*; Academic Press: New York, 1961.
3. Linhardt, K. Tenside 1972, *9*, 241.
4. Fernlay, G.W. J. Am. Oil Chem. Soc. 1978, *55*, 98–103.
5. Boiteux, J.P. Riv. Ital. Sostanze Grasse 1984, *61*, 491–495.
6. Larson, R.J. Appl. Env. Microbiol. 1979, *38*, 1153–1163.
7. Wentler, G.E.; McGrady, J.; Gosselink, E.P.; Cilley. W.A. EP 32 837 (Proctor & Gamble Co.).
8. Hein, H.; Jaroschek, H.J.; Melloh, W. Fette, Seifen, Anstrichm 1978, *80*, 448–453.
9. Takano, S.; Tsuji, K. J. Am. Oil Chem. Soc. 1983, *60*, 1807–1815.
10. Rieger, M.M. Cosmet. Toiletr. 1984, *99*, 61–67.
11. Watts, M.M. J. Am. Oil Chem. Soc. 1990, *67*, 993–995.
12. Schwarz, G.; Leenders, P.; Ploog, U. Fette, Seifen, Anstrichm 1979, *81*, 154–158.
13. Hein, H.; Jaroschek, H.J.; Melloh, W. Cosmet. Toiletr. 1980, *95*, 37–42.
14. Takano, S.; Tsuji, K. J. Am. Oil Chem. Soc. 1983, *60*, 1798–1806.
15. Zongshi, L.; Zhuangyu, Z. A Study on the Confirmation of the Structures of the Imidazoline Amphoteric Surfactant. 10th Symposium of the GDCh Section Detergents, Potsdam, 1993.
16. Palicka, J. Amphoterics in Household Detergents. Communicationes XXI Jornadas del Comite Espaniol de la Detergencia, Barcelona, 1990; 61–77.
17. Palicka, J. J. Chem. Tech. Biotechnol. 1991, *50*, 331–349.
18. Gericke, P. Parfuem. Kosmet. 1988, *69*, 130–132.
19. De Waart, J.; Van der Most, M.M. Int. Biodeterioration 1986, *22*, 113–120.
20. Eldib, L.A. Soap Chem. Spec. 1977, *41*, 77–80, 161, 163–165.
21. Takai, M.; Hidaka, H.; Ishikawa, S.; Takada, M.; Moriya, M. J. Am. Oil Chem. Soc. 1980, *57*, 183–188.
22. Gerhards, R.; Schulz, R. Tenside Surf. Det. 1999, *36*, 300–307.

21

Environmental Impact of Inorganic Detergent Builders

HARALD P. BAUER Clariant GmbH, Hürth, Germany

I. SODIUM TRIPOLYPHOSPHATE

A. Phosphate Ore Mining

The most important deposits of phosphate are in igneous rocks and sedimentary rocks [1]. Firstly there are granites, which contain on average 0.1% phosphorus by weight. The second are sandstones, with minor phosphorus contents of 0.05–0.1%, as well as phosphorites, which can contain up to 80% apatite.

Fluorapatite, which is the most frequently mined mineral, hydroxyapatite, carbonate apatite, as well as aluminum-containing types all belong to the family of apatites.

The largest phosphate deposits are the Bone Waley formation in Florida, which accounted for approximately 40% of world output in the early 1970s. Other deposits exist in the countries of the former USSR, central Morocco, and in the Middle East [2].

The mining of apatite ore is connected with the accumulation of overburden. For example, in the Florida production, for each ton of crude ore mined about one-third is apatite, one-third is sand, and another third is clay. The last is generated in a particle size of microns and forms a dilute (3–5% content of solids) colloidal suspension that is pumped into large settling areas [3]. The mining of phosphate ore in Florida generates nearly 100,000 tons of waste clays each day, and at the end of 1994 more than 131 square miles of clay-settling areas had accumulated in central Florida [3].

Weathering of phosphate deposits seems to be connected with the creation of new areal patterns of distribution and redistribution of major elements and fixation of minor elements. In particular a pronounced basal enrichment of uranium linked to radioactivity anomalies above the normal background was found [4].

B. The Wet Acid Process for Sodium Tripolyphosphate Production

For production of phosphoric acid and subsequently sodium tripolyphosphate, the most frequently used process worldwide is the wet acid process. Other routes, like electrothermal reduction, are less important in terms of volume [5].

The extracted rock (e.g., in Morocco) is calcined at temperatures of approximately 750°C [6] and transported to the phosphoric acid plant [7]. The phosphoric acid is digested via addition of mineral acids. Sulfuric acid is the most common, but hydrochloric or nitric acid is also applicable [5,8,9] and used mainly in fertilizer production. The "green" phosphoric acid is filtered and concentrated from 29% to 55% P_2O_5 by vaporization prior to shipment [6,7].

The residue from filtration is called *phosphogypsum*. In the Moroccan production it is filtered a second time and via a pipeline dicharged in slurry form into the Atlantic Ocean. The volumes of phosphogypsum generated that way are considerable. Every ton of P_2O_5 produced by the wet acid process implies 4.5–5.5 tons of impure calcium sulfate as by-product [3,5].

Concerns of at-sea dumping relate firstly to the large amounts of gypsum involved. The solubility of calcium sulfate in water appears to be high enough to guarantee its rapid dissolution, but silica and alumina as well as organic compounds remain undissolved.

The second item of concern is the content of heavy metals in phosphogypsum. The possibility of an enrichment of cadmium in the environment is seen critically [5].

Thirdly, a change in the pH of the seawater, owing to remains of phosphoric acid as well as hydrofluoric acid in the slurry, appears to be unlikely due to the buffering of seawater by its natural hydrogen carbonate content [5].

An alternative to sea dumping and the most common applied technology in terms of volume is dumping on land. Gypsum is slurried with water and pumped to settling vessels, and dried gypsum is stored in a dump or in an empty brown coal mine [5].

A phosphogypsum stack in Florida's phosphoric acid facilities was estimated to have to take up 50 million tons of gypsum for a plant producing 1 million t/a P_2O_5 during a 10-year life span. With a height of 200 ft, such a stack would occupy an area of 300 acres [3]. The primary problem in recultivating phosphogypsum stacks is its residual acidity. A further issue is the control of groundwater pollution from phosphoric acid waste gypsum stacks [10].

The efforts to find uses for Florida's phosphogypsum as a raw material have experienced a setback since the U.S. Environmental Protection Agency (U.S. EPA) in the early 1990s judged the radium/radon content in the waste to be dangerous. It effectively banned the use of phosphogypsum having a radium content higher than 10 pCi Ra/g for most applications, including agricultural use, construction, and research [3].

Efforts to reduce the environmental burden of the wet acid process were aimed at the use of sulfur rather than pyrites for sulfuric acid production and utilization of the double-contact sulfuric acid process [8].

Fluoride from the phosphate ore is distributed among the phosphoric acid, the phosphogypsum, and the facility's off-gases. The last are scrubbed in a wash, and the formed calcium fluoride or fluorosilicate are dumped [5].

The "green" phosphoric acid is subject to further purification from impurities in a countercurrent extraction of organic solvent. Frequently, alcohols, e.g., butanol and amylalcohol, are employed. The raffinate contains 90% of the heavy metal amount of the phosphate rock (the other approximately 10% are left in the phosphogypsum) and is reacted with powdered lime (raffinate solidification process) [6] and landfilled [7].

For production of pentasodium triphosphate, the phosphoric acid is reacted with alkali in the form of caustic soda or soda ash, yielding "ortho liquor" [5,6]. The soda ash has been produced by the Solvay process. The solution has a Na_2O-to-P_2O_5 molar ratio of 1.67 [11]. The solution is dried in one-step or two-step processes and subsequently calcined. Drying has to be achieved as fast as possible, e.g., by a spray dryer or in rotary dryers in which the salt solution is flash-dried on a bed of hot material. In most processes the bulk density of the product is fixed at the drying stage. Spray dryers give light products, drum dryers give medium-bulk density, and rotary dryers produce granular products of high density. The calcination is performed in rotary calciners being gas- or oil-fired at temperatures between 380 and 500°C [11]. Subsequently the product is cooled and seized to specified particle sizes. Partial prehydration of the sodium tripolyphosphate by addition of water result in a product of better hydration performance [11].

Criticism has been aimed at the energy-intensive steps of calcination of "green" phosphoric acid, purification, neutralization, spray-drying, and the use of soda ash produced by the Solvay process [7].

C. Life Cycle Assessment of Sodium Tripolyphosphate Production

Life cycle assessments are procedures to determine the environmental burdens caused, e.g., by chemical products. The assessment is structured first into a goal-definition phase, where reference quantities, system boundaries, and the level of detail are specified. Second, in the inventory-analysis phase, the streams of material and energy are included over the whole life cycle of the product. Third, in the impact-assessment phase, potential environmental impacts are determined for each item in the inventory analysis. Fourth, potential impact in the different environmental categories are evaluated in the impact-assessment stage. Finally, the ecologically most favorable variants are selected, existing systems are environmentally optimized, and new systems are developed [12–19].

Three major studies regarding sodium tripolyphosphate have been carried out. A cradle-to-grave inventory gathering data for life cycle assessments of detergent ingredients and their use in commercial laundries [6], a second study based on the delphi technique comparing sodium tripolyphosphate as a detergent builder with the alternative system of zeolite A–polycarboxylate [7], and a third study focusing on the environmental impact of sodium tripolyphosphate and zeolite A–polycarboxylate builder systems within the context of advanced wastewater treatment systems [20].

The cradle-to-grave study presents basic data regarding the energetic and substance process outputs of the manufacture of the most important detergent ingredients [6]. Accordingly, for sodium tripolyphosphate production 5.5 t/t feedstock resources, 16.9 m^3/t water, and 31.1 GJ/t of total primary energy are consumed and 0.6 t/t disposal, 2.8 t/t air pollutants, 45.5 m^3/t wastewater, and 4.1 t/t water pollutants are produced [6].

So far, no discussion relating to a comparison of the compiled data from the different builder substances has been published. But even if there should be a shift in the valuation, e.g., due to differences in the application performances, some items seem to be remarkable. Sodium tripolyphosphate production shows a significantly higher consumption in feed stock resources as well as total primary energy. Figure for disposal, air pollution, wastewater, and water pollution are considerably higher than those of the alternatives.

The Landbank study from 1994 cannot find significant differences between the environmental impacts of the two builder systems sodium tripolyphosphate and zeolite A–polycarboxylate but recommends efforts to reduce the at-sea dumping of phosphogypsum in the Moroccan production, reduce energy consumption, and address the problem of eutrophication, with the recommendation for a systematic program to recover and recycle phosphates from the wastewater.

D. Biogeochemical Cycles of Phosphates

A very slow primary inorganic geochemical cycle has its origin in the dissolution of calcium phosphate mineral deposits from the land by weathering [1,4]. Soluble inorganic phosphate is reprecipitated in the sea and the formed marine sediments uplifted in geological ages.

A second land-based biogeochemical cycle of phosphates starts in undisturbed ecosystems with the uptake of soluble, i.e. available, phosphate by plants. The phosphate is returned to the soil by decay of the plants or from excreta and debris of animals feeding on those plants. Generally the affinity of phosphate to the soil is high, but a small part is leached out by the rain and enters the surface waters.

Several types of phosphate can exist in aquatic systems. Though phosphorus also exists in natural phosphate rock in the form of the orthophosphate entity, orthophosphate in most aquatic systems has its main origin in man-made fertilizers [21]. Condensed phosphates come from detergents and are usually hydrolyzed over time to orthophosphate ions [17,22–26]. Phosphate can be absorbed onto particulate matter [21,27,28]; this seems to be the largest proportion because of the low solubility of organic and inorganic forms [21,28]. Organic bound phosphate is formed by biological processes when phosphate is assimilated by the organisms as structural and energy transfer compounds [21].

E. Input of Anthropogenic Phosphate to Aquatic Ecosystems

The input of phosphate into surface waters can be traced back to natural sources like groundwater [29] and the erosion of soil in the landscape and natural bedrock. In Europe, phosphate rock deposits are located in Scandinavia, mainly in Finland and Sweden. The phosphorus emissions from natural sources in European countries vary from 1% in the river Po in Italy (1986) to 53% in Swedish inland waters (data from 1986–1990). The figures refer to the total phosphorus emissions of a country [35]. Further phosphate input is due to runoff of fertilizer from agricultural land, milkhouse water, animal manure from animal farms, and drainage, to domestic wastewater, to industrial wastewater, and to domestic wastewater [25–27,29–50].

Phosphorus emissions of European countries into their surface waters from agricultural sources range between 17% into the Norwegian catchment of the North Sea (1990) and 42% in German surface waters (1989–1992) [35].

Domestic and industrial sources are usually summarized as point sources or selective phosphorus sources [29] insofar as the wastewater is collected in a sewerage system and discharged to surface water after treatment in a sewage treatment plant. Atmospheric deposition, runoff from agriculture, and erosion are attributed to the nonpoint or diffuse phosphorus sources [17,29,35,37,43,51].

In the European Union, the main contribution of phosphorus pollution (more than 50%) comes from point sources as, e.g., industrial and urban wastewater. In most parts of the European Union the anthropogenic phosphate input is far higher than from natural sources [35,37], e.g., by weathering of phosphate deposits, soil erosion, and from the athmosphere. Point sources, e.g., account for 18% of the total phosphorus emissions to Swedish inland waters (1986–1990) and up to 77% of the emissions to the German part of the Rhine catchment (1985) [35].

Phosphorus inputs from household activities separate into human metabolic waste, food residues, phosphate from laundry and automatic dishwasher detergents, as well as household cleaners, laundry soil, and phosphate from drinking water conditioning [25,26,29,30,32–34,38,41,42,44–46,49].

The contribution of detergent phosphates to the phosphorus load of domestic sewage was estimated in the 1960s. In Germany the total input of 3.35 g P per capita (per day) is divided into approx. 1.6 g P per capita from detergents and approx. 1.75 g P per capita from human metabolites (1965 data [44,45]). In the late 1970s the detergents were estimated to account for 40% of the domestic phosphate input [30].

In the United States, total phosphorus concentrations in raw wastewater effluent changed considerably from 11 mg total P per liter in the 1970s, which marked the height of phosphate detergent use, to 5 mg total P per liter in the 1990s [41].

At the end of the 1970s, an extensive study covered the pathways and fate of phosphorus in the Federal Republic of Germany, including the share of consumption of raw phosphate for detergent application. Accordingly, the total consumption of raw phosphate, phosphoric acid, and iron ore for detergents was found to be 69 kt P/a from 558 kt P/a phosphate consumption in total. The consumption for detergents divided into 60 kt P/a for household laundry detergents, 5 kt P/a for industrial and institutional (I&I) laundry detergents, and 4 kt P/a for automatic dishwasher detergents and cleaners (household and I&I) [32].

In the late 1980s to the early '90s, the amount of laundry wastewater was estimated to range between 8 and 16 liters per day and capita [25,52], 17 liters per day and capita [26,53], and 26 liters per day and capita [25,54]. These figures correspond to 4–8% to 13% of total sewage volume reaching a wastewater treatment plant.

In the period between 1975 and 1990, the phosphorus load from detergents in German domestic sewage declined considerably, from 3 grams per capita and day [2.25 from laundry detergents and 0.75 from automatic dishwasher detergents (ADDs) and household cleaners] to 0.45 grams per capita and day in 1989/1990 (0.3 from laundry detergents and 0.15 from ADDs and household cleaners) [29,38].

Owing to the use of phosphate in automatic dishwasher detergents, phosphate input by detergents increased in Germany in the 1990s from 5,100 t/a in 1994 to over 12,500 t/a in 1996 to 23,600 t/a in 1999 [46].

Studies focusing on the conditions in the United Kingdom in the late 1990s estimate that urban residents' discharge is between 2 and 3 g P per capita per day, which is composed of 1.2 g P per capita per day from human diet and 1.3–1.8 g P per capita per day from other household activities, including detergents [42].

In Australia, phosphate detergents in the late 1990s were seen as major contributors to nutrient load [49].

Surveys from European surface waters show that detergents account on average for 11% of total phosphorus input to European surface waters, with significant differences among the European countries, from 2% in Italy to 19% in the United Kingdom [55,56].

The raw effluent from the different sources just described is conducted via the municipal sewarage system to the wastewater treatment plants, where phosphorus elimination takes place. The phosphorus concentration in the effluent of the treatment plant depends upon whether a mechanical and biological treatment stage or an additional precipitation stage is installed. The effluent is discharged to surface waters.

F. Behavior of Sodium Tripolyphosphate in Municipal Wastewater Treatment

Early in the discussion of phosphorus inputs to surface waters and their eutrophication it was evident that point sources (e.g., municipal wastewater treatment plants and industrial discharges) were responsible for a major part of the phosphorus load and the source that could be influenced and controlled most easily. Detergent phosphate limitations were enacted to provide immediate relief, but it seemed obvious that additional actions for phosphorus removal in sewage treatment plants would have to be taken for a significant reduction in phosphorus load [32,33,37,40,43–45,50].

Scandinavian countries, e.g., Sweden, Finland, Denmark, and Norway, were early to adapt phosphorus removal in municipal wastewater treatment plants, as was Switzerland [20,29].

In some states of the United States and the provinces of Canada, the major municipal wastewater treatment plants of the Great Lakes Basin were to be improved as part of the Great Lake Water Quality Agreement [57–60].

In the European Community, the Urban Wastewater Treatment Directive [34,40,42,51,61–63] drives the improvement of existing sewage treatment plants or the installation of new ones, as a result of still-existing eutrophication in surface waters across Europe, irrespective of existing detergent bans. The Directive requires the collection of sewage and the removal of phosphate in sewage works for all agglomerations of more than 10,000-person equivalents (i.e., a population of 6,000–8,000 people) discharging into eutrophication-sensitive areas. The guidelines are maximum 2 mg P/L in effluents from sewage works serving agglomerations of 10,000–100,000 person equivalents, 1 mg P/L for sewage works serving more than 100,000 person equivalents, and a minimum phosphorus removal of 80% [34,42,62].

In Germany the guideline for minimum requirements on the discharge of sewages in waters (Allgemeine Rahmen-Verwaltungsvorschrift über Mindestanforderungen an das Einleiten von Abwasser in Gewässer—Rahmen Abw. VwV) [64] requires maximum 1 mg P/L for (large) sewage treatment works for more than 100,000 inhabitants, 2 mg P/L for (medium-size) sewage works for 20,000–100,000 inhabitants, minimum 90% phosphorus removal and no regulation for smaller sewage works [19,62].

About 5–10% of the incoming phosphorus can be removed in wastewater treatment plants via primary treatment (settling); 20–40% of the phosphorus can be removed in the biological treatment stage (secondary treatment) due to bacterial metabolic action [65]. Tertiary treatment by advanced biological treatment allows 40–

85% phosphorus removal. The achieved performance is strongly dependent on the process conditions. Tertiary chemical treatment allows removal of 95% or more phosphorus, depending on the quantity of calcium, aluminum, or iron salts added [29,31,48,56,66–77].

Several applications for the phosphate-containing sludge are in use. The largest portion (60%) is disposed via landfill [19]. Disposal via this route will have to be reduced in the European Union owing to the EU Landfill Directive [78], which poses a 65% reduction in biodegradable waste going to landfill over a horizon of 15–19 years [34].

Another 25% of sludge is distributed to agriculture as fertilizer [19,40,79–81]. The extent of this application seems to be limited because of the transportation costs of the bulky sludge and concerns about the heavy-metal content of the sludge [40,67].

A further 10% of the sludge is sent to incineration [19]. Products of concern in the ash are heavy metals and mercury (the EU Commission proposed much stricter controls on the composition of sludge used for spreading [40]) as well as dioxins and furans, acid gases, NO_x, and N_2O [40,82].

Other technologies to use the sludge have been developed, such as pyrolysis to yield oil, gas, and coke [67], composting the sludge [42], construction purposes [40,83], and calcium phosphate crystallization and recycling to phosphate production [79,42].

In any case, a possible renewed use of phosphate in detergents would raise the phosphorus loads as well as require higher quantities of precipitants and cause generation of more sludge [29,39].

G. Nutrients, Photosynthesis, and Biological Productivity in Aquatic Ecosystems

In aquatic environments, phosphate is an essential part of the photosynthetic process. Phosphate is assimilated by the organisms into structural and energy transfer compounds [21]. The basic equation of photosynthesis and respiration for algal protoplasm,

$$106CO_2 + 16NO_3^- + HPO_4^{2-} + 122H_2O$$

$$+18H^+ \underset{R}{\overset{S}{\rightleftharpoons}} C_{106}H_{263}O_{110}N_{16}P(I) + 138O_2$$

(S = sunlight, R = respiration), was formulated by Redfield [1,27,84–88]. So the elements carbon, nitrogen, and phosphorus are taken up during photosynthesis in the proportion of approximately C:N:P = 106:16:1 [1,84,89]. Sum formula (I) can also be seen as the composition of algal protoplasm.

The Redfield stochiometry has been subject to extensive studies [84,85,89–91]. Besides carbon, nitrogen, and others, phosphorus is a macronutrient in aquatic ecosystems. Silicon is attributed to the micronutrients [36].

Nutrients relate to the category resources [92] and are one of several abiotic factors driving biological productivity in surface waters [27]. They stimulate growth of green plants, including algae [57]. Nutrients can compensate, within limits, for physical limits such as light and temperature, and vice versa [89].

According to studies, basically all phosphorus compounds are biologically available, but orthophosphate is seen as a main source [93,94]. Laboratory studies support an effect of dissolved phosphorus compounds on the relative growth rate of algae following the enzyme kinetics after Michaelis–Menthen [93].

Other physical or abiotic factors belonging to the resources are the metorological influences of light and temperature [27,92,93]. The latter has an influence on the growth rate of the aquatic organisms, the sedimentation rate, the grazing activity, the stratification of the water, and its oxygen content [92].

Further physical factors are hydraulic properties of surface waters, like depth, flow speed, and retention time [27,93,95].

Besides the abiotic factors, there are also important biotic influences [27,95]. The primary productivity is determined by free-floating microscopic algae called *phytoplankton*, such as diatoms, green algae, and blue-green algae, by benthic algae (phytobentones), and by macroscopic plants (macrophytes) [27,36]. Blue-green algae belong to the bacteria and form metabolic products that can be toxic to fish and invertebrates [96]. The productivity of macroalgae is strongly influenced by temperature [21] and the macronutrients nitrogen and phosphorus, which regulate growth, reproduction, and the biochemistry of macroalgae [97]. They are the food basis for secondary consumers, like the zooplankton, which feeds on bacteria, phytoplankton, etc. Competition and grazing by zooplankton is an important growth factor for phytoplankton [27,93] as well as destructive processes. These are mineralization and decomposition of dead organisms by fungi and bacteria [27]. Zooplankton is consumed by many species of small fish.

H. Eutrophication Aspects of Groundwater

Usually phosphate is adsorbed to the soil. So its mobility is limited, and the concentration is low in the groundwater [98], typically in the range of a few micrograms per liter.

I. Eutrophication Aspects of Lakes and Reservoirs

In contrast to flowing surface waters, lakes have less turbulence through wind and water flow. In winter these influences are sufficient to achieve a vertical homogeneity in terms of nutrient distribution [1]. Algal activity is low because of insufficient temperature and sunlight. In spring, increasing hours of daylight and sunshine stimulate algal growth. The top layer of the lake is warmed by sunlight. In summer the natural turbulences are not intense enough to mix the warmer, less dense top layer (epilimnion) and the denser, colder bottom layer (hypolimnion). The water layers in the lake become stratified. Typically photosynthesis occurs in the top layer.

Dead algae being decayed and mineralized by microbes sediment to the bottom of the lake. Respiration and phosphate enrichment characterize the bottom layer, because mineralization consumes oxygen and phosphate is liberated. In autumn the boundaries merge and the top water layer is again supplied with nutrients [1].

Algae consume the nutrients nitrogen and phosphorus in approximately constant ratios of 16 to 1 by atoms and 7.2 to 1 by weight [99]. The practical range is 10–20 [99–101]. So if one of either nutrient falls short of this ratio, it determines algal growth: It is going to be the limiting nutrient. This correlation is described by *Liebig's law of the minimum* [44,84,102]. Aquatic environments with relative concentrations of nitrogen to phosphorus of less than 10 to 1 should be unable to support extensive algal growth. Nitrogen availability is the limiting factor in this situation. If the nitrogen-to-phosphorus ratio is in excess of 20, then phosphorus becomes the limiting nutrient for plant growth. At intermediate values, assessment is difficult because other limitations, like light etc., gain importance [99].

Productivity of lakes can be used as a basis for classification [19,57,92,103–108]. The least productive lakes are called oligotrophic (3–15 μg P/L), intermediate productive lakes mesotrophic (15–80 μg P/L), and most productive lakes eutrophic (20–300 μg P/L).

In the United States, the U.S. Environmental Protection Agency (U.S. EPA) established a recommended limit of 50 μg total phosphate/L in streams that enter lakes [41,98,109] and a water-quality endpoint of 30 μg P/L, which differentiates between "nonimpacted" lakes and "impacted" lakes in terms of eutrophication. In reservoirs, 50–150 μg total P/L is seen to be desirable, whereas 160–200 μg/L should be just tolerable [95].

Today there is strong support in the literature that for inland freshwater environment, e.g., lakes and reservoirs, phosphate is one of the macronutrients that determines algal growth [41,51,62,84,89,92,104,110,111].

In the 1960s, Vollenweider defined an empirical relationship between phosphorus loads to water bodies and algal productivity based on the findings of Atkins. Accordingly the biomass of plankton per unit volume of water shall be proportional to the initial mass of limited nutrient per unit volume. Although physical, chemical, and biological factors make each lake, reservoir, and watershed unique, the Vollenweider–OECD model was applied to a considerable number of lakes in most parts of the world, and in the meantime it is a tool for eutrophication-related water-quality management as well as for trophic-state classification of lakes [20,59,92,99,112–129].

Any excess in the limiting nutrient, phosphorus, can lead to excessive, accelerated growth of vegetation, e.g., microscopic floating plants like phytoplankton, macroalgae, and higher plants [19,34,48,56,103,130]. This is the meaning of the term eutrophication. Phytoplankton and macroalgal blooms occur more often, last longer, and extend over a greater area [21,99,131].

Reservoirs are particularly threatened. Reduced turbulences and mixing by the water flow make them vulnerable toward nutrient overenrichment.

Accelerated algal growth typically runs parallel with higher turbidity in the water. Extensive blooms of blue-green algae (cyanobacteria) [99] can result in the formation of toxins that are hazardous to animals [103]. This is the result of the appearance of toxic plankton surface scums [99] on lakes and reservoirs formed by blue-green algae, which have caused the deaths of a number of animals and cases of sickness in humans [7].

Even though macrocyte beds can be seen as important features of lakes, because they can act as spawning areas for fish and shellfish and may provide refuge for some fish against predator species [103], they can become a problem when they cover large areas of the lake surface. If they accumulate close to shore they can interfere with recreational activities and impede boat traffic and can create areas dangerous to fishermen and swimmers, respectively. Fishing can be hindered through the fouling of nets. When plants break off and are transported to water treatment facilities, they can clog water intakes and increase the cost of water treatment [21,30,103,132]. Blue-green algal strains are problematic and potentially toxic. Where they predominate they lower the water quality [7]. Decomposition of excessive macroalgal biomass can cause problems in drinking water, in terms of foul taste and odor [21,47]. Eutrophication generally is detrimental to drinking water production [19,30]. Treated drinking water taken from eutrophicated lakes and reservoirs can potentially also be contaminated with disinfection by-products, such as halomethanes, posing a threat to human health [99].

The green plants and algae, stimulated in growth by nutrients, die and are decomposed by bacteria. The largest part of this process occurs in the water zones close to the water surface [48]. Finally, the decaying organic matter sediments to the hypolimnion and the bottom and decays at the sediment–water interface [30,48,57,75,99]. Aerobic decomposition, which is the first step, is rather oxygen-consuming [19,48]. One milligram of phosphorus introduced into a stratified lake may lead to an oxygen consumption in the hypolimnion of 140 mg oxygen [84]. This may induce fluctuations in the oxygen balance [30,51] of the aquatic ecosystem. Over the day, photosynthesis results in an oversaturation of oxygen and a pH shift to the alkaline area. In the night, oxygen is consumed in the respiration phase [30]. This leads to decreased and low dissolved oxygen levels [19,21,47,48,57,75,99,103,133]. Changes in kind and diversity of species of algae, bottom-dwelling organisms, and fish are biological indicators of oxygen depletion [7,21,57,134].

In serious cases, the decomposition of excessive biomass will result in anoxia, a condition with complete absence of dissolved oxygen in the water. Oxygen depletion below or equal to 2 mg/L is called *hypnoxia* [1,21,60,135]. This will put severe stress on fish. The consequence is the loss of various bottom-dwelling plant and animal species as well as of seagrass [7,21,103,136,137].

J. Eutrophication Aspects of Rivers and Streams

Rivers and streams represent a diverse set of waterbody systems, ranging from cascading, cold-water mountain rivers to deep, slow-flowing streams [99]. Rivers are generally regarded as phosphate-limited with respect to their biological productivity [56,84,89,92,104,110,111]. Phosphorus concentrations in undisturbed rivers may vary from about 20 µg/L to hundreds of micrograms per liter total phosphorus [75,95]. The U.S. EPA established a recommended limit of 50 µg/L total phosphates in streams that enter lakes and reservoirs and 100 µg/L total phosphorus in flowing waters [41,98].

Running waters are not threatened by eutrophication as, e.g., lakes are [19]. This is because the turbulences of water flow result in an oxygen enrichment. Hence tissue of algae and plants can be decomposed without oxygen depletion [19]. The flow speed acts as a further limit on algal growth [19]. Because of all these factors and because of the limited residence time, rivers tolerate a higher phosphate load than lakes and reservoirs [19,95,138].

The situation can shift critically if the residence time is enhanced because of longer running distances in the lowlands and by extremely low river flow, if conditions get close to that of stagnant waters because of dams or weirs.

Fertilization of an aquatic system by nutrients results in enhanced primary production of algae and aquatic plants [130]. Potential nuisance effects of excessive phytoplankton and periphyton growth can affect the drinking water supply because of taste or odor problems or by blockage of intake screens and filters. Blue-green algae can produce toxins that can cause aquaculture fish deaths. In water treatment works, flocculation and chlorination processes can be disrupted. Because of reduced clarity and altered color the esthetic appeal can be reduced. Sloughed material can foul fishing lines and nets. Surface scums can appear (including "red tides") as well as floating mats. Dense mats covering the bed reduce intergravel flow and habitat quality for benthic invertebrates and fish spawning. Boating, swimming, water skiing, and other water-based recreation can be restricted or degraded. Shellfish contamination can result in human poisoning. Dense algal mats restrict invertebrates preferred as food by those fish targeted by sports fishermen. Fluctuations in pH and dissolved oxygen can eliminate sensitive species and release sediment phosphate. Reduced light protection may cause macrophyte decline [99,139].

K. Eutrophication Aspects of Estuaries

Marine waters are divided into estuarine, coastal, and oceanic waters [89]. The addressing of limiting factors seems to be less straightforward for estuarine waters than it is for inland surface waters. Large, real variability exists in physical and biological structure and processes between the systems. Hence a given input might result in different results. A quantitative prediction seems to be rarely if ever possible with acceptable accuracy.

For nutrients, various input sources, such as those from streams and rivers, from groundwater, runoff, tidal exchange, or atmospheric deposition, make it difficult to generalize what their effect on wetland quality is [99,140–142].

The impact of nutrients might be enhanced by the poor mixing of nutrient-rich coastal waters, accompanying hydrological stability, which results from salinity stratification [143]. So sources differ from a nitrogen limitation of estuarine phytoplankton [89], to phosphorus limitation of macroalgal growth in estuarine systems, to a possible fluctuation between phosphorus and nitrogen limitation in some situations [21,134,144].

In shallow, sheltered estuaries with long residence times, eutrophic conditions might cause excessive growth of green macroalgal blooms or development of epicytes that might interfere with the growth and productivity of wetland plants and microbes [51,99]. Localized overfertilization can result in algal blooms at the mouths of rivers and fjords [1]. Other consequences may be changes in plant species composition. Eutrophication may alter plant species dominance [99,145]. Changes in the trophic status may favor invasion by exotic plants [99,146].

L. Eutrophication Aspects of Coastal Waters

Nitrogen is generally believed to be the key nutrient limiting algal growth and eutrophication in coastal waters [21,51,89,147,148]. On the other hand, there seems to be evidence for exceptions like no strong limitation by nitrogen and phosphorus in some bays and coastal waters [89,149,150] and phosphorus limitation in Norwegian fjords and coastal waters [89,151].

Impacts of eutrophication in coastal waters are generally most evident in shallow waters, particularly in embayments, harbors, and lagoons with limited water exchange and that receive substantial inputs of nutrients from anthropological sources [143,150,152]. Most acute effects may be restricted to areas in the neighborhood to point sources of nutrients, but some impacts can also be documented over wider sea areas [143,153–155].

Consequences might be shifts in the relative abundance of key diatom species [143] as well as massive production of bentic macroalgae in coastal lagoons and shallow waters connected to the decline of sea grass [143] and the formation of dense mats of drifting benthic algae [143]. Dense algal growth and sedimentation of algae at the ocean bottom followed by decomposition [135] has severe effects. It results in the decrease of oxygen concentration in the water column, leading to more or less localized anoxia, which can have devastating impacts on the benthic community [143] and can result in the loss of many forms of aquatic life [47], such as massive fish kill and die-off of benthic organisms and production of hydrogen sulfide, which is toxic to marine life [35,143,150]. Draining of nutrients into oceans can also lead to an increase in the number of algal blooms, known as red tides, which make seafood unsafe to eat [47,143]. Red-tide dinoflagellates are capable of producing potent toxins [143].

M. Political Initiatives and Measures on Water-Quality Improvement

European studies suggested early in the discussion about detergent phosphate bans that eutrophication may be controlled only if nutrients of all sources were removed by improved wastewater treatment [7,20]. It pointed to phosphate removal programs in Sweden, Switzerland, and the Great Lakes region of North America where eutrophication should be controlled successfully.

In the period between 1990 and 1995, phosphorus discharges were reduced by 40–60% in the European Union. Nevertheless the problem of eutrophication of European rivers, lakes, reservoirs, estuaries, and coastal waters exists to date in a way the Dobris assessment described in 1991 [35]. In most parts of Europe, anthropogenic phosphate input appears to be substantially higher than that from natural sources. Action against eutrophication should include further reductions in phosphorus emissions from point as well as from diffuse sources [35].

Guided by this conclusion and by the fact that sources other than detergents are today the main drivers of phosphate input to European surface waters, European policies were introduced that pursued a number of initiatives and measures to reduce the problem of nutrient overenrichment of European surface waters, e.g., by intensified wastewater treatment.

The fifth Environmental Action Program led to a number of European directives in order to improve the ecological quality of European surface freshwaters [35]. Steps included Water Framework Directive 2000/60/EC [35,40,156], intended for the protection of freshwaters, estuaries, coastal waters, and groundwater, and Drinking Water Directive 1998/83/EC [30,35,157].

The Urban Wastewater Treatment Directive requires removal of phosphate in wastewater treatment plants for all agglomerations of more than 10,000 person equivalents discharging into "eutrophication-sensitive areas" and the installation of appropriate treatment for smaller communities by the deadline of Dec. 31, 1998 [34,56,62].

The Ecolabel Award Scheme, e.g., 1999/476/EC [158] and 1999/427/EC [159], established the criteria for the award of the Community Eco-Label to laundry and automatic dishwasher detergents. Other directives focusing on water quality are EC Freshwater Fish Directive 78/659/EEC [95,160] and EC Bathing Water Directive 76/160/EEC [161].

One of the international conventions to protect European marine ecosystems is the Paris and Oslo Convention to protect the North Sea and Atlantic Ocean (OSPAR)

[35,56,162,163]. The intergovernment body set the goal of reducing inputs of nitrogen and phosphorus by around 50% by 1995 [35].

The targets of the Helsinki Convention with the Baltic Sea Joint Comprehensive Environmental Action Program between 1993 and 2012 are to identify all major point sources of pollution and to initiate preventive and curative actions regarding them [35].

The Barcelona Convention is appointed to protect the Mediterranean Sea with all appropriate measures [35,56].

The Strategic Action Plan for the Rehabitation and Protection of the Black Sea aims at the reduction of nutrient discharges to rivers (mainly the Danube) and the reduction of inputs of insufficiently treated sewage by 2006 [35].

The Rhine Action Plan required a 50% reduction of phosphorus inputs to be achieved in the period 1985–1995 [35,40,162,164].

In the late 1960s, public concern arose regarding eutrophication of the Great Lakes, Lake Erie in particular. Municipal wastes containing phosphate detergents contributed 70% of the overall phosphorus input [60]. As a result the United States and Canada in 1972 signed the Great Lakes Water Quality Agreement (GLWQA [17,41,57,59,60,103]). The target was to reduce phosphorus inputs to the lakes in order to achieve a substantial elimination of nuisance algal growth in the lower lakes and the international section of the St. Lawrence River. Aerobic conditions should be reestablished in the hypolimnion of the Central Basin of Lake Erie. Lake Superior and Lake Huron should be maintained in the oligotrophic state [60]. This agreement required that all municipal discharges with flows in excess of 3800 m^3 per day had to be limited in their concentration of total phosphorus in the effluent to 1 mg/L on an annual average basis. Household detergents used in the Great Lakes Basin were limited in their phosphorus content. Industry was required to remove phosphorus from their discharges to a maximum extent practicable. Phosphorus loadings from agricultural activities should be controlled.

The agreement was revised in 1978 [59]. Effluents of the wastewater treatment plants discharging more than 1 million gallons per day to Lake Erie and Lake Ontario were not to exceed a phosphorus concentration of 0.5 mg P/L [59,165,166].

The result of the Soil and Water Environmental Enhancement Programme (SWEEP) carried out in Canada in the years 1985–1993 indicates that Canada met or exceeded its agricultural non–point source phosphorus loading reduction goals for Lake Erie.

The Federal Water Pollution Control Act of 1972 (Clean Water Act, CWA) and the Coastal Zone Management Act (CZMA) are aimed more at non point sources of nutrient pollutions [41,135].

In the second Chesapeake Bay Agreement, the federal government and the states in the Chesapeake drainage basin mandated in 1987 a 40% reduction from the controllable 1985 nutrient input level by the year 2000 [167].

In 1999, 22 U.S. states had numerical criteria for phosphorus, 12 states had narrative criteria, and 21 states had no water-quality standard for phosphorus [41,168].

An American study indicates that the extent of reduction of phosphate emissions attained by detergent phosphate bans in North America has not been shown to have an impact on eutrophication-related water quality [59]. Another American study [117] indicated that in cases of eutrophied waters where phosphate input by detergents is secondary, phosphate elimination from domestic wastewater can be an effective way to control phosphorus input.

N. Political Initiatives and Measures on Detergent Bans

Guided by the goal of an immediate reduction of phosphate inputs to surface waters, many countries issued regulations for the reduction of phosphate content in laundry detergents or entered into negotiations with the detergent industry for voluntary agreements.

Today we see regulations or voluntary agreements, e.g., in Austria, Brazil, Denmark, Finland, Ireland, Germany, Italy, The Netherlands, Norway, Sweden, Switzerland, the United States, Canada, and Japan [25,26,30,40]. In the meantime a complete shift from phosphate-containing laundry detergents to phosphate-free alternatives occurred in a large number of the industrialized countries, like Austria, Canada, Italy, Japan, Germany, The Netherlands, the Scandinavian countries, South Korea, Switzerland, and many states within the United States [26].

Authorities in Australia recommend that consumers in public education programs switch to household detergents without phosphates [49]. In Austria in 1994 the government imposed a regulatory ban of phosphates in household detergents [40,46].

In connection with the eutrophication problems of the Great Lakes, first the Canadian Province of Ontario adopted a limitation for phosphate in detergent formulas of 2.2% P by weight [17,59,60]. Today a federal limit of 2.2% phosphorus content is valid for household detergents in Canada [25].

In Denmark, a voluntary agreement exists that is in effect equivalent to a ban on phosphate in household detergents [40].

In Finland, an agreement between the government and the detergent industry is in force to limit the phosphate content to a maximum of 7% P [25].

In Germany, the Law of Environmental Compatibility of Laundry Detergents and Cleaners (Detergent Law) of 1975 [30,169] replaced the Detergent Law of 1961 [25,170]. The law is aimed at the protection of surface waters from detrimental effects by the application of household detergents and cleaners. It forms the basis for the phosphate regulation of 1980 and was revised in 1987 [43]. Phosphate reduction in Germany was enforced by the phosphate regulation of 1980 [25,26,30,31,40,48, 75,171]. The two-step decree established upper concentration limits for the phosphate content in the laundry liquor of the domestic laundry detergents. Limits for commercial detergents and cleaners were also set, but could be followed without replacement of phosphates [30]. The first step meant an overall decrease of 25% of the detergent phosphate, and it went into effect on October 1, 1981. The second step, on April 1, 1984, provided a 50% decrease, depending on the availability of environmentally sound phosphate alternatives.

The government of Ireland declared the reduction of phosphate inputs to waters as a key objective of their water pollution prevention objective. In 1999 a voluntary agreement was signed between the government and the Irish Detergents and Allied Products Association (IDAPA), which provides for an effective phasing-out of the marketing of phosphate-based domestic laundry detergent products [172]. It was said to be part of an incremental move to phosphate-free products by IDAPA members, with reduction targets of 55% by June 30, 2000, and 95% by December 31, 2002.

In Italy, a voluntary agreement exists that is in effect equivalent to a ban on phosphate in household laundry detergents. A first proposal for phosphate limits was introduced in March 1982. In negotiations with the industry in 1985, the content of sodium tripolyphosphate in household detergents was restricted to 4%. Finally, on

June 30, 1987, the phosphorus content in all laundry and dishwasher detergents was limited to 1% [25,40,46].

In Japan, detergent manufacturers voluntarily limited the phosphate level in laundry detergent powders to 10% or less in 1979. In the same year a regulation to prevent eutrophication of Lake Biwa was enacted and issued. Major detergent manufacturers launched phosphate-free synthetic detergents in 1980 [173]. In the meantime the use of sodium tripolyphosphate has been discontinued, and the industry made a total reformulation in favor of phosphate-free products [25,40].

In The Netherlands, a voluntary agreement has been negotiated with the industry to reduce phosphate levels in detergents [25,40].

In Norway, regulatory bans on phosphates in household detergents have been imposed. After January 1986, no household laundry detergent was permitted to contain more than 12% sodium tripolyphosphate [25,40,46].

In Poland, a voluntary agreement for phosphate restriction exists [46].

Sweden preferred the introduction of tertiary sewage treatment rather than a phosphate ban. In 1983 tertiary treatment served nearly 80% of the Swedish population [25].

In Switzerland, the Minister of State in 1977 ordered [30,174,175] limits for phosphates in laundry detergents that were stepwise tightened. In 1985 a ban of phosphates in laundry detergents was made obligatory, with a transitional phase of one year [25,40,46].

In the Czech Republic, a voluntary agreement is in effect for phosphate restriction [46].

As a result of the eutrophication problems encountered with the Great Lakes in the 1970s, the phosphorus content in household detergents used in the Great Lakes Basin was limited as a part of the Great Lakes Water-Quality Agreement [17,60]. Today a number of states of the United States have statewide limits of 0.5% phosphorus by weight in place: Arkansas, the District of Columbia (district-wide), Georgia, Indiana, Maine, Maryland, Massachusetts, Michigan, Minnesota, New Hampshire, New York, North Carolina, Oregon, Pennsylvania, South Carolina, Vermont, Virginia, Washington, and Wisconsin. In Connecticut and Florida, a limit of 8.7% phosphorus by weight is in force. In Idaho, Illinois, Missouri, Montana, Ohio, Rhode Island, and Texas, phosphate content in detergents is restricted by some cities and counties.

O. Ecotoxicological Aspects of Phosphates

Based on the LC_{50} value for the toxicity toward fish (1.65 g/L, golden orfe, determination after 48 h, in nonneutralized form; 2.4–4 g/L, limit of lethality, in nonneutralized form; 10–19 g/L, limit of lethality, at pH 7) [176] detergent phosphates can be classified as practically nontoxic in terms of acute aquatic toxicity.

The EC_{50} value (toxicity to *Daphnia*) was determined to be 1.154 g/L after 25 h and 1.089 g/L after 50 h [176].

The elimination from aquatic environment was determined in the OECD confirmatory test to be 30%, in mechanical-biological treatment plants 20–30%, and in tertiary treatment (phosphate elimination) 80–90% [176].

In the case of phosphate, the oxygen depletion is important. The EC_{10} value is 1.0–1.5 g/L (0.5 h, *Pseudomonas putida*) [176].

In the context of the ecological criteria for the award of the European community Eco-label to laundry detergents and automatic dishwasher detergents, the EU Directives 1999/427/EC and 1999/476/EC indicate in the DID lists a long-term effect (LTE) of 1,000 and a loading factor (LF) of 0.6 [158,159].

II. SODIUM SILICATES

A. Manufacture of Sodium Silicates

When considering the use of sodium silicates for detergents applications, one has to differentiate between amorphous and crystalline sodium silicates. Amorphous sodium silicates are hydrous sodium silicates (spray-dried powders and granules) and anhydrous silicates (sodium silicate lumps). The latter are possible precursors of aqueous sodium silicate solutions (water glass). Finally, there exist crystalline sodium silicates: sodium metasilicate in anhydrous form or as hydrate and crystalline layered sodium disilicates.

Amorphous silicates are differentiated by the ratio of SiO_2 to Na_2O on a molar basis or on a weight basis. This ratio of SiO_2 to Na_2O affects the properties of sodium silicates considerably [177]. Silicates of molar ratios between 0.5 and 4 are commercially available [24].

The anhydrous sodium silicates are commercially available in weight ratios SiO_2 to Na_2O in the range of 2.0 up to 3.3 [178,179]. They are prepared from quartz sand in a particle size range of 0.1–0.5mm and sodium carbonate. The sand is extracted from selected deposits that do not contain too many impurities. The most prominent contaminants to be avoided are organic components (in the hydrothermal process) and clay minerals [180]. The soda ash is produced by the ammonia-soda process (Solvay process) [181,182].

The mixture of quartz sand and soda ash is continuously fed into a furnace by means of a cooled screw conveyor [178]. The furnace is of a Siemens-Martin regenerative type and is heated by oil or gas [178,183,184]. The fusion of both components occurs at a temperature in the range of 1000°C up to 1500°C [6,178,183,185–188]. The melt of sodium silicate is removed continuously and cast into blocks, which are cooled. The blue to green or yellow-brown cullets are stored for subsequent production steps, like dissolution in water and milling to special product grades. Usually, the silicate lumps of a molar rtio of 3.33 [177] are the raw material for production of sodium water glass.

In the subsequent dissolution step, the solid silicate is dissolved in water in stationary or rotated vessels at temperatures of 100°C up to 150°C [178,183]. The pressure may be up to 5 bar. The solutions are clarified by settling at elevated temperatures or by filtration [178,183].

Finally, the solution may be confectioned by further concentration or by adjusting the molar ratio by addition of the other water glass types or sodium hydroxide solution [178,183]. The ratio of liquid sodium silicates can be adjusted to values ranging from 1.6 up to 3.75.

The hydrothermal route is an economic alternative for the production of water glass solutions with a ratio below 2.5 [178,185,186]. Quartz sand is dissolved in caustic soda in an autoclave [6,177,185–188]. The temperature may range from 130 to 200°C, and the pressure can be up to 20 bar. Thus a water glass solution of up to 48% solids by weight can be obtained in one production step. Finally, impurities are filtered from the sodium silicate solution [6,183,187,188].

The caustic soda is usually produced by alkali-chloride electrolysis [189]. The mercury cell process and the diaphragm process are commonly applied, but the membrane process, which is more economical and environmentally friendly, is gaining in importance [189–192].

Liquid sodium silicates are of complex composition made up from sodium and hydroxide ions as well as orthosilicate ions and a variable proportion of polysilicate anions [177,178]. The distribution of the constituents depends on the molar ratio of SiO_2 to Na_2O and the overall concentration of the solution.

Hydrous sodium polysilicates are amorphous silicates of SiO_2-to-Na_2O ratios between approximately 2 and 3 [185,186,193–195]. The moisture content of 5–20% determines the dissolution rate of the products in water [6,178,187,188,193,194]. They are typically obtained from liquid sodium silicates of adjusted molar ratio by drying in a spray tower or a drum drier [6,185–188]. The products are commercially available in this spray-dried powder form or as compacted granulate [185,186,193–195].

Sodium orthosilicate, sodium sesquisilicate, and sodium metasilicate are crystalline sodium silicates with SiO_2-to-Na_2O ratios of 0.5, 0.67, and 1.0, respectively [185,186]. The last is available either in anhydrous or in several hydrate forms [177].

Anhydrous metasilicate can be obtained by direct fusion of soda ash and sand at 1088°C [178] or by a solid-state reaction in a rotary kiln or in a fluidized bed at temperatures of 650–850°C [178,185,186].

For production of metasilicate pentahydrate, the molar ratio of water glass is adjusted with caustic soda and the heated solution crystallized in a fluidized bed or a drum granulator [178,196] and confectioned [6,187,188].

Sodium layered disilicate SKS-6 is a phyllosilicate with a crystal lattice characterized by silicate layers separated by sodium ions [197]. Spray-dried sodium disilicate is crystallized at 600–800°C in a rotary kiln [23,180,193,194,197,198]. The prepared layered silicate is finished to yield powder or granular product grades.

B. Life Cycle Assessment of Sodium Silicate Production

Life cycle assessments (LCAs) have gained importance as a model to gather environment-related data in a harmonized form. The environmental impact, e.g., of chemical products can be determined, a set of products can be compared, and proposals for improvements can be made [12–19]. Data regarding the energy and raw material consumption, the arising wastes, and the emissions into the air, water, and soil are recorded for the whole life cycle (cradle-to-grave assessment).

In the mid-1990s, a first German life cycle assessment dealt with the ecological aspects of the laundry wash process [181]. In a product line analysis (Produktlinienanalyse Waschen und Waschmittel), ecological data were gathered covering the process chain of production of detergent raw materials, the production of laundry detergents, and the domestic laundry washing process. In particular, the environmental impact of anhydrous sodium silicate of the composition $Na_2O \cdot 4\ SiO_2$ (solid) was estimated [181]. The production of sodium silicate consumes approximately 14 GJ/t primary energy and 1.7 t/t raw materials such as limestone for the production of soda ash by the Solvay process, sodium chloride, and sand and generate 1.9 t/t wastes and emissions.

In a later study, LCA data of detergent ingredients like sodium tripolyphosphate, zeolite (powder and slurry), sodium silicate (hydrothermal and furnace liquor, spray powder and furnace lumps, crystalline sodium metasilicate pentahydrate, and layered sodium disilicate SKS-6) were harmonized and partly newly compiled [6,187,188].

The production of sodium silicate spray powder (80% active matter, molar ratio $SiO_2:Na_2O = 2.0$) consumes 0.8 t/t resources of feedstock, 5.9 m^3/t water, and 18 GJ/t total primary energy, and 0.04 t/t disposal, 0.9 t/t air pollutants, 0.5 m^3/t wastewater, and 0.04 t/t water pollutants are produced [6].

The production of layered sodium disilicate is connected with the consumption of 1 t/t resources of feedstock, 4.3 m^3/t water, and 24.1 GJ/t total primary energy, and 2.2 kg/t recoverable wastes, 0.03 t/t disposal, 1.31 t/t air pollutants, 0.4 m^3/t wastewater, and 0.04 t/t water pollutants are produced [6].

The study was conducted to update the product line analysis of household laundry and detergents [181] and to provide data for the ecobalance of commercial laundry (UBA project "Life cycle assessments of detergent ingredients and their use in commercial laundries" of the Öko Institut e.V., Freiburg, Germany).

A comparison of selected data of ingredients that can actually be called detergent builders (sodium tripolyphosphate, zeolite A, sodium silicate spray powder, and layered sodium disilicate SKS-6) reveals the economy of sodium silicate production in terms of consumption of resources of feedstock and water and the production of solid waste and wastewater as well as air and water pollutants.

C. Biogeochemical Cycles and Nutritional Aspects of Silicates

The biogeochemical cycles of phosphorus, nitrogen, sulfur, and silicon are closely connected in freshwaters as well as in oceanic waters [84,199–203]. The main reservoirs in the silicon biogeochemical cycles are mineral silica and silicates, dissolved silica, sorbed silica, and biogenous silica.

Dissolved silica is released to the environment by weathering processes [89,202–204]. The average value of soluble silica in groundwater in the United States is 17 ppm [202,203,205]. Rivers and streams contain the highest levels of silica of the surface waters. The average of soluble silica in rivers worldwide is 13.1 ppm [202,203,206]. United States streams contain 14 ppm silica on average [203,205]. Public water supplies of the 100 biggest cities in the United States contain a mean value of 7.1 ppm of dissolved silica [203,207]. Lakes, seas, and oceans show the lowest concentrations of soluble silica. The average concentration of dissolved silica is about 6 ppm, or about 70 μM/L [202,203,208–210].

The solubility of amorphous silica in water appears to be defined by an equilibrium between the solid phase and a silica monomer, probably monosilicic acid [84,211]. Under normal environmental conditions, in terms of pH (pH 9) dissolved silica exists exclusively as monosilicic acid [203,212]. Also, the dissolved silica of ocean water seems to be mainly undissociated monosilicic acid [209].

Another closed biogeochemical cycle connects dissolved silica with silica that is adsorbed on suspended soil particles and sediments by adsorption and desorption processes [202,203,206,213]. This buffering mechanism appears to be responsible for the amount of soluble silica in natural waters [202].

The important biogeochemical cycle moves on from dissolved silica via assimilation processes to biogenous silica. There it is split up via dissolution processes in a short circuit back to dissolved silica and further via sedimentation and mineralization mechanisms back to mineral silica.

This large biogeochemical cycle is responsible for an annual turnover of 6.7 gigatons of silicon in marine biosystems [209,210]. The reason for this huge annual flux is the important skeletal role it fulfils in organisms [214]. Therefore silica is taken up and metabolized, e.g., by diatom algae [209,215–221]. Diatoms, silicoflagellates, radiolarians, and sponges are marine organisms that contain silica skeletons [209]. Diatoms have a dry weight of up to 50% silica [202,222]. Also, lower plants, such as grasses, are very rich in silica. In particular, wetland species typically contain the highest amount [203,223].

In organisms, biogenic silica may occur in a gel form, as soluble silica, or in animals bound to organic molecules such as polysaccharides, glycoproteins, and glycosaminoglycans [203,209,219,224,225]. The cell wall of diatoms (frustule) is built of amorphous silica with nanosize structures in the range of 2–50 nm [209,215]. In *Navicula pelliculosa*, the cell wall is composed of a silica shell and an organic skin [214,226]. In most diatoms the cell body is enclosed completely by the silica shell. If grown in nutrient media, the addition of silicon to *Navicula* induces synchronous silicon uptake, wall formation, and cell separation [214,227].

Diatom biomass production can be limited by the availability of dissolved silica [228]. At concentrations of less than 0.1 ppm, dissolved silica appears to be a limiting nutrient for diatoms and other algal species [95,202,203,229–231]. Ocean waters have lower nitrogen-to-phosphorus and silicon-to-phosphorus ratios than do river waters [89]. Measurements of dissolved inorganic nitrogen, phosphorus and silicon indicate that dissolved silicate controls the magnitude of diatom production during the spring bloom [228]. There is experimental evidence that diatom dominance occurs if silicate exceeds a concentration of approximately 2 μM [232]. Once the limiting concentration has been exceeded, diatom growth is independent of silica [202,203,233,234]. Diatoms are seen to be the most desirable phytoplankton in coastal waters because they are important in marine food chains, they do not form noxious surface blooms, they are not toxic, and they have no offensive visual or olfactory properties [89,202,235,236]. If depletion of dissociated silicate relative to nitrogen and phosphorus occurs and silica concentration drops below the limiting concentration for diatoms, this can favor nonsiliceous flora and the shift from a diatom-dominated population to one dominated by cyanobacteria and green algal species [89,202,203,228,231,237–240]. Decline of dissolved silica content, particularly of the surface water, occurs in eutrophic waters due to large phosphate input [203,237]. Diatoms are not so strongly associated with undesirable effects of eutrophication as are flagellate communities [232,236].

Changes in silicate retention in upstream sections of the Rhine River show that the pre-Alpine lakes have contributed to nondiatom phytoplankton blooms in receiving waters [199]. Eutrophication and construction of artificial lakes may have led to a decreased silicate input to the central North Sea and Elbe-influenced wates [232,241]. So it seems to be important to maintain the soluble silica levels to support diatom populations rather than green and blue-green algal species [202]. Also, this is because available silica can channel nitrogen and phosphorus into diatoms and away from less desirable species [89].

Domestic wastes typically have low contents of silicon in comparison to nitrogen or phosphorus [89,235]. Commercial sodium silicate once dissolved and diluted with water is indistinguishable from silica from natural sources. This is because polysilicate anions are depolymerized very rapidly upon dilution with water [203,242]. There have been efforts to calculate the impact of silicate input by use of detergents on silicate concentrations of surface waters [22,243]. It can be estimated that a linear relation exists between detergent use and silicate concentration in sewage entering municipal wastewater treatment plants. Detergent consumption (per capita per day) is an important parameter, as is the silicate content in the detergent and the sewage amount. The silicate concentration in tap water can be excluded for an instantaneous examination. With the assumption that silicate may not be detained in the sewage plant and ideal mixing may occur in the receiving river water, it can be found that the silicate concentration by detergent use depends on the dilution ratio of the river water flow to the sewage works effluent volume. As a result the silicate input is small in comparison to the natural silicate background level. To simulate silicate conditions in rivers, a plug flow model was developed on the basis of sky irradiance, light attenuation in the water, and integrating photosynthetic rates determined in the laboratory. Fluctuations in the silicate concentration were simulated and found to match observed changes in the Rhine River in The Netherlands [244].

D. Ecotoxicological Aspects of Sodium Silicates

No ecologically problematic properties of sodium silicates are known [245,246]. The aquatic toxicity of sodium silicates toward annelids was determined with 210–250 gram-atoms per liter [203,234], toward mosquitofish 2,320–3,200 ppm [203,247], toward *Daphnia magna* 247 ppm, toward snail eggs 632 ppm, and toward Amphipoda 160 ppm [203,248]. The LC_0 value was determined to be in excess of 1,000 g/L, and fish are affected only via the pH value [176].

In wastewater treatment plants, no elimination was found [176].

Regarding the ecological criteria for the award of the European Community Eco-label to laundry detergents and automatic dishwasher detergents, the EU Directives 1999/427/EC and 1999/476/EC indicate in the DID lists an EC_{50} value of more than 1,000 mg/L, a long-term effect of 1,000, and a loading factor of 0.8 [158,159].

III. ZEOLITE A

A. Manufacture of Zeolite A

Zeolite A for detergent builder use is most frequently manufactured by the aluminum silicate hydrogel route. Here a sodium aluminum silicate hydrogel is obtained from sodium silicate and sodium aluminate and converted to zeolite A by crystallization.

Sodium silicate production starts with sodium chloride mining and production of caustic soda by electrolysis of the sodium chloride brine. Quartz sand is dissolved in caustic soda to obtain sodium silicate solution.

The first step of sodium aluminate production is bauxite mining, followed by extraction of aluminum trihydroxide by the Bayer process. Alumina is converted to sodium aluminate by dissolving it in caustic soda.

Sodium chloride can be extracted from salt deposits either by rock salt mining or by solution mining, where water is pumped into a natural salt deposit. The resulting saturated brine solution is pumped back to the surface [249].

Sodium hydroxide solution is obtained by electrolysis of the brine by the amalgam cell, by diaphragm, or by the membrane cell process [189–192,250]. Sodium silicate solution is obtained in a concentration of approximately 42% after quartz sand is dissolved hydrothermally in caustic soda [6,178,251].

The world's largest producer of bauxite [251] is Australia, where the predominant mineral is gibbsite [252].

The Bayer process starts with the blending of the crushed and ground bauxite for uniform composition [181,251,252]. In most plants bauxite is ground while suspended in a portion of the process solution. This slurry is mixed with additional heated caustic soda and then treated in a digester vessel under conditions above atmospheric pressure [252,253]. Thus, the aluminum-containing minerals, such as boehmite, gibbsite, and diaspore, are dissolved under formation of sodium aluminate solution.

Most of the other components are insoluble under process conditions. Siliceous minerals react to soluble silicates that subsequently react with the constituents of the process solution to a series of poorly soluble precipitates having zeolite structure. The presence of seed particles is important [252].

Present phosphate affects clarification adversely. To avoid this, calcium hydroxide is added to precipitate calcium phosphate. Calcium hydroxide for the Bayer process is obtained by quarrying limestone. This is crushed, ground, sieved, dried, and finally calcined in a kiln [251].

The digestion process leaves the bauxite residue solids, such as aluminum oxide, titanium oxide, iron oxide, silicates, and calcium phosphates, as waste products (red mud). These are separated from the sodium aluminate solution and washed and discarded. The red mud has a pH of 7–9 [251], and its disposal represents the most important environmental problem of the Bayer process [252].

Previously, red mud was disposed of to the marine environment, but today this is unlikely to be permitted for new refineries.

In Australia and Africa, it is neutralized in big ponds, which may be recultivated after a few years. In Europe the red mud is diluted and filter-pressed into lagoons or shut-down mines. Therefore a convenient valley may be dammed or retaining dikes may be built on the flat land to form residue disposal areas.

For dry-stacking, the thixotropic residue is concentrated to 35–50 wt% solids. The slurry is agitated to reduce its viscosity and is pumped to the disposal area. The desilication products in the residue have ion exchange properties, and endrained fine residues do not have enough strength to support buildings or equipment even after years of consolidation. Caustic soda losses are mainly with the clay and iron oxide particles [253]. So the solution left with the residue is still very alkaline and may not make contact with the groundwater [252].

Solubility of aluminum hydroxide depends on temperature. Therefore the process solution is cooled and seeded with fine crystals of aluminum hydroxide. After half of the aluminum hydroxide content has been crystallized, the suspension of solids and solution is classified. Coarse particles are washed and calcined [252].

The main commercial process for the manufacture of sodium aluminate for zeolite production is to dissolve aluminum trihydroxide from the Bayer process in 10–

30% caustic soda at a temperature near the boiling point [254]. This affords a stable, concentrated solution of sodium aluminate [181,185,186,251].

In the hydrogel zeolite production route, sodium silicate and sodium aluminate are mixed together with sodium hydroxide to yield an amorphous sodium alumino-silicate gel [181,251,255]. This is stirred until homogeneous and crystallized at temperatures of 70–110°C for 1–8 hours [181,185]. The rate of zeolite formation is affected by the reactivity of the gel, the pH, and the temperature of the process solution, and the addition of seeds [256]. The crystals are separated from the mother liquor and the wet filter cake dried, e.g., by spray-drying [181,185,186,251].

B. Life Cycle Assessment of Zeolite A Production

Life cycle assessments are procedures to gather environmental related data, e.g., on a chemical product, in harmonized and peer-reviewed form. Typically, the environmental data cover raw material and energy consumption as well as the generated wastes and emissions into the air, water, and soil over the whole life cycle (cradle to grave) [12–19,257,258].

Three major life cycle assessments regarding zeolite A have been conducted. In 1993 an LCA study on the average situation of zeolite A production in Europe in 1993 was carried out by the EMPA Institute in St. Gallen, Switzerland, and commissioned by ZEODET (Association of Detergent Zeolite Producers) [6,251,257,258]. It was a cradle-to-factory-gate study excluding subsequent use of zeolite in detergent powders and its final fate.

In a later study the zeolite data were compiled together with data on other detergent ingredients, e.g., sodium silicates and sodium tripolyphosphate [6]. For the production of zeolite A, 1.5 t/t feedstock resources, 19 m^3/t water, and 26.5 GJ/t total primary energy are consumed, and 50.5 kg/t recoverable wastes, 0.3 t/t disposal, 1.5 t/t air pollutants, 3 m^3/t wastewater, and 0.6 t/t water pollutants are produced [6].

It was the starting point for the ecobalance of commercial laundry (UBA project "Life cycle assessments of detergent ingredients and their use in commercial laundries" of the Öko Institut e.V., Freiburg, Germany).

The later Landbank Phosphate Report [7] came to the main conclusion that the phosphate builder system is equivalent to the zeolite–polycarboxylate builder system. Later critics commented that the study had not complied with standardized requirements of LCAs, e.g., clear definition of the functional unit, incorporation of pre-combustion energy, and peer review [258].

A third study developed a product line analysis covering the process chain of production of detergent raw materials, the production of laundry detergents, and the domestic laundry washing process (Produktlinienanalyse Waschen und Waschmittel) [181]. For zeolite A the study cited a primary energy consumption of 41.1 GJ/t, raw materials consumption of 1.9 t/t (e.g., limestone, sodium chloride, sand, bauxite), and waste and emissions output of 2.8 t/t [181].

C. Behavior of Zeolite A in Municipal Wastewater Treatment

With the market introduction of zeolite A, extensive studies have been carried out to demonstrate its environmental safety. One of the key issues of concern was its behavior in the sewerage system. Calculations had shown that the mineral content in the sewage would increase by 40% in the domestic sewage and by 29% in the collector sediment [259]. Two field studies in Germany and one in Belgium have shown that no sludge deposits accumulated or solidified over a prolonged usage period of zeolite-based detergents, either in domestic sewers or in the main drains [260–262]. Zeolite causes no sediment deposits and hence clogging in domestic and municipal sewerage systems [26,246,255,259,260–266].

The major part (i.e., 21–80%) of the incoming zeolite is removed in the mechanical stage of sand trap and primary settler [26,176,246,255,259,263].

The remaining portion is subjected to biological treatment. After passing through the biological stage and the second settler, up to 95% of the zeolite has been eliminated, becoming part of the activated sludge [26,246,255,259–261,263,267].

The whole picture is that sodium calcium zeolite A is present in the washing liquor in typical concentrations of 150–200 mg/L. In the sewer and raw sewage, the concentration of calcium sodium zeolite is 15–30 mg/L. In the sewage treatment sludge, 90% of the origin zeolite is enriched, causing a concentration of 50–150 g/kg of calcium aluminum silicate phosphate. In the sewage treatment plant effluent, 10% of the origin zeolite is present, responsible for 1.5–3.0 mg/L of calcium aluminum silicate phosphate and calcium sodium zeolite A. In the receiving river, the concentration of calcium aluminum silicate phosphate and calcium sodium zeolite A is 0.1–0.5 mg/L [260], depending on the dilution factor.

Studies have shown that zeolite is transferred under wastewater treatment plant conditions together with calcium and phosphate ions to an extremely insoluble amorphous complex of basic aluminum silicate phosphate [26,257,258].

Further issues address the possible impact of zeolite on the operation and performance of the biological treatment stage. Laboratory and field tests indicated that zeolite has no detrimental effect on the aerobic and anaerobic treatment, even at high concentrations and enrichment in the sludge [26,176,255,259–264,268,269].

In some tests, positive effects of zeolite on biological treatment were found [264,270], such as a higher biodegradation of surfactants [255,271]. Other positive effects are improvements in sedimentation and dehydration properties of the sludge [260,267] and advantages with the nitrification [255,272,273].

The volume of sewage sludge is increased due to the enrichment with zeolite [260,263,265,274].

Other issues have been the uptake and release of heavy-metal ions, with impacts on their remobilization, e.g., from river and lake sediments [264,275,276]. Besides calcium and magnesium, lead, copper, silver, cadmium, zinc, and mercury ions are incorporated. Tests using pretreated municipal sewage showed a significant uptake of added heavy-metal ions and their immobilization [269,276].

Further questions address the fate of zeolite during sludge treatment [260,277]. During sludge incineration, the zeolite structure is completely destroyed and a mixture of aluminosilicates remains. No formation of quartz or cristobalite has been detected. The disposal of the sludge ash should pose no hazard [258,260].

In cases of sludge distribution to farmland, no disadvantage has yet been discovered. No remobilization of heavy metals occurred [258]. Studies have shown that possible effects on plants are determined more by the remaining composition of sludge than by the presence of zeolite [260,265,266].

An alternative possibility for sludge disposal is landfill [260].

In studies, zeolite A did not affect the phosphorus elimination with chemicals. The chemicals used enhanced removal of zeolite [260,261].

D. Behavior of Zeolite A Toward Aquatic and Terrestrial Ecosystems

Depending on the conditions, zeolite A is hydrolyzed to sodium aluminate and sodium silicate or converted to amorphous aluminosilicate, respectively [176,202,246,260, 263,278]. The half-life of hydrolysis is 1–2 months under conditions of natural waters [176,202,279]. In calcium- and phosphate-containing water with a half-life of 12–30 days, poorly soluble calcium aluminosilicate phosphates were formed having no ion exchange properties and therefore no ability to contribute to the remobilization of heavy metals or to lead to an increase in the level of soluble aluminium compounds [260].

Tests on the uptake and release of heavy-metal ions by zeolite A have shown that in synthetic tap water and river water (Rhine) no remobilization of sediment-bound heavy metals occurs [255,264,270,276].

Model calculations show that the contribution to the total environmental silicon burden from use of zeolite-containing detergents is almost negligible [260].

A study undertaken in 1979 by the German Federal Environmental Agency concluded that zeolite A does not display a harmful influence on aquatic organisms. It was not found to be toxic toward algae-like diatoms nor found to promote their growth; i.e., it does not have nutritional properties. The results were corroborated by later studies [255,260,264–266,280]. In lakes and rivers, no harmful effects on phytoplankton, zooplankton, and fish have been observed in 10 years of experience with zeolite-based detergents [260,281]. A growth-inhibiting effect on algae was observed at concentrations in excess of 10 mg/L in nutrient-poor culture media [26]. Other studies were carried out regarding the toxicity of zeolite toward bacteria, daphnia, and fish, which, besides algae, are important links in the food chain [270,282].

From the point of view of an acute toxicity toward fish (LC_{50} > 16,500 mg/L, golden orfe) and daphnia (EC_{50} > 1,000 mg/L) [176], zeolite A is not considered to be harmful. The acute and long-term toxicity of zeolite toward daphnia and fish is assessed as very low [26,176,255,264,270,283,284]. The EC_0 was determined to be larger than 250 mg/L [26,270,283].

For *Daphnia magna*, NOEC values of 10 mg/L [283] up to 37 mg/L (21-d reproducibility test) [285] were determined in long-term toxicity tests. Long-term ecotoxicology data from fish (early-life-stage test) revealed NOEC values of 87 mg/L [26]. In tests of commercial detergents based on phosphate and zeolite, there are indications that zeolite-based detergents could be somewhat more toxic to the tested fish species (*Poecilia reticulata*) than phosphate-based ones [284,42]. One possible reason may be the insolubility of zeolite particles, which could, after some time, accumulate in fish gills and cause their dysfunction. On the other hand, dissolved aluminum has been found to be toxic to fish [202,286]. Solubilized aluminum is able to bind to the epithelium of the fish's gill and causes loss of osmoregulatory functions [287]. These toxic effects seem to be blocked if there is a molar excess of silicon to aluminum, which makes the formation of a hydroxyaluminosilicate species possible. The latter is believed to limit the bioavailability of soluble aluminum [202].

Further investigations have been targeted on the effects of zeolite toward important benthic organisms, such as shellfish and river worms (*tubifex*). River worms and shellfish did not accumulate zeolite A under different test conditions (e.g., laboratory river model) during a 4-week exposure to 250 mg/L [255,264,270].

The zeolite elimination in mechanical-biological wastewater treatment plants was 96% [176].

The EU Directives 1999/427/EC and 1999/476/EC quote in their DID lists a NOEC level of 120 mg/L, a long-term effect of 120, and a loading factor of 0.05 for zeolite [158,159].

Issues of concern regarding the toxicity of zeolite toward the terrestrial environment may be due to the mobilization of aluminum, e.g., by acid rain under very acid conditions. Aluminum is toxic to the root hairs of tree roots and the mycorrhizal complex surrounding the roots [287,288,289].

PNEC values for zeolite A can be derived from NOEC values [36,163,290]. A PNEC for aquatic organisms of 3.7 mg/L is derived from a NOEC of 37 mg/L, which is the most sensitive value [283,285,290,291] from a chronic test data set on algae, invertebrates, and fish. Calculations of the PEC leads to river water concentrations of 150–300 µg/L, further decreasing to 57 µg/L if a half-life of 1–2 months is taken into account [26,263,292]. Since these PEC values are far below the PNEC, this is a strong rationale for the environmental safety of zeolite A.

A PNEC for terrestrial organisms of 20 mg/kg is derived from a NOEC of 1000 mg/kg [290], which is obtained from data on plants.

A PNEC for microorganisms in wastewater treatment plants of 33 mg/L is derived from a NOEC of 330 mg/L from an EC_{10} value referring to *Pseudomonas putida* [290,293].

Finally, a PNEC for sediment organisms of 2.3 mg/kg is derived from an aquatic NOEC [290].

ACKNOWLEDGMENTS

I should like to thank all individuals who provided their expertise. My gratitude to Catharine Dryden for proofreading the manuscript. I am indebted to Clariant GmbH/ Functional Chemicals Division for making infrastructure available to me. My heartfelt thanks go to my dear wife, Susanne Lang-Bauer, and my children, Eva-Maria Bauer and Claus-Jürgen Bauer, for their great patience, forbearance, and mental support while I was working on this chapter.

REFERENCES

1. Emsley, J.; Hall, D. *The Chemistry of Phosphorus*; Harper & Row: London, 1976.
2. Notholt, A.J.G.; Hartley, K. *Phosphate Rock: A Bibliography of World Resources*; Mining Journal Books, London: London, 1983.
3. Phosphate waste: Florida's dilemma. Phosphorus Potassium 1997, *211*, 38–44.
4. Altschuler, Z.S. In *Environmental Phosphorus Handbook*; Griffith, E.J., et al., Eds.; Wiley: New York, 1973.
5. Schrödter, K.; Bettermann, G.; Staffel, T.; Hofmann, T. In *Phosphoric acid and phosphates. Ullmann's Encyclopedia of Industrial Chemistry*, 5th Ed.; Elvers, B., Hawkins, S., Schulz, G., Eds.; VCH: Weinheim, 1991; Vol. A19, 465–503.

6. Dall'Acqua, S.; Fawer, M.; Fritschi, R.; Allenspach, A. *Life Cycle Inventories for the Production of Detergent Ingredients*; EMPA: St. Gallen, Switzerland, 1999.

7. Wilson, B.; Jones, B. *The Phosphate Report*; Landbank Environmental Research and Consulting, London: London, 1994.

8. NPK production and the environment. Phosphorus Potassium 1997, *209*, 42–51.

9. Raytheon's isothermal reactor process. Phosphorus Potassium 1997, *211*, 45–50.

10. Wissa, A.E.Z.; Fuleihan, N.F. Control of groundwater pollution from phosphoric acid waste gypsum stacks. In *Sulfuric/Phosphoric Acid Plant Operations*; Englund, S.W., Schneider, R.T., Eds.; American Institute of Chemical Engineers, New York: New York, 1982.

11. Hudson, R.B.; Dolan, M.J. Phosphoric acids and phosphates. In *Kirk-Othmer Encyclopedia of Chemical Technology*; Mark, H.F. Othmer, D.F., Overberger, C.G., Seaborg, G.T., Grayson, M., Eckroth, D., Eds.; Wiley: New York, 1982; Vol. 17.

12. Thomé-Kosmiensky, K.J.; Willnow, S.; Fleischer, G.; Schmidt, W.P.; Christ, C.; Menges, G.; Bilitewski, B.; Loll, U.; Gromotka, H.; Amsoneit, N.; Baerns, M.; Majunke, F.; Ehrig, H.J.; Schneider, H.J.; Gossow, V. Waste. In *Ullmann's Encyclopedia of Industrial Chemistry*, 5th Ed.; Elvers, B., Hawkins, S., Russey, W., Eds.; VCH: Weinheim, 1995; Vol. B8, 559–761.

13. DIN Deutsches Institut für Normung e.V., ed. ISO 14040: 1997 Environmental management—life cycle assessment—principles and framework. Berlin: Beuth Verlag, 1997.

14. DIN Deutsches Institut für Normung e.V., ed. ISO 14041: 1998, Environmental management—life cycle assessment—goal and scope definition and life cycle inventory analysis. Berlin: Beuth Verlag, 1998.

15. DIN Deutsches Institut für Normung e.V., ed. ISO 14042: 2000, Environmental management—life cycle assessment—life cycle impact assessment. Beuth Verlag: Berlin, 2000.

16. DIN Deutsches Institut für Normung e.V., ed. ISO 14043: 2000, Environmental management—life cycle assessment—life cycle interpretation. Berlin: Beuth Verlag, 2000.

17. Baird, C. *Environmental Chemistry*, 2nd Ed., Freeman: New York, 2000.

18. Troge, A. Ökobilanzen und produktbezogene Umweltpolitik. Ökobilanzen und Produktverantwortung, 2000.

19. Wagner, G. *Waschmittel*, 2nd Ed.; Ernst Klett Verlag: Stuttgart, 1997.

20. Wilson, B.; Jones, B. *The Swedish Phosphate Report*; Landbank Environmental Research and Consulting, London: London, 1995.

21. Nicholls, D.J. Eutrophication and excessive macroalgal growth in Lake Macquarie, New South Wales. Thesis, University of New South Wales, 1999.

22. Bauer, H.; Schimmel, G. Neue Aspekte bei der Anwendung des Builder-Silikates SKS-6. Tenside Surf. Det. 1997, *34*, 425–429.

23. Bauer, H.; Schimmel, G.; Jürges, P. The evolution of detergent builders from phosphates to zeolites to silicates. Tenside Surf. Det. 1999, *36*, 225–229.

24. Gorlin, P.A.; Dixit, N.; Lai, K.-Y. Liquid automatic dishwasher detergents. In *Liquid Detergents*; Lai, K.-Y., Ed.; Marcel Dekker: New York, 1997; 325–375.

25. Jakobi, G.; Löhr, A.; Schwuger, M.J.; Jung, D.; Fischer, W.K.; Gerike, P.; Künstler, K. Detergents. In *Ullmann's Encyclopedia of Industrial Chemistry*, 5th Ed.; Gerhartz, W., Yamamoto, Y.S., Kaudy, L., Pfefferkorn, R., Rounsaville, J.F., Eds.; VCH Verlagsgesellschaft: Weinheim, 1987; Vol. A8.

26. Smulders, E.; Hähse, W.; Rybinski, W.; Steber, J.; Sung, E.; Wiebel, F. *Laundry Detergents*; Wiley-VCH: Weinheim, 2002; 165–203.

27. Nusch, E.A.; Friedrich, G.; Davis, J.; Fischer, W.R.; Frank, C.; Hamm, A.; Heckman, C.; Herbst, V.; Kopf, W.; Lenhart, B.; Müller, D.; Pinter, J.; Schilling, N.; Schindele, X.; Schulte-Wülwer-Leidig, A.; Steinberg, C. Eutrophierung gestauter und freifließender Gewässer. In *Studie über die Wirkungen und Qualitätsziele von Nährstoffen in Fließgewässern*; Hamm, Alfred, Ed.; Sankt Augustin, Academia Verlag, 1991.

28. Flanagan, P.J. Parameter of Water Quality. *Interpretation and Standards*, 2nd Ed.; Environmental Research Unit: Dublin, 1990.

29. Metzner, G. Tenside Surf. Det. 2001, *38*, 360–367.

30. Au, I.; Leschber, R. Seifen—Öle—Fette—Wachse 1979, *105*, 337–340.

31. Beier, E. *Umweltlexikon für Ingenieure und Techniker*; VCH: Weinheim, 1994.

32. Bernhardt, H., Ed.; *Phosphor—Wege und Verbleib in der Bundesrepublik Deutschland*; Verlag Chemie: Weinheim, 1978.

33. Berth, P.; Jakobi, G.; Schmadel, E.; Schwuger, M.J.; Krauch, C.H. Angewandte Chemie 1975, *87*, 115–142.

34. Detergent phosphates: a sustainable detergent component, URL: Ceep-phosphates.org/ conf2001/Detergent_phosphates_engl.pdf.

35. Europe's Environment: The Second Assessment. Luxembourg: European Environmental Agency, Office for Official Publications of the European Communities, Luxembourg, 1998.

36. Fränzle, O.; Stražkraba, M; Jørgensen, S.E. Ecology and ecotoxicology. In *Ullmann's Encyclopedia of Industrial Chemistry*, 5th Ed.; Elvers, B., Hawkins, S., Russey, W., Eds.; VCH: Weinheim, 1995; Vol. B7, 19–154.

37. Frede, H.G.; Bach, A. Stoffbelastungen aus der Landwirtschaft. In Belastungen der Oberflächengewässer aus der Landwirtschaft. Frankfurt am Main: BLV-Verlags-Gesellschaft, 1993.

38. Hamm, A.; Gleisberg, D.; Hegemann, W.; Krauth, K.H.; Metzner, G.; Sarfert, F.; Schleypen, P. Stickstoff- und Phosphoreintrag in Oberflächengewässer aus punktförmigen Quellen. In *Studie über die Wirkungen und Qualitätsziele von Nährstoffen in Fließgewässern*; Hamm, A., Ed.; Academia Verlag: Sankt Augustin, Germany, 1991.

39. Hamm, A. Phosphate-free or phosphate-containing detergents. Tenside Surf. Det. 1991, *28*, 476–481.

40. Köhler, J. Detergent phosphates and detergent ecotaxes: a policy assessment. Brussels: Centre Européen d'Etudes des Polyphosphates CEEP, 2001.

41. Litke, D.L. Review of Phosphorus Control Measures in the United States and Their Effects on Water Quality, Water-Resources Investigations Report 99-4007; U.S. Geological Survey, Information Services: Denver, 1999.

42. Phosphate removal and recovery from wastewaters. Phosphorus Potassium 1998, *213*, 30–39.

43. Staffel-Schierhoff, U. Schadstoffgruppenorientierte Aktionskonzepte am Beispiel der Nährstoffreduzierung. In *Der Wassersektor in Deutschland, Methoden und Erfahrungen*; Rudolph, K.U., Block, T., Eds.; Berlin–Bonn: Bundesministerium für Umwelt, Naturschutz und Reaktorsicherheit, 2001.

44. Sprenger, F.J. Untersuchungen über die Phosphor- und Stickstoffbelastungen der Zuflüsse eines Sees. Vom Wasser 1967, *34*, 146.

45. Sprenger, F.J. Untersuchungen über die Phosphor- und Stickstoffbelastungen der Zuflüsse eines Sees. II. Teil. Vom Wasser 1968, *35*, 137–149.

46. Umweltbundesamt. Wasch- und Reinigungsmittel—Trends auf dem Wasch-und Reinigungsmittelmarkt, URL: http://www.umweltbundesamt.de/uba-info-daten/daten/ wasch/trends.htm, Umweltbundesamt, 2002.

47. United Nations E/CN.17/1997/9, Economic and Social Council, Commission on sustainable development, Fifth session, 7–25 April 1997.

48. Umwelt und Chemie von A - Z, Verband der Chemischen Industrie, 10th Ed, Freiburg im Breisgau: Verlag Herder, 1996.

49. Water facts, River and estuary pollution. East Perth: Water and Rivers Commission, Level 2, East Perth, WA, 1997.

50. Mohaupt, V.; Herata, H.; Mach, M.; Behrendt, H.; Fuchs, S. Kläranlagen saniert— Woher kommen Gewässerbelastungen heute? Wasser, Berlin, 2000.

51. Aquatic eutrophication in England and Wales, Environmental Issues Series. Almondsbury, Bristol BS32 4DU 1998. Environment Agency, 1998.

52. Schulze-Rettmer, R. GDCh-Fortbildungskurs Roetgen. 1980; 174–184.

53. Böhnke, B. Gutachten zur Belastung der Gewässer durch Waschmittelinhaltsstoffe aus Regenüberläufen und kommunalen Kläranlagen mit alleiniger mechanischer Reinigung. Frankfurt am Main: Industrieverband Körperpflege- und Waschmittel (IKW), 1991.

54. Krüssmann, H.; Hloch, H.G. Seifen Öle Fette Wachse 1981, *107*, 436–442.

55. Morse, G.K.; Lester, J.N.; Perry, R. *The Economic and Environmental Impact of Phosphorus Removal from Wastewater in the European Community*; Selper Publications: London, 1993.

56. Implementation of the 1991 EU urban wastewater treatment Directive and its role in reducing phosphate discharges, Summary of the report, SCOPE Newsletter no. 34. Brussels: CEEP, 1999.

57. *The Great Lakes: An Environmental Atlas and Resource Book*, 3rd Ed.; Government of Canada, Toronto, Ontario and United States Environmental Protection Agency, Great Lakes National Program Office: Chicago, 1995.

58. Protocol Amending the 1978 Agreement Between the United States of America and Canada on Great Lakes Water Quality, as Amended on 16. 10. 1983, Reaffirming their commitment to achieving the purpose and objectives of the 1978 Agreement between the United States of America and Canada on Great Lakes Water Quality, as amended on October 16, 1983.

59. Lee, G.F.; Jones, R.A. The North American Experience in Eutrophication Control Through Phosphorus Management. Proc Int Conf Phosphate, Water and Quality of Life, Paris, 1988.

60. Neilson, M.; L'Italien, S.; Glumac, V.; Williams, D. Nutrients: Trends and System Response, SOLEC Working Paper presented at State of the Lakes Ecosystem Conference, EPA 905-R-95-015. U.S. Environmental Protection Agency: Chicago, 1995.

61. Council Directive 91/271/EEC concerning urban wastewater treatment (The Urban Wastewater Treatment Directive, Official Journal L 135, 30.05.91), as amended by Commission Directive 98/15/EC (Official Journal L 67, 07.03.98).

62. Hamm, A. Problembereich Nährstoffe aus wasserwirtschaftlicher Sicht. In *Belastungen der Oberflächengewässer aus der Landwirtschaft*; Thoroe, C., Frede, H.G., Langholz, H.J., Schumacher, W., Werner, W., Eds.; Wissenschaftliche Arbeitstagung, Bonn, 1993.

63. Environmental Risk Assessment of Detergent Chemicals, Proceedings of the A.I.S.E / CESIO Limette III Workshop, 1993.

64. Allgemeine Rahmen-Verwaltungsvorschrift über Mindestanforderungen an das Einleiten von Abwasser in Gewässer—Rahmen Abw. VwV vom 25.11.1992, Anhang 1, Bundesanzeiger, Jg. 44, Nr. 233b.

65. Boughton, W.H.; Gottfried, R.J.; Sinclair, N.A.; Yall, I. Metabolic Factors Affecting Enhanced Phosphorus Uptake by Activated Sludge. Appl. Microbiol. 1971, *22*, 571–577.

66. Weitergehende Reinigung kommunaler Abwässer insbesondere zur Phosphatelimination. Hoechst-Symposium im Werk Knapsack, 1982.

67. Kandler, J. Phosphates as detergent builders. In *Formulation Technology: Builder Systems in Detergent Products*, Baldwin, A.R., Ed.; Proceedings of the Second World Conference on Detergents; American Oil Chemists Society.

68. Imhoff, K.; Imhoff, K.R. *Taschenbuch der Stadtentwässerung*, 28th Ed.; Oldenburg Verlag: Munich, 1993.

69. Seyfried, C.F. Fällungs- und Flockungsverfahren in der Abwasserreinigung, Hoechst Symposium, Frankfurt/Neu-Isenburg, 1986.

70. Simmler, W.; Mann, T.; Berger, M.; Kern, G.; Lemke, J.; Klockner, D.; Neuber, E.; Hebbel, G.H.; Werthmann, U.; Klinsmann, G.; Lawson, J.F.; Meyer, H.G.; Müller, M.;

Balser, K.; Maier, W.; Frieser, J.; Thüer, M.; Malaszkiewicz, J.; Aivasidis, A.; Koglin, B.; von Kienle, H.; Wegmann, U.; Weisbrodt, W.; Moldenhauer, W.; Mischer, G. Water. In *Ullmann's Encyclopedia of Industrial Chemistry*, 5th Ed.; Elvers, B., Hawkins, S., Russey, W., Eds.; VCH: Weinheim, 1995; Vol. B8, 1–152.

71. Jardin, N. Einfluss der biologischen P—Elimination auf die Schlammbehandlung. Mitt. Oswald-Schulze-Stiftung 1993; *16*, 9/1–9/34.

72. Cech, J.S.; Hartmann, P.; Wanner, J. Competition between polyP and non-polyP bacteria in an enhanced phosphate removal system. Water Environ. Res. 1993, *65*, 690–692.

73. Nultsch, W. *Allgemeine Botanik*, 8th Ed.; Thieme Verlag: Stuttgart, 1986.

74. Bever, J.; Stein, A.; Teichmann, J. *Weitergehende Abwasserreinigung*, 2nd Ed.; Oldenburg Verlag: Munich, 1993.

75. Streit, B. *Lexikon Ökotoxikologie*, 1st ed.; VCH: Weinheim, 1991.

76. Woods, N.C.; Sock, S.M.; Daigger, G.T. Phosphorus recovery technology modelling and feasibility evaluation for municipal wastewater treatment plants. Environ. Technol. 1999, *20*, 663–679.

77. Ramanathan, M. Water Pollution. In *Kirk-Othmer Encyclopedia of Chemical Technology*; Mark, H.F. Othmer, D.F., Overberger, C.G., Seaborg, G.T., Grayson, M., Eckroth, D., Eds.; Wiley: New York, 1984; Vol. 24.

78. Council Directive 1999/31/EC of 26 April 1999 on the landfill of waste. Official Journal L 182, 16/07/1999, 1–19.

79. Phosphates recovery for recycling from sewage and animal wastes. Phosphorus Potassium 1998, *216*, 17–21.

80. Towers, W.; Horne, P. Sludge recycling to agricultural land: the environmental scientist's perspective. J. CIWEM 1997.

81. Edge, D. Perspectives for nutrient removal from sewage and implication for sludge strategy. Environ. Technol. 1999, *20*, 759–763.

82. Werther, J.; Oganda, T. Sewage sludge combustion, progress in energy and combust. Science 1999, *25*, 55–116.

83. Guardian, Sludge on farms row, Guardian, 26th Feb. 1998.

84. Stumm, W.; Morgan, J.J. *Aquatic Chemistry—Chemical Equilibria and Rates in Natural Waters*, 3rd Ed.; Wiley: New York, 1996.

85. Redfield, A.C. In *The Sea*; Hill, M.N., Ed.; Wiley: New York, 1966; Vol. II.

86. Stumm, W.; Morgan, J.J. *Aquatic Chemistry*, 2nd Ed.; Wiley Interscience: New York, 1981.

87. Vollenweider, R.A.; Harris, G.P. Elemental ratios in marine and freshwater plankton. In *Phytoplankton Ecology. Structure, Function and Fluctuation*; Harris, G.P., Ed.; Chapman and Hall: London, New York, 1986.

88. Redfield, A.C. The biological control of chemical factors in the environment. Am. Sci. 1958, *46*, 205–222.

89. Hecky, R.E.; Lilham, P. Limnol. Oceanogr. 1986, *33*, 796–822.

90. Redfield, A.C. On the Proportions of organic derivatives in seawater and their relation to the comparison of plankton. In *James Johnson Memorial Volume*; Daniel, R.J., Ed.; Liverpool University Press: Liverpool, 1934.

91. Copin-Montegut, C.; Copin-Montegut, G. Stochiometry of carbon, nitrogen, and phosphorus in marine particulate matter. Deep Sea Res. 1983, *30*, 31–46.

92. Horn, H.; Paul, L.; Horn, W. Increase of phytoplankton production with decline of phosphorus. GWF, Wasser/Abwasser 2001, *142*, 268–278.

93. Poehlmann, W.; Kaul, U.; Kopf, W. Eutrophication of running waters. Part 1: Effect of temperature, phosphorus, silicate, carbon dioxide, ammonia, nitrate, and complexing agents on the growth of phytoplankton (laboratory studies). Z Wasser Abwasser Forsch 1989, *22*, 187–195.

94. Aidar, E.; Sigaud-Kutner, T.C.S.; Nishihara, L.; Schinke, K.P.; Braga, M.C.C.; Farah,

R.E.; Kutner, M.B.B. Marine phytoplankton assays: effects of detergents. Mar. Environ. Res. 1996, *43*, 55–68.

95. Hamm, A., Ed. Studie über die Wirkungen und Qualitätsziele von Nährstoffen in Fliessgewässern. Academia Verlag: Sankt Augustin, Germany, 1991.

96. URL: http://www.aquaristik-hilfe.de/system03.htm.

97. DeBoer, J.A. Nutrients. In *The Biology of Seaweeds*; Lobban, C.S., Wynne, M.J., Eds.; Blackwell Scientific: Oxford, 1981; 356–392.

98. Nolan, B.T.; Stoner, J.D. Nutrients in groundwaters of the conterminous United States 1992–1995. Environ. Sci. Technol. 2000, *34*, 1156–1165.

99. *National Nutrient Assessment Strategy: An Overview of Available Endpoints and Assessment Tools.* U.S. Environmental Protection Agency: Washington, DC, 1998.

100. Ghiaudani, G.; Vighi, M. The N:P ratio and tests with selenastrum to predict eutrophication in lakes. Water Res. 1974, *8*, 1063–1069.

101. Thomann, R.V.; Mueller, J.A. *Principles of Surface Water Quality Modelling and Control*; Harper and Row: New York, 1987.

102. Carpenter, S.R.; Cottingham, K.L.; Schindler, D.E. Biotic feedbacks in lake phosphorus cycles. TREE 1992, *7*, 332–336.

103. Dinar, A.; Seidl, P.; Olem, H.; Jordan, V.; Duda, A.; Johnson, R. *Restoring and Protecting the World's Lakes and Reservoirs*; World Bank, 1995.

104. Klapper, H. *Eutrophierung und Gewässerschutz*; Gustav Fischer Verlag: Jena, Stuttgart, 1982.

105. Uhlmann, D.; Hrbácek, J. Kriterien der Eutrophie stehender Gewässer. Limnologia (Berlin) 1976, *10*, 245–253.

106. DIN Deutsches Institut für Normung e. V., ed. DIN 4049-1 Hydrology; basic terms. Beuth Verlag: Berlin, 1992 (equivalent to ISO/DIS 772 (1993-05)).

107. DIN Deutsches Institut für Normung e. V., ed. DIN 4049-2 Hydrology; terms relating to quality of waters. Beuth Verlag: Berlin, 1990 (equivalent to ISO 772 DAM 1 (2000-04)).

108. DIN Deutsches Institut für Normung e. V., ed. DIN 4049-3 Hydrology—Part 3: Terms for the Quantitative Hydrology. Beuth Verlag: Berlin, 1994 (equivalent to ISO 772 (1996-04)).

109. Quality criteria for water 1986. U.S. Environmental Protection Agency Report 440/5-86-001. U.S. Environmental Protection Agency, Office of Water: Washington, DC, 1986.

110. Smith, V.H. In *Successes, Limitations and Frontiers in Ecosystem Science*; Pace, M.L. Groffmann, P.M. Eds.; Springer Verlag: New York, 1998; 7–49.

111. Schindler, D.W. Science 1977, *195*, 260.

112. Vollenweider, R.A. Advances in defining critical loading levels for phosphorus in lake eutrophication. Mem. 1st Ital. Idrobiol. 1976, *33*, 53–83.

113. Schindler, D.W. Predictive eutrophication models. Limnol. Oceanogr. 1978, *23*, 1080–1081.

114. Benndorf, J. A contribution to the phosphorus loading concept. Int. Revue Ges. Hydrobiol. 1979, *64*, 177–188.

115. Eutrophication of waters: monitoring, assessment and control. Organisation for Economic and Cooperative Development; OECD: Paris, 1982.

116. Zimmermann, U.; Foster, R.; Sontheimer, H. Long-term changes of water quality in three Swiss lakes (Lake Zürich, Zürichobersee, and Lake Walenstadt). Zurich Water Supply Authority: Zurich, 1991.

117. Lee, G.F.; Jones, R.A. Detergent phosphate bans and eutrophication. Environ. Sci. Technol. 1986, *20*, 330–331.

118. Atkins, W.R. The phosphate content of fresh and salt waters in its relationships to the growth of the algal population. J. Mar. Biol. Assu. UK 1923, *13*, 119–150.

119. Maki, A.W.; Porcella, D.; Wendt, R.H. The impact of detergent phosphorus bans on receiving water quality. Water Res. 1984, *18*, 893–903.

120. Vollenweider, R.A. Scientific Fundamentals of the Eutrophication of Lakes and Flowing Waters with Particular Reference to Nitrogen and Phosphorus as Factors in Eutrophication, Technical Report DAS/CSI/68. OECD: Paris, 1968.

121. Vollenweider, R.A. Input–output models with special reference to phosphorus loading concept in limnology. Schweiz A Hydrol. 1975, 37, 53–84.

122. Rast, W.; Lee, G.F. Summary Analysis of the North American (US Portion) OECD Eutrophication Project: Nutrient Loading - Lake Response Relationships and Trophic State Indices, U.S. EPA, Ecol Res Ser EPA 600/3-78-008. Corvallis, OR: US EPA, 1978.

123. Lee, G.F.; Rast, W.; Jones, R.A. Eutrophication of waterbodies: insights for an Age-old Problem. Environ. Sci. Technol. 1978, 12, 297–301.

124. Jones, R.A. Lee, G.F. Recent advances in assessing the impact of phosphorus loads on eutrophication-related water quality. Water Res. 1982, 16, 503–515.

125. Jones, R.A.; Lee, G.F. Eutrophication modelling for water quality management: an uptake of the Vollenweider–OECD Model. World Health Organization's Water Quality Bulletin 1986, 11, 67–74.

126. Lee, G.F.; Jones, R.A.; Rast, W. Availability of phosphorus to phytoplankton and its implication for phosphorus management strategies. *Phosphorus Management Strategies for Lakes*; Ann Arbor Press: Ann Arbor, MI, 1980; 259–308.

127. Welch, E.B.; Horner, R.R.; Patmont, C.R. Prediction of nuisance perphytic biomass: a management approach. Water Res. 1989, 23, 401–405.

128. Hutchinson, G.E. Eutrophication. The scientific background of a contemporary practical problem. American Sci. 1973, 61, 269–279.

129. Wetzel, R.G. *Limnology*, 2nd Ed.; Saunders College: Fort Worth, TX, 1983.

130. European Eco-label Revision of Eco-label criteria on laundry detergents, Report for the Commission of the European Union, DG XI, Berlin: UBA—German Federal Environmental Agency, 1998.

131. Ryther, J.H.; Dunstan, W.M. Nitrogen, phosphorus, and eutrophication in the coastal marine environment. Science 1971, 171, 1008–1013.

132. Fletcher, R.L.; Cuomo, V.; Palomba, I. The "green tide" problem, with particular reference to the Venice Lagoon. Brit. Phycolog. J. 1971, 25, 87.

133. Dibravko, J.; Rabalais, N.N.; Tyrner, R.E.; Wiseman, W.J. Seasonal coupling between riverborne nutrients, net productivity and hypnoxia. Mar. Pollut. Bull. 1993, 26, 184–189.

134. Kennish, M.J. *Ecology of Estuaries: Antropogenic Effects*; CRC Press: London, 1992.

135. Boesch, D.F.; Borroughs, R.H.; Baker, J.E.; Mason, R.P.; Rowe, C.L.; Siefert, R.L. *Marine Pollution in the United States*; New Oceans Commission: Arlington, VA, 2001.

136. Howard, R.K.; Edgar, G.J.; Hutchings, P.A. Faunal assemblages of seagrass beds. *Biology of Seagrasses*; Larkum, A.W.D., McComb, A.J., Sheperd, S.A., Eds.; Elsevier: Amsterdam, 1989; 536–564.

137. Neverauskas, V.P. Monitoring seagrass beds around a sewage sludge outfall in South Australia. Mar. Pollut. Bull. 1987, 18, 158–164.

138. Nusch, E.A. Zur Frage kritischer Nährstoffbelastung gestauter Fliessgewässer. Z Wasser- und Abwasser-Forsch 1982, 15, 102–112.

139. Quinn, J.M. Guidelines for the Control of Undesirable Biological Growths in Water. Consultancy Report No. 6213/2. Water Quality Centre: Hamilton, New Zealand, 1991.

140. Johnston, C.A. Sediment and nutrient retention by freshwater wetlands: effects on surface water quality. CRC Crit Rev Environ Control; University of Minnesota, Natural Resources Research Institute: Duluth MN, 1991; 491–565.

141. Leibowitz, N.C.; Brown, M.T. Indicator strategy for wetlands. *Environmental Monitoring and Assessment Program, Ecological Indicators*; EPA/600/390/060. Environmental Protection Agency: Washington, DC, 1990; 5.1–5.15.

142. Nixon, S.W.; Lee, V. Wetlands and Water Quality: A Regional Review of Recent Research in The United States on the Role of Freshwater and Saltwater Wetlands as Sour-

ces, Sinks and Transformers of Nitrogen, Phosphorus, and Various Heavy Metals. Technical Report Y 86 2, Final Report. Rhode Island University, Kinston Graduate School of Oceanography: Kinston, RI, 1986.

143. Johnston, P.; Santillo, D.; Stringer, R.; Ashton, J.; McKay, B.; Verbeek, M.; Jackson, E.; Landman, J.; van den Broek, J.; Samsom, D.; Simmonds, M. Report on the World's Oceans. Greenpeace Research Laboratories Report, 1998.

144. Laponite, B.E. Strategies for pulsed supply to Gracilaria cultures in the Florida Keys: interactions between concentration and frequency of nutrient pulses. J. Exp. Mar. Biol. Ecol. 1985, *93*, 211–222.

145. Howard-Williams, C. Cycling and retention of nitrogen and phosphorus in wetlands: a theoretical and applied perspective. Freshwater Biol. 1985, *15*, 391–431.

146. Kadlec, R.H. The Bellaire wetland: wastewater alteration and recovery. Wetlands 1983, *3*, 44.

147. Yentsch, C.M.; Yentsch, C.S.; Strube, L.R. Variations in ammonium enhancement, an indication of nitrogen deficiency in New England coastal phytoplankton populations. J. Mar. Res. 1977, *35*, 537–555.

148. Nixon, S.W.; Philson, M.E.Q. Nitrogen in estuarine and coastal ecosystems. In *Nitrogen in the Marine Environment*; Carpenter, E.J., Capone, D.G., Eds.; Academic Press: Orlando, FL, 1983.

149. McCarthy, J.J. Uptake of major nutrients by estuarine plants. In *Estuaries and nutrients*; Neilson, B.J., Cronin, L.E., Eds.; Humana Press: Totowa, NJ, 1981; 139–164.

150. Abdelmoati, M.A.R. Eutrophication in the coastal waters of Alexandria following the increase in phosphorus load. Fresenius Environ. Bull. 1996, *5*, 172–177.

151. Sakshaug, E.; Olsen, Y. Nutrient status of phytoplankton blooms in Norwegian waters and an algal strategy for nutrient competition. Can J. Fish Aquat. Sci. 1986, *43*, 389–396.

152. Webber, D.F.; Roff, J.C. Influence of Kingston Harbor on the plankton community of the nearshore Hellshire Coast, Southeast Jamaica. Bull. Mar. Sci. 1996, *59*, 245–258.

153. Lancelot, C.; Billen, G.; Sournia, A.; Weisse, T.; Colijn, F.; Veldhuis, M.J.W.; Davies, A.; Wassmann, P. Phaeocystis blooms and nutrient enrichment in the continental coastal zones of the North Sea. Ambio 1987, *16*, 38–46.

154. Aure, J.; Danielssen, D.; Saetre, R. Assessment of eutrophication in Skagerrak coastal waters using oxygen consumption in fjordic basins. ICES J. Mar. Sci. 1996, *53*, 589–595.

155. Lohrenz, S.E.; Fahnenstiel, G.L.; Redalje, D.G.; Lang, G.A.; Chen, X.G.; Dagg, M.J. Variations in primary production of northern Gulf of Mexico continental shelf waters linked to nutrient inputs from the Mississippi River. Mar. Ecol. Progress Series 1997, *155*, 45–54.

156. Directive 2000/60/EC of the European Parliament and of the Council of 23 October 2000 establishing a framework for Community action in the field of water policy (Water Framework Directive, Official Journal L 327, 22.12.2000, p. 1).

157. Council Directive 98/83/EC on the quality of water intended for human consumption (The Drinking Water Directive, Official Journal L 330, 05.12.98).

158. Commission Decision of 10 June 1999 Establishing the Ecological Criteria for the award of the Community Eco-label to Laundry Detergents (Official Journal L 187, 20.07.1999, p. 52).

159. Commission decision of 28 May 1999 establishing the ecological criteria for the award of the Community eco-label to detergents for dishwashers, Official Journal of the European Communities, 02.07.1999, 1999/427/EC.

160. EC Freshwater Fish Directive: Council Directive 78/659/EEC on the quality of fresh waters needing protection or improvement in order to support fish life (Official Journal L 222, 14.08.78).

161. Council Directive 76/160/EEC concerning the quality of bathing water (The Bathing Water Directive, Official Journal L 131, 05.02.76).

162. Lübbe, E. Politische Vorgaben und rechtliche Instrumentarien zur Problemlösung: eine Übersicht. In Belastungen der Oberflächengewässer aus der Landwirtschaft. Frankfurt am Main: BLV-Verlags-Gesellschaft, 1993.

163. Scailteur, V.; Pölloth, C.; Jassogne, C. "Hera": the human and environmental risk assessment industry initiative for chemicals used in household detergent and cleaning products. SÖFW J. 2000, *126*, 13–15.

164. Van der Kleij, M.; Dekker, R.H.; Kersten, H.; Wit, J.A.W. Water management of the River Rhine: past, present and future. European Water Pollution Control 1991, *1*, 9–15.

165. DePinto, J.V.; Young, T.C.; McIlroy, L.M. Great Lakes Water Quality Improvement. Environ. Sci. Technol. 1986, *20*, 752–759.

166. 1987 Report on Great Lakes Water Quality. Windsor, Ontario, Canada: International Joint Commission (IJC), 1987.

167. Gottlieb, S.J. Ecological role of atlantic menhaden (Brevoortia tyrranus) in Chesapeak Bay and implications for management of the fishery. Thesis, University of Maryland: College Park, MD, 1998.

168. Parry, R. Agricultural phosphorus and water quality—a U.S. Environmental Protection Agency perspective. J. Environ. Qual. 1998, *27*, 258–261.

169. Gesetz über die Umweltverträglichkeit von Wasch- und Reinigungsmitteln (Waschmittelgesetz) vom 20.08.1975 (DGBl. I, p. 2255).

170. Gesetz über Detergentien in Wasch- und Reinigungsmitteln vom 05.09.61 (DGBl. I, p. 1653).

171. Verordnung über Höchstmengen für Phosphate in Wasch- und Reinigungsmittel (Phosphathöchstmengenverordnung, PHöchstMeng V) vom 04.06.1980 (BGBl. I, pp 664-665).

172. Doyle, F., Ed. Environmental Awareness Section, Department of the Environment and Local Government, Custom House Dublin 1.

173. Japan Soap and Detergent Association (JSDA), 3-13-11, Nihonbashi, Chuo-ku, Tokyo Japan 103-0027, URL: http://www.jsda.org/e_history.html.

174. Müller, E. 10 Jahre Phosphatverbot für Textilwaschmittel in der Schweiz Erfahrungen und Auswirkungen. SEPAWA-Kongress, Bad Dürkheim, 1997.

175. Verordnung über Wasch-, Spül- und Reinigungsmittel (Waschmittelverordnung) vom 13.06.1977 (Sammlung des Bundesrechts, Jahrgang 1977; 1138 pp.

176. Schöberl, P.; Huber, L. Ökologisch relevante Daten von nichttensidischen Inhaltsstoffen in Wasch-und Reinigungsmitteln. Tenside, Surfactants, Detergents 1988, *25*, 99–107.

177. Falcone, J.S.; Spencer, R.W. Silicates expand role in waste treatment, bleaching, deinking. Pulp Paper 1975, *12*, 114–117.

178. Lagaly, G.; Klose, D.; Minihan, A.; Lovell, A. Silicates. In *Ullmann's Encyclopedia of Industrial Chemistry*, 5th Ed.; Elvers, B., Hawkins, S., Russey, W., Schulz, G., Eds.; VCH: Weinheim, 1993; Vol. A23, 661–719.

179. Barby, D.; Griffiths, T.; Jacques, A.R.; Pawson, D. The Modern Inorganic Chemicals Industry; Chemical Society: London, 1977; 320–352.

180. Mühlenkamp, S. Glasklare Sache. Process 1998, *11*, 158–159.

181. Griesshammer, R.; Bunke, D.; Gensch, C.O. Produktlinienanalyse Waschen und Waschmittel; Öko-Institut e.V: Freiburg, 1996.

182. Thieme, C. Sodium Carbonates. In *Ullmann's Encyclopedia of Industrial Chemistry*, 5th Ed.; Elvers, B., Hawkins, S., Russey, W., Schulz, G., Eds.; VCH: Weinheim, 1993; Vol. A24, 299–316.

183. Kuhr, W. Henkel Referate 1998, *34*, 7–13.

184. De Jong, B.H.W.S. Glass. In *Ullmann's Encyclopedia of Industrial Chemistry*, 5th Ed.; Elvers, B., Hawkins, S., Ravenscroft, M., Rousaville, J.F., Schulz, G., Eds.; VCH: Weinheim, 1989; Vol. A12, 365–432.

185. Schweiker, G.C. Sodium silicates and sodium aluminosilicates. J. Am. Oil Chem. Soc. 1978, *55*, 36–40.

186. Schweiker, G.C. Sodium silicates and sodium aluminosilicates, a worldwide update. In Proc 2nd World Conf on Dets AOCS; Baldwin, A.R., Ed.; 1987; 63–68.

187. Fawer, M. Life Cycle Inventories for the Production of Sodium Silicates, Report No. 241. St. Gallen, Switzerland: EMPA Eidgenössische Materialprüfungs- und Forschungsanstalt, 1997.

188. Fawer, M.; Concannon, M.; Rieber, W. Life cycle inventories for the production of sodium silicates. Int. J LCA 1999, *4*, 207–212.

189. Schmittinger, P.; Curlin, L.C.; Asawa, T.; Kotowski, S.; Beer, H.B.; Greenberg, A.M.; Zelfel, E.; Breitstadt, R. Chlorine. In *Ullmann's Encyclopedia of Industrial Chemistry*, 5th Ed.; Gerhartz, W., Yamamoto, Y.S., Campbell, T., Pfefferkorn, R., Rounsaville, J.F., Eds.; VCH: Weinheim, 1986; Vol. A6.

190. Bergner, D. Entwicklungsstand der Alkalichlorid-Elektrolyse Teil 1: Zellen, Membranen, Elektrolyte. Chemie Ingenieur Technik 1994, *66*, 783–791.

191. Bergner, D. Entwicklungsstand der Alkalichlorid-Elektrolyse Teil 2: Elektrochemische Grössen, Wirtschaftliche Fragen. Chemie Ingenieur Technik 1994, *66*, 1026–1033.

192. Bergner, D. 20 Jahre Entwicklung einer bipolaren Membranzelle für die Alkalichlorid-Elektrolyse vom Labor bis zur weltweiten Anwendung. Chemie Ingenieur Technik 1997, *69*, 438–445.

193. Upadek, U.; Kottwitz, B.; Schreck, B. Tenside Surf. Det. 1996, *33*, 385–392.

194. Upadek, U.; Kottwitz, B.; Schreck, B. Henkel Referate 1997, *33*, 57–63.

195. Weldes, H.H. Soap Cosmetics Chemical Spec. 1972, *48*, 72–96.

196. Bean, S.L.; Mouton, A.W. US 3,748,103, 24.07.1973.

197. Dany, F.J.; Gohla, W.; Kandler, J.; Rieck, H.P.; Schimmel, G. SÖFW 1990, *20*, 805–808.

198. de Lucas, A.; Rodriguez, L.; Sánchez, P.; Lobato, J. Ind. Eng. Chem. Res. 2000, *39*, 1249–1255.

199. Admiraal, W.; Breugem, P.; Jacobs, D.M.L.H.A.; de Ruyter van Stevenick, E.D. Biogeochem. 1990, *9*, 175–185.

200. Schindler, D.W. Interrelations between the cycles of elements in freshwater ecosystems. In *Some Perspectives of the Major Biogeochemical Cycles*; Likens, G.E., Ed.; Wiley: New York, 1981; 113–123.

201. Wollast, R. Interaction between major biogeochemical cycles in marine ecosystems. In *Some Perspectives of the Major Biogeochemical Cycles*; Likens, G.E. Ed.; Wiley: New York, 1981; 125–143.

202. Falcone, J.S.; Blumberg, J.G. In *The Handbook of Environmental Chemistry*; Hutzinger, O., Ed.; Springer-Verlag: Berlin, 1992; Vol. 3, 367–382.

203. Schleyer, W.L.; Blumberg, J.G. Health, safety, and environmental aspects of soluble silicates. J. Am. Chem Soc. 1982; 51–69.

204. Boyle, J.R.; Voight, G.K. Biological weathering of silicate minerals. Plant Soil 1973, *38*, 191–201.

205. Davis, S.N. Am. J. Sci. 1964, *262*, 870.

206. Edwards, A.M.C.; Liss, P.S. Nature 1973, *243*, 341.

207. Anon. Public Water Supplies of the 100 Largest Cities in the United States. U.S.G.S. Paper No. 1812, 1962.

208. Kido, K. Mar. Chem. 1974, *2*, 277–286.

209. Tacke, R. Meilensteine in der Biochemie des Siliciums: von der Grundlagenforschung zu biotechnologischen Anwendungen. Angew. Chem. 1999, *111*, 3197–3200.

210. Tréguer, P.; Nelson, D.M.; Van Bennekom, A.J.; DeMaster, D.J.; Leynaert, A.; Quéguiner, B. Science 1995, *268*, 375–379.

211. Alexander, G.B.; Heston, W.M.; Iler, R.K. J. Phys. Chem. 1954, *58*, 453–455.

212. Stumm, W.; Morgan, J.J. *Aquatic Chemistry*; Wiley: New York, 1970.

213. Oehler, J.H. In *Biogeochemical Cycling of Mineral-Forming Elements*; Trudinger, P.A., Swaine, D.J., Eds.; Amsterdam: Elsevier, 1979; 467–483.

214. Allison, A.C. Silicon compounds in biological systems. Proc. Roy Soc, Ser. B 1968, *171* (1022), 19–30.

215. Schuffenhauer, C. Phyllokieselsäuren: Darstellung, Charakterisierung und Intercalationsreaktionen mit Polyhydroxyverbindungen. Diplomarbeit, Universität Konstanz, 1998.

216. Mann, S., Webb, J., Williams, R.J.P., Eds.; *Biomineralization—Chemical and Biochemical Perspectives*; VCH: Weinheim, 1989.

217. Hildebrand, M.; Volcani, B.E.; Gassmann, W.; Schroeder, J.I. Nature 1997, *385*, 486.

218. Hildebrand, M.; Dahlin, K.; Volcani, B.E. Mol. Gen. Genet. 1998, *260*, 480–486.

219. Volcani, B.E. Role of silicon in diatom metabolism and silicification. Nobel Symp. 1978, *40*, 177–204.

220. Werner, D. Regulation of metabolism by silicate in diatoms. Nobel Symp. 1978, *40*, 149–176.

221. Werner, D. Silicate metabolism. Bot. Monogr. (Oxford) 1977, *13*, 110–149.

222. Soukup, M. U Mass Water Resources Res. Ctr. Pub. 1974, *39*, 67.

223. D'Hoore, J.; Coulter, J.K. *Solids of the Humid Tropics*; Nat Acad Sci: Washington, DC, 1972; 163–173.

224. Schwarz, K. Trace Elements Metab Anim, Proc Int Symp 2nd; 1974; 355 pp.

225. Simpson, T.L., Volcani, B.E., Eds.; Silicon and Siliceous Structures in Biological Systems; Springer: New York, 1981.

226. Reiman, B.E.F.; Lewin, J.C.; Volcani, B.E. J. Phycol. 1966, *2*, 74–84.

227. Coombs, J.; Halicki, P.J.; Holm-Hansen, O.; Volcani, B.E. Expl. Cell. Res. 1967, *47*, 315–328.

228. Conley, D.J.; Malone, T.C. Annual cycle of dissolved silicate in Chesapeake Bay: implications for the production and fate of phytoplankton biomass. Mar. Ecol: Prog. Ser. 1992, *81*, 121–128.

229. Kilham, S.S. J. Phycol. 1975, *11*, 396.

230. Klaveness, D.; Guillard, R.R.L. J. Phycol. 1975, *11*, 349–355.

231. Kilham, P. A hypothesis concerning silica and the freshwater planktonic diatoms. Limnology Oceanography 1971, *16*, 10–18.

232. Egge, J.K.; Aksnes, D.L. Silicate as regulating nutrient in phytoplankton competition. Mar. Ecol: Prog. Ser. 1992, *83*, 281–289.

233. Jorgensen, E. Dansk Botanisk Arkiv 1957, *18*, 1.

234. Schwartz, A.M. Interim Report for Environmental Protection Agency Contract FWQA 14-12-875. Environmental Protection Agency: Washington, DC, 1972.

235. Ryther, J.H.; Officer, C.B. Impact of nutrient enrichment on water uses. In *Estuaries and Nutrients*; Neilson, B.J., Cronin, L.E., Eds.; Humana Press: Totowa, NJ, 1981.

236. Officer, C.B.; Ryther, J.H. The possible importance of silicon in marine eutrophication. Mar. Ecol. Progr. Ser. 1980, *3*, 83–91.

237. Schelske, C.L.; Stoermer, E.F. Eutrophication, silica depletion and predicted changes in algal quality in Lake Michigan. Science 1971, *173*, 423–424.

238. Malone, T.C.; Garside, C.; Neale, P.J. Effects of silicate depletion on photosynthesis by diatoms in the plume of the Hudson River. Mar. Biol. 1980, *58*, 197–204.

239. Malone, T.C.; Ducklow, H.W.; Peele, E.R.; Pike, S.E. Picoplankton carbona flux in Chesapeake Bay. Mar. Ecol. Progr. Ser. 1991, *78*, 11–22.

240. Schleske, C.L.; Stoermer, E.F. Symposium on nutrients and eutrophication. Am. Soc. Limnology and Oceanography, 1971.

241. van Bennekom, A.J.; Salomons, W. Pathways of nutrients and organic matter from land through rivers. SCOR Working Group No. 46, Proc. of Review Workshop. River inputs to ocean systems; United Nations: New York, 1981.

242. O'Connor, T.L. J. Phys. Chem. 1961, *65*, 1.

243. Hauthal, H.G. 12. Vortragstagung der Fachgruppe Waschmittelchemie der Gesellschaft Deutscher Chemiker. SÖFW 1997, *123*, 708–722.

244. Admiraal, W.; Mylius, S.D.; de Ruyter van Steveninck, E.D.; Tubbing, D.M.J. J. Plankton Res. 1993, *15*, 659–682.

245. Blumberg, J.G.; Schleyer, W.L. Current Regulatory Status of Soluble Silicates. In *Soluble Silicates*; Falcone, J.S. Ed.; American Chemical Society: Washington, DC, 1982.

246. Schulze, C. Modelling and evaluating the aquatic fate of detergents, University of Osnabrück, Germany, 2001.

247. Reish, D.J. Water Res. 1970, *4*, 721.

248. Dowden, B.F.; Bennett, H.J. J. Water Pollut. Control Fed. 1965, *37*, 1308.

249. Westphal, G.; Kristen, G.; Wegener, W.; Ambatiello, P.; Geyer, H.; Epron, B.; Bonal, C.; Seebode, K.; Kowalski, U. Sodium chloride. In *Ullmann's Encyclopedia of Industrial Chemistry*, 5th Ed.; Elvers, B., Hawkins, S., Russey, W., Schulz, G., Eds.; VCH: Weinheim, 1993; Vol. A24, 317–339.

250. Minz, F.R. Sodium hydroxide. In *Ullmann's Encyclopedia of Industrial Chemistry*, 5th Ed.; Elvers, B., Hawkins, S., Russey, W., Schulz, G., Eds.; VCH: Weinheim, 1993; Vol. A24, 345–354.

251. Fawer, M. Life Cycle Inventory for the Production of Zeolite A for Detergents, St. Gallen, Switzerland: EMPA, 1996.

252. Hudson, L.K. Aluminum oxide. In *Ullmann's Encyclopedia of Industrial Chemistry*, 5th Ed.; Gerhartz, W., Yamamoto, Y.S., Campbell, F.T., Pfefferkorn, R., Rounsaville, J.F., Eds.; VCH: Weinheim, 1985; Vol. A1, 557–594.

253. Chemlink Pty Ltd, 1997. URL: www.chemlink.com.au.

254. Helmboldt, O.; Hudson, L.K.; Stark, H.; Danner, M. Aluminum compounds, inorganic. In *Ullmann's Encyclopedia of Industrial Chemistry*, 5th Ed.; Gerhartz, W., Yamamoto, Y.S., Campbell, F.T., Pfefferkorn, R., Rounsaville, J.F., Eds.; VCH: Weinheim, 1985; Vol. A1, 527–541.

255. Berth, P. Tenside Det. 1978, *15*, 176–180.

256. Roland, E.; Kleinschmidt, P. Zeolites. In *Ullmann's Encyclopedia of Industrial Chemistry*, 5th Ed.; Elvers, B., Hawkins, S., Eds.; VCH: Weinheim, 1996; Vol. A28, 475–504.

257. Hauthal, H.G. 43. SEPAWA-Jahrestagung 1996. SÖFW J. 1996, *122*, 912.

258. Hauthal, H.G. Detergent zeolites in an ecobalance spotlight. SÖFW J. 1996, *122*, 899–911.

259. Graffmann, G.; Roland, W.A.; Schmid, R.D.; Smolka, H.G.; Schneider, J.; Vogg, H. Chemiker Zeitung 1979, *103*, 123–129.

260. Zeolites for Detergents. Brussels: ZEODET Association of Detergent Zeolite Producers, 2000.

261. King, J.E.; Hopping, W.D.; Holman, W.F. Treatability of type A zeolite in wastewater. Part I. J. Water Pollut. Control Fed. 1980, *52*, 2875–2886.

262. Roland, W.A.; Graupner, W.; Holtmann, W. Sodium aluminum silicates in detergents—practical tests on sedimentation behavior in the sewerage. GWF, Gas- Wasserfach: Wasser/Abwasser 1979, *120*, 55–61.

263. Kurzendörfer, C.; Kuhm, P.; Steber, J. Zeolites in the environment. In *Detergents in the Environment*; Schwuger, M.J., Ed.; Marcel Dekker: New York, 1997.

264. Berth, P. J. Am. Oil Chem. Soc. 1978, *55*, 52–57.

265. Umweltbundesamt, ed. Materialien 4/79, Die Prüfung des Umweltverhaltens von Natrium-Aluminium-Silicat Zeolith A als Phosphatersatzstoff in Wasch- und Reinigungsmitteln. F Schmidt Verlag: Berlin, 1979.

266. Christophliemk, P.; Gerike, P.; Potokar, M. Zeolites. In *The Handbook of Environmental Chemistry Detergents*; DeOude, N.T., Ed.; Springer Verlag: Berlin, 1992; Vol 3F.

267. Holman, W.F.; Hopping, W.D. Treatability of type A zeolite in wastewater. Part II. J. Water Pollut. Control Fed. 1980, *52*, 2887–2905.

268. Wagener, R. GWF-Wasser/Abwasser 1978, *119*, 235–242.

269. Roland, W.A.; Schmid, R.D. Sodium aluminum silicates in detergents—model experi-

ments for the behavior of anaerobic alkaline-sludge digestion. Vom Wasser 1978, *50*, 177–190.

270. Fischer, W.K.; Gerike, P.; Kurzyca, G. Tenside Det. 1978, *15*, 60–64.
271. Pollution by Detergents: Determination of the biodegradability of anionic synthetic surface active agents. OECD: Paris, 1971.
272. Fischer, W.K.; Gerike, P.; Holtmann, W. Water Res. 1975, *9*, 1131.
273. Roland, W.A. Sodium aluminosilicates for detergents—phosphate successor in the environmental test. Forum Staedte-Hyg. 1979, *30*, 131–137.
274. Scherb, K. Results of studies of the determination of the ecological behavior of sodium aluminum silicate (SASIL) in sewage treatment plants. Muench Beitr. Abwasser-Fisch-Flussbiol. 1979, *31*, 179–193.
275. Roland, W.A.; Schmid, R.D. Sodium aluminum silicates in detergents—experiments with the ion exchange behavior towards heavy metals in waste waters. Tenside Deterg. 1978, *15*, 281–285.
276. Kurzendoerfer, C.P.; Schwuger, M.J.; Smolka, H.G. Use of sodium aluminum silicates in detergents. Part V. Model studies of heavy-metal ion exchange under wastewater and river water conditions. Tenside Deterg. 1979, *16*, 123–129.
277. Falke, J. Phosphate, Zeolite and Citrate in Detergents—Technical and Environmental Aspects of Detergent Builder Systems, Report No. 95002/06. MFG, Environmental Research Group: Gillelle, Denmark, 1996.
278. Cook, T.E. Environmental Sci. Technol. 1982, *16*, 344.
279. Llenado, R.A. In Proceedings of the Sixth International Zeolite Conference; Olson, D., Ed.; Butterworths: Guildford, UK, 1984; 947 pp.
280. Nusch, E.A. Study results on estimating the ecological effect of potential phosphorus substitutes. Vom Wasser 1980, *54*, 37–49.
281. EAWAG news, 42 D. EAWAG: Dübendorf, Switzerland, 1996.
282. Fischer, W.K.; Gode, P. The testing of sodium aluminum silicates as detergent additives for toxicity to water organisms. Vom Wasser 1977, *49*, 11–26.
283. Canton, J.H.; Slooff, W. Substitutes for phosphate-containing washing products: their toxicity and biodegradability in the aquatic environment. Chemosphere 1982, *11*, 891–907.
284. Koci, V. 12th Regional Conference IUAPPA and 4th International Conference on Environmental Impact Assessment, Praha, 2000.
285. Maki, A.W.; Macek, K.J. Aquatic environmental safety assessment for a nonphosphate detergent builder. Environ. Sci. Technol. 1978, *12*, 573–580.
286. Witters, H.E. Aquatic Tox 1986, *8*, 197.
287. Cross, D. The eco-toxicology of aluminium: an overview. Conference on the Camelford Aluminium Sulfate incident, Plymouth University, UK: 1998, URL: http://www.doublef.co.uk/texts/al/alecotox.htm.
288. Edwards, G.S.; Sherman, R.E.; Kelly, J.M. Red Spruce and Loblolly Pine Nutritional Responses to Acidic Precipitation and Ozone. Environmental Pollution; Elsevier Science: Amsterdam, 1995; 9–15.
289. Wellburn, A. *Air Pollution and Climate Change: The Biological Impact*; Longman Scientific and Technical: Essex, UK, 1994; 253 pp.
290. HERA Risk Assessment of Sodium Aluminum Silicate, Human and Environmental Risk Assessment on Ingredients of European Household Cleaning Products, 2001, URL: http://www.heraproject.com/files/HERA_Zeolite_Final_Version_03_2002.pdf.
291. Fischer, W.K.; Gode, P. Die Prüfung von Natriumaluminiumsilikat als Waschmittelzusatz auf Toxizität gegenüber Wasserorganismen. Vom Wasser 1977, *49*, 11–26.
292. Allen, H.E.; Cho, S.H.; Neubecker, T.A. Water Res. 1983, *17*, 1871–1879.
293. TNO report IMW-R 92/126, Henkel R9700476. TNO Institute of Environmental Sciences, 1992.

22

Environmental Impact of Aminocarboxylate Chelating Agents

OTTO GRUNDLER, HANS-ULRICH JÄGER, and HELMUT WITTELER BASF Aktiengesellschaft, Ludwigshafen, Germany

I. INTRODUCTION

Aminocarboxylates have been used as complexing agents in industrial applications for more than 60 years. Of all the aminocarboxylates, NTA and EDTA are used in the largest quantities as chelating agents. It is therefore understandable that most ecological and toxicological studies have been concerned with these two substances. Accordingly, the usual abbreviated names in the scientific literature, i.e., NTA (Trilon® A), EDTA (Trilon® B), DTPA (Trilon® C), HEDTA (Trilon® D), and MGDA (Trilon® M), IDS, and EDDS are used here (Fig. 1).

Annual world-wide consumption in 1998 was estimated to be ~200,000 tons (calculated as 100% free acid). NTA accounts for ca. 25% of this quantity, EDTA accounts for ca. 55%, and DTPA accounts for ca. 25%. Besides MGDA, the chelating agents EDDS and IDS were introduced into the market. Approximately ca. 30% of all chelating agents is consumed in Europe, and the remaining 70% is consumed mainly in North America, Japan, and Southeast Asia. NTA is employed chiefly in detergents and cleaners, whereas EDTA is a very versatile product used in a variety of different branches of the industry. No statistics are available on the consumption of chelating agents in all of their applications. However, sales of EDTA and DTPA by all manufacturers and importers in western Europe have been recorded by the CEFIC 1986 according to country and branch of industry, and comparable figures for NTA are available since 1990. These data, the best currently available, can be used to predict future trends, such as the concentration of EDTA in the Rhine River.

By its very nature the assessment of the environmental compatibility of the chelating agents under consideration here does not relate to special commercial products but fundamentally to the generic substances. NTA and EDTA are the chelating agents that have so far achieved special importance in terms of volume and hence also in terms of their relevance to the environment. For that reason, ecologi-

Nitrilotriacetic acid and salts (NTA)

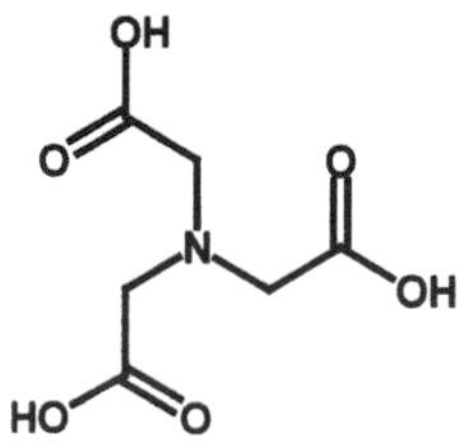

NTA (Trilon® A)

Ethylenediaminetetraacetic acid and salts (EDTA)

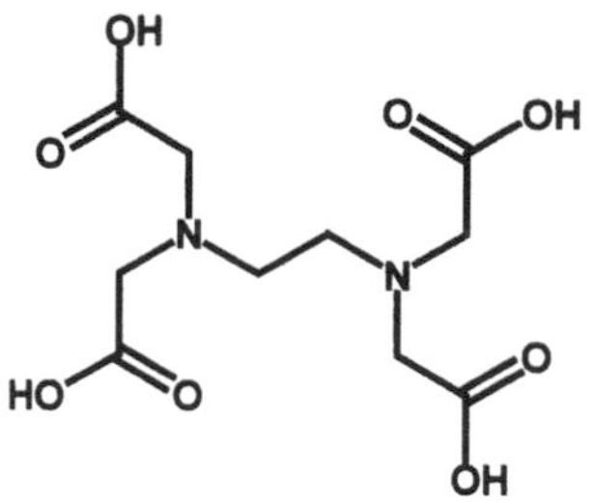

EDTA (Trilon® B)

Diethylenetriaminepentacetic acid and salts (DTPA)

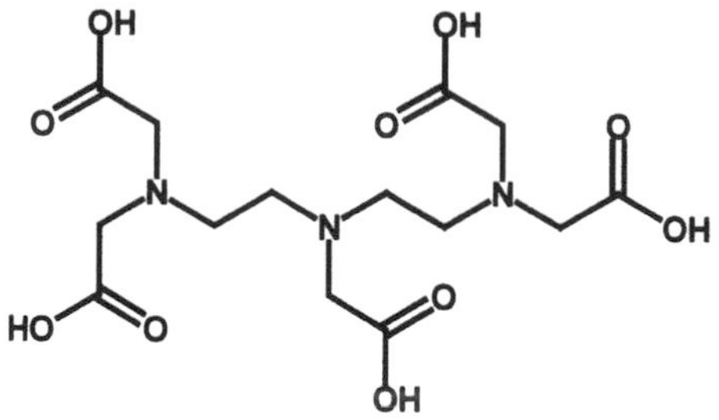

DTPA (Trilon® C)

FIG. 1 Structure of aminocarboxylates.

cal and toxicological studies have for the most part been conducted on these two representatives [1–3].

II. CONCENTRATION IN THE ENVIRONMENT

Chelating agents or their salts and metal complexes are very soluble in water and have little tendency to adsorb. After use they usually end up in sewage systems. Accordingly, residual quantities of chelating agents can be detected in surface waters, and advances in analytical methods are steadily making it easier to detect the presence of chelating agents in environmental samples.

Hydroxyethylenediaminetetraacetic acid and salts (HEDTA)

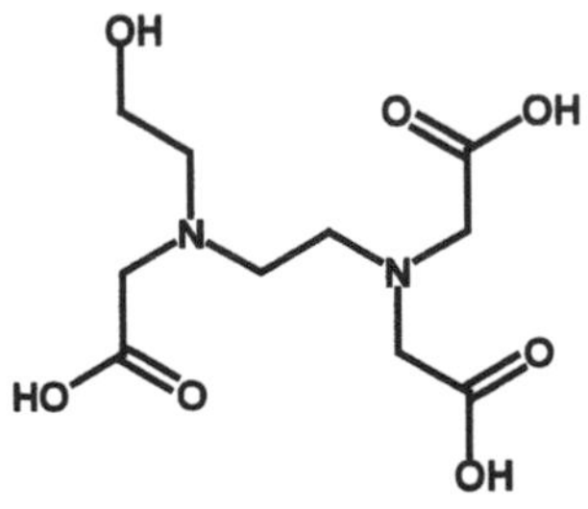

HEDTA (Trilon® D)

Methylglycinediacetic acid and salts (MGDA)

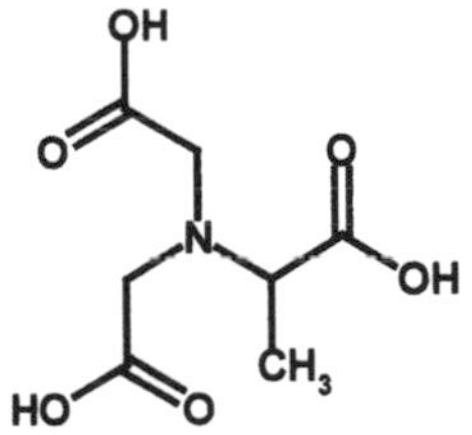

MGDA (Trilon® M)

Iminodisuccinic acid and salts (IDS)

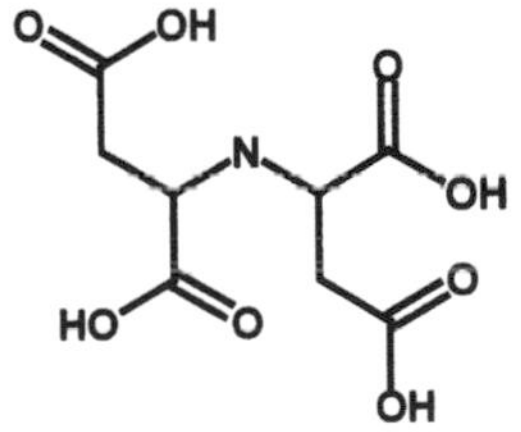

IDS

Ethylenediaminedisuccinic acid and salts (EDDS)

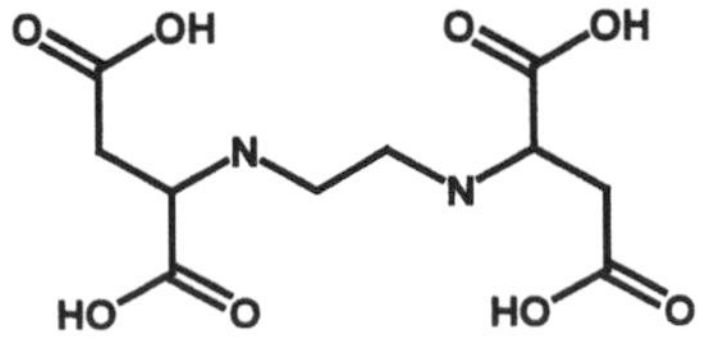

EDDS

FIG. 1 Continued.

For example, photometric and polarographic methods used earlier allowed the detection of NTA and EDTA only at concentrations above about 100 μg/l. However, new gas chromatography methods have limits of detection of 1–2 μg/L for NTA and EDTA. Much lower values are mentioned in the literature. The limits of detection stated, however, are attained at the cost of expensive means of sample preparation for separating out interfering substances present in waters, including adsorption on an ion exchanger, desorption using formic acid, and esterification with *n*-butanol. The problems in the method arise from the need to adhere strictly to the specified working conditions, for reproducibility is otherwise severely impaired.

The Rhine River, as the largest central river in Germany, is a good example for the assessment of concentration levels of chelating agents in the environment. At the measuring station in Wittlaer it has covered a large part of its course, which passes through heavily populated and highly industrialized regions. Here it contains the waters (and effluents) from most of its tributaries, such as the Neckar, Main, Ruhr, and Emscher.

Systematic determinations of the concentrations of NTA and EDTA in surface waters have been carried out in Germany since the 1980s. Figure 2 shows the 90th percentile figures for NTA and EDTA in the Rhine River and the Main from 1992 to 1997 [4].

Discussions of the concentrations of EDTA in the environment have been influenced in Germany in recent years by the "Declaration on the Reduction of Water Pollution by EDTA" of July 31, 1991. This agreement was reached voluntarily by the German Federal Environment Office (UBA), the water utilities, the manufacturers of EDTA, and the trade associations of the industries using EDTA. The aim of this agreement is to achieve in the medium term a 50% reduction in the levels of EDTA in surface waters by means of substitute products or other measures (Fig. 3).

At the measuring stations in Koblenz and Düsseldorf, the 50% cut aspired to was achieved as early as 1997.

BASF has also already fulfilled the obligation taken in the EDTA agreement to reduce emissions from production and in-house uses and has reduced its emissions by 75%. However, the search for possibilities for further reductions continues.

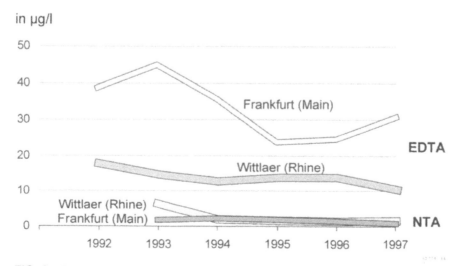

FIG. 2   EDTA and NTA concentrations in the Rhine and Main rivers.

Annual average, in ton/year (expressed as 100% free acid)

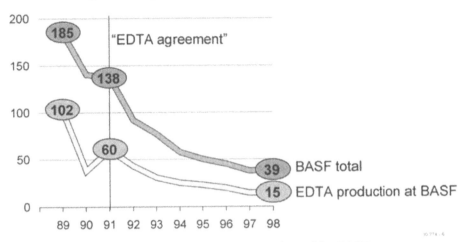

FIG. 3   Levels of EDTA in the treated effluent discharged by BASF.

Due to its ready biodegradability, NTA gets into surface water courses only in very low concentrations, and even there it is broken down further. This would be the case even if the volume used were to undergo a distinct rise.

An example from Switzerland demonstrates this very clearly. NTA was used as an alternative there in the mid-1980s, following the ban on the use of phosphates in detergents. This increased the consumption of NTA by about a factor of four without any significant change in NTA concentrations in surface water, as the long-term analytical study of Swiss waters from 1985 to 1987 shows (Table 1). Thus, increased consumption volumes of NTA, e.g., for use in detergents or cleaners, does not result in a corresponding rise in the concentrations in water courses.

Nowadays the major proportion of drinking water is obtained from ground-water, wells, and lakes and to a significant extent in many areas from bank filtration on rivers. Drinking water experts have tracked the effectiveness of treatment measures with regard to NTA and EDTA in a waterworks on the Rhine [26] (Fig. 4). The concentration of NTA was reduced by over 90% and that of EDTA by about

TABLE 1   Concentration of NTA in Swiss Waters, in μg/L (Selected Data)

	1985	1986	1987
Aare	2	1–3	1–9
Glatt	2–11	3–16	5–14
Limmat	2–4	2–5	2–5
Rhine	1–4	1–3	1–4
Rhône	1–7	1–5	1–4
Lake Constance	<0.2–0.6	<0.2–0.5	<0.2–0.3

Source: Ref. 5.

30% by bank filtration alone. This reveals the ready biodegradability of NTA by comparison with EDTA. The EDTA concentrations were reduced in the following treatment steps, especially in ozone treatment. Afterwards it was no longer possible to detect NTA, while at the same time the concentration of EDTA has fallen by 90% to about 3 µg/L. However, there are also reports according to which only small amounts of EDTA were eliminated under different treatment conditions. The NTA and EDTA concentrations found in the µg/l range in drinking water are safe in toxicological terms (Table 2). The acceptable daily intake (ADI) value is calculated with incorporation of a safety factor of 100 from the NOAEL (no-observed-adverse-effect level), i.e., from the dosage that exhibited no negative effects in investigations of lifelong administration (see also Section V). The toxicologically safe concentration in drinking water is for a body weight of 70 kg and a daily intake of 2 L of drinking

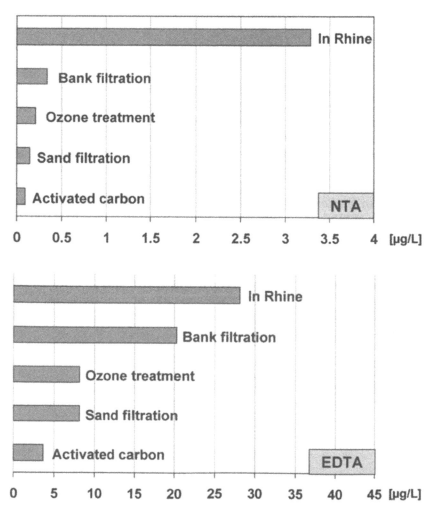

FIG. 4 Removal of NTA and EDTA during drinking water purification. (From Ref. 6.)

TABLE 2 Acceptable Daily Intake Values and Safe Concentration of NTA and EDTA in Drinking Water

	Accetable daily intake (mg/kg × d)	Safe concentration in drinking water (µg/L)
NTA	0.15	5,200
EDTA	2.5	875,000

Source: Ref. 7.

water. The concentrations of NTA and EDTA in drinking water are more than 1,000 times lower than these limiting values.

III. BIODEGRADABILITY AND ELIMINATION

For many years there was a certain level of uncertainty as to whether NTA really biodegrades, since completely contradictory study results had been published. Now, on the basis of numerous laboratory and field investigations, it has been unequivocally proven: NTA is readily biodegradable under environmental conditions. The same applies to MGDA. On the other hand, EDTA, DTPA, and HEDTA are broken down only slowly.

In the modified OECD screening test, a standard method for evaluating biodegradability, the NTA degradation curve reaches a value of >70% DOC after an adaptation time of a few days [8], which is required to qualify for ready biodegradability. At the same time, NTA is completely mineralized without the occurrence of measurable concentrations of metabolites. Thus, persistent degradation products are not produced. The biodegradation route is shown in Figure 5.

A number of publications also provide reports on the biodegradability of EDTA [3]. The prerequisite in all cases is for favorable adaptation conditions, e.g., relatively long residence times and concentrations, that are not too low. Conditions of this kind are fulfilled, for example, in test filters as specified by Sontheimer. Figure 6 shows that after adaptation in this system EDTA is completely biodegraded in 38 hours. In the case of NTA this occurs in 5 hours under the same conditions.

Nörtemann et al. (TU Braunschweig) have succeeded in isolating microorganisms that are capable of completely breaking down EDTA in the laboratory. Nörtemann's team is also working on the development of an industrial process for breaking down EDTA biologically. Removal rates for EDTA of greater than 90% have been achieved with pilot equipment.

Since EDTA is fundamentally biodegradable and is not persistent, there need be no fears of its continuous accumulation in the environment. In municipal treatment plants, however, EDTA is hardly broken down at all, but it can be eliminated to a certain extent in treatment plants using phosphate precipitants. Good rates of biodegradation can be achieved in industrial treatment plants in which there are favorable adaptation conditions [10,11].

Akzo have published the results of experiments that suggest that EDTA is completely biodegradable in sewage treatment plants in the slightly alkaline pH range.

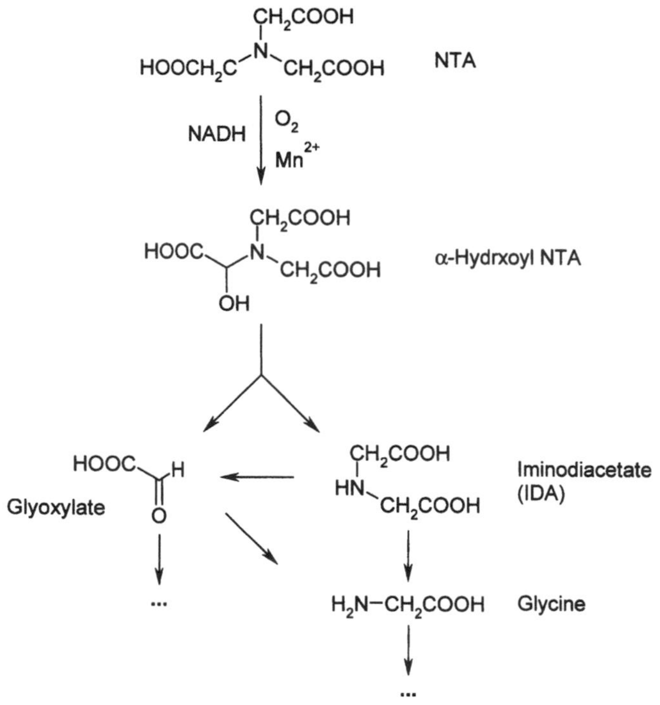

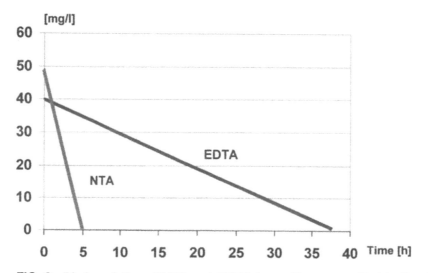

FIG. 5 Biodegredation pathway of NTA in *Pseudomonas* sp. (From Ref. 9.)

FIG. 6 Biodegradation of NTA and EDTA in test filters as specified by Sontheimer.

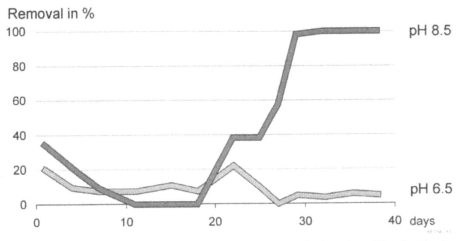

FIG. 7 EDTA degradation at elevated pH values, SCAS plant (van Ginkel, Akzo).

Removal rates for EDTA of greater than 90% were obtained for effluent from a dairy containing ca. 30 mg/L of EDTA at a pH of 7.5–8.5 [10] (Fig. 7).

A study of the effluent treatment plant at a paper mill has also been made by BASF (Table 3). The untreated wastewater from the paper mill contained an average of 23.8 mg/L EDTA, and a removal rate of ca. 76%, or 5.8 mg/L, was achieved. The degradability of EDTA was reproduced in the laboratory with adapted sludge from the effluent treatment plant [11].

These examples illustrate that it is possible to remove EDTA from wastewater by biotic means under the conditions that exist in effluent treatment plants. There is a large potential for reducing EDTA emissions into receiving waters, especially for large-volume consumers of EDTA with their own effluent treatment plants.

The mechanism by which EDTA is broken down has been a topic of discussion in the literature for many years. The first step is the formation of ED3A, which leads to the formation of a molecule that has been shown to be broken down quickly (Fig. 8).

Over the past few years, Ternes et al. of the Institute for Water Research and Water Technology (ESWE—*Institut für Wasserforschung und Wassertechnologie GmbH, Wiesbaden*) have published the results of their work on the metabolites of EDTA and DTPA, principally ketopiperazines. These are cyclic compounds formed

TABLE 3 Removal of EDTA in an Industrial Effluent Treatment Plant (Paper Mill)

H_4EDTA	Influent		Effluent		Removal Rate
	mg/L	mmol/L	mg/L	mmol/L	
Mean	23.8	0.0814	5.8	0.02	76%
Standard deviation	2.3		2.07		
Maximum	28.0	0.096	9.0	0.031	68%
Minimum	18.0	0.062	3.0	0.01	83%

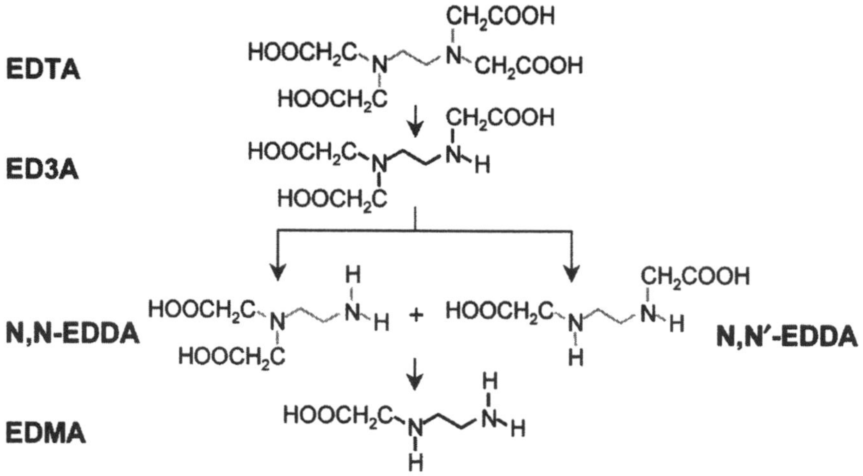

FIG. 8 Degradation of EDTA.

from ED3A. The impact of ketopiperazines on the environment has not yet been conclusively assessed, but joint studies by the ESWE and industry, with the participation of the German Federal Environment Agency, are planned in order to clarify the situation. Work by Van Ginkel at Akzo on the degradability of DTA in a semicontinuous activated-sludge plant has shown that ketopiperazines are biodegradable. As a result, more than 60% in terms of oxygen consumption was obtained in the OECD 301 D test for ready biodegradability. It has been shown that ketopiperazine diacetate is completely mineralized [23].

In the European Union, EDTA under the rules on existing chemicals is a top-priority substance. The EU authorities made an extensive risk assessment on EDTA and its sodium salt under the European Existing Substances Regulation for the protection of human health and the environment. The risk assessment report was discussed and endorsed by the technical experts of EU member states. The risk assessment report concluded that, regarding human health, no concern for consumers in any application, no concern for workers, and no concern for the public who may be exposed via the environment is expected. Aquatic toxicity is low. There is no risk of EDTA to the aqueous environment below the predicted no-effect concentration (PNEC) of 2.2 mg/L and no risk to the aqueous environment due to the influence of EDTA on the mobility of heavy metals, eutrophication, or nutrient deficiency. Thus, EDTA does not need to be classified and labeled with a specific environmental symbol or risk phrase. EDTA is not persistent in the aqueous environment. Biodegradability of EDTA depends on its concentration, the pH, and the species of complexed ions. Fe(III), Co(III), and Mn(II) complexes of EDTA are degraded photochemically. The metabolites are nontoxic to the aquatic environment. For risk characterization, the risk assessment report uses a worst-case assumption of "no biodegradation." For EDTA, no bioaccumulation in living organisms through the food chain has been observed or should be expected.

Risk from EDTA for the local aqueous environment is expected to occur only in some extreme cases at sites where a large, highly concentrated effluent—lacking adequate treatment—is connected to a small surface water recipient. In such a situation the concentration of EDTA in the local aqueous environment near the effluent release point could exceed the established no-effect concentration. In those cases appropriate measures should be considered to reduce the effluent concentrations. To limit risks associated with high release rates of EDTA into the environment, the technical experts of the EU member states indicated the following uses:

- Use in industrial detergents
- Use by paper mills
- Use by circuit board producers (metal plating)
- Recovery of EDTA-containing wastes

The biodegradability of NTA in treatment plants has been conclusively proven in numerous studies under various conditions. Like NTA, MGDA is readily bidegradable (>90%) after OECD 301E.

Results on NTA degradation from studies in the municipal treatment plant in Zurich [12] are presented in Figure 9. Despite low wastewater temperatures, large fluctuations in the concentration of NTA in the inflow, and relatively high levels of pollution, the degradation of NTA is not impaired. In spite of high removal rates, NTA can still be detected in surface waters, but it continues to be broken down. Nevertheless, biodegradation continues in the watercourse. Since this is difficult to check in an actual river because the distances between different waste water discharges are usually too short for appreciable effects, degradation was examined in an artificial watercourse [13]. The transit time was 10 days, which corresponds to the flow times of many river systems in central Europe. Figure 10 shows that when NTA is introduced into this watercourse at the rate of 100–200 µg/L, only about 10% of the initial quantity is present after 3 days. An equally rapid drop in concentration was also found in the seepage of river water into groundwater (Fig. 11) [14]. The NTA concentration dropped so sharply after a flow time of just a few days that NTA could no longer be detected quantitatively. NTA is biodegradable in groundwater not only under aerobic conditions; biodegradation in anaerobic systems has also been documented.

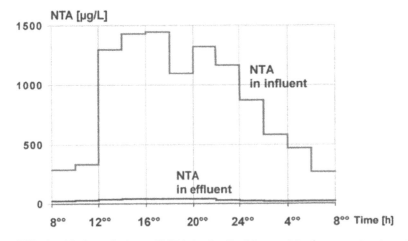

FIG. 9 Biodegradation of NTA in the Zurich municipal sewage treatment plant.

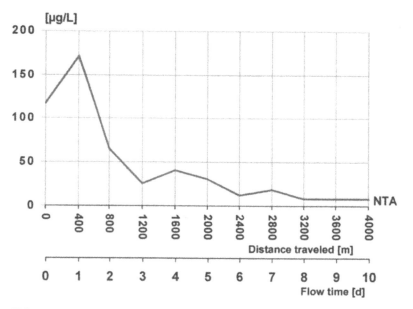

FIG. 10 Biodegradation of NTA in an artificial watercourse. (From Ref. 12.)

The elimination of NTA, EDTA, and other chelating agents in waters can also ensue in principle as a result of photochemical degradation. This is less important in the case of NTA because of its high biodegradability. For EDTA the situation is different. Here it has been scientifically proved that at least a proportion is degraded in this way. In comparison with the volumes used, very low concentrations are measured in waters. In particular, the iron complexes of the chelating agents decompose very

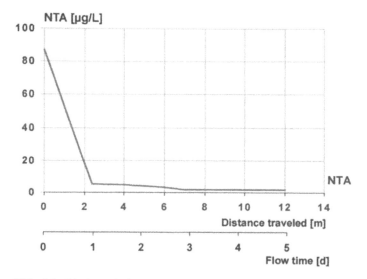

FIG. 11 Biodegradation of NTA during bank filtration. (From Ref. 14.)

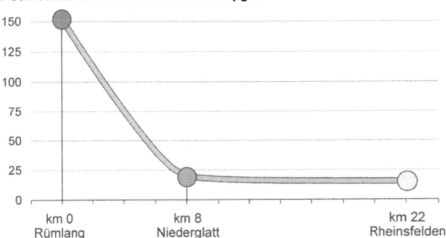

FIG. 12 Degradation of Fe[EDTA] by photolysis in the Glatt River. (From Ref. 15.)

rapidly in sunlight. However, since the penetration of light through the surface of a body of water depends on very many factors that are difficult to measure, the percentage of photochemical degradation is very hard to estimate in practice [15] (Figure 12).

IV. EFFECTS IN THE ENVIRONMENT

The acute toxicity of the chelating agents NTA, EDTA, DTPA, HEDTA, and MGDA to aquatic lifeforms is low, as Table 4 shows.

The testing of aquatic toxicity, with the aid of the usual test methods, and the evaluation of the results are not without problems in the case of chelating agents and can yield meaningful conclusions only when their chelating properties are taken into account. This is because, due to their ability to form complexes, chelating agents can alter the conditions specifically laid down for the test, such as pH, water hardness, and the presence of trace elements in the test medium, and in this way they have a significant effect on the results, regardless of their ecological properties, as the following examples demonstrate.

In the determination of the acute toxicity of H_4EDTA to fish, serious differences were obtained as a function of water hardness. In soft water it was 41 mg/L, in

TABLE 4   Acute Toxicity to Fish and Daphnia

	NTA	EDTA	DTPA	HEDTA	MGDA
Fish, LC_{50} (96 h) [mg/L]	>100	>100	>100	>100	>100
Daphnia, EC_{50} (48 h) [mg/L]	>100	>100	>100	>100	>100

water of medium hardness 150 mg/L, and in hard water 532 mg/L [3]. These results are not attributable to the specific toxicity of the substance but rather to the chelating properties of EDTA. This example also makes it clear how important it is to make sure that test conditions are identical when comparing literature data.

In the case of EDTA, the standardized test method for determining its aquatic toxicity to algae does not yield valid results because EDTA interferes with the test system. In consequence, (high) toxicity values were found that are not attributable to the ecotoxicological properties of EDTA but rather only to its general complex-forming properties [3]. The cause of this is that EDTA complexes the essential trace elements in the test medium and these are, accordingly, no longer present in adequate quantity for the growth of the algae.

The chelating agents under discussion here are highly polar substances that are readily soluble in water and hence as a rule do not accumulate in water-based organisms. The bioaccumulation characteristics of EDTA were examined in experiments, with the result that bioaccumulation was ruled out [3].

The question of the remobilization of heavy metals in treatment plants and sediments by chelating agents regularly gives rise to controversies. In connection with this, it can be concluded that chelating agents occurring in low concentrations in surface waters are not capable of leaching out heavy metal ions from sediments. On the contrary, they are present in the water as dissolved metal complexes of ions such as nickel, copper, and zinc and are thus inactive. Moreover, the concentration of chelating agents in waters is so low that they cannot substantially affect the balance of heavy metals. Exchange of the metal ions in the complex with heavy metal ions in the sediment is unlikely and could not be confirmed experimentally under conditions in practice [16].

V. TOXICITY TO MAMMALS

When the chelating agents are administered in the form of the metal complexes, the toxicity of the metal increasingly comes to the fore. Thus, for example, the LD_{50} (rats, oral) of the Ca-NTA and Ca-EDTA complexes is more than 10,000 mg/kg, while an LD_{50} of 800–900 mg/kg has been determined for the corresponding Cu complexes. It may be assumed that long-term studies on heavy metal complexes are substantially determined by the toxicity of the metal ion. Accordingly, it would be more in keeping with actual conditions to administer the chelating agents in the form of their Na salts or Ca salts. In this way, for example, insights may be obtained into the effect of the chelating agents on the metabolism of metals in the organism under study, unaffected by the toxic properties of the metals.

Alongside the results of subacute and subchronic studies that have been carried out many times on NTA, the NOAEL (no-observed-adverse-effects level) during lifelong administration is of particular importance. In these chronic studies with NTA on rats a daily intake of 15 mg/kg was found to have no toxic effects [17]. In a corresponding feeding experiment with EDTA, the daily intake value was 250 mg/kg [18]. In the case of NTA, higher dosages caused increasing damage to kidneys; in the case of EDTA, diarrhea was the main symptom observed.

NTA and EDTA are not metabolized in animals, and any quantity resorbed is excreted unchanged via the kidneys [17,19].

NTA is not teratogenic [17]. On the other hand, high dosages of EDTA via certain exposure or application routes can have a teratogenic effect. This applies only for dosages that are toxic for the dam. However, the same effects were also observed in animal experiments with a zinc-free diet. In the experiments it was confirmed that when EDTA is administered and the amount of zinc in the feed is simultaneously increased, malformations were completely suppressed [20]. EDTA concentrations found in practice cannot result in teratogenic effects, since adequate amounts of Zn are always available. NTA, upon 4 weeks of administration, does not increase the 8-OH-guanosine in kidney DNA, in contrast to FeNTA. This is true even for oral administration. There is no indication that Fe or FeNTA mediates the effects of NTA in the target tissue. The earlier hypothesis that Zn and/or ZnNTA is the essential cycotoxic species for a focal RTC toxiticity is further supported. Critical-dose levels of NTA cause enhanced lipid peroxidation and increased cell proliferation rates in RTC; these effects were also shown to be threshold related. Innocuos exposure levels of NTA may be defined via noninvasive methods. No such threshold has been observed as yet for FeNTA [24].

Neither NTA nor EDTA is mutagenic in the Ames test. Furthermore, numerous other tests for mutagenicity in different test systems have shown that NTA is not mutagenic [17]. No gene mutations or chromosomal aberrations have been recorded in studies in vitro on cell cultures, fungi, and bacteria or in studies in vivo on insects and rodents. Only in some systems were weak effects observed that were attributed to the complete sequestration of polyvalent cations in the culture medium or in the cells. Such weakly mutagenic effects are observed more frequently in the case of EDTA but also only at very high concentrations. Here, too, it may be assumed that they are not based on the direct action of EDTA on DNA but rather are a consequence of the sequestration of bivalent metal ions. Bivalent metal ions have a coenzyme function, for example, or like Zn, for instance, they are themselves components of proteins that interact with DNA.

The results from the aforementioned studies (Table 5) show that neither NTA nor EDTA is genotoxic. For NTA this means that there is a threshold concentration below which no toxic effects occur and hence no tumors. The trisodium salt of EDTA was studied for carcinogenicity in a 2-year experiment [21]. Even at high doses there were no indications of carcinogenic effects or other chronic toxic symptoms. Several long-term/carcinogenicity studies have shown that NTA administered at very high doses generates tumors in the urogenital system (kidneys, bladder, urinary tract). In the study of the National Cancer Institute, observations led to comparable results [22].

TABLE 5 Basic Toxicological Data for Complexing Agents (40% in Water)[a]

	NTA	EDTA	DTPA	HEDTA	MGDA
LD_{50} (rats, oral) (mg/kg)	3,900	3,200	3,500	8,000	>2,000
Skin irritation	n.i.	n.i.	n.i.	n.i.	n.i.
Mucous membrane	i.	i.	i.	i.	n.i.

n.i. = no irritation; i = irritant.
[a] In order to allow direct comparison, data for the trisodium and tetrasodium salts in 40% aqueous solution are assembled.

Incidence of tumor formation [%]

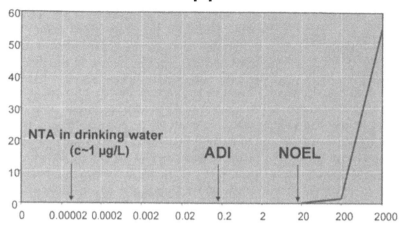

FIG. 13 Incidence of tumors as a function of NTA dosage.

The relationship between dosage and effect clearly reveals a disproportionate rise at higher doses, since the exceeding of a renotoxic dose is the precondition for the carcinogenic effect. Accordingly, a threshold value for tumor formation by NTA can be assumed below which no damage and hence no tumor formation occurs. This has been confirmed by toxicologists worldwide. Extrapolation of risk to low concentrations, therefore, does not seem appropriate. As can be seen from Figure 13, a carcinogenic risk to humans from NTA may be effectively ruled out at the levels of possible exposure. Intravenous administration of CaNa$_2$EDTA is standard therapy for treating cases of poisoning by Pb or other heavy metals. Following recent studies, the Zn-Ca complex is toxicologically more suitable and pharmacologically more effective. EDTA cannot be employed in cases of Cd poisoning.

VI. CONCLUSIONS ON THE ENVIRONMENTAL CHARACTERISTICS OF AMINOCARBOXYLATE CHELATING AGENTS

The ecological and toxicological properties of the aminocarboxylates NTA and EDTA have been established in an extensive series of studies. These are substantially determined by their ability to complex metals, and the effects observed are greatly dependent on concentration. At high concentrations, effects have been found under laboratory conditions that do not occur at the actual concentrations measured in the environment. From the ecological and toxicological perspectives, the use of Trilon® types is safe. Nevertheless, their discharge into the environment should be kept to the lowest possible level. On account of their ready biodegradability, the use of NTA (Trilon® A) and MGDA (Trilon® M liquid) is preferred as long as they fulfill the range of technical requirements.

REFERENCES

1. Brouwer, N.M.; Terpstra, P.M.J. Tenside Surf. Det. 1995, *32*, 4.
2. Kiessling, D.; Kaluza, U. In *Detergents in the Environment*; Schwuger, M.J., Ed.; Marcel Dekker: New York, 1996; 265 pp.
3. EDTA-BUA-Stoffbericht 168, S. Hirzel, Wissenschaftliche Verlagsgesellschaft, 1996.
4. 54th Annual Report of the Arbeitsgemeinschaft Rheinwasserwerke e.V. (ARW) [Rhine Water Works Working Group], 1997.
5. Houriet, J.-P. BUS/Bull. 1988, 42.
6. Brauch, H.J.; Schullerer, S. Vom Wasser 1987, *69*, 155.
7. Guidelines for Drinking-Water Quality. 2nd Ed.; Recommendations, WHO: Geneva, 1993; Vol. 1.
8. Canton, J.H.; Sloof, W. Chemosphere 1982, *11*, 891.
9. Firestone, M.K.; Tiedje, J.M. Appl. Environ. Microbiol. 1978, *35*, 956.
10. van Ginkel, C.G.; van den Grauke, K.K. Bioresource Technol. 1997, *59*, 151.
11. Kaluza, U.; Klingelhöfer, P.; Taeger, K. Water Res. 1998, *32*, 2843.
12. Siegrist, H.-J.; Alder, A.; Gujer, W.; Giger, W. Gas–Wasser–Abwasser 1988, *3*, 101.
13. Hansen, P.D.; Stehfest, H. NTA, Studie über die aquatische Umweltverträglichkeit von Nitrilotriacetat (NTA) [Study on the compatibility of nitrilotriacetate (NTA) with the aquatic environment]; Verlag Hans Richaz, 1984; 385 pp.
14. Kuhn, E.; van Loosdrecht, M.; Giger, W.; Schwarzenbach, R.P. Water Res. 1987, *21*, 1237.
15. Kari, F.G. PhD dissertation. ETH Zurich, No. 10698, 1994.
16. Potthoff-Karl, B.; Kiessling, D.; Gümbel, H.; Kaluza, U. Congress Proceedings, SEPAWA Annual Meeting, 1996; 26 pp.
17. Anderson, R.L.; Bishop, W.E.; Campbell, R.L. CRC Crit. Rev. Toxicol. 1985, *15*, 1.
18. Oser, B.; Oser, M.; Spencer, H.C. Tox. Appl. Pharmacol. 1963, *5*, 142.
19. Foreman, H.; Vier, M.; Magee, M.J. Biol. Chem. 1953, *203*, 1045.
20. Swenerton, H.; Hurley, L.S. Science 1971, *173*, 62.
21. National Technical Information Service (NTIS). Bioassay of Trisodium Ethylenediaminetetraacetate Trihydrate (EDTA) for Possible Carcinogenicity, PB-270 938, 1977.
22. National Cancer Institute (NCI). NCI Carcinogenesis, Technical Report No. 6, DHEW Publ. No. 77 (NIH), 1977.
23. Haberer, K.; Ternes, T.A. Wasser, Abwasser 1996, *137* (10), 573–578.
24. Leibold, E.; Deckardt, K.; Mellert, W.; Potthoff-Karl, B.; Grundler, O.; Jäckh, R. Hum. Experimental Toxicol. 2002, *21*, 445–452.

23

Environmental Impact of Bleaching Activators

VINCE CROUD* Warwick International Ltd., Holywell, Flintshire, United Kingdom

I. INTRODUCTION

Bleaching agents are included in cleaning formulations principally for the removal of stains such as tea, coffee, red wine and those arising from fruit and vegetable products. As well as removing colored ("bleachable") stains, they provide other important functions, such as antigreying and biocidal activity [1,2]. The total contribution that the bleach makes to the cleaning process has been reviewed elsewhere [2].

In the past, washing was typically undertaken using high temperatures (>60°C). Under such conditions, hydrogen peroxide, arising from the inclusion of persalts such as sodium perborate or sodium percarbonate in the formulation, is an effective bleach. The temperature also provides thermal disinfection.

Many modern fabrics and dyes, however, cannot be washed at high temperatures, and as a consequence wash temperatures have steadily decreased over the last two decades. Under low-temperature wash conditions, peroxide alone is ineffective as a bleach, and thermal disinfection is not possible. In order to overcome these problems, species known as *bleach activators* are incorporated into the formulation. These react with the persalt in the wash bath to produce stronger oxidants, typically peracid anions. Bleach activators can be described as acylating agents, and they are typically esters or amides of carboxylic acids.

Activators first found use in the early 1970s in Europe. They are now used on a global basis. In most cases their function is to provide both low-temperature bleaching and biocidal activity. However, in some instances they have been included solely to provide the latter.

Of all the activators researched and developed, only two (SNOBS and TAED) have met with commercial success. The reasons for this are given elsewhere, but in many cases it is the poor environmental and health and safety profile of the molecule that has contributed to the failure of the molecule to be commercialized [2].

**Current affiliation*: Antec International Ltd., Sudbury, Suffolk, United Kingdom.

SNOBS (sodium nonanoyloxybenzene sulfonate) is used principally in the Unites States and Japan, where the dilute, low-temperature, short-residence-time wash conditions used favor. Patent coverage restricts the use of SNOBS.

The activator with by far the greater commercial success is TAED (tetraacetyl ethylene diamine). Hence in any discussion of the environmental impact of activators it is TAED to which we should turn our greatest attention. Currently around 70,000 tons of TAED are used a year. Its present main end use is in fabric and dishwashing formulations and so-called "bleach boosters." It is, however, finding increasing uses in other cleaning and biocidal compositions, including hard-surface cleaners, medical sterilants, and food and agricultural cleaning applications.

II. ENVIRONMENTAL BENEFITS OF ACTIVATORS

It is typical for any review on the environmental impact of a chemical to focus solely on the chemistry of that molecule. Often the review will also look at the impact that the manufacturing process has. But to truly consider environmental impact one must also look at how the molecule is used and what impact that use has on the environment as a whole.

In the case of fabric washing, use of activators, as already indicated, facilitates low-temperature bleaching and biocidal action (up to sterilization if necessary). This results in energy savings as compared with high-temperature washing. It was calculated, in 1990, that the switch by 78% of the population from "boil" washing to washing at 60°C in the United Kingdom saved an estimated 2,200 GWh in a year. This represented nearly 3% of yearly domestic consumption in the United Kingdom, or the electricity consumed in 2 months by the entire City of Manchester (UK) and its suburbs. A further 940 GWh per year would have been saved if they had all switched to 40°C washing (Personal Communication, AJ Gradwell, Warwick International Ltd.).

Since activators provide for effective low-temperature bleaching and biocidal activity, garments are cleaned to a level satisfactory to the consumer in a single wash, obviating the need for further washing. This saves on energy, water, and detergent. Low-temperature washing also prolongs garment lifetime since less fiber degradation occurs in the wash and color and finish are better maintained as compared with high-temperature washing. As a consequence, the frequency of garment replacement is decreased, helping to conserve resources. Disinfection is also important, in terms of the operation of the washing machine. If bacteria are allowed to grow, slime builds up in the machine. This can impair the action of the machine, leading to inefficient operation, and can also cause mechanical failure. In addition, bacterial buildup is not only a potential health hazard but also undesirable, in terms of the generation of odors and potential staining of clothing.

In the case of ware washing, the advent of lower-alkalinity nonchlorine formulations incorporating activators and enzymes has led to less degradation of glaze, pattern, and glass, prolonging ware lifetimes, again decreasing replacement frequency.

In the area of industrial and institutional (I&I laundry), activators have been used to increase biocidal activity in rinse water in tunnel washers so that this water can be recycled back to the start of the tunnel and hence give rise to savings in water usage. In the area of I&I laundry as a whole, such as OPLs (on-premise laundries), use of lower-temperature washing helps extend fabric lifetime. Fabric replacement costs are second only to labor costs in this industry (labor is 36–55%, linen replacement 13–

25%). One I&I product based on activator technology has won an "EPA Design for the Environment Award" in the United States on the basis of its positive environmental profile. This is awarded on the basis of the reduction in chemical usage it facilitates (it is a single-shot product, replacing a four-stage process involving caustic detergent, chlorine bleach, sour, and then softener) and its energy reduction potential. Its use also gives rise to a reduction in effluent (compared with prior processes) and in the frequency of linen replacement (c.f. chlorine).

Hard-surface cleaners using peroxide and activators would be expected to give less surface and sealant damage than chlorine-based ones. They also permit effective combined cleaning and disinfection/sanitization, due to better compatibility of actives as compared with chlorine-based systems. This is especially so in the case of liquid products (a two-compartment system in the case of a TAED-based cleaner). Hence, rather than use a two-stage process, involving cleaning first and disinfection afterwards (typically a high-alkalinity cleaning stage followed by chlorine use as a biocide), a single-stage process can be used, saving water and decreasing effluent. This has applications in dairy and food preparation cleaning.

There are, of course, other examples. But in the cases cited, use of activators has a positive environmental impact through savings in either energy, water, or chemicals and in decreasing replacement frequency as compared with prior or alternative systems. In many ways activators can be thought of as enabling processes that have a positive environmental impact as compared with alternative technologies.

III. TAED

A. Introduction

TAED is a white, free-flowing powder. It is usually granulated (e.g., with CMC) when formulated into a powder, for reasons of handling, storage stability, and wash performance. It is typically present at a level of 4–6% in compact fabric washing detergents (with, e.g., 8–12% sodium perborate monohydrate, PBS1), and at 2.0–2.5% in fluffy powders. Occasionally it is found at lower levels when its prime function is to be disinfection only. Compacts are usually dosed, in Europe, at 5 g/L and fluffy powders at 7.5 g/L. Outside of Europe, detergent dosages are lower.

Under alkaline conditions, one mole of TAED reacts with two moles of peroxide anion to produce two moles of peracetic anion (Fig. 1). The reaction proceeds via the production of the short-lived intermediate tri-acetyl ethylene diamine (tri-AED) to give the residue di-acetyl ethylene diamine (DAED). The two remaining acetyl groups on DAED cannot be displaced by peroxide in the wash bath. The reaction can be

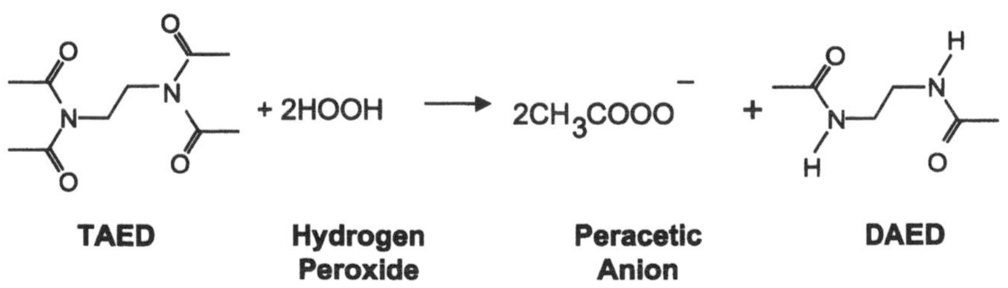

| **TAED** | **Hydrogen Peroxide** | **Peracetic Anion** | **DAED** |

FIG. 1 Perhydrolysis reaction of TAED.

considered to go to completion under typical wash conditions. As a consequence, the principal species being discharged to drain are DAED and acetate (the by-product of peracetic anion reaction or decomposition). Also present may be small quantities of unreacted TAED and peracetic anion. In the case of SNOBS, the three species of interest are phenyl sulfonate, unreacted pernonanoic anion, and nonanoate. SNOBS also generates a diacyl peroxide in the wash bath, but this can be considered to have a short lifetime and hence will not go to drain.

With TAED, any surviving peracetic anion will rapidly decompose on contact with the wastewater (being consumed either via oxidation of soils or via transition-metal-catalyzed decomposition) to give acetate. Acetate is readily biodegradable and of low toxicity, and its environmental impact can hence be ignored. Accordingly, the only environmental impact we need consider is that from the small quantities of unreacted TAED and from DAED.

TAED will hydrolyze to DAED, albeit slowly, in water of neutral pH or below. It will react more quickly under alkaline conditions and in the presence of strong nucleophiles.

The environmental impact of TAED has been reviewed previously [3,4].

B. The Biodegradability of TAED and DAED

The definition of ready biodegradability depends on the method used. For methods where direct measurement of DOC (dissolved organic carbon) is taken, the pass level is 70% loss of DOC within 28 days. For experiments that follow the decomposition by generation of CO_2, 60% of the theoretical yield must be produced within 28 days. In both cases a 10-day window is applied. This condition means that the pass level of decomposition (60% or 70%) must be achieved within 10 days of the attainment of 10% degradation. Only if both of these criteria are met can a material be defined as readily biodegradable.

The tests used in the work presented here are designed to show ready biodegradability. For ultimate degradation tests, different conditions are required. Using the test methods in this work, it can be estimated whether a material is likely to be ultimately biodegradable.

The studies of TAED and DAED were performed using a modified OECD (Organization for Economic Cooperation and Development) screening test (OECD procedure No 301E). The OECD test relies on the measurement of the DOC, i.e., a direct measurement of the levels of material in the test solution. It is important for the test material to be soluble in the test solvent (a buffered mineral salts medium), and this is tested by the preparation of standard solutions. A standard of sodium benzoate is used to ensure that the sewage sludge is active and to validate the test method. Preliminary studies can also be performed using sodium benzoate and test material to ensure that there is no interference.

The graph showing the biodegradation of TAED is given in Figure 2. This shows that TAED is biodegradable under the conditions set down for the test procedure used. There is good evidence that TAED is biodegradable under natural conditions.

The biodegradation of DAED is shown in Figure 3. After an acclimatization period, the decomposition of DAED was very rapid, with >80% decomposition occurring between days 7 and 14. Thus DAED is classed as being readily biodegradable under these conditions.

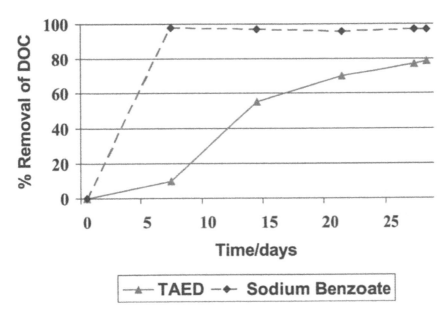

FIG. 2 Biodegradation of TAED.

It has previously been reported [5] that, using the modified Sturm test (which measures CO_2 evolution), TAED achieves 100% degradation under the conditions of the test (i.e., readily biodegradable).

Not all activators have the same positive biodegradability as TAED. For example, using the Sturm test methodology (OECD method 301B), the activator DADHT (1,5-diacetyl-2,4-dioxohexahydro-1,3,5-triazine), which appeared in the early 1990s

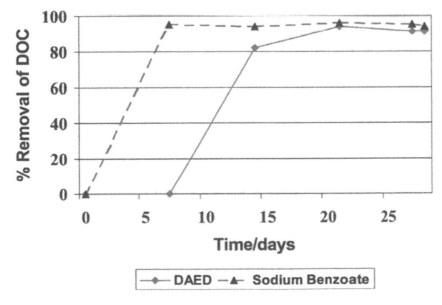

FIG. 3 Biodegradation of DAED.

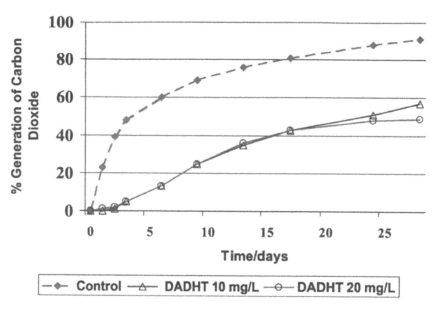

FIG. 4 Biodegradation of DADHT.

[16], did not reach 60% of theoretical in 28 days (Fig. 4). Hence it is less biodegradable than TAED and does not meet the test criteria. Its reaction residue, DHT (2,4-dioxohydro-1,3,5-triazine), managed to generate only ca. 8% of the theoretical CO_2 in 28 days, c.f. DAED's 94% (Fig. 5).

Preformed peracids have also been promoted for use in detergent formulations. They are attractive, in that they save formulation volume by avoiding the need to

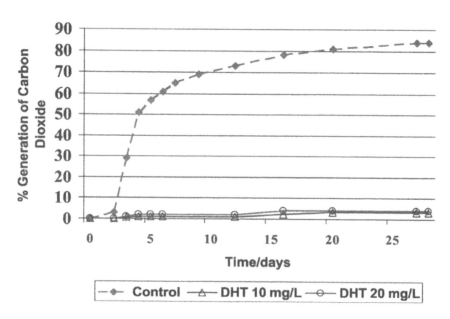

FIG. 5 Biodegradation of DHT.

include a persalt and activator, and, in theory, have benefits in terms of the bleaching species being instantly available rather than trying to optimize in situ generation. In practice the often low water solubility and the need for significant storage protection removes the latter advantage. One such species that has been promoted in the past is PAP (phthalimidoperoxycaproic acid). The biodegradability of the material has been studied using the modified Sturm test. Over a period of 28 days, PAP managed to generate only 3% of the theoretical CO_2 (Fig. 6). Hence DADHT, its reaction residue DHT, and PAP show very poor biodegradability.

C. Mechanisms of Biodegradation of TAED and DAED [5]

TAED and DAED are both acetylated ethylenediamine derivatives (AEDs). A potential mechanism for the biodegradation of these materials is via successive hydrolysis to yield ethylenediamine (as shown in Fig. 7). This is of possible concern because ethylenediamine is a chelating agent and could therefore lead to mobilization of heavy metals from the sediment. It was therefore decided to investigate the mechanism of degradation of TAED and DAED over an extended period of time to try and identify the intermediates of the process. This was achieved using a procedure based on the modified OECD screening test mentioned earlier.

For the first 28 days, the test procedure was as laid out in OECD 301E. After this period there was an extended period, to 307 days, where a semicontinuous batch replacement of spent media was performed on a regular basis. This procedure allowed for the ultimate biodegradability of TAED and DAED to be assessed. Samples were periodically removed from the test solutions and assayed for TAED, Tri-AED, DAED, and ethylenediamine. Measurements of the total nitrogen concentrations were also performed.

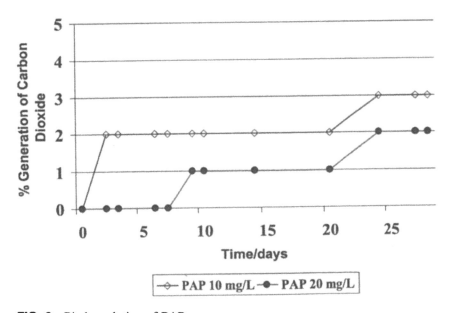

FIG. 6 Biodegradation of PAP.

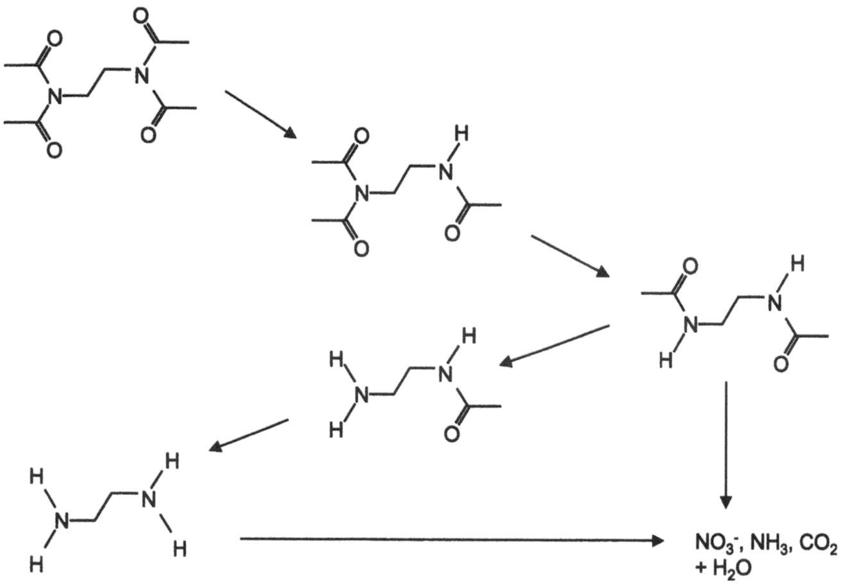

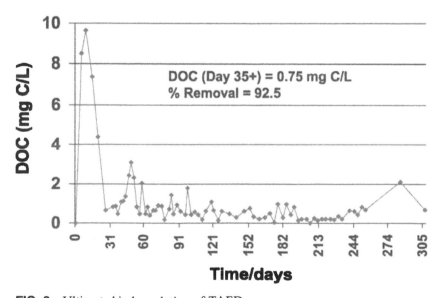

FIG. 7 Possible mechanism for the biodegradation of TAED.

During the test with acclimatized organisms, a consistently high removal of TAED and DAED was noticed. Mean levels of removal were found to be 89–93% in both cases (Figs. 8 and 9).

Analysis for AEDs was performed by HPLC during the first 26 days of the trial. The loss of DOC measurements are similar to those given in Figure 2. These results are given in Figure 10.

DOC (Day 35+) = 0.75 mg C/L
% Removal = 92.5

FIG. 8 Ultimate biodegradation of TAED.

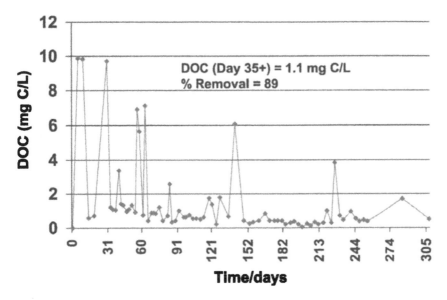

FIG. 9 Ultimate biodegradation of DAED.

The initial measurements show only TAED to be present. Over the next 9 days, deacetylation of TAED proceeds with little reduction in DOC and total AED concentrations. After this, the reduction of DOC and AED was more rapid. After 26 days a 94% decrease in DOC had occurred and no measurable quantities of AED were found. In the DAED sample, there was no appreciable loss of intensity during the first 9 days. By day 14, however, no DAED was detectable in the sample. The test solutions were assayed for ethylenediamine, but none of this material could be found.

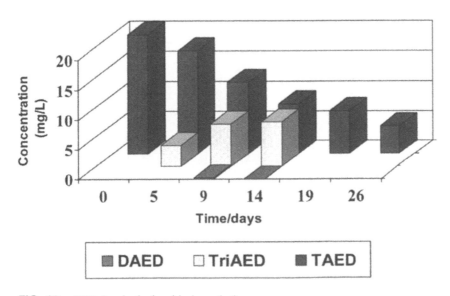

FIG. 10 AED levels during biodegradation.

Thus the levels of ethylenediamine in solution, if any, were less than 0.05 mg/L, the detection limit by HPLC.

The figures for determination of the nitrogen levels obtained toward the end of the experiment are given in Figure 11.

Samples were analyzed on three occasions (days 240, 244, and 285). The results were compared statistically using ANOVA and/or Student's t-test. The results were compared to those of a standard containing sodium benzoate (Refs. 1 and 2 in Fig. 11). The nitrate and ammonia levels were found to be significantly different in the AED samples. There was no significant difference between the samples in terms of nitrite and organic nitrogen.

The foregoing observations imply that the mechanism of degradation is deacetylation of TAED to DAED via Tri-AED, as expected. DAED, however, does not seem to decompose via a deacetylation route, because no evidence for MAED (monoacetyl ethylenediamine) or ethylenediamine formation was noticed. The nitrogen portions of TAED and DAED degrade ultimately to nitrate and ammonia.

There is no evidence for the formation of ethylenediamine during the decomposition process.

D. Toxicity of TAED and DAED

There is much information already in the public domain on the toxicity of TAED and to a lesser extent of DAED and Tri-AED [3,4].

From the studies conducted in can be concluded that.

- The aquatic toxicity of TAED, Tri-AED, and DAED to species representing the three trophic levels in the aquatic environment (algae, invertebrates, and fish) is remarkably low. The studies produced insufficient mortalities to allow calculation

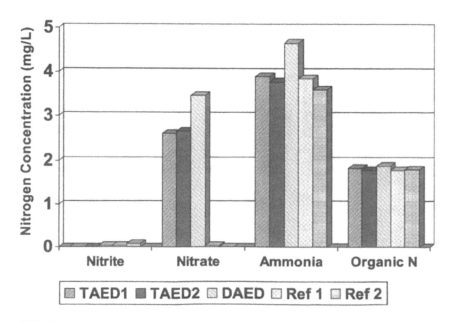

FIG. 11 Nitrogen concentrations (day 285).

of the EC_{50}. Hence EC_{50} for TAED was >500 mg/L for algae, >800 mg/L for invertebrates, and, for example, >2500 mg/L for goldfish.

- With respect to mammalian toxicity, TAED and DAED have low toxicity and are nonmutagenic, nonsensitizing with low irritancy potential.

There are now, however, products on the market that make a biocidal claim based on the incorporation of TAED. In addition, TAED is finding increasing use in formulations that are designed for sanitization (e.g., endoscope sterilization) rather than cleaning. Given the advent of the European Biocides Product Directive (BPD) and the requirements of the U.S. EPA and FDA for products making a biocidal claim, further toxicity testing has been required beyond that already reported.

These studies are:

- Avian Acute Oral toxicity Test in Bobwhite Quail (*Colinus virginianus*): The results of the 14-day acute oral toxicity test revealed the acute oral median dose (LD_{50}) to be greater than 2000 mg/kg. Based on the U.S. EPA Hazard Evaluation Division Standard Evaluation Procedure for an Avian Single Dose Oral LD_{50} (EPA-540/9-85-007, OPPTS 850.2100) TAED is classified practically nontoxic.
- Acute Oral Toxicity to Rainbow Trout (*Oncorhynchus mykiss*) Under Flow-through Conditions: Based on the results of this study, the 96-hour exposure to a concentration of 140 mg/L of TAED had no adverse effect on rainbow trout, and the LC_{50} was determined to be >140mg/L. The no-observed-effect-concentration (NOEC) was determined to be 140 mg/L. No further testing was required since TAED was not toxic at the maximum dissolved concentration (i.e. 100 mg/L) required under the existing guidelines. Based on these results and on criteria published by the U.S. EPA (1994, OPPTS Draft Guideline 850.1075), TAED would be classified as practically nontoxic to *Oncorhynchus mykiss*.
- Prenatal Development Toxicity Study in Rats via Oral Route (Teratogenesis study): TAED (0, 40, 200, and 1000 mg/kg/day) was given by the oral route to pregnant rats. No clinical signs, behavioral changes, death, or abortion were noted in any group. No embryotoxic effects were found, and no significant changes were found in the frequency of malformed fetuses. Based on these findings, the NOEL for pregnant females was 40 mg/kg/day and for fetuses 200 mg/kg/day [17].

Further genetic, acute and subchronic studies on TAED and DAED have also been conducted. The results are summarized, along with prior data, in Tables 1–3. The studies further confirm that TAED, DAED, and Tri-AED will not have any adverse effects on the environment or in man.

E. *n*-Octanol/Water Partition Coefficient

The octanol/water partition coefficient ($P_{o/w}$) of a compound is an indication of how the compound would distribute itself between the lipophilic tissue of an organism and the aqueous environment. It has also been shown that $P_{o/w}$ is a useful parameter in the prediction of adsorption on soil and sediments and for establishing quantitative structure–activity relationships for a wide range of biological effects. Octanol is used as the reference organic phase because it has a similar carbon-to-oxygen ratio as lipid materials in animal fats.

TABLE 1 Genetic Toxicity Studies

Study	Test material	Metabolic activation	Control	Vehicle	Strains	Result
A: Gene mutations in bacteria						
Ames test [7]	TAED	w/wo	+MNNG	DMSO	TA1535	Negative
			+9AA	DMSO	TA1537	Negative
			+4NOP	DMSO	TA1538	Negative
			+4NOP	DMSO	TA98	Negative
			+MNNG	DMSO	TA100	Negative
Ames test [3]	TAED	w/wo	Not reported	Not reported	TA 1535	Negative
			Not reported	Not reported	TA1537	Negative
			Not reported	Not reported	TA1538	Negative
			Not reported	Not reported	TA98	Negative
			Not reported	Not reported	TA100	Negative
Ames test [8]	DAED	w	2AA	NaCl	TA1535	Negative
			2AA	NaCl	TA1537	Negative
			2AA	NaCl	TA98	Negative
			2AA	NaCl	TA100	Negative
		wo	Na azide	NaCl	TA1535	Negative
			9AA	NaCl	TA1537	Negative
			2NF	NaCl	TA98	Negative
			Na azide	NaCl	TA100	Negative
E. coli [8]	DAED	w/wo	1-ethyl-3-nitro-1-nitroso-guanidine	NaCl	WP2	Negative
B: DNA damage/chromosomal abberations						
SCE [3]	TAED	w/wo	Not reported		CHO V9	No SCE, no chromosomal abberations

Key: DMSO = dimethyl sulfoxide; MNNG = *N*-methyl-*N'*-nitro-*N*-nitrosoguanide; 9AA = 9 amino acridine; 4NOP = 4-nitro-*ortho*-phenylenediamine; 2NF = 2-nitrofluorene; 2AA = 2-aminoanthracene; SCE = sister chromatid exchange (Chinese hamster V79 cells); w/wo = with/without.

TAED has a log $P_{o/w}$ of 0.06 or a $P_{o/w}$ of 0.876 (Protocol No. 1110598/OECD/EC/OPPTS/705). The low value for log $P_{o/w}$ indicates that TAED will not bioaccumulate through the food chain.

F. Soil Sorption Coefficient (K_{oc})

The method used is intended for the determination of the soil/water sorption coefficient, K_d, on some standard well-defined soils (in this case Whimpole, Bromsgrove, and Elmton soils in the United Kingdom). The technique is applicable to all pesticides and other organic compounds that can be analyzed by either liquid chromatography or a gas chromatographic method.

Estimates of the sorption coefficient to the organic matter or (carbon) component of the soil K_{om} (or K_{oc}) can be made knowing K_d and the organic matter content and assuming sorption solely by the organic matter component. The method was used in accordance with OECD test guideline method 106 (May 1981). The performance of the test was verified by the reference item, 2,4-dichlorophenoxy acetic acid (2,4-D). Three representative soils were chosen and the pedological parameters were analyzed independently from the GLP lab conducting the determination.

TABLE 2 Acute Studies

Study	Species	Test material	Number of animals	Result
Oral LD$_{50}$ [9]	Rat	TAED	24/sex	LD$_{50}$ 7.94 (6.46–9.77)
Acute dermal [10]	Rabbit	TAED	6	Mild irritant
Acute eye irritation [11]	Rabbit	TAED	6	Nonirritant
Skin sensitization [12]	Guinea pig	TAED	26	0% sensitization. Weak grade I sensitizer
Acute oral-limit test [13]	Rat	DAED	5/sex	No deaths at 2g/kg
Acute dermal [20]	NZ white rabbit	TAED	—	Limit dose to 2000 mg/kg produced no acute toxicity
Acute dermal [21]	NZ white rabbit	TAED	3	4-h application of 500 mg to shaved backs. No effect on mortality and no dermal irritation
Sensitization [3]	Guinea pig	DAED	Not reported	Nonsensitizer

Measurement of the sorption coefficients of TAED by the 1981 OECD test method no. 106 on three types of soils gave values of k_d = 1.40 (Whimpole), 0.60 (Bromsgrove), and 2.99 (Elmton). Relating these values to the organic carbon content of the soils to give an organic carbon sorption coefficient, K_{oc}, gave a mean value of ~ 65. This value puts TAED into the high-mobility class [6].

IV. TAED MANUFACTURE

TAED is manufactured via succesive acetylation of ethylenediamine, first with acetic acid and then with acetic anhydride. Excess acetic acid and acetic anhydride are recovered by distillation, separated, and reused within the process. Following crystallization, isolated TAED solid is washed with recovered distillates and then dried to remove any remaining residual wash materials.

TABLE 3 Subchronic Studies

Study	Species	Dose (mg/kg/day)	Test material	NOEL
90-day oral [14]	Rat	0, 25, 500, 1000	TAED	25 mg/kg/day
90-day dermal [15]	Rat	0, 20, 200, 2000	TAED	≥200 mg/kg/day
13-week feeding study [3]	Rat	Not reported	DAED	5.7 g bw/day

Utilizing outputs from the initial TAED process, two further process recovery stages are carried out to ensure maximum utilization of raw materials.

Final TAED purity from all elements of the process is in excess of 99%. Waste products are minimized by conversion of excess acetic acid back to acetic anhydride, which is then reused within the manufacturing process.

Small amounts of residue that cannot be utilized within the manufacturing process are diluted with water and used as a mixed fuel to generate process steam.

The TAED manufacturing process is thus efficient, with minimal environmental impact.

V. TAED ANALYSIS

A reliable analytical method for the determination of the concentration of a species in wastewater streams is clearly desirable in order to evaluate its environmental impact.

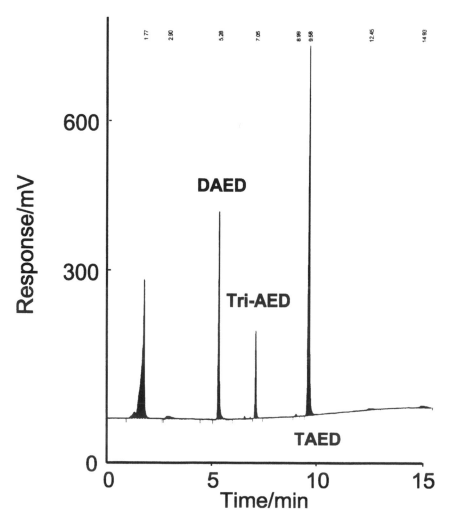

FIG. 12 Chromatogram of AEDs in wastewater.

TAED can be detected in wastewater streams using high-performance liquid chromatography (HPLC).

A representative sample of the water is filtered through a 0.45-μ 25-mm PTFE Acrodisc, and 20 μL is injected on to a Hypersil 5-μ C18 BDS 25 cm × 4.6 mm HPLC column. Separation of the TAED from any other detectable species in the sample is facilitated by the use of a gradient solvent program of 0–100% acetonitrile (100–0% water) over 20 minutes at a flow rate of 1.5 mL min^{-1}.

At near neutral pH, any TAED will remain in the water mainly intact, but movement of the pH away from neutral will have an effect on the stability of the molecule. If hydrolysis has occurred, then both the hydrolysis products, DAED and Tri-AED, could also be present in the sample.

The column is connected to a ultraviolet detector, which for the purpose of total AED species monitoring is set to 205 nm due to the poor chromophore on the DAED molecule. If quantitative analysis of TAED is required in the sample, then the wavelength should ideally be changed to 215 nm.

A typical chromatogram of a wastewater sample (Fig. 12) shows three major peaks, corresponding to DAED (t_R 5.28 min), TriAED (t_R 7.05), and TAED (t_R 9.58).

Depending on the nature of the waste stream, detection of TAED can be effective down to ppm levels.

VI. SUMMARY

Use of activators can have many positive benefits, in terms of enabling processes with lower environmental impact than alternatives. These include saving energy and water and extending material lifetimes.

TAED is the most widely used activator on a global basis. It has been determined that:

- TAED is completely and ultimately biodegradable.
- DAED is readily, completely and ultimately biodegradable.
- The ultimate biodegradation products of TAED and DAED are carbon dioxide, water, nitrate, and ammonia. There is no evidence for the formation of ethylenediamine as an intermediate in the process.
- TAED and its decomposition products, Tri-AED and DAED, have very low aquatic, avian, and mammalian toxicity.
- TAED and DAED are nonmutagenic and nonsensitizing and have low irritancy potential.
- TAED does not bioaccumulate or accumulate in soils in the environment.

It can thus be safely concluded the use of TAED in cleaning and hygiene formulations will have no adverse effect on either the environment or man.

ACKNOWLEDGMENTS

The author acknowledges SRS International, Washington, DC, for the tabulation of data in Table 1; Trevor Tommy, Warwick International Ltd. for information on TAED manufacture; and Paul Shrimpton, Warwick International Ltd. for information on TAED analysis. The author would also like to acknowledge input from Dr. S.J.Tompsett, previously at Warwick International.

REFERENCES

1. Croud, V.B. J. Soc. Dyers. Colour 1996, *112* (4), 117–122.
2. Croud, V.B. In *Handbook of Detergents, Part F: Properties*; Broze, Guy., Ed.; Marcel Dekker: New York, 1999, 597–617.
3. Gilbert, P.A. In *Handbook of Environmental Chemistry*; Hutzinger, O., Ed.; Springer-Verlag: Berlin, 1992; Vol 3 Part F, 319–328.
4. Reinhardt, G.; Schuler, W.; Quack, J.M. Comun. J. Com. Esp. Deterg. 1989, *20*, 165–179.
5. Tompsett, S.J. Mechanisms of biodegradation of TAED and DAED. 36th International wfk Detergency Conference, Krefeld: Germany, 1994.
6. Swann, R.L.; Laskowski, D.A.; McCall, P.J. Residue Rev. 1983, *85*, 17–28.
7. Todhunter, J.A. The *Salmonella typhymurium* Reverse Mutation Assay Testing TAED. Unpublished study by SRS International Corp, on behalf of Warwick International, Ltd., 2000.
8. Kight, V. *Salmonella typhimurium* and *E. coli* Reverse Mutation Assay—ICH. Unpublished study by Toxicon Labotratories, on behalf of Warwick International Ltd., 2000.
9. Collier, T.A. Acute Oral LD50 Determination of TAED in the Rat. Unpublished study by Safepharm Laboratories, on behalf of Warwick International Ltd., 1982.
10. Collier, T.A. Acute Dermal Irritation of TAED in the Rabbit. Unpublished study by Safepharm Laboratories, on behalf of Warwick International Ltd., 1982.
11. Collier, T.A. Acute Etye Irritation of TAED in the Rabbit. Unpublished study by Safepharm Laboratories, on behalf of Warwick International Ltd., 1982.
12. Collier, T.A. Skin Sensitization of TAED in the Guinea Pig. Unpublished study by Safepharm Laboratories, on behalf of Warwick International Ltd., 1982.
13. Tay, C. Acute Oral Toxicity Test (OECD) with DAED. Unpublished study by Toxicon Laboratories, on behalf of Warwick International Ltd., 2000.
14. Wolfe, G; Borst, S. 90-day Oral Toxicity Study of TAED in Sprague Dawley Rats. Unpublished study by TherImmune Research, on behalf of Warwick International Ltd., 2000.
15. Wolfe, G; Okoth, A. 90-day Oral Toxicity Study of TAED in Sprague Dawley Rats. Unpublished study by TherImmune Research, on behalf of Warwick International Ltd., 2000.
16. Porsch, M.; Tohomeczek, M.; Kaufman, D.; Bergk, K.-H. Seifen-Ole-Fette-Wasche 1990, *166*, 77–84.
17. Pedersen, C. A. Avian Acute Oral Toxicity Test with Tetracetylyethylenediamine in Bobwhite Quail. Unpublished study by Bio-Life Associates, on behalf of Warwick International Ltd., 2000.
18. Sousa, J.V. Tertracetylethylenediamine—Acute Toxicity to Rainbow Trout (*Oncohynchus mykiss*) Under Flow-Through Conditions. Unpublished study by Springborn Laboratories, on behalf of Warwick International Ltd., 2000.
19. Bussi, R. Prenatal Development Toxicity Study of Tetracetylethylenediamine in Rats via Oral Route. Unpublished study by Instituto Di Richerche Biomediche "Antoine Marxer" RBM S.p.A Dede Legale e Labor, 1994.
20. Acute Dermal (Limit Testing) in New Zealand White Rabbits. Therimune Research Corp. Report No. 1152-100, on behalf of Warwick International Ltd., 2000.
21. Acute Dermal Irritation study (Limit Testing) in New Zealand White Rabbits. Therimune Research Corp. Report No. 1152-101, on behalf of Warwick International Ltd., 2000.

24

Perborates

The Environmentally Problematic Bleaching Agents

J. TARCHITZKY   Israeli Ministry of Agriculture, Bet-Dagan, Israel

Y. CHEN   The Hebrew University of Jerusalem, Rehovot, Israel

I. INTRODUCTION

Boron (B) is widely distributed in nature and can be found in oceans, sedimentary rocks, and surface water and groundwater. Borates, boron–oxygen substances, are used extensively in industrial materials and end-use products. The extent and nature of their utilization determine the concentration of B discharged into water streams, lakes, or sewage water. Boron is taken up by plants and consumed by animals (aquatic and terrestrial). It is one of the essential microelements for crops, but only a narrow range separates deficiency from toxicity. Its availability to plants is affected by several factors, including soil solution pH, and soil texture, moisture, temperature, oxide content, carbonate content, and clay mineralogy. Boron toxicity to humans and animals has been only rarely reported. Its sorption and desorption from soil adsorption sites regulate the concentration of B in the soil solution. It is added to agricultural soils via the irrigation water. In arid and semiarid regions, the use of reclaimed wastewater for irrigation is on the rise, and it is expected to become the main irrigation water source. Therefore, B concentration could become a main limiting factor in wastewater recycling. During domestic, industrial, and agricultural use of freshwater, the concentrations of B increase. Sources are: human and animal excretions, detergents, laundry powders, paper mills, metal-coating processes, among others. Thus, recycling wastewater requires that special attention be paid to the prevention of B discharge into sewers.

II. CHEMICAL PROPERTIES

Boron is the first member (atomic number 5) of the metalloid family of elements, which includes silicon and germanium; it is the only nonmetal of the group IIIA elements. It has an atomic weight of 10.811, a melting point of 2,075°C, its specific gravity in crystal form is 2.34 and of the amorphous variety 2.37, and it has a valence

645

of 3. Although B compounds have been known for thousands of years, Gay-Lussac and Thenard did not discover the element until 1808. The pure element is not found free in nature: It usually occurs as orthoboric acid in certain volcanic spring waters and as borates in borax and colemanite [1]. Boron compounds make up about 0.001% of the earth's crust. It exists naturally as 19.78% ^{10}B isotope and 80.22% ^{11}B isotope.

III. NATURAL SOURCES

More than 200 minerals contain boric oxide, but only a few are of commercial significance (Table 1). Four minerals comprise almost 90% of the borates used by industry: borax and kernite, both of which are sodium borates; colemanite, which is a calcium borate; and ulexite, which is a calcium–sodium borate [2]. The United States, Turkey, Russia, Argentina, and Chile are the world's largest producers of B. World production of B minerals in 1999 was 4.37×10^6 tons, with about 29% (1.22×10^6 tons) of this total produced by the United States [2].

IV. USE OF BORON MINERALS AND PRODUCTS

Major uses for B include insulation- and textile-grade fiberglass, laundry bleach (sodium perborate), borosilicate glass, fire retardants, agricultural fertilizers and herbicides (as a trace element), and enamels, frits, and ceramic glazes. Glass fiber insulation accounts for about 30% of the total annual world consumption of borates. Quantitative distribution of its other uses is as follows: textile glass fiber (18%); soaps, detergents, and bleaches (10%); glass (10%); agriculture (7%); flame retardants (6%); enamels and glazes (3%); miscellaneous (16%) [3]. Some of the main industrial applications and functions of B compounds are presented in Table 2.

V. BORON COMPOUNDS IN DETERGENTS

Boron remains in a soluble form in fertilizers and detergents, but it is strongly bound into the structural components of ceramics, glass, and insulation products during the

TABLE 1 Boron Minerals of Commercial Importance

Mineral	Chemical composition	Boron oxide equivalent (B_2O_3) (wt%)	B (wt%)
Boracite (stassfurite)	$Mg_6B_{14}O_{26}C_{12}$	62.2	19.3
Colemanite	$Ca_2B_6O_{11} \cdot 5H_2O$	50.8	15.8
Datolite	$CaBSiO_4(OH)$	24.9	7.7
Hydroboracite	$CaMgB_6O_{11} \cdot 6H_2O$	50.5	15.7
Kernite (rasorite)	$Na_2B_4O_7 \cdot 4H_2O$	51.0	15.8
Priceite (pandermite)	$CaB_{10}O_{19} \cdot 7H_2O$	49.8	15.4
Probertite (kramerite)	$NaCaB_3O_{19} \cdot 5H_2O$	49.6	15.4
Sassolite (natural boric acid)	H_3BO_3	56.3	17.5
Szaibelyite (ascharite)	$MgBO_2(OH)$	41.4	12.8
Tincal (natural borax)	$Na_2B_4O_7 \cdot 10H_2O$	36.5	11.3
Tincalonite (mohavite)	$Na_2B_4O_7 \cdot 5H_2O$	47.8	14.8
Ulexite (boronatrocalcite)	$NaCaB_5O_9 \cdot 8H_2O$	43.0	13.3

TABLE 2 Some Industrial Applications and Functions of B Compounds

Industrial application	Function
Adhesives	Controls set time, viscosity, and solubility
Coatings	Corrosion inhibitor, adhesion promoter, flame retardant, and char former
Corrugated boxes	Used in adhesives
Detergents	Deodorant, buffer, enzyme stabilizer, builder, water softener, oxidizing bleaching agents
Diamond Cutting	Buffer and enhanced cleaning properties
Dyes	Raw material, buffer catalyst for organic dye intermediates
Enamels	Raw material, glass forming, chemical and physical durability, shine
Fiberglass, textile and insulation	Raw material batch ingredient; glass forming, chemical and physical durability
Flame retardants	Suppresses flame, controls afterglow, and promotes char formation
Guar gum	Viscosity control, cross-linking agent
Hand cleaners	Abradant, pH buffer, soil removal
Ink	pH buffer and fixative
Jewelry manufacturing	Flux, cleaning agent, buffer
Laundry additives	Alkalinity control, water conditioner, and stabilizer
Leather thinning	pH buffer, neutralizer
Lubricants	Corrosion inhibitor
Metal cleaners	Corrosion inhibitor and abrasive
Metal-polishing compounds	Buffer, corrosion inhibitor, and abrasive
Paint	Corrosion inhibitor, adhesion promoter, flame retardant
Paper	Flame retardant, filler, raw material for bleaching, starch adhesives
Rust inhibitors	Corrosion inhibitor and antioxidant
Shampoo	pH buffering agent, neutralizer, stabilizer
Soaps	Alkalinity control, conditioner and abradant, soil removal
Starch adhesives	Solubility and viscosity control, increases tack

Source: Ref. 4.

production phase. Process losses from industry and the use of detergents represent flow to aqueous systems of approx 5.3×10^7 kg yr^{-1} [5].

For centuries, soap has been manufactured by means of the reaction between alkaline products and animals fats or vegetable oil. This process is known as *saponification*, and in a modern soap-making plants is carried out by mixing together the oils and the alkali and then heating them under pressure to around 130°C. Detergents can be classified into two major groups. Light-duty detergents, designed for lightly soiled items or delicate fabrics, are used mainly for washing by hand and usually consist of a blend of surfactants, perfume, coloring, and preservatives and none of the bleach ingredients. Heavy-duty detergents, however, are designed to provide high quality cleaning results in any kind of washing machine by removing all types of dirt from a whole range of modern fabrics, and therefore, they do contain bleaching ingredients.

To meet all these needs, detergents products will include a number of ingredients: surfactants (surface-active agents), builders, antiredeposition agents, corrosion inhibitors, bleaching agents, bleach activator (TAED—tetraacetyl ethylene diamine), fluorescers, lather control agents, enzymes, fragrances, and preservatives.

Chemical bleaches help the cleaning because they remove stains, such as tea, coffee, and fruit juice, which is very difficult or impossible to do with surfactants alone. Several bleaching agents can be used: percarbonates (not stabilized), peracetic acid, hydrogen peroxide, persulfates, percarbonates (stabilized with strong complex builders), sodium dithionite, chlorisocyanurates, hypochlorite, and perborates.

Sodium perborate ($NaBO_3$) exists in the tetrahydrate, trihydrate, monohydrate, and anhydrous forms. The main B compounds used as bleaching agents in detergent powders are sodium perborate tetrahydrate ($NaBO_3 \cdot 4H_2O$) and sodium perborate monohydrate ($NaBO_3 \cdot H_2O$). Typically a detergent will contain up to 15 wt% of the tetrahydrate and/or up to 10 wt% of the monohydrate. Sodium perborate tetrahydrate is produced commercially from the oxidation of aqueous sodium metaborate by hydrogen peroxide. When dissolved in water, sodium perborate tetrahydrate releases back hydrogen peroxide; its aqueous solution practically performs like a solution of hydrogen peroxide. The bleaching activity is based on a chemical oxidation of stains, by means of the hydrogen peroxide, which destroys chromophore molecules [6]. Sodium perborate monohydrate is prepared by the partial dehydration of sodium perborate tetrahydrate [7]. The chemical behavior is the same as that of perborate tetrahydrate; when dissolved in water, it also releases hydrogen peroxide to the solution. In comparison with the tetrahydrate molecule, the monohydrate has a higher content of active oxygen and a higher rate of dissolution in water. Due to these properties the product is particularly appreciated in the manufacturing of concentrated ("compact") detergents and of detergents for low-temperature washing [8].

The dosage of laundry detergents depends on water hardness, because detergents contain chemical components that soften the water. A significant difference in detergent activity is observed between hard and soft water. The European average consumption of detergents in 1998 was 10 kg person^{-1} year^{-1} (about 27 g person^{-1} day^{-1}). Some countries have significantly lower per capita detergent usage (Finland, Sweden, Norway, Iceland, Denmark, The Netherlands, Austria, Switzerland) than the European average. Countries with higher detergent consumption exceed the European average by a factor of about 1.3 (France, Portugal, Spain, Italy) [9]. The per capita B usage in some European countries was: 0.119 g person^{-1} day^{-1} in Germany (decrease from 0.235 g person^{-1} day^{-1} in 1996), 0.148 g person^{-1} day^{-1} in The Netherlands, 0.22 g person^{-1} day^{-1} in the United Kingdom, and 0.30–0.33 g person^{-1} day^{-1} in Italy, [9]. The maximum value of per capita domestic water consumed in Europe is 450 liter person^{-1} day^{-1} [10]. From these data, the B concentration added to the sewage will be in the range of 0.26–0.73 mg B L^{-1}. In the United States, the increase of B concentration during freshwater usage is considered to be in the range of 0.2–0.4 mg B L^{-1} [11]. Per capita B loads reaching 48 sewage treatment works in Europe were determined via monitoring data. These were compared with the per capita input predicted from B in detergents, as determined from detergent product sales data. The resulting distribution of the ratios of measured to predicted B has a 90th percentile of less than 1.5 [9]. Values higher than 1 represents B input from sources other than detergents.

VI. DISCHARGE OF BORATES TO THE ENVIRONMENT

The discharge of inorganic B to the environment can arise from their anthropogenic use in industry, agriculture, and domestic utilization of end products. In general, industrial and agricultural B sources can be defined as point sources, i.e., high concentrations originating from one industrial process or on animal farms.

A. Industrial Discharges

The sources of industrial B can be divided into two main categories based on the potential solubility of B from the product: (1) products in which most of the borate is fixed into a water-insoluble matrix and only small quantities are discharged in water flows to the environment (glass wool, borosilicate glass, enamels, ceramics, and ferrous alloys); and (2) products from which most of the borate can be solubilized in water and almost all of it is discharged to the aquatic environment (from starch adhesives in paper-recycling plants, borax cleaning solutions in metal-coating plants, detergents, disposal of pulverized fuel ash from power stations, and others).

B. Agricultural Discharges

High B concentrations can also be observed in sewage water of agricultural origin, mainly animal husbandry farms. At lower water B concentrations (<30 mg L^{-1}), feed is a major source of B in cattle [12]. Most ingested B is rapidly absorbed and excreted in the urine. A concentration of 3 mg B L^{-1} was found in the urine of milk-producing cows [13]. Sheep trials have also indicated that a significantly large proportion of B is excreted in the urine. Komor [14] found a range of 0.4–8.2 mg B L^{-1} in hog manure in 23 farms in south central Minnesota.

C. Domestic Discharges

As previously explained, during the domestic use of freshwater, B is incorporated into the wastewater stream mainly via the contribution of detergents and laundry powders but also via excretions. Excessive B is removed via similar mechanisms in humans and animals. Removal of borates from the blood is largely by excretion of $>90\%$ of the administered dose via the urine [15]. The concentration of B in human urine was found to be in the range of 0.040–7.8 mg kg^{-1} [16], with a median of 0.75 mg B kg^{-1} [17] or 1.80 mg B kg^{-1} [18]. Higher values have been reported for farm animals (see Ref. 13).

VII. BORON IN THE AQUATIC ENVIRONMENT

A. Seawater

Most of the B in seawater is present in the form of boric acid. The average concentration of borate in all oceans is 4.6 mg B kg^{-1} water, but it can vary from 0.5 mg B kg^{-1} in the Baltic Sea to 9.6 mg B kg^{-1} in the Mediterranean Sea. Boron is continuously being added to the sea by the weathering of rocks and soils, by rivers, springs, marine volcanoes, rain, and dust eruptions [7]. Seawater desalination processes are usually based on the rejection ratio of salt needed to meet the level of chloride concentration for drinking purposes. However, Magara et al. [19] reported that the rejection ratio of B ranges from only 43% to 78%, and therefore the filtrate can contain high concentrations of B.

B. Freshwater

The concentration of B in freshwater depends on factors such as the geological properties of the area, distance to sea coastal regions, and contamination from municipal, agricultural, and industrial sewage discharges. Concentrations of B in various sources of freshwater are presented in Table 3.

C. Groundwater

Boron concentrations in a variety of groundwater resources are given in Table 4. In groundwaters, B concentration can be influenced by seawater intrusion into coastal aquifers because of the high concentration in seawater, or by contamination with anthropogenic B added to the soil via irrigation with treated wastewater.

TABLE 3 Concentrations of B in Fresh Surface Water

Area	B concentration (mg L^{-1})
United States	Median = 0.076; 90th percentile = 0.387
Drainage basins, United States	0.019–0.289
Coastal drainage waters, California	15 (boron-rich deposits)
Lakes, California	157–360 (boron-rich deposits)
Ontario, Canada	0.029–0.086
Cold river drainage, western Canada	0.0627
United Kingdom	0.046–0.822
Italy	0.4–1.0
Italy	<0.1–0.5
Sweden	0.013 (0.001–1.046)
Germany	0.02–2.0
The Netherlands	Range of medians = 0.09–0.145
Rivers, Austria	<0.02–0.6 (background level)
River Neva, Russia	0.01–0.02
Degh Nala, Pakistan	<0.01–0.46 (near effluent discharges)
Simav River, Turkey	<0.5 (uncontaminated) 4 (maximum 7) (contaminated with B mine waste)
Rio Arenales, Argentina	<0.3
	6.9 (near borate plant)
Loa River Basin, Chile	3.99–26 (soil rich in minerals and natural salts; low rainfall)
Japan (River Asahi)	0.009–0.0117
South Africa	0.02–0.33
Lake Kinneret, Israel	0.08

Source: Modified from Ref. 15.

TABLE 4 Concentration of B in Groundwater

Country	Area	No. of samples	B concentration (mg L^{-1})
Denmark		525	≤0.3 (92.2%)
			≥0.3 (7.4%)
			≥1.0 (0.4%)
France		716	≤0.3 (99.5%)
			≥0.3 (0.5%)
Germany	Baden-Wurttemberg	2574	≤0.1 (89%)
			≥0.1 (10.7%)
			≥1.0 (0.3%)
	Lower Saxony	188	≤1.0 (96%)
			≥1.0 (4%)
Greece	Patras	10	≤0.1 (100%)
	Halkidiki	3	2.3–5.4
Israel		994	<0.1 (61.3%)
			0.1–<0.3 (21.9%)
			0.3–0.5 (5.7%)
			>0.5 (11.1%)
Italy	North of Rome	423	Mean = 1.0
	Sicily	18	Mean = 1.5
	Paglia	102	Mean = 0.75
Netherlands	Inland		0.08–0.6
Spain	Valencia	21	Mean = 0.64
	Almeria	17	Mean = 0.98
	Murcia	15	Mean = 0.51
United Kingdom	London	21	0.02–0.54
	Northumbria	164	Mean = 0.31
	Dumfries and Galloway		Mean = 0.04
	Permo-Triassic (Scotland)		Mean = 0.04
United States	Minnesota, sand Plain aquifer	22	Mean = 0.04

Source: Modified from Ref. 15.

D. Sewage Waters

As previously explained, high quantities of B are discharged into the environment via sewage. Table 5 presents the concentration of B in several sources of sewage water.

E. Effects of Boron on Aquatic Organisms

Butterwick et al. [20] discuss the toxicological effect of B on aquatic freshwater life and show that the toxicity expressed by the LC50 data is in the range of 12–90 mg B L^{-1}. The early life stages of rainbow trout appear to be the most sensitive to B, with a lowest-observable-effect concentration (LOEC) of 0.1 mg B L^{-1} in reconstituted water [21]. In natural water exposures, the LOEC was found to be 1.0 mg L^{-1}. In the wild, healthy rainbow trout have been found in surface waters containing up to 13 mg B L^{-1} [22]. The WHO [15] presents data on the median response concentrations of aquatic invertebrates exposed to B compounds. The LC$_{50}$ range was between <52.0

TABLE 5 Concentration of B in Sewage Water

Area/source	B concentration (mg L^{-1})
United States: industrial waste discharge	0.4–1.5 (maximum 4.05)
Europe: domestic and industrial	2 (maximum 5)
Egypt: sewage water	0.32–0.38
Israel: domestic and industrial	0.46–1.74 (range of means)
Sweden: effluent	0.34–0.436
Spain, Alicante: industrial wastewater	1.45
Spain, Elche: industrial waste	3
United Kingdom: municipal	1.21–3.96 (range of means)

Source: Modified from Ref. 15.

mg B L^{-1} for *Daphnia magna* and 1376 mg B L^{-1} for *Chironomus decorus*. In marine fish, the range of LC$_{50}$ values was 5–3000 mg B L^{-1}.

VIII. BORON IN TERRESTRIAL ENVIRONMENTS

Boron is found in the earth's crust at an estimated level average of 10 mg kg^{-1}, depending on the rock type [23]. Total B content in the soil ranges widely from 2 to 250 mg kg^{-1}; however, in most soils it ranges from 20 to 50 mg kg^{-1}. Less than 5% of the total B content in the soil is available to plants [24]. High levels of B in the soil or aquifers are either natural occurrences or a result of fertilization, mainly through irrigation water [25]. Significant concentrations of B are also found in areas with little or no drainage. Since B adsorbs to various soil components, it takes more water to leach it from the soil than to leach chlorides [26].

Boron in the soil can be classified into three groups [26]:

1. Boron in primary minerals, the most common of which is tourmaline
2. Boron adsorbed to clay minerals, oxides, and organic matter (OM)
3. Boron in the soil solution

Solubilization of B from the earth crust's rocks usually occurs as borosilicate. Due to its high ionization potential, B forms mainly covalent bonds in complexes. The electron structure of B is $2s^2 2p$; nevertheless it forms a triple covalent bond with the hybrid sp^2 in a geometric trigonal. Boron compounds reach an octet configuration by behaving like Lewis acids, via the hybridization of sp^3 with Lewis bases [27].

Boron oxide (B_2O_3) reacts with water to create boric acid:

$$B_2O_3 - 3H_2O \longrightarrow 2H_3BO_3$$

Boric acid is a weak acid that acts like a Lewis acid when it receives a hydroxyl ion, creating a borate anion. The hydrolysis reaction is

$$B(OH)_3 + 2H_2O \Leftrightarrow B(OH)_4 + H_3O^+$$

The formation of a borate ion is spontaneous, and the pK of the reaction is 9.23. Boric acid exhibits medium solubility in water, and its solubility rises significantly as the temperature increases. Only two B species are present in solution at B concentrations lower than 0.025 M: (1) boric acid $B(OH)_3$ and (2) borate $B(OH)_4^-$ [21].

Boron adsorption controls the amount of soluble B available to crops. Boron toxicity is determined mainly by its concentration in the soil solution, rather than its total amount in the soil [28,29]. Boron adsorption sites in the soil serve as reservoirs, and they release B into the soil solution according to the various factors that influence B availability [26]. The main factors influencing B availability are: concentration, pH, clay content, iron and aluminum oxides, OM content, wetting and drying cycles, and temperature [26].

The amount of B adsorbed to soil components is, of course, related to the concentration of B in the soil solution and to the soil solution pH. The amount of adsorbed B increases as the pH rises, up to a maximum value in the range of pH 11. Any further increase in pH causes a decrease in the amount of adsorbed B [26]. The distribution of B species found in a solution varies according to its pH. At pH values of 7 or less, most B will be present in the form of boric acid [$B(OH)_3$], which has a relatively low adsorption affinity. An increase in the pH up to 9 results in an increase in the concentration of the borate ion, which exhibits high affinity for adsorption. At this level, the hydroxyl ion concentration is still low enough to prevent competition, allowing rapid and high B adsorption. Any further rise in the pH increases the concentration of hydroxyl ions, creating competition for the adsorption sites between the hydroxyl ions and the borate ions and resulting in a decrease in B adsorption [24,26,30,31].

Wetting and drying cycles increase B adsorption to the soil. During the drying period, the B availability decreases and there is a noticeable deficiency in the crops [26]. The influence of temperature on B adsorption has not yet been determined. It seems to cause changes in the dissociation constants, in ion activity, and in the solvent–solute interaction, which increases B fixation [24,30].

A. Adsorption of Boron to Soil Surfaces

1. Aluminum, Iron, and Magnesium Oxides

The relationship between B concentration in the soil solution and that of B adsorbed to soil particles is influenced by the composition of the adsorbing complex (clay minerals, metal oxides and hydroxides, and OM) [24,26,31,32].

Iron oxides play an important role in determining the distribution of B in its different phases: total, soluble, and adsorbed [33]. Boron adsorption has been found to be higher on aluminum oxides than on iron oxides [34]. This is explained by the fact that the aluminum oxide has a larger specific surface area than iron oxide, since adsorption on equally sized surface areas of the two oxides is identical [34]. Boron also adsorbs well to the broken edges of clay minerals. Adsorption to oxides is greatly influenced by the structural interactions between clays and oxides. Oxides intercalated between layers of clay are not available for adsorption, in contrast to free minerals. The mechanism of B adsorption to oxides involves ligand exchange with reactive hydroxyl grops [35,36]. This specific adsorption causes a change in the point zero charge (PZC) of the mineral to more acidic pH levels [37]. This mechanism of exchange between the hydroxyls and ligands is an acceptable description of ion adsorption to mineral surfaces [26].

Studies on sorption kinetics have shown that B adsorption occurs as a complex on the inner layer of aluminum oxides via ligand exchange of borate with a hydroxyl group on the surface [38]. Spectral FTIR studies have shown that B adsorption to iron oxide and amorphous aluminum oxides takes place via ligand exchange with $B(OH)_4^-$

and $B(OH)_3^0$ species [39]. Potentially competing ions, such as hydroxil silicate, sulfate, phosphate, and oxalate, reduce B adsorption to oxides [40]. Sulfate ions have less impact than phosphate ions. The ability of competing anions to accelerate the release of adsorbed B from oxides increases in the follow order: chloride < sulfate < arsenate < phosphate. Boron adsorption to iron oxide was found to depend only slightly on the ion strength of the solution, whereas B adsorption to aluminum oxide decreases as the ionic strength increases [37].

2. Clay Minerals

The silicate layers in clay minerals are important to the adsorption of B on soils. In general, for clay minerals, any increase in B concentration along with a pH increase results in an adsorption increase until a maximum is reached. A solution pH of 8–10 leads to adsorption maxima. When pH levels exceed 10, a decrease in adsorption is observed. The adsorption capacity differs among various clay minerals. The following is the adsorption capacity of three major clay minerals for arid and semiarid soils: illite > montmorillonite > kaolinite [41]. Despite the higher adsorption of B by illite, the overall specific surface area of montmorillonite is much larger. Assuming that (1) most of the B is adsorbed to illite or montmorillonite on fractured edges and (2) the surface of the fractured edges is about 8 m^2 g^{-1}, the B concentrations on montmorillonite minerals (based on fractured edges only) would be 6.6×10^{-5} mol g^{-1} clay. This is also the approximate value for illite (1.5×10^{-5} mol g^{-1}). Boron adsorption is greatly influenced by the size of the clay platelets; per clay unit mass, it increases with decreases in platelet size. Keren and Talpaz [42] showed that when ultrasonic forces are applied to montmorillonite suspensions, B adsorption increases. This increase results from the facts that B adsorbs to the edges of the clay platelets and that the area of the exposed edges increased when the clay platelets were broken into smaller ones. Some researchers have described B adsorption as a two-step process: (1) B adsorbs to the edges; and (2) B diffuses into the crystal and an exchange of Al^{3+} and Si^{4+} at the tetrahedral sites inside the crystal structure takes place [43,44].

Charges of cations such as Al^{3+} and Fe^{3+} that are part of the edge structure of clays are not totally compensated by the ions in the lattice, and therefore the electric charge on these surfaces is negative or positive, depending on the pH. The chemical and structural similarity between clay edges and oxides strengthens the theory that the adsorptive action in clays and oxides is similar [26,36,42,45,46]. Boron adsorption to clay minerals increases with the ionic strength of the solution [36,45,47]. Different ions compete for B adsorption sites. Chloride, nitrate, and sulfate have little effect on B adsorption; phosphate, on the other hand, reduces it. Boron adsorption to clay minerals depends on the clay's cation exchange capacity and on the adsorbed cations [41,47,48].

A decrease in water content reduces B adsorption to clays. Wetting and drying cycles increase B fixation on clays [37,48]. Calcium clays adsorb more B than sodium or potassium clays, whereas clay minerals adsorb considerably less B, per unit weight, than most oxides [41,47,48].

3. Organic Matter

Organic matter (OM) is an important soil component that influences B availability to plants in the soil, since it adsorbs much more B than soil minerals [41,48,49]. Boron

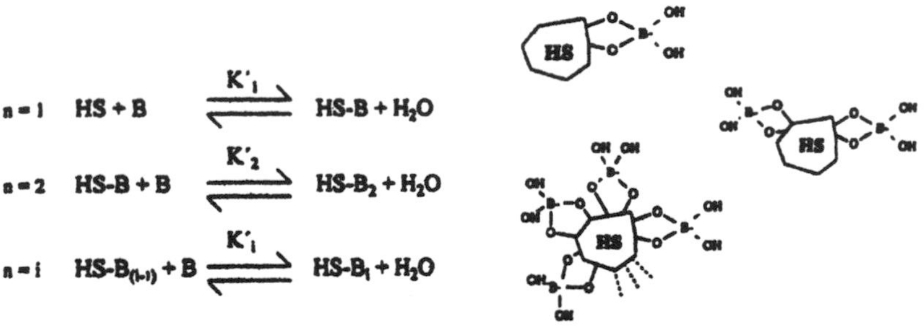

FIG. 1 Complexing of borate by HS (three stages). *Stage 1*: complexing of one borate ion (HS-B). *Stage 2*: formation of a diester (HS-B$_2$). *Stage 3*: continued complexation with borate ion. (Reprinted with permission from Ref. 50. Copyright 1998 American Chemical Society.)

adsorption to OM in soils and compost increases with increasing soil pH. Maximum absorption to humic acid (HA) is reached at pH = 9. Boron adsorption to OM is fast, and equilibrium is reached after about 3 h. The adsorption increases in proportion to the concentration of ions in solution [32].

Boric acid, and the ion borate, $B(OH)_4^-$, form complexes with diol groups of compatible spatial geometry that are found in cyclic and noncyclic polyhydroxyl [50,51], whereas complexes of B with compounds that contain pentagonal ring structures exhibit a 1,2-diol (n − 0) geometry. Complexes with compounds containing hexagonal ring structures exhibit a 1,3-diol geometry.

Complexes with dicarboxylic acids and hydrocarboxylic acids can be obtained, depending on the pH of the reaction solutions. Complexes formed with boric acid are less stable at high pH levels than those formed with borate ions [50,51].

Humic substances (HS) contain a wide range of alcoholic, phenolic, carboxyl, and hydroxyl groups. The formation of B-HS complexes can be described by a number of reactions between HS and borate ions, as shown in Figure 1. HS saturated with protons bind twice as much B as Na-HS. The complexes between the diol groups, which are present mainly in the OM and in B, are responsible for B binding and as such are not leached from soils that contain high levels of OM. Boron bound to OM can be released during microbiological activity, oxidation, and changes in B activity in the solutions [24].

B. Effects of Boron on Plant Growth

Although B is an essential element for normal plant growth, its role varies. It is needed for cell division and root-tissue elongation [52]. It plays an important role in sugar transport. Complexation with several hydroxyl-carrying molecules enables its involvement in several metabolic pathways [53]. Boron interacts with several enzymes, causing an increase in, stabilization of, or inhibition of their activity [54]. Inhibition of enzymatic activity is caused by the occupation of the active sites by boric acid or borate ion, resulting in repellence of the substrate needed for normal enzyme function. Boron also influences protein level and amino acid composition in plants. Boron deficiency does not affect the amino acid sequence in proteins, but it does affect their number. Boron also plays a role in normal membrane activity. Its deficiency causes a weakening of membrane functions, a reduction in its ability to transfer nutritional and metabolic elements. Boron presence accelerates the transport of sugars through the cell membrane via the formation of complexes. Boron plays an important part in cell elongation as well as in the stabilization and formation of cell walls [54].

The normal functioning of cell walls requires a constant supply of B. Brown and Hu [55] showed that most of the B in cell walls is found complexed to wall's pectin component. This component can produce very strong complexes with borate.

Boron deficiency causes anatomical, physiological, and biochemical changes in plants. These changes occur well in advance of any visual signs. The visual expression of B deficiency will first show up in the older leaves of the plant, and typical symptoms usually include growth inhibition, meristem necrosis, and sprouts, shortening of nodes, and excessive branching, giving the plant a bushy appearance. Under serious B deficiency, chlorosis develops as well as distorted and broken leaves, splitting of stems, a change in the color of the storage-tissue roots, for example, and decomposition. In some fruit trees, the fruits become distorted and changes in color are observed [26].

Toxic symptoms of B in most plant species are similar, with the older leaves exhibiting them first. Initial visible symptoms of B toxicity include chlorosis accompanied by necrosis of the edges and tips of the plant leaves. In some instances, chlorosis of the areas between the leaf veins occurs. Severe toxicity causes leaf shedding and, eventually, plant death [56].

Boron uptake by plants occurs only via the soil solution [28,29], through roots, and whether it is active or passive is a controversial issue [52]. The passive approach maintains that B is taken up through the formation of B-polysaccharide complexes and that the process is regulated by the gradient of B concentration [57]. This is corroborated by the fact that changes in soil solution pH cause a sharp decrease in B uptake, while changes in temperature and metabolic inhibitors have no effect on uptake level. The active approach maintains that uptake takes place via metabolic activity rather than via specific active passage through root-cell membranes. Uptake takes place via the formation of a complex ester of $B(OH)_3^0$ and polyol. Merging the two, i.e., passive passage of $B(OH)_3^0$ followed by the formation of diol complexes as a combined approach to uptake, could explain much of the data obtained on B uptake.

Boron mobility into the plant varies dramatically from one plant species to another [58]. A general differentiation can be made between plants with low and high B mobility. Transport of B from roots to foliage occurs through the xylem cells and is almost totally governed by the water gradient caused by transpiration. Hence, B transport in this system is directed to high transpiration areas, namely, leaf surfaces. These sites do not necessarily have the highest nutritional needs. In contrast, long-range transport through the phloem occurs in all directions, both upwards and downwards. Areas with high nutritional needs, such as young leaves, fruits, and seeds, namely, organs that do not lose water easily, are supplied through the phloem, which does not require transpiraton. There is a substantial exchange of elements from one transport system to the other. Plants in which sugars such as sorbitol, mannitol, and dolcitol occur as initial products of photosynthesis exhibit rapid B conduction due to the efficiency of these sugars at creating B complexes [55].

Boron concentration in leaves has been found to be in direct proportion to the B concentration in the soil solution. Moreover, strong correlation has been found between crop yield and B content in plant leaves [24,28,29]. Crops respond differently to B concentration, both in irrigation water and in soil solution. Accordingly, tables defining plant sensitivity to B in the irrigation water and soil solution have been published [59–61]. The threshold value is the maximal B concentration that does not cause a yield decrease or visible symptoms of damage to plants [26]. These threshold values were determined in sand using freshwater. Maas [60] classified the crops into groups according to their tolerance to B concentration in the soil solution (Table 6).

IX. BORON-CONTAINING DETERGENT FORMULATIONS IN ISRAEL: A CASE STUDY

The combination of severe water shortage, contamination of water resources, densely populated urban areas, and highly intensive irrigated agriculture leads Israel to optimize the treatment of wastewater and to reuse it for agriculture. Projects aiming to achieve this goal are placed high on the national list of priorities. Effluents are the most readily available and cheapest source of additional water, and they provide a partial solution to the water scarcity problem. To date, the rate of effluent reuse in Israel is the highest in the world. National policy calls for the gradual replacement of freshwater allocations to agriculture by reclaimed effluents. In the year 2000, treated wastewater constituted about 17% of consumption by the agricultural sector. It is estimated that effluents will constitute 50% of the water supplied to agriculture in 2020.

Average values for B concentration in wastewater in Israel were determined in 1997 [13]. The results (Table 7) show that although the concentration in tap water is in generally low (0.08–0.2 mg L^{-1}), wastewater sources are relatively enriched in B (0.46–1.41 mg L^{-1}). Water quality in some 120 effluent reservoirs used for irrigation is tested annually by the Israeli Ministry of the Environment in the months of July–August in order to characterize effluent water quality during the irrigation season. The results of water monitoring in 117 reservoirs, which supplied 95 million cubic meters (MCM) of effluents for irrigation in 2001, revealed that in 55% of the water, B concentrations exceeded 0.6 mg L^{-1} [62].

According to that document [62], detergents contribute up to 80–90% of the total added B to municipal sewage. The concentration of B in some detergents marketed in Israel was determined by our group [63]. The results revealed a concentration of 6,500–8,700 mg kg^{-1} in laundry powders (heavy-duty detergents), 16 mg kg^{-1} in concentrated ("compact") detergents, and 1–3 mg kg^{-1} in light-duty detergents [63].

Since B, like most other inorganic ions, is not removed during conventional sewage treatment, it is present in treated wastewater used for irrigation. To deal with the problem, a new Israeli environmental standard (IS 438) on labeling requirements for washing powders was published in 1999, which replaces a standard first published in 1982 (Table 8). According to the standard, industry will be required to reduce B concentrations by 60% in 4 years and 94% in up to 8 years. The new standard is innovative not only in Israel but in the world as well, although some applicability difficulties are known to exist (especially with respect to B). Recent surveys have confirmed the anticipated reduction in B concentration as a result of the implementation of the standard. The B concentration in treated effluents expected in 2008 is 0.2–0.3 mg L^{-1} [64].

TABLE 6 Crop Tolerance to B Concentration in the Soil Solution

Plant	Threshold value (mg L^{-1})
Very sensitive	
Lemon (*Citrus lemon*)	< 0.5
Blackberry (*Rubus* sp.)	
Sensitive	0.5–1.0
Avocado (*Persea americana*)	
Orange (*Citrus sinensis*)	
Grapefruit (*Citrus paradisi*)	
Apricot (*Prunus armeniaca*)	
Peach (*Prunus persica*)	
Plum (*Prunus domestica*)	
Persimmon (*Diosyros kaki*)	
Fig (*Ficus carica*)	
Grape (*Vitis vinifera*)	
Garlic (*Allium sativum*)	
Onion (*Allium cepa*)	
Kidney bean (*Phaseolus vulgaris*)	
Moderately sensitive	
Broccoli (*Brassica oleracea botrytis*)	1.0–2.0
Red pepper (*Capsicum annum*)	
Carrot (*Daucus carota*)	
Radish (*Raphanus sativus*)	
Potato (*Solanum tuberosum*)	
Cucumber (*Cucumis sativus*)	
Moderately tolerant	
Lettuce (*Lactuca sativa*)	2.0–4.0
Cabbage (*Brassica oleracea*)	
Corn (*Zea mays*)	
Celery (*Cepium graveolens*)	
Barley (*Hordeum vulgare*)	
Squash (*Cucurbita pepo*)	
Tolerant	
Sorghum (*Sorghum bicolor*)	4.0–6.0
Alfalfa (*Medicago sativa*)	
Oat (*Avena vulgare*)	
Parsley (*Petroselinum crispum*)	
Red beet (*Beta vulgaris*)	
Tomato (*Lycopersicon esculentum*)	
Sugarbeet (*Beta vulgaris*)	
Cotton (*Gossypium hirsutum*)	6.0–10.0
Asparagus (*Asparagus officinalis*)	10.0–15.0

Source: Adapted with permission from Ref. 60. Copyright CRC Press, Boca Raton, Florida.

TABLE 7 Boron Concentration and Pickup in Several Sewage Sources in Israel

| Site | Units | Hadera | | | | | | | | | | | | | |
		City	Paper factory	Ramla	Lod	Netania	Raanana	Kfar Saba	Ein Kerem	Emek Refaim	Nahal Og	Kiryat Gat	Ashdod	Dan region
Average	mg B L^{-1}	0.47	1.41	0.72	0.93	0.61	0.54	0.53	0.66	0.46	0.62	1.74	0.71	0.75
Boron pickup[a]	mg B L^{-1}	0.2–0.6	—	0.3–0.9	0.3–1.0	0.3–0.8	0.2–0.7	0.3–0.6	0.3–0.9	0.3–0.5	0.1–0.8	—	0.2–0.8	0.4–0.9

[a] Boron pickup: increase of B concentration during the usage of freshwater (domestic, industrial, or agriculture usage).
Source: Ref. 13.

TABLE 8 Boron in the Israeli Standard for Washing Powders in Comparison to the Previous Standard

Previous standard (1982)	New standard and timetable (IS438, 1999)
12% B, equaling 8.4 g B kg^{-1} product	8.4 g B kg^{-1} product max, until 6/30/1999 7.0 g B kg^{-1} product max, from 7/1/1999 5.6 g B kg^{-1} product max, from 7/1/2000 4.2 g B kg^{-1} product max, from 1/1/2002 3.5 g B kg^{-1} product max, from 1/1/2003 0.5 g B kg^{-1} product max, from 1/1/2008

Source: Ref. 64.

X. CONCLUDING COMMENTS

1. Excessive B in water used for irrigation in agriculture or in parks and nature reserves can pose a severe degree of risk to crops or to natural flora, even when present at relatively low concentrations.
2. Elimination or reduction of the B problem: Risks of pollution sources have to be identified and eliminated. In addition, regular determinations of B concentration in the source waters need to be performed.
3. Organic matter can reduce B hazards to crops; however, this should not be considered a long-term solution.
4. In order to reduce B hazard it is essential:

    a. To reduce the use of end products with relatively high concentrations of B (mainly detergents and laundry powders).
    b. To outline regulations for low B concentrations in detergents and laundry powders. This matter has already been dealt with in Israel.
    c. That industries with high-B discharges treat their effluents for B removal on site. New methods of B removal need to be developed.

5. Boron is only partially removed during reverse-osmosis desalination of seawater and saline water. Consequently, countries or regions in which a high proportion of the potable water supply originates from desalination will have to develop an additional treatment scheme that will remove B from the desalinized water.

ACKNOWLEDGMENT

The authors wish to thank the joint German-Israeli Water Technology Program, the BMBF-MOS cooperation, and GIF, the German-Israeli joint Research Fund for the support provided to this research work.

REFERENCES

1. Weast, R.C., Ed.; *CRC Handbook of Chemistry and Physics,* 68th Ed.; CRC Press: Boca Raton, FL, 1998 B–10.
2. Lyday, P.A. Boron, U.S. Geological Survey Minerals Yearbook; U.S. Department of Interior: Reston, VA, 1999; 13.1–13.8.

3. ChemExpo. ChemExpo Chemical profile: Borate. http://www.chemexpo.com, 2000.
4. U.S. Borax Inc. http://www.borax.com, 1999.
5. Argust, P. Biol. Trace Elem. Res. 1998, *66*, 131–143.
6. Ausimont USA. Sodium Perborate Tetrahydrate. Technical sheet. http://www.ausimont/docs/oxy_tetra.html, 2000.
7. ECETOC (European Center for Ecotoxicology and Toxicology of Chemicals). Ecotoxicology of Some Inorganic Borates. Special report No. 11. Brussels, 1997.
8. Ausimont USA. Sodium Perborate Monohydrate. Technical sheet. http://www.ausimont/docs/oxy_mono.html, 2000.
9. Fox, K.K.; Cassani, G.; Facchi, A.; Schroeder, F.R.; Poelloth, C.; Holt, M.S. Chemosphere 2002, *47*, 499–505.
10. European Space Agency (ESA). http://www.eso.org/seaspace/water/water3.html, 2000.
11. Metcalf; Eddy; Burton, F.L. *Wastewater Engineering*: Treatment, Disposal and Reuse, 3rd Ed.; Tchobanoglous, G., Ed.; Irwin McGraw-Hill: Boston, MA, 1991.
12. Weeth, H.J.; Speth, C.F.; Hanks, D.R. Am. J. Vet. Res. 1981, *42*, 474–477.
13. Tarchitzky J.R; Bar-Hai M; Keren R; Chen Y. Boron and salinity in wastewater: a survey. Field Service for Soil and Water. Extension Service. Ministry of Agriculture. Submitted to "Mekorot"-Israel Water Company 1997. *in Hebrew*.
14. Komor, S.C. J. Environ. Qual. 1997, *26*, 1212–1222.
15. WHO (World Health Organization). Environmental Health Criteria 204; Boron. World Health Organization: Geneva, 1998; 31–50.
16. Downing, R.G.; Strong, P.L.; Hovanec, B.M.; Northington, J. Biol. Trace Elem. Res. 1998, *66*, 3–21.
17. Imbus, H.R.; Cholak, J.; Miller, L.H.; Sterling, T. Arch. Environ. Health 1963, *6*, 286.
18. Minola, C.; Sabbioni, E.; Apostoli, P.; Pietra, R.; Pozzoli, L.; Gallorini, M. Sci. Total Environ. 1990, *95*, 89–105.
19. Magara, Y.; Tabata, A.; Kohki, M.; Kawasaki, M.; Hirose, M. Desalination 1998, *118*, 25–34.
20. Butterwick, L.; de Oude, N.; Raymond, K. Ecotoxicol Environ. Safety 1989, *17*, 339–371.
21. Birge, W.J; Black, J.A. Environmental Protection Agency, Office of Toxic Substances. Washington DC: Report No. EPA 560/1-76-008, 1977.
22. F.T. Bingham. Unpublished paper. Riverside, CA (Cited in: ECETOC-European Center for Ecotoxicology and Toxicology of Chemicals, 1995), 1982.
23. Krik-Othmer Encyclopedia of Chemical Technology. 4th Ed; Wiley: New York, 1992; Vol. 4, 67–110.
24. Yermiyaho, U. Boron sorption by soil and uptake by plants as affected by organic matter. MSc thesis, The Hebrew University of Jersualem, Rehovot, Israel, 1988. *in Hebrew*.
25. Nable, R.O.; Bañuelos, G.S.; Paull, J.G. Plant Soil 1997, *198*, 81–198.
26. Keren, R.; Bingham, F.T. Adv. Soil Sci. 1985, *1*, 229–276.
27. Cotton, F.A.; Wilkinson, G. In *Advanced Inorganic Chemistry*, 4th Ed.; Wiley: New York, 1980; 289–325.
28. Keren, R.; Bingham, F.T.; Rohades, J.D. Soil Sci. Soc. Am. J. 1985, *49*, 1466–1470.
29. Keren, R.; Bingham, F.T.; Rohades, J.D. Soil Sci. Soc. Am. J. 1985, *49*, 297–302.
30. Goldberg, S., Gupta, U.C., Ed.; In *Boron and Its Role in Crop Production*; CRC Press: Boca Raton, FL, 1993; 3–44.
31. Yermiyaho, U.; Keren, R.; Chen, Y. Soil Sci. Soc. Am. J. 1995, *59*, 405–409.
32. Yermiyaho, U.; Keren, R.; Chen, Y. Soil Sci. Soc. Am. J. 1988, *52*, 1035–1037.
33. Elrashidi, M.A.; O'Connor, G.A. Soil Sci. Soc. Am. J. 1982, *46*, 27–31.
34. Goldberg, S.; Glaubig, R.A. Soil Sci. Soc. Am. J. 1988, *52*, 87–91.
35. Keren, R.; Gast, R.G.; Bar Yossef, B. Soil Sci. Soc. Am. J. 1981, *45*, 45–48.
36. Goldberg, S.; Foster, H.S.; Heick, E.L. Soil Sci. Soc. Am. J. 1993, *57*, 704–708.
37. Goldberg, S. Plant Soil 1997, *193*, 35–48.

38. Toner, C.V.; Sparks, D.L. Soil Sci. Soc. Am. J. 1995, *59*, 395–404.
39. Su, C.; Suarez, D.L. Soil Sci. Soc. Am. J. 1997, *61*, 69–77.
40. Bloesh, P.M.; Bell, L.C.; Hughes, J.D. Aust. J. Soil Res. 1987, *25*, 377–390.
41. Keren, R.; Mezuman, U. Clays Clay Miner 1981, *29*, 128–204.
42. Keren, R.; Talpaz, H. Soil Sci. Soc. Am. J. 1984, *48*, 555–559.
43. Couch, E.L.; Grim, R.E. Clays Clay Miner 1968, *16*, 237–277.
44. Hingston, F.J. Aust. J. Soil Res. 1964, *2*, 83–95.
45. Keren, R.; Sparks, D.L. Soil Sci. Soc. Am. J. 1994, *58*, 1095–1100.
46. Keren, R.; Grossl, P.R.; Sparks, D.L. Soil Sci. Soc. Am. J. 1994, *58*, 1116–1122.
47. Keren, R.; O'Connor, G.A. Clays Clay Miner 1982, *30*, 341–346.
48. Keren, R.; Gast, R.G. Soil Sci. Soc. Am. J. 1983, *47*, 1116–1121.
49. Gu, B. Can. J. Soil Sci. 1990.
50. Schmitt-Kopplin, P.H.; Hertkon, N.; Garrison, A.W.; Fretag, A.W.; Kettrup, A. Anal. Chem. 1998, *70*, 3798–3808.
51. Shao, C.; Miyazaki, Y.; Matsuoka, S.; Yoshimura, K.; Sakashita, H. Macromolecules 2000, *33*, 19–25.
52. Shelp, B.J. Gupta, U.C., Ed.; In *Boron and Its Role in Crop Production*; CRC Press: Boca Raton, FL, 1993; 53–104.
53. Dugger, W.M., Lauchli, A., Bieleski, R.L., Eds.; *Inorganic Plant Nutrition Encyclopedia of Plant Phisiology*; Springer-Verlag: Heidelberg, 1983; 626–650.
54. Power, P.P.; Woods, W.G. Plant Soil 1997, *193*, 1–13.
55. Brown, P.H.; Hu, H. Physiol. Plant 1994, *91*, 435–441.
56. Gupta, U.C. In *Boron and Its Role in Crop Production*; Gupta, U.C., Ed.; CRC Press: Boca Raton, FL, 1993; 147–156.
57. Bingham, F.T.; Elseewi, A.; Oertli, J.J. Soil Sci. Soc. Am. Proc. 1970, *34*, 613–617.
58. Brown, P.H.; Shelp, B.J. Plant Soil 1997, *193*, 85–101.
59. US Salinity Laboratory Staff. Diagnosis and Improvement of Saline and Alkali Soils. Handbook No. 60; US Government Printing Office: Washington, DC, 1954.
60. Maas, E.V., Christie, B.R., Ed.; In *CRC Handbook of Plant Science in Agriculture*; CRC Press: Boca Raton, FL, 1986; 57–75.
61. Feigin, A.; Ravina, I.; Shalhevet, Y. *Irrigation with Treated Sewage Effluent: Management for Environmental Protection*; 1st Ed.; Springer-Verlag: Berlin, 1991.
62. Israeli Ministry of Environment. The Environment in Israel; Jerusalem Ministry of Environment, 2002.
63. Chen, Y.; Tarchitzky, J.; Benny, N.; Keren, R. The Fate of Boron in Reclaimed Wastewater and in Soils Irrigated with Them and Its Effects on Plants. Chief Scientist, Ministry of Agriculture and Rural Development of Israel, Project 821-0052, 1999. *in Hebrew*.
64. Weber, B.; Shilhav, U. Water Irrigation 2002, *427*, 14–17. *in Hebrew*.

25
Toxicology and Ecotoxicology of Minor Components in Personal Care Products

LOUIS HO TAN TAI   Consultant, Lambersart, France

I. INTRODUCTION

A. General

Personal care products are designed to be applied directly to parts of the human body, such as the skin (soaps, shower gels, bath foams) and the hair (shampoos, hair conditioners), but also, for some of them, on more intimate parts, such as the mouth mucous membranes (toothpastes). This is why personal care products should be considered with much care in terms of safety. To some extent we believe that they should be regarded as medicines. This is why the toxicology of personal care products (and by extension of the raw materials they are made of, including minor ingredients) is important.

Moreover, as for other categories of detergent products, their impact on the environment should be minimized as much as possible. For ecotoxicology also, every single ingredient plays its part, even minor ones. I might even say especially the minor ones, the risk being that because of their low levels of incorporation in personal care products, the formulators might be tempted not to pay much attention to their ecotoxicological properties.

B. Limitation

Many different products are available as personal care products. This chapter deals only with the most representative products on the market:

- Skin care products: soaps, toilet bars, shower gels, bath foams, and skin creams
- Hair care products: shampoos and conditioners
- Oral care products: toothpastes
- Deodorants and antiperspirants (aerosols)

Some specific cosmetic products have been excluded from this consideration.

A wide range of minor components may be used to obtain a similar functionality; only the most commonly or widely used are dealt with in this chapter.

C. Definition of Minor Components

Minor does not mean "useless." *Minor* just means that the ingredient is used at a low/very low concentration in the products; if it is an active ingredient, its activity may be very high (e.g., antibacterials). If it is a secondary ingredient, small quantities are usually sufficient to get the desired effect in the product (e.g., perfumes, colorants, fluorescers, opacifiers).

II. PERSONAL CARE PRODUCTS: FUNCTIONS AND FORMULATIONS

Before going further into the toxicological and ecotoxicological properties of the minor ingredients, we present a short summary of the different types of personal care products, together with their main functions and some examples of formulations currently used.

A. Soaps and Personal Wash Bars

Main function: cleaning of the skin.
Secondary functions: for specific soaps—antimicrobial effects.
Formulations: examples are given in Tables 1 [1–3] and 2 [4].

B. Shampoos and Hair Conditioners

Main function: cleaning of the hair.
Secondary functions: care of hair (these products must leave the hair supple, soft, easy
     to comb, and with less static).
Formulations: examples are given in Tables 3, 4, and 5 (see pp. 666 and 667) [5].

C. Toothpastes

Main function: cleaning of the teeth.
Secondary functions: the formulations must contain ingredients capable of getting rid
     of tartar, dental plaque, stains (abrasives), and bad breath, along with other
     types of ingredients to avoid caries and to protect gums.
Formulations: examples are given in Table 6 (see p. 668) [6–8].

D. Shower Gels and Bath Foams

Main function: cleaning of the skin.
Secondary functions: these products should leave the skin soft (nonaggressive formu-
     lations) and slightly perfumed.
Formulations: examples are given in Table 7 (see p. 669) [9,10].

E. Skin Creams

Main function: to take care of the skin.
Secondary functions: protection of the skin, slightly perfumed products.
Formulations: examples are given in Table 8 (see p. 670).

TABLE 1 Formulations of Classical and Germicidal Soaps

	Nonsuperfatted soap (%)	Superfatted soap (%)	Antibacterial soap (TCC*) (%)	Antibacterial soap (Irgasan) (%)
Nominal composition of fats	80–20 (palm/palm kernel)	65–35 (tallow/ coconut)	80–20 or 75–25 (palm/ coconut)	80–20 or 75–25 (tallow/coconut) (palm/palm kernel)
Ingredients				
Na soap	83–88	80–85	80–85	85–88
Free fatty acids	—	4–6		
Preservatives				
Na EDTA	0.015–0.030	0.015–0.030	0.01–0.05	0.02–0.05
EHDP	0.010–0.025	0.010–0.025	0.01–0.05	0.02–0.05
Orthophosphoric acid	0.1–0.2	0.1–0.2	0.005–0.015	0.1–0.2
Colorants	+	+	+	+
Antibacterials				
TCC	—	—	0.03–0.1	—
Irgasan DP 300	—	—	—	0.1–0.5
Opacifier (titanium oxide)	0.1–0.7	0.1–0.7	+	+
Brighteners	+	+	+	+
Perfume	+	+	+	+
Water, salts	Balance	Balance	Balance	Balance

Abbreviations: Sodium EDTA: ethylenediaminetetraacetic acid sodium salt; TCC: 3,4,4' trichlorocarbanilide; EHDP: ethanehydroxydiphosphonate.

TABLE 2 Sample Formulations of Detergent Bars (Unilever patents EP 0249474 A2)

Ingredients	%
Sodium cocoyl isethionate	44–60
Sodium alkylbenzenesulfonate	0–2
Anhydrous soap	7–8
Sodium isethionate	2
Stearic acid	15–19
Sodium sulfate	5
Antioxidant + sequestring agent[a]	+
Titanium oxide	0.2
Water, perfume	Balance

[a] Sodium EDTA (ethylenediaminetetraacetic acid sodium salt); EHDP (ethanehydroxydiphosphonate).

TABLE 3 Formulations of Shampoos

Ingredients	For normal hair, transparent (%)	For normal hair opaque (%)	For greasy hair (%)	For dry hair (%)
Surfactant (LES)	10–15	10–15	8–14	8–14
Cosurfactant (CAPB)	2–4	2–4	2–4	2–4
Mono- or diethanolamide	0–1	0–1	−/+	−/+
Polyoxyethylated sorbitan ester				0–1
Opacifiers (GMC or EGDS)	0	0.5–2	0–2	0–2
Protein hydrolyzates, egg			0.05–0.1	0.05–0.1
Olive oil, almond oil				0.05–0.1
Fatty alcohols (cetyl or stearyl)				0.1–0.4
Vitamins				0–0.2
Antioxidant (BHT)	+	+	+	+
Preservative (Bronopol)	+	+	+	+
Viscosity regulators	−/+	−/+		
Ingredients to adjust pH	−/+	−/+		
Perfume, colorants, water	Balance	Balance	Balance	Balance

Abbreviations: LES, lauryl ether sulfate; CAPB, cocamidopropyl betaine; GMS, glycol monostearate; EGDS, ethylene glycol distearate; BHT, butylated hydroxytoluene.

F. Deodorants/Antiperspirant Products

Main functions:
- *Deodorants*: inhibit malodor of perspiration by suppressing bacterial growth, or cover the malodor with a more pleasant odor.
- *Antiperspirants*: inhibit the flow of perspiration.
- *Formulations*: examples are given in Tables 9 and 10 (see p. 671).

III. TOXICOLOGY AND ECOTOXICOLOGY OF MINOR COMPONENTS IN PERSONAL CARE PRODUCTS

This chapter deals with only the most common "minor use" ingredients in personal care products: perfumes, antimicrobials, antioxidants, preservatives, fluorescers, and colorants.

TABLE 4 Formulations of Classical Hair Conditioners

A. Conditioning agents		B. Thickening and pearlescent agents	
Ingredients	%	Ingredients	%
Cetyltrimethylammonium chloride (CTAC)	0.5–1.2	Monostearate glycerol	0.5–1
Cetyl/stearyl alcohol	1.5–3	Stearyl stearate	0.3–0.7
Poly(dimethylsiloxane) (>5000 cps)	0.5–2.5	Cetyl palmitate	0.3–0.7
Preservatives[a]	+	Paraffin	0.5–1.5
		Hydroxyethylcellulose	0.7–1.5
		Preservatives[a]	+

[a] For example: DMDM hydantoin (1,3-dimethylol-5,5-dimethyl hydantoin).

TABLE 5 Formulations for Intensive Care Conditioners

Ingredients	Formulation A (%)	Formulation B (%)
CTAC	0.8–1.2	1–1.5
Cetyl/stearyl alcohol	2–3	2–4
Cetyl alcohol	0.5–1.5	
Paraffin		0.5–1.5
Stearyl stearate	1.5–3	
Hydroxyethyl cellulose	1–2	
Hydroxypropylmethyl cellulose		1–2
Poly(dimethylsiloxane) (> 5000 cps)	0.5–1.5	
Preservatives[a]	+	+
Perfume	+	+
Colorants	+	+
Water	Balance	Balance

Abbreviation: CTAC, cetyltrimethylammonium chloride.
[a] For example: DMDM hydantoin (1,3-dimethylol-5,5-dimethy hydantoin).

A. Perfumes

1. Toxicology

As we saw in the preceeding section, perfumes are minor ingredients incorporated in most personal care products. Perfumes can contain many different substances, each of which contributes to the global toxicity of a given fragrance.

The allergenic properties of perfumes have been known for a long time (they may cause allergic contact dermatitis). The International Fragrance Association (IFRA) was created to make recommendations concerning perfumes in the interests of public health. A system of self-regulation was established to which all fragrance suppliers adhere. Guidelines have been published that contain recommendations either not to use an ingredient or to limit its use [11].

IFRA collaborates with the Research Institute for Fragrance Materials (RIFM). In 1966 RIFM started a safety program for fragrance ingredients, in two steps:

1. To collect existing safety data
2. To get additional studies carried out by independant laboratories.

In total the program focused on approximately 1500 materials. If RIFM concludes that the available data are insufficient to show that the use of any ingredient is safe, then IFRA has to choose one of the following options:

1. Propose a guideline to restrict the use.
2. Ask for additional safety testing (in order to determine the "no effect" level).
3. Search for other qualities of the fragrance ingredient that could lead to a guideline based on purity specifications.
4. Decide not to use the ingredient.

IFRA employs the following process to evaluate any new ingredient:

- The ingredients must always conform to the relevant legislation and regulations of the countries where they are to be used.

TABLE 6 Toothpaste Formulations

Ingredients	Opaque toothpaste formulation with Na fluoride (%)	Transparent toothpaste formulation (%)	Opaque toothpaste formulation with two antimicrobials (%)
Glycerine (99.5%)		9.95	
Sodium fluoride	0.22	0.243	
Zinc chloride	2		
Zinc citrate dihydrate			1
Sorbitol (70%)	35	33.88	27
Glycerol	10		
Aluminium trihydrate			50
Sodium PAS		1.5	1.88
Sodium LAS			0.63
SCMC		0.4	0.8
Carrageenin		0.4	
Triclosan		0.3	0.5
Sodium monofluorophosphate			0.85
Hydrated silica	23		
N-Ethyl cocoyl laurate	3.75		
Xanthan gum	1		
Hydroxyethylcellulose	1		
Sodium gluconate	0.80		
Titanium oxide	0.80		
Sodium saccharinate	0.7	0.3	0.18
Poly(vinyl ether/maleic anhydride)		2	
Caustic soda (50%)		0.6	
Precipitated silica		22	
Saccharin	0.10		
Sodium benzoate	0.20		
Flavor	1.3	1	1.2
Formaldehyde			0.04
Demineralized water	Balance	Balance	Balance

Abbreviations: PAS, primary alcohol sulfate; LAS, linear alkylbenzenesulfonate; SCMC, sodium carboxymethylcellulose.

- Data enabling a proper toxicological evaluation to be made should be obtained from (1) the literature or available databases, (2) comparison with the legislation on health and safety at work, and (3) a consideration of structure–activity relationships based on known effects of other substances.

Systematic toxicity is considered in relation to the quantities used and to the quantities likely to enter the human body. Because fragrances frequently come in contact with the skin, the most important reasons for restricting use, appearing in the IFRA guidelines are:

1. *Skin sensitization* (appearing in 66 guidelines out of a total of 93). The ingredients are tested in a human repeated insult patch test (HRIPT), the concen-

TABLE 7 Formulations of Shower Gels/Bath Foams

Ingredients	Nonsoap shower gel/bath foam		Shower gel/bath foam with soap	
	Formulation A (%)	Formulation B (%)	Formulation C, transparent base (%)	Formulation D, opaque base (%)
Myristic acid			5–8	5–8
Lauric acid			5–8	5–8
Oleic acid			2–4.5	2–4.5
Glycerine			10–15	10–15
Sodium hydroxide 45%			7–10	7–10
Sodium EDTA			0.05–0.15	0.05–0.15
BHT			0.02–0.07	0.02–0.07
Sodium Cl			+	+
Sodium isethionate	9	5		
Lauryl ether sulfate		2		
Coco betaine	6			
Cocamidopropyl betaine		8	12–20	12–20
Silicone oil	5	5		2–6
Jaguar C-13-S	0.1	0.1		
Preservative	+	+	0.05–0.25 (formaldehyde)	0.05–0.25 (formaldehyde)
Colorants	+	+	+	+
Perfume, water	Balance	Balance	Balance	Balance

Abbreviation: BHT, butylated hydroxytoluene.

tration normally used is 10 times the concentration in leave-on-skin products. If there is evidence of sentitization, then the "no-effect" concentration (for induction of sensitization) is determined, and the maximum level permitted in the IFRA guideline is calculated as the value corresponding to one-tenth of the "no-effect" level.

2. Phototoxicity (in 14 guidelines).
3. Systemic toxicity (in 10 guidelines).
4. Photosensitization (in 6 guidelines).

At present, IFRA recommends that 35 ingredients not be used (e.g., Musk Ambrette), the use of 12 phototoxic ingredients be limited (e.g., Bergamot oil), and the use of 29 other ingredients be limited because they show a sensitisizing potential (e.g., trans-2-hexenal and isoeugenol). In addition, specific purity criteria or use instructions are recommended for 14 materials (the use instructions are, for example, not to use the material if other materials are present; and, if necessary, to add antioxidants to avoid peroxide formation).

IFRA guidelines have certainly reduced the number of allergies because of specific fragrances, but in the meantime the use of perfumed products in general has increased a lot.

TABLE 8 Formulations of Skin Care Creams

Ingredients	Day cream, Formulation A (%)
Ethoxylated fatty alcohol ester	3–7
Cetyl alcohol	0.25–0.80
Ceteth-20 (and) glyceryl stearate (and) PEG-6 stearate (and) steareth-20	3–6
Squalan	0.7–1.2
Paraffin oil	1–2.5
Petrolatum	4–8
Propylene glycol	4–6
DMDM Hydantoin (preservative)	0.1–0.2
Perfume	0.1–0.3
Water	0.15–0.45
	Balance

Ingredients	Day cream, Formulation B (%)
Glyceryl stearate (and) ceteareth-20 (and) ceteareth-12 (and) cetearyl alcohol (and) cetyl palmitate	3–7
Cetyl alcohol	0.8–2.2
Stearic acid	0.8–2
Isopropyl myristate	3–7
Octyldodecanol	2–6
Mineral oil	3–7
Cetyl acetate (and) acetylated lanolin alcohol	1–2.5
Sorbitol 70%	2.5–4.5
DMDM hydantoin (and) iodopropynyl butylcarbamate (preservative)	0.05–0.15
Perfume	0.25–0.55
Water	Balance

Ingredients	Night cream (%)
Polyglyceryl-3 dimerate	1.5–5
Cetearyl octanoate (and) isopropyl myristate	9–15
Isopropyl myristate	4–9
Beeswax white	0.7–1.5
Hydrogenated castor oil	0.3–0.8
Glycerin 86%	2–5
Magnesium sulfate	0.7–1.5
DMDM hydantoin (preservative)	0.05–0.15
Perfume	0.25–0.55
Water	Balance

TABLE 9 Formulations of Deodorant Aerosols

Ingredients	%
Alcohol (denatured)	35–45
Isopropyl myristate	0.5–1.5
Perfume	0.5–1.5
Triclosan	0.2–0.5
Butane, isobutane, propane	Balance

Recently groups of European dermatologists have called for the introduction of a new system in which persons already sensitive to certain ingredients can be informed about their presence in cosmetic products by listing them on the label [12]:

Amylcinnamic aldehyde (not limited by IFRA)
Amylcinnamic alcohol
Anis alcohol (not limited by IFRA)
Benzyl alcohol (not limited by IFRA)
Benzyl benzoate (not limited by IFRA)
Benzyl cinnamate (not limited by IFRA)
Benzyl salicilate (not limited by IFRA)
Cinnamic alcohol (currently limited by IFRA to 0.8% in the final product)
Cinnamic aldehyde (current IFRA guideline requires "quenching agents" to be present)
Citral (current IFRA guideline requires "quenching agents" to be present)
Citronellol (not limited by IFRA)
Coumarin (not limited by IFRA)
Eugenol (not limited by IFRA)
Farnesol (not limited by IFRA)
Geraniol (not limited by IFRA)
Hexylcinnamic aldehyde (not limited by IFRA)
Hydroxycitronellal
Isoeugenol (currently limited by IFRA to 0.2% in the final product)
Lilial (not limited by IFRA)
d-Limonene (in nearly all naturals)

TABLE 10 Formulations of Antiperspirant Products

Ingredients	Aerosol (%)	Spray (%)
Aluminium chlorhydrate	3–7	28–35
Cyclomethicone	5–7.5	45–55
Disteardimonium hectorite	0.25–0.75	
Quaternium-18 hectorite		2–4.5
Perfume	0.5–1	2.5–4
Isopropyl myristate		7–12
Silica		0.7–1.3
Butane, isobutane, propane	Balance	

Linalool (not limited by IFRA)
Lyral (not limited by IFRA)
"gamma-Methylionone" (not limited by IFRA)
Methyl heptine carbonate (currently limited by IFRA to 0.01% in the final product)

2. Ecotoxicology

Even if fragrance ingredients are used at low volumes (which could be a kind of "guarantee" against noticeable environmental effects), they are also known to be chemically stable on the one hand and to have a low solubility in water on the other hand.

Many fragrance ingredients do not meet the extremely demanding OECD criteria for ready biodegradability, and many do not even meet those for inherent biodegradability. In some cases their chemical stability may provide biochemical inertness. In other cases, their bioavailability is reduced by their low water solubility in standard tests. Their high lipophilicity can lead to their bioaccumulation and bioconcentration.

Under real conditions (in the environment), analysis has led to the conclusion that fragrance ingredients can sometimes be detected in aquatic organisms, surface water, and sewage sludge. These are the ingredients that are much used (in high volumes) and that show the higher lipophilicity and the higher stability. Among them, "nitro musks" (Musk xylene, Musk ketone) and "polycyclic musks" have been detected in European river life (eels and other fish), in the North Sea (mussels and crabs), and in Asia (shrimps). It can be anticipated that, as analytical chemistry techniques improve, many of the higher volume, biochemically stable, less water-soluble/highly lipophilic substances will be detected in these kinds of organisms (and possibly in other terrestrial species). But the presence of these chemicals in aquatic organisms does not prove that these ingredients manifest any really adverse environmental effects.

The fragrance industry (via RIFM) is taking an increasing role in investigating questions of persistence in the environment and accumulation in different organisms.

B. Antimicrobials

Two ingredients have been used as effective antiseptics in personal care products (especially soaps) since the 1960s. One is trichlorocarbanilide (TCC), or triclocarban, with the following formula: 3-(4-chlorophenyl)-1(3,4-dichlorophenyl) urea:

The other one is triclosan. Because it is the most widely used antimicrobial today in personal care products, we will detail the toxicology and ecotoxicology of triclosan.

Triclosan has been used in a wide variety of personal care products for more than 25 years: soaps and toilet bars, deodorants and antiperspirants, and, more recently, toothpastes (to prevent plaque formation). Triclosan is a 2,4,4'-tricloro-2'-hydroxydiphenyl ether with the following formula:

OH Cl

O

Cl Cl

1. Toxicology

(a) Acute Toxicity. Triclosan is not an oral toxicant: When administered orally to rats, the LD_{50} (dose of a substance that produces lethal effects in at least 50% of the test subject population) is greater than 5000 mg/kg [13].

Human safety studies have also been conducted using dental products containing triclosan in concentrations ranging from 0.06% to 0.6%. The duration of the testing ranged from less than 1 week to more than 12 weeks. No adverse effects were noted at any time period, in any product, at any triclosan concentration. Blood chemistry and hematologic measurements conducted during these studies showed no difference between control subjects and subjects using triclosan products.

Triclosan is not a dermal toxicant: Acute systemic toxicity was not observed when doses of up to 6000 mg/kg were applied to rabbits under an occluded patch test (the substance is placed directly in contact with the skin, and the passage of air to/from the covered area is blocked by covering it with a polymeric material) [14].

Dermal Toxicity. Acute dermal toxicity tests carried out on various products containing triclosan show that triclosan in concentrations found in these products does not produce acutely toxic effects. In these tests, 2g/kg of the individual products (shower gels, lotions, soaps) were applied to rabbits, which were observed for 14 days. The primary dermal index for all tested products was less than 5.0 (i.e., none of the products is considered dermally irritating) [14].

Human safety studies have also been conducted using dental products containing triclosan in concentrations ranging from 0.036% to 0.36%. The conclusion is that triclosan does not have a skin sensitizing or photosensitizing effect [15–17]. These tests were: (a) prophetic patch tests (24-hour application of triclosan to the skin) and (b) repeat insult patch tests (two or more applications to the same area and often over a longer period of time).

Eye Irritation. Triclosan induced a moderate eye irritation when 0.1 g was applied to the rabbit eye mucosa. But the amount applied was much greater than the possible human exposure concentrations: formulations containing triclosan do not irritate the eyes [13].

(b) Chronic Toxicology/Carcinogenicity. The long-term toxicity profile of triclosan has been established in a combined chronic toxicity/carcinogenicity study. It demonstrated the lack of carcinogenic activities and established a "no-effect" level for long-term toxicity [14]. Male and female rats were administered dietary concentrations of 0, 300, 1000, and 3000 ppm for 2 years. The following results were found:

- There were no treatment-related effects on mortality during the course of the study.
- Dose-related changes were observed in mean body weight gain.

- Observed toxicity was mainly hepatic in nature.
- No preneoplastic or neoplastic lesions were found in an increased incidence in any treated group.
- The "no-effect" level was considered to be 1000 ppm (corresponding to 52 mg/kg/day in males and 67 mg/kg/day in females).

Mutagenicity [13]. Triclosan showed no mutagenic activity in the following tests:

- Ames test
- Chromosome aberration in vitro using CHO cells
- Mouse lymphoma TK locus test
- In vivo mouse micronucleus test (chromosome aberration in vivo)
- Unscheduled DNA synthesis assay using adult rat hepatocyte primary cultures

2. Ecotoxicology [13]

(a) Environmental Fate.

Biodegradation in the Aquatic Environment. Under tests conditions using sewage sludge at environmentally relevant concentrations, it has been demonstrated that triclosan undergoes complete biodegradation. Almost complete removal of triclosan from wastewater and the absence of persistent metabolites have been confirmed in further sludge tests and continuous activated-sludge tests with radio-labeled material:

1. Removal of parent: >99.5%
2. Complete biodegradation: 78–88%
3. Primary biodegradation: 5–16%

Measurements were carried out in sewage plants monitoring projects in North America and Europe. In effluents from the most frequently used system (activated-sludge treatment plants), 0.1–0.5 µg/L of triclosan were found. Thus, triclosan shows a very good biodegradability under real-life conditions.

At in-use concentrations, triclosan does not affect the effectiveness of aerobic microbial treatment plants [an inhibition concentration IC_{50} (3h) of 20 mg/L was found in a respiratory inhibition test on activated sludge].

Triclosan does not meet the criteria set forth in several standard biodegradation tests (modified MITI I test—OECD 301C, modified MITI II test—OECD 302C, and coupled-unit test—OECD 303 A. These results are because of the high concentrations of test material used in screening-level tests and because of the antimicrobial properties of triclosan.

The BOD (biochemical oxygen demand) is 0 mg oxygen per gram substance. The COD (chemical oxygen demand) is 1116 mg oxygen per gram substance.

Biodegradation in the Terrestrial Environment. Traces of triclosan adsorbed onto activated sludge may reach the soil when sludge is used in agricultural soil enrichment. The substance remained in the top layer of soil (it was not leached into the groundwater). In soil mineralization tests, primary and ultimate biodegradation was observed in three standard soils (loss of 77–93%). The calculated half-life of triclosan in these tests was 15–35 days (depending on the nature of the soil). In several tests under anaerobic conditions, no concrete evidence for the biodegradation of triclosan was observed. An adsorption study was conducted using deactivated sludge solids as the sorbent matrix. The adsorption value (K_{OC} = 47,500 L/kg) suggested that 10–20% triclosan was sorbed onto solids in the effluent and receiving waters. Regarding photodegradation, when highly dilute aqueous solutions are exposed to sunlight, triclosan undergoes rapid photolysis. The aqueous photolysis half-life is less than 1 hour.

(b) Environmental Effects [13].

Toxicity to Aquatic Organisms. The results of a series of studies carried out on fish, daphnia, and algae can be summarized as follows: The toxicity of triclosan was influenced by pH. Under the test pH conditions, triclosan is available both under the ionized form and under the (more toxic) unionized form. Therefore, the lowest "no-observed-effect concentration" (NOEC) in algae of 0.69 µg/L has to be adjusted for ionization to a value of 0.5 µg/L. Using the lowest NOEC divided by a safety factor of 10, a predicted no-effect concentration (PNEC) of 0.05 µg/L can be calculated.

Toxicity to Terrestrial Organisms.

Earthworm studies: The acute toxicity EC50 (14 days) to earthworm (*Eisenia foetidia*), OECD 207/2, was found to be over 1026 mg/kg soil (dry weight).

Avian studies: The acute oral toxicity (14 days) to mallard duck was found to be over 2150 mg/kg and to bobwhite quail over 825 mg/kg. The acute dietary toxicity test in bobwhite quail resulted in an LD_{50} of over 5000 ppm and a NOEC of greater than 1250 ppm.

Terrestrial plants: The chronic toxicity of triclosan was evaluated in a 28-day seedling growth phototoxicity test on cucumber. For the parameters % emergence, shoot length, shoot weight, and root weight, a NOEC of over 1000 µg/kg soil (dry weight) was found.

C. Antioxidants

1. EDTA

EDTA (ethylenediaminetetraacetic acid) has the following chemical structure:

EDTA is incorporated into soap bars to chelate traces of metal (iron, copper, manganese) in order to prevent degradation due to metal-catalyzed oxidation and into shower gels/bath foams as a preservative enhancer.

(a) Toxicology.

Acute and Longer-Term Toxicology. Exposure to EDTA in drinking water was conducted primarily by the oral route. Because a large stoichiometric excess of calcium ions is likely to be present in surface waters, the calcium salt of EDTA should be used for safety assessment purposes.

Toxicity of EDTA to Mammals. Conversion from the tetrasodium salt to the calcium disodium salt greatly reduces the toxicity of EDTA.

Teratology. Studies involving EDTA have sometimes caused congenital abnormalities (3% EDTA in the diet produced 100% malformations). One hypothesis is that EDTA induced a zinc deficiency: No congenital abnormalities were found in the young of rats given a similar diet reinforced with 1000 ppm zinc in addition. Another possibility is that EDTA is complexed by the excess zinc and therefore is no longer available for chelating another essential element [18]. Subcutaneous administration of Ca EDTA/Zn EDTA or both to pregnant rats (from day 11 to day 15 of gestation) showed that Ca EDTA is teratogenic in rats (at concentrations producing no discernible toxicity in the dam other than reduced weight gain) [19].

Acceptable Daily Intake for Man.   Because EDTA has a low toxicity to man, its use in foods is allowed in several countries. Based on the studies in rats (level causing no toxicological effect in the rat = 250 mg/kg), the WHO calculated the level of EDTA as $CaNa_2$ EDTA that was an acceptable daily intake for man of 0–2.5 mg/kg (i.e., 1/100 of no-effect level), far above the quantity potentially found in drinking water [20]. (Drinking 2 liters of water per day would involve ingestion of up to 50 µg EDTA.) It is also to be noted that EDTA does not absorb through the skin.

(b)   Ecotoxicology.

Biodegradability.   In current laboratory tests, the results are poor: EDTA biodegrades at low levels only [21]. For example:

* In the AFNOR test using the effluent from a municipal sewage treatment plant as inoculum, no DOC (dissolved organic carbon) removal occurred within 42 days.
* In the OECD screening test (after 19 days incubation), 10% of DOC removal was found.

No degradation of any EDTA species by sewage bacteria was observed during the 72-day test period in a study of mineralization of EDTA (1–100 µg/L) in buffer media. But under aerobic conditions, EDTA displays some biodegradability. For example, $(2-^{14}C-)$ and (ethylene $^{14}C-)$ radiolabeled FeNH4 EDTA was found to degrade when the test mixture was inoculated with microorganisms from an aerated lagoon receiving industrial effluents containing EDTA [22].

Measurements in field conditions also show that EDTA is eliminated (either by biodegradation or by photodegradation). However, removal of EDTA during the biological treatment of sewage is far from being optimal, since the rate of biodegradation of EDTA is too low, but EDTA should not persist indefinitely in receiving soils or waters.

Photodegradation.   IronIII EDTA photodegrades quickly (calculated half-life is less than 1 day in shallow rivers and surface waters of lakes). NickelII, copperII, and zincII complexes of EDTA are inert to photodegradation. A significant level of photodegradation in surface waters will occur only if the ironIII chelate is present to a sufficient extent.

Toxicity to Aquatic Organisms.   Toxicity of EDTA to aquatic organisms depends on the form in which it is presented to them: The influence of pH and on free calcium ion concentration of the medium can explain the differences, as described in the following test (see Table 11): 96-hour LC_{50} to bluegill sunfish of EDTA tetrasodium salt in:

* Very soft water (61.2 mg/L)
* Medium-hard water (401.7 mg/L)
* Very hard water (807.3 mg/L)

TABLE 11 Ninety-Six-Hour LC_{50} of EDTA to Bluegill Sunfish

Form	LC_{50} (mg/L)	pH
EDTA—acid	159	3.7
EDTA—tetrasodium salt	486	8.9
EDTA—calcium disodium salt	2340	7.4

Concentrations of 95 and 190 mg/L in soft water (30–32 mg/L as $CaCO_3$) caused a significant decrease in the adsorption and an increase in the excretion of calcium by goldfish (this means that the basic requirement for calcium of the fish were not satisfied) [23].

In practical environmental conditions, a large stoichiometric excess of calcium ions will allow EDTA not to exert toxicity by affecting the calcium balance of aquatic organisms. Concentrations of EDTA in surface waters were much lower than those for which any adverse effects on aquatic life should occur.

Bioaccumulation. EDTA did not accumulate in the aquatic food chain (a whole-body bioconcentration factor of 1, with a half-life for depuration of 128–242 hours was observed for bluegill sunfish exposed for 28 days to radiolabeled EDTA (0.08 and 0.76 mg/L) [24]. The concentrations of EDTA in fish eaten by man will not be higher than the concentrations of EDTA in the water they came from.

2. EHDP

EHDP (ethane-1-hydroxy-1, 1-diphosphonic acid) is an antioxidant used mainly in soaps.

(a) Structure.

$$\text{H}_2\text{O}_3\text{P} - \underset{\underset{\text{PO}_3\text{H}_2}{|}}{\overset{\overset{\text{OH}}{|}}{\text{C}}} - \text{CH}_3$$

(b) Toxicology. In a pre- and postnatal study [25], pregnant female Long–Evans rats (20/group except 21/high-dose group) were orally exposed by gavage to EHDP (acid) at dose levels of 16.51, 110.09, and 330.28 mg/kg/day (the rats were given one-half doses, twice daily in deionized water, 10 mL/kg/day) on gestation day 15 through to the end of gestation and through the period of lactation. Surviving animals were sacrificed on lactation day 21.

Significant differences between treated and control animals were observed in the following:

- Decreased implantations/dam (high-dose group)
- Decreased live fetuses/total fetuses (low-dose group)
- Increased pup survival (high-dose group)
- Increased pup body weight (high- and mid-dose groups)
- Decreased number of females (mid-dose group)

No significant differences were observed in pregnancy rate, length of gestation period, number of litters delivered, maternal weight gain, mortality, life and necropsy observations, numbers of pups/dam, live or dead fetuses/litter, live fetuses/implantation, pup sex ratios, and gross pup necropsy observations.

Yet in a second study, the effects of EHDP on the fertility and reproductive behavior of male rats were evaluated in groups of 24 male Wistar rats administered 0, 15, 100, and 300 mg EHDP/kg of body weight by oral gavage daily for 10 weeks before mating, up to the end of the mating period (24 nontreated females/group). In the 300-mg/kg treatment group, the following findings were recorded:

- Reduced body weight during treatment
- Defects on the dental enamel (on the incisor of one-third of the males)
- A significant increase in the index of resorption in the laparotomized females
- Increased index of preimplantation loss in the laparotomized females
- Significantly reduced liver weights among male and female pups

The results of another study of tooth morphology and the mechanism of calcification suggests that EHDP has an influence on the dental tissue and on mice fetuses.

(c) Ecotoxicology. The main pathway for phosphonate removal in the environment is by adsorption onto surfaces. A study [26] has examined the effect of Ca, Cu, Zn, and FeIII on the adsorption of six phosphonates, including EHDP, onto iron (hydr)oxide goethite. When the molar concentration of Ca, Cu, Zn, and FeIII is equal to the concentration of the phosphonate, the effects on phosphonate adsorption were slight or negligible. When excess concentrations of Ca and Zn were added, adsorption increased considerably–presumably through tertiary surface complex formation and adsorption onto precipitated (hydr)oxides of zinc.

Optimal phosphonate removal can be expected in waters containing calcium and employing addition of iron salt for flocculation or phosphonate elimination.

3. BHT (Butylated Hydroxy Toluene)

(a) Chemical Names. 2,6-bis(1,1-Dimethyl-ethyl)-4-methyl phenol or 2,6-*tert*-butyl-*p*-cresol or 2,6-*tert*-butyl-4-methylphenol.

(b) Structure.

It is used as an antioxidant for soaps, shampoos, gel shower, etc.

(c) Toxicology. The LD_{50} orally in mice is 1040 mg/kg [58].

D. Preservatives

Preservatives are used in personal care products in order to prevent deterioration by microbial action.

1. Formaldehyde (or formol or formalin): HCHO

(a) Structure.

$$H - C \overset{\displaystyle O}{\underset{\displaystyle H}{\Big\langle}}$$

(b) Toxicology.

Exposition to Formaldehyde.   Individual reaction to formaldehyde depends on hereditary and lifestyle factors. Formaldehyde may cause allergenic reactions. Symptoms include minor respiratory irritation and watery eyes. Main health problems occur with direct exposure. Most adults react to formaldehyde at a level of 0.5–1.5 ppm in the air (symptoms: nausea, vomiting, diarrhea), but the prognosis for recovery is good unless the dosage is very high.

Studies requested by the FDA (Food and Drug Administation) in the United States were carried out by independant laboratories (in 1982), the results of which have led the FDA to propose a permissible exposure level (PEL) of 1 ppm (for 8 hours of exposure per day). Since 1992, the PEL has been decreased to 0.5 ppm in the United States. This limit concerns occupational exposure. For all other places (whatever the time of exposure), the FDA has recommended a maximum concentration below 0.3 ppm.

Adsorption.   The effect of subchronic exposure to formaldehyde on blood formaldehyde concentrations was studied in monkeys [27]. Young adult rhesus monkeys were exposed to 0–6 ppm formaldehyde vapor 6 hours per day, 5 days per week for 4 weeks. Blood samples were obtained after exposure for 7 minutes and 45 hours after the last exposure. The average blood formaldehyde concentrations obtained after 7 minutes was 1.84 µg/g and after 45 hours was 2.04 µg/g. For the controls, the value was 2.42 µg/g. None of the concentrations were statistically different from one another. Subchronic exposure to a relatively high concentration of formaldehyde did not significantly increase the blood formaldehyde concentration of rhesus monkeys.

Because formaldehyde is rapidly metabolized it does not accumulate in the blood or produce toxic effects at distant sites.

Potential Fatal Human Dose.   The approximate MLD (minimum lethal dose)—for a 150-lb man—was 30 mL [28].

There is limited evidence in humans for the carcinogenicity of formaldehyde. There is sufficient evidence in experimental animals for the carcinogenicity of formaldehyde. Overall, formaldehyde is probably carcinogenic to humans, according to the International Agency for Research on Cancer (IARC).

(c) Ecotoxicology.

Environmental Fate.

Terrestrial fate: When released on soil, aqueous solutions containing formaldehyde will leach through the soil. While formaldehyde is biodegradable under both aerobic and anaerobic conditions, its fate in soil is unknown.

Aquatic fate: When released into water, formaldehyde will biodegrade to low concentrations in a few days. Little adsorption to sediment would be expected to occur. In nutrient-enriched seawater, there is a long lag period (approximately 40 h) before measurable loss of added formaldehyde, presumably by biological processes [29].

Biodegradation. Formaldehyde in aqueous effluent is degraded by activated sludge and sewage in 48–72 hours. In a die-away test using water from a stagnant lake, degradation was complete in 30 hours under aerobic conditions and in 48 hours under anaerobic conditions [30].

2. Benzoic Acid/Sodium Benzoate [31] (Benzene Carboxylic Acid/Sodium Salt)

(a) Structure.

(b) Mode of Action. The antimicrobial activity of benzoic acid/sodium benzoate is based upon the ability of the undissociated acid to pass through the cell membrane and to interfere with enzymatic systems in the cell, resulting in inhibition of the nutrient transport activity of the cell membrane.

(c) Toxicology. Benzoic acid/sodium benzoate is widely used as a food preservative and is generally recognized as safe by the Food and Drug Administration (FDA). The Joint Expert Committee on Food Additives (JECFA) has assigned an acceptable daily intake of 0.5 mg/kg by weight. Contact urticaria and asthma have been reported as side effects of exposure in the literature. Most of these reports are associated with the ingestion of 50–500 mg of benzoic acid or sodium benzoate.

(d) Ecotoxicology. At environmentally relevant pH, benzoic acid will be present as the benzoate ion. At neutral pH it is unlikely to bioaccumulate in aquatic species and, in theory, would not be expected to reach the terrestrial environment via adsorption onto sewage sludge.

Benzoic acid is readily and ultimately biodegradable (by ring opening) and is of low acute toxicity to aquatic species. Current data on algae, daphnia, and fish are the results of nonstandard tests but do suggest that the lowest acute EC_{50} is likely to be above 10 mg/L.

3. Bronopol/Myacide BT [31] (1,3-Propanediol,2-Bromo-2-Nitro)

(a) Structure.

$$\underset{\underset{Br}{|}}{\overset{\overset{NO_2}{|}}{HOCH_2 - C - CH_2OH}}$$

Bronopol is the cosmetic grade and myacide BT is the technical grade of this preservative. They are effective against bacteria, especially *Pseudomonads* (and less effective against fungi).

(b) Mode of Action. This chemical reacts with enzyme thiol groups present in the cell membrane, thus inhibiting active transport across the membrane, inhibiting specific enzymes, and lyzing the cell membrane.

(c) Toxicology. The available toxicological evidence indicates that very low levels (i.e., 0.01% in leave-on and rinse-off personal products and 0.02% in liquid detergent products) present no undue health risks to the consumer with regard to acute toxicity, skin and eye irritation, skin sensitization, mutagenicity, carcinogenicity, or reproductive effects.

(d) Ecotoxicology. 1,3-Propanediol,2-bromo-2-nitro is unlikely to bioaccumulate in aquatic species and would not be expected to reach the terrestrial environment via adsorption to sewage sludge. Bronopol did not readily biodegrade in a closed-bottle test, probably because it inhibited the inoculum, but it has been shown to be inherently biodegradable in a modified Zahn–Wellens test. It is not expected to affect sewage treatment processes because it does not inhibit activated sludge at 50 ppm.

Levels reaching surface waters will be reduced both by photolysis and by hydrolysis. Hydrolysis occurs via four pathways, with the initial breakdown products being *tris*-(hydroxymethyl)-methane, 2-bromo-2-nitro-ethanol, and formaldehyde, all of which are expected to break down further to simple, naturally occurring substances that should not have an adverse effect on the environment.

Acute toxicity data for bronopol to fish, daphnia, and algae indicate that algae are the most sensitive species, with a 72-h EC_{50} of 0.2 mg/L. A chronic daphnia study gave a NOEC of 0.27 mg/L.

4. Glydant/Dantogard [31] (1,3-Dimethylol-5,5-Dimethyl Hydantoin (DMDM Hydantoin), or 1,3-bis (Hydroxymethyl)-5,5-Dimethyl Imidazolidine 2,4-Dione in 45% Water)

(a) Structure.

(b) Mode of Action. Glydant/Dantogard is an effective antibacterial agent, but its antifungal activity is weaker. Glydant is a formaldehyde donor; therefore it reacts with proteins, nucleic acids, cell walls, and cell membranes. It will react directly with genetic chromosomal material.

(c) Toxicology. The safe use of glydant has been assessed using the results of clinical and nonclinical testing, including studies of acute and subacute toxicity, irritation and skin sensitization, and photosensitization, potential, genotoxicity, and reproductive toxicicology. The Cosmetic Ingredient Review Expert Panel has produced a safety assessment (CIR, 1988) that concluded that DMDM hydantoin is safe as a cosmetic ingredient under present practices of use (generally up to 1%).

Human skin sensitization to glydant has been reported, and there is cross-reaction with formaldehyde. However, it is unlikely that any adverse skin reactions will be elicited in consumers when glydant is used as recommended.

(d) Ecotoxicology. DMDM hydantoin is unlikely to bioaccumulate in aquatic species and would not be expected to reach the terrestrial environment via adsorption to sewage sludge. DMDM hydantoin is biodegradable, although not readily, when tested in a sealed-vessel ready biodegradation test, below its inhibitory concentration. It is unlikely to inhibit the sewage treatment processes because the NOEC for the inhibition of *Pseudomonas putida* is 32 mg/L. Acute aquatic toxicity data for hydantoin mixture to fish, daphnia, and algae indicate that algae are the most sensitive species, with a 96-h EC_{50} of 4.2 mg/L.

E. Fluorescers

1. Structure

The two main fluorescers used in personal care products have the following structures.

"A" type: disodium 4,4'-bis(2-sulfostyryl)biphenyl (Tinopal CBS-X Ciba):

"B" Type: disodium 4,4'-bis(4-phenyl-1,2,3-triazol-2yl)stilbene-2,2'-disulfonate (Blankophor BHC, Bayer):

2. Toxicology

(a) Mammalian Toxicology. Oral intake of fluorescers from fish is negligible. The real concentrations found in drinking water of seven European countries are not detectable (at a detection limit of 0.01 ppb) [32], and in the United States the average concentration is less than 0.1 ppb [33,34]. These concentrations have no effect on rats in oral toxicity tests. Several studies showed no irritation to skin [35,36], no skin sensitization potential, no phototoxicity [37,38], and no photoallergy [39,40].

    Studies on animals showed:

• No skin effects
• No carcinogenicity
• No photocarcinogenicity [41]

(b) Acute Toxicity to Fish, Algae, and Daphnia. The LC_{50} values in Tables 12 and 13 [42] demonstrate that the fluorescers used in personal care products do not cause acute toxicity to water organisms at the concentrations in surface water (environmental concentrations are about 1000 times lower than the concentrations that produce mortality in rats).

TABLE 12 LC_{50} Values after 96-h Exposure (ppm)

	Catfish	Trout	Bluegill
"A"-type fluorescer	126	130	241
"B"-type fluorescer		221	689

3. Ecotoxicology

(a) Environmental Fate. Fluorescers enter the environment mainly through domestic wash water effluents. Field monitoring of sewage treatment plants in the United States showed that primary or mechanical treatment reduced the fluorescer concentration by more than 50% (certainly mainly by adsorption). Fluorescers were almost completely removed by secondary (biological) and tertiary (chemical) treatments [43].

Fluorescers accumulate strongly in sewage sludge (140–1080 ppm on a dry weight basis have been found in tests). When the sludge is used as a fertilizer, it is important to know whether or not the fluorescer adsorbed may be leached into groundwater (potentially used as a source of drinking water).

Experiments have shown that fluorecers are strongly adsorbed by the soil and that pollution of groundwater is unlikely [44,45].

(b) Biodegradation and Photodegradation. Laboratory experiments showed that fluorescers are not readily biodegradable. Fluorescers can be used as a substrate for bacterial growth, measured by the oxygen consumption over 5 days (BOD5). In these tests, the BOD5 of "A" type and "B" type is 0 (no evidence of biodegradation) [46]. No biodegradation was found in a river water die-away test (the CO_2 production was measured in solutions of maximum concentrations 1000 ppb over a period of 35 days) [47]. Table 14 shows that "A" type undergoes incomplete degradation (after an adaptation period) [48]. Therefore, in sewage treatment plants the elimination process will mostly be by adsorption. Photodegradation ("A" type) is rather fast: After 15 days, only small amounts of the initial product remain (test with 10-ppm solutions).

A bluegill sunfish test (with an accumulation period of 70 days) showed, for "A" type, detectable concentrations (<0.05 ppm), but below the concentration in the water (0.1 ppm) [49]. The fluorescer was quickly eliminated when the fish were transferred to clean water.

TABLE 13 LC_{50} Values of "A"-Type Fluorescer for Algae and Daphnia

	Concentration (ppm)	Time (h)
Algae	10	72
	8	96
Daphnia	>1000 (87% active substance)	24

TABLE 14 Biological Degradation of "A"-Type Fluorescer

	Percentage of degradation after:		
5 days	10 days	15 days	30 days
0	0	13	63
0	0	15	62

Radiolabeled "A" type was used in another experiment, where golden ides were exposed to 0.1 mg/L "A" type. A concentration of 6 ppm was found in the intestines (concentration factor of 60) [50]. An experiment using radiolabeled "B" type and golden ides showed that an equilibrium develops between uptake and elimination of the fluorescer within 1 week [51].

Experiments using a model aquatic food chain model showed that accumulation through the food chain is very slight (and no higher than for the direct uptake) [50]. Another study determined the uptake of radiolabeled "A" type in bean plants (plants grown in soil containing 17.5 ppm of "A" type). Over 90% of the uptake was found in the roots [52]. Other uptake experiments lead to similar findings. This indicates that the fluorescer present in the soil will probably not be translocated to the main stem or other parts of the plants.

F. Colorants

The main characteristics of a few representative dyes used in personal care products can be summed up as follows.

1. Pigment Yellow 1/CI 11680 [53]

(a) Structure: Azo Chromophore.

(b) Toxicology.

- In animal studies, the acute toxicity (LD_{50}) in rats (oral) is over 5000 mg/kg.
- Pigment yellow 1 was not irritant to rabbit skin and was not irritant to rabbit eyes.
- No adverse effects have been reported in man.

(c) Ecotoxicology.

Fish toxicity. $LC_{50} > 100$ mg/L (rainbow trout, 48 h)—Pigment yellow 1 was not toxic or harmful to aquatic organisms.

2. Acid Yellow 23 (Tartrazine)/CI 19140 [54]

(a) Structure: Azo Dye.

(b) Toxicology.

- Acute toxicity (LD50) was over 10,000 mg/kg (rats/oral).
- Acid yellow 23 was not irritant to rabbit skin and eyes.
- In use, cases of skin sensitization have been observed with this dye.

(c) Ecotoxicology.

Bioelimination. Acid yellow 23 was partially eliminated by adsorption on effluent treatment sludge (50–60% DOC analysis—using OECD method 302B).

3. Food Yellow 13 (Quinoline Yellow)/CI 47005 [55]

(a) Structure: Quinoline Dye.

(b) Toxicology.

- Acute toxicity (LD_{50}) was over 2000 mg/kg (rats/oral).
- Food yellow 13 was not irritant to rabbit skin or eyes.

(c) Ecotoxicology.

Bioelimination. Food yellow 13 is poorly eliminated by adsorption on effluent treatment sludge (10–25% DOC analysis—OECD method 302B). No disturbance of water treatment plants has been recorded under recommended application techniques.

4. Pigment Red 5/CI 12490 [56]

(a)   *Structure: Azo Chromophore.*

(b)   *Toxicology.*

*   Acute toxicity (LD_{50}) was over 5000 mg/kg (rats/oral).
*   Pigment red 5 was not irritant to rabbit skin or eyes.
*   No adverse effects in use have been reported in man.

(c)   *Ecotoxicology.*

*   Fish toxicity: LC_{50} > 100 mg/l (48 h—rainbow trout).
*   Pigment red 5 was not toxic or harmful to aquatic organisms.

5. Acid Blue 9/CI 42090 [57]

(a)   *Structure: Triarylmethane Dye.*

(b) Toxicology.

- Acute toxicity (LD_{50}) was over 10,000 mg/kg (rats/oral).
- Acid blue 9 was not irritant to rabbit skin or eyes.
- No adverse effects have been reported in man.

IV. CONCLUSIONS

In this chapter we have looked at a number of "minor ingredients" in personal care products. Because this list of chemicals is very long, we have not attempted to cover all the materials in use. We hope that the information provided is useful and meets the requirements of the reader. We believe that considerable resources are going to be deployed in the future to improve these products or their replacements significantly from both toxicological and environmental perspectives.

It should be noted that some companies have higher standards of safety and are more demanding than national and international regulations, in order to maximize safety for both consumers and the environment.

Finally, we have to mention that the information provided in this chapter is based on the present state of knowledge. It shows the uses to which the described ingredients can be put into formulations of personal care products.

ACKNOWLEDGMENTS

I would like to acknowledge the contributions of the following expert scientists to this chapter: Dr. A. Caby (Firmenich S.A. Geneva) for performing the unreferenced studies, Mr. M. Graff and Dr. B. Jones (Ciba Basel) for providing precious data, Mr. M. Capelle (ex. Lever France) for great assistance in preparing the manuscript.

REFERENCES

1. Ho Tan Tai, Louis. Formulating Detergents and Personal Care Products; AOCS Press: Champlain, IL, 2000; 235 pp.
2. Helliwel, J.F. Unilever, World Patent WO 9,403,151.
3. Cherrey, M.; Filiciano, D.; Wivell, S. Cheseborough Ponds, U.S. Patent US 5,441,671.
4. Unilever Patent EP 0249474 A2.
5. Ho Tan Tai, Louis. Formulating Detergents and Personal Care Products; AOCS Press, 2000; 255, 257, 258, 260, 261.
6. Asano, A.; Gaffer, M.C. Johnson & Johnson, European Patent EP 0,162,574-B1.
7. Collins, M.A.; Duckenfield. J.M. Colgate, European Patent EP 0,549,287-A1.
8. Roger, M.L. Unilever, U.S. Patent US 4,759,562.
9. Helliwell, J.F. Unilever, World Patent WO 9,403,151.
10. Ho Tan Tai, Louis. Formulating Detergents and Personal Care Products; AOCS Press: Champlain, IL, 2000; 24 pp.
11. Grundschober, Friedrich. The IFRA Guidelines; IFRA, 1998.
12. Committee on Cosmetic Products (SCCNF). Fragrance Allergy in Consumers. Opinion SCCNF p 0017/98 adopted on 8 December 1999 on fragrance allergens.
13. Ciba-Geigy. Irgasan DP 300—Irgacare MP—Toxicological and Ecological Data—Official Registrations 2521; additionnal reference: Barghana, H.; Leonard, P.A.; Triclosan: application and safety. AJIC 1996; *24* (3).

14. Kanetoshi, A.; Katsura, E.; Ogawa, H.; Ohyama, T.; Kaneshima, H.; Miura, T. Acute toxicity, percutaneous absorption and effects on hepatic mixed-function oxidase activity of 2,4,4′trichloro 2′-hydroxydiphenylether (Irgasan DP 300) and its chlorinated derivatives. Arch. Environ. Contam. Toxicol. 1992, *23*, 91–98.

15. Veronisi, S.; DePadova, M.P.; Vanni, D.; Melino, M. Contact dermatitis to triclosan. 1986, *15*, 257.

16. Black, J.G.; Howes, D.; Rutherford, T. Percutaneous absorption and metabolism of Irgasan DP 300. Toxicology 1975, *3*, 33–47.

17. Anderson, K.; Nilsson, C.A.; Rappe, C. Studies on chlorinated phenoxyphenols (predioxius). Environ. Qual. Saf. 1975, *3* (suppl), 731–733.

18. Swenerton, H.; Harley, L.A. Science 1971, *173*, 62.

19. Brownie, C.F.; Brownie, C.; Noden, O.; Krook, L.; Haluska, M.; Aronsen, A.L. Toxicol. Appl. Pharmacol. 1986, *82*, 426.

20. Toxicological evaluation of some food additives including anticaking agents, antimicrobials, antioxidants, emulsifiers and thickening agents. WHO Food Additive Series no. 5, WHO, Geneva, 1974.

21. Wolf, K.; Gilber, P.A. EDTA. In *The Handbook of Environmental Chemistry*; Hatzinger, O., Ed.; 1992; Vol 3, Part F, 248 pp.

22. Belly, R.T.; Lauff, J.J.; Goodhue, C.T. Appl. Microbiol. 1975, *29*, 787.

23. Berg, M. First Ital. Idrobiol. 1970, *26*, 257.

24. Bishop, W.E.; Maki, A.W. Proc 3rd Ann Symp on Aquat Tox ASTM STP 707, 1980.

25. Reports on the effects of EDITMPA in animals and humans, Vol 2, EPA/OTS; Doc #88-8400594, 1984.

26. Nowack, B.; Stone, A.T. The influence of metal ions on the adsorption of phosphonates onto goethite. Environ. Sci. Technol. 1999, *33*, 3627–3633.

27. Casanova, M., et al. Food Chem. Toxicol. 1988, *8*, 715–716.

28. Arena, J.M. *Poisoning: Toxicology, Symptoms, Treatments*, 4th Ed.; Charles C. Thomas: Springfield, IL, 1979; 97 pp.

29. Mopper, K.; Stahovec, W.L. Marine Chem. 1986, *19*, 305–321.

30. Kitchens, J.F., et al. Investigation of selected potential environmental contaminants: formaldehyde. U.S. EPA 560/2-76-009, 1976; 99–110.

31. Ho Tan Tai, L. Private document.

32. Anders, G. In 1975; 143 pp.

33. Procter & Gamble Co. Fluorescent whitening agents: occurrence in wastewater, river waters, and drinking waters, unpublished report cited in Ref. 34, 1974.

34. Burg, A.W.; Rohovsky, M.W.; Kensler, C.J. CRC Crit Rev Env Contr, 1977, *7*, 91.

35. GAF Corp. Unpublished data cited in Ref. 34.

36. Ciba-Geigy, A.G. Repeated insult patch test with compound "A" type, Food and Drug Research Laboratories, unpublished report cited in Ref. 34.

37. Forbes, P.D.; Urbach, F. In [5] 1975, 212 pp.

38. Ciba Geigy, AG. Acute study: human skin. Unpublished report by Urbach F cited in Ref. 34.

39. Verona Corp. Unpublished data cited in Ref. 34.

40. Griffith, J.F. Arch Dermatol. 1973, *107*, 728.

41. Kramer, J.B. Fluorescent whitening agents. In: *The Handbook of Environmental Chemistry*; Hutzinger, O., Ed.; Vol 3, Part F, 1992; 363 pp.

42. Kramer, J.B. Fluorescent whitening agents. In *The Handbook of Environmental Chemistry*; Hutzinger, O., Ed.; 1992; Vol. 3, (Part F) 365 pp.

43. Ganz, C.R.; Liebert, C.; Schulze, J.; Stensby, P.S. J. Water Pollut. Control Fed. 1975, *47*, 2834.

44. Esser, H.O. and collaborators. Personal communication cited in Ref. 45.

45. Zinkernagel R. In 1975; 129 pp.

46. Bayer, A.G.; Ciba-Geigy, A.G. Product information, 1990.
47. Procter & Gamble Co. Unpublished report, 1988.
48. Guglielmetti, I. In [5] 180 pp.
49. Ganz, C.R.; Schulze, J.; Stensby, P.S.; Lyman, F.L.; Macek, K. Env. Sc. ant Tech. 1975, *9*, 738.
50. Feron, J.-P.; Hitz, H.R. In: 1975; 157 pp.
51. Hamburger, B.; Maul, W.; Patzschke, K.; Theidel, H.; Wegner, L.-A. In 1975; 165 pp.
52. Muecke, W.; Dupuis, G. Hesser HO. In 1975; 174 pp.
53. Ciba-Geigy. Vibracolor Yellow PYE1-L Safety Data Sheet.
54. Ciba-Geigy. Puricolor Yellow AYE23 Safety Data Sheet.
55. Ciba-Geigy. Puricolor Yellow FYE13 Safety Data Sheet.
56. Ciba-Geigy. Vibracolor Red PRE5-L Safety Data Sheet.
57. Ciba-Geigy. Puricolor Blue ABL9 Safety Data Sheet.
58. J. Am. Pharm. Assoc. 1949, *38*, 366.

26

The Environmental Impact of Surfactant Ingredients in Pesticide Formulations— Special Focus on Alcohol Ethoxylates and Alkylamine Ethoxylates

KRISTINE A. KROGH and BENT HALLING-SØRENSEN
The Danish University of Pharmaceutical Sciences, Copenhagen, Denmark

BETTY B. MOGENSEN and KARL V. VEJRUP
National Environmental Research Institute, Roskilde, Denmark

I. INTRODUCTION

Surfactants are widely used as adjuvants in agrochemicals, e.g., pesticide formulations and pesticide tank mixing additives. Beside the solvents, surfactants make up the largest group of adjuvants [1,2]. Adjuvants are general added to enhance the effectiveness (bioavailability) of the active ingredients of the pesticide formulation, by enhancing the solubility, or the compatibility of the active ingredients. Other functions are enhancing adsorption, penetration, and translocation of the active ingredients into the target, increasing rain fastness, and altering selectivity of the active ingredient toward different plants [2–4]. Adjuvants in pesticides comprise a large and heterogeneous group of substances, which include solvents, surfactants, spreaders, dispersants, emulsifiers, antifoaming agents, wetting agents, antifreezing agents, and preservatives [2]. The major types of surfactants are listed in Table 1. The nonionic surfactants alkylphenol ethoxylates (APEOs), alcohol ethoxylates (AEOs) and alkylamine ethoxylates (ANEOs) are especially commonly employed in agrochemical formulations and as spray additives [5,6]. Figure 1 shows the chemical structure of the compounds.

The literature dealing with adjuvants focuses mostly on comparing the effectiveness of different adjuvant combinations, especially different surfactant mixtures [4]. The lack of information on the effect and toxicity as well as risk on the environment of

TABLE 1 Surfactants Used in Pesticide Formulations and Spray Additives

Nonionic surfactants	Anionic surfactants
Alcohol alkoxylates	Alkyl benzene sulfonates
Alkylamine ethoxylates	• Calcium dodecyl benzene sulfonates
• Tallow amine ethoxylates	Alkyl naphathalene sulfonates
Alkylphenol ethoxylates	• Sodium naphathalene sulfonates
• Nonyl- and octylphenol ethoxylates	Alkyl sulfonates
Castor oil ethoxylates	Fatty acid sulfonates
Ethylene oxide/propylene oxide block copolymers	Ligno sulfonates
Fatty acid ethoxylates	Dioctyle sulfosuccinates
Sorbitan ethoxylates	

Source: Refs. 5 and 18.

adjuvants has been noted previously [7]. One paper summarizing biodegradation data on ethoxylated surfactants (AEO and APEO) used in agrochemical formulations has been published [6], and a review has recently become available on environmental properties and effects on nonionic surfactant adjuvants (AEO and ANEO) [8]. Furthermore, during the last 10 years, several papers dealing with APEO and their degradation products concerning occurrence, biodegradation, and toxicity have been published [9–12]. In particular, papers demonstrating estrogenic responses of the APEO degradation products (alkylphenols) in fish have become available [13–15].

APEO are often identified as one of the most used groups of surfactants in pesticide formulations. However, due to their lack of biodegradation and adverse effects (degradation products of the alkylphenols induce estrogenic effects) [11], APEO are no longer used in Denmark [16]. Because of a voluntary agreement between the manufacturer of the pesticides and the Danish Environmental Protection Agency (Danish EPA) to phase out the APEO from the market, several pesticide formulations have been reformulated and others have not been reapproved. Nevertheless, APEO are of low cost as compared to the AEO; as a result these surfactants are still widely used in other countries.

This chapter will summarize the existing knowledge of exposure, fate, and effects of surfactants used as adjuvants in pesticide formulations, with special focus on the nonionic surfactants, AEO and ANEO. APEO will be discussed only very briefly.

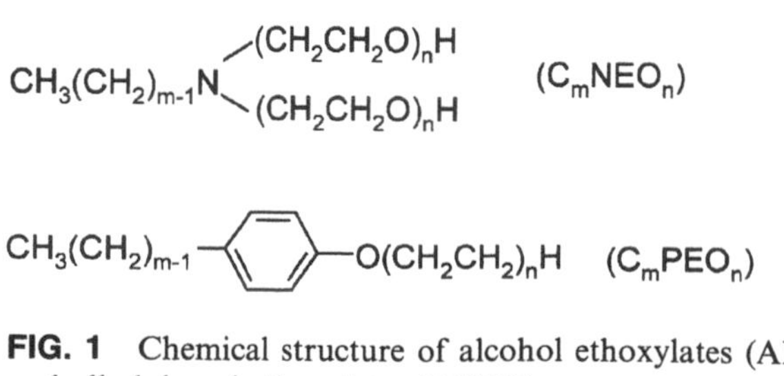

$$CH_3(CH_2)_{m-1}O(CH_2CH_2O)_nH \qquad (C_mEO_n)$$

$$CH_3(CH_2)_{m-1}N\begin{array}{c}(CH_2CH_2O)_nH\\(CH_2CH_2O)_nH\end{array} \qquad (C_mNEO_n)$$

$$CH_3(CH_2)_{m-1}-\langle\bigcirc\rangle-O(CH_2CH_2)_nH \qquad (C_mPEO_n)$$

FIG. 1 Chemical structure of alcohol ethoxylates (AEO), alkylamine ethoxylates (ANEO), and alkylphenol ethoxylates (APEO).

Due to the limited chapter size, the data presented are illustrated primarily by post-1995 publications. A more comprehensive overview is given in a 2003 review [8]. Furthermore, a primary attempt to suggest environmental hazard assessment has been proposed [17].

II. EXPOSURE OF SURFACTANT ADJUVANTS

Data on the consumption of adjuvants (including surfactants used as adjuvants) are given in Table 2. In 1992, the amount used as pesticide spray additives was estimated to make up 150,000 tons ($196 million). Of this, a large part consisted of the surfactants shown in Table 1 [18]. The mean Danish load of surfactants on farming areas coming from pesticide application is between 0.3 and 0.4 kg surfactants per ha per year [19]. However, the load depends on the crop, the frequency of pesticide treatment, and the formulation used.

Comparing the annual consumption of surfactants in Denmark, we find that 750–1000 tons originate from agrochemicals (1998) [19], 16,360 tons from household detergents (1998), and 2780 tons from industrial and institutional detergent products (1997) [20]. It becomes obvious that agrochemicals represent the smallest source. However, surfactants coming from pesticide applications are sprayed directly on humid or dry soil, whereas exposures from household, industrial, and institutional detergent products enter the environment via either untreated sewage waters, effluent, or sludge from sewage treatment plants that are discharged into the environment. Furthermore, degradation and adsorption conditions are different in soil and presumably not as optimal as in sludge and sewage water.

Runoff, leaching, adsorption, and biodegradation are some of the possible exposure routes and main fate processes. Potential transport routes via rainwater may be either horizontal, as runoff ending up in surface waters, or vertical, leaching to groundwater. Another fate route is adsorption to soil or biological material such as plants, plant roots, and plant debris or microorganisms in the soil. Additionally, the adjuvants may be biodegraded either aerobically or anaerobically.

III. ENVIRONMENTAL CONCENTRATION LEVELS OF SURFACTANT ADJUVANTS

Most data consider only environmental occurrence and effects related to sewage sludge, sewage water, and sediments of sewage-receiving waters [21–23]. Only few data

TABLE 2 Consumption of Adjuvants in Pesticide Formulations

| Region (year) | Total use of adjuvants | | Surfactants used as adjuvants | | Ref. |
	Cost ($)	Mass (tons)	Cost ($)	Mass (tons)	
USA (1992)	179×10^6	0.5×10^6	75×10^6		18
Worldwide (1994)				0.11×10^6 (nonionic)	115
Worldwide (1998)				0.18×10^6	116
Denmark (1998)		10×10^3		750–1000	19

on the occurrence of surfactant adjuvants originating from pesticide application on agricultural land exist. These data cover only the nonionic surfactants: AEO and ANEO. Concerning aquatic data, concentrations in groundwater (2–3 m below the groundwater table) were between 61 and 189 ng/L for the different AEO homologues $C_{12}EO_{3-9}$, giving a total concentration of the six homologues of 710 ng/L [24]. In soil interstitial water, concentrations were, respectively, 33 ng/L of $C_{12}EO_4$ and varied from 48 to 73 ng/L for $C_{12}EO_{3-5}$, with a total concentration in the deeper layers of 194 ng/L. In all of the samples, the AEO with the shortest ethoxy chain were detected in the highest concentrations. ANEO were not detected in any of the samples. In a recent published thesis, more aquatic data were given [17]. AEO were detected in soil interstitial water and groundwater at concentrations of the different homologues ranging between 31 and 157 ng/L, again with the highest concentration of $C_{12}EO_3$. Also for these aqueous samples, ANEO were not detected. The presented data indicated that the frequency of the detection as well as the concentrations are highly dependent on the cultivated crop and the type of pesticide applied as well as the application frequency. Surface water concentrations of the AEO originating from other sources are higher than the ones coming from agrochemicals. Concentrations of AEO as high as 6.5 μg/L [25] and 12.5–300 μg/L [26] have been detected in treated sewage, whereas 10–190 mg/kg has been found in sewage sludge [27].

In another paper, terrestrial data on the occurrence of nonionic surfactant adjuvants in agricultural soils have been presented [28]. Here the concentration of ANEO ($C_{16-18}NEO_{13-18}$) reached 218–524 μg/kg dry soil 14 days after the pesticide application of glyphosate formulations, whereas prior to the application the concentration levels were 97–170 μg/kg dry soil. The homologues detected at the highest concentration were the ANEO with the longest ethoxy chains as well as the alkyl chains. AEO were not detected in any of the samples. In agreement with the expected results, where only ANEO were detected in the fields sprayed with glyphosate formulations, since these formulations are generally known to contain the ANEO (tallow amine ethoxylates) [18].

Considering the available data on occurrence, it is important to bear in mind that the applied analytical methods quantify only selected surfactants of the homologous series. This implies that other homologues might be present. However, since the surfactants are added as technical mixtures consisting of various homologous series, the analytical methods are believed to include representatives of the homologous series.

IV. SORPTION BEHAVIOR OF THE AEO AND ANEO USED AS SURFACTANT ADJUVANTS

A. Possible Bonds to Soil Particles for the AEO and ANEO

Different binding possibilities expected for AEO and ANEO in soil are summarized in two schematic illustrations (see Fig. 2). The hydrophilic and hydrophobic parts of these surfactant molecules enable various binding types, e.g., hydrogen bonds and hydrophobic bonds to different soil components. For AEO, the hydrophobic alkyl chain can adsorb to organic matter by hydrophobic bonds (see Fig. 2a), whereas the hydrophilic ethoxy chain, containing ether-bound oxygen, might bind via hydrogen bonds to rather polar clay minerals in the soil. As described for AEO, ANEO can

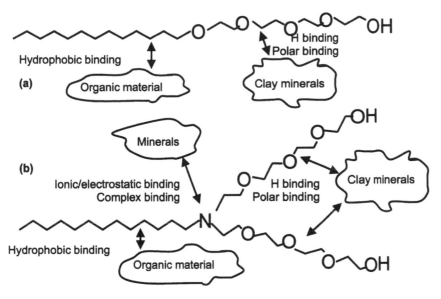

FIG. 2 Schematic illustrations summarizing binding possibilities between soil components and (a) AEO and (b) ANEO.

participate in the same kind of bonds to the soil components (see Fig. 2). Because ANEO consist of two ethoxy chains (see Fig. 2b), the polar bonds to clay minerals or other polar components are probably more pronounced for the ANEO than for the AEO. Additionally, for the ANEO the hydrophobic and hydrophilic moieties are connected with a nitrogen atom, unlike the AEO, where oxygen joins the two moieties. This nitrogen in the ANEO will become protonated, depending on the pH of the surrounding media and the pK_a value of the specific ANEO. The protonated form of the ANEO may then form ionic bonds, while the deprotonated form participates in complex bonds to minerals in the soil (see Fig. 2b).

Partition coefficients between soils, sediments, or plant material and water for the AEO and ANEO are shown in Table 3, together with data on the critical micelle concentration (CMC). CMC decreases with increasing alkyl chain length and decreases slightly with decreasing ethoxylate chain length. The partitioning coefficient between octanol and water (K_{ow}) decreased for the compounds with more ethoxylate units and with decreasing alkyl chain length [8]. On the other hand, K_{sw} increased with increasing ethoxy chain length (see Table 3). Alkyl chain length was shown to have a more dominant effect than the ethoxy chain length [29].

B. Adsorption to Soils and Sediments

AEO adsorb to clay. Swelling clay minerals and other oxygen-containing mineral surfaces, e.g., soil silicates such as vermiculite and montmorillonite, are especially prone to absorb the AEO in soil [30,31] and in sediments [32–34]. AEO homologues possessing longer ethoxy chains were especially found to have a higher affinity to soil components. The ethoxy part of the AEO molecule is identical to polyethoxylene

TABLE 3 Partition Coefficients and CMC of AEO and ANEO

Compound	Partition coefficient	Log CMC	Sorption media	Method	Conc.	Ref.
AEO						
$C_{10}EO_3$	$\log K_{sw} = 1.61$		Sediment	Analyzing both phases	<CMC	37
$C_{10}EO_8$	$\log K_{sw} = 2.10$		Sediment	Analyzing both phases	<CMC	37
$C_{10}EO_9$	$\log K_{om} = 3.18$	0.094 mol/m^3	O.c.n. sediment			37
$C_{12}EO_3$	$\log K_{sw} = 1.86$		Sediment	Analyzing sediment, container wall, water	<CMC	34
$C_{12}EO_3$	$\log K_{sw} = 2.41$		Sediment	Analyzing both phases	<CMC	37
$C_{12}EO_4$	$\log K_{sw} = 7.79$	−4,64 mol/L	Soil	Surface tension analyzed in water	<CMC	43
$C_{12}EO_5$	$\log K_{sw} = 2.86$		Sediment	Analyzing both phases	<CMC	37
$C_{12}EO_6$	$\log K_{sw} = 1.60–1.79$		EPA sediments	Analyzing sediment, container wall, water	<CMC	34
$C_{12}EO_8$	$\log K_{sw} = 3.09$		Sediment	Analyzing both phases	<CMC	37
$C_{12}EO_{8.5}$	$\log K_d = 1.99–2.21$		Different minerals, sand (desorption)	Analyzing both phases, using ^{14}C		72
$C_{12}EO_{8.5}$	$\log K_d = 2.13$		Humic acid (desorption)	Analyzing both phases, using ^{14}C		72
$C_{12}EO_9$	$\log K_{om} = 3.83$	−0.892 mol/m^3	O.c.n. sediment			37
$C_{12}EO_{23}$	$\log K_{sw} = 1.70–1.81$		Different EPA soils	Analyzing only aqueous phase	<CMC	31
$C_{13}EO_2$	$\log K_{om} = 3.75$	−1.707 mol/m^3	O.c.n. sediment			37
$C_{13}EO_3$	$\log K_{sw} = 2.04–2.70$		Different sediments	Analyzing both phases, using ^{14}C		32
$C_{13}EO_4$	$\log K_{om} = 3.84$	−1.615 mol/m^3	O.c.n. sediment			37

Compound	Partition coefficient	Concentration	Matrix	Remarks	Conc.	Ref.
$C_{13}EO_8$	$\log K_{om} = 4.09$	-1.431 mol/m^3	O.c.n. sediment			37
$C_{13.5}EO_9$	$\log K_{om} = 4.30$	-1.6315 mol/m^3	O.c.n. sediment			37
$C_{13}EO_9$	$\log K_{sw} = 2.04$–2.77		Different sediments	^{14}C-labeled analyzing both phases		32
$C_{14}EO_3$	$\log K_{sw} = 3.47$		Sediment	Analyzing both phases	<CMC	37
$C_{14}EO_5$	$\log K_{sw} = 3.54$		Sediment	Analyzing both phases	<CMC	37
$C_{14}EO_8$	$\log K_{sw} = 3.55$		Sediment	Analyzing both phases	<CMC	37
$C_{14}EO_9$	$\log K_{om} = 4.48$	-1.878 mol/m^3	O.c.n. sediment			37
$C_{15}EO_9$	$\log K_{sw} = 2.54$–3.23/ 2.67–3.32		Different sediments	Analyzing both phases, using ^{14}C/^{3}H	10 mg/L	32
$C_{16}EO_5$	$\log K_{sw} = 3.68$		Sediment	Analyzing both phases	<CMC	37
$C_{16}EO_8$	$\log K_{sw} = 3.79$		Sediment	Analyzing both phases	<CMC	37
$C_{16}EO_9$	$\log K_{om} = 5.14$	-2.864 mol/m^3	O.c.n. sediment			37
$C_{18}EO_{20}$	$\log K_{cw} = 1.92$; 1.94	0.0038 g/L	Cuticle (tomato; pepper)	Analyzed as radioactivity of the water	>CMC	117
ANEO						
$C_{18}NEO_{20}$	$\log K_{cw} = 1.70$; 1.66	0.002 g/L	Cuticle (tomato; pepper)	Sorption analyzed as radioactivity of the water	>CMC	117
APEOs						
$C_8PEO_{9.5}$	$\log K_{sw} = 0.86$		Soil	Surface tension analyzed in water	<CMC	43
C_8PEO_{12}	$\log K_{sw} = 0.058$		Soil	Surface tension analyzed in water	<CMC	43
$C_9PEO_{10.5}$	$\log K_{sw} = 0.41$		Soil	Surface tension analyzed in water	<CMC	43

Partitioning coefficients: K_{sw}: soil or sediment/water; K_{om}: organic carbon normalized (o.c.n.) sediment/water; K_{cw}: cuticle/water.

glycol (PEG). Thus, similar to the AEOs, longer PEG adsorbs more strongly to sediments as compared to shorter PEG. In addition, the content of expandable clay (vermiculite and montmorillonite) enhanced the adsorption of the PEG. PEG forms hydrogen bonds with the negatively charged sediment sites [35]. Likewise, polar binding forces, such as hydrogen bonds, between the ethoxy groups in the AEO and polar soil substituents, are probably the primary mechanism of adsorption to soil [31,36] (see Fig. 2).

Besides the length of the ethoxylate chain, increasing the alkyl chain length of AEO also increases adsorption [37]. The fact that an increase in the length of the hydrophobic alkyl chain increases the potential of hydrophobic binding as well as weakening the repulsion caused by the hydrophilic part can be anticipated from the chemical structure (see Fig. 2). Adsorption of AEO is proportional to the content of organic carbon in sediments [38]; however, the content of clay was not revealed in the paper. Relationships between the percentage of expandable clay minerals and adsorption can therefore not be discussed. Inconsistent with the findings described earlier, more studies [31–34] found no correlation between organic content and the amount of AEO adsorbed to different sediments and soils. Similarly, it was found that the organic content in sediments did not have any effect on the adsorption of PEG [35]. These inconsistencies may be explained by differences in sediment organic carbon content. Urano et al. [38] used a sediment with a organic carbon content ranging from 1% to 6%, while for the other studies sediments with lower (between 0.2 and 3%) sediment organic carbon content were applied. Even though only one paper found a correlation between adsorption and organic content, organic matter in the soil is expected to participate in binding to the hydrophobic part of the surfactant by hydrophobic bonds; however, other bonds might be more pronounced.

Variation in the cation exchange capacity of the soil did not correlate with the extent of adsorption of AEO [31]. The effect of pH was most pronounced for AEO with the longer EO chain, for which increased sorption to sediments was observed when lowering the pH [34]. In other soils, no effect of pH was found [31]. Concerning adsorption of ANEO, pH might be an important parameter because these compounds can become protonated at the nitrogen (see Fig. 2b). Typical pK_a values for ANEO are between 5.8 and 7.4 [39], so the nitrogen is partly protonated at normal soil pH, around 5 [40]. Complementary to the previously described binding abilities, the N atom in the ANEO enables ionic/electrostatic and complex bonds to minerals in the soil.

Additionally, surfactant concentrations, being below and above CMC, influence the sorption of an AEO [41–43]. When the concentration is above CMC, the surfactants will aggregate into micelles. The hydrophilic part of the molecules will probably then become exposed to aqueous phase or may bind to the matrix, e.g., soil particles.

An investigation of kinetics of the sorption behavior of an AEO showed that both sorption and desorption processes were rapid and reversible. Equilibrium was already reached after 24 hours [33]. Despite the addition of water to allow leaching, the major part of a nonionic surfactant (AEO) remains adsorbed to the upper 2 cm of the topsoil after application [42]. These findings indicate that AEOs bind to the topsoil shortly after application. As a consequence of sorption to the soil minerals (such as montmorillonite, illite, or humics), AEO both mineralized more slowly and to a lower extent than the nonadsorbed ones or the ones adsorbed to sand and kaolinite [44].

C. Influence of Surfactants on the Fate of Other Chemicals

Adsorption of surfactants onto soil particles is anticipated to influence both the physicochemical and the biological properties of the soil [45,46]. Key factors affecting the mobility of chemicals (pesticides or other pollutants) in soil is the concentration of the surfactant and the hydrophobic character of the chemicals [47–52]. At concentrations below and above CMC, the mobility in soil of the pesticide active ingredients (a.i.'s), diazinon, atrazine, and acephate, decreased in the presence of nonionic surfactants [51,52]. Likewise, the mobility of pesticide a.i. metachlor decreased at low surfactant concentrations, around 0.04 g/kg soil (equal to CMC), while at high concentrations (5–50 g/kg) the mobility increased [51]. Furthermore, high surfactant concentrations (50 g/L) in the leaching water increase the mobility of the pesticide a.i.'s. Addition of nonionic surfactants to soil decreases the mobility of the pesticides, which probably are adsorbed by the surfactants previously, adsorbed by the soil. On the other hand, the increasing mobility when surfactants are added to the leaching water might be due to the absorption in the micelles.

Beside mobility, degradation of pesticides is also affected by the occurrence of surfactants. The pesticide a.i.'s atrazine and coumaphos were degraded at a lower rate and to a lower extent in the presence of a nonionic surfactant [53].

V. DEGRADATION BEHAVIOR OF AEO AND ANEO USED AS SURFACTANT ADJUVANTS

Degradation of chemical compounds is a complex process that is affected by both intrinsic properties of the compound and properties of the receiving matrix. It is thus often necessary to distinguish between abiotic and biotic degradation. Among the different abiotic degradation processes occurring in the environment, photolysis is the most prevalent. Recently, investigations of photolytic degradation as a useful tool for AEO removal in water have been performed [54,55]. In the latter study, a technical mixture of AEO was 95% mineralized after 7 days using irradiation at 365 nm and photoinduced Fe(III).

The remaining part will deal with biotic degradation (biodegradation). Biodegradation pathways depend on the redox potential, i.e., oxidative or nonoxidative conditions as well as on the availability of the microorganisms in the soil media. In a 2003 review, the different biodegradation pathways for AEO and ANEO are outlined [8]. Here the effects of the chemical structure on biodegradation are considered in combination with available data on aerobic and anaerobic biodegradation in various matrices, such as soil, sediment, and waters.

It is always difficult to compare data on biodegradation kinetics, because the degradation rate often depends on several factors related to the experimental design and the analytical method applied. Concerning the experimental design, the following specifications are of utmost importance: test conditions of the microorganisms, acclimatization of the microorganisms to the test compound, sorption of the test compound to inoculum and test container, the source and quantity of inoculum, the diversity of the population of microorganisms, the presence or absence of nutrients in the media. The choice of method is linked to the available technique and the extent of degradation, i.e., the endpoint, which distinguish between primary and ultimate biodegradation (mineralization). In primary biodegradation, the characteristic properties of the compounds are lost. In ultimate biodegradation/mineralization, the test compound is decomposed into CH_4, CO_2, and H_2O and, for the ANEO, also NH_3.

As outlined in a 2003 review, most of the available data on the biodegradation of AEO and ANEO have been obtained from biodegradation studies in sludge and surface waters using cultures of sewage microorganisms [8]. The same conditions may not be present in agricultural fields, even though soil microorganisms may be amended with organic fertilizers. Principally data on biodegradations in soil and subsequently surface water and groundwater with the naturally occurring microorganisms are discussed.

A. Biodegradation and Chemical Structure

In 1955 it was already indicated that the molecular structure of both the hydrophobic alkyl chain and the hydrophilic ethoxylate group of the surfactants is of major importance for the susceptibility to biochemical oxidation [56]. Branching in the alkyl group of the AEO reduced the susceptibility to biodegradation; i.e., branched AEO degraded much more slowly and to a lower extent than the linear ones [56–58]. Other studies have weakened these findings, since the ultimate biodegradation of secondary AEO seemed to decrease only slightly as compared to the primary AEO, while slight methyl branching in a primary AEO appeared to have no significant effect [59]. Likewise, in a comparison of the ultimate aerobic biodegradation of linear and branched $C_{14-15}EO$, degradation rates were similar for the different isomers [60].

Primary and ultimate biodegradation of primary AEO were unaffected by variation in alkyl chain length [57,59]. In contrast to these findings, another study indicated that the removal rates increase slightly in proportion to the length of the alkyl chain of the AEO. Here the $C_{15}EO$ degraded faster than the $C_{14}EO$, $C_{13}EO$, and $C_{12}EO$ [61].

The ethoxylate chain length seems to affect the biodegradability of AEO [59,62]. Reduced biodegradation rates have been observed for AEO as the ethoxylate chain length increased [57,63,64]. Inconsistent findings were obtained in a study for $C_{17}EO_3$ to $C_{17}EO_{11}$, where no significant effects on the ultimate biodegradability were seen when the ethoxylate chain length increased [59]. However, AEO with EO greater than 20 show reduced biodegradability. Still another study showed that the percentage of mineralization increased with increasing EO chain length of the AEO [65].

B. Ultimate Biodegradation in Sewage Sludge, Soil, and Sediments

Biodegradation of a linear primary $C_{12}EO_n$ in an aerobic shake-flask experiment, was highly concentration dependent, comparing sub-CMC levels and supra-CMC levels [66]. Likewise, concentration dependence of degradation has been seen for a $C_{10}EO_8$, since when the concentration increases to above the CMC, less degradation was observed in a sequencing batch reactor study [67]. Surfactants in a micellar phase may have lower biodegradability than monomeric surfactant molecules, for example, because the structure hinders contact between the surfactant and the microorganisms or because the microorganisms become inactive. Additionally, the surfactants might destroy the cell membrane of the microorganisms.

Only in one study was the biodegradation of two ANEO investigated using a closed-bottle test [68]. Here the ANEO were found to be ready biodegradable; i.e., 60% tallowbis(2-hydroxyethyl)amine and 63% oleylbis(2-hydroxyethyl)amine were degraded after 28 days when silica gel was added to the test medium (sludge and sewage), whereas 52% and 13%, respectively, were degraded when no silica was added. By adding silica gel, the concentration of the test compound in the water phase was reduced, thereby reducing the toxicity of the ANEO to the microorganisms in the inoculum of the closed-bottle tests. ANEO containing 15 polyoxyethylene groups were not ready biodegradable in this test. Only 42% and 23% were degraded after 28 days for tallowamine EO_{15} and oleylamine EO_{15}. These results are inconsistent with what is anticipated, possibly because the adsorption and biodegradation were not separated in the experiment.

Investigating the effect of adsorption on degradation (CO_2 evolution), the highest extent of degradation was generally associated with soils having the lowest adsorption capacity for the nonionic surfactants [69]. Testing the biodegradation in 11 different soils, rapid mineralization without any lag period was observed for $C_{12}EO_{8-9}$ [70]. The soils consist of a wide range of different soil types (from sand, and loamy sand, and loam to silt). Investigating the influence of recent climatic events, i.e., cycles of wetting and drying, a significant increase in biodegradation was found for most soils when the soil was wetted and dried prior to the mineralization experiments. Degradation extent monitored after 38 days varied with soil type: 61% (sandy soil), 46–54% (loamy sand), 43% (sandy loam), 54–60% (loam), and 30–69% (silt loam). The same group of scientists has investigated the effect of soil depths on the mineralization of an AEO [71]. In the topsoil profile (above 2.5 m), $C_{12}EO_{8-9}$ degraded rapidly, while the degradation decreased dramatically below this level in the vadose zone. The observed degradation levels after 32 days were 35%, 22%, and 44% in soil collected at 1.89-, 8.2-, and 19.1-m depths, respectively.

Different soil constituents added to the soil–water mixture were found to influence the degradation rate and extent [72]. Initial mineralization rates for minerals (montmorillonite, kaolinite, illite, and sand) -bound AEO were $k_1 = 0.33 \pm 0.5$ d^{-1} and when bound to humic and fulvic acid $k_1 = 0.15$ and 0.09 d^{-1}, which are much slower than when no medium was added to the test system ($k_1 = 0.68$ d^{-1}). Binding of AEO to montmorillonite and fulvic acid seems more irreversible than binding to kaolinite, illite, sand, or humic acid. AEO sorbed to montmorillonite and fulvic acid were mineralized to a lower extent.

Comparing biodegradation of a $C_{12}EO_{8,9}$ in two subsurface sediments, we find that the sediments were obtained, respectively, from a control pond and a laundromat wastewater pond [73]. Degradation rate was more rapid in the laundromat (0.08–0.25) than in the control (0.01–0.20) ponds, which indicates that adaptation plays an important role in the mineralization of AEO. Federle and co-workers [74] used the same ponds to study mineralization of a ^{14}C-labeled $C_{12}EO_{8-9}$ by the microbiota of submerged plant detritus (submerged oak leaves). Only for the control pond was a lag period observed before degradation started, so prior exposure to surfactants seems to have a major effect on the ability of the detrital microorganisms to mineralize the AEO. Degradation levels after 56 day were 56% and 60%, respectively, for the control and the laundromat wastewater pond.

Surfactants might also become adsorbed on microorganisms [65]. Within 10 minutes after mixing, 73%, 32%, and 15% of the stearyl–AEO with EO_5, EO_{10}, and

EO_{17} were adsorbed to the inoculum consisting of a microorganism culture. The adsorption processes took place before any ultimate biodegradation was observed. Neither the presence of sterilized activated sludge nor anionic surfactants in the water media inhibited biodegradation of, respectively, a $C_{12}EO_{10}$ and a stearyl-AEO [65,75]. In fact, the presence of the anionic surfactants enhanced the extent of biodegradation of the AEO.

C. Ultimate Biodegradation in Waters

Adaptation was not found to be very important in affecting the degradation rates of an AEO using groundwater from an up-gradient and a down-gradient well in the vicinity of a field exposed to infiltrating wastewater [71]. Native microbes present in a soil/groundwater system simulating a shallow subsurface system have been applied in a microcosm experiment to study the biodegradation of a $C_{10-12}EO_5$ [76]. The study demonstrated that the rate of degradation was improved by adding nutrients, especially nitrogen and phosphorous.

The degradation of two AEO at environmentally relevant trace concentrations was investigated in estuarine water [77]. The average half-life for ultimate biodegradation was 2.3 days for the $*C_{16}EO_3$ (^{14}C-labeled in the alkyl chain), and the rate constant was independent of the concentration in the examined concentration range from 0.85 to 68 µg/L. The kinetic degradation of $C_{12}*EO_9$ (^{14}C-labeled in the ethoxy chain) was more complex. The half-life for ultimate biodegradation was 5.8 days at the concentration 0.42 µg/L, while at 3.9 µg/L the enzymatic activity approached saturation and much longer half-life was observed. The degradation pathway can explain the difference in degradation rate. After the ether cleavage, the degradation of the two chains diverges; therefore the observed difference in rate and enzyme saturation is observed.

Biodegradation of four different ^{14}C-labeled AEO has been examined in river water by measuring the $^{14}CO_2$ evolution in a closed-shake-flask system [78]. Only a little influence of the ethoxylate chain length on the degradation was indicated. Additionally, the rate constants were independent of the concentration when the concentration varied over a factor of 10 or 2 (from 1–10 or 50–100 µg/L). Degradation tests of an $C_{11.5}EO_{6.5}$ in river water indicated that degradation is proportionately enhanced at low concentrations in the ppb range [79]. Increased degradation rates followed increasing temperature in the range of 3–34°C. Inconsistent with these results, Ref. 80 found that mineralization of an AEO was unchanged or increased gradually during seasonal decreases in temperature from 26 to 2°C in a stream mesocosm study.

D. Ultimate Anaerobic Biodegradation

A $C_{10}EO_8$ was found to be readily biodegradable applying both DOC (modified OECD screening test) and BOD (closed-bottle test) using secondary effluent as source of inoculum [81]. Furthermore, this group demonstrated the importance of optimizing the test conditions in biodegradation tests in order to reduce sorption of the test compound to substrate and microorganisms and to avoid toxic effects on bacteria caused by the test compound. In the anaerobic biodegradation tests of the AEO, a lag period of 42 days was observed prior to observation of biodegradation, while after 56 days 47–82% of the theoretical biodegradation was achieved. In other investigations of anaerobic degradation using cultures inoculated with anoxic sewage sludge for three different AEO, $C_{12}EO_3$, $C_{12}EO_{23}$, and $C_{18}EO_{20}$, the ethoxylate moiety degraded while the respective fatty acids were not degraded further [82]. In an anaerobic study using a model digester, more than 80% of a ^{14}C-labeled stearyl-AEO was degraded to CH_4 and CO_2 within 4 weeks, while between 10% and 15% of the ^{14}C-labeled stearyl-AEO was sorbed to and/or incorporated into the anaerobic sludge [83].

Other studies have performed anaerobic biodegradation tests using different types of sediments. Thirteen% and 24–40% of a $C_{12}EO_{8-9}$ were degraded to CH_4 and CO_2 after 87 days when anaerobic sediments from control and wastewater ponds were used as test medium [84]. Higher extents of degradation (30–75%) were achieved for $C_{9-11}EO_8$ with marine sediments in a 56-day anaerobic biodegradation test monitoring CH_4 levels [85]. Applying the same test samples and test conditions except from the test medium that was replaced by freshwater swamp, more than 75% was degraded [85].

VI. TOXIC EFFECTS OF THE AOE AND ANEO USED AS SURFACTANT ADJUVANTS

In the past, a variety of test procedures was used to investigate the toxicity of chemicals. These procedures vary in test organisms, life stages of the test organism, test duration, and, thereby, endpoints. Tables 4 and 5 review the available literature published after 1995 on the acute and chronic toxicity data presented of the study compounds. Several species covering different trophic levels are included. Both direct and indirect effects causing disorder in the food chain might be consequences of toxicity. Examples of indirect effects may be changes in the soil surface due to adsorption of surfactants, effects on leaching, mobility, and persistence of pesticides and other pollutants in the soil as well as on mixture toxicity. More toxicity data on AEO is found in the literature than on ANEO. The pH of the test media might influence the toxicity of ANEO, since these compounds dissociate between a protonated cationic and a deprotonated neutral form. As seen for other chemicals, this feature may be important for the toxicity [86].

A. Single-Species Level

1. Cellular Targets

Cserhati [87] reviewed available data concerning the interaction of AEO with bioactive compounds and biological effects. AEO have been found to interact with proteins and membrane phospholipids by modifying their structure and permeability. These interactions may cause harmful or beneficial effects on the activity of various enzymes in biological systems.

2. Microorganisms

The ability of AEO to inhibit the mobility of protozean *Tetrahymena elliotti* was studied [88]. The minimal inhibitory concentration (MIC), which is the lowest concentration that inhibits movement of the protozean in cell suspension, was applied as the model for the surfactant–membrane relationships. The study was based on

TABLE 4 Acute Toxicity Data for AEO and ANEO (Published After 1995)

Compound[a]	Test organisms	Acute toxicity[b]	Test method and test period	Ref.
AEO				
Algae				
$C_{10}EO_{7-8}$ [br]	*Scenedesmus subspicatus CHODAT*	IC_{50} = 50 mg/L	Screening test (DIN 38412), 72 h	101
$C_{11-12}EO_{7-8}$ [br]	*S. subspicatus CHODAT*	IC_{50} = 5 mg/L	Screening test (DIN 38412), 72 h	101
$C_{12-14}EO_{7-8}$	*S. subspicatus CHODAT*	IC_{50} = 0.5 mg/L	Screening test (DIN 38412), 72 h	101
$C_{12-15}EO_7$	*Kirchneria subcapitata*	EC_{50} = 0.85 mg/L	Static toxicity test (ISO 8692), 72 h	91
$C_{12-15}EO_x$	*Selenastrum capricornutum*	EC_{50} = 0.5 mg/L	Static renewal test (ISO 8692), 72 h	118
$C_{13}EO_{7-8}$ [l]	*S. subspicatus CHODAT*	IC_{50} = 0.5 mg/L	Screening test (DIN 38412), 72 h	101
$C_{13}EO_{7-8}$ [br]	*S. subspicatus CHODAT*	IC_{50} = 0.5–5 mg/L	Screening test (DIN 38412), 72 h	101
$C_{13-15}EO_{7-8}$ [l/br]	*S. subspicatus CHODAT*	IC_{50} = 0.5 mg/L	Screening test (DIN 38412), 72 h	101
$C_{15}EO_{7-8}$ [l/br]	*S. subspicatus CHODAT*	IC_{50} = 0.05 mg/L	Screening test (DIN 38412), 72 h	101
Rotifers				
$C_{9-11}EO_{5,8}$ [br]	*Brachionus calyciflorus*	LC_{50} = 83.2–89.1 µM	Standard operational procedure by EPA	119
$C_{10}EO_{5-8}$ [hbr]	*B. calyciflorus*	LC_{50} = 166.0–239.9 µM	Standard operational procedure by EPA	119
$C_{11}EO_5$ [br]	*B. calyciflorus*	LC_{50} = 56.2–128.8 µM	Standard operational procedure by EPA	119
$C_{12}EO_{8,9,11}$ [br]	*B. calyciflorus*	LC_{50} = 63.1–69.2 µM	Standard operational procedure by EPA	119
$C_{12-14}EO_6$ [l]	*B. calyciflorus*	LC_{50} = 12.6–12.9 µM	Standard operational procedure by EPA	119
$C_{12-14}EO_{6,9,11}$ [br]	*B. calyciflorus*	LC_{50} = 21.9–22.9 µM	Standard operational procedure by EPA	119
$C_{12-15}EO_7$ [br]	*B. calyciflorus*	LC_{50} = 8.5 µM	Standard operational procedure by EPA	119
$C_{13}EO_6$ [hbr]	*B. calyciflorus*	LC_{50} = 28.2 µM	Standard operational procedure by EPA	119
$C_{13}EO_8$ [hbr]	*B. calyciflorus*	LC_{50} = 21.4–25.7 µM	Standard operational procedure by EPA	119
$C_{13}EO_{10}$ [hbr]	*B. calyciflorus*	LC_{50} = 18.6–25.7 µM	Standard operational procedure by EPA	119
$C_{14-15}EO_{11}$ [br]	*B. calyciflorus*	LC_{50} = 6.5 µM	Standard operational procedure by EPA	119
$C_{16-18}EO_{10}$ [l]	*B. calyciflorus*	LC_{50} = 2.1–2.3 µM	Standard operational procedure by EPA	119
$C_{16-18}EO_{10}$ [l, uns]	*B. calyciflorus*	LC_{50} = 3.3 µM	Standard operational procedure by EPA	119
$C_{18}EO_7$ [l, uns]	*B. calyciflorus*	LC_{50} = 2.8 µM	Standard operational procedure by EPA	119
$C_{18}EO_9$ [l, uns]	*B. calyciflorus*	LC_{50} = 2.0 µM	Standard operational procedure by EPA	119

Compound	Species	Value	Test method	Ref.
$C_{18}EO_{11}$ [l, uns]	B. calyciflorus	LC_{50} = 2.8 μM	Standard operational procedure by EPA	119
Crustaceans				
$C_{9-11}EO_5$ [br]	Thamnocephalus platyurus (freshwater shrimp)	LC_{50} = 17.8 μM	Standard operational procedure by EPA	119
$C_{9-11}EO_6$	Daphnia magna	EC_{50} = 5.3 mg/L	Static renewal test (USEPA-TSCA/ASTM), 48h	98
$C_{9-11}EO_6$	Hyalella azteca (amphipod)	LC_{50} = 14 mg/L	10-d flow-through laboratory exposure test	106
$C_{9-11}EO_6$	H. azteca	LC_{50} = 9.1 mg/L	10-d stream mesocosms	106
$C_{9-11}EO_8$ [br]	T. platyurus	LC_{50} = 30.9 μM	Standard operational procedure by EPA	119
$C_{10}EO_4$	Mysidopsis bahia (estuarinic)	LC_{50} = 5.5 mg/L	Static renewal test, 48 h	120
$C_{10}EO_{5-8}$ [hbr]	T. platyurus	LC_{50} = 125.9–166.0 μM	Standard operational procedure by EPA	119
$C_{10}EO_{7-8}$ [br]	D. magna	LC_{50} = 50 mg/L	Screening test (EG-Richtlinie 79/381/EWG), 48 h	101
$C_{11}EO_7$	D. magna	EC_{50} = 2.1 mg/L	Static renewal test (USEPA-TSCA/ASTM), 48 h	98
$C_{11}EO_{7-8}$ Oxo	D. magna	LC_{50} = 5 mg/L	Screening test (EG-Richtlinie 79/381/EWG), 48 h	101
$C_{11}EO_5$ [br]	T. platyurus	LC_{50} = 9.1–58.9 μM	Standard operational procedure by EPA	119
$C_{11}EO_9$	D. magna	EC_{50} = 6.7 mg/L	Static renewal test (USEPA-TSCA/ASTM), 48 h	98
$C_{11-12}EO_{7-8}$ [br]	D. magna	LC_{50} = 5 mg/L	Screening test (EG-Richtlinie 79/381/EWG), 48 h	101
$C_{12}EO_{8,9,11}$ [br]	T. platyurus	LC_{50} = 29.5–51.3 μM	Standard operational procedure by EPA	119
$C_{12-13}EO_{4.5-6}$	D. magna	EC_{50} = 0.59 mg/L	Static renewal test (USEPA-TSCA/ASTM), 48 h	98
$C_{12-13}EO_5$	D. magna	EC_{50} = 0.46 mg/L	Static renewal test (USEPA-TSCA/ASTM), 48 h	98
$C_{12-13}EO_{6.5}$	D. magna	EC_{50} = 0.74 mg/L	Static renewal test (USEPA-TSCA/ASTM), 48 h	98
$C_{12-14}EO_6$ [l]	T. platyurus	LC_{50} = 3.31–3.89 μM	Standard operational procedure by EPA	119
$C_{12-14}EO_{6,9,11}$ [br]	T. platyurus	LC_{50} = 12.3–13.8 μM	Standard operational procedure by EPA	119
$C_{12-14}EO_{7-8}$	D. magna	LC_{50} = 0.5 mg/L	Screening test (EG-Richtlinie 79/381/EWG), 48 h	101
$C_{12-15}EO_7$	D. magna	EC_{50} = 1.0–2.0 mg/L	Static toxicity test ISO 6341, 48 h	91
$C_{12-15}EO_7$ [br]	T. platyurus	LC_{50} = 3.5 μM	Standard operational procedure by EPA	119
$C_{12-15}EO_{12}$	D. magna	EC_{50} = 1.4 mg/L	Static renewal test (USEPA-TSCA/ASTM), 48 h	98
$C_{13}EO_6$ [hbr]	T. platyurus	LC_{50} = 20.9 μM	Standard operational procedure by EPA	119
$C_{13}EO_{7-8}$ [l]	D. magna	LC_{50} = 0.5 mg/L	Screening test (EG-Richtlinie 79/381/EWG), 48 h	101
$C_{13}EO_{7-8}$ [br]	D. magna	LC_{50} = 0.5–5 mg/L	Screening test (EG-Richtlinie 79/381/EWG), 48 h	101
$C_{13}EO_8$ [hbr]	T. platyurus	LC_{50} = 12.8–19.5 μM	Standard operational procedure by EPA	119
$C_{13}EO_{9.75}$	M. bahia	LC_{50} = 2.2 mg/L	Static renewal test, 48 h	120

TABLE 4 Continued

Compound[a]	Test organisms	Acute toxicity[b]	Test method and test period	Ref.
$C_{13}EO_{10}$ [hbr]	T. platyurus	$LC_{50} = 16.6–26.3$ μM	Standard operational procedure by EPA	119
$C_{13–14}EO_{7–8}$ [br]	D. magna	$LC_{50} = 0.5$ mg/L	Screening test (EG-Richtlinie 79/381/EWG), 48 h	101
$C_{13–15}EO_{7–8}$ [l]	D. magna	$LC_{50} = 0.5$ mg/L	Screening test (EG-Richtlinie 79/381/EWG), 48 h	101
$C_{13–15}EO_{7–8}$ Oxo	D. magna	$LC_{50} = 0.5$ mg/L	Screening test (EG-Richtlinie 79/381/EWG), 48 h	101
$C_{14}EO_1$	D. magna	$LC_{50} = 0.83$ ppm	Based on USEPA guidelines, 48-h	101
$C_{14–15}EO_{13}$	D. magna	$EC_{50} = 1.2$ mg/L	Static renewal test (USEPA-TSCA/ASTM), 48 h	98
$C_{14–15}EO_{11}$ [br]	T. platyurus	$LC_{50} = 3.7$ μM	Standard operational procedure by EPA	119
$C_{15}EO_{7–8}$ [l]	D. magna	$LC_{50} = 0.5$ mg/L	Screening test (EG-Richtlinie 79/381/EWG), 48 h	101
$C_{15}EO_{7–8}$ [br]	D. magna	$LC_{50} = 0.5$ mg/L	Screening test (EG-Richtlinie 79/381/EWG), 48 h	101
$C_{16–18}EO_{10}$ [l]	T. platyurus	$LC_{50} = 1.6–2.3$ μM	Standard operational procedure by EPA	119
$C_{16–18}EO_{10}$ [l, uns]	T. platyurus	$LC_{50} = 3.2$ μM	Standard operational procedure by EPA	119
$C_{18}EO_{7,9,11}$ [l, uns]	T. platyurus	$LC_{50} = 1.9–2.8$ μM	Standard operational procedure by EPA	119
Insects				
$C_{9–11}EO_6$	Chironomus tentans (midge)	$LC_{50} = 5.7$ mg/L	10-d flow-through laboratory exposure test	106
Fish				
$C_{9–11}EO_6$	Lepomis macrochirus	$LC_{50} = 6.3–6.77$ mg/L	30-d stream mesocosms	103,106
$C_{9–11}EO_6$	Pimephales promelas	$LC_{50} = 2.7$ mg/L	10-d flow-through laboratory exposure test	106
$C_{9–11}EO_6$	P. promelas	$LC_{50} = 5.5$ mg/L	30-d stream mesocosms	103,106
$C_{9–11}EO_6$	P. promelas	$LC_{50} = 8.5$ mg/L	Static renewal test (USEPA-TSCA/ASTM), 96 h	98

Compound	Organism	Value	Test	
$C_{9-11}EO_6$	*P. promelas* (fry, fry survival)	LC_{50} = 4.87 mg/L	28-d laboratory flow-through early-life-stage test	105
$C_{9-11}EO_6$	*P. promelas* (juveniles)	LC_{50} = 6.42 mg/L	10-d stream mesocosms	103
$C_{9-11}EO_6$	*P. promelas* (larvae)	LC_{50} = 2.65 mg/L	10-d flow-through laboratory exposure test	103
$C_{11}EO_7$	*P. promelas*	LC_{50} = 3.9 mg/L	Static renewal test (USEPA-TSCA/ASTM), 96 h	98
$C_{11}EO_9$	*P. promelas*	LC_{50} = 7.1 mg/L	Static renewal test (USEPA-TSCA/ASTM), 96 h	98
$C_{12-13}EO_{4.5-6}$	*P. promelas*	LC_{50} = 0.96 mg/L	Static renewal test (USEPA-TSCA/ASTM), 96 h	98
$C_{12-13}EO_5$	*P. promelas*	LC_{50} = 1.0 mg/L	Static renewal test (USEPA-TSCA/ASTM), 96 h	98
$C_{12-13}EO_{6.5}$	*L. macrochirus*	LC_{50} = 1.30 mg/L	30-d stream mesocosms	107
$C_{12-13}EO_{6.5}$	*P. promelas*	LC_{50} = 1.27 mg/L	30-d stream mesocosms	107
$C_{12-13}EO_{6.5}$	*P. promelas*	LC_{50} = 1.3 mg/L	Static renewal test (USEPA-TSCA/ASTM), 96 h	98
$C_{12-13}EO_{6.5}$	*P. promelas* (fry, fry survival)	LC_{50} = 2.39 mg/L	28-d laboratory flow-through early-life-stage test	105
$C_{12-15}EO_7$	*Brachdanio rerio*	LC_{50} = 1.0–2.0 mg/L	Static toxicity test (OECD test guidelines), 96 h	91
$C_{12-15}EO_{12}$	*P. promelas*	LC_{50} = 1.4 mg/L	Static renewal test (USEPA-TSCA/ASTM), 96 h	98
$C_{14-15}EO_7$	*L. macrochirus*	LC_{50} = 0.65; 0.56 mg/L	Flow-through laboratory exposure test, 96 h; 10 d	104
$C_{14-15}EO_7$	*P. promelas*	LC_{50} = 0.77; 0.69 mg/L	Flow-through laboratory exposure test, 96 h; 10 d	104
$C_{14-15}EO_7$	*P. promelas* (fry, fry survival)	LC_{50} = 1.02 mg/L	28-d laboratory flow-through early-life-stage test	105
$C_{14-15}EO_{13}$	*P. promelas*	LC_{50} = 1.0 mg/L	Static renewal test (USEPA-TSCA/ASTM), 96 h	98
ANEOs				
$C_{12-16}NEO_2$	Sludge bacteria	EC_{50} = 21.0 mg/L	Viable plate counting (pour plate method)	121
$C_{16-18}NEO_{16}$	Sludge bacteria	EC_{50} = 21.0 mg/L	Viable plate counting (pour plate method)	121

[a] br: branched; hbr: highly branched; l: linear; uns: unsaturated.

[b] EC_{50}: concentration at which 50% of the organisms are effected; IC_{50}: concentration with 50% cell reproduction; LC_0: concentration at which no effect can be seen; LC_{50}: concentration at which 50% of the organisms die.

TABLE 5 Chronic Toxicity Data for AEO and ANEO (Published After 1995)

Compound	Test organisms	Chronic toxicity (mg/L)		Test method and test period	Ref.
		NOEC	LOEC		
AEO					
Algae					
$C_{9-11}EO_6$	*Periphyton* (58 taxa)	5.7		30-d stream mesocosms	106
$C_{12-15}EO_7$	*K. subcapitata*	1.0		Static toxicity test (ISO 8692), 72 h	91
$C_{12-18}EO_9$	*S. subspicatus* (growth)	0.2	1.0	EG92/69 procedures, 72 h	122
$C_{13}EO_8$	*S. subspicatus* growth	0.32	1.0	EG92/69 procedures, 72 h	122
$C_{13-15}EO_7$	*S. subspicatus* growth	0.07	0.25	EG92/69 procedures, 72 h	122
$C_{14-15}EO_8$	*S. subspicatus* growth	0.48	0.15	EG92/69 procedures, 72 h	122
Crustaceans					
$C_{9-11}EO_6$	Copepoda and cladocera	2.0	4.35	30-d stream mesocosms	106,109
$C_{9-11}EO_6$	*D. magna*	2.77	5.57	Flow-through laboratory test	107
$C_{9-11}EO_6$	*H. azteca* (amphipod)	2.0		10-d stream mesocosms	106
$C_{12-13}EO_{6.5}$	Copepoda and cladocera	1.99	5.15	30-d stream mesocosms	107,110
$C_{12-13}EO_{6.5}$	*D. magna*	1.75	2.69	Flow-through laboratory test	107
$C_{12-18}EO_9$	*D. magna* (reproduction)	2.9	9.6	OECD 202 procedures, 21 d	122
$C_{13}EO_8$	*D. magna* (reproduction)	0.45	0.93	OECD 202 procedures, 21 d	122
$C_{13-15}EO_7$	*D. magna* (reproduction)	0.2	0.66	OECD 202 procedures, 21 d	122
$C_{14-15}EO_7$	*D. magna*	0.79	1.02	Flow-through laboratory test	107
$C_{14-15}EO_8$	*D. magna* (reproduction)	0.28	0.93	OECD 202 procedures, 21 d	122
Insects					
$C_{12-13}EO_{6.5}$	Simuliidae	0.32	0.32	30-d stream mesocosms	107,110
$C_{14-15}EO_7$	Simuliidae	0.08	0.16	28-d stream mesocosms	108

		NOEC	LOEC		
Fish					
$C_{9-11}EO_6$	*L. macrochirus* (survival, growth)	5.7	11.2	30-d stream mesocosms	106
$C_{9-11}EO_6$	*P. promelas*	1.01	3.15	Flow-through laboratory test	107
$C_{9-11}EO_6$	*P. promelas*	0.73–2.0	4.4	30-d stream mesocosms	103,106
$C_{9-11}EO_6$	*P. promelas*	1	3	10-d flow-through laboratory exposure test	103
$C_{9-11}EO_6$	*P. promelas*	1.01–10.27	1.01–10.27	28-d laboratory flow-through early-life-stage test	105
$C_{9-11}EO_6$	*P. promelas* (juvenile survival)	4.35	5.7	10-d stream mesocosms	103
$C_{12-13}EO_{6.5}$	*L. macrochirus* (survival; growth)	0.88; 0.88	1.99, > 0.88	30-d stream mesocosms	107
$C_{12-13}EO_{6.5}$	*P. promelas*	1.76	> 1.76	Flow-through laboratory test	107
$C_{12-13}EO_{6.5}$	*P. promelas*	1.76–8.06	> 1.76–8.06	28-d laboratory flow-through early-life-stage test	105
$C_{12-13}EO_{6.5}$	*P. promelas* (survival)	0.88	1.99	30-d stream mesocosms	107
$C_{12-18}EO_9$	*O. mykiss* (growth)	0.25	1.0	OECD 204 procedures, 28 d	122
$C_{13}EO_8$	*O. mykiss* (growth)	0.6	2.0	OECD 204 procedures, 28 d	122
$C_{13-15}EO_7$	*O. mykiss* (growth)	0.17	0.54	OECD 204 procedures, 28 d	122
$C_{14-15}EO_7$	*L. macrochirus* (survival)	0.16	0.46	10-d flow-through laboratory exposure test	104
$C_{14-15}EO_7$	*L. macrochirus*	> 0.33		30-d stream mesocosms	104
$C_{14-15}EO_7$	*P. promelas*	0.16–0.74	0.46–1.12	10-d flow-through laboratory exposure test	104,107
$C_{14-15}EO_7$	*P. promelas* (survival)	0.28	0.33	30-d stream mesocosms	104
$C_{14-15}EO_8$	*O. mykiss* (growth)	0.15	0.51	OECD 204 procedures, 28 d	122

NOEC: no-observed-effect concentration; LOEC: lethal-observed-effect concentration.

structure–activity relationships (SARs) and showed that an increase in the ethoxy chain length was associated with a decreasing inhibitory activity. The AEO with $C_{12}EO_n$ were found to be far more powerful inhibitors (MIC: 0.1 mM $C_{12}EO_{4,10}$; 1.2 mM $C_{12}EO_{20}$) than the corresponding AEO with 4, 8, 16, or 18 carbons in their alkyl chain (MICs range from 6.6 mM C_4EO_{20} to >128.9 mM $C_{18}EO_{20}$). In a 5-min Microtox® toxicity test using *Photobacterium phosphoreum*, EC_{50} were 8.1 and 11.4 mg/L for the branched $C_{12-15}EO_7$ and $C_{13}EO_7$, respectively, whereas EC_{50} was 1.5 mg/L for the linear $C_{12-15}EO_9$ [89].

3. Phytoplankton

In series of AEO it was found that with increasing hydrophobicity of the test compound, growth of the algae (*Chlamydomonas reinhardi*) decreased [90]. A freshwater microalga (*Kirchneria subcapitata*, previously called *Selenastrum capricornutum*) was used as test organism in toxicity tests comparing branched and linear AEO [89]. Branched AEO were less toxic by a factor of 10 (EC_{50} 7.5 and 10 mg/L for $C_{13}EO_7$ and $C_{12-15}EO_7$, respectively) than the linear ones (EC_{50} 0.7 mg/L for $C_{12-15}EO_9$). Similar toxicity levels (EC_{50} 0.9 mg/L) were obtained for a linear primary $C_{12-15}EO_7$ in a chronic static toxicity test (NOEC 1.0 mg/L) using the same test organism [91]. Comparable EC_{50} values (4.1–4.9 mg/L) for *S. capricornutum* were obtained with ANEO ($C_{14-18}NEO_{15}$) [92].

Comparing toxicity levels of $C_{14-15}EO_6$ to three individual algae (*S. capricornutum, Microcystis aeruginosa,* and *Navicula pelliculosa*), it was found that EC_{50} values (0.1–0.6 mg/L) were lower than the EC_{50} of lake photosynthesis studies (EC_{50} 2.1 ± 2.0 mg/L) [93]. Furthermore, testing the three freshwater planktonic algae species (*S. capricornutum, M. aeruginosa,* and *Nitzschia fonticola*) to $C_{12-14}EO_9$, it was found that *S. capricornutum* was the most sensitive, having an EC_{50} of 4–8 mg/L, which is why this green alga was recommended for standard algae tests [94].

In a stream mesocosm study assessing the effects of a $C_{14-15}EO_7$, no effects on the periphyton and macrophyte community were observed at concentrations up to 550 μg/L [95].

4. Plants

Root dry weight of barley plants (*Hordeum vulgare* L.) was significantly reduced and the porosity of the roots was reduced by the presence of an AEO (at concentrations above 40 ppm) [96]. At 100 ppm, the root dry weight decreased to 32% of the corresponding weight of the control plants.

5. Crustaceans

Several studies have tried to relate toxicity on crustaceans and the structural characteristics of AEO. With increasing ethoxylate chain length of a series of AEO, the acute toxicity to *Daphnia magna* [97,98] and to *Mysidopsis bahia* [99] decreased. A similar trend was observed for ANEO, for which the EC_{50} of daphnia increased from 1 to 10–30 mg/L for a C_xNEO_2 and a C_xNEO_{12-25}, respectively [100]. Linear AEO have been shown to be both more acutely and chronically toxic than branched AEO in *D. magna* toxicity tests [58,89], as indicated by these data: LC_{50} 1.3 mg/L, NOEC and LOEC 1.0 mg/l for the linear $C_{12-15}EO_9$, contra LC_{50} 11.6 mg/L, NOEC and LOEC 4.0 mg/L for the branched $C_{13}EO_7$ [58] in a 48-h toxicity test. Similar results of increasing toxicity with decreasing branching and increasing alkyl chain length have been confirmed by others [101]. By moving the branching from near the terminal end of the alkyl chain to the middle of the alkyl chain, the toxicity decreased. The overall results show that the more hydrophobic AEO are generally more toxic than the less hydrophobic ones.

6. Fish

Several toxicity tests of AEO have been performed using different test organisms and AEO (see Tables 4 and 5). LC_{50} values of one AEO and three ANEOs were determined for salmon (*Salmon salar* L.) [102]. LC_{50} (96-h) was 1.5 mg/L for the AEO, in contrast to the slightly lower LC_{50} (0.09–0.78 mg/L) of the ANEO. This study is the only one comparing the toxicity of the two types of nonionic surfactants. Unfortunately, no data on the pH of the test water is given in the bioassay. However, when considering the pk_a values of the ANEO the protonated fraction is the major fraction at typical freshwater pH. As expected from the structure, the ANEO were more toxic than the AEO. Only few papers dealing with the toxicity of ANEO have been found. Schöberl et al. [100] found lower toxicity for fish (LC_{50} 0.7 mg/L for C_xNEO_2 and 5–10 mg/L for C_xNEO_{12-25}) than Wildish [102].

So with the other test organisms, relationships between toxicity and the chemical structure of the AEO were indicated. Linear AEO were more toxic than the branched AEO in *Pimephales promelas* (fathead minnow) toxicity experiments [58,89]. LC_{50} levels of the *P. promelas* in a 7-day static renewal test were 4.6 and 1.3 mg/L for a branched $C_{12-15}EO_7$ and linear $C_{12-15}EO_9$, respectively [89]. The acute toxicity tended to increase with increasing alkyl chain length for fathead minnow [98,103,104], namely, a twofold toxicity increase per two-carbon addition [105]. Concerning the chronic toxicity, the effect concentrations were nearly independent of branching, since NOEC was 1.0 and LOEC 2.0 mg/L for a branched $C_{12-15}EO_7$, whereas the NOEC was 0.4–1.0 and LOEC 1.0–2.0 mg/L for a linear $C_{12-15}EO_9$ [89].

Attempts have been made to evaluate the toxicity data obtained in laboratory experiments with data from mesocosm studies. Fathead minnow were the most sensitive organism toward $C_{9-11}EO_6$ in a stream mesocosm study including periphyton, aquatic macrophytes, benthic invertebrates, insects, and two fish species [106]. With the same test compound and mesocosm facilities used to determine the 30-day survival LC_{50}, bluegill sunfish (LC_{50} 6.8 mg/L) were found less sensitive than fathead minnow (LC_{50} 5.5 mg/L, adult) [103]. Reproduction of the fathead minnow was a more sensitive response than growth or survival. Indication of reproductive effects was also found for $C_{14-15}EO_7$ [104]. Effects on reproduction were reversible, since fish exposed to the test compound showed increased egg production when the test compound was removed [103,106]. Additionally, the toxicity data obtained in the laboratory tests were remarkably similar to the data from the mesocosm study, so the authors suggest that a safety factor may not be necessary when going from the laboratory to the field [104].

Early-life-stage toxicity of fathead minnow exposed to three different AEO ($C_{9-11}EO_6$, $C_{12-13}EO_{6.5}$, and $C_{14-15}EO_7$) was determined in a 28-day flow-through laboratory toxicity test [105]. The results obtained for all three AEO were within a factor of 2–5 of the NOEC found for the respective AEOs in 10-day laboratory toxicity tests and 30-day stream mesocosm experiments [103,104,107]. Due to the similarity in the toxicity endpoints from the exposure experiments in laboratory and stream mesocosm, it might be possible to use laboratory data to predict "safe" environmental concentrations of these AEOs for fish [105].

7. Insects

Effects of different AEO on the aquatic invertebrates were tested in stream mesocosm studies [108–110]. The results collectively supported a hypothesis of increasing toxicity with increasing alkyl chain length of the AEO. Investigating the effect of the C_{14-15} EO_7 on more than 30 invertebrate taxa in a stream mesocosm, only for Simuliidae were negative effects seen at concentrations lower than 0.550 mg/L [95].

8. Earthworms

Three different species of earthworms have been tested for the effect of the $C_{14-15}EO_7$ [111]. LC_{50} values were, respectively, 1.0, 2.8, and 6.8 mg/L for *Dugesia* sp. (flatworm), *Dero* sp. (oligochaete), and *Rhabditis* sp. (nematode). These data are the only available on soil organisms.

B. Ecosystem Level

Aquatic microecosystem studies have shown that most ecosystems are triggered by lower concentrations of the AEO than observed for the classical chronic single-species tests [112]. In an allochthonous biocenosis, consisting of algae, protozoa, and small multiple-cell organisms, with a flow-through system over 3–5 weeks, the acute toxicity EC_0, ranged from 0.3 to 5 mg/L for three different AEO. For the corresponding single-species toxicity test of 24–96 hours, EC_0 was 0.3–30 mg/L for algae, LC_0 was 2–700 mg/L for *D. magna*, and LC_0 was 2.5–100 mg/L (48 h) for fish [112].

It is important to correlate the toxicity and biodegradation of the surfactants. The lower aquatic toxicity shown by the branched AEO is offset by their slower biodegradability; this makes these more harmful to ecosystems than the AEO with more linear hydrophobic chains [58]. Similarly, studies investigating the relationship between toxicity and biodegradation have been made with ANEO [68,92]. Here, the toxicity was coupled to a biodegradation test by investigating the toxicity as acute toxicity to *D. magna* and growth of the alga *S. capricornutum* during biodegradation. It was found that the initial toxicity of the parent compound was not observed for the water-soluble intermediate degradation products, which were nontoxic to both algae and *D. magna*. Thus, loss of toxicity correlates with loss of surfactant properties when the ANEO is cleaved. Furthermore, ANEO with a fewer number of ethoxylate groups were more toxic but were biodegraded faster than the ones with longer ethoxylate chains.

Response addition has been found to be the best model to predict toxicity of mixtures containing nonionic as well as anionic and cationic surfactants [113]. The response addition model assumes that each component in a mixture affects a different target organ of the organism or has a different mode of toxic action. In another study, the toxicity of mixtures has been investigated with mixtures of surfactants and mixtures of pesticides and surfactants [114]. In their study, the response addition model underestimated the toxicity, while concentration addition provides a balance of over- and underestimated mixture toxicity. The best predictions of combination effects of mixtures of pesticides and surfactants should therefore be obtained using concentration addition.

VII. DISCUSSION

Based on the comprehensive overview given in Krogh et al. [8], a preliminary attempt to make an environmental hazard assessment for AEO and ANEO was published [17]. As seen for Denmark, it has been possible to eliminate the APEO from both the pesticide formulation and spray additives. Consequently, since the reformulation of the pesticide products on the Danish market is already done, it should be possible for other countries to follow the example of Denmark and phase out these environmentally critical substances. Due to the fact that APEO have successfully been phased out of the pesticide formulations in Denmark, due to the environmental hazards of these compounds and, in particular the degradation products, these compounds are discussed only very briefly.

The environmental legislation of such compounds is indirectly included in EU directive 91/414/EEC. Here, environmental requirements, e.g., degradation and toxicity data, are given for the active ingredients and the pesticide formulations, while no specific requirements are set for the adjuvants.

Both AEO and ANEO are used as technical mixtures. This implies that they are not a single compound but a whole range of compounds present in different ratios with their own physicochemical behavior. Structurally both groups of substances have a mutual core with side chains of varying lengths (see Fig. 1). This means that each of these single compounds (homologues), besides having the overall ability to distribute between different phases, also possesses some single-compound behavior. Each of these single compounds has its own physicochemical behavior. Consequently, environmental assessment of such compounds is difficult and challenging because each of the single substances, even though present in a mixture in the pesticide formulation, will have slightly different behavior in the environment. This dilemma is reflected not only in the parameters describing the fate, e.g., distribution coefficient, leaching, runoff, adsorption to soil, degradation, and effects of these substances, but also in analyzing the substances in often-difficult matrices, such as soil, sediment, and sludge, because the ones with very short side chains will behave differently than the ones with longer side chains.

Adjuvants end up in the environment either after local application on agricultural fields or from point sources. The latter are, e.g., connected to the pesticide handling (preparation of spraying mixture and cleaning of spraying equipment) in the farmyard. Further, short-range drift of the pesticide under application may lead to minor distribution in the environment, while long-distance transport is not expected for these compounds.

For the ANEO (pK_a: 6–7), pH is also a critical factor and influences the fate of the compounds. At low pH the nitrogen of the ANEO become protonated, while at higher pH the compounds are neutral. The fate of the compound is thus matrix dependent. AEO in contrast have no pH-dependent forms.

Data on occurrence obtained from monitoring studies show that ANEO were not detected, whereas AEO were found in the ng/L range in the aquatic environment. Contrary to the aquatic data, only ANEO at concentrations as high as 218–524 µg/kg were detected, while AEO were not found in the terrestrial samples.

Considering ANEO, ultimate degradation half-lives are on the order of 20–40 days or even more in sludge and sewage, depending on the chemical structure. Corresponding data for AEO are around 30–40 days in soil. The degradation rates of AEO are highly dependent on soil type and soil depth. Likewise, ANEO degradation is expected to be matrix dependent because of the pH-related tendency to sorb and complex to soil and sediment matrices. However, compared to persistent organic pollutants (POPs), the AEO and ANEO should be considered easily degraded. This, together with the results of the monitoring studies, supports the fact that these compounds either become adsorbed to the soil matrix or are degraded. Further, these chemicals are not very mobile and therefore are not expected to leach through the soil column to the groundwater.

Compared to POPs, these compounds are generally relatively nontoxic to both aquatic and terrestrial organisms. EC_{50} values are in the low mg/L range for both ANEO and AEO. Toxicity generally increases with a decrease in the number of ethoxy groups and an increasing alkyl chain length. Important factors to investigate include if the combined toxicity of the surfactants, with the pesticide, follow concentration or response addition models. Considering the adsorption behavior of these compounds as well as the data published so far, we believe that environmental concentrations in soil are required to understand the fate of these compounds. More research is needed examining parameters affecting the adsorption as well as desorption of the ANEO to soil and sediments. These investigations for ANEO are highly relevant, considering the rising consumption of glyphosate pesticides, which often contain ANEO. Further, knowledge is needed concerning the influence of surfactants on the mobility of other chemicals in the soil. In total, risk assessment is highly relevant, considering the high volumes used.

REFERENCES

1. Underwood, A.K. In *Adjuvants for Agrochemicals*; Foy, C.L., Ed.; CRC Press: Boca Raton, FL, 1992; 489–501.
2. Foy, C.L. In *Pesticide Formulation and Adjuvant Technology*; Foy, C.L., Pritchard, D.W., Eds.; CRC Press: Boca Raton, FL, 1996; 323–352.
3. Foy, C.L. In *Adjuvants and Agrochemicals*; Chow, P.N.P., Grant, C.A., Hinshalwood, A.M., Simundson, E., Eds.; CRC Press: Boca Raton, FL, 1987; 1–15.
4. Foy, C.L. Pestic. Sci. 1993, *38*, 65–76.
5. Mulqueen, P.J. In *Industrial Applications of Surfactants II*; Karsa, D.R., Ed.; Royal Society of Chemistry, Cambridge, UK: 1990: 276–302.
6. White, G.F. Pestic Sci 1993, *37*, 159–166.
7. Chow, P.N.P.; Grant, C.A. In *Adjuvants and Agrochemicals*; Chow, P.N.P., Grant, C.A., Hinshalwood, A.M., Simundson, E., Eds.; CRC Press: Boca Raton, FL, 1987, 169–170.
8. Krogh, K.A.; Halling-Sørensen, B.; Vejrup, K.V.; Mogensen, B.B. Chemosphere 2003, *50*, 871–901.
9. Talmage, S.S. Environmental and Human Safety of Major Surfactants; Lewis FL, 1994; 3–188.
10. Scott, M.J.; Jones, M.N. Biochim. Biophys. Acta 2000, *1508*, 235–251.
11. Thiele, B.; Günther, K.; Schwuger, M.J. Chem. Rev. 1997, *97*, 3247–3272.
12. Ying, G.G.; Williams, B.; Kookana, R. Environment Int. 2002, *28*, 215–226.
13. Jobling, S.; Sumpter, J.P. Aquat. Toxicol. 1993, *27*, 361–372.
14. White, R.; Jobling, S.; Hoare, S.A.; Sumpter, J.P.; Parker, M.G. Endocrinology 1994, *135*, 175–182.
15. Jobling, S.; Sheahan, D.; Osborne, J.A.; Matthiessen, P.; Sumpter, J.P. Environ. Toxicol. Chem. 1996, *15*, 194–202.

16. Kudsk, P. In *Landbrugets Rådgivningscenter, Landskontoret for Planteavl*; Skejby: Planteværnsafdelingen, 1999.
17. Krogh, K.A. Nonionic surfactants as adjuvants in pesticides—methods of analysis and environmental significance. Ph.D. thesis, National Environmental Research Institute: Denmark, 2003; p.182.
18. Hochberg, E.G. In *Pesticide Formulation and Adjuvant Technology*; Foy, C.L., Pritchard, D.W., Eds.; CRC Press: Boca Raton, FL, 1996; 203–208.
19. Løkke, H. DJF rapport 2000, *23*, 79–85.
20. Boyd, H.B.; Nylén, D.; Pedersen, A.R.; Petersen, G.I.; Simonsen, F. Environmental and health assessment of substances in household detergents and cosmetic detergent products. Miljø- og Energiministeriet;. 2000, 1–36.
21. Ahel, M.; Schaffner, C.; Giger, W. Water Res. 1996, *30*, 37–46.
22. Ahel, M.; Molnar, E.; Ibric, S.; Giger, W. Water Sci. Technol. 2000, *42*, 15–22.
23. La Guardia, M.J.; Hale, R.C.; Harvey, E.; Mainor, T.M. Environ. Sci. Technol. 2001, *35*, 4798–4804.
24. Krogh, K.A.; Vejrup, K.V.; Mogensen, B.B.; Halling-Sørensen, B. J. Chromatogr. A 2002, *957*, 45–57.
25. Matthijs, E.; Holt, M.S.; Kiewiet, A.; Rijs, G.B.J. Environ. Toxicol. Chem. 1999, *18*, 2634–2644.
26. Castillo, M.; Martinez, E.; Ginebreda, A.; Tirapu, L.; Barceló, D. Analyst 2000, *200*, 1733–1739.
27. Petrovic, M.; Barceló, D. Anal. Chem. 2000, *72*, 4560–4567.
28. Krogh, K.A.; Mogensen, B.B.; Halling-Sørensen, B.; Cortés, A.; Vejrup, K.V.; Barceló, D. Anal. Bioanal. Chem. 2003, *376*, 1089–1097.
29. Müller, M.T.; Zehnder, A.J.B.; Escher, B.I. Environ. Toxicol. Chem. 1999, *18*, 2191 2198.
30. Platikanov, D.; Weiss, A.; Lagaly, G. Colloid Polym. Sci. 1977, *255*, 907–915.
31. Ching, Y.; Jafvert, C.T. J. Contam. Hydrol. 1997, *28*, 311–325.
32. Cano, M.L.; Dorn, P.B. Chemosphere 1996, *33*, 981–994.
33. Cano, M.L.; Dorn, P.B. Environ. Toxicol. Chem. 1996, *15*, 684–690.
34. Brownawell, B.J.; Chen, H.; Zhang, W.; Westall, J.C. Environ. Sci. Technol. 1997, *31*, 1735–1741.
35. Podoll, R.T.; Irwin, K.C.; Brendlinger, S. Environ. Sci. Technol. 1987, *21*, 562–568.
36. Law, J.P.; Bloodworth, M.E.; Runkles, J.R. Soil Sci. Soc. Am. Proc. 1966, *30*, 327–332.
37. Kiewiet, A.T.; de Beer, K.G.M.; Parsons, J.R.; Govers, H.A.J. Chemosphere 1996, *32*, 675–680.
38. Urano, K.; Saito, M.; Murata, C. Chemosphere 1984, *13*, 293–300.
39. Krogh, K.A.; Schilder, C.; Vejrup, K.V. Determination of the CMC and pKA values of three alkylamine ethoxylates, 2001. *unpublished work.*
40. Petersen, P.H.; Clowes, L.A. Vand Jord 2000, *4*, 151–155.
41. Miller, W.W.; Valoras, N.; Letey, J. Soil Sci. Soc. Am. Proc. 1975, *39*, 11–16.
42. Valoras, N.; Letey, J.; Osborn, J. Agron. J. 1976, *68*, 591–595.
43. Liu, Z.; Edwards, D.A.; Luthy, R.G. Water Res. 1992, *26*, 1337–1345.
44. Knaebel, D.B. Abstracts of the Annual Meeting of the American Society of Microbiology, Washington, DC, 90; 216–1990.
45. Bayer, D.E.; Foy, C.L. In *Adjuvants for Herbicides*; Hodgson, R.H., Maryland, F., Eds.; Weed Science Society of America: IL, 1982; 84–92.
46. Kuhnt, G. Environ. Toxicol. Chem. 1993, *12*, 1813–1820.
47. Huggenberger, F.; Letey, J.; Farmer, W.J. Proc. Soil Sci. Am. 1973, *37*, 215–219.
48. Aronstein, B.N.; Calvillo, Y.M.; Alexander, M. Environ. Sci. Technol. 1991, *25*, 1728–1731.
49. Laha, S.; Luthy, R.G. Biotechnol. Bioeng. 1992, *40*, 1367–1380.

50. Cesare, D.D.; Smith, J.A. Rev. Environ. Contam. Toxicol. 1994, *134*, 1–29.
51. Sanchez-Camazano, M.; Arienzo, M.; Sanchez-Martin, M.J.; Crisanto, T. Chemosphere 1995, *31*, 3793–3801.
52. Iglesias-Jimenez, E.; Sanchez-Martin, M.J.; Sanchez-Camazano, M. Chemosphere 1996, *32*, 1771–1782.
53. Mata-Sandoval, J.C.; Karns, J.; Torrents, A. J. Argic. Food Chem. 2001, *49*, 3296–33303.
54. Sherrard, K.B.; Marriott, P.J.; Amiet, R.G.; Colton, R.; Mccormick, M.J.; Smith, G.C. Environ. Sci. Technol. 1995, *29*, 2235–2242.
55. Brand, N.; Mailhot, G.; Bolte, M. Chemosphere 2000, *40*, 395–401.
56. Bogan, R.H.; Sawyer, C.N. Sewage and Industrial Wastes. Federation of Sewage and Industrial Wastes Associations, 1955; 917–928.
57. Patterson, S.J.; Scott, C.C.; Tucker, K.B.E. J. Am. Oil Chem. Soc. 1967, *44*, 407–412.
58. Kravetz, L.; Salanitro, J.P.; Dorn, P.B.; Guin, K.F. J. Am. Oil Chem. Soc. 1991, *68*, 610–618.
59. Sturm, R.N. J. Am. Oil Chem. Soc. 1973, *50*, 159–167.
60. Salanitro, J.P.; Diaz, L.A.; Kravetz, L. Chemosphere 1995, *31*, 2827–2837.
61. Tobin, R.S.; Onuska, F.I.; Brownlee, B.G.; Anthony, D.H.J.; Comba, M.E. Water Res. 1976, *10*, 529–535.
62. Patterson, S.J.; Scott, C.C.; Tucker, K.B.E. J. Am. Oil Chem. Soc. 1970, *47*, 37–41.
63. Nooi, J.R.; Testa, M.C.; Willemse, S. Tenside Deterg. 1970, *7*, 61–65.
64. Steber, J.; Wierich, P. Tenside Deterg. 1983, *20*, 183–187.
65. Neufahrt, A.; Lötzsch, K.; Gantz, D. Tenside Deterg. 1982, *19*, 264–268.
66. Zhang, C.L.; Valsaraj, K.T.; Constant, W.D.; Roy, D. J. Environ. Sci. Health Pt A: Toxic/Hazard Subst. Environ. Eng. 1998, *A33*, 1249–1273.
67. Figueroa, L.A.; Miller, J.; Dawson, H.E. Water Environ. Res. 1997, *69*, 1282–1289.
68. van Ginkel, C.G.; Stroo, C.A.; Kroon, A.G.M. Tenside Surfact. Deterg. 1993, *30*, 213–216.
69. Valoras, N.; Letey, J.; Martin, J.P.; Osborn, J. Soil Sci. Soc. Am. J. 1976, *40*, 60–63.
70. Knaebel, D.B.; Federle, T.W.; Vestal, J.R. Environ. Toxicol. Chem. 1990, *9*, 981–988.
71. Federle, T.W.; Ventullo, R.M.; White, D.C. Microb. Ecol. 1990, *20*, 297–314.
72. Knaebel, D.B.; Federle, T.W.; McAvoy, D.C.; Vestal, J.R. Environ. Toxicol. Chem. 1996, *15*, 1865–1875.
73. Federle, T.W.; Pastwa, G.M. Ground Water 1988, *26*, 761–770.
74. Federle, T.W.; Ventullo, R.M. Appl. Environ. Microbiol. 1990, *56* (Suppl), 333–339.
75. Kiewiet, A.T.; Weiland, A.R.; Parsons, J.R. Sci. Total Environment. 1993; (Suppl.), 417–422.
76. Ang, C.C.; Abdul, A.S. J. Hydrol. 1992, *138*, 191–209.
77. Vashon, R.D.; Schwab, B.S. Environ. Sci. Technol. 1982, *16*, 433–436.
78. Larson, R.J.; Games, L.M. Environ. Sci. Technol. 1981, *15*, 1488–1493.
79. Larson, R.J.; Perry, R.L. Water Res. 1981, *15*, 697–702.
80. Lee, D.M.; Guckert, J.B.; Belanger, S.E.; Feijtel, T.C. Chemosphere 1997, *35*, 1143–1160.
81. Madsen, T.; Damborg, A.; Rasmussen, H.B.; Seierø, C. Miljøstyrelsen 1994, 1–69.
82. Wagener, S.; Schink, B. Appl. Environ. Microbiol. 1988, *54*, 561–565.
83. Steber, J.; Wierich, P. Water Res. 1987, *21*, 661–667.
84. Federle, T.W.; Schwab, B.S. Water Res. 1992, *26*, 123–127.
85. Madsen, T.; Rasmussen, H.B.; Nilsson, L. Chemosphere 1995, *31*, 4243–4258.
86. Holten Lützhøft, H.-C.; Halling-Sørensen, B.; Jørgensen, S.E. Arch. Environ. Contam. Toxicol. 1999, *36*, 1–6.
87. Cserhati, T. Environ. Health Perspect. 1995, *103*, 358–364.
88. Baillie, A.J.; Al-Assadi, H.; Florence, A.T. Int. J. Pharm. 1989, *53*, 241–248.
89. Dorn, P.B.; Salanitro, J.P.; Evans, S.H.; Kravetz, L. Environ. Toxicol. Chem. 1993, *12*, 1751–1762.
90. Ernst, R.; Gonzales, C.J.; Arditti, J. Environ. Pollut. 1983, *31*, 159–175.

91. Madsen, T.; Petersen, G.; Seierø, C.; Tørsløv, J. J Am. Oil Chem. Soc. 1996, 72, 929–933.

92. van Ginkel, C.G.; Stroo, C.A.; Kroon, A.G. Sci. Total Environ. 1993; (Suppl 1), 689–697.

93. Lewis, M.A.; Hamm, B.G. Water Res. 1986, 20, 1575–1582.

94. Yamane, A.N.; Okada, M.; Sudo, R. Water Res. 1984, 18, 1101–1105.

95. Dorn, P.B.; Rodgers, J.H.; Dubey, S.T.; Gillespie, W.B.; Figueroa, A.R. Ecotoxicol. Environ. Saf. 1996, 34, 196–204.

96. Luxmoore, R.J.; Valoras, N.; Letey, J. Agron. J. 1974, 66, 673–675.

97. Maki, A.W.; Bishop, W.E. Arch. Environ. Contam. Toxicol. 1979, 8, 599–612.

98. Wong, D.C.; Dorn, P.B.; Chai, E.Y. Environ. Toxicol. Chem. 1997, 16, 1970–1976.

99. Hall, W.S.; Patoczka, J.B.; Mirenda, R.J.; Porter, B.A.; Miller, E. Arch. Environ. Contam. Toxicol. 1989, 18, 765–772.

100. Schöberl, P.; Bock, K.J.; Huber, L. Tenside Surfact. Deterg. 1988, 25, 86–98.

101. Kaluza, U.; Taeger, K. Tenside Surfact. Deterg. 1996, 33, 46–51.

102. Wildish, D.J. Water Res. 1974, 8, 433–437.

103. Harrelson, R.A.; Rodgers, J.H.; Lizotte, R.E.; Dorn, P.B. Ecotoxicology 1997, 6, 321–333.

104. Kline, E.R.; Figueroa, R.A.; Rodgers, J.H.; Dorn, P.B. Environ. Toxicol. Chem. 1996, 15, 997–1002.

105. Lizotte, R.E.; Wong, D.C.; Dorn, P.B.; Rodgers, J.H. Arch. Environ. Contam. Toxicol. 1999, 37, 536–541.

106. Dorn, P.B.; Rodgers, J.H.; Dubey, S.T.; Gillespie, W.B.; Lizotte, R.E. Ecotoxicology 1997, 6, 275–292.

107. Dorn, P.B.; Rodgers, J.H.; Gillespie, W.B.; Lizotte, R.E.; Dunn, A.W. Environ. Toxicol. Chem. 1997, 16, 1634–1645.

108. Gillespie, W.B.; Rodgers, J.H.; Crossland, N.O. Environ. Toxicol. Chem. 1996, 15, 1418–1422.

109. Gillespie, W.B.; Rodgers, J.H.; Dorn, P.B. Aquat. Toxicol. 1997, 37, 221–236.

110. Gillespie, W.B.; Rodgers, J.H.; Dorn, P.B. Ecotoxicol. Environ. Saf. 1998, 41, 215–221.

111. Lewis, M.A.; Suprenant, D. Ecotoxicol. Environ. Saf. 1983, 7, 313–322.

112. Guhl, W.; Gode, P. Tenside Surfact. Deterg. 1989, 26, 282–287.

113. Lewis, M.A.; Perry, R.L. ASTM STP 1981, 737, 402–418.

114. Altenburger, R.; Boedeker, W.; Faust, M.; Grimme, L.H. Food Chem. Toxicol. 1996, 34, 1155–1157.

115. Schulze, K. Tenside Surfact. Deterg. 1996, 33, 94–95.

116. Hewin International. Surfactants and Other Additives in Agricultural Formulations; Wiley: New York, 2000.

117. Chamel, A.; Gambonnet, B. Chemosphere 1997, 34, 1777–1786.

118. Anastacio, P.M.; Holten Lützhøft, H.-C.; Halling-Sørensen, B.; Marques, J.C. Chemosphere 2000, 40, 835–838.

119. Uppgård, L.-L.; Lindgren, Å.; Sjöström, M.; Wold, S. J. Surfact. Deterg. 2000, 3, 33–41.

120. Patoczka, J.B.; Pulliam, G.W. Water Res. 1990, 24, 965–972.

121. Halling-Sørensen, B. Inhibition of Aerobic Growth and Nitrification of Bacteria in Sewage Sludge by Alkyamine Ethoxylate (technical mixtures). 2002. *unpublished work*.

122. Scholz, N. Tenside Surfact. Deterg. 1997, 34, 229–232.

27

Biodegradation of Surface-Active Detergent Components in Sewage Treatment Plants

ZENON ŁUKASZEWSKI Poznan University of Technology, Poznan, Poland

I. INTRODUCTION

Detergents are complex systems consisting of a mixture of surfactants, builders, bleach, corrosion inhibitors, fillers, and numerous auxiliary additives, such as enzymes, optical brighteners, foam regulators, perfumes, and protective colloids. The concentration of each component in the environment depends mostly on the substance stream, the amount of water in the environment, and the biodegradation rate. Biodegradation in sewage treatment plants (STPs) or in surface water is a decisive factor in the total effect. Each organic compound of detergent composition may undergo biodegradation in STPs. However, observation of this process is difficult because of the limited potential of analysis. To date, this potential has been rather poor. The analytical success that has resulted in the development of a real analytical tool sooner or later leads to a series of papers concerning monitoring of the surfactant or its metabolite in STPs and in the aquatic environment. A surrogate measure of biodegradation is the testing of the biodegradation of detergent components under different test conditions.

II. DETERGENT COMPOSITION

Surfactants, because of their amphiphile structure, have found wide application as emulsifiers, wetting agents, softeners, antistatic agents, etc. However, the major application of surfactants is as detergents. Annual detergent consumption per person in developed countries varies from 5 to 13 kg and is dropping due to the use of more concentrated detergents [1]. Detergents are purchased in the form of powders, liquids, and tablets. Modern detergents are relatively complex multicomponent systems. Apart from surfactants, detergents contain builders, corrosion inhibitors, fillers, and numerous auxiliary additives, such as enzymes, optical brighteners, foam regulators, perfumes, and protective colloids. Additionally, heavy-duty detergent powders contain a bleach system.

The mixture of anionic and nonionic surfactants is most frequently used in detergent formulations. Cationic surfactants are used mostly as softeners and anti-static agents, though rarely in detergent formulation. Soap is the most popular surfactant [2]. In liquid detergent, soap concentration is from 4% to 22%, and in washing powder soap is a foam regulator (app. 1%) [1,3]. A mixture of anionic and nonionic surfactants is usually used in washing powder formulation: 10–25% of surfactants is used. Liquid detergents contain more surfactants: 7–25% of anionic surfactants and 6–30% of nonionic surfactants. Linear alkylbenzene sulfonates (LAS) comprise the major surfactant. However, they are partially substituted by alcohol sulfates (AS) in washing powders and by alcohol ether sulfates (AES) in liquid detergent formulations [4]. Alcohol ethoxylates (AE) are the major nonionic surfactants. Alkylglucosides are frequently used as cosurfactants. Application of alkylphenol ethoxylates (APE) is radically reduced for environmental reasons.

The main functions of a builder in a detergent formulation are to remove the effect of water hardness and to prevent incrustation. In classical formulations, soda ash and sodium polyphosphates were used, the most typical being sodium tripolyphosphate. An auxiliary builder is sodium methasilicate, or water glass, whose main role is corrosion inhibition. However, phosphates cause the eutrophication effect in surface water and therefore have been gradually withdrawn from detergent formulations. Nitrilotriacetic acid (NTA) as well as EDTA were introduced instead of polyphosphates, and more recently Zeolith A or P have been applied [1,4]. Polycarboxylates or citrates as cobuilders are used. Concentrations of 3–8% of cobuilder are used. Thus organic cobuilders are a major component of detergent in terms of organic matter to be biodegraded.

III. LINEAR ALKYLBENZENE SULFONATES (LAS)

To date, LAS are the major surfactant in household detergent formulations. Commercially available LAS are a mixture of homologues having different lengths of alkyl chain. A typical distribution is given by Schöberl [5]. Standard content of the homologue PhenylC10 in the mixture is 5–10%. The homologue PhenylC11 represents 40–45%; homologue PhenylC12 represents 35–40%; and the content of homologue PhenylC13 is 10–15%. LAS sodium salt is usually used. The environmental impact of LAS is reviewed by Jensen [6]. The no-effect concentration for LAS is between 100 and 350 ppb [7].

A. Analysis

Analysis is a crucial stage in the successful monitoring of LAS, their biointermediates in STPs, as well as in the aquatic environment. Since the mid-1980s, high-performance liquid chromatography (HPLC) has been used for the determination of LAS in the environment [8–23]. Previously, gas chromatography was used for this purpose [24–34]. HPLC determination in developed form [35] applies preconcentration in a C18 spe cartridge, elution with methanol and treatment in an anionite SAX column. The analyte is an eluted methanol hydrochloric acid mixture. An evaporated sample reconstituted to methanol is determined by the RP HPLC with fluorescence detection (excitation wavelength 230 nm and emission wavelength 290 nm). Four characteristic peaks appear on the chromatogram, each corresponding to a particular homologue. The detection limit for effluent and river water is 4 ppb and for raw sewage, 160 ppb. The determination recoveries are 77–83% for river water and effluent and 97% for raw sewage.

Apart from LAS, their main biodegradation intermediates, sulfophenyl carboxylates, may also be determined by HPLC [36–39] or HPLC-MS [40] as well as GC-MS [41].

Apart from specific methods, nonspecific methods for the determination of anionic surfactants can be used, especially in model investigations where a sole surfactant is tested. Also, when LAS are the major anionic surfactants on a particular market (e.g., Poland), the results for total anionic surfactant concentration reflect LAS concentration, at the initial approximation.

The main tool for the determination of the total concentration of anionic surfactants is the methylene blue–active substances method [42,43]. The results of this nonspecific determination are known as the MBAS concentration, and the principle of the procedure is the formation of ion pairs between anionic surfactants and cationic methylene blue. This ion pair is extracted with chloroform, and the concentration is validated spectrophotometrically ($\lambda = 650$ nm).

B. Model Biodegradation

A large number of works report on LAS biodegradation under different test conditions. These are broadly discussed in the Swisher monograph [44]. More recently, the question has been also discussed in the Scott and Jones review [45]. Three stages of biotransformation may be selected: shortening of the alkyl chain, desulfonation, and breakdown of the phenyl ring. Shortening of the straight alkyl chain follows the Heyman–Molof scheme [46], which starts from ω-oxidation of the alkyl chain:

$$CH_3\text{-}CH_2\text{-}CH(C_6H_4SO_3Na)\text{-}CH_2\text{-}CH_2\text{-}(CH_2)_n\text{-}CH_2\text{-}CH_3$$

$$\downarrow$$

$$CH_3\text{-}CH_2\text{-}CH(C_6H_4SO_3Na)\text{-}CH_2\text{-}CH_2\text{-}(CH_2)_n\text{-}CH_2\text{-}COOH$$

Sulfophenyl carboxylic acids are formed. The biotransformation then follows the β-oxidation pathway with shortening of the alkyl chain by two carbons in a single cycle.

Desulfonation may occur alternatively via hydrolysis or monooxygenase catalysis or via reductive desulfonation [47]. A broad discussion concerning potential pathways of LAS biodegradation is given in the Schöberl review [48].

Biodegradation of branched alkylbenzene sulfonates is significantly worse than that of LAS.

C. Biodegradation in Sewage Treatment Plants

Special attention is paid to LAS biodegradation in STPs, and numerous papers are devoted to this problem. This is due to the fact that LAS are the major surfactant. Studies performed before 1980 are broadly reviewed by Schöberl [48].

Highly significant investigation of LAS in STP and rivers receiving treated sewage was performed by an international group within the ERASM project [7]. The ERASM project—environmental risk assessment was established jointly by the AIS

(Association Internationale de la Savonnerieet de la Detergence) and CESIO (Comité Europeen des Agents de Surface et Intermediares Organiques). The aim of the project is to monitor major surfactants in the environment, i.e., LAS, alcohol ethoxylates (AE), alcohol ethoxylated sulfates (AES), alcohol sulfates (AS), secondary alkane sulfonates (SAS), and soap. Here LAS were the reference surfactant. One STP participating in the project was from Germany, and others were from the Netherlands, the United Kingdom, Spain, and Italy. The research program embraces two conventional activated-sludge STPs, a conventional activated-sludge STP with the application of trickling filters for polishing effluent (see also Ref. 35), a carrousel activated-sludge plant (see also Ref. 49), and a two-stage activated-sludge plant with phosphate removal. The hydraulic retention time of the investigated STP varied from 1.3 to 12 hours, and the sludge retention time varied from 5 to 19.8 days. A 7-day investigation was performed for each STP. LAS concentration in the raw sewage of these STPs ranged between 4.0 to 15.1 ppm, and in effluents ranged between 9 and 140 ppb. LAS removal varied from 98.5% to 99.9% and was worse for conventional STPs. LAS concentration in rivers, as receivers of treated sewage, varied between 2 and 130 ppb. The differences between LAS concentration below STP and above STP were not observed. LAS adsorption on activated sludge varied between 0.2 and 0.4 ppm, adsorption on digested sludge ranged between 6 and 9.4 ppm, adsorption on primary sludge varied between 4.3 and 8.3 ppm, and adsorption on river sediments was within 0.17–12.3 ppm.

Similar investigations of LAS biodegradation in STPs and LAS concentrations in associated rivers were performed by Schöberl [5,50]. Six German STPs and associated rivers receiving treated sewage were investigated. Five STPs had biological sewage treatment, and one had only mechanical treatment. A 7-day investigation was performed for each STP. Hydraulic retention time varied from 3 to 21 hours. LAS concentration in raw sewage varied from 1.9 to 5.08 ppm, while in treated sewage of biological plants LAS concentration varied from 2.6 to 42 ppb. In biological STPs, LAS was eliminated on average to 99.2%. LAS removal seems to be correlated with sludge age—the older the sludge, the better the LAS removal. However, in STPs without the biological step, LAS reduction was not observed.

Through comparison of LAS concentration in rivers upstream and downstream with respect to the STP, the effect of STP on LAS concentration in rivers was measured. No effect or negligible effect was observed in the cases of biological STPs (LAS concentration reduction 98% or more). In the case of STPs without the biological step, LAS concentration in an associated river grew from 63 ppb upstream to 623 ppb 2.5 km downstream. Reported LAS concentration in river water varied from 0.9 to 620 ppm.

Seven STPs in Madrid having a different flow and hydraulic retention times from 4 to 8 hours were investigated [36]. LAS concentration in the influent varied from 6.7 to 11.4 ppm, and in the effluent varied from 80 to 310 ppb. This corresponds to an LAS concentration reduction from 97.1% to 99.1%. LAS adsorption on dry sludge varied from 5.4 to 13.9 g/kg.

Six representative trickling-filter STPs in Yorkshire (UK) were investigated for LAS removal [51]. LAS concentration in raw sewage varied from 2.5 to 4.4 ppm. Five of the tested STPs showed an LAS concentration from 0.08 to 0.38 ppm and LAS removal between 91% and 97%. However, in one STP, LAS concentration in treated sewage was 0.76 ppm, with only 75% LAS concentration reduction.

LAS removal occurs through biodegradation and through adsorption on activated sludge. The LAS portion removed through adsorption depends on water hardness; the harder the water, the more LAS is adsorbed [52]. Part of LAS removed this way is reported as equal to 16% [52], 27% [53], or 37% [15] of total LAS and depends on water hardness. The significant role of LAS biodegradation in the sewer system was evidenced [11]. Almost 40% of LAS is biodegraded in the sewer system.

LAS concentration in STP effluent can be calculated using the STP mathematical model SIMPLETREAT [54] or WWTREAT [55]. On the basis of the experimental results of five STPs from different countries, it was shown that the SIMPLETREAT model underestimates results [7] while the WWTREAT model predicts data more precisely. The aim of the more sophisticated mathematical model GREAT-ER (Geography-Referenced Regional Exposure Assessment Tool for European Rivers) is to calculate predicted environmental concentrations [51,56].

Thus, the conclusion may be drawn that LAS are almost completely removed during treatment in the modern STP (above 99%) and treated sewage may not affect LAS concentration in rivers receiving treated sewage. Even STPs with trickling filters efficiently reduce LAS concentration. However, the STP without biological treatment has no cleaning effect in terms of LAS concentration. Treated sewage from such an STP significantly increases LAS concentrations in the receiving surface waters.

Unfortunately, these conclusions concern only primary biodegradation. Apart from LAS, sulfophenyl carboxylic acids (SPC), being the main soluble metabolite of LAS biodegradation [48], appeared in the effluent of the STPs. In seven STPs in Madrid, SPC concentrations in influents varied from 0.25 to 0.85 ppm, and in the effluent SPC concentrations ranged from 0.7 to 1.4 ppm; i.e., growth by 80% was observed [36]. Also SPC were monitored in the raw sewage and in the effluent of a lagoon treatment plant [57]. An SPC concentration of the order of 1 ppm was found both in raw and in treated sewage. It needs to be stressed that the concentration of SPC in the effluent was two times higher than the residual concentration of LAS.

To date, SPC monitoring is not included in large projects concerning STPs. However, an interest in SPC in the environment is growing. Tropical rivers with large volumes of discharged nontreated sewage are monitored [40,58], as are estuaries [59]. SPC concentrations of the order of 10 ppb were determined.

IV. BRANCHED ALKYLBENZENE SULFONATES (ABS)

Barely biodegradable ABS are banned in developed countries. However, they are still used in numerous countries. Poor ABS biodegradability was reconfirmed by Sengul [60] in the continuous-flow activated-sludge simulation test. LAS available on the Turkish market was biodegraded to 75%, while tetrapropylene benzene sulfonate was biodegraded to only 44%. Washing powders and dishwashing detergents available on the market were biodegraded to 30–60%.

In the Philippines, where ABS are still in use, ABS are detected in river water and the LAS/ABS ratio ranges from 0.9 to 7 [58].

V. OTHER ANIONIC SURFACTANTS

Very little is published on other anionic surfactants in STPs. Fatty alcohol sulfates (AS) and alcohol ether sulfates (AES) were monitored in STPs [61,62]. AES concentration in raw sewage was 3.5 ppm, and 99.99% surfactant was removed in STPs. AS concentration in raw sewage was 0.6 ppm, and surfactant is reported to be biodegraded to 99.9%. Under the conditions of OECD screening test, AES are biodegraded to 98–99%, AS are biodegraded to 99%, and secondary alkane sulphonates are biodegraded more than 90% [48]. Cleavage of the ether bond is a predominant pathway of AES [63].

In sewers and STPs, soap is eliminated as a precipitate of calcium salt and passes into the sludge [2]. During anaerobic sludge digestion, soap is biodegraded to 80% [64].

VI. OXYETHYLATED ALKYLPHENOLS

Alkylphenol ethoxylates (APE) are a major group of nonionic surfactants (NS) [65]; however, their application is slowly decreasing due to their unsatisfactory biodegradation. Nonylphenol ethoxylates and octylphenol ethoxylates are used, both having a branched alkyl chain. APE are banned in numerous countries because of this unsatisfactory biodegradation.

A. Analysis

Most attention has been paid to the determination of this class of nonionic surfactants because of their poor biodegradability. Additionally, due to the presence of the phenyl ring in the molecules, it was an easier task than in the case of other classes of nonionic surfactants. The presence of the phenyl ring facilitated the use of UV-spectrophotometric or fluorimetric detection. Only oxyethylated alkylphenols among nonionic surfactants possess this property. Ahel and Giger [66] developed an HPLC method with UV detection for the specific determination of oxyethylated alkylphenols. According to Wickbold [67], gas-stripping separation of nonionic surfactants was the first stage of separation, and an aluminum oxide column for additional cleaning of the sample was used. The method was used for the determination of oxyethylated alkylphenols in different environmental samples [68,69]. Application of fluorimetric instead of UV-spectrophotometric detection substantially improved the detection limit of the method [70–73]. Kubeck and Naylor [70] applied another separation scheme to the HPLC determination of oxyethylated alkylphenols. The sample was cleaned on the ion exchange column with the mixed bed and, in the solid phase extraction column with a C_{18} reverse phase. Hot methanol was used to wash the determined oxyethylated alkylphenols from the column, prior to separation on the HPLC column.

B. Model Biodegradation

A biodegradation pathway of APE is considered to have been discovered: Molecules undergo stepwise ω-oxidation of the oxyethylene chain leading to stepwise shortening of the oxyethylene chain (see Fig. 1). The final product of this stage of biodegradation is free alkylphenols (AP) and alkylphenol mono-and diethoxylate (APE1 and APE2, respectively). These products were identified and determined as a result of the method developed by Ahel and Giger [66]. AP, APE1, and APE2 are persistent in the biodegradation process [68,69] and are accumulated in activated sludge [74].

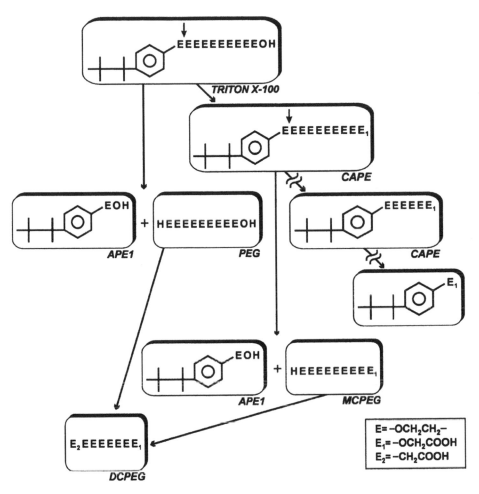

FIG. 1 Biodegradation pathways of Triton X-100. (Reprinted from Ref. 75 with very kind permission of the copyright owner.)

It was recently discovered that an alternative pathway for APE under the conditions of continuous-flow activated simulation test is central fission, leading to PEG formation [75] (see Fig. 1). A mixed pathway leads to the formation of monocarboxylated PEG and dicarboxylated PEG. A series of peaks recorded with the LC-ESI-MS technique provides strong evidence for such a pathway. Formation of mono- and dicarboxylated PEG cannot be explained by ω-oxidation of the oxyethylene chain pathway.

C. Biodegradation in Sewage Treatment Plants

Relatively little information concerning APE reduction in STP is published. Two STPs in High Point, North Carolina, were monitored for APE concentration [70]. NPE concentration in raw sewage ranged from 1.6 to 2.52 ppm, while NPE concentration in treated sewage ranged from 51 to 104 ppb. An STP in Thessaloniki, Greece, treats sewage containing 1.18–1.62 ppm APE and discharges treated sewage containing 35–130 ppb APE directly into Thermaikos Bay [76]. In this STP, 92–97% of APE is reduced.

There is more information concerning alkyl phenol and short-chained APE in the treated sewage of STPs [77–86]. The highest reported concentrations of free nonylphenol in treated sewage reached 350 ppb [86]. However, none of the 16 samples of German treated sewage exceeded the concentration of 1 ppb [81]. These differences probably reflect a divergent level of APE consumption. The highest reported concentration of free octylphenol is equal to 1.7 ppb [82]. Concentrations of short-chained APE in treated sewage are usually higher than those of free alkylphenol [77,83]. The problem of alkylphenols in the environment has recently been reviewed [87].

Many papers are devoted to the determination of free AP and short-chained APE in surface water [78,80,81,83,85,86,88–91]. The highest recorded concentration of free nonyl phenol is 644 ppb [86]. However, the highest NP concentration in the German samples was 0.13 ppb [81]. Concentrations of free octylphenol are reported as significantly lower than those of free nonylphenol.

VII. ALCOHOL ETHOXYLATES (AE)

Alcohol ethoxylates are the second largest group of surfactants in terms of manufacture (after linear alkylbenzenesulphonates) and the main group of nonionic surfactants (NS) [92]. The main area of AE application is in the manufacture of washing powders, liquids, etc. AE are classified as easily biodegradable surfactants [93], in contrast to the previously used oxyethylated alkylphenols or oxyethylene-oxypropylene block copolymers.

A. Analysis

Alcohol ethoxylates have no chromophoric groups. This gives rise to detection problems in HPLC. In order to solve the detection problem, Allen and Linder [94] derivatized oxyethylated alcohols with phenylisocyanate, converting them to a form detectable in UV-spectrophotometry. The preliminary separation scheme was partially adapted from the previous schemes concerning oxyethylated alkylphenols. Another scheme, developed by Schmitt et al. [95], proposes a sequential application of the column with macroreticular adsorbent Ameberlite XAD-2, liquid–liquid extraction with ethyl acetate, and ion exchange treatment prior to their derivatization with phenylisocyanate. HPLC separation with the normal phase (μBond-apak NH$_2$) leads to the separation of the treated mixture in accordance with the length of the oxyethylene chain, while separation on the reverse-phase column (μBond-apak C$_{18}$) accords with the length of the alkyl chain. More sensitive is the reversed-phase HPLC method with fluorescence detection [96]. Naphthyl isocyanate or naphtoyl chloride are used for derivatization. A graphitized carbon black spe cartridge is used for preconcentration. Apart from AE, PEG, being the main metabolite, can be also determined.

The liquid chromatography–electrospray ionization mass spectrometry technique shows great potential in AE and PEG determination [97,98].

Apart from specific methods, nonspecific methods of AE analysis can also be used, especially in model investigations where a sole surfactant is tested. Equally, when

AE are the major nonionic surfactants on a particular market, the results for total NS concentration reflect AE concentration in the initial approximation.

The bismuth active substances method (BiAS) is recommended in Europe for the determination of nonionic surfactants in aquatic environment components [67,99,100]. This method is also known as the Wickbold method or the method with modified Dragendorff reagent. A selective step of the BiAS method is the precipitation of the orange-colored compound of ethoxylates with Dragendorff reagent in the presence of barium(II) ions. The concentration of nonionic surfactants is determined indirectly by the determination of bismuth(III) in the precipitate.

The cobalt thiocyanate active substances method (CTAS) is recommended in the United States [101] for the determination of nonionic surfactants in environmental samples. CTAS is considered equivalent to the BiAS method and is an extraction-spectrophotometric method. Ethoxylates having more than five oxyethylene subunits form colored compounds with an anionic complex of cobalt(II) and thiocyanates. This colored compound is extracted to dichloromethane, and spectrophotometric measurement is performed. To eliminate interferences, a relatively complicated separation procedure is recommended [95].

The GC–hydrogen bromide cleavage method is based on the cleavage of the ethoxylene chain with hydrogen bromide and gas-chromatographic determination of dibromoethane, the product of the cleavage [102–106]. The final version of the method, developed by Wee [107] and recommended by Matthijs and Hennes [12], is preceded by a relatively complicated separation scheme.

The indirect tensammetric technique (ITT) is the technique specified for the determination of nonionic surfactants in the aquatic environment samples and is therefore much better adapted to this task than the majority of other tensammetric techniques. The lowering of the tensammetric peak of a monitoring substance, due to competitive adsorption of surfactants to be determined, is the analytical signal in the ITT. Anionic surfactants do not interfere with the determination of nonionic surfactants by the ITT, with ethyl acetate as the monitor [108]. Another important advantage of the ITT is the approximate additivity of the analytical signals of mixture components [109].

The method combining the BiAS separation procedure with the indirect tensammetric technique (BiAS-ITT) is more selective than the ITT, due to the additional step of selective precipitation of ethoxylates with modified Dragendorff reagent [110,111]. Nonionic surfactants are determined in the final stage of the BiAS-ITT method instead of bismuth(III), as in the classical version of the BiAS. Only ethoxylates having 4–30 oxyethylene subunits can be determined by the method, i.e., similar to the classical BiAS. NS having shorter oxyethylene chains, which remain in the solution, can be extracted and determined by the ITT. This is a fraction of short-chained ethoxylates.

Two tensammetric methods for the determination of PEG have been developed [112,113]. The simplest applied is the modified ITM [112]. The first stage of the procedure is the extraction of nonionic surfactants, which hinder the determination. The sequential extraction to chloroform causes PEG to be separated from the water matrix. PEG are determined using the ITT. However, ethoxylates having more than 30 oxyethylene subunits are also determined in this way [114].

The other tensammetric method for the determination of PEG is the modified BiAS-ITT [113–115]. The separation of PEG by sequential extraction with ethyl acetate and then chloroform is identical to the method described earlier. In the next stage, PEG are precipitated with the modified Dragendorff reagent. PEG are determined by the ITT in the dissolved precipitate. PEG determined by this method are long-chained PEG. PEG having shorter oxyethylene chains, which remain in the solution, can be extracted and determined by the ITT. This is a fraction of short-chained PEG.

B. Model Biodegradation

The most probable by-products of AE biodegradation are free alcohols (FA) and poly(ethylene glycols) (PEG) as a result of the "central fission" pathway [44]:

enzyme

$\Downarrow$

$CH_3\text{-}CH_2\text{-}(CH_2)_n\text{-}CH_2\text{-}O\text{-}CH_2\text{-}CH_2\text{-}(O\text{-}CH_2\text{-}CH_2)_m\text{-}OH$

$\Downarrow$

$CH_3\text{-}CH_2\text{-}(CH_2)_n\text{-}CH_2\text{-}OH \quad + \quad HO\text{-}CH_2\text{-}CH_2\text{-}(O\text{-}CH_2\text{-}CH_2)_m\text{-}OH$

or

enzyme

$\Downarrow$

$CH_3\text{-}CH_2\text{-}(CH_2)_n\text{-}CH_2\text{-}O\text{-}CH_2\text{-}CH_2\text{-}(O\text{-}CH_2\text{-}CH_2)_m\text{-}OH$

$\Downarrow$

$CH_3\text{-}CH_2\text{-}(CH_2)_n\text{-}CH_2\text{-}O\text{-}CH_2\text{-}CH_2\text{-}OH \quad + \quad HO\text{-}CH_2\text{-}CH_2\text{-}(O\text{-}CH_2\text{-}CH_2)_{m-1}\text{-}OH$

Free fatty alcohols (FFA) or short-chained ethoxylates (SCE) containing 1–3 oxyethylene subunits (EO) as well as PEG are formed.

Though the occurrence of the central fission pathway has been confirmed in several papers [44,116,117], the by-products of central fission are not as readily biodegraded as was judged previously. Further FFA biodegradation, formed from AE by central fission, was considered as fast and complete [44]. This opinion, affirmed by Patterson et al. [118], is based on the fact that FFA were not detected in AE biodegradation experiments.

Apart from the central fission pathway, the alternative pathways for ethoxylates have been published: ω-oxidation of alkyl moiety and ω-oxidation of oxyethylene moiety [119]. The first leads to the shortening of the alkyl moiety, while the other to shortening of the oxyethylene moiety. ω-Oxidation of the oxyethylene chain was found to be the dominating biodegradation pathway in the case of oxyethylated alkylphenols [68]. In accordance with Schöberl et al. [120], the fast ω-oxidation of branched alkyl chains of oxyethylated oxo-alcohols leads to the formation of dicarboxylated PEG. Carboxylated PEG may also be formed by the terminal oxidation of neutral PEG [44,117].

According to the most recent research [121–123] performed with several nonionic surfactants under the conditions of both the continuous-flow activated-sludge test as well as the die-away test, poly(ethylene glycols) (PEG) and short-chained ethoxylates or/and free alcohols are formed. This strongly supports the version of the central fission pathway. The presence of short-chained ethoxylates together with PEG presence is crucial evidence. The presence of short-chained ethoxylates in the die-away test of oxyethylated fatty alcohol C12E10 is shown in Figure 2 (curve b) [123]. The dashed curve b_c shows the concentration of short-chained ethoxylates calculated on the basis of the central fission of surfactant C12E10. Both curves coincide within the initial period of the test; partial biodegradation of short-chained ethoxylates is then observed. Even stronger evidence for the presence of short-chained ethoxylates is given in Figure 3, showing short-chained ethoxylate concentration in four separate continuous-flow activated-sludge simulation tests. In separate experiments, two fatty alcohol ethoxylates and two oxo-alcohol ethoxylates were tested. The investigated

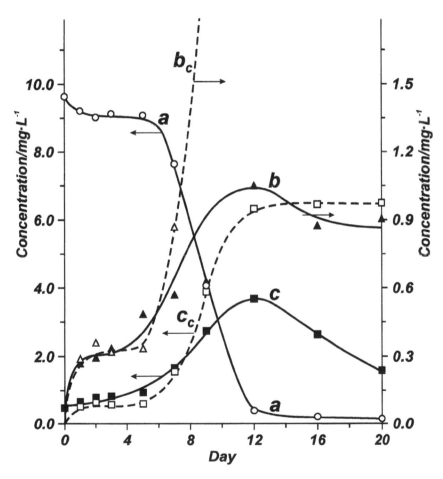

FIG. 2 Residual concentration of surfactant C12E10 (a), and short-chained ethoxylates (b) and total poly(ethylene glycol) (c) concentration during the test. Curves b_c and c_c correspond to the concentrations of the short-chained ethoxylates (b_c) and the total poly(ethylene glycols) (c_c) calculated on the basis of the central fission and curve a. (Reprinted from Ref. 130 with very kind permission of the copyright owner.)

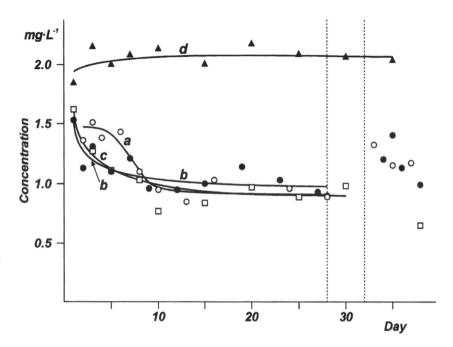

FIG. 3 Free alcohol concentration in treated sewage of oxo-alcohol ethoxylates (*a* and *b*) and fatty alcohol ethoxylates (*c* and *d*) under the conditions of continuous-flow activated-sludge. (a) (open circles) LIAL 125/14 BRD (oxo-C13.5); (b) (solid circles) LIAL 125/14 NRD (oxo-C13.5); (c) (open squares) C12E10 (fatty-C12); (d) (solid triangles) Marlipal 1618/25 (fatty-C16-18). Vertical dashed lines indicate the pause in surfactant supply (only for *a* and *b*). (Reprinted from Ref. 122 with very kind permission of the copyright owner.)

fatty alcohol ethoxylates differ from each other by the length of alkyl chain. The investigated oxo-alcohol ethoxylates represented broad- or narrow-distribution homologues. High concentrations of short-chained ethoxylates were observed during the test. This, jointly with PEG formation, is strong evidence that the central fission is the only dominant pathway. PEG formation is shown in Figure 2 as curve *c*, together with the dashed line c_c showing PEG concentrations calculated on the basis of the central fission of surfactant C12E10. It is worth stressing that the most rapid increase in PEG and short-chained ethoxylates corresponds to equally rapid primary biodegradation of the surfactant.

Apart from total PEG concentration in treated sewage of biodegradation tests, PEG distribution into two fractions (long-chained and short-chained) is also reported [121–123]. The long-chained PEG fraction contains oligomers that are precipitated with modified Dragendorff reagent, i.e., those having more than six oxyethylene subunits. The short-chained PEG fraction contains shorter oligomers. PEG distribution on long- and short-chained PEG provides information concerning the shortening of PEG chains during the test. Table 1 shows an example of concentration of PEG fractions in four continuous-flow activated-sludge simulation tests, discussed earlier, concerning four different alcohol ethoxylates. It may be observed that PEG formed by the biodegradation of oxyethylated oxo-alcohol having broad M.W. distribution are mostly short-chained, while PEG formed from oxyethylated oxo-alcohol having

TABLE 1 Residual Concentrations of Tested Oxyethylated Fatty and Oxo-Alcohols and Concentration of Metabolites of Their Biodegradation Under Conditions of the Test in the Classical Husmann Plant

	Residual surfactant (ppm)	Free alcohol (ppm)	Total PEG (ppm)	Long-chained PEG (ppm)	Short-chained PEG (ppm)	Ultimate biodegradation (%)
9* or 10 day						
C12E10	0.22	0.77	2.93	1.63	1.30	62.2
1618/25	0.04	2.14	2.32	0.62	1.70	57.1
125/14B	0.26	0.95	3.18	0.45	2.73	56.1
125/14N*	0.12	0.96	4.63	4.23	0.39	42.9
25* or 28 day						
C12E10*	0.28	0.89	1.80	1.49	0.31	72.3
1618/25*	0.034	2.09	1.42	0.53	0.89	66.6
125/14B	0.42	0.89	1.15	0.36	0.78	75.4
125/14N	0.14	0.90	1.89	1.42	0.47	70.7

Source: Data from Refs. 121 and 122.

narrow M.W. distribution are mostly long-chained. Further PEG biodegradation under test conditions can be described by the following reaction:

$$HO\text{-}CH_2\text{-}CH_2\text{-}(O\text{-}CH_2\text{-}CH_2)_m\text{-}OH$$
$$\Downarrow$$
$$nH\text{-}(O\text{-}CH_2\text{-}CH_2)_p\text{-}OH$$

where $p = 1$–6. An additional conclusion from Table 1 is the ready primary biodegradation of tested alcohol ethoxylates. This fact has already been shown in several papers [124–126].

Determination of the adsorption of surfactant C12E10 (alcohol ethoxylate) and its metabolites on alive and dead activated sludge is reported [127]. Concentration of surfactant adsorbed on activated sludge is essential because it is a measure of the amount of surfactant removed with activated sludge. A very small concentration of surfactant C12E10 on alive activated sludge was determined as well as high concentrations of adsorbed metabolites: short-chained ethoxylates, long- and short-chained PEG. Higher adsorption on dead activated sludge than on alive sludge is reported. These facts are evidence of the very fast fission of the surfactant on alive activated sludge. Concentration of surfactant C12E10 on alive activated sludge ranged from 3 to 16 ppm; however, on dead activated sludge it ranged from 43 to 58 ppm. Concentration of short-chained ethoxylates on alive activated sludge ranges from 340 to 460 ppm, of long-chained PEG ranges from 80 to 140 ppm, and of short-chained PEG from 40 to 75 ppm. The slope of the Henry isotherm calculated from the data of this experiment yields $0.19 \pm 0.05 \, l\,g^{-1}$ for surfactant C12E10, $0.25 \pm 0.05 \, l\,g^{-1}$ for short-chained ethoxylates, $0.12 \pm 0.02 \, l\,g^{-1}$ for long-chained PEG, and $0.08 \pm 0.02 \, l\,g^{-1}$ for short-chained PEG. The higher the hydrophobicity of species being adsorbed, the steeper the slope of Henry's isotherm.

TABLE 2 Concentration of NS in Raw and Treated Sewage of STPs in Wielkopolska (Poland)

STP	Influent (ppm)	Effluent (ppm)	Removal (%)
Mosina	6.45	0.83	87.1
Pleszew	5.94	0.46	92.2
Goluchów	4.20	0.97	76.9
Jarocin	6.89	0.40	94.2
Wronki (I)	5.67	0.60	89.4
Ostroróg	8.82	0.49	94.4
Szamotuly	7.69	0.61	92.1
Kiekrz	5.13	0.39	92.4
Kozieglowy	9.65	0.54	94.4
Kórnik	3.89	0.38	90.2
Average	6.43	0.57	91.1

Source: Data from Ref. 128.

C. Biodegradation in Sewage Treatment Plants

Very little is published concerning the removal of alcohol ethoxylates in STPs, despite the fact that these surfactants were to be included in the ERASM project (Environmental Risk Assessment) established jointly by AIS (Association Internationale de la Savonnerieet de la Detergence) and CESIO (Comité Europeen des Agents de Surface et Intermediares Organiques) with the aim of monitoring major surfactants in the environment. An attempt to determine AE in raw and treated sewage was performed by Schmitt et al. [95]. AE concentration ranging from 0.67 to 1.31 ppm in raw sewage was determined. CTAS determination corresponding to total NS concentration gave 3.22 ppm. AE concentration in treated sewage ranged from 9 to 20 ppb, while the CTAS value was 320 ppb.

TABLE 3 Concentrations of AS, NS, and PEG in Raw and Treated Sewage of STPs in Wrzesnia and Gniezno (Poland)

	Influent (ppm)	Effluent (ppm)	Reduction (%)
STP Wrzesnia			
AS	4.2	0.15	96.5
NS_{ITM}	4.7	1.15	75.5
NS_{BiAS}	3.4	0.4	88.2
PEG_{total}	1.7	1.6	7.7
PEG_{BiAS}	1.0	0.75	25.7
$PEG_{sh.ch.}$	0.45	0.7	Incr.55
STP Gniezno			
AS	6.5	0.15	97.5
NS_{ITM}	7.9	1.1	85.5
PEG_{total}	1.45	1.2	16.5
PEG_{BiAS}	0.95	0.55	42.1
$PEG_{sh.ch.}$	0.45	0.7	Incr. 49.2

Source: Data from Ref. 128.

Some data are available from tensammetric measurements reported in academic works. Eleven biological STPs located in Wielkopolska (Poland) were tested [128]. The results, shown in Table 2, represent the total concentration of nonionic surfactants and therefore cannot be strictly considered as concerning oxyethylated alcohols. The results represent, rather, the total concentration of alcohol ethoxylates and alkylphenol ethoxylates. Nonionic surfactant concentration in raw sewage varied from 3.9 to 9.65 ppm. Treated sewage contains from 0.39 to 0.97 ppm of NS, which corresponds to NS removal between 76.9% and 94.4%. Two other biological STPs of medium size, located in Wielkopolska (Poland), were investigated more thoroughly [128]. The results, given in Table 3, show a high reduction of anionic surfactants, as can be expected, and the removal on NS to 75–85%. It is worth stressing that NS reduction

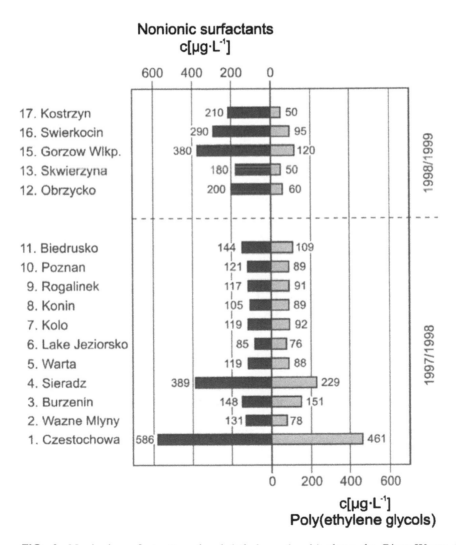

FIG. 4 Nonionic surfactants and poly(ethylene glycols) along the River Warta (autumn–winter periods 1997–1999). (Reprinted from Ref. 131.)

measured with the application of specific precipitation with Dragendorff reagent is higher (88%). Significant PEG concentrations were determined both in raw and in treated sewage and small PEG reduction. The presence of PEG in raw sewage is evidence of preliminary NS biodegradation in the sewer system. Low reduction of PEG concentration in STPs is caused by equally fast PEG formation as well as their biodegradation. Moreover, the concentration of long-chained PEG is reduced, but the concentration of short-chained PEG increases. This is evidence for the fragmentation of PEG in the treatment process. Similar concentrations of total short- and long-chained PEG were determined in two other STPs in Wielkopolska (Poland) [129].

No research has been done showing the impact of alcohol ethoxylates on surface water. Generally, even the total concentration of NS in surface water is rarely published. Recently, concentrations of NS along the River Warta have been published, showing a range between 85 and 586 ppb [130], with peaks of NS concentration related to industrial centers. The central fission of NS causes PEG to appear in river water. PEG concentrations along the River Warta are shown in Figure 4 [131]. The range of indicated PEG concentration ranges between 50 and 461 ppb. It is clear that a link exists between PEG and NS concentration in river water. Fractional concentrations of long- and short-chained PEG in the Rivers Warta and Odra were published [129,132].

REFERENCES

1. Smulders, E.; Krings, P.; Verbeek, H. Tenside Surf. Det. 1997, *34*, 386–392.
2. Moreno, A.; Bravo, J.; Ferrer, J.; Bengoechea, C. Tenside Surf. Det. 1996, *33*, 479–482.
3. Fell, B. Tenside Surf. Det. 1991, *28*, 385–395.
4. Zweig, J.E. Tenside Surf. Det. 1993, *30*, 306–309.
5. Schöberl, P. Tenside Surf. Det. 1997, *34*, 233–237.
6. Jensen, J. Sci. Total Environ. 1999, *226*, 93–111.
7. Waters, J.; Feijtel, T.C.J. Chemosphere 1995, *30*, 1939–1956.
8. Matthijs, E.; Stalmans, M. Tenside Surf. Det. 1993, *30*, 29–33.
9. Matthijs, E.; De Henau, H. Tenside Surf. Det. 1987, *24*, 193–199.
10. Castles, M.A.; Moore, B.L.; Ward, S.R. Anal. Chem. 1989, *61*, 2534–2540.
11. Moreno, A.; Ferrer, J.; Berna, J.L. Tenside Surf. Det. 1990, *27*, 312–315.
12. Matthijs, E.; Hennes, E.C. Tenside Surf. Det. 1991, *28*, 22–27.
13. Sanchez Leal, J.; Garcia, M.T.; Ferrer, J.; Bongoechea, C. Tenside Surf. Det. 1994, *31*, 253–256.
14. Schöberl, P.; Klotz, H.; Spilker, R.; Nitschkke, L. Tenside Surf. Det. 1994, *31*, 243–252.
15. Prats, D.; Ruiz, F.; Vazquez, B.; Rodriguez-Pastor, M. Water Res. 1997, *31*, 1925–1930.
16. Marcomini, A.; Capri, S.; Giger, W. J. Chromat. 1987, *403*, 243.
17. Di Corcia, A.; Marchetti, M.; Marcomini, A.; Samperi, R. Anal. Chem. 1991, *63*, 1179–1182.
18. Crescenzi, C.; Di Corcia, A.; Marchiori, E.; Samperi, R.; Marcomini, A. Water Res. 1996, *30*, 722–730.
19. Nakae, A.; Tsuji, K.; Yamanaka, M. Anal. Chem. 1980, *52*, 2275–2277.
20. Marcomini, A.; Giger, W. Anal. Chem. 1987, *59*, 1709–1715.
21. De Henau, H.; Matthijs, E.; Hopping, W.D. J. Environ. Anal. Chem. 1989, *26*, 279–282.
22. Holt, M.S.; Matthijs, E.; Waters, J. Water Res. 1989, *23*, 749–759.

23. Waters, J.; Holt, M.S.; Matthijs, E. Tenside Surf. Det. 1989, *26*, 109.
24. Fairing, J.D.; Short, F.R. Anal. Chem. 1956, *28*, 1827–1834.
25. Association of American Soap and Glycerine Producers. Analytical Subcommittee. Anal. Chem. 1956, *28*, 1822.
26. Webster, H.L.; Halliday, J. Analyst. 1959, *84*, 552.
27. Ogden, C.P.; Webster, H.L.; Halliday, J. Analyst. 1961, *86*, 22–29.
28. Setzkorn, E.A.; Carel, A.B. J. Am. Oil Chem. Soc. 1963, *40*, 57.
29. Frazee, C.D.; Osburn, Q.W.; Crisler, R.O. J. Am. Oil Chem. Soc. 1964, *41*, 808.
30. Simko, J.P.; Blank, Emery E.W. J. Am. Oil Chem. Soc. 1965, *42*, 627.
31. Swisher, R.D. J. Am. Oil Chem. Soc. 1966, *43*, 137–140.
32. Sullivan, W.T.; Swisher, R.D. Envir. Sci. Technol. 1969, *3*, 481.
33. Waters, J.; Garrigan, J.T. Water Res. 1983, *17*, 1549–1562.
34. Osburn, Q.W. J. Am. Oil Chem. Soc. 1986, *63*, 257–263.
35. Holt, M.S.; Waters, J.; Comber, M.H.I.; Armitage, R.; Morris, G.; Newbery, C. Water Res. 1995, *29*, 2063–2070.
36. Berna, J.L.; Moreno, A.; Ferrer, J. Tenside Surf. Det. 1993, *30*, 217–222.
37. Crescenzi, C.; Di Corcia, A.; Passariello, G.; Turnes Carou, M.I. J. Chromatogr. A 1996, *733*, 41–55.
38. Hrsak, D. Environ. Poll. 1995, *89*, 285–292.
39. Leon, V.M.; Gonzales-Mazo, E.; Gomez-Parra, A. J. Chromatogr. A 2000, *889*, 211–219.
40. Eichhorn, P.; Rodrigues, S.V.; Baumann, W.; Knepper, T.P. Sci. Total. Environ. 2002, *284*, 123–134.
41. Wang-Hsien, Ding; Jun-Hui, Lo; Shin-Haw, Tzing J. Chromatogr. A 1998, *818*, 270–279.
42. Longwell, J.; Menice, W.D. Analyst. 1955, *80*, 167.
43. Abbot, D.C. Analyst. 1962, *87*, 286–293.
44. Swisher, R.D. *Surfactants Biodegradation*, 2nd Ed.; Marcel Dekker: New York, 1987.
45. Scott, M.J.; Jones, M.N. Biochim. Biophys. Acta. 2000, *1508*, 235–251.
46. Heyman, J.J.; Molof, A.H. J. Water. Poll. Contr. Fed. 1967, *39*, 50–62.
47. Hashim, M.A.; Kulandai, J.; Hassan, R.S. J. Chem. Tech. Biotechnol. 1992, *54*, 207–214.
48. Schöberl, P. Tenside Surf. Det. 1998, *26*, 86–94.
49. Feijtel, T.C.J.; Matthijs, E.; Ruiters, A.; Rijs, G.B.J.; Klewiet, A.; de Nijs, A. Chemosphere 1995, *30*, 1053–1066.
50. Schöberl, P. Tenside Surf. Det. 1995, *32*, 25–35.
51. Holt, M.S.; Fox, K.K.; Burford, M.; Daniel, M.; Buckland, H. Sci. Total Environ. 1998, *210/211*, 255–269.
52. Berna, J.L.; Ferrer, J.; Moreno, A. Tenside Surf. Det. 1989, *26*, 101–107.
53. Giger, W.; Alder, A.C.; Brunner, P.H.; Marcomini, A.; Siegrist, H. Tenside Surf. Det. 1989, *26*, 95–100.
54. Struijs, J.; Stoltenkamp, J.; van de Meent, D. Water Res. 1991, *25*, 891–900.
55. Cowan, C.E.; Larson, R.J.; Feijtel, T.C.J.; Rapaport, R.A. Water Res 1993, *27*, 561–573.
56. Schröder, F.R. Tenside Surf. Det. 1997, *34*, 442–445.
57. Moreno, A.; Ferrer, J.; Bevia, F.; Prats, D.; Vazquez, B.; Zarzo, D. Water Res. 1994, *28*, 2183–2189.
58. Eichhorn, P.; Flavier, M.E.; Paje, M.L.; Knepper, T.P. Sci. Total Environ. 2001, *269*, 75–85.
59. Leon, V.M.; Saez, M.; Gonzalez-Mazo, E.; Gomez-Parra, A. Sci. Total Environ. 2002, *288*, 215–226.
60. Sengul, F. Tenside Surf. Det. 1991, *28*, 356–361.
61. Schröder, F.R. Tenside Surf. Det. 1995, *32*, 492–497.
62. Schröder, F.R.; Schmitt, M.; Reichensperger, U. Waste Manage. 1995, *19*, 125–131.

63. Hales, G.S.; Watson, G.K.; Dodgson, K.S.; White, G.F. J. Gen. Microbiol. 1986, *132*, 953–963.

64. Hales, G.S.; Watson, G.K.; Dodgson, K.S.; White, G.F. J. Gen. Microbiol. 1986, *132*, 953–963.

65. Information of Association of the Industry Producing Chemicals for Textile, Leather and Surfactant Plant TEGEWA, Germany, 2001.

66. Ahel, M.; Giger, W. Anal. Chem. 1985, *57*, 2584–2590.

67. Wickbold, R. Tenside Deterg. 1972, *9*, 173–177.

68. Ahel, M.; Giger, W.; Koch, M. Water Res. 1994, *28*, 1131–1142.

69. Ahel, M.; Giger, W.; Schaffner, C. Water Res. 1994, *28*, 1143–1152.

70. Kubeck, E.; Naylor, C.G. J. Am. Oil Chem. Soc. 1990, *67*, 400–405.

71. Kudoh, M.; Ozwa, H.; Fudano, S.; Tsuji, K. J. Chromatog. 1984, *287*, 337.

72. Holt, M.S.; McKerrell, E.H.; Perry, J.; Watkinson, R.J. J. Chromatog. 1986, *362*, 419.

73. Yoshimura, K. J. Am. Oil Chem. Soc. 1986, *63*, 1590–1596.

74. Brunner, P.H.; Capri, S.; Marcomini, A.; Giger, W. Water Res. 1988, *22*, 1465–1472.

75. Franska, M.; Franski, R.; Szymanski, A.; Lukaszewski, Z. Water Res. 2003, *37*, 1005–1014.

76. Fytianos, K.; Pegiadou, S.; Raikos, N.; Eleftheriadis, I.; Tsoukali, H. Chemosphere 1997, *35*, 1423–1429.

77. Ahel, M.; Giger, W. Anal. Chem. 1985, *57*, 1577–1583.

78. Bleckburn, M.A.; Waldock, M.J. Water Res. 1955, *29*, 1623–1629.

79. Di Corcia, A.; Samperi, R. Environ. Sci. Technol. 1994, *28*, 850–858.

80. Isobe, T.; Nishiyama, H.; Nakashima, A.; Takada, H. Environ. Sci. Technol. 2001, *35*, 1041–1049.

81. Kuch, H.M.; Ballschmitter, K. Environ. Sci. Technol. 2001, *35*, 3201–3206.

82. Lee, H.B.; Peart, T.E. Anal. Chem. 1995, *67*, 1976–1980.

83. Naylor, C.G.; Mieure, J.P.; Adams, W.J.; Weeks, J.A.; Castaldi, F.J.; Ogle, L.D.; Romero, R.R. J. Am. Oil Chem. Soc. 1992, *69*, 695–703.

84. Rudel, R.A.; Melly, S.J.; Geno, P.W.; Sun, G.; Brody, J.G. Environ. Sci. Technol. 1998, *32*, 861–869.

85. Snyder, S.A.; Keith, T.L.; Verbrugge, D.A.; Snyder, E.M.; Gross, T.S.; Kannan, K.; Giesy, J.P. Environ. Sci. Technol. 1999, *33*, 2814–2820.

86. Sole, M.; Lopez de Alda, M.J.; Castillo, M.; Porte, C.; Ladegaard-Pedersen, K.; Barcelo, D. Environ. Sci. Technol. 2000, *34*, 5076–5083.

87. Guang-Guo, Ying; Williams, B.; Kookana, R. Chemosphere 2002, *28*, 215–226.

88. Bennie, D.T.; Sulivan, C.A.; Lee, H.B.; Peart, T.E.; Maguire, R.J. Environ. Sci. Technol. 1997, *193*, 1623–1629.

89. Ahel, M.; Molnar, E.; Ibric, S.; Giger, W. Water Sci. Technol. 2000, *42*, 15–22.

90. Tabata, A.; Kashiva, S.; Ohnishi, Y.; Ishikawa, H.; Miyamoto, N.; Itosh, M.; Magara, Y. Water Sci. Technol. 2001, *43*, 109–116.

91. Ferguson, P.L.; Iden, C.R.; Brownwell, B.J. Environ. Sci. Technol. 2001, *35*, 2428–2435.

92. H. Palicka, Development and Trends in the European Laundry Detergent Market, III International Symposium "Chemistry Forum'97," Warsaw, April 1997.

93. Janicke, W. Tenside Surf. Det. 1988, *25*, 345–355.

94. Allen, M.C.; Linder, D.E. J. Am. Oil Chem. Soc. 1981, *58*, 950.

95. Schmitt, T.M.; Allen, M.C.; Brain, D.K.; Guin, K.F.; Lemmel, D.E.; Osburn, Q.W. J. Am. Oil Chem. Soc. 1990, *67*, 103–109.

96. Zanette, M.; Marcomini, A.; Marchiori, E.; Samperi, R. J. Chromatogr. A 1996, *756*, 159–174.

97. Crescenzi, C.; Di Corcia, A.; Marcomini, A.; Samperi, R. Anal. Chem. 1995, *67*, 1797–1804.

98. Castillo, M.; Riu, J.; Venture, F.; Boleda, R.; Scheding, R.; Schröder, H.Fr.; Nistor, C.; Emneus, J.; Eichhorn, P.; Knepper, Th.P.; Jonkers, C.C.A.; de Voogt, P.; Gonzalez-Mazo, E.; Leon, V.M.; Barcelo, D. J Chromatogr A 2000, 889, 195–209.

99. Waters, J.; Garrigan, J.T.; Paulson, A.M. Water Res. 1986, 20, 247–253.

100. Woltering, D.M.; Larson, R.J.; Hopping, W.D.; Jamieson, R.A. Tenside Surf. Det. 1988, 25, 116.

101. American Public Health Association. Standard Methods for the Examination of Water and Wastewater, 18th Ed.; 1992, 5540 Surfactants.

102. Mathias, A.; Mellor, N. Anal. Chem. 1966, 38, 472–477.

103. Slagt, C.; Daemen, J.M.H.; Dankelman, W.; Sipman, W.A. Z. Anal. Chem. 1973, 264, 401–406.

104. Tsuji, K.; Konishi, K. J. Am. Oil Chem. Soc. 1974, 51, 55–60.

105. Kaduji, I.I.; Stead, J.B. Analyst. 1976, 101, 728–731.

106. Tobin, R.S.; Onuska, F.I.; Brownlee, B.G.; Anthony, D.H.J.; Comba, M.E. Water Res. 1976, 10, 529–535.

107. Wee, V.T. Determination of linear alcohol ethoxylates in waste- and surface water. In Advances in the Identification and Analysis of Organic Pollutants in Water; Keith, L.H., Ed.; Ann Arbor Science: Ann Arbor, MI, 1981.

108. Szymanski, A.; Lukaszewski, Z. Anal. Chim. Acta 1994, 293, 77–86.

109. Szymanski, A.; Lukaszewski, Z. Anal. Chim. Acta 1993, 273, 313–321.

110. Wyrwas, B.; Szymanski, A.; Lukaszewski, Z. Talanta 1994, 41, 1529–1535.

111. Wyrwas, B.; Szymanski, A.; Lukaszewski, Z. Talanta 1995, 42, 1251–1258.

112. Szymanski, A.; Lukaszewski, Z. Analyst. 1996, 121, 1897–1901.

113. Lukaszewski, Z.; Szymanski, A.; Wyrwas, B.; Tomaszewski, K. Determination of poly(ethylene glycols) by the indirect tensammetric method combined with the BiAS separation scheme. Deauville Conference: 4th Symposium on Analytical Sciences, Brussels, 1996.

114. Lukaszewski, Z.; Szymanski, A.; Wyrwas, B. Application of tensammetry for the monitoring of metabolic products of biodegradation of oxyethylated alcohols. 6th European Conference on ElectroAnalysis, Durham, England, 1996.

115. Lukaszewski, Z.; Szymanski, A.; Wyrwas, B. Poly(ethylene glycols) as metabolites of nonionic surfactants biodegradation in the aquatic environment. 9th International Conference on Surface and Colloid Science, Sofia, Bulgaria, 1997.

116. Steber, J.; Wierich, P. Tenside Deterg. 1983, 20, 183–187.

117. Kravetz, L.; Chung, H.; Guin, K.F.; Shebs, W.T.; Smith, L.S. Tenside Deterg. 1984, 21, 1–6.

118. Patterson, S.J.; Scott, C.C.; Tucker, K.B.E. J. Am. Oil Chem. Soc. 1970, 47, 37–41.

119. Burczyk, B. In Encyclopedia of Surfactants and Colloid Science; Hubbard, A.T., Ed.; Marcel Dekker: New York, 2002, p 724–752.

120. Schöberl, P.; Kunkel, E.; Espeter, K. Tenside Deterg. 1981, 18, 64–72.

121. Szymanski, A.; Wyrwas, B.; Swit, Z.; Jaroszynski, T.; Lukaszewski, Z. Water Res. 2000, 34, 4101–4109.

122. Szymanski, A.; Wyrwas, B.; Bubien, E.; Kurosz, T.; Hreczuch, W.; Zembrzuski, W.; Lukaszewski, Z. Water Res. 2002, 36, 3378–3386.

123. Szymanski, A.; Bubien, E.; Kurosz, T.; Wolniewicz, A.; Lukaszewski, Z. Polish J. Environ. Stud. 2002, 11, 429–433.

124. Battersby, N.S.; Sherren, A.J.; Bumpus, R.N.; Eagle, R.; Molade, I.K. Chemosphere 2001, 45, 109–121.

125. Matthijs, E.; Holt, M.S.; Kiewiet, A.; Rijs, G.B.J. Tens. Surf. Det. 1997, 34, 238–241.

126. McAvoy, D.C.; Dyer, S.D.; Fendinger, N.J.; Eckhoff, W.S.; Lawrence, D.L.; Begley, W.M. Environ. Toxicol. Chem. 1998, 17, 1705–1711.

127. Szymanski, A.; Wyrwas, B.; Lukaszewski, Z. Water Res. 2003, 37, 281–288.

128. Traczyk, L. Biodegradation of surfactants in sewage treatment plant. Ph.D. dissertation, Poznan University of Technology, Faculty of Chemical Technology, to be defended.

129. Szymanski, A.; Wyrwas, B.; Szymanowska, M.; Lukaszewski, Z. Water Res. 2002, *35*, 3599–3604.

130. Szymanski, A.; Wyrwas, B.; Jesiolowska, A.; Przybysz, T.; Grodecka, J.; Lukaszewski, Z. Polish J. Environ. Stud. 2001, *10*, 371–377.

131. Sniadek, E. Biodegradation of poly(ethylene glycols). PhD dissertation, Poznan, University of Technology, Faculty of Chemical Technology, 2003.

132. Tomaszewski, K.; Szymanski, A.; Lukaszewski, Z. Talanta 1999, *50*, 299–306.

28

Surfactant Biodegradation: Sugar-Based Surfactants Compared to Other Surfactants

I. J. A. BAKER   Kodak (Australasia) Pty Ltd., Coburg, Victoria, Australia

C. J. DRUMMOND   CSIRO Molecular Science, Clayton South, Victoria, Australia

D. N. FURLONG   RMIT University, Bundoora, Victoria, Australia

F. GRIESER   University of Melbourne, Melbourne, Victoria, Australia

I. SUGAR-BASED SURFACTANTS

Surfactants can be produced from sugar or starch in combination with fatty acids. Products of this type have been known for as long as other "synthetic" surfactants, formed from petrochemical-derived materials. There has been a recent trend toward surfactants based on renewable resources [1]. This trend has been supported by several commercial plants opened by Cognis (formerly Henkel), Akzo Nobel, BASF, ICI (now Uniqema and Huntsman), SEPPIC, Dai ichi-Kogyo Seiyaku, Kao, and others [2]. The driving force behind the trend is the number of benefits gained by sugar-based surfactants. Desirable properties of these surfactants, in addition to being produced from renewable resources, are that they are readily biodegradable [3,4], they have low toxicity, they are produced from and degrade into materials of low toxicity, they are mild on the eyes and skin [4,5], and some can perform synergistically with other surfactants [6,7]. About 150,000 tons of sugar-derived surfactants are produced annually worldwide [2,8].

The major classes of surfactants produced from sugars are alkyl polyglycosides, fatty acid glucamides, and sucrose or sorbitan esters of fatty acids [2]. Examples of the structures of these surfactants are shown in Figure 1. The sugar groups form the hydrophilic portion of the surfactant, and the fatty alkyl chains form the hydrophobic portion. Surfactants with different properties can be obtained by incorporating fatty alkyl chains of different lengths or varying the number of

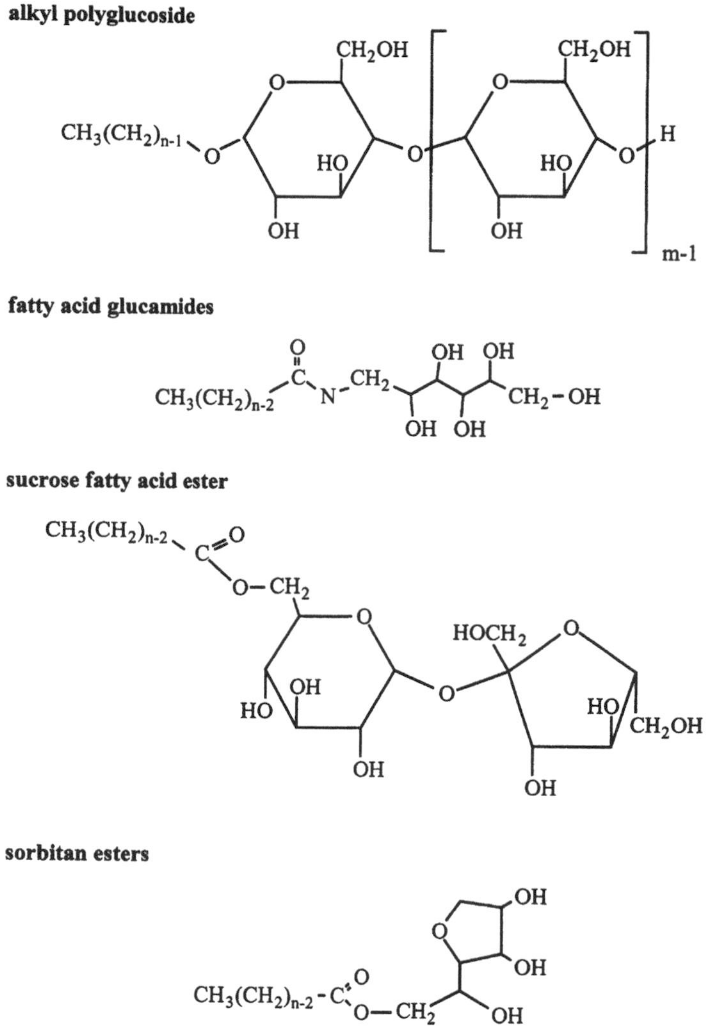

alkyl polyglucoside

fatty acid glucamides

sucrose fatty acid ester

sorbitan esters

FIG. 1 Examples of sugar-based surfactants. These types of surfactants commonly contain fatty acyl chains with 8–18 carbon atoms ($n = 8\text{--}18$) and 1–3 sugar units ($m = 1\text{--}3$).

sugar units attached. Generally chain lengths of between 8 and 18 carbon atoms are used in combination with 1–3 sugar units. These are nonionic surfactants. Anionic analogues can be formed by incorporation of groups such as a sulfonate or a carboxylate. The properties of the surfactants may also be varied by incorporation of polyethylene oxide functionality in the sugar headgroup.

A. Alkyl Polyglycosides

The major class of sugar-based surfactants in the market are alkyl polyglycosides [2]. The first glycosylation procedures were performed more than 100 years ago. Emil Fischer produced an alkyl glycoside material with surface-active properties in 1893 [9]. Fischer glycosylation consists of reacting a sugar (or other carbohydrate) with a fatty alcohol in the presence of an acid catalyst, to form an ether link between the fatty alkyl chain and the saccharide, as depicted in Figure 2. The reaction results in a complex equilibrium mixture of isomers and oligomers. The number of glycoside units attached varies and is referred to as the degree of polymerization; it usually averages around 1.4–1.7 [6]. In contrast to the normal distribution of ethylene oxide units in more conventional nonionic surfactants, equilibrium polyglycoside mixtures contain an asymmetric distribution of oligomers dominated by monoglycosides, according to Flory statistics [9]. In addition to direct glycosidation, transglycosidation may be performed to replace a short-chain by a long-chain alkyl group.

In 1901 Konigs and Knor achieved a more stereoselective synthesis by forming a polyacetylated bromoglucose intermediate that was reacted with fatty alcohol under base catalysis [9]. This is not suitable for large-scale production, however, and Fischer glycosylation is used for commercial production of alkyl polyglycosides.

Alkyl polyglycosides show foaming and surface-tension-lowering properties between those of alcohol ethoxylates and alkylbenzene sulfonates. A distinct advantage they have over alcohol ethoxylates is that their aqueous solutions do not have cloud points. Another advantage is that they show a marked synergism in combination with other surfactants [6]. This means that they can be used to replace a portion of a more harsh surfactant to produce a milder formulation that is equally effective. Alkyl polyglycosides have been found to be readily biodegradable under both aerobic and anaerobic conditions [1,3]. They are used in detergents, toiletries, and cosmetics [8]. The synthesis, properties, and applications of alkyl polyglycosides are described more fully elsewhere [2].

B. Sugar Fatty Acid Esters

In the early 1950s the Sugar Research Foundation sponsored Hass and Snell to investigate non-food uses of sucrose. This was at the time when researchers were looking for biodegradable alternatives to branched alkylbenzene sulfonate surfactants. Sucrose ester–based surfactants offered an attractive possibility. Sucrose esters

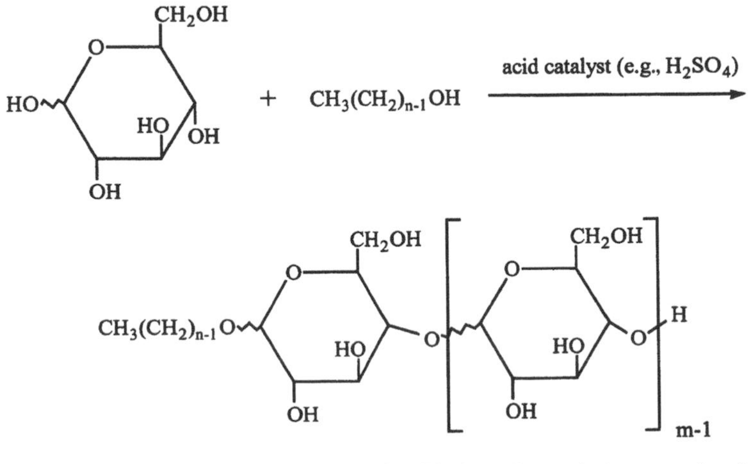

FIG. 2 Schematic representation of a Fischer glycosylation reaction between glucose and a fatty alcohol.

were first produced by base-catalyzed transesterification between a triglyceride and sucrose in a mutual solvent, such as dimethylformamide, at 90°C [10]. This results in a mixture of glycerides, sucrose esters, and unreacted starting materials. Purification of the mixed product is expensive; this means that sucrose ester use is economically viable only in food and cosmetic applications that can cover the cost of production.

Methyl fatty acid esters may be used in the transesterification process in place of triglycerides, as shown in Figure 3 [11]. Methanol produced in the reaction can be continuously removed under reduced pressure. This results in a more complete reaction but also increases the proportion of di- and triesters formed. Esterification of sucrose occurs preferentially at the primary 6 and 6' hydroxyl groups; however, it also occurs at the 1' and the secondary hydroxyl groups to a lesser extent. This process was first used commercially by the Dai Nippon Sugar Manufacturing Co. (now Mitsubishi-Kasei Food Corporation (MFC)). Sugar esters were approved as food additives in Japan in 1959 and were approved worldwide by 1983. For food applications, solvents such as dimethylformamide must not be present in the product at levels greater than 50 ppm.

Alternative methods of synthesizing sugar esters remain a subject of research. Tate and Lyle developed a solventless process in which sucrose is transesterified with triglyceride in the presence of a base catalyst at 125°C [10]. The product is a complex mixture with surface-active properties. It has not been commercially successful. Another solventless process has been suggested in which transesterification of molten sucrose and fatty acid esters occurs at temperatures between 170 and 187°C in the presence of a sodium or potassium salt as a solubilizer and catalyst [12]. This has the disadvantage that extensive discoloration occurs at these temperatures, and subsequent purification is not straightforward. Other methods involve emulsions of the

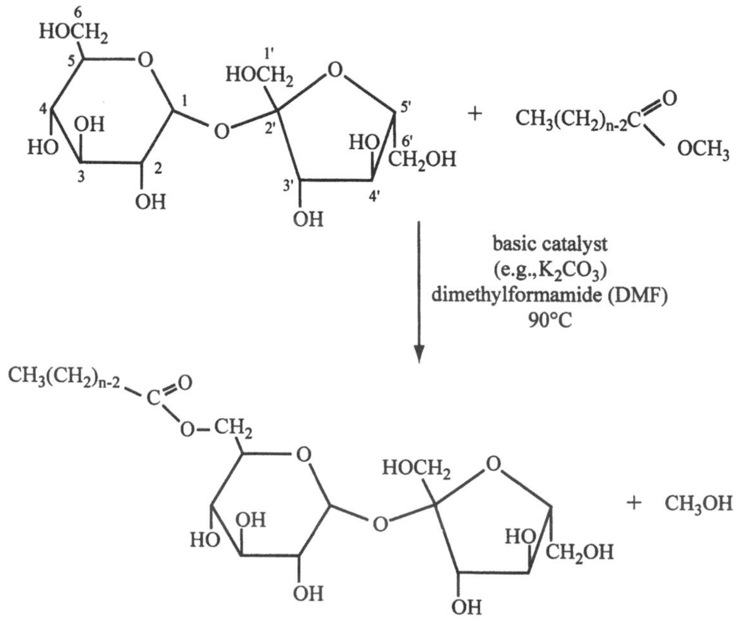

FIG. 3   Transesterification of a methyl fatty acid ester with sucrose to produce sucrose ester.

reactants and catalyst in either water or propylene glycol [13,14]. More recent approaches have employed enzymes, with the intention of obtaining more stereo-selectivity [15,16].

The results of several investigations have shown that sugar ester surfactants are very mild, being tasteless, odorless, nonirritant to skin, nontoxic to most plants and animals, and readily biodegradable [10,11,17]. The properties of these surfactants are affected by their headgroup size and by the length and number of alkyl chains they contain. Some of these structure–property relationships have been studied [18–20]. Sugar ester surfactants are compatible with formulation components such as builders and fillers. When built into heavy-duty cleaning formulations, they show detergency equivalent to that of alkylbenzene sulfonate formulations [11]. They are moderate-to-low foamers and have surface-tension-lowering properties between those of alcohol ethoxylates and alkylbenzene sulfonates [11,17,21]. These nonionic surfactants do not show a decrease in water solubility with increased temperature, which is in contrast to the cloud point behavior displayed by ethoxylated nonionic surfactants [11]. They can act as viscosity modifiers and are excellent emulsifiers, particularly when combined with a more lyophilic material [11,21].

The properties of sugar ester surfactants make them suited to many applications. They are widely used in food, cosmetic, and pharmaceutical formulations [10,16,17]. They are not widely used in household and industrial cleaning applications due to their high production costs, which are not covered by the market value of these products. Another factor limiting their application in cleaning products is their susceptibility to alkaline hydrolysis.

Sugar ester surfactants with long alkyl chains (greater than 16 carbons) or more than one alkyl chain have low water solubilities. Surfactants with greater water solubility have been obtained by incorporating a sulfonate group adjacent to the ester bond. This is achieved by esterifying a sugar with an α-sulfonyl fatty acid or by transesterification with the corresponding methyl ester [22]. In addition to being more water soluble, these sulfonated sugar esters have greater stability toward acid or alkaline hydrolysis than their unsulfonated analogues. They do not precipitate in the presence of doubly charged metal cations [22].

As already touched upon, the ready biodegradability of sugar-based surfactants is one of their major attractions. Since the environmental acceptability of surfactants has risen to importance on a par with their performance in products [16], surfactants derived from sugar and fatty acids have attracted renewed attention. This may also be attributed in part to the fact that they are derived from renewable resources (although the latter point is subject to debate [23,24]).

II. ENVIRONMENTAL IMPORTANCE OF SURFACTANT BIODEGRADABILITY

After use in a wide range of applications (cleaning formulations, mineral recovery, bitumen emulsions and cement, manufacturing, wool scouring, photographics, paints, plastics, paper, adhesives, pesticide formulations, pharmaceutical preparations, cosmetics, and foods), surfactants can enter waterways via industrial waste treatment processes, via municipal sewage treatment plants, or in some cases directly. This represents a considerable load on the environment.

In 1946 a synthetic surfactant, tetrapropylenebenzene sulfonate (TBS), was first used to replace soap in cleaning formulations in the United States [25]. TBS was the most commonly used of the alkylbenzene sulfonate (ABS) class of surfactants. It soon became the major surfactant in formulations around the world wherever chemical technology made it possible. The rapid takeover by TBS was due to several factors, including better performance, since it does not precipitate with calcium and magnesium ions present in hard water in the way that soap does, and low cost of production. In addition, the availability of suitable fats and oils from which to manufacture soap had become limited during the First and Second World Wars [26]; this stimulated the move toward alternatives, which were manufactured from petroleum-based raw materials.

The accumulation of TBS caused large mountains of foam to build up in rivers, canals and water treatment plants. Examples of this occurred in 1953 on the Ohio River, which receives effluent from Chicago wastewater treatment plants [25], and in Germany during the drought summers of 1959 and 1960 [27]. TBS undergoes biodegradation relatively slowly; in addition it is not removed from waters by precipitation to the same extent that soaps are. These factors, together with its low foaming threshold of about 1 part per million (ppm) [25], led to the problems described earlier. Levels of ABS in sewage effluents ranged up to about 5–10 ppm in the early 1960s. In addition to the nasty appearance of foam on waterways, the concentrations of surfactant were in the range that is toxic to some aquatic life forms [27]. The dramatic effect when TBS accumulates in waterways initiated the growth of concern about the biodegradability of chemicals discharged into the environment. This has now become a high priority to consumers of all types of products disposed of to the environment.

Very soon after TBS was identified as the cause of foaming problems in waterways, laws were put in place requiring detergents to be biodegradable [27]. Soon an alternative class of surfactants, linear alkylbenzene sulfonates (LAS), was developed that are more readily biodegraded while offering the performance and cost advantages of TBS. The success of using more readily biodegradable surfactants is exemplified by the drop in alkylbenzene sulfonate levels observed in the Pekin–Peoria stretch of the Illinois River. This averaged around 0.5 ppm during 1959–1965 and dropped to 0.05 ppm by 1968 [25].

Since synthetic surfactants were first produced, the variety available has expanded rapidly to meet the specific requirements of their diverse applications. They are commonly classified according to the nature of the hydrophilic group, which may be anionic, nonionic, cationic, or zwitterionic. Common hydrophilic groups are sulfonate ($-SO_3^-$), sulfate ($-OSO_3^-$), carboxylate ($-CO_2^-$), polyoxyethylene ($-(OCH_2CH_2)_nOH$), polyglucoside ($-(O-C_6H_7O(OH)_3-O)_n-H$), sucrose ($-O-C_6H_7O(OH)_3-O-C_6H_7O(OH)_4$), polypeptide ($-NH-CHR-CO-NHCHR-COOH$), quaternary ammonium ($R_4N^+$), and betaines ($RN^+(CH_3)_2CH_2CO_2^-$). Hydrophilic groups can be linked to a range of hydrophobic groups derived from either petroleum or animal or vegetable sources.

Simultaneous with the expansion in the range of surfactants available, a considerable amount of research has been performed on the biodegradability of surfactants so as to minimize detrimental effects on the environment [25,28–30]. Much is known regarding both the aerobic and anaerobic biodegradabilities of conventional surfactants. It has been found that the biodegradability of surfactants is largely controlled by the structure of the hydrophobic group and the nature of the link between the hydrophilic and hydrophobic groups. The structure of the hydrophobic group varies depending on whether it is derived from petrochemical or oleochemical raw materials.

The significance of any surfactant accumulation that may occur depends on the effects that the surfactant has on the environment. Surfactants are known to have low human toxicities but fairly high toxicities toward aquatic organisms [27,31]. By comparing reported toxicity data and environmental concentration data, the risk that surfactants pose to the environment may be assessed. Such a comparison shows that for certain surfactants there is a zone of uncertainty where the environmental concentrations and the effect concentrations overlap, leading to some risk of detrimental effects. For example, the lowest no-observed-effect concentration (NOEC) for anionics was 0.1 mg/L, while the highest environmental concentration was 0.3 mg/L. Similarly, the lowest NOEC for nonionics is 0.18 mg/L, while concentrations up to 0.8 mg/L have been measured in the environment. However, it must be realized that concentrations in the environment greater than the effect levels are exceptional, occurring only in the worst-case situations. In the majority of cases, environmental concentrations are well below those at which adverse effects occur.

Clearly the faster surfactants are degraded, both aerobically and anaerobically, the more their environmental impact is minimized. Generally a material that undergoes 80% biodegradation within 21 days is considered readily biodegradable and unlikely to accumulate in the environment. In general it has been found that the rate at which common surfactants degrade in most sewage treatment plants is sufficient to ensure that concentrations do not accumulate to detrimental levels in the environment. However, in cases where effluent dilution is low or treatment plants are not operating optimally, the safety margin between environmental concentrations and effect concentrations is less than desirable. For this reason there are advantages to be gained from the use of surfactants whose environmental safety is more certain, particularly in situations where waste treatment is least efficient and effluent dilution factors are low.

III. BIODEGRADABILITY OF COMMON AND SUGAR-BASED SURFACTANTS

A. Common Surfactants

Many laboratory studies have been performed to determine the biodegradabilities of a wide range of surfactants [25,27,28]. The results are reported as either primary degradation, referring to the initial oxidation or other alteration of the molecule so that characteristic properties, such as foaming, are destroyed; or ultimate degradation, which is the complete conversion of the molecule into CO_2, water, inorganic salts, and metabolic products or biomass. Many international and national standard methods have been established to measure these stages of biodegradation [28,32,33]. The test methods are classified as screening tests—used to determine readily biodegradable materials under harsher than real conditions; confirmatory tests—used to confirm the biodegradability of materials under more realistic conditions, and simulation tests, which simulate the real environmental conditions. The aerobic biodegradabilities of a selection of surfactants determined in studies employing screening tests are shown in Table 1 and those determined in confirmatory or simulation tests in Table 2.

Most of the surfactants listed in Table 1 and Table 2 were found to pass the tests for ready biodegradability. The exceptions are the anionic branched alkylbenzene sulfonates, the nonionic alcohol polypropoxy-polyethoxylates and branched alkylphenol ethoxylates, the cationic didodecyl dimethylammonium chloride, and the zwitterionic surfactants (the last show ready primary but not ultimate biodegradation). Branched alkylbenzene sulfonates and alcohol polypropoxy-polyethoxylates are also found not to pass the confirmatory test. Branched alkylphenol ethoxylates are generally found to pass the confirmatory test but show incomplete biodegradation in simulation tests for ultimate biodegradability. The cationics are shown to be effectively removed during simulation tests. More detailed investigation of the behavior of cationic surfactants in simulation tests has, however, shown that their initial removal is due mainly to adsorption to sludge. Monoalkyl trimethylammonium chlorides and alkylbenzyl dimethylammonium chlorides extensively biodegraded after adsorption to sludge [28]. In the case of dialkyl dimethylammonium chloride, biodegradation after adsorption to sludge is much slower. The zwitterionic alkylbetaines were found to be extensively removed in the coupled units simulation test [28].

Compared with the amount of information available regarding the aerobic biodegradability of surfactants, there is little regarding their anaerobic biodegradation. Some studies of anaerobic biodegradation have been performed [25,28], the findings are summarized in Table 3.

As well as laboratory studies of biodegradability, some observations of surfactant biodegradation in actual sewage treatment plants and in natural waterways have been made. The most famous of these was performed in Luton, England, in the early 1960s. Over a wide area around Luton all stocks of branched alkylbenzene sulphonate–based detergents were replaced by products formulated with an essentially linear alkylbenzene sulphonate; this resulted in a dramatic decrease in the ABS content of Luton's treated sewage [25]. In other field studies, surfactants have been found to be significantly removed from sewage, by combined adsorption and biodegradation processes in the sewer, before they reach treatment plants [25,34–36], with up to 35% of this removal being accounted for by primary biodegradation [25].

The in situ removal of surfactants in sewage treatment plants has also been investigated. It has been found that activated-sludge sewage treatment processes with aerobic sludge digestion are the most effective and that trickling-filter systems are least effective [37]. Activated-sludge sewage treatment with aerobic sludge digestion resulted in 94–99% removal of LAS. Essentially all of this removal was accounted for by primary biodegradation [35–38]. Activated-sludge treatment with anaerobic sludge digestion gave 96–99% LAS removal, with 68–91% removal accounted for by primary biodegradation. In contrast, LAS was only 77% removed in a trickling-filter treatment plant, with 33% accounted for by primary biodegradation. A better result of 96% removal was achieved in a trickling-filter plant by recycling the effluent [39]. Nonylphenol ethoxylate was 84–98% removed in an activated-sludge treatment plant [38], and alkyl sulfate was more than 90% removed in a trickling-filter plant [40].

TABLE 1 Surfactant Biodegradabilities Determined in Screening Tests

Surfactants	Primary biodegradation (%)	Ultimate biodegradation (%)	Ref.
Anionic surfactants			
Sodium laurate		100	28
Sodium stearate (soap)		85–100	28
Linear alkylbenzene sulfonates	90–95	65, 73	27
Branched alkylbenzene sulfonates	8–25	0–13	27
Secondary alkyl sulfonates (C13–18)[a]	97–98	73, 80	27
Linear primary alkyl sulfates (C16–18)	99	91, 88	27
α-Sulfonyl fatty acid methyl esters (C16–18)	99	76	27
Sucrose α-sulfonyl fatty acid esters (C12–16)		68–98	51
Nonionic Surfactants			
Linear alcohol ethoxylate (C12E8)[b]	99–100	77–100	25
Linear alcohol ethoxylates (C16–18E14)	99	80, 86	27
Linear alcohol ethoxylates (C12–14E30)	99	27	27
Linear alcohol polypropoxy-polyethoxylates (C12–18P8E5)[c]	70	15	25
Linear alkylphenol ethoxylates (C8–10E9)	84	29	27
Branched alkylphenol ethoxylate (C9E9)	6–78	5–17	27
Sucrose fatty acid esters (C12–16)		99–100	51
Alkyl polyglycoside (C12/14)		72–88	45
Cationic surfactants			
Dodecyl trimethylammonium chloride	100	63, 90	25
Didodecyl dimethylammonium chloride	25–75	0	25
Dodecyl benzyl dimethylammonium chloride	100	97, 95	25
Zwitterionic surfactants			
Alkylbetaine (C12)	100	55	28
Sulfobetaines	90–97	25–40	28

[a] C13–18 represents surfactants with alkyl chain lengths of 13–18 carbon atom.
[b] E8 represents surfactants with 8 ethoxylate units.
[c] P8 represents surfactants with 8 propoxylate units.

TABLE 2 Surfactant Biodegradabilities Determined in Model Sewage Treatment Plants

Surfactants	Primary biodegradation (%)	Ultimate biodegradation (%)	Ref.
Anionic surfactants			
Linear alkylbenzene sulfonates	90–95	73 ± 6 (C)	27
Branched alkylbenzene sulfonates	36	41 ± 9 (COD)	27
Secondary alkane sulfonates (C13–18)[a]	97–98	93 ± 5	27
Linear primary alkane sulfates (C16–18)	100	100	25
α-Sulfonyl fatty acid methyl esters (C16–18)	95	98 ± 6	27
Nonionic surfactants			
Linear alcohol ethoxylate (C12E8)[b]	100		25
Linear alcohol ethoxylates (C16–18E10)	98	62 ± 28 (3-h retention time) 90 ± 16 (6-h retention time)	27
Linear alcohol ethoxylates (C12–14E30)	98	59 ± 20 (C)	27
Linear alcohol ethoxypropoxylates (C11–15E8P5)[c]		24 ± 5	25
Linear alkylphenol ethoxylates (C8–10E9)	96	68 ± 3	27
Branched alkylphenol ethoxylate (C9E9)	97	48 ± 6	27
Sucrose stearate	100		4
Cationic surfactants			
Octadecyl trimethylammonium chloride	98		25
Dioctadecyl dimethylammonium chloride	91–93	108 ± 9	25
Dodecyl benzyl dimethylammonium chloride	96	83 ± 7	27
Zwitterionic surfactants			
Alkylbetaines (C8–18)		95	28

[a] C13–18 represents surfactants with alkyl chain lengths of 12–18 carbon atoms.
[b] E8 represents surfactants with 8 ethoxylate units.
[c] P5 represents surfactants with 5 propoxylate units.

TABLE 3 Anaerobic Biodegradability of Some Common Surfactants

Surfactants	Anaerobically biodegradable
Anionic surfactants	
Soaps	Yes
Alkylbenzene sulfonates	No
Secondary alkane sulfonates	No
Linear primary alkane sulfates	Yes
α-Sulfonyl fatty acid methyl esters	No
Nonionic surfactants	
Linear alcohol ethoxylates	Yes
Alkylphenol ethoxylates	Partially
Sugar esters	Yes
Alkyl polyglycosides	Yes
Cationic surfactants	
Alkyl trimethylammonium chlorides	No

Source: Refs. 25 and 28.

Some studies of biodegradation have also been made in marine and estuarine conditions. Schimp [41] has studied the biodegradation of ^{14}C-labeled linear alkylbenzene sulfonate in samples of estuarine water from pristine and sewerage-effluent-exposed sites. It was demonstrated that exposure to LAS present in sewerage effluent enables its biodegradation by estuarine microbial communities. LAS was mineralized 40–60% in 28 days in effluent-exposed samples, while only 10% mineralization occurred in pristine samples. The study showed that marine microbial communities are less active toward LAS degradation than freshwater communities and that bacterial acclimatization is more essential. The importance of adaptation of marine microbial communities to surfactants was also demonstrated by Quiroga and Sales [42] and Terzic et al. [43]. Stalmans et al. [44] measured concentrations of LAS in the North Sea and found that samples taken 15 km or more offshore contained less than 0.0005 mg/L (the limit of detection), while samples taken in the mouth of the Scheldt River contained up to 0.0094 mg/L. This decrease in concentration further offshore, greater than that expected due to dilution, could be attributed to either biodegradation or increased settling of suspended solids under marine conditions or a combination of both.

B. Sugar-Based Surfactants

From their first discovery, the ready biodegradability of sugar ester surfactants has been one of their main attractions. Alkyl polyglycosides have been shown to be readily and completely biodegradable in the OECD closed-bottle test [45]. Sugar ester surfactants have also generally been found to be readily biodegradable via initial ester hydrolysis [25]. Biodegradation of alkyl plyglycosides and sugar esters occurs in both aerobic and anaerobic conditions [2,25].

Primary and ultimate biodegradabilities of sugar esters have been monitored by surface tension measurements, BOD and CO_2 evolution. Early studies showed that water-soluble sucrose esters promoted the oxygen uptake by inoculated media at a rate similar to that of LAS and to a much greater extent than TBS [46,47]. Sucrose esters of cottonseed oil showed 65.8% of theoretical BOD in 250 hours. Less soluble esters promoted oxygen uptake to a smaller extent; mixed sucrose stearate and palmitate gave only 11–23% of theoretical BOD in 250 hours. However, the biodegradation of sucrose stearate in a model sewage treatment plant was found to occur readily. A laboratory-scale semicontinuous activated-sludge plant was fed daily with sucrose stearate, to give a total concentration of 200 ppm (BOD of 400–450 ppm). Throughout 2 months of operation, good-quality effluent was produced with a BOD consistently less than 10 ppm [4]. The disappearance of sucrose ester surfactants from river water samples was studied by measuring the consequent increase in surface tension. This showed 100% primary biodegradation within 2 days [48,49]. Complete primary biodegradation within 2 days was also observed under anaerobic conditions [48].

The biodegradation of a related surfactant formed by ethoxylation of sucrose tallowate has also been studied [50]. Surface tension measurements indicated complete biodegradation in an inoculated medium within 7 days. Gas chromatography was used to analyze samples, taken from the medium during biodegradation, for fatty acids content. This revealed that ester hydrolysis occurred almost instantaneously upon addition of the surfactant to the culture, releasing fatty acids. The subsequent removal of the fatty acids from solution was found to occur within about 7 days. Stearic acid was removed more rapidly than palmitic acid. The levels of myristic and lauric acid initially decreased but then increased. The observed increase in the levels of the shorter-chain fatty acids was attributed to the presence of partial degradation products of the longer-chain fatty acids.

A comprehensive study of the biodegradabilities of sugar esters was published in 2000 [51–53]. The International Standards Organization procedure [32] for determination of biodegradation by dissolved organic carbon (DOC) was used to investigate the ultimate aerobic biodegradabilities of a range of nonionic and anionic sugar ester surfactants. This is a nonspecific measure of ultimate aerobic biodegradability. It was observed that nonionic sugar esters, such as sucrose laurate, undergo rapid and complete aerobic biodegradation within a day. This is more rapid than most conventional surfactants mentioned earlier. Anionic sugar ester surfactants, such as sucrose sulfonyl laurate, with a sulfonyl group adjacent to the ester bond, were found to be degraded more slowly, reaching 90% degradation after 25 days. The biodegradation of sucrose sulfonyl laurate was very similar to that of LAS, and it would be classified as readily biodegradable. The results of this study are discussed in more detail in later sections.

IV. EFFECT OF STRUCTURE ON BIODEGRADABILITY

An understanding of the relationship between structure and biodegradability enables the design of biodegradable surfactants. Studies of structure–biodegradability relationships have been reviewed by Swisher [54], Gerike and Jasiak [55], and Kravetz [56].

A. Common Surfactants

The biodegradability of surfactants is governed mostly by the structure of the hydrophobe and the nature of the link between the hydrophobe and the hydrophile

[25,28], while the chemical nature of the hydrophilic group is of less importance. The main exception to this rule is that the biodegradability of ethoxylate surfactants increases with decreasing hydrophile chain length.

The two following general observations have been made:

- Surfactants with branched hydrophobes, particularly quaternary branching, are more resistant to biodegradation than linear ones.
- Increased distance between a sulfonate group and the far end of the alkyl chain increases the speed of primary biodegradation.

The latter effect is referred to as the distance principle. Both effects are presumed to reflect the relationship between surfactant structure and the geometry of the enzymes performing the biodegradation steps.

The distance principle holds for ABS surfactants with chains up to 12 carbons long. Longer-chain ABS may have inhibitory effects on bacteria; however, if sufficient time is allowed for acclimatization, the distance principle will still hold. For chain lengths above about 18, the distance principle breaks down, possibly because of changes in solubility or surfactancy [25]. The distance principal also applies to other types of surfactants to some extent.

With regard to the extent to which branching hinders biodegradation, it has been found that single-terminal methyl groups do not have any significant effect on biodegradation and that the effect of several methyl groups attached to different carbons is small. It is the presence of quaternary branching, especially at or near the far end of the chain, that has an inhibitory effect on biodegradation. If an unbranched chain is present, in addition to the chain terminated by a quaternary branch, biodegradation may occur rapidly via oxidation of this chain. Even where all the chains have terminal quaternary branching, acclimatization of the microbial population can eventually be achieved, allowing biodegradation to occur.

As already mentioned, the biodegradability of ethoxylated nonionic surfactants decreases with increasing ethoxylate chain length; this may be due to the concomitant reduction in lipophilicity, inhibiting the passage of the surfactants through biological membranes. At the same time, branching in the hydrophobe chains also imparts some resistance to primary alcohol ethoxylate biodegradation. A single branch in an otherwise linear chain has little or no effect, but a multiplicity of branching can greatly impair biodegradation. For alcohol ethoxylates it appears that increased carbon chain length gives increased degradability, in accordance with the distance principle. However, this effect may be masked by simultaneous inhibition due to the longer ethoxylate chain lengths incorporated to retain solubility and surfactancy properties of these materials. Secondary alcohol ethoxylates are slightly slower to degrade than primary alcohol ethoxylates.

The biodegradation of alkylphenol ethoxylates (APE) has given rise to a great deal of controversy. Linear alkylphenol ethoxylates, which are mixed isomers with phenols attached at all positions along the carbon chain, show relatively slow biodegradability. Phenol linkage at or near the center of the chain increases resistance to a much greater extent than for corresponding LAS isomers. Branched alkylphenol ethoxylates are very resistant, especially if acclimitization of the microbial population to the APE has not occurred.

It has been generalized that predominantly linear alcohol ethoxylates (AE) undergo primary and ultimate biodegradation more rapidly than linear or branched

APE [56–58]. The resistance of the latter is exacerbated in cool weather or under other conditions that result in stress on biological systems (for instance, high surfactant load).

Quaternary ammonium surfactants with single alkyl chains are the most readily biodegradable; greater resistance to biodegradation is imparted by extra chains, benzyl groups, pyridinium, and quinolinium groups, in that order. The distance principle holds for chains up to 10 carbons long; above this length, bacteriostatic properties mask further increase in degradability.

B. Sugar-Based Surfactants

The number of investigations into the effects of structure on the biodegradability of sugar-based surfactants is relatively few. In one early study the rates of primary biodegradation of sucrose laurate, palmitate, and stearate (indicated by surface tension increase) were found to differ slightly, increasing with alkyl chain length [49]. Another study measured the ultimate biodegradation of the same sucrose esters by measurement of CO_2 evolution [59]. They all gave greater than 70% biodegradation within 30 days. In this study the biodegradation rate (indicated by CO_2 evolution) was found to decrease with increasing chain lengths. The discrepancy between the findings of these two studies may be due to the different biodegradation methods and measurement techniques employed in each case. It has been demonstrated that the presence of extensive alkyl chain branching greatly reduces the biodegradability of sugar ester surfactants [25].

A systematic study of the impacts that structural changes have on the ultimate aerobic biodegradability of sugar ester surfactants has now been performed, to reveal some correlations between the structure and biodegradability of these surfactants [51–53]. An array of sugar ester surfactants whose molecular structures were systematically varied was studied. The structures of the sugar surfactants included in the study are shown in Figure 4. The size of the sugar headgroup was varied between a monosaccharide and a trisaccharide; the length of the alkyl chain was varied between 12 and 16 carbons; structures with two alkyl chains attached were studied, and structures where a sulfonate or alkyl side group had been attached in a position α to the ester bond were also included.

First sucrose laurate and its anionic analogue, sucrose sulfonyl laurate, where a sulfonate group is attached α to the ester bond, were considered. The biodegradation profiles for sucrose laurate and sucrose sulfonyl laurate, obtained in five separate biodegradation tests, are shown in Figure 5. There is some scatter in the results, which was also observed for standard surfactants included in the study; however, it is clearly seen that sucrose laurate is completely degraded within a day while the sucrose sulfonyl laurate degrades more slowly, reaching about 90% degradation after 25 days.

The ready biodegradability of sucrose esters, as observed for sucrose laurate in this study, has previously been ascertained. The biodegradation of α-sulfonated sugar esters has not previously been determined; however, that of other substances containing a sulfonyl group α to an ester bond have been reported. The primary biodegradation of methyl α-sulfonyl fatty acid esters is reported to be complete within 3–10 days under aerobic conditions [27,48,60–62]. The ultimate biodegradation is reported to reach about 80% within 20 days [27,63]. These results are similar

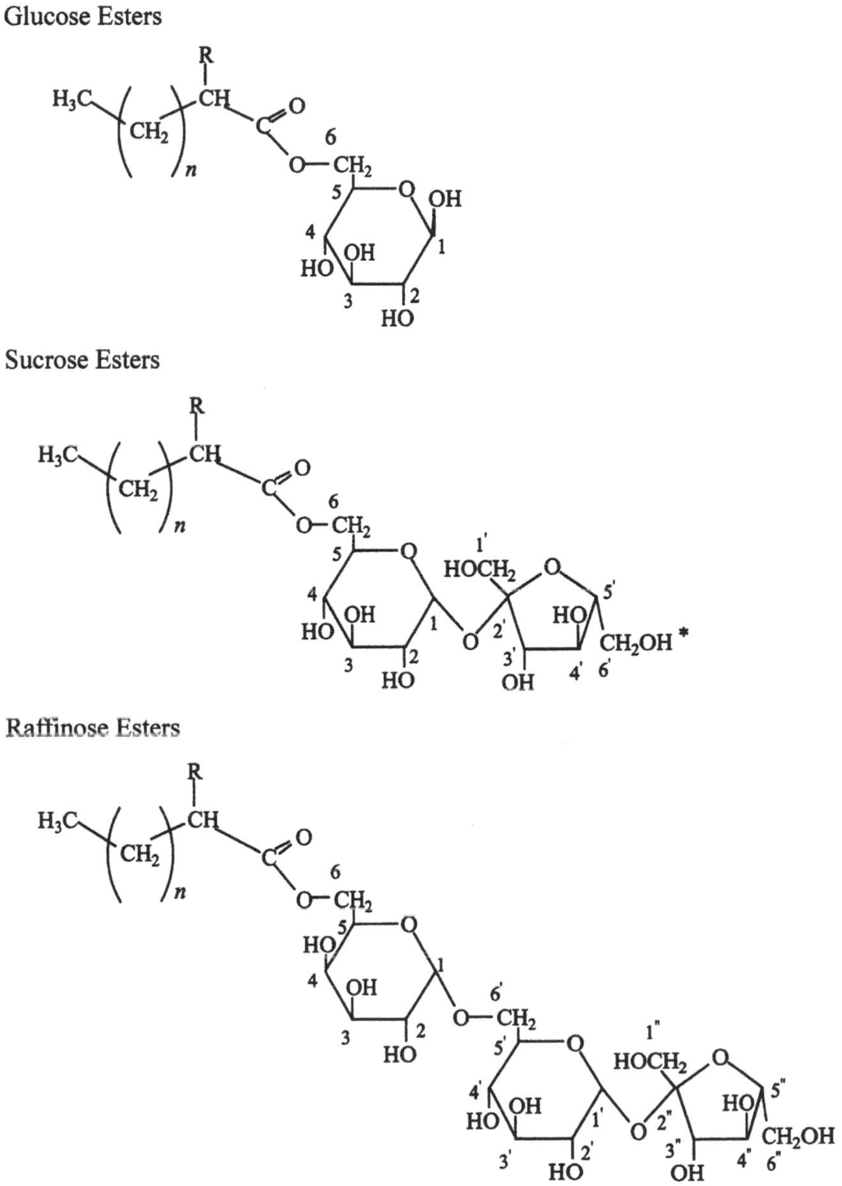

Glucose Esters

Sucrose Esters

Raffinose Esters

FIG. 4 Schematic of the sugar ester surfactants included in the study by Baker et al. (From Refs. 51–53.)

to those observed for α-sulfonated sugar esters. It is evident that the addition of the α-sulfonate group produces an ester that is biodegraded at a dramatically reduced rate compared with the unsulfonated analogue. It is also known that the stability of the ester bond toward chemical hydrolysis is increased by the presence of a sulfonate group in the α position [60,64,65].

The effect of changes in sugar headgroup size on the biodegradability of the unsulfonated and the sulfonated surfactants was not significant. In the case of the

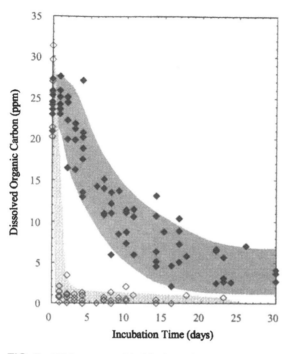

FIG. 5 Ultimate aerobic biodegradation profiles determined for sucrose laurate and sucrose sulfonyl laurate by nonspecific dissolved organic carbon analysis. (From Refs. 51–53. Reprinted with permission of AOCS Press.)

sulfonated surfactants, the presence of two alkyl chains, rather than one, was not observed to affect biodegradability significantly. Similarly, variations in the alkyl chain length were not found to have a significant effect. As already mentioned, previous studies have reported slight differences between the primary and ultimate biodegradabilities of sucrose laurate, palmitate, and stearate [49,59]. The previous investigations involved test methods that resulted in lower overall biodegradation potentials than that employed in the more recent study. The effect of this is that in the two earlier studies the biodegradation of the materials tested occurred over a longer time frame, such that small differences in biodegradability were observed with greater resolution. Thus it is possible that there may be slight differences in the biodegradation rates of sugar ester surfactants with different chain lengths, which have not shown up on the time scale of the test used in the recent study. Any such effects are unlikely to be of significance in real environmental situations, such as in sewage treatment plants, where the biodegradation potential is higher than that of these screening tests.

The results obtained for sucrose methyl laurate, with a methyl group α to the ester bond, and sucrose ethyl laurate, with an ethyl group α to the ester bond, indicated reduced biodegradability of these esters relative to the underivatized sucrose ester, but not as low as in the presence of a sulfonate group in the same position. This indicates that the inhibitory effect of the α-sulfonate group is unlikely to be due solely to its charge but is also a consequence of its size.

V. BIODEGRADATION PATHWAYS

In many instances it is helpful to know the sequence of chemical alterations that take place during the bacterial breakdown of a material, referred to as the biodegradation pathway. Such knowledge can help to explain the differences in biodegradation behavior between different chemical structures. It also helps to determine whether incomplete ultimate biodegradation of a substance is likely to result in the buildup of toxic intermediate metabolites. The biodegradation pathways of a number of common surfactants have been determined [25,28].

A. Common Surfactants

The biodegradation pathway of alkylbenzene sulfonates is depicted in Figure 6. The first phase of biodegradation is conversion of the terminal methyl group to carboxylate by ω-oxidation. This is analogous to the microbial attack that occurs in the case of linear hydrocarbons. This terminal β-oxidation step for the most part destroys the surfactancy properties of even the longest-chain species. The next phase is rapid β-oxidation of the carbon chain. In this process, acetate groups are

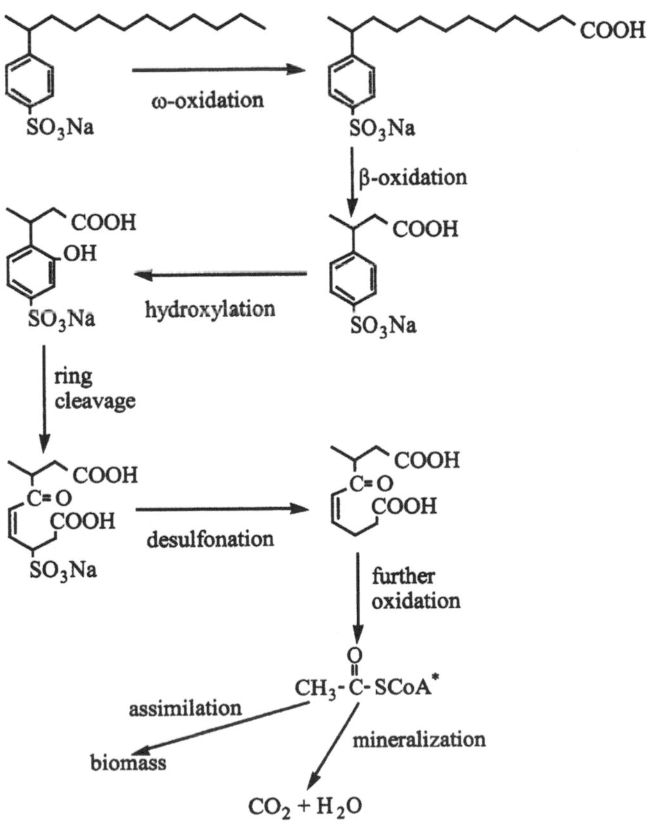

FIG. 6 Schematic diagram showing the biodegradation pathway of alkylbenzene sulfonates. (*CH₃COSCoA represents an acetyl group bound to coenzyme A.)

sequentially cleaved off, leaving carboxylates two carbons shorter than the original. This can continue until about four carbons remain in the chain. At this stage the sulfophenyl group appears to clash with the topography of the β-oxidation enzyme. Thus 3-sulfophenyl butyrate has been observed as an intermediate [25]. However, once acclimatization has occurred, the benzene ring disappears and inorganic sulfate is liberated. α-Oxidation may occur in place of β-oxidation, removing one carbon at a time. The presence of chain branching inhibits β-oxidation, but α-oxidation can achieve further breakdown albeit at a slower rate. Dicarboxylates may also be formed by oxidation of both ends of the alkyl chain.

The biodegradation of alcohol ethoxylate surfactants can occur via two alternative pathways. The most common route involves initial fission of the molecule at the ether linkage into hydrophobic and hydrophilic parts. The hydrophobic portion is then degraded by β-oxidation, while the hydrophilic portion is degraded by sequential cleavage of ethoxylate units from the chain. The alternative route is by simultaneous stepwise ethoxylate cleavage and slower β-oxidation of the alkyl chain without initial fission into two parts. In contrast, alkylphenol ethoxylate biodegradation proceeds via slow oxidation of the ethoxylate chain, while the rest of the molecule remains intact. Biodegradation via initial fission is in this case sterically blocked by the benzene group.

Alkyl trimethylammonium surfactants are thought to be degraded by initial fission of the C–N bonds, forming alkanals and trimethylamine. The alkanal portions are subsequently degraded by β-oxidation and trimethylamine by sequential removal of the methyl groups in the form of formaldehyde. The alternative route via initial ω- and β-oxidation of the alkyl chain has been found not to occur. The initial fission of the C–N bond is more sterically hindered in the case of di- and trialkylammoniums, which explains their increased resistance to biodegradation.

Alkyl betaines are degraded by initial ω- and β-oxidation of the alkyl chain. The resultant trimethyl glycine occurs naturally and is subsequently degraded by previously known metabolic paths.

B. Sugar-Based Surfactants

It has been known for some years that the biodegradation of sugar esters proceeds by initial hydrolysis of the ester bond. It has also been reported that α-sulfonyl methyl esters of fatty acids are degraded by initial oxidation of the alkyl chain followed by desulfonation, ester hydrolysis, and bioassimilation [63]. By analogy with α-sulfonyl methyl esters studied previously, the α-sulfonyl fatty acid esters of sugar could be expected to be degraded by initial oxidation of the alkyl chain. The pathways followed during biodegradation of three key sugar ester surfactants have been determined [51–53]. The three surfactants studied were sucrose laurate, sucrose sulfonyl laurate, and sucrose ethyl laurate, which have significantly different aerobic biodegradabilities.

Biodegradation pathways were determined by identification of intermediates by HPLC and NMR analysis of samples taken from the biodegradation test medium at intervals throughout the tests. Sucrose laurate was determined to be degraded by initial hydrolysis of the ester bond, forming free lauric acid and sucrose, consistent with previously reported findings. In contrast, the degradation of sucrose sulfonyl laurate shown in Figure 7 does not occur by initial hydrolysis but by a pathway

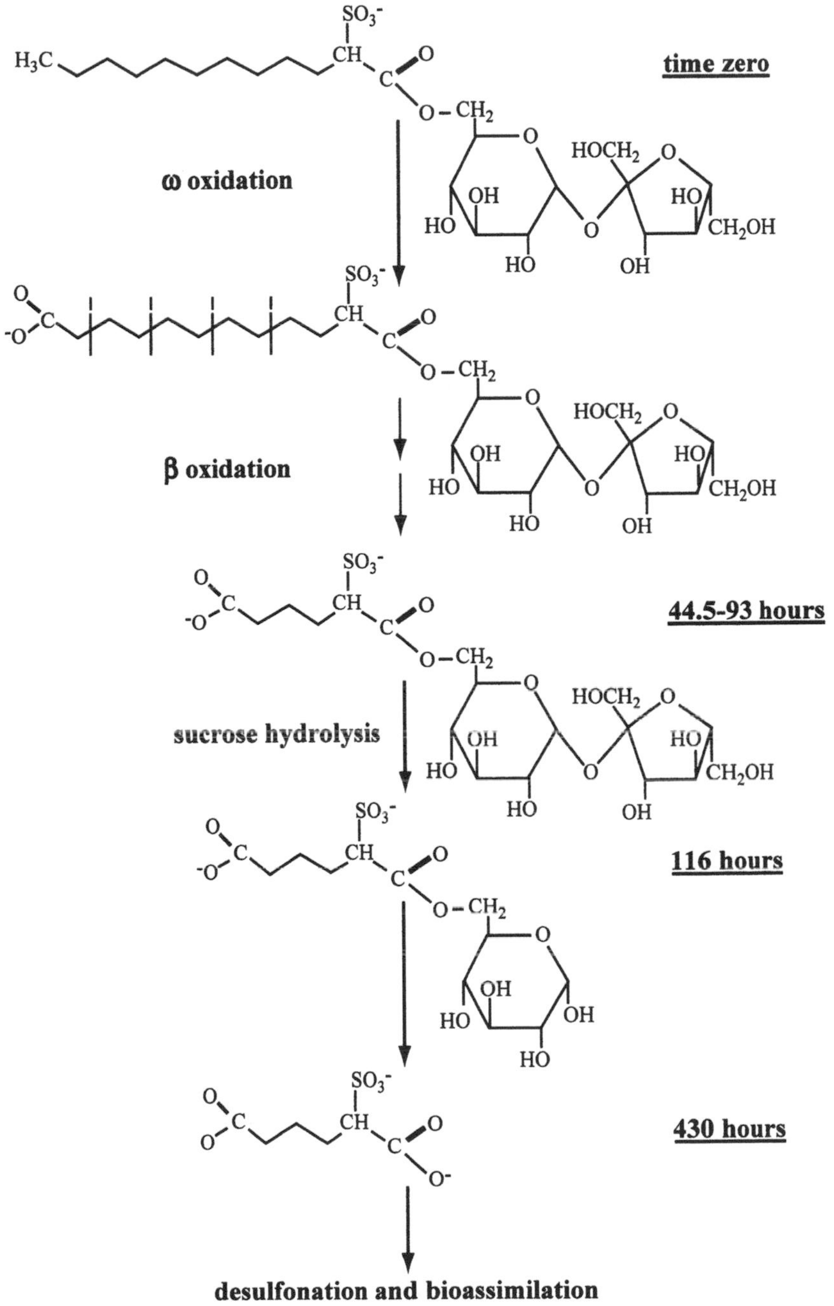

FIG. 7 Biodegradation pathway determined for sucrose sulfonyl laurate. (From Ref. 52. Reprinted with permission of AOCS Press.)

analogous to those previously reported for linear alkylbenzene sulfonates and α-sulfonyl methyl esters. The ultimate biodegradation behavior of sucrose sulfonyl laurate, shown in Figure 5, was very similar to that of linear alkylbenzene sulfonate, lending further support to the proposal that these two compounds are degraded by similar pathways involving initial oxidation of the alkyl chains.

The fact that sucrose sulfonyl laurate is not biodegraded via initial ester hydrolysis, as occurs in the case of sucrose laurate, indicates that the presence of the sulfonyl group adjacent to the ester bond inhibits ester hydrolysis. Consequently the sulfonyl ester cannot be degraded by initial ester hydrolysis but follows an alternative pathway involving initial alkyl chain oxidation. Because the latter process occurs more slowly than the former, this has the effect that the two compounds show different ultimate biodegradation rates as measured by dissolved organic carbon analyses. In the case of sucrose ethyl laurate, the presence of an ethyl group adjacent to the ester bond was found to inhibit ester hydrolysis so that biodegradation of sucrose ethyl laurate occurs, at least in part, via alkyl chain oxidation. Further studies suggested that steric effects are the major factors controlling the relative rates at which the three types of sugar esters are degraded.

The differences in the pathways followed during biodegradation of the three key sucrose ester surfactants explain the different rates at which they were found to biodegrade. The combination of HPLC and ^{1}H NMR analysis techniques provided a powerful method for determining biodegradation pathways.

VI. CONCLUSIONS

Surfactants formed from sugar and fatty acids have attracted attention from manufacturers and product developers for many decades because they are based on renewable resources, are readily biodegradable, and are very mild on the skin and eyes. The performance advantages of many sugar-based surfactants are well established. They are utilized in cleaning agents, pharmaceutics, cosmetics, personal care, and food products. There is a comprehensive understanding of the biodegradation of many sugar-based surfactants. They are generally found to be rapidly and completely biodegradable under aerobic and anaerobic conditions. There is also a good understanding of the effects that structural variations have on the performance and biodegradability of sugar esters, allowing the targeted desig of environmentally benign surfactants for specific applications.

REFERENCES

1. Hill, K. Pure Appl. Chem. 2000, 72 (7), 1255–1264.
2. Hill, K.; Rhode, O. Fett/Lipid 1999, 101 (1), S:25–33.
3. Matsumura, S.; Imai, K.; Yoshikawa, S.; Kawada, K.; Uchibori, T. J. Am. Oil Chem. Soc. 1990, 67 (12), 996–1001.
4. Isaac, P.C.G.; Jenkins, D. Conference of Biological Waste Treatment: Advances in Biological Waste Treatment; London: Macmillan, 1963 (Paper no. 5), 61–76.
5. Salka, B. Cosmetics Toiletries 1993, 108, 89–94.
6. Siracusa, P.A. happi Apr., 1992; 100–108.
7. Hauthal, H.G. Chim. Oggi May, 1992; 9–13.

8. Brancq, B. Proceedings of the 3rd CESIO International Surfactants Congress, London, Plenary Lectures, June 1992; Plenary Lectures, 60–66.

9. Schulz, P. Chim. Oggi Aug/Sep, 1992; 33–38.

10. Parker, K.J.; James, K.; Horford, J. In *Sucrochemistry*; Hickson, J.L., Ed.; Washington, DC, 1977; Chapter 7, 97–114.

11. Osipow, L.; Snell, F.D.; Marra, D.; York, W.C. Ind. Eng. Chem. 1956, *48* (9), 1462–1464.

12. Feuge, R.O.; Zeringue, H.J., Jr; Weiss, T.J.; Brown, M. J. Am. Oil Chem. Soc. 1957, *34*, 185–188.

13. Osipow, L.; Rosenblatt, W. Paper presented at the American Oil Chemists Society Meeting, Philadelphia, 1966.

14. British Patent 1332190 to Dai-Ichi Kyogo Seigako Co. Ltd.: Kyoto, Japan, 1973.

15. Seino, H.; Uchibori, T.; Nishitani, T.; Inamasu, S. J. Am. Oil Chem. Soc. 1984, *61* (11), 1761–1765.

16. Vulfson, E.N. In *Surfactants in Lipid Chemistry: Recent Synthetic, Physical and Biodegradative Studies*; Tyman, J.H.P., Ed.; The Royal Society of Chemistry: London, 1992; 16–37.

17. Ryoto Sugar Ester Technical Information. Nonionic Surfactant/Sucrose Fatty Acid Ester/Food Additive; Mitsubishi-Kasei Foods Corporation, 1989.

18. Söderbrg, I.; Drummond, C.J.; Furlong, D.N.; Godkin, S.; Matthews, B. Colloids Surf. A 1995, *102*, 91–97.

19. Godkin, S.; Furlong, D.N.; Drummond, C.J.; Matthews, B. Proceedings of the 3rd CESIO International Surfactants Congress, London, June 1992; Vol. C, 266 pp.

20. Herrington, T.M.; Midmore, B.R.; Sahi, S.S. In *Microemulsions and Emulsions in Foods*; El-Nokaly, M., Cornell, D., Eds.; American Chemical Society: Washington, DC, 1991.

21. Sethi, S.C.; Adyanthaya, S.D.; Deshpande, S.D.; Kelkar, R.G. J. Surf. Sci. Technol. 1986, *2* (2), 103–107.

22. Bistline, R.G.; Smith, F.D.; Weil, J.K.; Stirton, A.J. U.S. Patent 3,808,200, April 30, 1974.

23. Vogel, W.J.B. Paper presented at the 3rd CESIO International Surfactants Congress, London, June 1992.

24. Schirba, C.A. Int. News Fats Oils Relat. Mater. 1991, *2* (2), 1062–1072.

25. Swisher, R.D. In *Surfactant Biodegradation*, 2nd Ed.; Marcel Dekker: New York, 1987.

26. White, G.F.; Russell, N.J. In *Surfactants in Lipid Chemistry: Recent Synthetic, Physical and Biodegradative Studies*; Tyman, J.H.P., Ed.; Royal Society of Chemistry: London, 1992.

27. Gerike, P. In *Surfactants in Consumer Products: Theory, Technology and Applications*; Falbe, J., Ed.; Springer-Verlag: Heidelberg, 1987.

28. Karsa, D.R.; Porter, M.R., Eds. *Biodegradability of Surfactants*; Blackie Academic and Professional/Chapman and Hall: Glasgow, 1995.

29. Falbe, J., Ed.; *Surfactants in Consumer Products: Theory, Technology and Application*; Springer-Verlag: Heidelberg, 1987.

30. Pitter, P.; Chudoba, J. *Biodegradability of Organic Substances in Aquatic Environments*; CRC Press: Boca Raton, FL, 1990.

31. Sivak, A.; Goyer, M.; Perwk, J.; Thayer, P. In *Solution Behaviour of Surfactants*; Mittal, K.L., Fendler, E.J., Eds.; Plenum Press: New York, 1982; Vol. 1, 161–198.

32. Water Quality—Evaluation in an Aqueous Medium of the Ultimate Aerobic Biodegradability of Organic Compounds—Method by Analysis of Dissolved Organic Carbon (DOC), ISO 7827, 1984 (E).

33. Revised Guidelines for Tests of Ready Biodegradability and Revised Zahn–Wellens/EMPA Tests for Inherent Biodegradability, OECD Chemicals Group, Paris, 1993.

34. Sanchez Leal, J.; Ribosa, I.; Gonzalez, J.J.; Garcia, M.T. Eur. Water Pollut. Control 1991, *1* (4), 51–55.

35. Cavalli, L.; Gallera, A.; Landone, A. Environ. Toxicol. Chem. 1993, *12*, 1777–1788.

36. Berna, J.L.; Ferrer, J.; Moreno, A. Tenside Surfactants Deterg. 1989, *26* (2), 101–107.

37. McAvoy, D.C.; Eckoff, W.S.; Rapaport, R.A. Environ. Toxicol. Chem. 1993, *12*, 977–987.
38. Di Corcia, A.; Samperi, R.; Marcomini, A. Environ. Sci. Technol. 1994, *28*, 850–858.
39. Field, J.A.; Barber, L.B.; Thurman, E.M.; Moore, B.L.; Lawrence, D.J.; Peake, D.A. Environ. Sci. Technol. 1992, *26*, 1140–1148.
40. Fendinger, N.J.; Begley, W.M.; McAvoy, D.C.; Eckoff, W.S. Environ. Sci. Technol. 1992, *26*, 2493–2498.
41. Schimp, R.J. Tenside Surfactants Deterg. 1989, *26* (6), 390–393.
42. Quiroga, J.M.; Sales, D. Tenside Surfactants Deterg. 1991, *28* (3), 200–203.
43. Terzic, S.; Hrsak, D.; Ahel, M. Water Res. 1992, *26* (5), 585–591.
44. Stalmans, M.; Matthijs, E.; DeOude, N. T. Water Sci. Technol. 1991, *24* (10), 115–126.
45. von Rybinski, W.; Hill, K. Angew Chem. Int. Ed. 1998, *37*, 1328–1345.
46. Isaac, PC.G.; Jenkins, D. Chem. Ind. Aug. 1958, *2*, 976–977.
47. Isaac, P.C.G.; Jenkins, D.J. Proc. Inst. Sewage Purif. 1960; 314–329.
48. Wayman, C.H.; Roberts, J.B. Biotechnol. Bioeng. V, 1963; 367–384.
49. Ruiz Cruz, J.; Doborganes Garcia, M.C. Grasas Aceitas 1978, *29*, 1–8.
50. Brebion, G.; Cabridenc, R.; Lerenard, A. Rev. Fr. Corps Gras. 1964, *11*, 191–204.
51. Baker, I.J.A.; Matthews, B.; Suares, H.; Krodkiewska, I.; Furlong, D.N.; Grieser, F.; Drummond, C.J. J. Surfactants Deterg. 2000, *3* (1), 1–11.
52. Baker, I.J.A.; Willing, R.I.; Furlong, D.N.; Grieser, F.; Drummond, C.J. J. Surfactants Deterg. 2000, *3* (1), 12–27.
53. Baker, IJ.A.; Furlong, D.N.; Grieser, F.; Drummond, C.J. J. Surfactants Deterg. 2000, *3* (1), 29–32.
54. Swisher, R.D. J. Water Pollut. Control Fed. 1963, *35*, 877–892.
55. Gerike, P.; Jasiak, W. Tenside Deterg. 1986, *23* (6), 300–304.
56. Kravetz, L. J. Am. Oil Chem. Soc. 1981, *58*, 58A–65A.
57. Kravetz, L.; Salanitro, J.P.; Dorn, P.B.; Guin, K.F. J. Am. Oil Chem. Soc. 1991, *68* (8), 610–615.
58. Huddleston, R.L.; Allred, R.C. 4th International Congress on Surface-Active Substances, 1964; 871–882.
59. Sturm, R.N. J. Am. Oil Chem. Soc. 1973, *50*, 159–167.
60. Kapur, B.L.; Solomon, J.M.; Blustein, B.R. J. Am. Oil Chem. Soc. 1978, *55*, 549–557.
61. Stein, W.; Baumann, H. J. Am. Oil Chem. Soc. 1975, *52*, 323–329.
62. Weil, J.K.; Stirton, A.J. J. Am. Oil Chem. Soc. 1964, *41*, 355–358.
63. Masuda, M.; Odake, H.; Miura, K.; Ho, K.; Yamada, K.; Oba, K. J. Jpn. Oil Chem. Soc. (Yukagaku) 1993, *42* (ii), 21–25.
64. Lower, E.S. La Revista Delle Sostanze Grasse, 1986, *LXII*, 271–280.
65. Weil, J.K.; Bistline, R.G., Jr.; Stirton, A.J. J. Am. Chem. Soc. 1953, *75*, 4859–4860.

29

Biotreatment of Wastewater with a High Concentration of Surfactants

ODED VASHITZ and ESTER GORELIK   Zohar Dalia, Kibbutz Dalia, Israel

I. INTRODUCTION

The detergent industry has been transforming in the last 50 years from the use of soap to the use of synthetic surfactants as the main workhorse. Such transformation has encountered new problems in the biodegradation of surfactants by microbial cultures in wastewater. Soap is derived from natural sources—plant or animal. As such it is biodegradable by microbial cultures that were adapted to its components throughout thousands of years in use. The synthetic surfactants, produced from petrochemical sources by chemical processes, formed new, xenobiotic compounds. As such, the existing enzymatic systems of the microbial cultures could not recognize and degrade these compounds and they were defined as "hard" surfactants.

The history of synthetic surfactants is relatively short, but it is sufficient for various microbial cultures, mainly those that survived in environments with a high concentration of surfactants, to adjust their enzymatic systems and adapt to the synthetic compounds. Given sufficient time, microbial cultures mutate and adapt to degrade and utilize most of the compounds that form their environment. Such mutants are selected naturally by the rules of "survival of the fittest." Thus, surfactants that were formerly classified as "hard" will become gradually "softer" by natural genetic adaptation of the microbial system. The classification of a specific compound as "hard" or "soft" can therefore change with time.

In addition, the biodegradation of a chemical compound depends on its concentration. A specific microbial system that degrades a defined compound at low concentration may be inhibited by a higher concentration of the same compound. That compound will then be considered and treated as nonbiodegradable, or "hard." Biodegradation of contaminants at the high concentration that is typical of industrial wastewater is the main subject of this chapter.

Apart from biological treatment of wastewater, various physical, chemical, and physicochemical methods are used for treatment of highly concentrated surfactants in industrial wastewater streams. Various commercial methods will be mentioned here briefly as an alternative to biotreatment. Nonbiological treatment methods for the elimination of surfactants use either separation or decomposition technologies.

A. Separation Methods

1. Separation of anionic surfactants by adsorption to a specific water-insoluble polymer. The detergent–polymer sludge is separated from the effluents by a sedimentation unit. The toxic detergent–polymer concentrate is eventually sent to a hazardous waste treatment site at high cost or to some other disposal site according to local rules.
2. A similar method is based on a cationic flocullant. Anionic surfactants like LAS are precipitated as sludge and separated by filtration. The main disadvantages are the high cost of the flocullant and of the sludge disposal.
3. A process based on flocculation and sedimentation of detergents and other contaminants. An additional step of polishing by electrocatalytic method, by biotreatment, or by RO membranes is needed for proper removal of detergents. Daily adjustment of coagulants and flocculants is needed to achieve reasonable separation. Disposal of the concentrated sediment streams is expensive.
4. A one-stage membrane separation process composed of ultrafiltration membrane elements can achieve only partial removal of surfactants, to a concentration of 200–900 mg/L. For good surfactant removal, a second stage of reverse osmosis (RO) or nanofiltration is needed. The RO purified water can be recycled to production units. Frequent membrane cleaning and replacement is needed. The disposal of concentrate streams is expensive.
5. Methods for natural and induced evaporation or drying are used. Induced evaporation is expensive, and natural evaporation ponds have unpleasant ecological side effects. These methods are suitable for streams with a high concentration of solids.
6. Foam fractionation was suggested, by bubbling of air and collecting the separated foam. This method is limited by the critical micelle concentration to several hundreds of milligrams per liter of surfactant and is not efficient at higher concentrations.

An intrinsic disadvantage of all the separation methods is the need for disposal of the separated contaminants. Besides costs, the contaminants are not decomposed but accumulate and contaminate the global environment.

B. Decomposition Methods

1. Chemical decomposition of detergents by oxidation, using ozone or hydrogen peroxide. The use of ozone, produced onsite, is expensive, and therefore it is suitable mainly for low-contaminant-concentration, polishing applications of wastewater after pretreatment or in combination with other treatment methods.
2. *Low-pressure wet oxidation* was reported to be cheaper than the preceding methods. But this too is not a cheap solution. Tests with high-surfactant waste streams from a detergent plant were discontinued.
3. A similar solution is thermocatalytic oxidation of hazardous wastewater at 400–700°C. High oxidation efficiency was reported for nonionic detergents. Treatment cost is high.

None of the aforementioned methods is accepted as the industry standard, and specific solutions are tailored to the needs of each industrial source. Prevention of surfactant discharge by means of good manufacturing procedures in the production or processing site is of course the best and usually the cheapest solution. Dilution to low concentration is not acceptable legally, although this is a logical solution in certain applications.

The concept of sustainability forms an environmental management basis. One of the more widely quoted definitions is that of the "Bruntland report" [1], "Development that meets the needs of the present without compromising the ability of future generations to meet their own needs." On that basis, treatment of wastewater should not form waste that contaminates the global environment. The major surfactants in use, such as linear alkylbenzene sulfonates (LAS), are based on nonrenewable sources. Hence the best treatment approach should be that of decomposition of the surfactants, in contrast to separation and dumping or landfill of the contaminant concentrates. This approach is most important at the high concentration of industrial effluent sources.

It is well known in the industry that biodegradation is more cost effective than chemical decomposition. Advanced biodegradation methods should therefore be employed to mineralize such xenobiotic products at high concentration of industrial effluents and avoid contamination of the environment for future generations.

This chapter will describe various aspects of biodegradation of surfactants at high concentration in wastewater, defined as "hard" according to the foregoing concept, with an emphasis on practical solutions. In view of the ample literature on the biodegradation of surfactants at low concentration, such conditions will not be discussed thoroughly.

II. INDUSTRIAL VS. SANITARY WASTEWATER CONTAINING SURFACTANTS

Many publications analyze the biodegradation of surfactants in wastewater. Most of the published data about surfactants in wastewater concerns sanitary sewage in municipal treatment plants [2]. Activated-sludge plants are usually used for that purpose. The inlet concentration of anionic surfactants in such plants is lower then 40 mg/L. The microbial cultures in the sludge are adapted to that low concentration, and complete biodegradation of the surfactants can be achieved. An extensive Dutch environmental safety assessment [3] concluded that when treatment plants are operated properly, biodegradation of linear alkylbenzene sulfonate (LAS) is complete and that LAS under proper sewage treatment is safe for the environment. Extensive data and references about this subject can be found in the CLER homepage: http://www.cler.com/homea.html.

Industrial wastewater discharged from manufacturing and processing industries and from institutional consumers (e.g., commercial laundries, hospitals, hotels) is a totally different story. The concentration of surfactants in such wastewater can reach 500–10,000 mg/L. Natural microbial systems are not adapted to that concentration and are inactivated or inhibited. In addition, surfactant-concentrated effluents are highly foaming, which affects the choice of a technical solution. In most cases pretreatment of industrial effluents is needed before discharge to municipal treatment plants. Table 1 shows a comparison between industrial and sanitary wastewater.

TABLE 1 Comparison of Industrial and Sanitary Wastewater

Municipal wastewater	Industrial wastewater
High capacity: 10k–1M m³/day	Low capacity: 10–200 m³/day
Low biological load	High biological load
Biodegradable contaminants	Limited biodegradation
Low contaminant level: COD <200	High contaminant level: COD-500–100,000
Activated sludge	Pretreatment needed
Low cost per m³ Wastewater	High cost per m³ wastewater

About 57% of the surfactants are used in consumer products and are discharged to municipal systems at low concentration. About 37% are in technical and in industrial and institutional applications [4] that discharge wastewater with high surfactant concentration. Most of the large and medium-size industrial plants have pretreatment facilities. Several companies prefer to discharge untreated wastewater and pay a fee to central treatment facilities. Some of the smaller establishments discharge directly to municipal or public wastewater systems.

III. AEROBIC–ANAEROBIC BIODEGRADATION OF SURFACTANTS

The biodegradation of surfactants in wastewater is known to be more efficient under aerobic conditions. This is most pronounced for surfactants with lower biodegradability, such as LAS. Decomposition is faster, and the aerobic process can reach complete mineralization to CO_2, H_2O, and mineral salts. Under anaerobic conditions, decomposition is not complete and hydrocarbon intermediate products remain in the treated wastewater. Anaerobic biodegradation is also slower than aerobic, and a higher retention time is needed. Moreover, since anionic surfactants contain sulfur, anaerobic biodegradation products include reduced sulfo-organic compounds, such as mercaptanes, thio-compounds, and H_2S. Such compounds have disagreeable and irritating odors. Anaerobic biodegradation of anionic surfactants is avoided in most cases.

The anaerobic recalcitrance of LAS was demonstrated in a recent investigation by Federle and Schwab [5], who examined sediments of a wastewater pond that had been exposed to the discharge of wastewater from a local laundromat for more than 25 years. The authors reached similar conclusions by studying several anaerobic ponds of wastewater containing 3000–5000 mg/L LAS. Partial degradation was observed, with emission of H_2S and bad odors, possibly due to sulfur-reducing bacteria. The concentration had decreased slowly for 2–3 months and than stabilized at around 2000–3000 mg/L LAS. The slow and partial degradation observed can be related to slow subsurface oxygen diffusion in open ponds. This subject is also discussed in Section VII of this chapter.

The main problem with the aerobic biotreatment of surfactants is the production of foam, as a result of aeration in the treatment plant. The foaming problem can be solved by:

1. Addition of antifoam: At high surfactant concentration, a large quantity is needed, expressed by high cost. Antifoam products of organic source are usually

TABLE 2 Disadvantages of Aerobic and Anaerobic Biotreatment of LAS

Aerobic biotreatment	Anaerobic biotreatment
Foaming by aeration	Slow biodegradation
Sensitive to high concentration	Partial degradation
Higher sludge formation	Bad odors

decomposed or deactivated by the microbial cultures, so more antifoam has to be added, increasing the contamination load on the plant as well as the cost. Antifoam products that are not decomposed should be separated in a following step, which also adds to the cost.

2. Aeration by oxygen instead of air: The cost of oxygen is high, special aeration equipment and a dissolved oxygen control system is needed, so the overall cost is high.
3. Low-foaming aeration equipment based on a rotating biological contactor is a recommended solution for the problem [6].

Another problem encountered in aerobic biotreatment is inhibition and inactivation of the microbial cultures by a high concentration of surfactants. The selection of resistant and adaptable microbial cultures can offer a solution and will be discussed later.

Anaerobic treatment is used as a cost-effective solution for high organic load in wastewater due to minor production of sludge, but it is not a preferred solution for the major surfactants. A summary of the disadvantages of each method in existing biotreatment plants is given in Table 2.

Certain surfactants, such as aromatic and poly-aromatic surfactants, are degraded more efficiently by a combination of aerobic and facultative anaerobic cultures in the biotreatment process. Such a combination can be achieved in immobilized microbial environments composed of aerobic and anaerobic interconnecting zones. A high or low oxygen supply defines the aerobic or anaerobic nature of these internal zones. Subsequent aerobic and anaerobic processes are interconnected in an optimal degradation pathway. Such a process is more efficient than activated sludge, where biodegradation takes place in an aerobic vessel and biomass digestion proceeds in a separate, anaerobic vessel.

IV. LIMITATIONS ON DISCHARGE OF SURFACTANTS IN WASTEWATER

The limitations on discharge of surfactants to public treatment works are not consistent in different countries. Differences exist, even between states and districts. In the United States, EPA guidelines for discharge from industrial sources are based on the quantity of surfactants manufactured or processed. The rules are strict and definite but difficult to impose, because the concentration in wastewater is not defined. Local discharge permits are required.

Partial guidelines of pretreatment standards for discharge of LAS into publicly owned treatment works by a point source are summarized in Table 3. The guidelines

TABLE 3 U.S. Guidelines for LAS Discharge to Public Wastewater Treatment Plants

Manufacturing process	Surfactants (MBAS) (kg/1000 kg dry product)		COD (kg/1000 kg dry product)	
	Max. 1 day	Max. of daily average for 30 days	Max. 1 day	Max. of daily average for 30 days
Sulfonation/sulfation	0.36 (0.90)	0.18 (0.30)	1.10 (4.05)	0.55 (1.35)
Neutralization	0.04 (0.06)	0.02 (0.02)	0.10 (0.15)	0.05 (0.05)
Spray-drying	0.04 (0.06)	0.02 (0.02)	0.08 (0.15)	0.04 (0.05)
Liquid detergents (fast turnaround)	0.12 (0.43)	0.12 (0.43)	0.51 (1.95)	0.29 (1.2)

Values for best practicable control technology currently available (BPT) are given in parentheses.

define the degree of effluent reduction attainable by the application of the best available technology economically achievable for existing sources [7].

In the European Union there is at present no specific directive for maximum concentration of surfactants for discharge from an industrial point source to central treatment plants. Surfactants are controlled in several countries (e.g., Germany), within COD and BOD limitations. Problems linked to surfactant elimination are at present under study by the European Commission (JO S 165, 08/30/2000, p. 35). The use of LAS and especially APE is under debate and is gradually shrinking in favor of the more degradable surfactants, due to the "green" political bodies. European public opinion and the "green" pressure are forcing down the use of LAS. An example is Danish Executive Order number 823 [8]. The subject of this order is limitations on the concentration of LAS and APE in sludge from biotreatment plants, used as fertilizer in agriculture. The order led to a limitation by the municipalities on the use of LAS in consumer detergents and consequently came under fire by the detergent industry. Further studies led to less strict limitations, concluding that LAS of 10–50 kg/ha would be unlikely to cause any adverse effects to the soil ecosystem [9]. The concentration of LAS in sludge for "sludge-amended soils" was limited to 2600 mg/kg. Nevertheless, many municipalities in Denmark prevent the use of LAS to avoid problems with sludge from treatment plants. In other European countries and in the United States, LAS in sewage sludge is not regarded as a problem.

The use of APE in many European countries is voluntarily phased out, or its use was banned and strict limitations were imposed. In the United States, the limitations on APE are not so strict, and its use is growing, with an emphasis on effective elimination in treatment plants. APE and mainly NPE (nonylphenol ethoxylates) were considered difficult-to-biodegrade, or "hard," surfactants and are less biodegradable than LAS. Their biodecomposition products were suspected to be endocrine disrupters or, more specifically, pseudoestrogens. The observed decline of fish and of human male fertility was related to the existence of NP in rivers and in drinking water [10]. Other studies [11] show that nonoccupational exposure to NP does not pose an estrogenic health risk to humans. More recent studies [12] clear NPE and NP from suspected estrogenic effects at the actual low concentrations found in European water bodies.

As a result of the higher toxicity of APE and its biodegradation products, the limitations on APE in drinking water sources are around 1 ppb in most countries, compared to limitations on LAS of around 250 ppb. The concentration of NPE is limited to 5 or even 1 mg/L in the wastewater inlet to central activated-sludge plants in certain countries, compared to 40 mg/L (or no limitation) for LAS.

V. BIODEGRADABILITY OF SURFACTANT GROUPS

Biodegradability is the capacity of an organic compound to be degraded by microorganisms, usually bacteria or fungi. The extent and rate of biodegradability depends on different factors, such as the specific combination of environmental conditions, the specific microorganisms, and the interaction between the involved microorganism cultures. Anionic surfactants were defined as biodegradable, or "soft," if 90–95% of the parent compound was degraded [13,14]. Biodegradation of surfactants, like of any carbohydrate, is a source of carbon for energy and growth by the microorganisms in wastewater. Various surfactants are considered more difficult to degrade. These include anionic surfactants such as LAS and nonionic surfactants such as alkylphenol ethoxylates (APE). The main inhibitor for biodegradation of both is the aromatic ring that induces xenobiotic nature to the compounds. Both surfactants are the most cost effective in their categories and are used in large quantities. Linear nonionic ethoxylates are readily biodegradable and replace APE, although they are less cost effective.

Several cationic surfactants are bactericides and at high concentration may inhibit or destroy biotreatment systems. Cationic surfactants are used, however, at relatively low quantities, and at low concentration they have negligible effect on such systems. Amphoterics and other categories of surfactants are biodegradable and are generally not regarded as a problem.

Our discussion will concentrate on LAS and APE, as the surfactants that are more difficult to biodegrade.

VI. PRIMARY AND SECONDARY BIODEGRADATION OF LAS AND APE

Primary biodegradation can be defined as partial degradation to a degree that the surfactant loses its basic hydrophilic–hydrophobic structure. At each stage the microbial cultures degrade the most accessible parts of the surfactant molecule. Primary biodegradation can be achieved by slicing the alkyl chain, reducing the hydrophobic section and the detergency of the molecule. Partially degraded intermediates are further degraded in consecutive steps to reach complete or ultimate mineralization of the surfactants. The terms *ultimate biodegradation* and *mineralization* are identical and were defined [15] as "complete biodegradation of an organic compound to methane and/or carbon dioxide, water, minerals and new biomass."

Standard analytical methods such as the methylene blue–active substance (MBAS) method (ASTM D 1681) [16] determine the concentration of anionic surfactants (as LAS). This method identifies only the nondegraded LAS and not the intermediate biodegradation products. Therefore it is a simple method to estimate primary biodegradation.

TABLE 4 Comparison of Primary and Complete Biodegradation of LAS and NPE

	LAS (mg/L)			NPE (mg/L)		
	Start	48 hours	Conversion	Start	48 hours	Conversion
Surfactant	2100	<10	>99%	1000	<10	>99%
COD	15680	1180	92.5	3280	1290	60.5%
TOC	20360	470	97.5	793	330	58.5%

The same logic applies also to nonionic surfactants. Analytical methods such as CTAS (cobalt thiocyanate-active substance) [17] are used. The decomposition products of APE contain benzene derivatives with short alkyl chains that can be more toxic than the original surfactant. It should be noted that the definition of "complete" mineralization is changing with the introduction of new and improved analytical methods.

Ultimate degradation or mineralization can be determined by COD (chemical oxygen demand) and TOC (total organic carbon) analysis and by gas chromatography (GCMS). As shown in Table 4, a consistent reduction of surfactant, COD, and TOC levels indicate complete mineralization of the surfactant. An inconsistent degradation level in APE indicates partial degradation. The data indicates that LAS was completely mineralized in 48 hours of aerobic biotreatment, while NPE (nonylphenol ethoxylate) was mineralized only ca. 60%, in spite of complete primary degradation (>99%). NPE is defined as a relatively "hard" surfactant due to the low secondary biodegradation.

Biodegradability of surfactants is determined by the difference between primary and complete biodegradation. Ultimate biodegradation of LAS and NPE can be based on analytical parameters such as COD and BOD (biological oxygen demand) [18,13]. A common criterion for biodegradability is the ratio between COD and BOD [19]. A ratio higher than 5 indicates generally low biodegradability. In our opinion this is not a consistent criterion, since BOD is not a good indicator for xenobiotic compounds, including surfactants. The BOD test depends on the source of biomass used in the test and on the degree of its adaptation to the specific surfactants tested. Use of adapted biomass can show considerably different results than nonadapted biomass, and the results may be misleading. It should be noted that commercial LAS have non-homogenous composition. The rate and extent of primary biodegradation depends on the alkyl chain length and the position of the sulfophenyl group of its homologues and isomers.

VII. MECHANISMS OF BIODEGRADATION: ANIONIC SURFACTANTS

LAS is biodegradable at low concentrations (mostly below 40 mg/L) but is "harder" to degrade as compared to other anionic surfactant groups, such as fatty alcohol sulfates (FAS) and fatty alcohol ethoxy sulfates (FAES). Therefore we use it as a model for biodegradation of anionic surfactants.

Branched alkanes are generally less susceptible to biodegradation then *n*-alkanes. Specifically, a quaternary carbon structure inhibits microbial attack of the alkyl group. Branched surfactants were traditionally considered partially biodegradable [20–24]. Branched alkyl benzene sulfonates (BABS) were degraded only 20–45% and were defined as nonbiodegradable, or "hard," surfactants [18]. In the late 1990s a strain of *Pseudomonas aeruginosa* that can mineralize BABS by 70% and LAS up to 100% was reported [25,26].

The mechanism and pathways for biodegradation of LAS were studied at a low concentration of about 10 mg/L, but basically the same mechanism applies at higher concentration. Several mechanisms were proposed. Several authors [18,27–30] suggested the more widely accepted mechanism. This mechanism involves three bacterial groups, or tiers. The first is involved in omega oxygenation at the end of the C-12 chain and partial beta-oxidation of the alkyl chain of LAS, to form sulfophenyl carboxylates (SPC). The second converts the sulfophenyl carboxylates formed, by further beta-oxidation of the alkyl chain, to short-chain sulfophenyl carboxylates (sc-SPC). The third group mineralizes the sc-SPC to CO_2, H_2O, and sulfate via 4-sulfocatechol, ortho ring cleavage and desulfonation. The basic degradation steps described are possible only under aerobic conditions.

Anaerobic LAS biodegradation is not well understood. It is based on strict anaerobic sulfate-reducing bacteria that reduce SO_4^{-2} to H_2S [31], on methanogenic bacteria, or on nitrate-reducing facultative anaerobic bacteria [32]. It is performed as well by consortiums of microorganisms. The resistance of aromatic sulfonated surfactants to degradation under strict anaerobic conditions is related [33,34] to the dependence on molecular O_2 to initiate biodegradation, either by hydrophyl removal or for alpha-oxidation of the alkyl chain. Bruce et al. [35] give additional evidence that LAS is not biodegradable anaerobically in practice. Data by Swisher [18] suggest as well poor LAS primary biodegradation under anaerobic conditions. Most of the degradation of surfactants is by aerobic processes.

Earlier publications [36–38] suggest that the structure of the LAS molecule offers three potential sites where the initial enzymatic attack could take place. Each of the potential models can take place depending on the microorganisms involved and the different environmental conditions:

1. Oxidation of the terminal methyl groups of the alkyl chain and subsequent degradation of the chain, yielding short-chain sulfophenyl alkanoic acids.
2. Oxidative cleavage of the aromatic ring, yielding sulfonate-substituted dicarboxylic acids.
3. Desulfonation of the ring degradation products, yielding the keto-unsaturated dicarboxylic acids plus inorganic sulfate. These aliphatic intermediates are further decomposed by beta-oxidation and other enzymatic reactions. Further catabolism is performed via general metabolic routes.

VIII. MECHANISMS OF BIODEGRADATION: NONIONIC SURFACTANTS

APE and specifically NPE (nonylphenol ethoxylates) were considered difficult-to-biodegrade, or "hard," surfactants and are less biodegradable than LAS. Linear nonionic ethoxylates are readily biodegradable and replace APE in use, although they are more expensive.

The biodegradability of NPE has been a source of controversy and disagreement. It is agreed now that APE undergo nearly complete primary biodegradation, given sufficient acclimation. Primary degradation of NPE proceeds rapidly through slicing of the ethoxy chain and of their carboxylates as intermediates [39–44].

Further evidence was provided by the isolation of bacterial cultures able to grow on NPE. Frassinetti et al. [45] isolated three gram-negative bacteria that can individually attack NPE and affect its primary biodegradation. *Pseudomonas* sp. strains identified by Maki et al. [46] and John and White [47] were unable to mineralize NPE but were able to degrade its ethoxy chain. The resulting dominant intermediate was an NP ethoxylate with two ethoxy units.

All commercial NPE contain isomers of NP with a branched nonyl chain. The branched alkyl group, especially the quaternary carbon structure at the end of the alkyl group, inhibits biological attack and biotransformation of the alkyl group [18]. Transformation of compounds with quaternary carbon chain is said to be possible [48]. Only one report describes a bacterial culture attacking branched NP with a fission of the phenol ring and preferred use of the para isomer and not the ortho isomer [49]. Primary degradation intermediates include benzene derivatives.

IX. BACTERIAL CONSORTIUMS IN BIOTREATMENT OF SURFACTANTS

Wastewater is an open environmental system that always contains diverse microbial communities. Microorganisms in wastewater react as in any aqueous system. The contaminants are used as nutrients, and the microbes behave according to the laws developed for interacting microbial communities. Various behavior patterns were analyzed [50]. The most common is *competition*, or "survival of the fittest." Such a pattern is typical of a system that contains simple carbohydrates readily accessible to enzymatic degradation by the microorganisms. Such a system can be used for the primary degradation of surfactants, where readily accessible fractions of the carbohydrate chains are decomposed by the enzymatic systems of certain bacterial cultures.

For biodegradation of LAS and APE, a pattern of commensalism is preferred. In such patterns, a consortium of bacterial cultures is coordinating in a feed chain, where the molecule degradation is in a stepwise metabolic pathway. The degradation products of one bacterial culture at each step are nutrients for a second culture at the following step, and so on. A mixed culture is essential for biodegradation, in the case when a single culture is missing from the community, the biodegradation process is inhibited [28]. A product of the first step can catabolically inhibit the culture that is performing the first step, so the whole chain is inhibited [50]. A xenobiotic contaminant is usually not decomposed by a single culture but by a consortium of bacterial cultures [51–53]. Mixed cultures of microorganisms are needed for complete mineralization of surfactants [54]. Some members of the consortium will be responsible for primary biodegradation, while others utilize the intermediate products to achieve complete biodegradation.

Practically, in industrial wastewater, a commensal pattern for a contaminant-specific microbial consortium always coexists with a competition pattern for microorganisms that are not members of that consortium.

In spite of the importance of these processes, there is limited understanding of the relationship between microbial community structure and function [55]. Studies have generally demonstrated that the dominant members of aerobic reactors treating municipal wastewater are from the proteobacteria [56–58]. Manz et al. [57] reported that the dominant members of an industrial treatment facility are from the Cytophaga–Flavobacterium–Bacteriodes divisions. Most of the reported microorganisms are inhibited by surfactant concentrations above 40–60 mg/L.

The authors found [52,53] that in industrial wastewater from a detergent plant, the main bacterial groups are *Citrobacter* and *Aeromonas* spp. These bacterial cultures were used to construct a bacterial consortium for biodegradation of surfactants at high concentration, discussed later in this chapter.

Jimenez et al. [59] constructed a four-member consortium with three *Pseudomonas* spp. and *Aeromonas* sp. that was adapted to mineralization of LAS. Excluding any single species from the consortium, mineralization was not achieved, although three isolates could carry out primary degradation. This study confirmed earlier works [18,60,61], indicating that consortia of bacteria were involved in LAS biodegradation in natural environments. The presence of a microbial consortium is necessary to obtain complete mineralization of surfactants.

Various experiments were performed using a single bacterial culture [27,28]. This experimental technique is recommended for studying the metabolic pathways [30] but not for practical use. It was confirmed that the complete aerobic degradation of LAS involves communities and not single cultures.

Genetically modified (GM) bacteria, which can induce all the enzymatic systems and perform single-handedly all the biodegradation steps of LAS or APE, have been considered. In the authors' opinion, such a strategy is problematic. Since wastewater treatment is an open system, the GM bacteria will have to compete with natural bacterial cultures that are better adjusted to the environment and grow faster. In addition, in an open system the GM bacteria will be mutated soon and lose their special enzymatic advantages, unless original GM bacteria will be added to the wastewater constantly. Change in wastewater composition, which happens frequently in industrial effluents, may inhibit or deactivate the GM bacteria.

X. BIOTREATMENT TECHNIQUES FOR WASTEWATER WITH SURFACTANTS

Bioreactors are commonly used to treat wastewater containing soluble and particulate organic material. They support mixed consortia of microorganisms that can simultaneously convert a broad spectrum of compounds into biomass, innocuous byproducts, carbon dioxide, and water.

Aerobic and anaerobic techniques and combinations of both are used for urban wastewater treatment. Standard methods of aerobic biotreatment include activated sludge, trickling filters, aerated lagoons, rotating biological contactors (RBC), and membrane bioreactors (MBR). A description of these techniques can be found in textbooks [62,63].

The activated-sludge process is applied mostly to large-scale wastewater aerobic treatment plants. It is used mostly for municipal wastewater treatment with high flow and low organic load, as shown in Table 1. The process involves two distinct steps: aeration and settling. In the first step the organic components in the wastewater are metabolized to end products and to new biomass (microorganisms) under aeration and mixing. In the settling step the biomass is allowed to settle as sludge, also containing untreated suspended solids and adsorbed contaminants. A portion of the sludge is recycled to the aeration step, and the rest is discharged to further anaerobic processing. Anaerobic treatment of the sludge reduces its volume and stabilizes it. Treatment and discharge of the sludge is a major cost of the process.

Reduction of LAS in urban activated-sludge treatment plants, with inlet concentration below 40 mg/L, was reported to reach up to 99% [2].

In trickling filters, wastewater is sprayed on a fixed bed with a high surface area, where the biomass is attached in a film to the bed particles' surface. Wastewater trickles down the bed, and the microorganisms in the film metabolize the contaminants. Aerobic conditions are maintained by flow of air through the bed. Removal of LAS is less efficient in trickling filters, at about 77% in urban plants [2].

Both methods are sensitive to foam formation and are not recommended for surfactants at high concentration.

Rotating biological contactors (RBC) use rotating discs or wheels as aerators and as attachment surface for biomass film growth. The rotating contactors are submerged 40–90% in the treated wastewater. Aeration is performed by exposing the biofilm surface periodically to air and wastewater or by special design of the discs to induce air into the wastewater. RBC are less sensitive to foam formation and are more suitable for wastewater with surfactants, depending on the specific design. RBC units are suitable for medium- or small-scale plants. Good reduction efficiency of 96–99% for LAS was reported in urban wastewater.

Aerated lagoons were used historically in places with a large, low-cost free area. This technique is suitable only for wastewater with low surfactant concentration or when high dilution is possible. High concentration results in oxygen depletion, anaerobic conditions, low efficiency, and bad odor.

Membrane bioreactors combine biotreatment and submerged hollow-fiber membranes, where air streams prevent clogging of the hollow fibers. This method is limited for use with surfactants, due to foam formation.

Biotreatment of surfactants at high concentration was optimized [52,53] by the use of two consecutive treatment stages. This can be explained by a better adaptation of the bacterial mix to the narrow concentration range. A full-scale industrial wastewater treatment facility, consisting of seven reactors operated in series, was described [64], with thermophilic (45–65°C) stages and lower temperatures (25–35°C) stages.

The treated sludge can be used for composting and as fertilizer in agriculture. Up to 30% of the LAS remain in the sludge [65,2]. Toxic levels of nonionic and anionic surfactants were reported, but other reports suggest that long composting process can decompose the surfactants [66]. LAS degraded very slowly in anaerobic storage of sludge, but under aerobic stabilization it decomposed 90% in 3–9 hours. Surfactants in anaerobic sludge from activated-sludge plants had been the source for Danish Executive Order number 823, mentioned earlier. Elimination of LAS during tunnel composting of sludge was reported [67]. In 40 days of the described process, nearly 100% degradation was achieved and the half-life was 7.5–9.5 days. Such a solution is suitable for solid or semisolid treated sludge from an activated-sludge process.

XI. FREE AND IMMOBILIZED MICROORGANISMS

Immobilized bacteria have been widely used in wastewater treatment. Many reports have been published on the use of immobilized microorganisms to degrade aromatic and xenobiotic compounds [68–70]. Various methods of immobilization were described [71]: covalent coupling, adsorption, and entrapment in a three-dimensional polymer network or within a semipermeable membrane. In many cases, immobilized cells have advantages over free cells. Catalytic stability is higher and they can tolerate higher concentrations of toxic compounds and higher concentration fluctuations [68,72,73]. Immobilized cells are more protected from pH and temperature peaks and from toxic nutrients [70,74]. By attachment to inert support material it is possible to retain high cell density, even under washout conditions. For cells with a low specific growth rate, immobilization can improve the reactor performance in comparison to free-cell systems.

On the other hand, oxygen and nutrient transfer rate often is the rate limiting factor in the performance of aerobic, immobilized culture due to film formation by the dense biomass or by the immobilization medium. Aeration and mass transfer limitations must be considered in bioreactor design using immobilized aerobic cultures.

Immobilization techniques are most suitable for biotreatment of surfactants at high concentration, where toxic conditions prevail and strong aeration should be avoided to prevent foam. The low growth in such adverse conditions becomes an advantage due to low sludge formation. Sludge formation is lower due to the longer retention time of the biomass and to extended digestion. Sludge reduction is a very important factor in biotreatment plant design.

Different carriers are used for support of the cells. Immobilized microorganisms in trickling filters or RBC units tend to be packed in thick films and are inhibited by slow diffusion of oxygen, nutrients, and biodecomposition products. The authors found that a porous structure is more efficient (with suitable design) than a flat surface of rotary discs or a gel-immobilized cell system due to high surface area and low resistance to mass transfer. Biotreatment of industrial wastewater, using different carriers, including porous polymer, was reported (patents RU 2039013 and JP 4187086). Aeration techniques are a very important consideration in immobilized bioreactor design.

XII. EXAMPLE OF A BIOTREATMENT PROCESS FOR INDUSTRIAL WASTEWATER: ZOHAR BIO-D PROCESS

It is generally agreed that biotreatment is the most cost-effective method for industrial wastewater treatment. Biotreatment technology that is effective at high surfactant concentration could have been the best solution from the economic and ecological viewpoints.

An example of such a technology that was developed in Zohar Dalia as a solution for wastewater of a detergent factory [6,52,53] is presented. The process was developed on the principles discussed earlier to achieve high performance. It is based on three principles:

- Selection and adaptation of a bacterial consortium that can achieve complete mineralization of the contaminant surfactants without being inhibited by high concentration of up to 10% anionic surfactants
- Immobilization of the consortium on a porous media with high surface area to achieve good contact and mass transfer between microorganisms, oxygen, and contaminants
- Equipment design for efficient aeration while avoiding foam formation

A. Selection and Adaptation

The bacterial consortium was selected by inoculating samples of wastewater and soil of a detergent factory into minimal medium. LAS basal mineral solution was added daily to the inoculated samples, increasing gradually the LAS concentration in the feed up to 2000 mg/L. Surviving cultures were isolated and screened for LAS degradation. None of the pure cultures showed significant degradation of LAS [52]. Consecutive transfers of the selected cultures were employed in a similar procedure to construct a bacterial consortium.

Using an adaptation procedure such as shown in Figure 1, an optimal consortium was selected. It consisted of rod- shaped gram-negative aerobic and anaerobic facultative bacteria identified as genera *Citrobacter, Aeromonas,* and *Pseudomonas.* With any single member missing, biodegradation was reduced to an insignificant level. The consortium was named BIO-V.

The concentration of LAS is gradually increased to adapt the consortium to a high concentration. The adaptation procedure is essential for adjusting the bacterial consortium to high surfactant concentration. Various enzymatic systems should be induced to support the biodegradation metabolic pathways.

B. Immobilization

This biodegradation process was shown to be more efficient when the cells of bacterial consortium BIO-V were immobilized on a porous media. A spongelike matrix with open pore structure was used. An advantage of this porous carrier is the excellent mass transfer between the external and internal fluid phases, by internal convection through the pore system. Another advantage is the large surface area exposed to nutrients and to aeration. In contrast, in the gel-immobilized cell systems, as well as in the thick films of a trickling filter, the nutrient and oxygen transport to the immobilized cells is mainly by diffusion.

The cells were immobilized via enhanced adsorption. Binding of the consortium cells to the carrier surface is enhanced by the specific surface properties of the porous

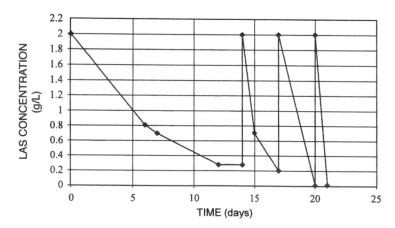

FIG. 1 Effect of the adaptation of consortium BIO-V on the decomposition of LAS in the wastewater of a detergent plant.

material. The effect of immobilization, as shown in Figure 2, is considerable, and more than 10-fold improvement was recorded. The immobilized consortium degrades LAS completely in 1 day, while free consortium cells degrade it to 80% in 12 days. Such an effect can be explained by several of the factors mentioned earlier, such as high surface area, high biomass concentration, and high mass transfer rate. Another factor could be the effect of close contact between the consortium members, enhancing the multistep enzymatic degradation by good transfer of intermediate product–nutrient chain.

C. Description of the Biotreatment Plant

The consortium BIO-V is composed of aerobic and anaerobic-facultative bacteria. Optimal biodegradation of surfactants is achieved at aerobic conditions. High aeration is difficult to achieve in conventional plants, without considerable foam formation. To maximize the oxygen supply to the biomass while maintaining minimal foam formation, a special design of a modified rotating biological contactor was developed. A frame carrying porous elements rotates in a large vessel that contains the wastewater, in such a way that the elements are submerged and aerated periodically. The benefits of this specific contactor design are high mass transfer rate, good oxygen supply, high stability of the biomass, low sensitivity to concentration variations, and fast recovery from concentration shocks. As a result, the effectiveness of biodegradation is highly increased.

In addition, a high biomass concentration of immobilized cells is achieved, and most of the biodegradation activity is in the porous substrate. Hence the mild but efficient internal aeration in the porous substrate is very important. Controlling the rotation speed of the frame in the wastewater vessel solves the foaming problem. The internal aeration also contributes to foam reduction. Low sludge is produced due to low free-cell concentration. The operation of the plant is continuous—with a constant controlled flow of wastewater in and out, or batchwise—where wastewater is filled and

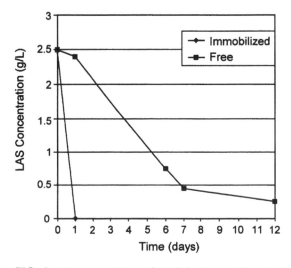

FIG. 2 Decomposition of LAS by immobilized and free cells.

discharged periodically. The flow rate or the quantity of water to be replaced in every cycle is controlled by the surfactant concentration. A typical period is 1 day, but other values can be used. The biotreatment plant can be built of one or two consecutive units.

The effect of LAS concentration on biodegradation is demonstrated in Figure 3. Raw wastewater at "inlet concentration" is filled to the vessel. After dilution with the treated wastewater retained in the porous carrier, the concentration reduces to the "start concentration." With an inlet concentration of 1.5 g/L, the start concentration is 0.7–0.8 g/L and the final concentration after 24 hours is less than 10 mg/L, so one treatment stage is sufficient. An inlet concentration of 4 g/L results in a start concentration of 2–2.5 g/L, and the final concentration after 24 hours is increasing up to 0.5 g/L. At such a concentration, a second treatment stage is needed to reduce the final concentration to zero.

The number of units depends on the inlet surfactant concentration and on the outlet purification requirements. A single-stage plant can degrade up to 800 mg/L LAS to less than 40 mg/L in 24 hours. A two-stage plant can degrade 2000–2500 mg/L LAS to a concentration below 40 mg/L. Each stage lasts 24 hours.

APE was degraded by the same process, from 1000 mg/L to less than 10 mg/L in 48 hours (primary biodegradation).

D. Comparison of Batch and Continuous Biotreatment

Large-scale municipal biotreatment systems operate inherently in continuous mode. Small or medium-size industrial systems can sometimes be more efficient when operating in batch mode. The authors found that in the biotreatment of wastewater that contains high concentrations of anionic surfactants, the biodegradation of surfactants can be more efficient and faster in batch as compared to continuous flow. Figure 4 shows the LAS concentration at start time, after first stage biotreatment of

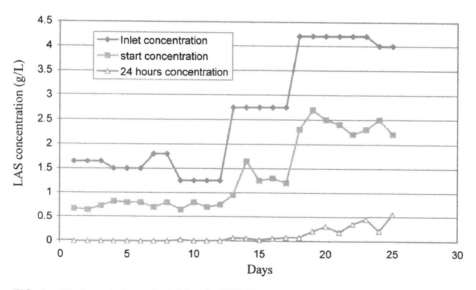

FIG. 3 Biodegradation of LAS by the BIO-V process.

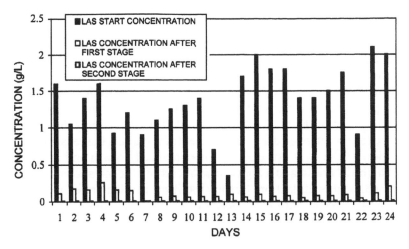

FIG. 4 Biodegradation of LAS in a two-consecutive-units batch plant.

24 hours, and after a second stage of an additional 24 hours in a batch two-stage process. Figure 5 shows similar data from continuous operation. The continuous flow rate is only 70% of the full capacity achieved in batch operation.

That result may be related to an inhibition by a constant high level of intermediate decomposition products of LAS. Another explanation can be that the microorganisms are more efficient in starvation–abundance cycles.

Comparing the batch and continuous modes, the degradation results are similar. The difference is in the capacity of the biotreatment plant. In the same pilot units, batch operation can treat 100 L in 24 hours and continuous operation only 70 L in 24 hours.

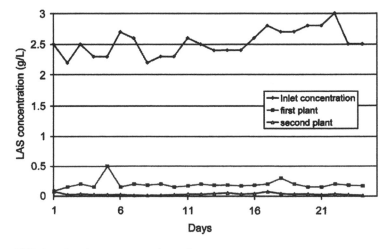

FIG. 5 Continuous operation of a two-stage plant.

E. Special Features

Figure 6 shows data of a 10,000-liter unit that has been in operation for more than 24 months without failure of the bacterial biomass. Data is for the first stage; a second stage reduces LAS concentration to zero.

One of the main advantages of the process described is the low sludge formation. Sludge formation depends on biomass retention time and on the wastewater composition. A long retention time of the immobilized cells (sludge age) reduces the sludge level. Wastewater with a low nitrogen derivative concentration limits the cells growth. A balanced proportion of the C/N ratio leads to maximal cell growth and biomass production. With a high C/N ratio and a high excess of carbon, the metabolism of the cells is optimized for efficient N utilization. Carbon is utilized at a high rate for energy and for maintenance of the cells, with low efficiency. This phenomenon is known in fermentation processes [75,76]. Control of the C/N ratio, which is normally high in wastewater of detergent manufacturing plants, leads to low biomass production and low sludge formation.

The sludge after the BIO-D process was about 10–20% of the sludge separated from the same wastewater by a parallel ultrafiltration process. This is compared to the conventional doubling or tripling of the sludge in activated-sludge plants due to biomass growth. Most of the carbon was released as CO_2, and TOC was reduced up to 97.5% of the start value in two stages of treatment, with a total retention time of 48 hours.

Figure 7 shows the dependence of the rate of surfactant degradation on the surfactant concentration. The rate increases up to around of 1800 mg/L (start concentration of LAS) and then tends to slow down, possibly due to inhibition by too high a concentration. Such behavior depends on the specific local wastewater composition.

A comparison of the operating costs of biotreatment and nonbiological treatment methods shows that biotreatment is generally more cost effective. A comparison of costs for several methods is shown in Table 5.

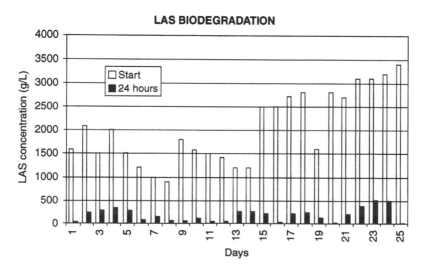

FIG. 6   Data from a 10,000-L biotreatment unit.

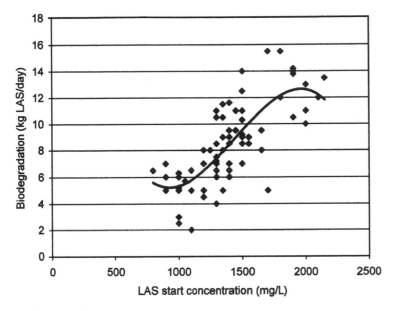

FIG. 7 LAS biodegradation rate.

The consortium degrades high concentrations of surfactants, including LAS and other anionic surfactants, and APE and other nonionic surfactants at high, industrial concentrations. The consortium is active in a wide range of conditions and is adaptable to a wide range of contaminants and concentrations.

XIII. SUMMARY

Most surfactants are discharged to natural water systems after varying levels of treatment. A vast amount of research has been reported on the biodegradation of LAS, and its final degradation products are considered safe to the environment after more than 40 years of use. Sewage treatment plants can remove 99% of the major surfactants. The removal efficiency depends on the treatment plant's design.

TABLE 5 Comparison of Relative Operating Costs in Different Treatment Methods, for Wastewater with a Similar Composition Profile

Process	LAS—inlet concentration (mg/L)	LAS—outlet concentration (mg/L)	Relative operating cost	Comments
BIO-D	300–800	<40	1	1 stage
BIO-D	2000–2500	<40	2	2 stages
BIO-D	3000	<40	2.5	2 stages
UF membranes	300–3000	200–900	3.5–4	Off-limits product
UF + RO membranes	300–3000	<40	6–7.5	
Evaporation/drying	300–3000	<40	6–7.5	

The case of industrial wastewater that contains surfactant at high concentration is different. Detergent manufacturers face considerable economic and ecological constraints with their effluents. LAS and APE are relatively more resistant to biodegradation than fatty alcohol sulfates and "soft" nonionics. Wastes streams with LAS concentration of 100 mg/L or higher are usually toxic to the biomass in standard wastewater biotreatment plants. The upper limit for acceptance in municipal treatment plants is around 40 mg/L LAS. Wastewater streams from industrial laundries contain 100–600 mg/L LAS, and untreated wastewater from detergent manufacturing plants can reach more than 5000 mg/L LAS. Most biotreatment plant operators request that such wastewater be pretreated before accepting it in conventional municipal plants. Several methods are used, with aerobic biotreatment considered as the most cost effective.

Similar problems are encountered in cosmetics processing, textile, leather, metal machining, and other industries that use surfactants and discharge toxic and foaming wastewater.

REFERENCES

1. WCED. Our Common Future, World Commission on Environment and Development; Oxford University Press: New York, 1987.
2. McEvoy, D.C.; Eckhoff, W.S.; Rapaport, R.A. Fate of linear alkylbenzene sulfonate in the environment. Environ. Toxicol. Chem. 1993, *12*, 977–987.
3. Feijtel, T.C.J.; van de Plassche, E.J. Environmental Risk Characterization of four Major Surfactants used in the Netherlands. National Institute of Public Health and Environmental Protection and Dutch Soap Association. The Netherlands, Sept, 1995. Report no. 679101 025.
4. Dolkemeyer, W. Surfactants on the eve of the third millenium—challenges and opportunities. Proceedings of the 5th World Surfactants Congress (CESIO). Florence, 2000.
5. Federle, T.W.; Schwab, B.S. Mineralization of surfactants in anaerobic sediments of a laundromat wastewater pond. Water Res. 1992, *26*, 123–127.
6. Gorelik, E.; Sivan, D.; Weiss, D.; Ben-Neria, T.; Vashitz, O. Biotreatment of alkylbenzene sulfonate and other surfactants in industrial concentrated effluents of a detergent factory. In Proceedings of the 4th World Conference on Detergents; Cahn, Arno., Ed.; AOCS Press: Champaign, IL, 1999; 346–348.
7. EPA 40 CFR part 417, 1995.
8. Heinze, J.E. Science versus politics in the environmental regulatory process. In Proceedings of the 4th World Conference on Detergents: Strategies for the 21st Century, Cahn, A., Ed.; AOCS Press: Champaign, IL, 1999; 83–88.
9. SPT Workshop in coordination with the Danish EPA. Linear Alkyl Benzene Sulphonate (LAS) Risk Assessment for Sludge-Amended Soils. Copenhagen, April 19–20 1999.
10. Facemire, C.F.; Groos, T.S.; Guillette, L.J. Jr. Reproductive impairment in the Florida panther: nature or nurture? Environ. Health Perspect. 1995, *103*, 79–86.
11. Muller, S, et al. Evaluation of the estrogenic potency of nonylphenol in non-occupationally exposed humans. Environ. Toxicol. and Pharmacol. 1998, *6*, 27–33.
12. Basler; Lebsanft. Environ. Sci. Pollut. Res. 1999, *6* (1), 44–48.
13. Schoberl, P. Basic principles of LAS biodegradation. Tenside Surfac. Deterg. 1989, *26*, 86–94.
14. Painter, H.A.; Zabel, T. The behavior of LABS in sewage treatment. Tenside Surfac. Deterg. 1989, *26*, 108–115.
15. Gilbert, P.A.; Watson, G.K. Tenside Surfac. Deterg. 1977, *14*, 171–177.

16. Milwidsky, B.M.; Holzmann, S. Rapid method to test soaps/syndets. Soap Chemical Specialties, May 1966; (+ 154), 83–86.

17. Milwidsky, B.M. Analist 1969, *94*, 377–386.

18. Swisher, R.D. Surfactant Biodegradation; Marcel Dekker: New York, 1987.

19. Schoberl, P.; Bock, K.J.; Huber, L. Oekologish relevante daten von Tensiden in wasch und reinigungs mitteln. Tenside Surfac. Deterg. 1988, *25*, 86–98.

20. Alexander, M. Nonbiodegradable and other recalcitrant molecules. Biotechnol. Bioeng 1973, *15*, 611–647.

21. Cain, R.B. Biodegradation of anionic surfactants. Biochem. Soc. Trans. 1987, *15* (suppl), 7–22.

22. Cain, R.B. Biodegradation of detergents. Curr. Opin. Biotechnol. 1994, *5*, 266–274.

23. Greek, B.F. Sales of detergents growing despite recession. Chem. Eng. News 1991, *69*, 25–52.

24. Swisher, R.D. Biodegradation of ABS in relation to chemical structure. J. Water Pollut. Control Fed. 1963, *35*, 877–892.

25. Soberon-Chavez, G.; Haidour, A.; Campos, J.; Ortigoza, J. Selection and preliminary characterization of a *Pseudomonas aeruginosa* strain mineralizing some isomers in a branched-chain dodecylbenzene sulfonate mixture. World J. Microbiol. Biotechnol. 1996, *12*, 362–367.

26. Campos-Garcia, J.; Esteve, A.; Vazquez-Duhalt, R.; Ramos, J.L.; Soberon-Chavez, G. The branched-chain dodecylbenzene sulfonate degradation pathway of *Pseudomonas aeruginosa* W51D involves a novel route for degradation of the surfactant lateral alkyl chain. Appl Environ. Microb. 1999, *65* (8), 3730–3734.

27. Cook, A.M. Tenside Surfac. Deterg. 1998, *35*, 52–56.

28. Cook, A.M.; Hrsak, D. The complete degradation of LAS is becoming better understood by pure cultures of bacteria. Proceedings of the 5th World Surfactants Congress (CESIO), Florence, 2000; 1387–1396.

29. Falbe, J. Surfactants in consumer products: theory, technology and application; Springer Verlag: Berlin, 1987.

30. Hrsak, D.; Grbic-Galic, D.; Prehrambeno. Tehnol. Biotehnol. Rev. 1993, *31* (1), 7–14.

31. Huxtable, R.J. Biochemistry of Sulfur; Plenum Press: New York, 1986.

32. Berna, J.L.; Battersby, N.; Cavali, L.; Fletcher, R.; Guldner, A.; Schowanek, D.; Steber, J. Assessment of the environmental relevance of anaerobic biodegradation of surfactants. Proceedings of the 5th World Surfactants Congress (CESIO), Florence, 2000; 1414–1427.

33. Wood, A.A.; Claydon, M.B.; Finch, J. Synthetic detergents: some problems. Water Pollut. Control 1970, *69*, 675–683.

34. McEvoy, J.; Giger, W. Determination of linear alkylbenzene sulfonates in sewage sludge by high-resolution gas chromatography mass spectrometry. Environ. Sci. Technol. 1986, *20*, 376–383.

35. Bruce, M.A.; Swanwick, J.D.; Ownsworth, R.A. Synthetic detergents and sludge digestion: some plant observations. J. Proc. Inst. Sewage Purif. Pt. 1966, *5*, 427–447.

36. Willets, A.J. Microbial aspects of the biodegradation of synthetic detergents: a review. Int. Biodetn. Bull. 1973, *9*, 3–10.

37. Cain, R.B.; Willets, A.J.; Bird, J.A. In *Biodeterioration of materials*, 2nd; Walters, A.H., Van der Plas, Hueck, Eds.; Int. Biodet. Symp. Lunteren; Wiley: New York, 1971; Vol. 2, 136–144.

38. Schoberl, P.; Kunkel, E. Tenside Surfac. Deterg. 1977, *14*, 293–296.

39. Kravetz, L.; Chung, H.; Guin, K.F.; Shebs, W.T.; Smith, L.S. Ultimate biodegradation of an alcohol ethoxylate and a nonylphenol ethoxylate. Household Personal Prod. Ind. Mar, Apr, 1982, *48–72*, 62–70.

40. Rudling, L.; Solyom, P. The investigation of biodegradability of branched nonyl phenol ethoxylates. Water Res. 1974, *8*, 115–119.

41. Ahel, M.; Giger, W.; Koch, M. Behavior of alkylphenol polyethoxylate surfactants in the aquatic environment. I. Occurrence and transformation in sewage treatment. Water Res. 1994, *28*, 1131–1142.

42. Ahel, M.; Giger, W.; Koch, M. Behavior of alkylphenol polyethoxylate surfactants in the aquatic environment. II. Occurrence and transformation in rivers. Water Res. 1994, *28*, 1143–1152.

43. Espadaler, I.J.; Caixach, J.; Om, F.; Ventura, F.; Cortina, M.; Paune, F.; Rivera, J. Identification of organic pollutants in Ter River and its system of reservoirs supplying water to Barcelona (Catalonia, Spain): a study by GC/MS and FAB/MS. Water Res. 1997, *31*, 1996–2004.

44. Osburn, O.W.; Benedict, J.H. Polyethoxylated alkyl phenols: relationship of structure to biodegradation mechanism. J. Am. Oil Chem. Soc. 1966, *43*, 141–146.

45. Frassinetti, S.; Isopps, A.; Corti, A.; Vallini, G. Bacterial attack of nonionic aromatic surfactants: comparison of degradative capabilities of new isolates from nonylphenol polyethoxylate–polluted wastewaters. Environ. Technol. 1996, *17*, 199–205.

46. Maki, H.; Masuda, N.; Fujiwara, Y.; Ike, M.; Fujita, M. Degradation of alkylphenol ethoxylates by *Pseudomonas* sp. strain TR01. Appl. Environ. Microboiol. 1994, *60*, 2265–2271.

47. John, D.M.; White, G.F. Mechanism for biotransformation of nonylphenol polyethoxy-lates to xenoestrogens in *Pseudomonas* putida. J. Bacteriol. 1998, *180*, 4332–4338.

48. Ball, H.A.; Reinhard, M.; McCarty, P.L. Biotransformation of halogenated and nonhalogenated octylphenol polyethoxylate residues under aerobic and anaerobic conditions. Environ. Sci. Technol. 1989, *23*, 951–961.

49. Tange, T.; Dhooge, W.; Verstraete, W. Isolation of a bacterial strain able to degrade branched nonylphenol. Appl. Environ. Microbiol. 1999, *65* (8), 746–751.

50. Bailey, J.E.; Ollis, D.F. Biochemical Engineering Fundamentals. 2nd ed. McGraw-Hill: New York, 1986; 854–902.

51. Knackmuss, H.J. Basic knowledge and perspectives of bioelimination of xenobiotic compounds. Environ Biotechnol Workshop. Technion, Haifa, Nov 1995.

52. Gorelik, E.; Ben-Neria, T.; Vashitz, O. Biodegradation of alkylbenzene sulphonate in concentrated effluents by bacterial immobilized consortiums. Proc. 7th Int. Cong. Bact. Appl. Micr., IUMS, 1984.

53. Gorelik, E.; Sivan, D.; Ben-Neria, T.; Shimon, E.; Vashitz, O. Biodecomposition of alkylbenzene sulphonate, alkyl phenol ethoxylates and other surfactants in industrial wastewater by bacterial immobilized consortium. Proceedings of the 4th World Surfactants Congress (Cesio), Barcelona, 1996.

54. Van Ginkel, C.G.; Kroon, A.G.M. Metabolic pathway for the biodegradation of octadecylbis (2-hydroxyethyl) amine. Biodegradation 1993, *3*, 435–443.

55. Lapara, T.M.; Nakatsu, C.H.; Pantea, L.; Alleman, J.E. Appl. Environ. Microbiol. 2000, *66* (9), 3951–3959.

56. Bond, P.L.; Hugenholtz, P.; Keller, J.; Blackall, L.L. Bacterial community structure of phosphate removing and non-phosphate-removing activated sludges from sequencing batch reactors. Appl. Environ. Microbiol. 1995, *61*, 1910–1916.

57. Manz, W.; Wanger, M.; Amann, R.; Schleifer, K.H. In situ characterization of the microbial consortia in two wastewater treatment plants. Water Res. 1994, *28*, 1715–1723.

58. Snaidr, J.; Amann, R.; Huber, I.; Ludwig, W.; Schleifer, K.H. Phylogenetic analysis and in situ identification of activated sludge. Appl. Environ. Microbiol. 1997, *63*, 2884–2896.

59. Jimenez, L.; Breen, A.; Thomas, N.; Federle, T.W.; Sayler, G.S. Mineralization of linear alkylbenzene sulfonate by a four-member aerobic bacterial consortium. Appl. Environ. Microbiol. 1991, *57*, 1566–1569.

60. Hrsak, D.; Bosnjak, M.; Johanides, V. J. Appl. Bacteriol. 1982, *53*, 413.

61. Sigoillot, J.C.; Nguyen, M.H. Complete oxidation of linear alkylbenzene sulfonate by

bacterial communities selected from coastal seawater. Appl. Environ. Microbiol. 1992, *58*, 1308–1312.

62. Droste, R.L. Theory and Practice of Water and Wastewater Treatment; Wiley: New York, 1997.

63. Metcalf and Eddy, Inc. Rev. by G. Tchobanoglous. Wastewater Engineering Treatment, Disposal, Reuse. 2nd ed. McGraw-Hill: New York, 1979.

64. Eckenfelder, W.W.; Musterman, J.L. Activated-sludge treatment of industrial wastewater; Technomic: Lancaster, PA, 1995.

65. Matthijs, E.; de Henau, H. Adsorption and desorption of LAS. Tenside Surf. Det. 1985, *22*, 299–304.

66. Madsen, T.; Winther-Nielsen, M.; Rasmussen, D. Studies of the fate of linear alkylbenzene sulphonates (LAS) in sludge and sludge-amended soil. Proceedings of the 5th World Surfactants Congress, 2000, 1428–1432.

67. Prats, D.; Rodrigues, M.; Muela, M.A.; Llamas, J.A.; Moreno, A.; De Ferrer, J.; Berna, L. Tenside Surf. Deterg. 1999, *36*, 294–298.

68. Dwyer, D.F.; Krumme, M.L.; Boyd, S.A.; Tiedje, J.M. Kinetics of phenol biodegradation by an immobilized methanogenic consortium. Appl. Environ. Microbiol. 1986, *52*, 345–351.

69. Klein, J.; Hackel, U.; Wangner, F. Phenol degradation by *Candida tropicalis* whole cells entrapped in polymeric ionic network. In *Immobilized Microbial Cells*; Venkatasubramanian, K., Ed.; American Chemical Society: Washington, DC, 1979, 101–118.

70. Heinze, U.; Rehm, H.-J. Appl. Microbiol. Biotechnol. 1993, *40*, 158–184.

71. Mattiasson, B. Immobilization Methods. In *Immobilized Cells and Organelles*; Mattiasson, B., Ed.; CRC Press: Boca Raton, FL, 1983; Vol I, 3–26.

72. Klein, J.; Kluge, M. Immobilization of microbial cells in polyurethane matrices. Biotechnol. Lett. 1981, *3*, 65–70.

73. Westmeier, F.; Rehm, H.J. Appl. Microbiol. Biotechnol. 1985, *22*, 301–305.

74. Balfanz, J.; Rehm, H.-J. Biodegradation of 4-chlorophenol by adsorptive immobilized *Alcaligenes* sp. A7-2 in soil. Appl. Microbiol. Biotechnol. 1991, *35*, 662–668.

75. Vashitz, O.; Ulitzur, S.; Sheintuch, M. Mass transfer, batch and continuous kinetics, in a luminous strain of *Xantomonas campestris*. Chem. Eng. Sci. 1988, *43* (8), 1883–1890.

76. Vashitz, O.; Ulitzur, S.; Sheintuch, M. Analysis of polymer synthesis rates during steady-state growth of *X. campestris*. Biotechnol. Bioeng. 1990, *37*, 383–385.

30

Science Versus Politics in the Environmental Regulatory Process

JOHN E. HEINZE Environmental Health Research Foundation, Manassas, Virginia, U.S.A.

I. INTRODUCTION

Detergent products are commonly used worldwide for cleaning laundry, washing dishes, and other household and industrial cleaning applications. Because detergent products are used daily by most people, major detergent ingredients, such as surfactants and cleaning agents, are produced in large volume. After use these materials are typically discharged into the aquatic environment, e.g., rivers, lakes, and the oceans, or, in more developed countries, into public sewers. In North America and western Europe, sewers are typically connected to municipal sewage treatment plants or publicly owned treatment works (POTW). Extensive research conducted around the world over the past 15 years or more has documented that the major surfactants used in laundry and cleaning products are effectively removed in POTW so that the residual levels of these materials in wastewater discharges are low and pose little risk to the environment [1–4].

Environmental regulations on detergent ingredients have been in force since 1973, when the European Union enacted Council Directive 73/404/EEC [5] to ensure that only biodegradable surfactants were used in laundry detergents. This legislation followed and codified the detergent industry's shift from reliance on poorly biodegradable branched alkylbenzene sulfonates (ABS) as the major laundry detergent cleaning agent. ABS were the first fully synthetic surfactants whose use in laundry detergents led to major improvements in cleaning performance as compared to soap, which readily forms unsightly scums in the presence of water-hardness ions, calcium and magnesium. Because of the highly branched alkyl carbon side chain, ABS were not easily biodegraded, leading to problems of excessive foaming in treatment plants and in streams and rivers receiving treated and untreated wastewater. When this environmental problem began to be recognized in the mid-1960s, the detergent and surfactant industry in North America and western Europe developed replacement surfactants, linear alkylbenzene sulfonates (LAS), which were rapidly biodegradable and easily removed in POTW using biological (secondary) wastewater treatment [6]. In recent years there has been a renewed interest in environmental regulation and environmental labeling of detergent surfactants.

This chapter focuses on is the process by which environmental regulations are enacted and the sometimes-conflicting roles of politics and science in the regulatory process. The role of science is crucial for effective regulatory decisions, especially those involving human health and the environment. The scientific process is the only objective method to organize and evaluate technical information to produce an impartial assessment of health and environmental safety based on the facts. To ignore science in creating regulations governing human health or the environment is to impose risks (inadequate regulation) or costs (excessive regulation) on society due to the lack of adequate information for decision making.

This review of recent environmental regulations on detergent ingredients, especially surfactants, reveals two types of common failures in the regulatory processes used: (1) inadequate stakeholder involvement and (2) failure to make use of independent scientific review. Either or both of these lapses can lead to a regulatory process that is based primarily on politics and often is in conflict with science.

If the scientific basis for the regulatory decision and if the best available science has not been used as a basis for a regulation, then the affected stakeholders, including industry, must become advocates for the legitimate role of science in the regulatory process. Otherwise, the acquiescence of stakeholders will be interpreted as agreement with the regulation and the process by which it was derived. In recent years the process used in the development of regulations affecting the detergent and cleaning products industry has demonstrated this pattern.

This chapter will consider regulations on sewage sludge disposal in Denmark and the United States of America (U.S.), the City of Toronto (Canada) sewer discharge law, and two types of environmental labeling regulations, the U.S. Federal Trade Commission guidelines on environmental marketing claims and the Nordic Swan Eco-label program.

II. DENMARK SEWAGE SLUDGE DISPOSAL REGULATION, EXECUTIVE ORDER NUMBER 823

The bio-solids, or sludge, from municipal sewage treatment plants has been used for many years as a low-cost fertilizer and soil conditioner. The European Union Council Directive of June 12, 1986, set limits on the concentrations of heavy metals (cadmium, chromium, copper, lead, mercury, nickel, and zinc) in sludge that can be used for agricultural purposes [7]. Danish Executive Order Number 823 sets maximum permissible levels, or "cutoff values," on four organic compounds in sewage sludge used for agricultural purposes in Denmark [8]. The four compounds regulated under the executive order are: LAS, NPE, DEHP, and total PAH. LAS, as previously indicated, is linear alkylbenzene sulfonate, the major surfactant used in laundry detergents and cleaning products in Denmark and throughout the world.

NPE is defined as nonylphenol plus the one- and two-mole ethoxylates of nonylphenol. These are biodegradation intermediates of commercial nonylphenol ethoxylates (NPE), which are highly effective surfactants that have been widely used for more than 50 years in a number of industrial sectors, including textiles, pulp and paper, paints, adhesives, resins, and protective coatings. NPE are also used in a variety of detergents and cleaning products for home and institutional use.

DEHP is di(2-ethylhexyl)phthalate, a softening agent, or plasticizer, used to impart flexibility, specifically, in the production of vinyl plastic.

Total PAH is defined as the sum of nine specific polycyclic aromatic hydrocarbons [acenaphthene, benzo(a)pyrene, benzo(ghi)perylene, fluoranthene, fluorene, indeno(1,2,3-cd)pyrene, phenanthrene, pyrene, and three benzofluoranthenes (b + j + k)]. These materials are by-products of combustion or incineration.

Executive Order Number 823 set cutoff limits for these four materials in sludge used for agricultural purposes that went into effect July 1, 1997. These limits were lowered further by Denmark, and new cutoffs became effective July 1, 2000.

The stated purpose of the executive order is to "ensure elimination of concentrations of substances that over a period of time and by repeated application may have damaging effects on farmland or, due to unacceptable loading to subsurface waters, crops, or domestic animals, on human beings" [8]. The justification is based on the mere *possibility* of environmental and human health effects. The executive order does not consider whether environmental concentrations are at levels that cause any harm or pose actual risk.

In fact, the executive order is not based on an environmental safety assessment or an overall risk-to-benefit balance of these chemicals for society. Instead, it seems to be based on an interpretation of the *precautionary principle*. The precautionary principle [9] states: "In order to protect the environment, the precautionary approach should be widely applied by states according to their capabilities. When there are threats of serious or irreversible damage, lack of full scientific certainty shall not be used for postponing cost-effective measures to prevent environmental degradation."

However, in the case of LAS and NPE, there are extensive scientific data, including environmental monitoring studies, to demonstrate that the residual levels present in sludge-amended soil rapidly biodegrade and do not accumulate [10–17]. Furthermore, this extensive data set indicates that current levels of LAS and NPE in Europe and North America pose minimal risk to the environment [3,18–24]. Consequently, the sludge limits set in Danish Executive Order Number 823, at least for the detergent-related materials, cannot be justified based on the precautionary principle.

The follow-up to the executive order has been a series of campaigns by the government of Denmark to encourage consumers not to purchase specific detergent products. Avoiding products containing LAS and NPE would provide no real benefit to the environment in Denmark, as is known from the extensive environmental information cited earlier, but *is* likely to lead to higher prices or poorer performance for laundry detergents and cleaning products.

A further implication of the executive order illustrates how environmental regulations can have unintended consequences that actually produce an adverse impact on the environment. Existing Danish regulations allow sludge application to soil only in the spring and autumn of the year, and one of the provisions of Danish Executive Order Number 823 requires each sewage treatment plant providing fertilizer to have enclosed facilities adequate to store its entire production of sludge between soil applications. Thus, treatment plants may have to store sludge up to 7 months before it can be applied to the soil. Traditional sludge storage practices consisted of placing sludge in open fields and covering it with straw [25]. Thus this provision of the executive order appears to mandate improved storage of sludge that would reduce possible runoff and leaching of sludge components and reduce odors associated with storage of mounds of sewage sludge in open fields.

The capital costs of building the required sludge storage facilities greatly exceed the cost of alternative sludge disposal methods, including construction of incinerators to combust the sludge [26]. As a result of this provision of the executive order, the uncertainties that sewage sludge will meet organic compound cutoff limits, and the continuing concerns of farmers regarding the safety of sewage sludge, the percentage of municipalities choosing to apply sludge to soil has been reduced from 70% in 1994, before the executive order was enacted, to 50% in 1997 and was expected to fall to 25% by the year 2000 [27]. By 2001, less than 8% of municipal sludge was applied to soil in Denmark [59].

The Danish EPA has publicly stated that the intention of the executive order was to reassure farmers in Denmark that sewage sludge was safe for application to agricultural soil and thus to maintain this environmentally preferable option for sludge disposal [28]. How did the Danish government enact an executive order that has had exactly the opposite impact?

The short answer is that the regulatory process was not open to stakeholders such as farmers, the municipalities (the operators of the POTW in Denmark), or the detergent and surfactant industry [29]. Furthermore, the regulation was not subject to independent scientific review. Thus this closed regulatory process and the lack of adequate scientific review led to a regulation that in fact is in conflict with the available scientific information.

It is instructive to compare this regulation with those developed under a more open regulatory process, such as the one used by the U.S. Environmental Protection Agency (EPA) to develop sludge disposal regulations.

III. EPA SLUDGE DISPOSAL REGULATIONS AND THE REGULATORY PROCESS USED

U.S. EPA regulations on the application of sewage biosolids to land were enacted in 1992 and amended in October 1995 [30]. Under the regulation, certain "organic pollutants" are eligible for a "removal credit" (U.S. government funds) if sewage treatment equipment is installed to remove the organics before sludge disposal. The amount of credit depends on the method of disposal (land application, landfill, or incineration). The organics listed include: DEHP, the PAH benzo(a)pyrene, and various others (aldrin/dieldrin, benzene, chlordane, DDD, DDE, DDT, 2-4-dichloro-phenoxy acetic acid, heptachlor, hexachlorobenzene, hexachlorbutadiene, lindane, malathion, N-nitrosodimethylamine, pentachorophenol, phenol, polychlorinated biphenyls, toxaphene, and trichloroethylene). However, LAS and NPE are not listed, and no surfactant or detergent ingredient is listed.

Consequently, the EPA regulation on biosolids disposal to soil is very different from the Danish regulation:

1. The EPA regulation does not regulate highly biodegradable detergent ingredients but instead focuses on compounds with evidence of persistence in the environment.
2. The EPA regulation does not place restrictions on the levels of targeted organics in biosolids but encourages their reduction through the incentive of offering financial credits for installing sewage treatment equipment that will reduce the level of the targeted organics.

How did the United States end up with such a different regulation on the organic components of biosolids? The U.S. EPA followed a rigorous policy of scientific assessment and outside scientific review (summarized in Table 1) that revealed no risk to human health or the environment from the presence of organic substances in biosolids intended for land application [31]. Because some organics found in biosolids were identified as having persistence in the environment, EPA developed an incentive program to reduce the levels of the substances in biosolids through improved sewage treatment.

Even after the regulation was enacted, the U.S. EPA invited periodic outside review by the U.S. National Academy of Sciences [32]. These reviews ensure that the

TABLE 1 U.S. EPA Biosolids Risk Assessment and Rule Development Process

Developmental step	Result
1. EPA task force formed to study biosolids	Comprehensive plan developed for management and regulation of biosolids.
2. EPA initial identification of pollutants	200 pollutants identified based on toxicity.
3. Expert panel selection of pollutants for further study	Selection based on exposure and toxicity data.
4. Exert panel identification of appropriate exposure pathways	Exposure pathways detailed for each pollutant and biosolids disposal option.
5. EPA development of initial risk assessments	Assessment based on pollutant toxicity and concentration in biosolids and worst-case exposures
6. EPA development of detailed risk assessments on selected pollutants	Pollutants selected based on initial risk assessments
7. EPA Science Advisory Board (SAB) review	This provides independent scientific review of risk assessment methods.
8. EPA revision of risk assessments	EPA revises risk assessments based on SAB review.
9. EPA publication of proposed regulation for biosolids	Proposed rule, based on risk assessments, published for public and scientific comment.
10. EPA revision of proposed regulation for biosolids	EPA revises based on public and scientific comment.
11. National Sewage Sludge Survey (NSSS)	EPA conducts a statistically representative survey of pollutants in biosolids and biosolids use and disposal practices.
12. EPA publication of NSSS results and revised proposed biosolids rule	Proposed biosolids regulation revised based on NSSS and published for public and scientific comment.
13. EPA final revision of biosolids regulation	Final revisions conducted with internal and external experts.
14. Final regulation published.	Final regulation subject to challenge and review under U.S. law.

Source: Ref. 31.

regulation takes into account the most recent available health and environmental information and uses the most advanced scientific methods to evaluate safety.

Because the regulation is based on a rigorous evaluation of the science and an open process of outside scientific review, EPA could be assured that the regulation would be protective of human health and the environment and that public funds spent on improved sewage treatment would actually benefit the environment by reducing the levels of persistent substances in biosolids.

IV. CITY OF TORONTO, CANADA SEWER DISCHARGE LAW

In June 2000, the City of Toronto, Ontario, Canada, adopted a municipal regulation prohibiting the discharge into city sewers of wastewater exceeding prescribed levels of BOD, oil and grease, phenolics, total nitrogen, suspended solids, three salts, 18 metals, and 27 organic compounds, including the detergent ingredients commercial NPE and the compound from which they are made, nonylphenol (Table 2) [33]. The stated purpose of the regulation is to allow the city to sell its biosolids to a private company to make fertilizer. Previously, Toronto had been incinerating its biosolids, but the incinerator was being closed because it was too expensive and because of concerns regarding air pollution [34].

The regulation applies to all businesses in the city with 10 or more employees. Automobile, medical, printing, and transportation businesses were given until December 31, 2001, to verify compliance with the regulation or submit a pollution prevention plan to meet the discharge requirements; all other businesses had until June 30, 2002.

Because analysis of wastewater for the substances listed in Table 2 would be quite expensive, especially for small or medium size businesses, the most practical method to ensure that the sewer discharge requirements are met is for each business to ensure that they do not discharge to the sewer any products containing the listed substances. Indeed, this seems to be the major purpose of the pollution prevention plan required for businesses that do not meet the discharge limits [35].

While the use of biosolids as fertilizer is environmentally preferred to incineration because nutrient recycling returns the substance to the earth, scientifically there are a number of flaws in the approach taken by the City of Toronto:

1. No consideration was given to the biodegradability of the listed organics during sewage treatment.
2. The sewer discharge limits were not based on any finding of harm to human health or the environment from existing levels of any of the substances in City of Toronto biosolids.
3. No safety assessment was conducted to determine maximum safe levels for the listed substances in biosolids. Indeed, the regulated levels were set based on the lowest levels that could be reliably detected by the analytical methods recommended [33].

The approach used by the City of Toronto indicates that the unstated purpose of the regulation is to ensure that businesses do not discharge to city sewers prod-

TABLE 2 Regulated Substances for Discharge to Toronto Sewers

Compounds	Compounds
Biochemical oxygen demand	Organics
Oil and grease—animal and vegetable	Benzene
Oil and grease—mineral and synthetic	Chloroform
Phenolics (4AAP)	1,2-Dichlorobenzene
Suspended solids (total)	1,4-Dichlorobenzene
Total Kjeldahl nitrogen	cis-1,2-Dichloroethylene
	trans-1,3-Dichloropropylene
Salts	Ethyl benzene
Cyanide (total)	Methylene chloride
Fluoride	1,1,2,2-Tetrachloroethane
Phosphorus (total)	Tetrachloroethylene
	Toluene
Metals	Trichloroethylene
Aluminum (total)	Xylenes (total)
Antimony (total)	Di-n-butyl phthalate
Arsenic (total)	Bis(2-ethylhexyl) phthalate
Cadmium (total)	Nonylphenols
Chromium (hexavalent)	Nonylphenol ethoxylates
Chromium (total)	Aldrin/dieldrin
Cobalt (total)	Chlordane
Copper (total)	DDT
Lead (total)	Hexachlorobenzene
Manganese (total)	Mirex
Mercury (total)	PCBs
Molybdenum (total)	3,3'-Dichlorobenzidine
Nickel (total)	Hexachlorocyclohexane
Selenium (total)	Pentachlorophenol
Silver (total)	Total PAHs
Tin (total)	
Titanium (total)	
Zinc (total)	

Source: Ref. 35.

ucts containing *any* amount of the listed substances. Since many consumer, institutional, and industrial products containing these substances are intended for discharge to sewers, the discharge regulations are in effect restrictions on the sale and use of these products.

The lack of scientific basis for the regulation is particularly striking for commercial NPE and nonylphenol (NP) because of the considerable data available on these compounds.

1. NP and NPE are effectively removed from wastewater in well-functioning sewage treatment plants, such as those operated by the City of Toronto, with only a small fraction of the NP and NPE remaining in the biosolids [16,36,37].
2. NP and NPE in soil have been shown to undergo rapid and extensive biodegradation [15,17].

3. Because of this biodegradation, the levels of NP remaining in soil as a result of biosolids application are below the level that would pose a risk to typical soil organisms, such as earthworms [38,39].
4. Laboratory experiments indicate that NP and NPE in soils amended with biosolids do not leach into lower soil layers towards groundwater [40].
5. Uptake of NP by crop plants from biosolids applied to soil is unlikely because of the rapid biodegradation of NP and NPE in soil and the typical fallow period between the application of biosolids and planting [41].

Consequently, the Toronto sewer discharge regulation provides no safety benefit to the environment, but imposes costs on local businesses to identify alternative or replacement products for the many consumer, institutional, and industrial products using commercial NPE or other listed substances. In addition the regulation imposes costs to manufacturers and suppliers in Toronto and beyond to reformulate or supply alternative products for customers in Toronto. For businesses unable to identify alternative or replacement products, the costs include preparation of a pollution prevention plan and verification, by analytical chemistry testing, of compliance with the discharge limits.

When environmental regulations are not based on science, the unfortunate result is a sewer discharge regulation that imposes costs on the public without demonstrable benefit to the environment. Here, as with the sludge regulation in Denmark, stakeholder comments on the lack of a scientific basis for the regulation [42,43] were ignored, and the regulation was not subject to independent scientific review.

V. ENVIRONMENTAL LABELING REGULATIONS

Other regulations on detergent products have focused on environmental labeling. Two very different types of regulations have been considered: One focuses on the types of environmental marketing claims that can be made about a detergent or other product, and the other provides a mechanism for environmentally preferred labeling ("Ecolabeling") of detergents products.

A. Environmental Marketing Claims

In August, 1992, the U.S. Federal Trade Commission (FTC) published guidelines for environmental labeling of products and packages, including detergent products, that applied to all forms of marketing these products to the public [44]. Although the guidelines were labeled as "voluntary," the FTC advised that:

> Conduct inconsistent with the positions articulated in these guidelines may result in corrective action by the Commission under section 5 [of the FTC Act] if, after investigation, the Commission has reason to believe that the behavior falls within the scope of conduct declared unlawful [e.g., "unfair or deceptive acts or practices"] by the statute [44, pp. 36364, 36366].

In other words, the guidelines may not have the force of law but do provide guidance as to how the FTC will choose to enforce the law against unfair or deceptive advertising.

The guidelines are composed of general principles and specific guidance on the use of environmental claims. The four general principles that apply to all environmental claims are:

1. Qualifications and disclosures related to any claims should be clear and prominently visible.
2. For each claim, it should be clear whether it applies to the product, the package, or a component of the product or the package.
3. Environmental claims should not be overstated.
4. Comparative claims should be clear and understandable.

The guidelines also address 10 commonly used environmental terms: *degradable, biodegradable, photodegradable, compostable, recyclable, recycled content, source reduction, refillable, ozone safe,* and *ozone friendly.* The guidance on each specific term includes examples that illustrate the FTC's concern regarding an unfair or deceptive environmental claim.

For instance, regarding *biodegradable,* the FTC guidelines [44, pp. 36365–36366] state:

> An unqualified claim that a product or package is biodegradable should be substantiated by competent and reliable scientific evidence that the entire product or package will completely break down and return to nature, i.e., decompose into elements found in nature within a reasonably short period of time after customary disposal. Claims of biodegradability should be qualified to the extent necessary to avoid consumer deception about:
>
> 1. The product or package's ability to degrade in the environment where it is customarily disposed; and
> 2. The rate and extent of degradation.
>
> *Example:* A soap or shampoo product is advertised as "biodegradable," with no qualification or other disclosure. The manufacturer has competent and reliable scientific evidence demonstrating that the product, which is customarily disposed of in sewage systems, will break down and decompose into elements found in nature in a short period of time. The claim is not deceptive.

In short, the FTC guidelines call for environmental claims to be factual and based on appropriate scientific data. Furthermore, environmental claims must be specific and not overstate the findings or relevance of the data.

How did the FTC manage to enact guidelines for environmental claims that make so much sense and are based on science? Several factors can be cited:

1. The guidelines were the result of a petition filed by manufacturers and retailers seeking uniform national standards for environmental claims on consumer products. Public hearings were held on the guidelines, along with a 90-day public comment period [44]. Consequently, the process of development of the guidelines was open to stakeholder involvement from the beginning.
2. Furthermore, the guidelines were developed in a cooperative effort with the U.S. EPA and the White House Office of Consumer Affairs to broaden the base of scientific and technical input into the process [45].

3. Once the guidelines were developed, the FTC published a brochure for consumers to explain what was meant by overly broad or vague environmental
   claims and why such claims had little meaning [45].
4. When the FTC issued the guidelines in 1992, it included a provision that, three
   years after adoption, it would seek public comment on "whether and how the
   guides need to be modified in light of ensuing developments" [44, p. 36364]. The
   FTC then issued a call for public comment in July 1995, held more public hearings
   in December 1995, and issued revised guidelines in 1996 and in 1998 [46].
   Consequently, the guidelines have been open to further refinement based on public
   and scientific comment since they were issued.

As a result of these factors, the FTC guidelines on environmental claims are
based on science and are clearly understandable to the detergent industry and the
public.

B. Eco-Label Programs

A number of organizations have taken a quite different approach to environmental
claims, providing a single, easily recognized symbol, or Eco-label, to denote environmentally superior products among those in a particular category of products, such as
laundry detergents [47,48].

While the various Eco-label programs differ in the products covered and the
specific criteria used, the Nordic Swan, the official Eco-label in Denmark, Finland,
Iceland, Norway, and Sweden [49], will be examined in more detail because it is one of
the oldest Eco-labels (it was initiated in 1989) and because the approach taken by the
Nordic Swan is typical of other Eco-label programs.

Moreover, the Nordic Swan program has probably had more influence on
laundry detergent products than any other Eco-label program. This is because retailers
in Sweden have insisted that only laundry products displaying the Nordic Swan may
be sold in their stores. In Denmark, the EPA initiated a series of campaigns to
encourage consumers not to purchase laundry products unless they carried the Eco-
label. Both cases raise troubling questions regarding the supposed voluntary nature of
the program and focus renewed attention on the scientific basis and adequacy of
stakeholder involvement.

The Nordic Swan currently provides Eco-label criteria for 48 types of products,
including durable goods (dishwashers, washing machines, refrigerators, etc.), household products, including laundry detergents, home and garden products, paper
products, office equipment and supplies, and a number of consumer and industrial
products and services (Table 3).

According to the Nordic Eco-labeling Board, which chooses the products and
decides the final criteria, the objective of the Nordic Swan is "to provide information
to consumers to enable them to select products that are the least harmful to the
environment" and "to stimulate environmental concern in product development" [50].

The difficulty of this task can be demonstrated by considering the Nordic Swan
criteria for laundry detergents. Manufacturers of laundry detergents seeking to earn
the ecolabel are required to:

1. Meet all general restrictions, summarized in Table 4, on the product and on all
   components, including packaging.

TABLE 3 List of Products for Which Nordic Swan
Criteria Are Available

Durable goods
  Audiovisual equipment
  Dishwashing machines
  Refrigerators, freezers
  Washing machines
Household products
  All-purpose cleaners
  Detergents for sanitary facilities
  Dishwasher detergents
  Floor care products
  Hand dishwashing detergents
  Shampoo and soap
  Textile (laundry) detergents
Home and garden products
  Building board: chip-, fiber-, and gypsum board
  Closed toilet systems
  Composters
  Flooring materials
  Lawnmowers
  Oil burner/boiler combinations
  Oil burners
  Small heat pumps
  Wall coverings
  Windows
Paper products
  Coffee filters
  Grease-proof paper
  Paper envelopes
  Printed matter
  Printing paper
  Tissue paper
Office equipment and supplies
  Copying machines
  Faxes and printers
  Personal computers
  Toner cartridges
  Writing instruments
Miscellaneous consumer and industrial products and services
  Adhesives
  Batteries, primary
  Batteries, rechargeable
  Car care products
  Diapers, disposable
  Diapers, washable
  Female sanitary products
  Hotels
  Ice combating agents
  Industrial cleaning and degreasing agents
  Light sources
  Lubricating oils
  Marine engines
  System for towels in dispensers
Textiles
Wooden furniture and fittings

Source: Ref. 49.

TABLE 4 Restrictions on Laundry Detergents in the Nordic Swan Program

Product must not be classified as "very toxic," "toxic," "corrosive," or "harmful."

The pH of the laundry detergent solution prepared in accordance with the directions on the package must not exceed 11.0.

Products formed during decomposition must not be classified environmentally harmful by Nordic or EU criteria.

Components that have an environmental hazard classification of R50 + R53, R51 + R53 or R52 + R53 in accordance with any Nordic or EU Directive are limited to a maximum of 0.12 g/wash. Components classified as R50 (not in combination with other classifications) are limited to a maximum of 7.5 g/wash.

Swan-labeled detergents containing more phosphorus than permitted by Norwegian regulations may not be sold in Norway.

The total content of phosphonates and NTA must not exceed 0.5 g/wash.

Surfactants must be readily biodegradable and anaerobically degradable.

Components considered carcinogens, genotoxic, or teratogen in accordance with the relevant regulations in Finland, Iceland, Norway, or Sweden must not be intentionally added.

Enzymes must be encapsulated or mixed into slurry before they are added to the laundry detergent and must not contain residues of microorganisms originating from production.

Only fragrances that are recommended by the International Fragrance Association Code of Practice (IFRA) are permitted.

The following components may not be added: alkylphenol ethoxylates, reactive chlorine compounds (e.g., sodium hypochlorite or organic chlorine compounds), EDTA, optical whiteners, or dyes.

All plastic must be labeled. PVC or other chlorinated plastics must not be included either in the package or in the label.

Source: Ref. 50.

2. Meet all specific restrictions on the maximal amounts of ingredients per wash load and score at least 42 of a possible 84 points by further limiting the amounts of these ingredients per load according to a scoring system summarized in Table 5, including documentation of all results.

3. Document satisfactory washing performance in a standardized test with "normally soiled laundry."

4. Provide all required consumer labeling information plus amounts of phosphates and recommendations to use water softeners in areas where hard water is commonly found.

5. Document that the detergent manufacturing plant meets all environmental rules in the country of manufacture and that the manufacturer follows all national rules for recycling of products and packaging in the Nordic countries.

6. Set up and administer an environmental management system, directed by a single contact person, to ensure that all Eco-label and environmental rules are followed.

7. Provide a plan for educating retailers regarding the requirements of the Eco-label program for this product category, and share in the responsibility for following the program rules and principles.

8. Pay for costs of analysis of random samples of product taken from retail stores at the discretion of any of the Nordic country Eco-labeling bodies.

In addition to the obvious complexity of such a system, a number of scientific and practical difficulties are apparent. First and most importantly, the criteria are not

TABLE 5 Maximal Amounts of Ingredients Permitted per Wash Load and Mandatory Scoring System for Ingredients in Laundry Detergents for Nordic Swan

Parameter (g/wash)	Highest permissible quantity	Points, P, calculated in accordance with:	Weighting factor, WF	Highest permissible partial sum WF $\times$ P
Total quantity of chemicals, K (g/wash)	60	$12-K/5$	3	12
Toxicity and degradability, GN (L/wash)	14,000	$4.7-GN/2985$	8	32
Phosphorus quantity, P (g/wash)	4	$4-P$	3	12
Poorly soluble inorganic compounds, SOO (g/wash)	15	$6-SOO/2.5$	0.5	2
Easily soluble inorganic compounds, LOO (g/wash)	35	$7-LOO/5$	0.5	2
Not ultimately degradable compounds, IPN (g/wash)	4	$4-IPN$	1	4
Not anaerobically degradable compounds, IAN (g/wash)	4	$4-IAN$	2	8
Quantity of organic compounds, TOC (g/wash)	10	$2/3(10-TOC)$	2	8
Packaging: weight/ utility, VNF (g/dose)	6	$6-VNF$	1	4
Sum				At least 42 points required of a possible 84 points

Source: Ref. 50.

based on a scientific determination of health and environmental safety—no justification whatsoever is provided for any of the restrictions listed in Table 4. In fact, when the available science is examined, it does not support the specific restrictions in the Nordic Swan.

One example is the restriction on the use of alkylphenol ethoxylate (APE) surfactants. Environmental monitoring studies have demonstrated that levels of nonylphenol ethoxylates, the principle APE surfactant, and their biodegradation intermediates, are generally found at low concentrations in the environment and pose little risk to fish or other aquatic organisms [3]. Exceptions have been noted, as might be expected, where untreated or partially treated wastewater is discharged. Since there are numerous potentially harmful substances in untreated or partially treated waste-water, the available information provides no scientific justification for the restrictions on the use of APE surfactants in laundry detergents.

Another important example where the available science does not support the restrictions in the Nordic Swan is the requirement in Table 4 that all surfactants used must be anaerobically degradable, that is, undergo biodegradation in the absence of oxygen.

An expert panel of the Environmental Risk Assessment Steering Management (ERASM), a cooperative effort of the International Association for Soaps, Detergents, and Maintenance Products (AISE, Association Internationale de la Savonnerie de la Détergence et des Produits d'Entretien) and the European Committee of Surfactants and Their Organic Intermediates (CESIO, Comité Européen des Agents de Surface et leurs Intermédiares Organiques), reviewed the environmental relevance of anaerobic biodegradation of surfactants [51]. This extensive review of the available scientific evidence concludes:

> The majority of surfactants entering the environment will be exposed to and degraded under aerobic conditions, and only less than 20% will potentially reach anaerobic environmental compartments. In all but a few cases their presence in these will not be permanent. A systematic evaluation of the risk to the structure and function of these compartments due to the presence of undegraded surfactants led to the conclusion that, in contrast to the adverse effects observed in the absence of aerobic degradation, the lack of anaerobic biodegradation does not seem to be correlated with any apparent environmental problem for most compartments. Consequently it is concluded that anaerobic biodegradability does not have the same environmental relevance as aerobic. Anaerobic biodegradability should not, therefore, be used as a pass/fail property for the environmental acceptability of surfactants that are readily biodegradable under aerobic conditions.

Furthermore, the Nordic Swan classification of surfactants that do not undergo anaerobic biodegradation is outdated. Recent studies indicate that LAS, which are listed as not anaerobically biodegradable, in fact *do* undergo anaerobic biodegradation [52–56].

Because the available science indicates that the restrictions on surfactants in the Nordic Swan are not based on science, the Eco-label provides no guarantee that laundry detergents meeting the restrictions are any better for the environment than products that simply meet all legal requirements.

The same concern applies to the scoring system for ingredients (Table 5). No scientific justification has been provided for the scoring criteria. Neither impact nor risk to the environment has been demonstrated for laundry detergents exceeding the permitted levels of any of the ingredients. Consequently, the Eco-label provides no guarantee that products meeting the scoring criteria are any better for the environment than products that do not.

This lack of scientific basis for the eco-label criteria leads to a second difficulty with such schemes—the apparent unfair advantage that Eco-labeled products have over their competitors. This concern arises because only "satisfactory" cleaning performance is required by the Nordic Swan. As a result, consumers may have to use more of an Eco-labeled laundry detergent to achieve the same cleaning as products optimized for cleaning performance. Consumer use of an Eco-labeled product at doses exceeding the recommended amounts provides an unfair advantage to that product because Eco-label criteria are based on recommended doses, not the doses actually used by consumers.

A third difficulty with Eco-label criteria is that, once adopted, they tend to be rigid and to stifle innovation and product improvement. Since Eco-labels are based on pass/fail criteria; there is little incentive, once all the criteria have been met, for additional reformulations even if these would improve the environmental performance of the product.

The Nordic Swan program, to its credit, has a procedure for periodic updating of the criteria, which should allow for the input of new scientific developments and new production methods. The Nordic Swan follows the principle that "interested parties should be given the opportunity to participate and account should be taken of their comments" [50]. Despite these procedures, it is apparently difficult for the Eco-labeling board to agree to loosen criteria even when overwhelming scientific evidence is presented, e.g., the ERASM report on anaerobic biodegradation criteria [57].

A final difficulty with the Nordic Swan is that it does not educate consumers regarding the environmental properties of the products they purchase and use. This is because the focus of the program, apart from some concern to reduce packaging, is on reducing inputs of detergent ingredients to the sewers without consideration of the fact that improving wastewater treatment systems is recognized as the best method for reducing the impact of sewage discharges on the environment. The Nordic Swan criteria document for laundry detergents acknowledges that consumer choices such as selection of water temperature, presorting and adjusting detergent use by degree of soiling, and use of prewashing for heavily soiled laundry may provide opportunities for major environmental gains but indicates these choices are difficult to guide by Eco-labeling.

Furthermore, the criteria document does not educate, and may mislead, consumers on the true environmental properties of laundry detergents because it does not discuss the scientific basis for the criteria, which in fact are not based on science.

In short, the Nordic Swan program on laundry detergents, despite its appearance of objectivity, is not based on science and thus provides no demonstrable environmental benefit. Moreover, the program focuses on a limited set of properties of laundry detergents and largely ignores consumer decisions on how to use detergents that are major determinants of their environmental impacts.

These limitations of the Nordic Swan Eco-label on laundry detergents are particularly disappointing because the Nordic Swan is one of the oldest Eco-label programs and the criteria on laundry detergents are typical of other Eco-labels.

Comparison of the Nordic Swan program with science-based guidelines for environmental marketing claims, such as the FTC guidelines discussed in the previous section, reveals why major detergent manufacturers strongly prefer environmental marketing guides to Eco-label programs [58]. In contrast to the process used to develop the FTC guidelines, Eco-label programs typically have limited stakeholder involvement, especially scientific input. A further key limitation of Eco-label programs is the absence of independent scientific review. Given the very unscientific nature of the

Eco-label process, this review should begin with establishing a scientific approach to supporting a claim that Eco-labeled products are "least harmful to the environment."

VI. CONCLUDING REMARKS

In this chapter we have examined the ongoing policy debate regarding the best method for setting environmental and health regulations using detergent surfactants as a vehicle of the discussion. This analysis has pointed out that the choice is not between stakeholder involvement or science—both must be involved. Stakeholder involvement is critical for ensuring that the regulations have the intended effect and are based on science. When scientific input is not considered, regulations will inevitably impose public costs without reducing health or environmental impacts or risks. Independent review by scientific experts is a very effective tool, when employed, to ensure a sound scientific basis for environmental and health regulations.

Lack of independent scientific review and inadequate stakeholder involvement, as shown by examples presented in this chapter, have led to environmental regulations that are poor science and poor public policy. The best approach for environmental and health regulations is not science versus politics, as has all too often been the case, but science *and* politics, including stakeholder involvement, working together for the benefit of an improved environment for everyone.

REFERENCES

1. Feijtel, T.C.J.; Struijs, J.; Matthijs, E. Environ. Toxicol. Chem. 1999, *18*, 2645–2652.
2. Matthijs, E.; Holt, M.S.; Kiewiet, A.; Rijs, G.B.J. Environ. Toxicol. Chem. 1999, *18*, 2634–2644.
3. Staples, C.A.; Williams, J.B.; Naylor, C.G. An environmental risk assessment of the biodegradation intermediates of nonylphenol ethoxylates. Proceedings of the 4th World Conference on Detergents: Strategies for the 21st Century, Montreux, Switzerland, 1998, 298–303.
4. van de Plassche, E.J.; de Bruijn, J.H.M.; Stephenson, R.R.; Marshall, S.J.; Feijtel, T.C.J.; Belanger, S.E. Environ. Toxicol. Chem. 1999, *18*, 2653–2662.
5. Off J. Eur. Com. 1973, *L347*, 51.
6. Swisher, R.D. Surfactant Biodegradation. 2nd ed.; Marcel Dekker: New York, 1987.
7. Off. J. Eur. Com. 1986, *L181*, 6–12.
8. Lorenzen, I.M. Draft for a New Executive Order on the Utilization of Waste Products for Agricultural Purposes; Danish Ministry for Energy and the Environment, June 1, 1996.
9. Principle 15 (The Precautionary Principle), Agenda 21: The United Nations Program of Action from Rio; United Nations: New York, 1992.
10. Berna, J.L.; Ferrer, F.; Moreno, A.; Prats, D.; Bevia, F.R. Tenside Surfac. Det. 1989, *32*, 101–107.
11. Figge, K.; Schöberl, P. Tenside Surfac. Det. 1989, *26*, 122–128.
12. Holt, M.S.; Matthijs, E.; Waters, J. Water Res. 1989, *23*, 749–759.
13. Marcomini, A.; Capel, P.D.; Lichtensteiger, T.; Brunner, P.H.; Giger, W. J. Environ. Qual. 1989, *18*, 523–528.
14. Prats, D.; Ruiz, F.; Váquez, B.; Zarzo, D.; Berna, J.L.; Moreno, A. Environ. Toxicol. Chem. 1993, *12*, 1599–1608.
15. Hughes, A.I.; Fisher, J.; Brumbaugh, E. Biodegradation of NPE in soil. Proceedings of the 4th World Surfactants Congress, Barcelona, 1996; Vol. 4, 365–372.
16. Maguire, R.J. Water Qual. Res. J. Canada 1996, *34*, 37–38.

17. Topp, E.; Starratt, A. Environ. Toxicol. Chem. 2000, *19*, 313–318.
18. Mieure, J.P.; Waters, J.; Holt, M.S.; Matthijs, E. Chemosphere 1990, *21*, 251–262.
19. Holmstrup, M.; Krogh, P.H. Environ. Toxicol. Chem. 1996, *15*, 1745–1748.
20. Kloepper-Sams, P.; Torfs, F.; Feijtel, T.; Gooch, J. Sci. Total Environ. 1996, *185*, 171–185.
21. de Wolf, W.; Feijtel, T. Chemosphere 1998, *36*, 1319–1343.
22. Cavalli, L.; Valtora, L. Tenside Surfac. Det. 1999, *36*, 22–28.
23. Solbé, J. CLER Rev 1999, *5*, 24–60.
24. Elsgaard, L.; Petersen, S.O.; Debosz, K.; Kristiansen, I.B. Tenside Surfac. Det. 2001, *38*, 94–97.
25. Madsen C.S. Utilization of Sludge on Farmland in Denmark. The Association of Danish Cosmetics, Toiletries, Soap and Detergent Industries (SPT), September 18, 1996.
26. Jørgensen L. Ingeniøren, January 3, 1997.
27. Jørgensen L.; Weitling, H. Ingeniøren, January 3, 1997.
28. Lynghus J. New Draft on Limit Values for LAS Released, Denmark Faces Toughest Sludge Legislation in the World, Summary of Danish Press Reports. Mannov Consult, Rodovre, May 6, 1996.
29. Heinze, J.E. Science versus politics in the environmental regulatory process. Proceedings of the 4th World Conference on Detergents: Strategies for the 21st Century, Montreux, Switzerland, 1998; 83–88.
30. Fed. Reg. 1995, *60*, 54763–54770.
31. A Guide to the Biosolids Risk Assessments for the EPA Part 503 Rule, Report EPA832-B-93-005. Environmental Protection Agency, Washington, DC, 1995.
32. Renner, R. Environ. Sci. Technol. 2001, *35*, 183A.
33. New Sewer Use By-Law, Consultation Feedback Before June 1999, Industry Issues/Comments List, Up To June 1, 1999. Works & Emergency Services, City of Toronto, 1999.
34. Mittelstaedt M. Globe and Mail, June 3, 1999.
35. Municipal Code, Chapter 681, Sewers. City of Toronto, 2000. Available at: www.city.toronto.on.ca/legdocs/municode/ch681.pdf.
36. Lee, H.B.; Peart, T.E. Anal. Chem. 1995, *67*, 1976–1980.
37. Naylor, C.G. Textile Chem. Colorists 1995, *27*, 29–33.
38. Krogh, P.H.; Holmstrup, M.; Jensen, J.; Petersen, S.O. 1996. Priority Substance List Assessment Report, Nonylphenol and Its Ethoxylates, Draft for Public Comment. Environment Canada, Health Canada, 2000; 41 pp.
39. Jensen, J.; Krogh, P.H. Ecotoxicological assessment of sewage sludge application. Specialty Conference, Management and Fate of Toxic Organics in Sludge Applied to Land, Copenhagen, 1997.
40. Kujawa, M.; Schnaak, S.; Kuchler, T. Occurrence of organic pollutants in sewage sludge and influence of surfactants on their mobility in amended soils. Specialty Conference, Management and Fate of Toxic Organics in Sludge Applied to Land, Copenhagen, 1997.
41. Priority Substance List Assessment Report, Nonylphenol and Its Ethoxylates, Draft for Public Comment, March 2000. Environment Canada, Health Canada, 2000; 41–69 pp. Available at: www.ec.gc.ca/cceb1/eng/public/npe_e.pdf.
42. Fensterheim, R. Letter to Mr. Victor Lim, Manager of Industrial Waste and Storm Water Quality, City of Toronto. Alkylphenols and Ethoxylates Research Council, Washington, DC, 1999.
43. Fensterheim, R. Letter to the Honorable Mel Lastman, Mayor, City of Toronto. Alkylphenols and Ethoxylates Research Council, Washington, DC, 2000.
44. Fed. Reg. 1992, *57*, 36363–36369.
45. Green Advertising Claims, EPA530-F-92-024. RCRA Information Center, U.S. Environmental Protection Agency, Washington, DC, 1992.
46. Fed. Reg 1998, *63*, 24240–24250.
47. Off. J. Eur. Com. 1992, *L99*, 1.

48. Off. J. Eur. Com. 1995, *L217*, 14–30.
49. Welcome to the Nordic Swan, at www.ecolabel.no/Engelsk/main.html, and Criteria for the Nordic Swan Label at www.ecolabel.no/Engelsk/Criteria.html, 2001.
50. Ecolabeling of Laundry Detergents, Criteria Document 16 June 1995 to 15 June 2002, Version 3.8. Nordic Ecolabeling Board, 2000.
51. Berna, J.L.; Battersby, N.; Cavalli, L.; Fletcher, R.; Guldner, A.; Schowanek, D.; Steber, J. Tenside Surfac. Det. 2001, *38*, 86–93.
52. Denger, K.; Cook, A.M. J. Appl. Microbiol. 1999, *86*, 165–168.
53. Angelidaki, I.; Haagensen, F.; Ahring, B.K. Anaerobic transformation of LAS in continuous stirred-tank reactors treating sewage sludge. Proceedings of the 5th World Surfactants Congress, Florence, 2000; 1551–1557.
54. Ferrer, J.; Moreno, A.; Berna, J.L.; Sanz, J.L.; Rodriguez, N. Evaluation of the inhibition potential of LAS (linear alkylbenzene sulfonate) to the methanogenic process. Proceedings of the 5th World Surfactants Congress, Florence, 2000; 1603–1608.
55. León, V.M.; González-Mazo, E.; Forja, J.M.; Gómez-Parra, A. Identification of linear alkylbenzene sulfonate biodegradation intermediates in anoxic marine coastal sediments. Proceedings of the 5th World Surfactants Congress, Florence, 2000; 1638–1643.
56. Prats, D.; Rodriguez, M.; Llamas, J.M.; De La Muela, M.A.; Ferrer, J.; Moreno, A.; Berna, J.L. The use of specific analytical methods to assess the anaerobic biodegradation of LAS. Proceedings of the 5th World Surfactants Congress, Florence, 2000; 1655–1658.
57. Phosphate and LAS ecoprofiles under siege in Scandinavia. CMR Online, June 11, 2001. Available at www.chemexpo.com/cmronline/stories/06_11_01/42_06_11_01.cfm.
58. De Henau, H. EU ecolabeling scheme for laundry detergents: criteria and relevance to environmental progress. Proceedings of the 4th World Surfactants Congress, Barcelona 1996; Vol. 3, 9
59. Affaldsstastistik 2002, Orientering fra Miljøstyrelsen Nr. 6, Danish Ministry for Energy and the Environment, 2003. Available at www.mst.dk.

31

The Environmental Impact of Detergents: Which Way Is the Wind Blowing?

URI ZOLLER   Haifa University–Oranim, Kiryat Tivon, Israel

I. INTRODUCTION: ENVIRONMENTAL IMPACT

Environmental impact (EI) is a complex, multidimensional systemic issue that may be approached from various perspectives, points of view, and predetermined goals by using different criteria for its evaluation in different contexts leading, in turn, to different and even contradictory conclusions. The product(s) of any EI assessment (EIA) of any environmental process that occurred in the past, is occurring at present, or is expected/predicted to occur in the future is, ultimately, being evaluated and translated into action in the context of science–technology–environment–society interfaces by relevant community and local, regional, or international policy-making bodies at different levels. The problem is that, regardless of the level of sophistication, accuracy, appropriateness, and comprehensiveness of the science-and-technology-based models, methodologies, data analysis and criteria applied, both science and technology or each separately may at best tell us what can be done. But neither one nor both can tell us what *should* be done in the socioeconomic environmental context [1]. The latter requires *evaluative* thinking, focusing on the economic feasibility, sustainability, and expected systemic short- and long-term outcomes of any action taking, rather than on the technological/engineering feasibility of the remediation process(es) and/or the detection limits/precision levels of the analytical methods employed in the determination of contaminants in the various environmental compartments studied. As far as detergents are concerned, the expected EI of each of their components reaching the environment, e.g., surfactants, builders, chelating agents, bleaches, coupling agents (such as ethylene glycol), and additives, should be assessed, separately and synergistically with other components of detergent formulations as well as with other pollutants present in receiving soils, surface water, and groundwater. For obvious reasons, the long-term EIA is almost always preferred, since it facilitates long-term planning and implementation of appropriate, feasible, long-term environmental policy, management remediation processes (if needed), and prevention program(s).

In response to the question posed in the title of this last chapter, the essence of selected relevant recent papers published during 2003 is presented and briefly discussed. Since surfactants constitute qualitatively and, in most cases, quantitatively too the leading active organic components in detergent formulations, their persistence and their metabolites' persistence in the natural environment and in sewage treatment plants, their biodegradability, their toxicity/ecotoxicity, and their bioavailability constitute areas of study. It is not surprising, therefore, that most detergent-related environmental studies deal with surfactants and/or with environmental concerns pertinent to surfactants and their EI.

II. OCCURRENCE OF SURFACTANTS

The main problem associated with the occurrence of surfactants in the environment is that, although more than 90–95% of them can be eliminated by conventional, mainly activated-sludge-based biological wastewater treatment, lower-molecular-weight recalcitrant metabolites are formed out of the parent surfactants. In the case of the nonionic alkylphenol ethoxylates [2] (APEOs), the shorter-chain alkylphenol ethoxylates formed, are more resistant to further degradation and more toxic than the parent compounds. Since in reality no operating sewage treatment plants may completely eliminate the original APEOs and/or their metabolites from the wastewater, they ultimately find their way into both surface water and groundwater [3] and, consequently, as potential endocrine disrupters (in this case), have their EI in these and related soil environmental compartments. Recently reported concentrations of the anionic linear alkylbenzene sulfonates (LASs), alcohol ether sulfates (AESs), alcohol ethoxylates (AEOs), the nonionic APEOs and the quaternary cationic surfactants in the *effluents* of wastewater treatment plants (WWTPs) in selected representative western countries are given in Table 1 [2]. These concentrations may provide an initial point of origin for the estimation—taking into account the biodegradation rate of these surfactants (in different aquatic environments, under different conditions) as well as the dilution effect—of the concen-

TABLE 1 Reported Concentrations of LASs, EASs, AEOs, APEOs and Cationic Surfactants in WWTP Effluents

Country	Surfactant	Concentration	Refs.
United States	LAS	<1–1500	4, 2
	AES	4–164	4, 2
	AEO	8–509	4, 2
Germany	LAS	65–115	5, 2
	Cationics	~0.4	6, 2
The Netherlands	LAS	19–71	7, 2
	AES	3–11.5	7, 2
	AEO	22–13	7, 2
	Cationics	1.2–9.1	8, 2
Switzerland	APEOs[a]	~1–20	9, 2

[a] NPEOs: Sum of NP, NP_1EO, and NP_2EO.

trations of these surfactants in the environmental receiving water bodies, i.e., rivers, water reservoirs, lakes, and/or seas/oceans.

Since the APEOs are substantially more biodegradation resistant ("hard")—both in WWTPs and in the natural environment—compared with the anionic surfactants ("soft"), it is not surprising that their and their metabolites' occurrence/survival in the receiving aquatic environment in affecting concentrations is expected to have an impact on the environment. This is particularly so because they are known to be endocrine-disrupting chemicals (EDCs) [10].

Three 2003 studies report the concentrations of APEOs and their metabolites in U.S. rivers [11], in river water entering waterworks in Spain [12], and in Israel surface water and groundwater [13]. Thus, the occurrence of APEOs, their metabolites, and nonylphenol (NP) was determined over a 74-mile length of the Cuyahoga River in Ohio. As one might expect, the concentrations were found to be higher in the more urbanized downstream section of the river, with the maximum values in the river water just downstream of the local WWTP and in the sediment at the most downstream site. The concentration ranges were 5.16–5.29 µg/L for the river water just downstream of the WWTP and 507–541 µg/kg dry wt for the river sediment in that site [11]. The range of values found in the river water was within the range of values found in other rivers in the United States [14], whereas that of the sediment was moderate as compared to published data for sediments throughout the United States and Canada. Much higher levels have been reported in U.S. upper midwestern rivers [15]. While the concentrations of the short-chain nonylphenol ethoxylates and carboxylates in raw water entering waterworks, taken from the Llobregat River in Spain, ranged from 8.3 to 22 µg/L of NP_1EO, NP_2EO, NP_1EC (nonylphenol monocarboxylate), and NP_2EC (NP-ethoxycarboxylate), they do not include the higher homologues of APEOs. This means not only a significantly higher concentration of APEOs in the river studied, but also an overall significantly higher concentration than that found in the U.S. study [11].

APEOs are components of aircraft deicer and antiicer fluids. A recent preliminary study was conducted at the international airport in Milwaukee to identify and quantify the NPEOs in the airport runoff. The concentrations of the NPEOs were found to be up to 1190 µg/L in airport outfall samples, up to 77 µg/L in samples from the receiving stream, and less than 5 µg/L in samples from the upstream reference [16]. The contribution of these NPEOs to the EI is apparent.

The essence of a relevant study, published in Israel in 2003, is presented here as an illustrative comprehensive case study [13]. The country is located in a semiarid region, thus experiencing an extreme shortage of water supplies. Of about 5×10^8 m^3 of the annually produced sewage—containing ca. 9–12 mg/L of anionic (mainly LABS) and 1–3 mg/L of nonionic (mainly APEO) surfactants (approximately 85:15 ratio)—about 27% and 45% of the total quantity is used following secondary treatment or directly, for aquifer recharge and agricultural irrigation, respectively. About 15% of the total amount is disposed into the Mediterranean Sea. Since (1) only secondary treatment is available for sewage effluents in the country, (2) APEOs are quite biodegradation resistant, and (3) about two-thirds of the nonionic surfactants used in the country are of the "hard" APEO type, these nonionic surfactants and/or their metabolites reach surface water and groundwater. As a result, the nonionic alkylphenol ethoxylate (APEO) surfactant concentration profiles of Israel's rivers, groundwater, and coastal water of the eastern Mediterranean Sea were found to be within the ranges of 12.5–74.6, trace–20.2, and 4.2–25.0 µg/L, respectively [3,13].

III. BIODEGRADATION AND REMOVAL

The occurrence/survival of surfactants in the aquatic environment constitutes a determining factor that affects their overall EI, particularly of those that have a toxic or ecotoxic potential, like the "hard" APEOs. However, not less important in this respect are the ultimate environmental biodegradation of each surfactant type, its biodegradation rate and pathway, as well as the percentage of its removal in WWTPs of different types and operating conditions. In a critical review published in 2003 that focuses on the treatment/removal of the endocrine- disrupting APEOs in activated-sludge treatment works [17], the authors conclude that:

> This example of the fate of APEOs. . . [in activated-sludge-based WWTPs]. . . is a classic example of the activated-sludge process with regard to some organic contaminants. On a basic level, it has successfully eliminated the parent compound, but incomplete biodegradation and sorption release breakdown products into the effluent, potentially more harmful to the environment than the original compound.

The implications of this to the EI and its assessment, should be considered not only in view of the unremoved parent APEOs and their metabolites that are discharged (via the effluents) into receiving waters, but also in view of the "hard" biodegradation products that distribute into the sludge fraction [17], where they persist [18] and therefore may later reach agricultural soils and, eventually, groundwater.

Perchlorination in river water used in a drinking water treatment plant (DWTP) reduced the concentrations of short-chain ethoxy NPEC, NPEO, and NP by about 25–35% and ~90%, respectively, partly due to their transformation to halogenated derivatives. Settling and flocculation followed by rapid sand filtration, ozonization, GAC filtration, and final chlorination resulted in 96–98% overall elimination of nonylphenolic compounds, a few of which were still detected in the drinking water. This is an additional manifestation of the EI of surfactants in aquatic environmental compartments that should be taken into consideration.

In a related investigation of LAS, NP, and NP_2EO biodegradation in activated-sludge WWTPs [19], it was found that for LAS, less than 1% was in the treated water, 84% was biodegraded, and 15% was in the sludge, compared with 4% and 2% of the entering NP and NP_2EO found in the effluent, 80% that was biodegraded for both, and 16% and 18%, respectively, that were in the sludge. The simple model developed based on these results allows the inclusion of complex, alternatively operated WWTPs in risk assessment tools and thus contributes to the assessment of one of the important components of the relevant EIs.

IV. TOXICITY/ECOTOXICITY

Numerous studies have focused on the toxicity and/or ecotoxicity component(s) of the EI associated with the constituents of detergent formulation, particularly the surfactants, the main concern being the potential risks to human health. A substantial portion of these studies is summarized in the relevant chapters of this volume. At any rate, the guiding perception appears to be that our modern, technology-based life exposes us to a "cocktail" of anthropogenic chemicals, some of which are toxic and therefore act, on their own or synergistically with others, directly or indirectly (e.g., via the food chain), on one or more components of exposed biological systems, thereby causing some form of malfunction and/or damage to these systems and, ultimately, health problems.

Following our use of the case of the APEOs in the environment in the preceding sections of this chapter, the toxicity/ecotoxicity issue in the EI context will be dealt with here, using the APEOs EDCs as an illustrative case study. A 2003 comprehensive study of endocrine disruptors in aquatic environments was aimed at establishing whether the estrogenicity so evident in certain UK rivers is characteristic of other European aquatic environments and at identifying the principal ingredients, understanding their behavior and distribution in the environment, and looking for evidence of impacts on aquatic wildlife [17]. In concluding this study the investigators claim that theirs and related toxicological studies did demonstrate the deleterious effects and, therefore, the EI of the EDCs studied on fish exposed to environmentally realistic levels of estrogenic substances (such as APEOs and their metabolites). However, the question of deleterious impacts of estrogenic effluents on fish *populations* is one of the most important that is still to be answered. The results of a functional reproductive capacity may be more sensitive than gross morphological endpoints (condition, ovosomatic index sex ratio) in zebrafish exposed to zenoestrogens during sexual differentiation and early gametogenesis [20].

In taking all of the foregoing into consideration, as well as the results of a study on the environmental occurrence/survival and homologic distribution of APEOs and their metabolites [13], it was concluded that there is an APEO-related potential environmental and health risk problem. Consequently:

1. A total ban on the use of the "hard" APEOs in detergent formulations should be seriously considered, i.e., *prevention.*
2. Appropriate and effective sewage treatment facilities should immediately be constructed and operated in order to effectively remove the "hard" nonionic APEO from domestic and industrial wastewaters.
3. A switch to a research-based management of our water resources, groundwater in particular, should immediately be implemented [13].

The overall implications—concerning the estimation of the EI—are: (1) Measurement of chemical targets alone is not sufficient; estimation of total estrogenic activity (of the APEOs) in the effluents or receiving water is needed; (2) combining analytical chemistry and bioassay is needed when setting environmental studies; and (3) the issue of the effects of mixtures of chemicals, particularly at low doses, requires much more research. [17].

V. EI: WHICH WAY IS THE WIND BLOWING?

Attaining sustainability in the environmental context in light of the ever-increasing EI as a result of human activity is an enormous challenge facing modern society. Several recently published papers provide a relevant response to the question of which way the wind is blowing or should be blowing, in the context of EI, the EI of detergents included. All of them emphasize the importance of reconceptualization in the context of the environmental system's challenge that we are confronting:

1.   Given that the public, many policymakers, and some environmental professionals believe that science and technology can solve most pollution problems, prevent future (undesirable) environmental impacts, and pave the way for sustainable development, a claim is made that since science and technology alone cannot meet the challenge of sustainable development or protect the environment now or in the future, we should all recognize the limits (reconceptualize the problem-solving capability) of environmental science and technology. Furthermore, we must address the root causes of environmental problems: our society's preoccupation with economic expansion. Thus, long-term protection of the environment and sustainable lifestyles are primarily not scientific-technological or economic problems, but rather social-behavioral and moral ones [21]. It is therefore the responsibility of each one of us, to get involved in a process of self-evaluative thinking and reconceptualization of the issues related to the EI of human activity, to be followed by an in accord responsible action [22].

2.   It is claimed that since major environmental pollutants are coming under the control of regulatory authorities, the relevant part of ecotoxicology is more or less completed. Yet there is still work remaining to be done that is not expected to call for major scientific innovation and discovery. It is concluded that the merger of ecotoxicology and ecology would give rise to a new science, denoted as *stress ecology*, at the crossroads of ecology, genomics, and bioinformatics [23]. The expected contribution of such a development to the scientific basis of the EIA of detergents—surfactants in particular—is apparent. Since demonstrating effects of toxic chemicals (e.g., EDCs) at the *population level* (ecotoxicology) is highly important, molecular techniques such as genetic sex probes and the use of population genetics are strongly recommended in this respect [17].

3.   In view of the increasing worldwide attention to environmental impact assessment in the context of which the consequences of defined actions are identified, quantified, evaluated, and predicted with a broad remit to socioeconomic developments, it is claimed [24] that *genomic* tools provide an option for analyzing basic resources at the nucleic acid level of an organism or the population of microorganisms [25]. Since genomics tools can be used for handling *complex* systems and for helping to identify the real stressors that cause impacts and observed perturbations [26], subtle impacts caused by industrial development that may have long-term effects can be monitored at the microbial level [24]. Such a contribution to EIA is highly desirable and needed.

4.   In response to the question of whether regulations will protect more people more often if new *economics* is applied, a new research initiative into environmental health effects is being advanced—the cost effectiveness analysis (CEA) [27]. Analysts use CEA to consider all the factors, measured in terms of costs per unit of health benefit gains associated with regulation. Although the relevance of the CEA to EIA is apparent, it does not tell regulators whether a rule is worth pursuing; rather, it creates a common metric (a dollar value in this case) needed to achieve a certain benefit, as gauging the price of good health [28].

In summary: The EI of detergents, the subject of this volume, is the core issue in the context of detergent production, formulation, applications/usage, and discharge. This impact should be assessed via appropriate EIA methodologies, both qualitatively and, more important, quantitatively to serve as a scientifically based rationale for environmental regulation and policy targeted at sustainable development as well as for evaluative thinking–based EIA by all involved—scientists, engineers, policymakers, economists, politicians, and the public at large, to be followed by responsible action.

REFERENCES

1. Zoller, U.; Scholz, R.W. Environ. Sci. Technol. 2003. *submitted.*
2. Petrovic, M.; Barceló, D. In *Comprehensive Analytical Chemistry XL*; Knepper, T.P., Barceló, D., de Voogt, P., Eds.; Elsevier Science: Amsterdam, The Netherlands, 2003; 655–673.
3. Zoller, U. J. Environ. Eng. 2003. *in press.*
4. McAvoy, D.C.; Dyer, S.D.; Fendinger, N.J.; Eckhoff, W.S.; Lawrence, D.L.; Begely, W.M. Environ. Toxical. Chem. 1998, *27*, 1705.
5. Li, H.-Q.; Jiku, F.; Schroeder, H.Fr. J. Chromatogr. A 2000, *889*, 155.
6. Radke, M.; Berhrends, T.; Foster, J.; Herrman, R. Anal. Chem. 1999, *71*, 5362.
7. Feijtel, T.C.; Struijs, J.; Matthijs, E. Environ. Toxicol. Chem. 1999, *18*, 2645.
8. Waters, J.; Lee, K.S.; Perchard, V.; Flanagan, M.; Clarke, P. Tenside Surfac. Det. 2000, *37*, 161.
9. Ahel, M.; Giger, W. Am. Chem. Soc. Natl. Meeting Extended Abst. 1999, *38*, 276.
10. Pikering, A.S.; Sumpter, J.P. Environ. Sci. Technol. 2003, *37* (17), 331A–336A.
11. Price, C.P.; Schmitz-Afonso, I.; Loyo-Rosales, I.E.; Link, E.; Thoma, B.; Fay, L.; Alfater, D.; Camp, M.J. Environ. Sci. Technol. 2003, *37* (19), 3747–3754.
12. Petrovic, M; Diza, A; Ventura, F; Barceló, D. Environ. Sci. Technol. 2003, *37* (19), 4442–4448.
13. Zoller, U.; Plaut, I.; Hushan, M. Water Sci. Technol. 2003. *submitted.*
14. Berber, L.B.; Brown, G.K.; Zaugg, S.D. Analysis of environmental endocrine disruptors. In *Analytical Method for Endocrine Disrupters*; Keith, L.H., Jones-Lepp, T.L., Nedham, L.L., Eds.; Oxford University Press: Washington, D.C., 2000; 97–123.
15. Zinteck, L.B.; Parachuri, B.; Wesolowski, D.J. Society of Enviroinmental Toxicology and Chemistry (SETAC 22nd Annual Meeting), Abstract Book. 2001; 131.
16. Corsi, S.R.; Zitomer, D.H.; Field, J.A.; Canciilla, D.A. Environ. Sci. Technol. 2003, *33* (18), 4031–4037.
17. Johnson, A.C.; Sumpter, J.P. Environ. Sci. Technol. 2003, *35* (24), 4697–4703.
18. Ejlertsson, L.; Nilsson, M.-L.; Kilin, H.; Bergman, A.; Karlson, L.; Oquist, M.; Svensson, B.H. Environ. Sci. Technol. 2003, *35* (24), 4697–4703.
19. Fauser, P.; Vikelsöe, J.; Sörensen, P.B.; Carlson, L. Water Res. 2003, *17*, 1288–1295.
20. Hill, R.L., Jr.; Janz, D.M. Aquat. Toxicol. 2003, *63*, 417–429.
21. Huesemann, M.H. Environ. Sci. Technol. 2003, *37* (13), 259A–261A.
22. Zoller, U.; Scholz, R.W. Water Sci. Technol. 2003. *submitted.*
23. Van Straaleno, N.M. Environ. Sci. Technol. 2003, *37* (17), 324A–330A.
24. Purohit, H.I.; Raie, D.V.; Kapley, A; Padmanbhan, R. Environ. Sci. Technol. 2003, *37* (19), 356A–363A.
25. Newman, D.K.; Banfield, I.E. Science 2002, *296*, 1071–1076.
26. Strogatz, S.H. Nature 2002, *410*, 268–276.
27. Cooney, C.M. Environ. Sci. Technol. 2003, *37* (15), 272A–273A.
28. Conney, C.M. Environ. Sci. Tech. 2003, *37* (19), 365A–368A.

Index